Problem-Solving Strategies and Model Boxes

Note for users of the three-volume edition:
Volume 1 (pp. 1–600) includes Chapters 1–21.
Volume 2 (pp. 601–1062) includes Chapters 22–36.
Volume 3 (pp. 1021–1240) includes Chapters 36–42.

Chapters 37–42 are not in the Standard Edition.

Brief Contents

PHYSICS

FOR SCIENTISTS AND ENGINEERS **A STRATEGIC APPROACH** 4/E

VOLUME TWO

RANDALL D. KNIGHT

California Polytechnic State University
San Luis Obispo

PEARSON

Editor-in-Chief:	Jeanne Zalesky
Acquisitions Editor:	Darien Estes
Project Manager:	Martha Steele
Program Manager:	Katie Conley
Senior Development Editor:	Alice Houston, Ph.D.
Art Development Editors:	Alice Houston, Kim Brucker, and Margot Otway
Development Manager:	Cathy Murphy
Program and Project Management Team Lead:	Kristen Flathman
Production Management:	Rose Kernan
Compositor:	Cenveo® Publisher Services
Design Manager:	Mark Ong
Cover Designer:	John Walker
Illustrators:	Rolin Graphics
Rights & Permissions Project Manager:	Maya Gomez
Rights & Permissions Management:	Rachel Youdelman
Photo Researcher:	Eric Schrader
Manufacturing Buyer:	Maura Zaldivar-Garcia
Executive Marketing Manager:	Christy Lesko
Cover Photo Credit:	Thomas Vogel/Getty Images

Library of Congress Cataloging-in-Publication Data On File

ISBN 10: 0-134-11066-8; ISBN 13: 978-0-134-11066-0 (Volume 2)

www.pearsonhighered.com   1 2 3 4 5 6 7 8 9 10—**V011**—18 17 16 15

About the Author

Randy Knight taught introductory physics for 32 years at Ohio State University and California Polytechnic State University, where he is Professor Emeritus of Physics. Professor Knight received a Ph.D. in physics from the University of California, Berkeley and was a post-doctoral fellow at the Harvard-Smithsonian Center for Astrophysics before joining the faculty at Ohio State University. It was at Ohio State that he began to learn about the research in physics education that, many years later, led to *Five Easy Lessons: Strategies for Successful Physics Teaching* and this book, as well as *College Physics: A Strategic Approach,* co-authored with Brian Jones and Stuart Field. Professor Knight's research interests are in the fields of laser spectroscopy and environmental science. When he's not in front of a computer, you can find Randy hiking, sea kayaking, playing the piano, or spending time with his wife Sally and their five cats.

Detailed Contents

Volume 1 contains chapters 1–21; Volume 2 contains chapters 22–36; Volume 3 contains chapters 36–42.

OVERVIEW

Forces and Fields

Amber, or fossilized tree resin, has long been prized for its beauty. It has been known since antiquity that a piece of amber rubbed with fur can attract feathers or straw—seemingly magical powers to a pre-scientific society. It was also known to the ancient Greeks that certain stones from the region they called *Magnesia* could pick up pieces of iron. It is from these humble beginnings that we today have high-speed computers, lasers, and magnetic resonance imaging as well as such mundane modern-day miracles as the lightbulb.

The basic phenomena of electricity and magnetism are not as familiar as those of mechanics. You have spent your entire life exerting forces on objects and watching them move, but your experience with electricity and magnetism is probably much more limited. We will deal with this lack of experience by placing a large emphasis on the *phenomena* of electricity and magnetism.

We will begin by looking in detail at *electric charge* and the process of *charging* an object. It is easy to make systematic observations of how charges behave, and we will consider the forces between charges and how charges behave in different materials. Similarly, we will begin our study of magnetism by observing how magnets stick to some metals but not others and how magnets affect compass needles. But our most important observation will be that an electric current affects a compass needle in exactly the same way as a magnet. This observation, suggesting a close connection between electricity and magnetism, will eventually lead us to the discovery of electromagnetic waves.

Our goal in Part VI is to develop a theory to explain the phenomena of electricity and magnetism. The linchpin of our theory will be the entirely new concept of a *field*. Electricity and magnetism are about the long-range interactions of charges, both static charges and moving charges, and the field concept will help us understand how these interactions take place. We will want to know how fields are created by charges and how charges, in return, respond to the fields. Bit by bit, we will assemble a theory—based on the new concepts of electric and magnetic fields—that will allow us to understand, explain, and predict a wide range of electromagnetic behavior.

The story of electricity and magnetism is vast. The 19th-century formulation of the theory of electromagnetism, which led to sweeping revolutions in science and technology, has been called by no less than Einstein "the most important event in physics since Newton's time." Not surprisingly, all we can do in this text is develop some of the basic ideas and concepts, leaving many details and applications to later courses. Even so, our study of electricity and magnetism will explore some of the most exciting and important topics in physics.

These bright loops above the surface of the sun—called *coronal loops*—are an extremely hot gas ($>10^6$ K) of charged particles moving along field lines of the sun's magnetic field.

22 Electric Charges and Forces

Electricity is one of the fundamental forces of nature. Lightning is a vivid manifestation of electric charges and forces.

IN THIS CHAPTER, you will learn that electric phenomena are based on charges, forces, and fields.

What is electric charge?

Electric phenomena depend on charge.

- There are two kinds of charge, called positive and negative.
- Electrons and protons—the constituents of atoms—are the basic charges of ordinary matter.
- Charging is the transfer of electrons from one object to another.

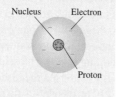

How do charges behave?

Charges have well-established behaviors:

- Two charges of the same kind repel; two opposite charges attract.
- Small neutral objects are attracted to a charge of either sign.
- Charge can be transferred from one object to another.
- Charge is conserved.

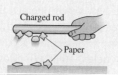

What are conductors and insulators?

There are two classes of materials with very different electrical properties:

- Conductors are materials through or along which charge moves easily.
- Insulators are materials on or in which charge is immobile.

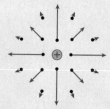

What is Coulomb's law?

Coulomb's law is the fundamental law for the electric force between two charged particles. Coulomb's law, like Newton's law of gravity, is an inverse-square law: The electric force is inversely proportional to the *square* of the distance between charges.

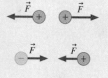

≪ LOOKING BACK Sections 3.2–3.4 Vector addition
≪ LOOKING BACK Sections 13.2–13.4 Gravity

What is an electric field?

How is a long-range force transmitted from one charge to another? We'll develop the idea that charges create an electric field, and the electric field of one charge is the agent that exerts a force on another charge. That is, charges interact via electric fields. The electric field is present at all points in space.

Why are electric charges important?

Computers, cell phones, and optical fiber communications may seem to have little in common with the fact that you can get a shock when you touch a doorknob after walking across a carpet. But the physics of electric charges—how objects get charged and how charges interact with each other—is the foundation for all modern electronic devices and communications technology. Electricity and magnetism is a very large and very important topic, and it starts with simple observations of electric charges and forces.

22.1 The Charge Model

You can receive a mildly unpleasant shock and produce a little spark if you touch a metal doorknob after walking across a carpet. Vigorously brushing your freshly washed hair makes all the hairs fly apart. A plastic comb that you've run through your hair will pick up bits of paper and other small objects, but a metal comb won't.

The common factor in these observations is that two objects are *rubbed* together. Why should rubbing an object cause forces and sparks? What kind of forces are these? Why do metallic objects behave differently from nonmetallic? These are the questions with which we begin our study of electricity.

Our first goal is to develop a model for understanding electric phenomena in terms of *charges* and *forces*. We will later use our contemporary knowledge of atoms to understand electricity on a microscopic level, but the basic concepts of electricity make *no* reference to atoms or electrons. The theory of electricity was well established long before the electron was discovered.

Experimenting with Charges

Let us enter a laboratory where we can make observations of electric phenomena. The major tools in the lab are plastic, glass, and metal rods; pieces of wool and silk; and small metal spheres on wood stands. Let's see what we can learn with these tools.

Discovering electricity I

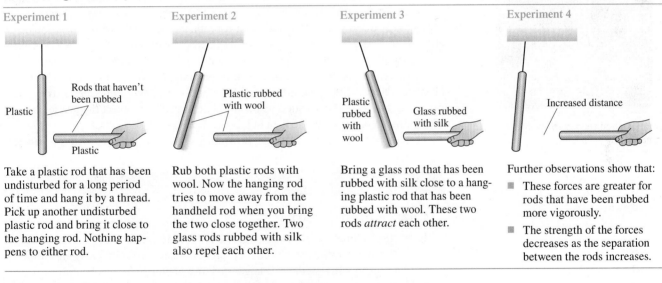

Experiment 1	Experiment 2	Experiment 3	Experiment 4

Take a plastic rod that has been undisturbed for a long period of time and hang it by a thread. Pick up another undisturbed plastic rod and bring it close to the hanging rod. Nothing happens to either rod.

Rub both plastic rods with wool. Now the hanging rod tries to move away from the handheld rod when you bring the two close together. Two glass rods rubbed with silk also repel each other.

Bring a glass rod that has been rubbed with silk close to a hanging plastic rod that has been rubbed with wool. These two rods *attract* each other.

Further observations show that:
- These forces are greater for rods that have been rubbed more vigorously.
- The strength of the forces decreases as the separation between the rods increases.

No forces were observed in Experiment 1. We will say that the original objects are **neutral.** Rubbing the rods (Experiments 2 and 3) somehow causes forces to be exerted between them. We will call the rubbing process **charging** and say that a rubbed rod is *charged*. For now, these are simply descriptive terms. The terms don't tell us anything about the process itself.

Experiments 2 and 3 show that there is a *long-range repulsive force,* requiring no contact, between two identical objects that have been charged in the *same* way. Furthermore, Experiment 4 shows that the force between two charged objects depends on the distance between them. This is the first long-range force we've encountered since gravity was introduced in Chapter 5. It is also the first time we've observed a repulsive force, so right away we see that new ideas will be needed to understand electricity.

Experiment 3 is a puzzle. Two rods *seem* to have been charged in the same way, by rubbing, but these two rods *attract* each other rather than repel. Why does the outcome of Experiment 3 differ from that of Experiment 2? Back to the lab.

Discovering electricity II

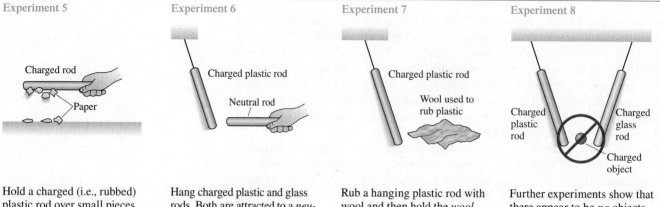

Experiment 5	Experiment 6	Experiment 7	Experiment 8

Hold a charged (i.e., rubbed) plastic rod over small pieces of paper. The pieces of paper leap up and stick to the rod. A charged glass rod does the same. However, a neutral rod has no effect on the pieces of paper.

Hang charged plastic and glass rods. Both are attracted to a *neutral* (i.e., unrubbed) plastic rod. Both are also attracted to a *neutral* glass rod. In fact, the charged rods are attracted to *any* neutral object, such as a finger or a piece of paper.

Rub a hanging plastic rod with wool and then hold the *wool* close to the rod. The rod is weakly *attracted* to the wool. The plastic rod is *repelled* by a piece of silk that has been used to rub glass.

Further experiments show that there appear to be *no* objects that, after being rubbed, pick up pieces of paper and attract *both* the charged plastic and glass rods.

A comb rubbed through your hair picks up small pieces of paper.

Our first set of experiments found that charged objects exert forces on each other. The forces are sometimes attractive, sometimes repulsive. Experiments 5 and 6 show that there is an attractive force between a charged object and a *neutral* (uncharged) object. This discovery presents us with a problem: How can we tell if an object is charged or neutral? Because of the attractive force between a charged and a neutral object, simply observing an electric force does *not* imply that an object is charged.

However, an important characteristic of any *charged* object appears to be that **a charged object picks up small pieces of paper.** This behavior provides a straightforward test to answer the question, Is this object charged? An object that passes the test by picking up paper is charged; an object that fails the test is neutral.

These observations let us tentatively advance the first stages of a **charge model.**

MODEL 22.1

Charge model, part I

1. Frictional forces, such as rubbing, add something called **charge** to an object or remove it from the object. The process itself is called *charging.* More vigorous rubbing produces a larger quantity of charge.
2. There are two and only two kinds of charge. For now we will call these "plastic charge" and "glass charge." Other objects can sometimes be charged by rubbing, but the charge they receive is either "plastic charge" or "glass charge."
3. Two **like charges** (plastic/plastic or glass/glass) exert repulsive forces on each other. Two **opposite charges** (plastic/glass) attract each other.
4. The force between two charges is a long-range force. The size of the force increases as the quantity of charge increases and decreases as the distance between the charges increases.
5. *Neutral* objects have an *equal mixture* of both "plastic charge" and "glass charge." The rubbing process somehow manages to separate the two.

Postulate 2 is based on Experiment 8. If an object is charged (i.e., picks up paper), it always attracts one charged rod and repels the other. That is, it acts either "like plastic" or "like glass." If there were a third kind of charge, different from the first two, an object with that charge should pick up paper and attract *both* the charged plastic and glass rods. No such objects have ever been found.

The basis for postulate 5 is the observation in Experiment 7 that a charged plastic rod is attracted to the wool used to rub it but repelled by silk that has rubbed glass. It appears that rubbing glass causes the silk to acquire "plastic charge." The easiest way to explain this is to hypothesize that the silk starts out with equal amounts of "glass charge" and "plastic charge" and that the rubbing somehow transfers "glass charge" from the silk to the rod. This leaves an excess of "glass charge" on the rod and an excess of "plastic charge" on the silk.

While the charge model is *consistent* with the observations, it is by no means proved. We still have some large unexplained puzzles, such as why charged objects exert attractive forces on neutral objects.

Electric Properties of Materials

We still need to clarify how different types of materials respond to charges.

Discovering electricity III

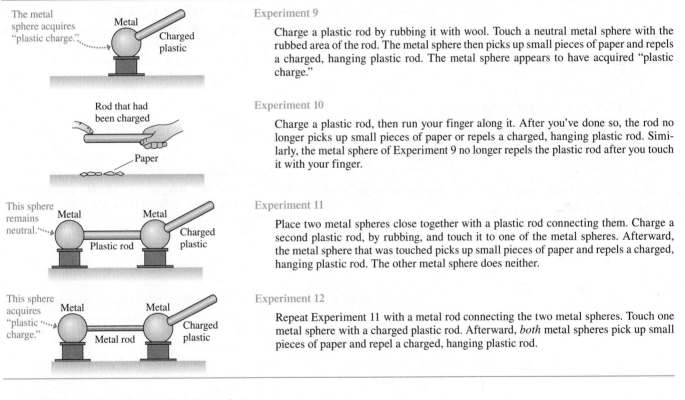

Experiment 9

Charge a plastic rod by rubbing it with wool. Touch a neutral metal sphere with the rubbed area of the rod. The metal sphere then picks up small pieces of paper and repels a charged, hanging plastic rod. The metal sphere appears to have acquired "plastic charge."

Experiment 10

Charge a plastic rod, then run your finger along it. After you've done so, the rod no longer picks up small pieces of paper or repels a charged, hanging plastic rod. Similarly, the metal sphere of Experiment 9 no longer repels the plastic rod after you touch it with your finger.

Experiment 11

Place two metal spheres close together with a plastic rod connecting them. Charge a second plastic rod, by rubbing, and touch it to one of the metal spheres. Afterward, the metal sphere that was touched picks up small pieces of paper and repels a charged, hanging plastic rod. The other metal sphere does neither.

Experiment 12

Repeat Experiment 11 with a metal rod connecting the two metal spheres. Touch one metal sphere with a charged plastic rod. Afterward, *both* metal spheres pick up small pieces of paper and repel a charged, hanging plastic rod.

Our final set of experiments has shown that

- Charge can be *transferred* from one object to another, but only when the objects *touch*. Contact is required. Removing charge from an object, which you can do by touching it, is called **discharging.**
- There are two types or classes of materials with very different electric properties. We call these *conductors* and *insulators*.

Experiment 12, in which a metal rod is used, is in sharp contrast to Experiment 11. Charge somehow *moves through* or along a metal rod, from one sphere to the other, but remains *fixed in place* on a plastic or glass rod. Let us define **conductors** as those materials through or along which charge easily moves and **insulators** as those materials on or in which charges remain immobile. Glass and plastic are insulators; metal is a conductor.

This information lets us add two more postulates to our charge model:

MODEL 22.1

Charge model, part II

6. There are two types of materials. Conductors are materials through or along which charge easily moves. Insulators are materials on or in which charges remain fixed in place.
7. Charge can be transferred from one object to another by contact.

NOTE Both insulators and conductors can be charged. They differ in the *mobility* of the charge.

We have by no means exhausted the number of experiments and observations we might try. Early scientific investigators were faced with all of these results, plus many others. How should we make sense of it all? The charge model seems promising, but certainly not proven. We have not yet explained how charged objects exert attractive forces on *neutral* objects, nor have we explained what charge is, how it is transferred, or *why* it moves through some objects but not others. Nonetheless, we will take advantage of our historical hindsight and continue to pursue this model. Homework problems will let you practice using the model to explain other observations.

EXAMPLE 22.1 | **Transferring charge**

In Experiment 12, touching one metal sphere with a charged plastic rod caused a second metal sphere to become charged with the same type of charge as the rod. Use the postulates of the charge model to explain this.

SOLVE We need the following postulates from the charge model:

7. Charge is transferred upon contact.
6. Metal is a conductor, and charge moves through a conductor
3. Like charges repel.

The plastic rod was charged by rubbing with wool. The charge doesn't move around on the rod, because it is an insulator, but some of the "plastic charge" is transferred to the metal upon contact. Once in the metal, which is a conductor, the charges are free to move around. Furthermore, because like charges repel, these plastic charges quickly move as far apart as they possibly can. Some move through the connecting metal rod to the second sphere. Consequently, the second sphere acquires "plastic charge."

STOP TO THINK 22.1 To determine if an object has "glass charge," you need to

a. See if the object attracts a charged plastic rod.
b. See if the object repels a charged glass rod.
c. Do both a and b.
d. Do either a or b.

22.2 Charge

As you probably know, the modern names for the two types of charge are *positive charge* and *negative charge*. You may be surprised to learn that the names were coined by Benjamin Franklin.

So what is positive and what is negative? It's entirely up to us! Franklin established the convention that **a glass rod that has been rubbed with silk is *positively* charged.** That's it. Any other object that repels a charged glass rod is also positively charged. Any charged object that attracts a charged glass rod is negatively charged. Thus **a plastic rod rubbed with wool is negative.** It was only long afterward, with the discovery of electrons and protons, that electrons were found to be attracted to a charged glass rod while protons were repelled. Thus *by convention* electrons have a negative charge and protons a positive charge.

Atoms and Electricity

Now let's fast forward to the 21st century. The theory of electricity was developed without knowledge of atoms, but there is no reason for us to continue to overlook this important part of our contemporary perspective. **FIGURE 22.1** shows that an atom consists of a very small and dense *nucleus* (diameter $\sim 10^{-14}$ m) surrounded by much less massive orbiting *electrons*. The electron orbital frequencies are so enormous ($\sim 10^{15}$ revolutions per second) that the electrons seem to form an **electron cloud** of diameter $\sim 10^{-10}$ m, a factor 10^4 larger than the nucleus.

Experiments at the end of the 19th century revealed that electrons are particles with both mass and a negative charge. The nucleus is a composite structure consisting of *protons,* positively charged particles, and neutral *neutrons.* The atom is held together by the attractive electric force between the positive nucleus and the negative electrons.

One of the most important discoveries is that **charge, like mass, is an inherent property of electrons and protons.** It's no more possible to have an electron without charge than it is to have an electron without mass. As far as we know today, electrons and protons have charges of opposite sign but *exactly* equal magnitude. (Very careful experiments have never found any difference.) This atomic-level unit of charge, called the **fundamental unit of charge,** is represented by the symbol e. **TABLE 22.1** shows the masses and charges of protons and electrons. We need to define a unit of charge, which we will do in Section 22.4, before we can specify how much charge e is.

The Micro/Macro Connection

Electrons and protons are the basic charges of ordinary matter. Consequently, the various observations we made in Section 22.1 need to be explained in terms of electrons and protons.

NOTE Electrons and protons are particles of matter. Their motion is governed by Newton's laws. Electrons can move from one object to another when the objects are in contact, but neither electrons nor protons can leap through the air from one object to another. An object does not become charged simply from being close to a charged object.

Charge is represented by the symbol q (or sometimes Q). A macroscopic object, such as a plastic rod, has charge

$$q = N_p e - N_e e = (N_p - N_e)e \tag{22.1}$$

where N_p and N_e are the number of protons and electrons contained in the object. An object with an equal number of protons and electrons has no *net* charge (i.e., $q = 0$) and is said to be *electrically neutral.*

NOTE *Neutral* does *not* mean "no charges" but, instead, no *net* charge.

A charged object has an unequal number of protons and electrons. An object is positively charged if $N_p > N_e$. It is negatively charged if $N_p < N_e$. Notice that an object's charge is always an integer multiple of e. That is, the amount of charge on an object varies by small but discrete steps, not continuously. This is called **charge quantization.**

In practice, objects acquire a positive charge not by gaining protons, as you might expect, but by losing electrons. Protons are *extremely* tightly bound within the nucleus and cannot be added to or removed from atoms. Electrons, on the other hand, are bound rather loosely and can be removed without great difficulty. The process of removing an electron from the electron cloud of an atom is called **ionization.** An atom that is missing an electron is called a *positive ion*. Its *net* charge is $q = +e$.

Some atoms can accommodate an *extra* electron and thus become a *negative ion* with net charge $q = -e$. A saltwater solution is a good example. When table salt (the chemical sodium chloride, NaCl) dissolves, it separates into positive sodium ions Na^+ and negative chlorine ions Cl^-. **FIGURE 22.2** shows positive and negative ions.

FIGURE 22.1 An atom.

The nucleus, exaggerated for clarity, contains positive protons.

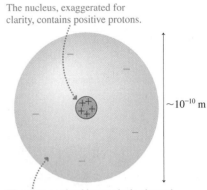

$\sim 10^{-10}$ m

The electron cloud is negatively charged.

TABLE 22.1 Protons and electrons

Particle	Mass (kg)	Charge
Proton	1.67×10^{-27}	$+e$
Electron	9.11×10^{-31}	$-e$

FIGURE 22.2 Positive and negative ions.

Positive ion   Negative ion

The atom has lost one electron, giving it a net positive charge.   The atom has gained one electron, giving it a net negative charge.

FIGURE 22.3 Charging by friction usually creates molecular ions as bonds are broken.

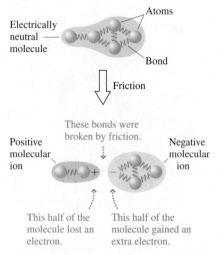

All the charging processes we observed in Section 22.1 involved rubbing and friction. The forces of friction cause molecular bonds at the surface to break as the two materials slide past each other. Molecules are electrically neutral, but FIGURE 22.3 shows that *molecular ions* can be created when one of the bonds in a large molecule is broken. The positive molecular ions remain on one material and the negative ions on the other, so one of the objects being rubbed ends up with a net positive charge and the other with a net negative charge. This is the way in which a plastic rod is charged by rubbing with wool or a comb is charged by passing through your hair.

Charge Conservation and Charge Diagrams

One of the most important discoveries about charge is the **law of conservation of charge:** Charge is neither created nor destroyed. Charge can be transferred from one object to another as electrons and ions move about, but the *total* amount of charge remains constant. For example, charging a plastic rod by rubbing it with wool transfers electrons from the wool to the plastic as the molecular bonds break. The wool is left with a positive charge equal in magnitude but opposite in sign to the negative charge of the rod: $q_{wool} = -q_{plastic}$. The *net* charge remains zero.

Diagrams are going to be an important tool for understanding and explaining charges and the forces on charged objects. As you begin to use diagrams, it will be important to make explicit use of charge conservation. The net number of plusses and minuses drawn on your diagrams should *not* change as you show them moving around.

STOP TO THINK 22.2   Rank in order, from most positive to most negative, the charges q_a to q_e of these five systems.

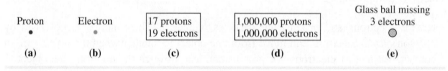

22.3 Insulators and Conductors

FIGURE 22.4 A microscopic look at insulators and conductors.

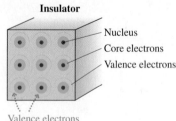

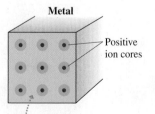

You have seen that there are two classes of materials as defined by their electrical properties: insulators and conductors. FIGURE 22.4 looks inside an insulator and a metallic conductor. The electrons in the insulator are all tightly bound to the positive nuclei and not free to move around. Charging an insulator by friction leaves patches of molecular ions on the surface, but these patches are immobile.

In metals, the outer atomic electrons (called the *valence electrons* in chemistry) are only weakly bound to the nuclei. As the atoms come together to form a solid, these outer electrons become detached from their parent nuclei and are free to wander about through the entire solid. The solid *as a whole* remains electrically neutral, because we have not added or removed any electrons, but the electrons are now rather like a negatively charged gas or liquid—what physicists like to call a **sea of electrons**— permeating an array of positively charged **ion cores.**

The primary consequence of this structure is that electrons in a metal are highly mobile. They can quickly and easily move through the metal in response to electric forces. The motion of charges through a material is what we will later call a **current,** and the charges that physically move are called the **charge carriers.** The charge carriers in metals are electrons.

Metals aren't the only conductors. Ionic solutions, such as salt water, are also good conductors. But the charge carriers in an ionic solution are the ions, not electrons. We'll focus on metallic conductors because of their importance in applications of electricity.

Charging

Insulators are often charged by rubbing. The charge diagrams of **FIGURE 22.5** show that the charges on the rod are on the surface and that charge is conserved. The charge can be transferred to another object upon contact, but it doesn't move around on the rod.

FIGURE 22.5 An insulating rod is charged by rubbing.

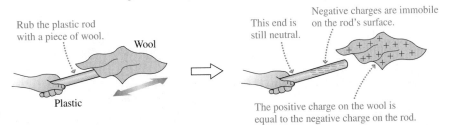

Rub the plastic rod with a piece of wool.

Wool

Plastic

This end is still neutral.

Negative charges are immobile on the rod's surface.

The positive charge on the wool is equal to the negative charge on the rod.

Metals usually cannot be charged by rubbing, but Experiment 9 showed that a metal sphere can be charged by contact with a charged plastic rod. **FIGURE 22.6** gives a pictorial explanation. An essential idea is that **the electrons in a conductor are free to move.** Once charge is transferred to the metal, repulsive forces between the negative charges cause the electrons to move apart from each other.

Note that the newly added electrons do not themselves need to move to the far corners of the metal. Because of the repulsive forces, the newcomers simply "shove" the entire electron sea a little to the side. The electron sea takes an extremely short time to adjust itself to the presence of the added charge, typically less than 10^{-9} s. For all practical purposes, a conductor responds *instantaneously* to the addition or removal of charge.

Other than this very brief interval during which the electron sea is adjusting, the charges in an *isolated* conductor are in static equilibrium. That is, the charges are at rest (i.e., static) and there is no net force on any charge. This condition is called **electrostatic equilibrium.** If there *were* a net force on one of the charges, it would quickly move to an equilibrium point at which the force was zero.

Electrostatic equilibrium has an important consequence:

> In an isolated conductor, any excess charge is located on the surface of the conductor.

To see this, suppose there *were* an excess electron in the interior of an isolated conductor. The extra electron would upset the electrical neutrality of the interior and exert forces on nearby electrons, causing them to move. But their motion would violate the assumption of static equilibrium, so we're forced to conclude that there cannot be any excess electrons in the interior. Any excess electrons push each other apart until they're all on the surface.

FIGURE 22.6 A conductor is charged by contact with a charged plastic rod.

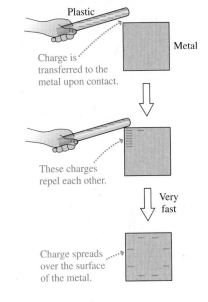

Plastic

Metal

Charge is transferred to the metal upon contact.

These charges repel each other.

Very fast

Charge spreads over the surface of the metal.

EXAMPLE 22.2 | Charging an electroscope

Many electricity demonstrations are carried out with the help of an *electroscope* like the one shown in **FIGURE 22.7**. Touching the sphere at the top of an electroscope with a charged plastic rod causes the leaves to fly apart and remain hanging at an angle. Use charge diagrams to explain why.

MODEL We'll use the charge model and the model of a conductor as a material through which electrons move.

VISUALIZE **FIGURE 22.8** on the next page uses a series of charge diagrams to show the charging of an electroscope.

FIGURE 22.7 A charged electroscope.

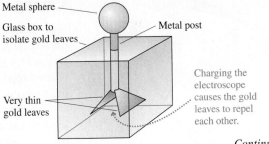

Metal sphere

Glass box to isolate gold leaves

Metal post

Very thin gold leaves

Charging the electroscope causes the gold leaves to repel each other.

Continued

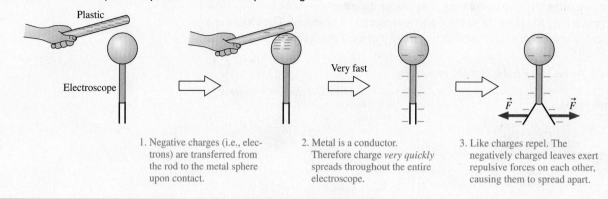

FIGURE 22.8 The process by which an electroscope is charged.

Plastic

Electroscope

Very fast

$\vec{F}$ $\vec{F}$

1. Negative charges (i.e., electrons) are transferred from the rod to the metal sphere upon contact.

2. Metal is a conductor. Therefore charge *very quickly* spreads throughout the entire electroscope.

3. Like charges repel. The negatively charged leaves exert repulsive forces on each other, causing them to spread apart.

FIGURE 22.9 Touching a charged metal discharges it.

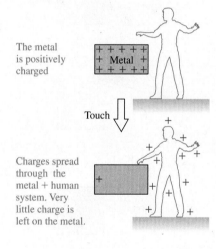

The metal is positively charged

Metal

Touch

Charges spread through the metal + human system. Very little charge is left on the metal.

FIGURE 22.10 A charged rod held close to an electroscope causes the leaves to repel each other.

The electroscope is polarized by the charged rod. The sea of electrons shifts toward the positive rod.

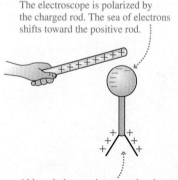

Although the net charge on the electroscope is still zero, the leaves have excess positive charge and repel each other.

Discharging

The human body consists largely of salt water. Pure water is not a terribly good conductor, but salt water, with its Na^+ and Cl^- ions, is. Consequently, and occasionally tragically, humans are reasonably good conductors. This fact allows us to understand how it is that *touching* a charged object discharges it, as we observed in Experiment 10. As **FIGURE 22.9** shows, the net effect of touching a charged metal is that it and the conducting human together become a much larger conductor than the metal alone. Any excess charge that was initially confined to the metal can now spread over the larger metal + human conductor. This may not entirely discharge the metal, but in typical circumstances, where the human is much larger than the metal, the residual charge remaining on the metal is much reduced from the original charge. The metal, for most practical purposes, is discharged. In essence, two conductors in contact "share" the charge that was originally on just one of them.

Moist air is a conductor, although a rather poor one. Charged objects in air slowly lose their charge as the object shares its charge with the air. The earth itself is a giant conductor because of its water, moist soil, and a variety of ions. Any object that is physically connected to the earth through a conductor is said to be **grounded.** The effect of being grounded is that the object shares any excess charge it has with the entire earth! But the earth is so enormous that any conductor attached to the earth will be completely discharged.

The purpose of *grounding* objects, such as circuits and appliances, is to prevent the buildup of any charge on the objects. The third prong on appliances and electronics that have a three-prong plug is the ground connection. The building wiring physically connects that third wire deep into the ground somewhere just outside the building, often by attaching it to a metal water pipe that goes underground.

Charge Polarization

One observation from Section 22.1 still needs an explanation. How do charged objects exert an attractive force on a *neutral* object? To begin answering this question, **FIGURE 22.10** shows a positively charged rod held close to—but not touching—a *neutral* electroscope. The leaves move apart and stay apart as long as you hold the rod near, but they quickly collapse when it is removed.

The charged rod doesn't touch the electroscope, so no charge is added or removed. Instead, the metal's sea of electrons is attracted to the positive rod and shifts slightly to create an excess of negative charge on the side near the rod. The far side of the electroscope now has a deficit of electrons—an excess positive charge. We say that the electroscope has been *polarized*. **Charge polarization** is a slight separation of the positive and negative charges in a neutral object. Because there's no net charge, the electron sea quickly readjusts when the rod is removed.

Why don't *all* the electrons rush to the side near the positive charge? Once the electron sea shifts slightly, the stationary positive ions begin to exert a force, a restoring force, pulling the electrons back to the right. The equilibrium position for the sea of electrons shifts just enough that the forces due to the external charge and the positive ions are in balance. In practice, the displacement of the electron sea is usually *less than 10^{-15} m!*

Charge polarization is the key to understanding how a charged object exerts an attractive force on a neutral object. FIGURE 22.11 shows a positively charged rod near a neutral piece of metal. Because the electric force decreases with distance, **the attractive force on the electrons at the top surface is slightly greater than the repulsive force on the ions at the bottom.** The net force toward the charged rod is called a **polarization force.** The polarization force arises because the charges in the metal are separated, *not* because the rod and metal are oppositely charged.

A negatively charged rod would push the electron sea slightly away, polarizing the metal to have a positive upper surface charge and a negative lower surface charge. Once again, these are the conditions for the charge to exert a *net attractive force* on the metal. Thus our charge model explains how a charged object of *either* sign attracts neutral pieces of metal.

The Electric Dipole

Polarizing a conductor is one thing, but why does a charged rod pick up paper, which is an insulator? Consider what happens when we bring a positive charge near an atom. As FIGURE 22.12 shows, the charge polarizes the atom. The electron cloud doesn't move far, because the force from the positive nucleus pulls it back, but the center of positive charge and the center of negative charge are now slightly separated.

FIGURE 22.12 A neutral atom is polarized by and attracted toward an external charge.

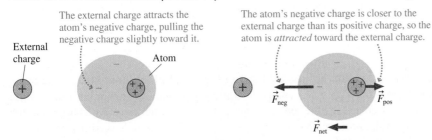

Two opposite charges with a slight separation between them form what is called an **electric dipole.** (The actual distortion from a perfect sphere is minuscule, nothing like the distortion shown in the figure.) The attractive force on the dipole's near end *slightly* exceeds the repulsive force on its far end because the near end is closer to the external charge. The net force, an *attractive* force between the charge and the atom, is another example of a polarization force.

An insulator has no sea of electrons to shift if an external charge is brought close. Instead, as FIGURE 22.13 shows, all the individual atoms inside the insulator become polarized. The polarization force acting *on each atom* produces a net polarization force toward the external charge. This solves the puzzle. A charged rod picks up pieces of paper by

■ Polarizing the atoms in the paper,
■ Then exerting an attractive polarization force on each atom.

This is important. Make sure you understand all the steps in the reasoning.

FIGURE 22.11 The polarization force on a neutral piece of metal is due to the slight charge separation.

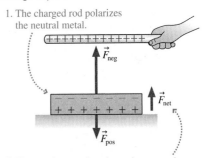

1. The charged rod polarizes the neutral metal.

2. The nearby negative charge is attracted to the rod more strongly than the distant positive charge is repelled, resulting in a net upward force.

FIGURE 22.13 The atoms in an insulator are polarized by an external charge.

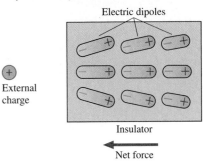

STOP TO THINK 22.3 An electroscope is positively charged by *touching* it with a positive glass rod. The electroscope leaves spread apart and the glass rod is removed. Then a negatively charged plastic rod is brought close to the top of the electroscope, but it doesn't touch. What happens to the leaves?

a. The leaves get closer together.
b. The leaves spread farther apart.
c. One leaf moves higher, the other lower.
d. The leaves don't move.

Charging by Induction

Charge polarization is responsible for an interesting and counterintuitive way of charging an electroscope. **FIGURE 22.14** shows a positively charged glass rod held near an electroscope but not touching it, while a person touches the electroscope with a finger. Unlike what happens in Figure 22.10, the electroscope leaves hardly move.

FIGURE 22.14 A positive rod can charge an electroscope negatively by induction.

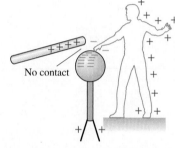

1. The charged rod polarizes the electroscope + person conductor. The leaves repel slightly due to polarization.

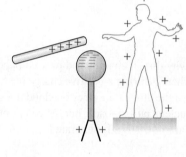

2. The negative charge on the electroscope is isolated when contact is broken.

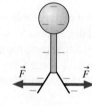

3. When the rod is removed, the leaves first collapse as the polarization vanishes, then repel as the excess negative charge spreads out.

Charge polarization occurs, as it did in Figure 22.10, but this time in the much larger electroscope + person conductor. If the person removes his or her finger while the system is polarized, the electroscope is left with a net *negative* charge and the person has a net positive charge. The electroscope has been charged *opposite to the rod* in a process called **charging by induction.**

22.4 Coulomb's Law

The first three sections have established a *model* of charges and electric forces. This model has successfully provided a qualitative explanation of electric phenomena; now it's time to become quantitative. Experiment 4 in Section 22.1 found that the electric force decreases with distance. The force law that describes this behavior is known as *Coulomb's law.*

Charles Coulomb was one of many scientists investigating electricity in the late 18th century. Coulomb had the idea of studying electric forces using the torsion balance scheme by which Cavendish had measured the value of the gravitational constant G (see Section 13.4). This was a difficult experiment. Despite many obstacles, Coulomb announced in 1785 that the electric force obeys an *inverse-square law* analogous to Newton's law of gravity. Today we know it as **Coulomb's law.**

Coulomb's law

1. If two charged particles having charges q_1 and q_2 are a distance r apart, the particles exert forces on each other of magnitude

$$F_{1 \text{ on } 2} = F_{2 \text{ on } 1} = \frac{K|q_1||q_2|}{r^2} \qquad (22.2)$$

where K is called the **electrostatic constant.** These forces are an action/reaction pair, equal in magnitude and opposite in direction.

2. The forces are directed along the line joining the two particles. The forces are *repulsive* for two like charges and *attractive* for two opposite charges.

We sometimes speak of the "force between charge q_1 and charge q_2," but keep in mind that we are really dealing with charged *objects* that also have a mass, a size, and other properties. Charge is not some disembodied entity that exists apart from matter. Coulomb's law describes the force between charged *particles,* which are also called **point charges.** A charged particle, which is an extension of the particle model we used in Part I, has a mass and a charge but has no size.

Coulomb's law looks much like Newton's law of gravity, but there is one important difference: The charge q can be either positive or negative. Consequently, the absolute value signs in Equation 22.2 are especially important. The first part of Coulomb's law gives only the *magnitude* of the force, which is always positive. The direction must be determined from the second part of the law. **FIGURE 22.15** shows the forces between different combinations of positive and negative charges.

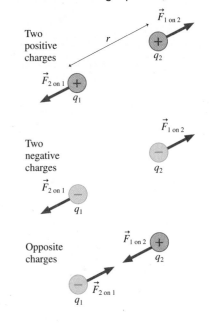

FIGURE 22.15 Attractive and repulsive forces between charged particles.

Units of Charge

Coulomb had no *unit* of charge, so he was unable to determine a value for K, whose numerical value depends on the units of both charge and distance. The SI unit of charge, the **coulomb** (C), is derived from the SI unit of *current,* so we'll have to await the study of current in Chapter 27 before giving a precise definition. For now we'll note that the fundamental unit of charge e has been measured to have the value

$$e = 1.60 \times 10^{-19} \text{ C}$$

This is a very small amount of charge. Stated another way, 1 C is the net charge of roughly 6.25×10^{18} protons.

NOTE The amount of charge produced by friction is typically in the range 1 nC (10^{-9} C) to 100 nC. This is an excess or deficit of 10^{10} to 10^{12} electrons.

Once the unit of charge is established, torsion balance experiments such as Coulomb's can be used to measure the electrostatic constant K. In SI units

$$K = 8.99 \times 10^9 \text{ N m}^2/\text{C}^2$$

It is customary to round this to $K = 9.0 \times 10^9 \text{ N m}^2/\text{C}^2$ for all but extremely precise calculations, and we will do so.

Surprisingly, we will find that Coulomb's law is not explicitly used in much of the theory of electricity. While it *is* the basic force law, most of our future discussion and calculations will be of things called *fields* and *potentials.* It turns out that we can make many future equations easier to use if we rewrite Coulomb's law in a somewhat more complicated way. Let's define a new constant, called the **permittivity constant** ϵ_0 (pronounced "epsilon zero" or "epsilon naught"), as

$$\epsilon_0 = \frac{1}{4\pi K} = 8.85 \times 10^{-12} \text{ C}^2/\text{N m}^2$$

Rewriting Coulomb's law in terms of ϵ_0 gives us

$$F = \frac{1}{4\pi\epsilon_0} \frac{|q_1||q_2|}{r^2} \tag{22.3}$$

It will be easiest when using Coulomb's law directly to use the electrostatic constant K. However, in later chapters we will switch to the second version with ϵ_0.

Using Coulomb's Law

Coulomb's law is a force law, and forces are vectors. It has been many chapters since we made much use of vectors and vector addition, but these mathematical techniques will be essential in our study of electricity and magnetism.

There are two important observations regarding Coulomb's law:

1. **Coulomb's law applies only to point charges.** A point charge is an idealized material object with charge and mass but with no size or extension. For practical purposes, two charged objects can be modeled as point charges if they are much smaller than the separation between them.
2. **Electric forces, like other forces, can be superimposed.** If multiple charges 1, 2, 3, ... are present, the *net* electric force on charge j due to all other charges is

$$\vec{F}_{net} = \vec{F}_{1 \text{ on } j} + \vec{F}_{2 \text{ on } j} + \vec{F}_{3 \text{ on } j} + \cdots \tag{22.4}$$

where each of the $\vec{F}_{i \text{ on } j}$ is given by Equation 22.2 or 22.3.

These conditions are the basis of a strategy for using Coulomb's law to solve electrostatic force problems.

PROBLEM-SOLVING STRATEGY 22.1 (MP)

Electrostatic forces and Coulomb's law

MODEL Identify point charges or model objects as point charges.

VISUALIZE Use a *pictorial representation* to establish a coordinate system, show the positions of the charges, show the force vectors on the charges, define distances and angles, and identify what the problem is trying to find. This is the process of translating words to symbols.

SOLVE The mathematical representation is based on Coulomb's law:

$$F_{1 \text{ on } 2} = F_{2 \text{ on } 1} = \frac{K|q_1||q_2|}{r^2}$$

- Show the directions of the forces—repulsive for like charges, attractive for opposite charges—on the pictorial representation.
- When possible, do graphical vector addition on the pictorial representation. While not exact, it tells you the type of answer you should expect.
- Write each force vector in terms of its x- and y-components, then add the components to find the net force. Use the pictorial representation to determine which components are positive and which are negative.

ASSESS Check that your result has correct units and significant figures, is reasonable, and answers the question.

Exercise 26

EXAMPLE 22.3 | Lifting a glass bead

A small plastic sphere charged to -10 nC is held 1.0 cm above a small glass bead at rest on a table. The bead has a mass of 15 mg and a charge of $+10$ nC. Will the glass bead "leap up" to the plastic sphere?

MODEL Model the plastic sphere and glass bead as point charges.

VISUALIZE FIGURE 22.16 establishes a y-axis, identifies the plastic sphere as q_1 and the glass bead as q_2, and shows a free-body diagram.

FIGURE 22.16 A pictorial representation of the charges and forces.

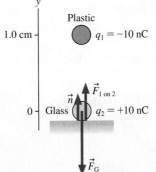

accelerate upward from the table. Using the values provided, we have

$$F_{1\,on\,2} = \frac{K|q_1||q_2|}{r^2}$$

$$= \frac{(9.0 \times 10^9 \, \text{N m}^2/\text{C}^2)(10 \times 10^{-9}\text{C})(10 \times 10^{-9}\text{C})}{(0.010 \, \text{m})^2}$$

$$= 9.0 \times 10^{-3} \, \text{N}$$

$$F_G = m_{\text{bead}}g = 1.5 \times 10^{-4} \, \text{N}$$

$F_{1\,on\,2}$ exceeds $m_{\text{bead}}g$ by a factor of 60, so the glass bead will leap upward.

SOLVE If $F_{1\,on\,2}$ is less than the gravitational force $F_G = m_{\text{bead}}g$, then the bead will remain at rest on the table with $\vec{F}_{1\,on\,2} + \vec{F}_G + \vec{n} = \vec{0}$. But if $F_{1\,on\,2}$ is greater than $m_{\text{bead}}g$, the glass bead will

ASSESS The values used in this example are realistic for spheres ≈ 2 mm in diameter. In general, as in this example, electric forces are *significantly* larger than gravitational forces. Consequently, we can neglect gravity when working electric-force problems unless the particles are fairly massive.

EXAMPLE 22.4 | The point of zero force

Two positively charged particles q_1 and $q_2 = 3q_1$ are 10.0 cm apart on the *x*-axis. Where (other than at infinity) could a third charge q_3 be placed so as to experience no net force?

MODEL Model the charged particles as point charges.

VISUALIZE FIGURE 22.17 establishes a coordinate system with q_1 at the origin and q_2 at $x = d$. We have no information about the sign of q_3, so apparently the position we're looking for will work for either sign. Suppose q_3 is off the *x*-axis, such as at point A. The two repulsive (or attractive) electric forces on q_3 cannot possibly add to zero, so q_3 must be somewhere on the *x*-axis. At point B, outside the two charges, the two forces on q_3 will always be in the same direction and, again, cannot add to zero. The only possible location is at point C, on the *x*-axis *between* the charges where the two forces are in opposite directions.

FIGURE 22.17 A pictorial representation of the charges and forces.

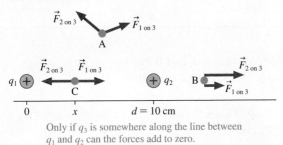

Only if q_3 is somewhere along the line between q_1 and q_2 can the forces add to zero.

SOLVE The mathematical problem is to find the position for which the forces $\vec{F}_{1\,on\,3}$ and $\vec{F}_{2\,on\,3}$ are equal in magnitude. If q_3 is distance x from q_1, it is distance $d - x$ from q_2. The *magnitudes* of the forces are

$$F_{1\,on\,3} = \frac{Kq_1|q_3|}{r_{13}^2} = \frac{Kq_1|q_3|}{x^2}$$

$$F_{2\,on\,3} = \frac{Kq_2|q_3|}{r_{23}^2} = \frac{K(3q_1)|q_3|}{(d-x)^2}$$

Charges q_1 and q_2 are positive and do not need absolute value signs. Equating the two forces gives

$$\frac{Kq_1|q_3|}{x^2} = \frac{3Kq_1|q_3|}{(d-x)^2}$$

The term $Kq_1|q_3|$ cancels. Multiplying by $x^2(d-x)^2$ gives

$$(d-x)^2 = 3x^2$$

which can be rearranged into the quadratic equation

$$2x^2 + 2dx - d^2 = 2x^2 + 20x - 100 = 0$$

where we used $d = 10$ cm and x is in cm. The solutions to this equation are

$$x = +3.66 \text{ cm and } -13.66 \text{ cm}$$

Both are points where the *magnitudes* of the two forces are equal, but $x = -13.66$ cm is a point where the magnitudes are equal while the directions are the same. The solution we want, which is between the charges, is $x = 3.66$ cm. Thus the point to place q_3 is 3.66 cm from q_1 along the line joining q_1 and q_2.

ASSESS q_1 is smaller than q_2, so we expect the point at which the forces balance to be closer to q_1 than to q_2. The solution seems reasonable. Note that the problem statement has no coordinates, so "$x = 3.66$ cm" is *not* an acceptable answer. You need to describe the position relative to q_1 and q_2.

EXAMPLE 22.5 | Three charges

Three charged particles with $q_1 = -50$ nC, $q_2 = +50$ nC, and $q_3 = +30$ nC are placed on the corners of the 5.0 cm $\times$ 10.0 cm rectangle shown in FIGURE 22.18. What is the net force on charge q_3 due to the other two charges? Give your answer both in component form and as a magnitude and direction.

MODEL Model the charged particles as point charges.

VISUALIZE The pictorial representation of FIGURE 22.19 establishes a coordinate system. q_1 and q_3 are opposite charges, so force vector $\vec{F}_{1 \text{ on } 3}$ is an attractive force toward q_1. q_2 and q_3 are like charges, so force vector $\vec{F}_{2 \text{ on } 3}$ is a repulsive force away from q_2. q_1 and q_2 have equal magnitudes, but $\vec{F}_{2 \text{ on } 3}$ has been drawn shorter than $\vec{F}_{1 \text{ on } 3}$ because q_2 is farther from q_3. Vector addition has been used to draw the net force vector $\vec{F}_3$ and to define its angle ϕ.

FIGURE 22.18 The three charges.

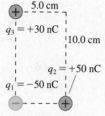

FIGURE 22.19 A pictorial representation of the charges and forces.

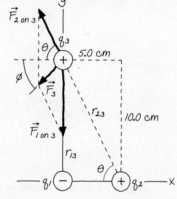

SOLVE The question asks for a *force*, so our answer will be the *vector* sum $\vec{F}_3 = \vec{F}_{1 \text{ on } 3} + \vec{F}_{2 \text{ on } 3}$. We need to write $\vec{F}_{1 \text{ on } 3}$ and $\vec{F}_{2 \text{ on } 3}$ in component form. The magnitude of force $\vec{F}_{1 \text{ on } 3}$ can be found using Coulomb's law:

$$F_{1 \text{ on } 3} = \frac{K|q_1||q_3|}{r_{13}^2}$$
$$= \frac{(9.0 \times 10^9 \text{ N m}^2/\text{C}^2)(50 \times 10^{-9} \text{ C})(30 \times 10^{-9} \text{ C})}{(0.100 \text{ m})^2}$$
$$= 1.35 \times 10^{-3} \text{ N}$$

where we used $r_{13} = 10.0$ cm.

The pictorial representation shows that $\vec{F}_{1 \text{ on } 3}$ points downward, in the negative y-direction, so

$$\vec{F}_{1 \text{ on } 3} = -1.35 \times 10^{-3} \hat{\jmath} \text{ N}$$

To calculate $\vec{F}_{2 \text{ on } 3}$ we first need the distance r_{23} between the charges:

$$r_{23} = \sqrt{(5.0 \text{ cm})^2 + (10.0 \text{ cm})^2} = 11.2 \text{ cm}$$

The magnitude of $\vec{F}_{2 \text{ on } 3}$ is thus

$$F_{2 \text{ on } 3} = \frac{K|q_2||q_3|}{r_{23}^2}$$
$$= \frac{(9.0 \times 10^9 \text{ N m}^2/\text{C}^2)(50 \times 10^{-9} \text{ C})(30 \times 10^{-9} \text{ C})}{(0.112 \text{ m})^2}$$
$$= 1.08 \times 10^{-3} \text{ N}$$

This is only a magnitude. The *vector* $\vec{F}_{2 \text{ on } 3}$ is

$$\vec{F}_{2 \text{ on } 3} = -F_{2 \text{ on } 3} \cos \theta \hat{\imath} + F_{2 \text{ on } 3} \sin \theta \hat{\jmath}$$

where angle θ is defined in the figure and the signs (negative x-component, positive y-component) were determined from the pictorial representation. From the geometry of the rectangle,

$$\theta = \tan^{-1}\left(\frac{10.0 \text{ cm}}{5.0 \text{ cm}}\right) = \tan^{-1}(2.0) = 63.4°$$

Thus $\vec{F}_{2 \text{ on } 3} = (-4.83\hat{\imath} + 9.66\hat{\jmath}) \times 10^{-4}$ N. Now we can add $\vec{F}_{1 \text{ on } 3}$ and $\vec{F}_{2 \text{ on } 3}$ to find

$$\vec{F}_3 = \vec{F}_{1 \text{ on } 3} + \vec{F}_{2 \text{ on } 3} = (-4.83\hat{\imath} - 3.84\hat{\jmath}) \times 10^{-4} \text{ N}$$

This would be an acceptable answer for many problems, but sometimes we need the net force as a magnitude and direction. With angle ϕ as defined in the figure, these are

$$F_3 = \sqrt{F_{3x}^2 + F_{3y}^2} = 6.2 \times 10^{-4} \text{ N}$$
$$\phi = \tan^{-1}\left|\frac{F_{3y}}{F_{3x}}\right| = 38°$$

Thus $\vec{F}_3 = (6.2 \times 10^{-4}$ N, $38°$ below the negative x-axis).

ASSESS The forces are not large, but they are typical of electrostatic forces. Even so, you'll soon see that these forces can produce very large accelerations because the masses of the charged objects are usually very small.

STOP TO THINK 22.4 Charged spheres A and B exert repulsive forces on each other. $q_A = 4q_B$. Which statement is true?

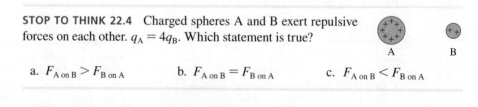

a. $F_{A \text{ on } B} > F_{B \text{ on } A}$ b. $F_{A \text{ on } B} = F_{B \text{ on } A}$ c. $F_{A \text{ on } B} < F_{B \text{ on } A}$

22.5 The Electric Field

Electric and magnetic forces, like gravity, are *long-range* forces; no contact is required for one charged particle to exert a force on another. But this raises some troubling issues. For example, consider the charged particles A and B in FIGURE 22.20. If A suddenly starts moving, as shown by the arrow, the force vector on B must pivot to follow A. Does this happen *instantly*? Or is there some *delay* between when A moves and when the force $\vec{F}_{A \text{ on B}}$ responds?

Neither Coulomb's law nor Newton's law of gravity is dependent on time, so the answer from the perspective of Newtonian physics has to be "instantly." Yet most scientists found this troubling. What if A is 100,000 light years from B? Will B respond *instantly* to an event 100,000 light years away? The idea of instantaneous transmission of forces had become unbelievable to most scientists by the beginning of the 19th century. But if there is a delay, how long is it? How does the information to "change force" get sent from A to B? These were the issues when a young Michael Faraday appeared on the scene.

Michael Faraday is one of the most interesting figures in the history of science. Because of the late age at which he started his education—he was a teenager—he never became fluent in mathematics. In place of equations, Faraday's brilliant and insightful mind developed many ingenious *pictorial* methods for thinking about and describing physical phenomena. By far the most important of these was the field.

The Concept of a Field

Faraday was particularly impressed with the pattern that iron filings make when sprinkled around a magnet, as seen in **FIGURE 22.21.** The pattern's regularity and the curved lines suggested to Faraday that the *space itself* around the magnet is filled with some kind of magnetic influence. Perhaps the magnet in some way alters the space around it. In this view, a piece of iron near the magnet responds not directly to the magnet but, instead, to the alteration of space caused by the magnet. This space alteration, whatever it is, is the *mechanism* by which the long-range force is exerted.

FIGURE 22.22 illustrates Faraday's idea. The Newtonian view was that A and B interact directly. In Faraday's view, A first alters or modifies the space around it, and particle B then comes along and interacts with this altered space. The alteration of space becomes the *agent* by which A and B interact. Furthermore, this alteration could easily be imagined to take a finite time to propagate outward from A, perhaps in a wave-like fashion. If A changes, B responds only when the new alteration of space reaches it. The interaction between B and this alteration of space is a *local* interaction, rather like a contact force.

Faraday's idea came to be called a **field.** The term "field," which comes from mathematics, describes a function that assigns a vector to every point in space. When used in physics, a field conveys the idea that the physical entity exists at every point in space. That is, indeed, what Faraday was suggesting about how long-range forces operate. The charge makes an alteration *everywhere* in space. Other charges then respond to the alteration at their position. The alteration of the space around a mass is called the *gravitational field.* Similarly, the space around a charge is altered to create the **electric field.**

> **NOTE** The concept of a field is in sharp contrast to the concept of a particle. A particle exists at *one* point in space. The purpose of Newton's laws of motion is to determine how the particle moves from point to point along a trajectory. A field exists simultaneously at *all* points in space.

Faraday's idea was not taken seriously at first; it seemed too vague and nonmathematical to scientists steeped in the Newtonian tradition of particles and forces. But the significance of the concept of field grew as electromagnetic theory developed during the first half of the 19th century. What seemed at first a pictorial "gimmick" came to be seen as more and more essential for understanding electric and magnetic forces.

The Field Model

The basic idea is that **the electric field is the agent that exerts an electric force on a charged particle.** Or, if you prefer, that charged particles interact via the electric field. We postulate:

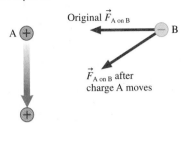

FIGURE 22.20 If charge A moves, how long does it take the force vector on B to respond?

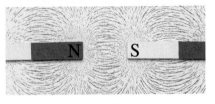

FIGURE 22.21 Iron filings sprinkled around the ends of a magnet suggest that the influence of the magnet extends into the space around it.

FIGURE 22.22 Newton's and Faraday's ideas about long-range forces.

In the Newtonian view, A exerts a force directly on B.

In Faraday's view, A alters the space around it. (The wavy lines are poetic license. We don't know what the alteration looks like.)

Particle B then responds to the altered space. The altered space is the agent that exerts the force on B.

1. Some set of charges, which we call the **source charges,** alters the space around them by creating an electric field $\vec{E}$ at all points in space.
2. A separate charge q *in* the electric field experiences force $\vec{F} = q\vec{E}$ exerted *by the field.* The force on a positive charge is in the direction of $\vec{E}$; the force on a negative charge is directed opposite to $\vec{E}$.

Thus the source charges exert an electric force on q through the electric field that they've created.

We can learn about the electric field by measuring the force on a *probe charge q.* If, as in **FIGURE 22.23a**, we place a probe charge at position (x, y, z) and measure force $\vec{F}_{\text{on }q}$, then the electric field at that point is

FIGURE 22.23 Charge q is a probe of the electric field.

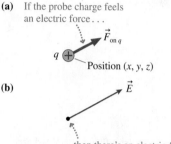

(a) If the probe charge feels an electric force...

$\vec{F}_{\text{on }q}$

q

Position (x, y, z)

(b)

$\vec{E}$

... then there's an electric field at this point in space.

$$\vec{E}(x, y, z) = \frac{\vec{F}_{\text{on }q} \text{ at } (x, y, z)}{q} \qquad (22.5)$$

We're *defining* the electric field as a force-to-charge ratio; hence the units of electric field are newtons per coulomb, or N/C. The magnitude E of the electric field is called the **electric field strength.**

It is important to recognize that probe charge q allows us to *observe* the field, but q is not responsible for creating the field. The field was already there, created by the source charges. **FIGURE 22.23b** shows the field at this point in space after probe charge q has been removed. You could imagine "mapping out" the field by moving charge q all through space.

Because q appears in Equation 22.5, you might think that the electric field depends on the size of the charge used to probe it. It doesn't. Coulomb's law tells us that $\vec{F}_{\text{on }q}$ is proportional to q, so the electric field defined in Equation 22.5 is *independent* of the charge that probes it. The electric field depends only on the source charges that create it.

We can summarize these important ideas with the **field model** of charge interactions:

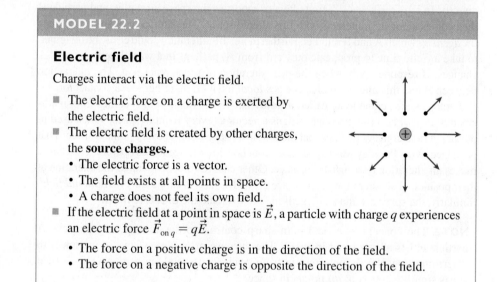

MODEL 22.2

Electric field

Charges interact via the electric field.

- The electric force on a charge is exerted by the electric field.
- The electric field is created by other charges, the **source charges.**
  - The electric force is a vector.
  - The field exists at all points in space.
  - A charge does not feel its own field.
- If the electric field at a point in space is $\vec{E}$, a particle with charge q experiences an electric force $\vec{F}_{\text{on }q} = q\vec{E}$.
  - The force on a positive charge is in the direction of the field.
  - The force on a negative charge is opposite the direction of the field.

EXAMPLE 22.6 | **Electric forces in a cell**

Every cell in your body is electrically active in various ways. For example, nerve propagation occurs when large electric fields in the cell membranes of neurons cause ions to move through the cell walls. The field strength in a typical cell membrane is 1.0×10^7 N/C. What is the magnitude of the electric force on a singly charged calcium ion?

MODEL The ion is a point charge in an electric field. A singly charged ion is missing one electron and has net charge $q = +e$.

SOLVE A charged particle in an electric field experiences an electric force $\vec{F}_{\text{on }q} = q\vec{E}$. In this case, the magnitude of the force is

$$F = eE = (1.6 \times 10^{-19} \text{ C})(1.0 \times 10^7 \text{ N/C}) = 1.6 \times 10^{-12} \text{ N}$$

ASSESS This may seem like an incredibly tiny force, but it is applied to a particle with mass $m \sim 10^{-26}$ kg. The ion would have an unimaginable acceleration ($F/m \sim 10^{14}$ m/s^2) were it not for resistive forces as it moves through the membrane. Even so, an ion can cross the cell wall in less than 1 μs.

FIGURE 22.26 Using the unit vector $\hat{r}$.

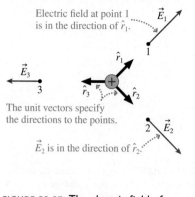

The unit vectors specify the directions to the points.

$\vec{E}_2$ is in the direction of $\hat{r}_2$.

FIGURE 22.27 The electric field of a negative point charge.

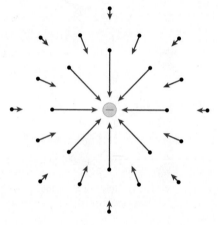

expressing certain directions—namely, the unit vectors $\hat{\imath}$, $\hat{\jmath}$, and $\hat{k}$. For example, unit vector $\hat{\imath}$ means "in the direction of the positive x-axis." With a minus sign, $-\hat{\imath}$ means "in the direction of the negative x-axis." Unit vectors, with a magnitude of 1 and no units, provide purely directional information.

With this in mind, let's define the unit vector $\hat{r}$ to be a vector of length 1 that points from the origin to a point of interest. Unit vector $\hat{r}$ provides no information at all about the distance to the point; it merely specifies the direction.

FIGURE 22.26 shows unit vectors $\hat{r}_1$, $\hat{r}_2$, and $\hat{r}_3$ pointing toward points 1, 2, and 3. Unlike $\hat{\imath}$ and $\hat{\jmath}$, unit vector $\hat{r}$ does not have a fixed direction. Instead, unit vector $\hat{r}$ specifies the direction "straight outward from this point." But that's just what we need to describe the electric field vector, which is shown at points 1, 2, and 3 due to a positive charge at the origin. No matter which point you choose, the electric field at that point is "straight outward" from the charge. In other words, the electric field $\vec{E}$ points in the direction of the unit vector $\hat{r}$.

With this notation, the electric field at distance r from a point charge q is

$$\vec{E} = \frac{1}{4\pi\epsilon_0} \frac{q}{r^2} \hat{r} \quad \text{(electric field of a point charge)} \quad (22.8)$$

where $\hat{r}$ is the unit vector from the charge toward the point at which we want to know the field. Equation 22.8 is identical to Equation 22.7, but written in a notation in which the unit vector $\hat{r}$ expresses the idea "away from q."

Equation 22.8 works equally well if q is negative. A negative sign in front of a vector simply reverses its direction, so the unit vector $-\hat{r}$ points *toward* charge q. **FIGURE 22.27** shows the electric field of a negative point charge. It looks like the electric field of a positive point charge except that the vectors point inward, toward the charge, instead of outward.

We'll end this chapter with three examples of the electric field of a point charge. Chapter 23 will expand these ideas to the electric fields of multiple charges and of extended objects.

EXAMPLE 22.7 | Calculating the electric field

A -1.0 nC charged particle is located at the origin. Points 1, 2, and 3 have (x, y) coordinates $(1\text{ cm}, 0\text{ cm})$, $(0\text{ cm}, 1\text{ cm})$, and $(1\text{ cm}, 1\text{ cm})$, respectively. Determine the electric field $\vec{E}$ at these points, then show the vectors on an electric field diagram.

MODEL The electric field is that of a negative point charge.

VISUALIZE The electric field points straight *toward* the origin. It will be weaker at $(1\text{ cm}, 1\text{ cm})$, which is farther from the charge.

SOLVE The electric field is

$$\vec{E} = \frac{1}{4\pi\epsilon_0} \frac{q}{r^2} \hat{r}$$

where $q = -1.0\text{ nC} = -1.0 \times 10^{-9}\text{ C}$. The distance r is $1.0\text{ cm} = 0.010\text{ m}$ for points 1 and 2 and $(\sqrt{2} \times 1.0\text{ cm}) = 0.0141\text{ m}$ for point 3. The *magnitude* of $\vec{E}$ at the three points is

$$E_1 = E_2 = \frac{1}{4\pi\epsilon_0} \frac{|q|}{r_1^2}$$

$$= \frac{(9.0 \times 10^9\text{ N}\,\text{m}^2/\text{C}^2)(1.0 \times 10^{-9}\text{ C})}{(0.010\text{ m})^2} = 90,000\text{ N/C}$$

$$E_3 = \frac{1}{4\pi\epsilon_0} \frac{|q|}{r_3^2}$$

$$= \frac{(9.0 \times 10^9\text{ N}\,\text{m}^2/\text{C}^2)(1.0 \times 10^{-9}\text{ C})}{(0.0141\text{ m})^2} = 45,000\text{ N/C}$$

Because q is negative, the field at each of these positions points directly at charge q. The electric field vectors, in component form, are

$$\vec{E}_1 = -90,000\hat{\imath}\text{ N/C}$$

$$\vec{E}_2 = -90,000\hat{\jmath}\text{ N/C}$$

$$\vec{E}_3 = -E_3\cos 45°\hat{\imath} - E_3\sin 45°\hat{\jmath} = (-31,800\hat{\imath} - 31,800\hat{\jmath})\text{ N/C}$$

These vectors are shown on the electric field diagram of **FIGURE 22.28**.

FIGURE 22.28 The electric field diagram of a -1.0 nC charged particle.

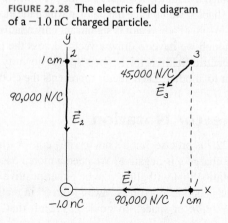

EXAMPLE 22.8 | The electric field of a proton

The electron in a hydrogen atom orbits the proton at a radius of 0.053 nm.

a. What is the proton's electric field strength at the position of the electron?

b. What is the magnitude of the electric force on the electron?

SOLVE a. The proton's charge is $q = e$. Its electric field strength at the distance of the electron is

$$E = \frac{1}{4\pi\epsilon_0} \frac{e}{r^2} = \frac{1}{4\pi\epsilon_0} \frac{1.6 \times 10^{-19}\,\text{C}}{(5.3 \times 10^{-11}\,\text{m})^2} = 5.1 \times 10^{11}\,\text{N/C}$$

Note how large this field is compared to the field of Example 22.7.

b. We could use Coulomb's law to find the force on the electron, but the whole point of knowing the electric field is that we can use it directly to find the force on a charge in the field. The magnitude of the force on the electron is

$$F_{\text{on elec}} = |q_e| E_{\text{of proton}}$$
$$= (1.60 \times 10^{-19}\,\text{C})(5.1 \times 10^{11}\,\text{N/C})$$
$$= 8.2 \times 10^{-8}\,\text{N}$$

STOP TO THINK 22.6 Rank in order, from largest to smallest, the electric field strengths E_a to E_d at points a to d.

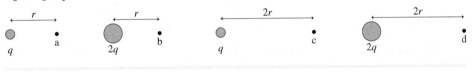

CHALLENGE EXAMPLE 22.9 | A charge in static equilibrium

A horizontal electric field causes the charged ball in FIGURE 22.29 to hang at a 15° angle, as shown. The spring is plastic, so it doesn't discharge the ball, and in its equilibrium position the spring extends only to the vertical dashed line. What is the electric field strength?

FIGURE 22.29 A charged ball hanging in static equilibrium.

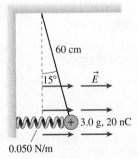

MODEL Model the ball as a point charge in static equilibrium. The electric force on the ball is $\vec{F}_E = q\vec{E}$. The charge is positive, so the force is in the same direction as the field.

VISUALIZE FIGURE 22.30 is a free-body diagram for the ball.

SOLVE The ball is in equilibrium, so the net force on the ball must be zero. With the field applied, the spring is stretched by $\Delta x = L \sin\theta = (0.60\,\text{m})(\sin 15°) = 0.16\,\text{m}$, where L is the string length, and exerts a pulling force $F_{\text{Sp}} = k\,\Delta x$ to the left. Newton's first law, with $\vec{a} = \vec{0}$ for equilibrium, is

$$\sum F_x = F_E - F_{\text{Sp}} - T\sin\theta = 0$$

$$\sum F_y = T\cos\theta - F_G = T\cos\theta - mg = 0$$

FIGURE 22.30 The free-body diagram.

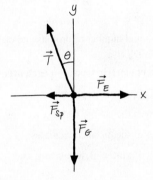

From the y-equation,

$$T = \frac{mg}{\cos\theta}$$

The x-equation is then

$$qE - k\,\Delta x - mg\tan\theta = 0$$

We can solve this for the electric field strength:

$$E = \frac{mg\tan\theta + k\,\Delta x}{q}$$

$$= \frac{(0.0030\,\text{kg})(9.8\,\text{m/s}^2)\tan 15° + (0.050\,\text{N/m})(0.16\,\text{m})}{20 \times 10^{-9}\,\text{C}}$$

$$= 7.9 \times 10^5\,\text{N/C}$$

ASSESS We don't yet have a way of judging whether this is a reasonable field strength, but we'll see in the next chapter that this is typical of the electric field strength near an object that has been charged by rubbing.

SUMMARY

The goal of Chapter 22 has been to learn that electric phenomena are based on charges, forces, and fields.

GENERAL PRINCIPLES

Coulomb's Law

The forces between two charged particles q_1 and q_2 separated by distance r are

$$F_{1 \text{ on } 2} = F_{2 \text{ on } 1} = \frac{K|q_1||q_2|}{r^2}$$

The forces are repulsive for two like charges, attractive for two opposite charges.
To solve electrostatic force problems:

MODEL Model objects as point charges.

VISUALIZE Draw a picture showing charges and force vectors.

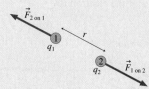

SOLVE Use Coulomb's law and the vector addition of forces.

ASSESS Is the result reasonable?

IMPORTANT CONCEPTS

The Charge Model

There are two kinds of charge, positive and negative.

- Fundamental charges are protons and electrons, with charge $\pm e$ where $e = 1.60 \times 10^{-19}$ C.

- Objects are charged by adding or removing electrons.

- The amount of charge is $q = (N_p - N_e)e$.

- An object with an equal number of protons and electrons is **neutral,** meaning no *net* charge.

Charged objects exert electric forces on each other.

- Like charges repel, opposite charges attract.

- The force increases as the charge increases.

- The force decreases as the distance increases.

There are two types of material, insulators and conductors.

- Charge remains fixed in or on an insulator.

- Charge moves easily through or along conductors.

- Charge is transferred by contact between objects.

Charged objects attract neutral objects.

- Charge polarizes metal by shifting the electron sea.

- Charge polarizes atoms, creating electric dipoles.

- The **polarization** force is always an attractive force.

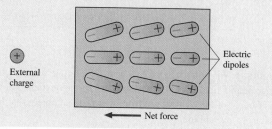

External charge

Electric dipoles

Net force

The Field Model

Charges interact with each other via the **electric field** $\vec{E}$.

- Charge A alters the space around it by creating an electric field.

- The field is the agent that exerts a force. The force on charge q_B is $\vec{F}_{\text{on B}} = q_B\vec{E}$.

An electric field is identified and measured in terms of the force on a **probe charge** q:

$$\vec{E} = \vec{F}_{\text{on } q}/q$$

- The electric field exists at all points in space.

- An electric field vector shows the field only at one point, the point at the tail of the vector.

The electric field of a **point charge** is

$$\vec{E} = \frac{1}{4\pi\epsilon_0}\frac{q}{r^2}\hat{r}$$

Unit vector $\hat{r}$ indicates "away from q."

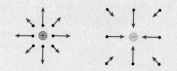

TERMS AND NOTATION

neutral	electron cloud	electrostatic equilibrium	coulomb, C
charging	fundamental unit of charge, e	grounded	permittivity constant, ϵ_0
charge model	charge quantization	charge polarization	field
charge, q or Q	ionization	polarization force	electric field, $\vec{E}$
like charges	law of conservation of charge	electric dipole	source charge
opposite charges	sea of electrons	charging by induction	electric field strength, E
discharging	ion core	Coulomb's law	field model
conductor	current	electrostatic constant, K	field diagram
insulator	charge carrier	point charge	

CONCEPTUAL QUESTIONS

1. Can an insulator be charged? If so, how would you charge an insulator? If not, why not?

2. Can a conductor be charged? If so, how would you charge a conductor? If not, why not?

3. Four lightweight balls A, B, C, and D are suspended by threads. Ball A has been touched by a plastic rod that was rubbed with wool. When the balls are brought close together, without touching, the following observations are made:
   - Balls B, C, and D are attracted to ball A.
   - Balls B and D have no effect on each other.
   - Ball B is attracted to ball C.

   What are the charge states (glass, plastic, or neutral) of balls A, B, C, and D? Explain.

4. Charged plastic and glass rods hang by threads.
   a. An object repels the plastic rod. Can you predict what it will do to the glass rod? If so, what? If not, why not?
   b. A different object attracts the plastic rod. Can you predict what it will do to the glass rod? If so, what? If not, why not?

5. A lightweight metal ball hangs by a thread. When a charged rod is held near, the ball moves toward the rod, touches the rod, then quickly "flies away" from the rod. Explain this behavior.

6. A plastic balloon that has been rubbed with wool will stick to a wall. Can you conclude that the wall is charged? If so, where does the charge come from? If not, why does the balloon stick?

7. Suppose there exists a third type of charge in addition to the two types we've called glass and plastic. Call this third type X charge. What experiment or series of experiments would you use to test whether an object has X charge? State clearly how each possible outcome of the experiments is to be interpreted.

8. The two oppositely charged metal spheres in FIGURE Q22.8 have equal quantities of charge. They are brought into contact with a neutral metal rod. What is the final charge state of each sphere and of the rod? Use both charge diagrams and words to explain.

FIGURE Q22.8   FIGURE Q22.9

9. Metal sphere A in FIGURE Q22.9 has 4 units of negative charge and metal sphere B has 2 units of positive charge. The two spheres are brought into contact. What is the final charge state of each sphere? Explain.

10. A negatively charged electroscope has separated leaves.
    a. Suppose you bring a negatively charged rod close to the top of the electroscope, but not touching. How will the leaves respond? Use both charge diagrams and words to explain.
    b. How will the leaves respond if you bring a positively charged rod close to the top of the electroscope, but not touching? Use both charge diagrams and words to explain.

11. Metal spheres A and B in FIGURE Q22.11 are initially neutral and are touching. A positively charged rod is brought near A, but not touching. Is A now positive, negative, or neutral? Use both charge diagrams and words to explain.

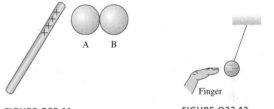

FIGURE Q22.11   FIGURE Q22.12

12. If you bring your finger near a lightweight, negatively charged hanging ball, the ball swings over toward your finger as shown in FIGURE Q22.12. Use charge diagrams and words to explain this observation.

13. Reproduce FIGURE Q22.13 on your paper. Then draw a dot (or dots) on the figure to show the position (or positions) where an electron would experience no net force.

FIGURE Q22.13

14. Charges A and B in FIGURE Q22.14 are equal. Each charge exerts a force on the other of magnitude F. Suppose the charge of B is increased by a factor of 4, but everything else is unchanged. In terms of F, (a) what is the magnitude of the force on A, and (b) what is the magnitude of the force on B?

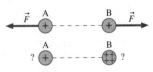

FIGURE Q22.14

15. The electric force on a charged particle in an electric field is F. What will be the force if the particle's charge is tripled and the electric field strength is halved?

EXERCISES AND PROBLEMS

Problems labeled ▨ integrate material from earlier chapters.

Exercises

Section 22.1 The Charge Model

Section 22.2 Charge

1. | A glass rod is charged to +8.0 nC by rubbing.
   a. Have electrons been removed from the rod or protons added?
   b. How many electrons have been removed or protons added?
2. | A plastic rod is charged to −12 nC by rubbing.
   a. Have electrons been added to the rod or protons removed?
   b. How many electrons have been added or protons removed?
3. | A plastic rod that has been charged to −15 nC touches a metal sphere. Afterward, the rod's charge is −10 nC.
   a. What kind of charged particle was transferred between the rod and the sphere, and in which direction? That is, did it move from the rod to the sphere or from the sphere to the rod?
   b. How many charged particles were transferred?
4. | A glass rod that has been charged to +12 nC touches a metal sphere. Afterward, the rod's charge is +8.0 nC.
   a. What kind of charged particle was transferred between the rod and the sphere, and in which direction? That is, did it move from the rod to the sphere or from the sphere to the rod?
   b. How many charged particles were transferred?
5. ‖ What is the total charge of all the electrons in 1.0 L of liquid water?
6. ‖ What mass of aluminum has a total nuclear charge of 1.0 C? Aluminum has atomic number 13.
7. | A chemical reaction takes place among 3 molecular ions that have each lost 2 electrons, 2 molecular ions that have each gained 3 electrons, and 1 molecular ion that has gained 2 electrons. The products of the reaction are two neutral molecules and multiple molecular ions that each have a charge of magnitude e. How many molecular ions are produced, and are they charged positively or negatively?
8. | A *linear accelerator* uses alternating electric fields to accelerate electrons to close to the speed of light. A small number of the electrons collide with a target, but a large majority pass through the target and impact a *beam dump* at the end of the accelerator. In one experiment the beam dump measured charge accumulating at a rate of −2.0 nC/s. How many electrons traveled down the accelerator during the 2.0 h run?

Section 22.3 Insulators and Conductors

9. | Figure 22.8 showed how an electroscope becomes negatively charged. The leaves will also repel each other if you touch the electroscope with a positively charged glass rod. Use a series of charge diagrams to explain what happens and why the leaves repel each other.
10. | Two neutral metal spheres on wood stands are touching. A negatively charged rod is held directly above the top of the left sphere, not quite touching it. While the rod is there, the right sphere is moved so that the spheres no longer touch. Then the rod is withdrawn. Afterward, what is the charge state of each sphere? Use charge diagrams to explain your answer.
11. ‖ You have two neutral metal spheres on wood stands. Devise a procedure for charging the spheres so that they will have like

charges of *exactly* equal magnitude. Use charge diagrams to explain your procedure.

12. ‖ You have two neutral metal spheres on wood stands. Devise a procedure for charging the spheres so that they will have opposite charges of *exactly* equal magnitude. Use charge diagrams to explain your procedure.

Section 22.4 Coulomb's Law

13. | Two 1.0 kg masses are 1.0 m apart (center to center) on a frictionless table. Each has +10 μC of charge.
    a. What is the magnitude of the electric force on one of the masses?
    b. What is the initial acceleration of this mass if it is released and allowed to move?
14. ‖ Two small plastic spheres each have a mass of 2.0 g and a charge of −50.0 nC. They are placed 2.0 cm apart (center to center).
    a. What is the magnitude of the electric force on each sphere?
    b. By what factor is the electric force on a sphere larger than its weight?
15. ‖ A small glass bead has been charged to +20 nC. A small metal ball bearing 1.0 cm above the bead feels a 0.018 N downward electric force. What is the charge on the ball bearing?
16. | Two protons are 2.0 fm apart.
    a. What is the magnitude of the electric force on one proton due to the other proton?
    b. What is the magnitude of the gravitational force on one proton due to the other proton?
    c. What is the ratio of the electric force to the gravitational force?
17. ‖ What is the net electric force on charge A in **FIGURE EX22.17**?

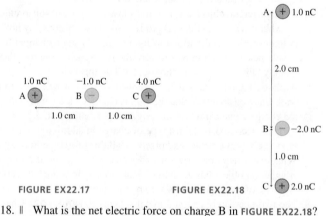

FIGURE EX22.17   FIGURE EX22.18

18. ‖ What is the net electric force on charge B in **FIGURE EX22.18**?
19. ‖ What is the force $\vec{F}$ on the 1.0 nC charge in **FIGURE EX22.19**? Give your answer as a magnitude and a direction.

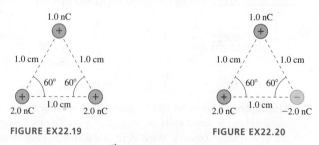

FIGURE EX22.19   FIGURE EX22.20

20. ‖ What is the force $\vec{F}$ on the 1.0 nC charge in **FIGURE EX22.20**? Give your answer as a magnitude and a direction.

21. | Object A, which has been charged to +4.0 nC, is at the origin. Object B, which has been charged to −8.0 nC, is at $(x, y) = (0.0 \text{ cm}, 2.0 \text{ cm})$. Determine the electric force on each object. Write each force vector in component form.

22. | A small plastic bead has been charged to −15 nC. What are the magnitude and direction of the acceleration of (a) a proton and (b) an electron that is 1.0 cm from the center of the bead?

23. | A 2.0 g plastic bead charged to −4.0 nC and a 4.0 g glass bead charged to +8.0 nC are 2.0 cm apart and free to move. What are the accelerations of (a) the plastic bead and (b) the glass bead?

24. ‖ Two positive point charges q and $4q$ are at $x = 0$ and $x = L$, respectively, and free to move. A third charge is placed so that the entire three-charge system is in static equilibrium. What are the magnitude, sign, and x-coordinate of the third charge?

25. ‖ A massless spring is attached to a support at one end and has a 2.0 μC charge glued to the other end. A −4.0 μC charge is slowly brought near. The spring has stretched 1.2 cm when the charges are 2.6 cm apart. What is the spring constant of the spring?

Section 22.5 The Electric Field

26. | What are the strength and direction of the electric field 1.0 mm from (a) a proton and (b) an electron?

27. ‖ The electric field at a point in space is $\vec{E} = (400\,\hat{i} + 100\,\hat{j})$ N/C.
    a. What is the electric force on a proton at this point? Give your answer in component form.
    b. What is the electric force on an electron at this point? Give your answer in component form.
    c. What is the magnitude of the proton's acceleration?
    d. What is the magnitude of the electron's acceleration?

28. | What are the strength and direction of the electric field 4.0 cm from a small plastic bead that has been charged to −8.0 nC?

29. ‖ What magnitude charge creates a 1.0 N/C electric field at a point 1.0 m away?

30. ‖ What are the strength and direction of an electric field that will balance the weight of a 1.0 g plastic sphere that has been charged to −3.0 nC?

31. ‖ The electric field 2.0 cm from a small object points away from the object with a strength of 270,000 N/C. What is the object's charge?

32. ‖ A +12 nC charge is located at the origin.
    a. What are the electric fields at the positions $(x, y) = (5.0 \text{ cm}, 0 \text{ cm})$, $(−5.0 \text{ cm}, 5.0 \text{ cm})$, and $(−5.0 \text{ cm}, −5.0 \text{ cm})$? Write each electric field vector in component form.
    b. Draw a field diagram showing the electric field vectors at these points.

33. ‖ A −12 nC charge is located at $(x, y) = (1.0 \text{ cm}, 0 \text{ cm})$. What are the electric fields at the positions $(x, y) = (5.0 \text{ cm}, 0 \text{ cm})$, $(−5.0 \text{ cm}, 0 \text{ cm})$, and $(0 \text{ cm}, 5.0 \text{ cm})$? Write each electric field vector in component form.

34. | A 0.10 g honeybee acquires a charge of +23 pC while flying.
BIO
    a. The earth's electric field near the surface is typically (100 N/C, downward). What is the ratio of the electric force on the bee to the bee's weight?
    b. What electric field (strength and direction) would allow the bee to hang suspended in the air?

Problems

35. ‖ Pennies today are copper-covered zinc, but older pennies are 3.1 g of solid copper. What are the total positive charge and total negative charge in a solid copper penny that is electrically neutral?

36. ‖ Two 1.0 g spheres are charged equally and placed 2.0 cm apart. When released, they begin to accelerate at 150 m/s². What is the magnitude of the charge on each sphere?

37. ‖ The nucleus of a ^{125}Xe atom (an isotope of the element xenon with mass 125 u) is 6.0 fm in diameter. It has 54 protons and charge $q = +54e$.
    a. What is the electric force on a proton 2.0 fm from the surface of the nucleus?
    b. What is the proton's acceleration?
    Hint: Treat the spherical nucleus as a point charge.

38. ‖ A *Van de Graaff generator* is a device that accumulates electrons on a large metal sphere until the large amount of charge causes sparks. As you'll learn in Chapter 23, the electric field of a charged sphere is exactly the same as if the charge were a point charge at the center of the sphere. Suppose that a 25-cm-diameter sphere has accumulated 1.0×10^{13} extra electrons and that a small ball 50 cm from the edge of the sphere feels the force $\vec{F} = (9.2 \times 10^{-4}$ N, away from sphere$)$. What is the charge on the ball?

39. ‖ A smart phone charger delivers charge to the phone, in the form of electrons, at a rate of −0.75 C/s. How many electrons are delivered to the phone during 30 min of charging?

40. ‖ Objects A and B are both positively charged. Both have a mass of 100 g, but A has twice the charge of B. When A and B are placed 10 cm apart, B experiences an electric force of 0.45 N.
    a. What is the charge on A?
    b. If the objects are released, what is the initial acceleration of A?

41. ‖ What is the force $\vec{F}$ on the −10 nC charge in **FIGURE P22.41**? Give your answer as a magnitude and an angle measured cw or ccw (specify which) from the +x-axis.

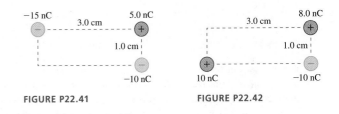

FIGURE P22.41   **FIGURE P22.42**

42. ‖ What is the force $\vec{F}$ on the −10 nC charge in **FIGURE P22.42**? Give your answer as a magnitude and an angle measured cw or ccw (specify which) from the +x-axis.

43. ‖ What is the force $\vec{F}$ on the 5.0 nC charge in **FIGURE P22.43**? Give your answer as a magnitude and an angle measured cw or ccw (specify which) from the +x-axis.

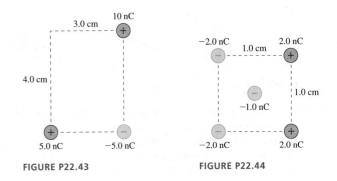

FIGURE P22.43   **FIGURE P22.44**

44. ‖ What is the force $\vec{F}$ on the −1.0 nC charge in the middle of **FIGURE P22.44** due to the four other charges? Give your answer in component form.

45. ‖ What is the force $\vec{F}$ on the 1.0 nC charge at the bottom in **FIGURE P22.45**? Give your answer in component form.

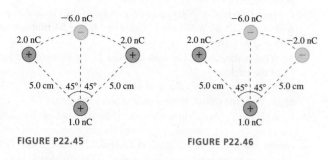

FIGURE P22.45   **FIGURE P22.46**

46. ‖ What is the force $\vec{F}$ on the 1.0 nC charge at the bottom in **FIGURE P22.46**? Give your answer in component form.

47. ‖‖ A +2.0 nC charge is at the origin and a −4.0 nC charge is at $x = 1.0$ cm.
   a. At what x-coordinate could you place a proton so that it would experience no net force?
   b. Would the net force be zero for an electron placed at the same position? Explain.

48. ‖ The net force on the 1.0 nC charge in **FIGURE P22.48** is zero. What is q?

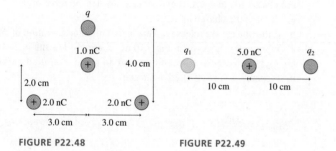

FIGURE P22.48   **FIGURE P22.49**

49. ‖ Charge q_2 in **FIGURE P22.49** is in equilibrium. What is q_1?

50. ‖ A positive point charge Q is located at $x = a$ and a negative point charge $-Q$ is at $x = -a$. A positive charge q can be placed anywhere on the y-axis. Find an expression for $(F_{net})_x$, the x-component of the net force on q.

51. ‖ **FIGURE P22.51** shows four charges at the corners of a square of side L. What is the magnitude of the net force on q?

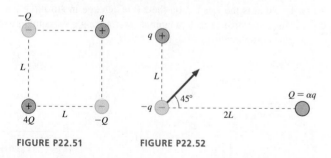

FIGURE P22.51   **FIGURE P22.52**

52. ‖ **FIGURE P22.52** shows three charges and the net force on charge $-q$. Charge Q is some multiple α of q. What is α?

53. ‖‖ Suppose the magnitude of the proton charge differs from the magnitude of the electron charge by a mere 1 part in 10^9.
   a. What would be the force between two 2.0-mm-diameter copper spheres 1.0 cm apart? Assume that each copper atom has an equal number of electrons and protons.
   b. Would this amount of force be detectable? What can you conclude from the fact that no such forces are observed?

54. ‖ In a simple model of the hydrogen atom, the electron moves in a circular orbit of radius 0.053 nm around a stationary proton. How many revolutions per second does the electron make?

55. ‖ You have two small, 2.0 g balls that have been given equal but opposite charges, but you don't know the magnitude of the charge. To find out, you place the balls distance d apart on a slippery horizontal surface, release them, and use a motion detector to measure the initial acceleration of one of the balls toward the other. After repeating this for several different separation distances, your data are as follows:

Distance (cm)	Acceleration (m/s²)
2.0	0.74
3.0	0.30
4.0	0.19
5.0	0.10

Use an appropriate graph of the data to determine the magnitude of the charge.

56. ‖ A 2.0 g metal cube and a 4.0 g metal cube are 6.0 cm apart, measured between their centers, on a horizontal surface. For both, the coefficient of static friction is 0.65. Both cubes, initially neutral, are charged at a rate of 7.0 nC/s. How long after charging begins does one cube begin to slide away? Which cube moves first?

57. ‖ Space explorers discover an 8.7×10^{17} kg asteroid that happens to have a positive charge of 4400 C. They would like to place their 3.3×10^5 kg spaceship in orbit around the asteroid. Interestingly, the solar wind has given their spaceship a charge of −1.2 C. What speed must their spaceship have to achieve a 7500-km-diameter circular orbit?

58. ‖ Two equal point charges 2.5 cm apart, both initially neutral, CALC are being charged at the rate of 5.0 nC/s. At what rate (N/s) is the force between them increasing 1.0 s after charging begins?

59. ‖ You have a lightweight spring whose unstretched length is 4.0 cm. First, you attach one end of the spring to the ceiling and hang a 1.0 g mass from it. This stretches the spring to a length of 5.0 cm. You then attach two small plastic beads to the opposite ends of the spring, lay the spring on a frictionless table, and give each plastic bead the same charge. This stretches the spring to a length of 4.5 cm. What is the magnitude of the charge (in nC) on each bead?

60. ‖ An electric dipole consists of two opposite charges $\pm q$ separated by a small distance s. The product $p = qs$ is called the *dipole moment*. **FIGURE P22.60** shows an electric dipole perpendicular to an electric field $\vec{E}$. Find an expression in terms of p and E for the magnitude of the torque that the electric field exerts on the dipole.

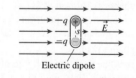

FIGURE P22.60   Electric dipole

61. ‖ You sometimes create a spark when you touch a doorknob after shuffling your feet on a carpet. Why? The air always has a few free electrons that have been kicked out of atoms by cosmic rays. If an electric field is present, a free electron is accelerated until it collides with an air molecule. Most such collisions are elastic, so the electron collides, accelerates, collides, accelerates, and so on, gradually gaining speed. But if the electron's kinetic energy just before a collision is 2.0×10^{-18} J or more, it has sufficient energy to kick an electron out of the molecule it hits. Where there was one free electron, now there are two! Each of these can then accelerate,

hit a molecule, and kick out another electron. Then there will be four free electrons. In other words, as **FIGURE P22.61** shows, a sufficiently strong electric field causes a "chain reaction" of electron production. This is called a *breakdown* of the air. The current of moving electrons is what gives you the shock, and a spark is generated when the electrons recombine with the positive ions and give off excess energy as a burst of light.

a. The average distance between ionizing collisions is 2.0 μm. (The electron's mean free path is less than this, but most collisions are elastic collisions in which the electron bounces with no loss of energy.) What acceleration must an electron have to gain 2.0×10^{-18} J of kinetic energy in this distance?

b. What force must act on an electron to give it the acceleration found in part a?

c. What strength electric field will exert this much force on an electron? This is the *breakdown field strength*. **Note:** The measured breakdown field strength is a little less than your calculated value because our model of the process is a bit too simple. Even so, your calculated value is close.

d. Suppose a free electron in air is 1.0 cm away from a point charge. What minimum charge is needed to cause a breakdown and create a spark as the electron moves toward the point charge?

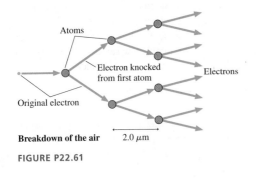

Atoms

Electron knocked from first atom

Electrons

Original electron

Breakdown of the air   2.0 μm

FIGURE P22.61

62. ‖ Two 5.0 g point charges on 1.0-m-long threads repel each other after being charged to +100 nC, as shown in **FIGURE P22.62.** What is the angle θ? You can assume that θ is a small angle.

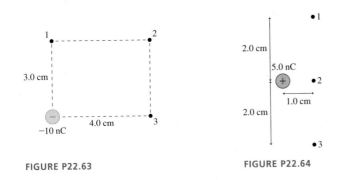

1.0 m θ θ 1.0 m

5.0 g (+)   (+) 5.0 g

FIGURE P22.62   100 nC   100 nC

63. ‖ What are the electric fields at points 1, 2, and 3 in **FIGURE P22.63**? Give your answer in component form.

1   2

3.0 cm

-10 nC   4.0 cm   3

FIGURE P22.63

64. ‖ What are the electric fields at points 1, 2, and 3 in **FIGURE P22.64**? Give your answer in component form.

• 1

2.0 cm

5.0 nC
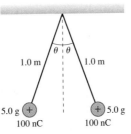
(+)   • 2

1.0 cm

2.0 cm

• 3

FIGURE P22.64

65. ‖ A 10.0 nC charge is located at position $(x, y) = (1.0$ cm, 2.0 cm). At what (x, y) position(s) is the electric field

a. $-225,000\,\hat{\imath}$ N/C?

b. $(161,000\,\hat{\imath} + 80,500\,\hat{\jmath})$ N/C?

c. $(21,600\,\hat{\imath} - 28,800\,\hat{\jmath})$ N/C?

66. ‖ Three 1.0 nC charges are placed as shown in **FIGURE P22.66.** Each of these charges creates an electric field $\vec{E}$ at a point 3.0 cm in front of the middle charge.

$q_1 = 1.0$ nC (+)

1.0 cm

$q_2 = 1.0$ nC (+) - - - - - - 3.0 cm - - - - - - •

1.0 cm

$q_3 = 1.0$ nC (+)

FIGURE P22.66

a. What are the three fields $\vec{E}_1$, $\vec{E}_2$, and $\vec{E}_3$ created by the three charges? Write your answer for each as a vector in component form.

b. Do you think that electric fields obey a principle of superposition? That is, is there a "net field" at this point given by $\vec{E}_{net} = \vec{E}_1 + \vec{E}_2 + \vec{E}_3$? Use what you learned in this chapter and previously in our study of forces to argue why this is or is not true.

c. If it is true, what is $\vec{E}_{net}$?

67. ‖ An electric field $\vec{E} = 100,000\,\hat{\imath}$ N/C causes the 5.0 g point charge in **FIGURE P22.67** to hang at a 20° angle. What is the charge on the ball?

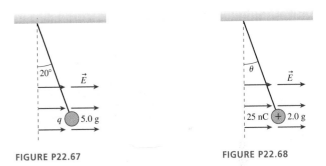

20° $\vec{E}$

q (●) 5.0 g

FIGURE P22.67

θ $\vec{E}$

25 nC (+) 2.0 g

FIGURE P22.68

68. ‖ An electric field $\vec{E} = 200,000\,\hat{\imath}$ N/C causes the point charge in **FIGURE P25.68** to hang at an angle. What is θ?

In Problems 69 through 72 you are given the equation(s) used to solve a problem. For each of these,

a. Write a realistic problem for which this is the correct equation(s).

b. Finish the solution of the problem.

69. $\dfrac{(9.0 \times 10^9 \,\text{N m}^2/\text{C}^2) \times N \times (1.60 \times 10^{-19}\,\text{C})}{(1.0 \times 10^{-6}\,\text{m})^2}$

$= 1.5 \times 10^6$ N/C

70. $\dfrac{(9.0 \times 10^9\,\text{N m}^2/\text{C}^2)q^2}{(0.0150\,\text{m})^2} = 0.020$ N

71. $\dfrac{(9.0 \times 10^9\,\text{N m}^2/\text{C}^2)(15 \times 10^{-9}\,\text{C})}{r^2} = 54,000$ N/C

72. $\sum F_x = 2 \times \dfrac{(9.0 \times 10^9\,\text{N m}^2/\text{C}^2)(1.0 \times 10^{-9}\,\text{C})q}{((0.020\,\text{m})/\sin 30°)^2} \times \cos 30°$

$= 5.0 \times 10^{-5}$ N

$\sum F_y = 0$ N

Challenge Problems

73. ‖ Two 3.0 g point charges on 1.0-m-long threads repel each other after being equally charged, as shown in FIGURE CP22.73. What is the magnitude of the charge q?

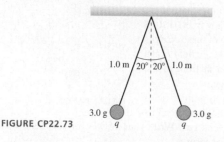

FIGURE CP22.73

74. ‖ Three 3.0 g balls are tied to 80-cm-long threads and hung from a *single* fixed point. Each of the balls is given the same charge q. At equilibrium, the three balls form an equilateral triangle in a horizontal plane with 20 cm sides. What is q?

75. ‖ The identical small spheres shown in FIGURE CP22.75 are charged to $+100$ nC and -100 nC. They hang as shown in a 100,000 N/C electric field. What is the mass of each sphere?

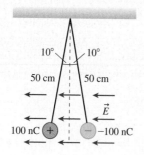

FIGURE CP22.75

76. ‖ The force on the -1.0 nC charge is as shown in FIGURE CP22.76. What is the magnitude of this force?

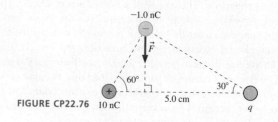

FIGURE CP22.76   10 nC

77. ‖ In Section 22.3 we claimed that a charged object exerts a net attractive force on an electric dipole. Let's investigate this. FIGURE CP22.77 shows a *permanent* electric dipole consisting of charges $+q$ and $-q$ separated by the fixed distance s. Charge $+Q$ is distance r from the center of the dipole. We'll assume, as is usually the case in practice, that $s \ll r$.

   a. Write an expression for the net force exerted on the dipole by charge $+Q$.
   b. Is this force toward $+Q$ or away from $+Q$? Explain.
   c. Use the *binomial approximation* $(1 + x)^{-n} \approx 1 - nx$ if $x \ll 1$ to show that your expression from part a can be written $F_{net} = 2KqQs/r^3$.
   d. How can an electric force have an inverse-cube dependence? Doesn't Coulomb's law say that the electric force depends on the inverse square of the distance? Explain.

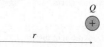

FIGURE CP22.77

STOP TO THINK 22.5 An electron is placed at the position marked by the dot. The force on the electron is

a. Zero.
b. To the right.
c. To the left.
d. There's not enough information to tell.

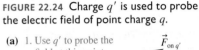

The Electric Field of a Point Charge

We will begin to put the definition of the electric field to full use in the next chapter. For now, to develop the ideas, we will determine the electric field of a single point charge q. FIGURE 22.24a shows charge q and a point in space at which we would like to know the electric field. To do so, we use a second charge q' as a probe of the electric field.

For the moment, assume both charges are positive. The force on q', which is repulsive and directed straight away from q, is given by Coulomb's law:

$$\vec{F}_{\text{on } q'} = \left(\frac{1}{4\pi\epsilon_0} \frac{qq'}{r^2}, \text{ away from } q \right) \tag{22.6}$$

It's customary to use $1/4\pi\epsilon_0$ rather than K for field calculations. Equation 22.5 defined the electric field in terms of the force on a probe charge; thus the electric field at this point is

$$\vec{E} = \frac{\vec{F}_{\text{on } q'}}{q'} = \left(\frac{1}{4\pi\epsilon_0} \frac{q}{r^2}, \text{ away from } q \right) \tag{22.7}$$

The electric field is shown in FIGURE 22.24b.

NOTE The expression for the electric field is similar to Coulomb's law. To distinguish the two, remember that Coulomb's law has a product of two charges in the numerator. It describes the force between *two* charges. The electric field has a single charge in the numerator. It is the field of *a* charge.

If we calculate the field at a sufficient number of points in space, we can draw a **field diagram** such as the one shown in FIGURE 22.25. Notice that the field vectors all point straight away from charge q. Also notice how quickly the arrows decrease in length due to the inverse-square dependence on r.

Keep these three important points in mind when using field diagrams:

1. The diagram is just a representative sample of electric field vectors. The field exists at all the other points. A well-drawn diagram can tell you fairly well what the field would be like at a neighboring point.
2. The arrow indicates the direction and the strength of the electric field *at the point to which it is attached*—that is, at the point where the *tail* of the vector is placed. In this chapter, we indicate the point at which the electric field is measured with a dot. The length of any vector is significant only relative to the lengths of other vectors.
3. Although we have to draw a vector across the page, from one point to another, an electric field vector is *not* a spatial quantity. It does not "stretch" from one point to another. Each vector represents the electric field at *one point* in space.

Unit Vector Notation

Equation 22.7 is precise, but it's not terribly convenient. Furthermore, what happens if the source charge q is negative? We need a more concise notation to write the electric field, a notation that will allow q to be either positive or negative.

The basic need is to express "away from q" in mathematical notation. "Away from q" is a *direction* in space. To guide us, recall that we already have a notation for

FIGURE 22.24 Charge q' is used to probe the electric field of point charge q.

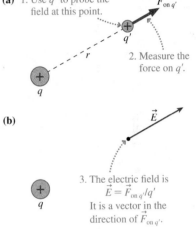

(a) 1. Use q' to probe the field at this point.

2. Measure the force on q'.

(b)

3. The electric field is $\vec{E} = \vec{F}_{\text{on } q'}/q'$
It is a vector in the direction of $\vec{F}_{\text{on } q'}$.

FIGURE 22.25 The electric field of a positive point charge.

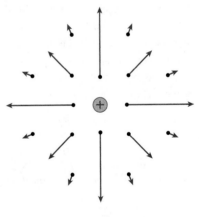

23 The Electric Field

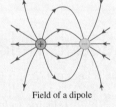

In a plasma ball, electrons follow the electric field lines outward from the center electrode. The streamers appear where gas atoms emit light after the high-speed electrons collide with them.

IN THIS CHAPTER, you will learn how to calculate and use the electric field.

Where do electric fields come from?

Electric fields are created by charges.

- **Electric fields add.** The field due to several point charges is the sum of the fields due to each charge.
- **Electric fields are *vectors*.** Summing electric fields is vector addition.
- Two equal but opposite charges form an electric dipole.
- Electric fields can be represented by electric field vectors or electric field lines.

Field of a dipole

« LOOKING BACK Section 22.5 The electric field of a point charge

What if the charge is continuous?

For macroscopic charged objects, like rods or disks, we can think of the charge as having a continuous distribution.

- A charged object is characterized by its charge density—the charge per length, area, or volume.
- We'll divide objects into small point charge-like pieces ΔQ.
- The summation of their electric fields will become an integral.
- We'll calculate the electric fields of charged rods, loops, disks, and planes.

Total charge Q

ΔQ

L

Linear charge density $\lambda = Q/L$

What fields are especially important?

We will develop and use four important electric field models.

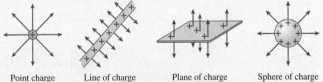

Point charge   Line of charge   Plane of charge   Sphere of charge

What is a parallel-plate capacitor?

Two parallel conducting plates with equal but opposite charges form a parallel-plate capacitor. You'll learn that the electric field between the plates is a uniform electric field, the same at every point. Capacitors are also important elements of circuits, as you'll see in Chapter 26.

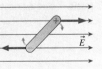

How do charges respond to fields?

Electric fields exert forces on charges.

- **Charged particles *accelerate*.** Acceleration depends on the charge-to-mass ratio.
- A charged particle in a uniform field follows a parabolic trajectory.
- A dipole in an electric field feels a torque that aligns the dipole with the field.

$\vec{E}$

« LOOKING BACK Section 4.2 Projectiles

23.1 Electric Field Models

Chapter 22 introduced the key idea that **charged particles interact via the electric field.** In this chapter you will learn how to calculate the electric field of a set of charges. We will start with discrete point charges, then move on to calculating the electric field of a continuous distribution of charge. The latter will let you practice the mathematical skills you've been learning in calculus. Only at the end of the chapter will we look at what happens to charges that find themselves *in* an electric field.

The electric fields used in science and engineering are often caused by fairly complex distributions of charge. Sometimes these fields require exact calculations, but much of the time we can understand the physics by using simplified models of the electric field. A *model*, you'll recall, is a highly simplified picture of reality, one that captures the essence of what we want to study without unnecessary complications.

Four common electric field models are the basis for understanding a wide variety of electric phenomena. We present them here together as a reference; the first half of this chapter will then be devoted to justifying and explaining these results.

MODEL 23.1

Four key electric fields

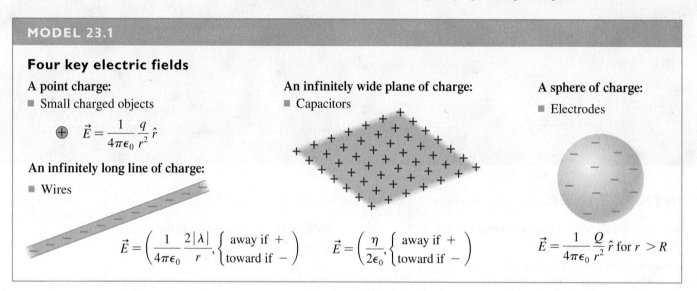

A point charge:
- Small charged objects

$$\vec{E} = \frac{1}{4\pi\epsilon_0}\frac{q}{r^2}\hat{r}$$

An infinitely long line of charge:
- Wires

$$\vec{E} = \left(\frac{1}{4\pi\epsilon_0}\frac{2|\lambda|}{r}, \left\{ \begin{array}{l} \text{away if } + \\ \text{toward if } - \end{array} \right. \right)$$

An infinitely wide plane of charge:
- Capacitors

$$\vec{E} = \left(\frac{\eta}{2\epsilon_0}, \left\{ \begin{array}{l} \text{away if } + \\ \text{toward if } - \end{array} \right. \right)$$

A sphere of charge:
- Electrodes

$$\vec{E} = \frac{1}{4\pi\epsilon_0}\frac{Q}{r^2}\hat{r} \text{ for } r > R$$

23.2 The Electric Field of Point Charges

Our starting point, from « Section 22.5, is the electric field of a point charge q:

$$\vec{E} = \frac{1}{4\pi\epsilon_0}\frac{q}{r^2}\hat{r} \qquad \text{(electric field of a point charge)} \qquad (23.1)$$

where $\hat{r}$ is a unit vector pointing away from q and $\epsilon_0 = 8.85 \times 10^{-12}$ C^2/Nm^2 is the permittivity constant. **FIGURE 23.1** reminds you of the electric fields of point charges. Although we have to give each vector we draw a length, keep in mind that each arrow represents the electric field *at a point*. The electric field is not a spatial quantity that "stretches" from one end of the arrow to the other.

▶ **FIGURE 23.1** The electric field of a positive and a negative point charge.

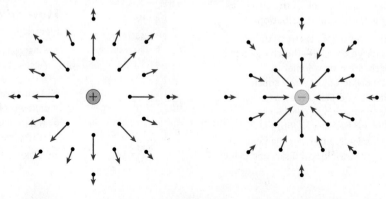

Multiple Point Charges

What happens if there is more than one charge? The electric field was defined as $\vec{E} = \vec{F}_{\text{on } q}/q$, where $\vec{F}_{\text{on } q}$ is the electric force on charge q. Forces add as vectors, so the net force on q due to a group of point charges is the vector sum

$$\vec{F}_{\text{on } q} = \vec{F}_{1 \text{ on } q} + \vec{F}_{2 \text{ on } q} + \cdots$$

Consequently, the net electric field due to a group of point charges is

$$\vec{E}_{\text{net}} = \frac{\vec{F}_{\text{on } q}}{q} = \frac{\vec{F}_{1 \text{ on } q}}{q} + \frac{\vec{F}_{2 \text{ on } q}}{q} + \cdots = \vec{E}_1 + \vec{E}_2 + \cdots = \sum_i \vec{E}_i \qquad (23.2)$$

where $\vec{E}_i$ is the field from point charge i. That is, **the net electric field is the *vector sum* of the electric fields due to each charge.** In other words, electric fields obey the *principle of superposition*.

Thus vector addition is the key to electric field calculations.

PROBLEM-SOLVING STRATEGY 23.1   (MP)

The electric field of multiple point charges

MODEL Model charged objects as point charges.

VISUALIZE For the pictorial representation:

- Establish a coordinate system and show the locations of the charges.
- Identify the point P at which you want to calculate the electric field.
- Draw the electric field of each charge at P.
- Use symmetry to determine if any components of $\vec{E}_{\text{net}}$ are zero.

SOLVE The mathematical representation is $\vec{E}_{\text{net}} = \sum \vec{E}_i$.

- For each charge, determine its distance from P and the angle of $\vec{E}_i$ from the axes.
- Calculate the field strength of each charge's electric field.
- Write each vector $\vec{E}_i$ in component form.
- Sum the vector components to determine $\vec{E}_{\text{net}}$.

ASSESS Check that your result has correct units and significant figures, is reasonable (see **TABLE 23.1**), and agrees with any known limiting cases.

TABLE 23.1 Typical electric field strengths

Field location	Field strength (N/C)
Inside a current-carrying wire	10^{-3}–10^{-1}
Near the earth's surface	10^2–10^4
Near objects charged by rubbing	10^3–10^6
Electric breakdown in air, causing a spark	3×10^6
Inside an atom	10^{11}

EXAMPLE 23.1 | The electric field of three equal point charges

Three equal point charges q are located on the y-axis at $y = 0$ and at $y = \pm d$. What is the electric field at a point on the x-axis?

MODEL This problem is a step along the way to understanding the electric field of a charged wire. We'll assume that q is positive when drawing pictures, but the solution should allow for the possibility that q is negative. The question does not ask about any specific point, so we will be looking for a symbolic expression in terms of the unspecified position x.

VISUALIZE **FIGURE 23.2** shows the charges, the coordinate system, and the three electric field vectors $\vec{E}_1$, $\vec{E}_2$, and $\vec{E}_3$. Each of these fields points *away from* its source charge because of the assumption that q is positive. We need to find the vector sum $\vec{E}_{\text{net}} = \vec{E}_1 + \vec{E}_2 + \vec{E}_3$.

Before rushing into a calculation, we can make our task *much* easier by first thinking qualitatively about the situation. For example, the fields $\vec{E}_1$, $\vec{E}_2$, and $\vec{E}_3$ all lie in the xy-plane, hence we can conclude without any calculations that $(E_{\text{net}})_z = 0$. Next, look

FIGURE 23.2 Calculating the electric field of three equal point charges.

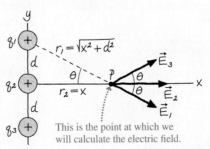

This is the point at which we will calculate the electric field.

at the y-components of the fields. The fields $\vec{E}_1$ and $\vec{E}_3$ have equal magnitudes and are tilted away from the x-axis by the same angle θ. Consequently, the y-components of $\vec{E}_1$ and $\vec{E}_3$ will *cancel* when added. $\vec{E}_2$ has no y-component, so we can conclude that $(E_{\text{net}})_y = 0$. The only component we need to calculate is $(E_{\text{net}})_x$.

Continued

SOLVE We're ready to calculate. The x-component of the field is

$$(E_{net})_x = (E_1)_x + (E_2)_x + (E_3)_x = 2(E_1)_x + (E_2)_x$$

where we used the fact that fields $\vec{E}_1$ and $\vec{E}_3$ have *equal* x-components. Vector $\vec{E}_2$ has *only* the x-component

$$(E_2)_x = E_2 = \frac{1}{4\pi\epsilon_0}\frac{q_2}{r_2^2} = \frac{1}{4\pi\epsilon_0}\frac{q}{x^2}$$

where $r_2 = x$ is the distance from q_2 to the point at which we are calculating the field. Vector $\vec{E}_1$ is at angle θ from the x-axis, so its x-component is

$$(E_1)_x = E_1\cos\theta = \frac{1}{4\pi\epsilon_0}\frac{q_1}{r_1^2}\cos\theta$$

where r_1 is the distance from q_1. This expression for $(E_1)_x$ is correct, but it is not yet sufficient. Both the distance r_1 and the angle θ vary with the position x and need to be expressed as functions of x. From the Pythagorean theorem, $r_1 = (x^2 + d^2)^{1/2}$. Thus

$$\cos\theta = \frac{x}{r_1} = \frac{x}{(x^2 + d^2)^{1/2}}$$

By combining these pieces, we see that $(E_1)_x$ is

$$(E_1)_x = \frac{1}{4\pi\epsilon_0}\frac{q}{x^2 + d^2}\frac{x}{(x^2 + d^2)^{1/2}} = \frac{1}{4\pi\epsilon_0}\frac{xq}{(x^2 + d^2)^{3/2}}$$

This expression is a bit complex, but notice that the dimensions of $x/(x^2 + d^2)^{3/2}$ are $1/\text{m}^2$, as they *must* be for the field of a point charge. Checking dimensions is a good way to verify that you haven't made algebra errors.

We can now combine $(E_1)_x$ and $(E_2)_x$ to write

$$(E_{net})_x = 2(E_1)_x + (E_2)_x = \frac{q}{4\pi\epsilon_0}\left[\frac{1}{x^2} + \frac{2x}{(x^2 + d^2)^{3/2}}\right]$$

The other two components of $\vec{E}_{net}$ are zero, hence the electric field of the three charges at a point on the x-axis is

$$\vec{E}_{net} = \frac{q}{4\pi\epsilon_0}\left[\frac{1}{x^2} + \frac{2x}{(x^2 + d^2)^{3/2}}\right]\hat{i}$$

ASSESS This is the electric field only at points *on the x-axis*. Furthermore, this expression is valid only for $x > 0$. The electric field to the left of the charges points in the opposite direction, but our expression doesn't change sign for negative x. (This is a consequence of how we wrote $(E_2)_x$.) We would need to modify this expression to use it for negative values of x. The good news, though, is that our expression *is* valid for both positive and negative q. A negative value of q makes $(E_{net})_x$ negative, which would be an electric field pointing to the left, toward the negative charges.

Limiting Cases

There are two cases for which we know what the result should be. First, let x become really small. As the point in Figure 23.2 approaches the origin, the fields $\vec{E}_1$ and $\vec{E}_3$ become opposite to each other and cancel. Thus as $x \to 0$, the field *should* be that of the single point charge q at the origin, a field we already know. Is it? Notice that

$$\lim_{x \to 0}\frac{2x}{(x^2 + d^2)^{3/2}} = 0 \tag{23.3}$$

Thus $E_{net} \to q/4\pi\epsilon_0 x^2$ as $x \to 0$, the expected field of a single point charge.

Now consider the opposite situation, where x becomes extremely large. From very far away, the three source charges will seem to merge into a single charge of size $3q$, just as three very distant lightbulbs appear to be a single light. Thus the field for $x \gg d$ *should* be that of a point charge $3q$. Is it?

The field is zero in the limit $x \to \infty$. That doesn't tell us much, so we don't want to go *that* far away. We simply want x to be very large in comparison to the spacing d between the source charges. If $x \gg d$, then the denominator of the second term of $\vec{E}_{net}$ is well approximated by $(x^2 + d^2)^{3/2} \approx (x^2)^{3/2} = x^3$. Thus

$$\lim_{x \gg d}\left[\frac{1}{x^2} + \frac{2x}{(x^2 + d^2)^{3/2}}\right] = \frac{1}{x^2} + \frac{2x}{x^3} = \frac{3}{x^2} \tag{23.4}$$

Consequently, the net electric field far from the source charges is

$$\vec{E}_{net}(x \gg d) = \frac{1}{4\pi\epsilon_0}\frac{(3q)}{x^2}\hat{i} \tag{23.5}$$

As expected, this is the field of a point charge $3q$. These checks of limiting cases provide confidence in the result of the calculation.

FIGURE 23.3 is a graph of the field strength E_{net} for the three charges of Example 23.1. Although we don't have any numerical values, we can specify x as a multiple of the charge separation d. Notice how the graph matches the field of a single point charge when $x \ll d$ and matches the field of a point charge $3q$ when $x \gg d$.

FIGURE 23.3 The electric field strength along a line perpendicular to three equal point charges.

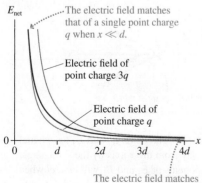

The electric field matches that of a single point charge q when $x \ll d$.

Electric field of point charge $3q$

Electric field of point charge q

The electric field matches that of point charge $3q$ when $x \gg d$.

The Electric Field of a Dipole

Two equal but opposite charges separated by a small distance form an *electric dipole*. **FIGURE 23.4** shows two examples. In a *permanent electric dipole,* such as the water molecule, the oppositely charged particles maintain a small permanent separation. We can also create an electric dipole, as you learned in Chapter 22, by polarizing a neutral atom with an external electric field. This is an *induced electric dipole.*

FIGURE 23.5 shows that we can represent an electric dipole, whether permanent or induced, by two opposite charges $\pm q$ separated by the small distance s. The dipole has zero net charge, but it *does* have an electric field. Consider a point on the positive y-axis. This point is slightly closer to $+q$ than to $-q$, so the fields of the two charges do not cancel. You can see in the figure that $\vec{E}_{\text{dipole}}$ points in the positive y-direction. Similarly, vector addition shows that $\vec{E}_{\text{dipole}}$ points in the negative y-direction at points along the x-axis.

Let's calculate the electric field of a dipole at a point on the axis of the dipole. This is the y-axis in Figure 23.5. The point is distance $r_+ = y - s/2$ from the positive charge and $r_- = y + s/2$ from the negative charge. The net electric field at this point has only a y-component, and the sum of the fields of the two point charges gives

$$(E_{\text{dipole}})_y = (E_+)_y + (E_-)_y = \frac{1}{4\pi\epsilon_0}\frac{q}{(y - \frac{1}{2}s)^2} + \frac{1}{4\pi\epsilon_0}\frac{(-q)}{(y + \frac{1}{2}s)^2}$$

$$= \frac{q}{4\pi\epsilon_0}\left[\frac{1}{(y - \frac{1}{2}s)^2} - \frac{1}{(y + \frac{1}{2}s)^2}\right] \tag{23.6}$$

Combining the two terms over a common denominator, we find

$$(E_{\text{dipole}})_y = \frac{q}{4\pi\epsilon_0}\left[\frac{2ys}{(y - \frac{1}{2}s)^2(y + \frac{1}{2}s)^2}\right] \tag{23.7}$$

We omitted some of the algebraic steps, but be sure you can do this yourself. Some of the homework problems will require similar algebra.

In practice, we almost always observe the electric field of a dipole at distances $y \gg s$—that is, for distances much larger than the charge separation. In such cases, the denominator can be approximated $(y - \frac{1}{2}s)^2(y + \frac{1}{2}s)^2 \approx y^4$. With this approximation, Equation 23.7 becomes

$$(E_{\text{dipole}})_y \approx \frac{1}{4\pi\epsilon_0}\frac{2qs}{y^3} \tag{23.8}$$

It is useful to define the **dipole moment** $\vec{p}$, shown in **FIGURE 23.6**, as the vector

$$\vec{p} = (qs, \text{ from the negative to the positive charge}) \tag{23.9}$$

The direction of $\vec{p}$ identifies the orientation of the dipole, and the dipole-moment magnitude $p = qs$ determines the electric field strength. The SI units of the dipole moment are $C\,m$.

We can use the dipole moment to write a succinct expression for the electric field at a point on the axis of a dipole:

$$\vec{E}_{\text{dipole}} \approx \frac{1}{4\pi\epsilon_0}\frac{2\vec{p}}{r^3} \qquad \text{(on the axis of an electric dipole)} \tag{23.10}$$

where r is the distance measured from the *center* of the dipole. We've switched from y to r because we've now specified that Equation 23.10 is valid only along the axis of the dipole. Notice that the electric field along the axis points in the direction of the dipole moment $\vec{p}$.

A homework problem will let you calculate the electric field in the plane that bisects the dipole. This is the field shown on the x-axis in Figure 23.5, but it could equally well be the field on the z-axis as it comes out of the page. The field, for $r \gg s$, is

$$\vec{E}_{\text{dipole}} \approx -\frac{1}{4\pi\epsilon_0}\frac{\vec{p}}{r^3} \qquad \text{(bisecting plane)} \tag{23.11}$$

FIGURE 23.4 Permanent and induced electric dipoles.

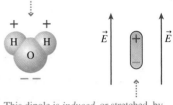

A water molecule is a *permanent* dipole because the negative electrons spend more time with the oxygen atom.

This dipole is *induced,* or stretched, by the electric field acting on the + and − charges.

FIGURE 23.5 The dipole electric field at two points.

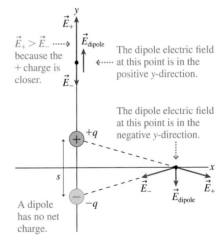

$\vec{E}_+ > \vec{E}_-$ because the + charge is closer.

The dipole electric field at this point is in the positive y-direction.

The dipole electric field at this point is in the negative y-direction.

A dipole has no net charge.

FIGURE 23.6 The dipole moment.

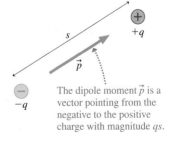

The dipole moment $\vec{p}$ is a vector pointing from the negative to the positive charge with magnitude qs.

This field is *opposite* to $\vec{p}$, and it is only half the strength of the on-axis field at the same distance.

> **NOTE** Do these inverse-cube equations violate Coulomb's law? Not at all. Coulomb's law describes the force between two *point charges,* and from Coulomb's law we found that the electric field of a *point charge* varies with the inverse square of the distance. But a dipole is not a point charge. The field of a dipole decreases more rapidly than that of a point charge, which is to be expected because the dipole is, after all, electrically neutral.

EXAMPLE 23.2 | **The electric field of a water molecule**

The water molecule H_2O has a permanent dipole moment of magnitude 6.2×10^{-30} C m. What is the electric field strength 1.0 nm from a water molecule at a point on the dipole's axis?

MODEL The size of a molecule is ≈ 0.1 nm. Thus $r \gg s$, and we can use Equation 23.10 for the on-axis electric field of the molecule's dipole moment.

SOLVE The on-axis electric field strength at $r = 1.0$ nm is

$$E \approx \frac{1}{4\pi\epsilon_0} \frac{2p}{r^3} = (9.0 \times 10^9 \, \text{N m}^2/\text{C}^2) \frac{2(6.2 \times 10^{-30} \, \text{C m})}{(1.0 \times 10^{-9} \, \text{m})^3}$$

$$= 1.1 \times 10^8 \, \text{N/C}$$

ASSESS By referring to Table 23.1 you can see that the field strength is "strong" compared to our everyday experience with charged objects but "weak" compared to the electric field inside the atoms themselves. This seems reasonable.

Electric Field Lines

We can't see the electric field. Consequently, we need pictorial tools to help us visualize it in a region of space. One method, introduced in Chapter 22, is to picture the electric field by drawing electric field vectors at various points in space.

Another common way to picture the field is to draw **electric field lines.** As FIGURE 23.7 shows,

- Electric field lines are *continuous* curves tangent to the electric field vectors.
- Closely spaced field lines indicate a greater field strength; widely spaced field lines indicate a smaller field strength.
- Electric field lines start on positive charges and end on negative charges.
- Electric field lines never cross.

The third bullet point follows from the fact that electric fields are created by charge. However, we will have to modify this idea in Chapter 30 when we find another way to create an electric field.

FIGURE 23.8 shows three electric fields represented by electric field lines. Notice that the electric field of a dipole points in the direction of $\vec{p}$ (left to right) on both sides of the dipole, but points opposite to $\vec{p}$ in the bisecting plane.

FIGURE 23.7 Electric field lines.

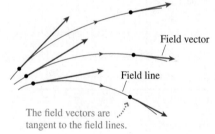

FIGURE 23.8 The electric field lines of (a) a positive point charge, (b) a negative point charge, and (c) a dipole.

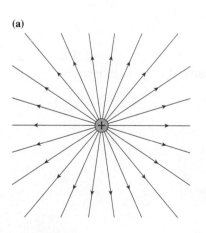

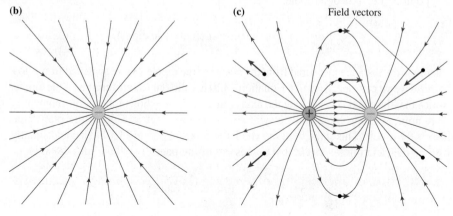

Neither field-vector diagrams nor field-line diagrams are perfect pictorial representations of an electric field. The field vectors are somewhat harder to draw, and they show the field at only a few points, but they do clearly indicate the direction and strength of the electric field at those points. Field-line diagrams perhaps look more elegant, and they're sometimes easier to sketch, but there's no formula for knowing where to draw the lines. We'll use both field-vector diagrams and field-line diagrams, depending on the circumstances.

STOP TO THINK 23.1 At the dot, the electric field points

a. Left.
b. Right.
c. Up.
d. Down.
e. The electric field is zero.

23.3 The Electric Field of a Continuous Charge Distribution

Ordinary objects—tables, chairs, beakers of water—seem to our senses to be continuous distributions of matter. There is no obvious evidence for atoms, even though we have good reasons to believe that we would find atoms if we subdivided the matter sufficiently far. Thus it is easier, for many practical purposes, to consider matter to be continuous and to talk about the *density* of matter. Density—the number of kilograms of matter per cubic meter—allows us to describe the distribution of matter *as if* the matter were continuous rather than atomic.

Much the same situation occurs with charge. If a charged object contains a large number of excess electrons—for example, 10^{12} extra electrons on a metal rod—it is not practical to track every electron. It makes more sense to consider the charge to be *continuous* and to describe how it is distributed over the object.

FIGURE 23.9a shows an object of length L, such as a plastic rod or a metal wire, with charge Q spread uniformly along it. (We will use an uppercase Q for the total charge of an object, reserving lowercase q for individual point charges.) The **linear charge density** λ is defined to be

$$\lambda = \frac{Q}{L} \tag{23.12}$$

Linear charge density, which has units of C/m, is the amount of charge *per meter* of length. The linear charge density of a 20-cm-long wire with 40 nC of charge is 2.0 nC/cm or 2.0×10^{-7} C/m.

NOTE The linear charge density λ is analogous to the linear mass density μ that you used in Chapter 16 to find the speed of a wave on a string.

We'll also be interested in charged surfaces. **FIGURE 23.9b** shows a two-dimensional distribution of charge across a surface of area A. We define the **surface charge density** η (lowercase Greek eta) to be

$$\eta = \frac{Q}{A} \tag{23.13}$$

Surface charge density, with units of C/m^2, is the amount of charge *per square meter*. A 1.0 mm × 1.0 mm square on a surface with $\eta = 2.0 \times 10^{-4}$ C/m^2 contains 2.0×10^{-10} C or 0.20 nC of charge. (The volume charge density $\rho = Q/V$, measured in C/m^3, will be used in Chapter 24.)

FIGURE 23.9 One-dimensional and two-dimensional continuous charge distributions.

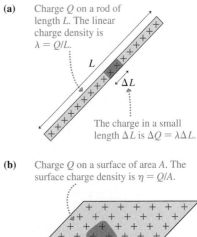

(a) Charge Q on a rod of length L. The linear charge density is $\lambda = Q/L$.

L

ΔL

The charge in a small length ΔL is $\Delta Q = \lambda \Delta L$.

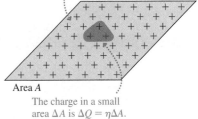

(b) Charge Q on a surface of area A. The surface charge density is $\eta = Q/A$.

Area A

The charge in a small area ΔA is $\Delta Q = \eta \Delta A$.

Figure 23.9 and the definitions of Equations 23.12 and 23.13 assume that the object is **uniformly charged,** meaning that the charges are evenly spread over the object. We will assume objects are uniformly charged unless noted otherwise.

NOTE Some textbooks represent the surface charge density with the symbol σ. Because σ is also used to represent *conductivity,* an idea we'll introduce in Chapter 27, we've selected a different symbol for surface charge density.

STOP TO THINK 23.2 A piece of plastic is uniformly charged with surface charge density η_a. The plastic is then broken into a large piece with surface charge density η_b and a small piece with surface charge density η_c. Rank in order, from largest to smallest, the surface charge densities η_a to η_c.

Integration Is Summation

Calculating the electric field of a continuous charge distribution usually requires setting up and evaluating integrals—a skill you've been learning in calculus. It is common to think that "an integral is the area under a curve," an idea we used in our study of kinematics.

But integration is much more than a tool for finding areas. More generally, **integration is summation.** That is, an integral is a sophisticated way to add an infinite number of infinitesimally small pieces. The area under a curve happens to be a special case in which you're summing the small areas $y(x)\,dx$ of an infinite number of tall, narrow boxes, but the idea of integration as summation has many other applications.

Suppose, for example, that a charged object is divided into a large number of small pieces numbered $i = 1, 2, 3, \ldots, N$ having small quantities of charge $\Delta Q_1, \Delta Q_2, \Delta Q_3, \ldots, \Delta Q_N$. Figure 23.9 showed small pieces of charge for a charged rod and a charged sheet, but the object could have any shape. The total charge on the object is found by *summing* all the small charges:

$$Q = \sum_{i=1}^{N} \Delta Q_i \tag{23.14}$$

If we now let $\Delta Q_i \rightarrow 0$ and $N \rightarrow \infty$, then we *define* the integral:

$$Q = \lim_{\Delta Q \to 0} \sum_{i=1}^{N} \Delta Q_i = \int_{\text{object}} dQ \tag{23.15}$$

That is, integration is the summing of an infinite number of infinitesimally small pieces of charge. This use of integration has nothing to do with the area under a curve.

Although Equation 23.15 is a formal statement of "add up all the little pieces," it's not yet an expression that can actually be integrated with the tools of calculus. Integrals are carried out over coordinates, such as dx or dy, and we also need coordinates to specify what is meant by "integrate over the object." This is where densities come in.

Suppose we want to find the total charge of a thin, charged rod of length L. First, we establish an x-axis with the origin at one end of the rod, as shown in **FIGURE 23.10**. Then we divide the rod into lots of tiny segments of length dx. Each of these little segments has a charge dQ, and the total charge on the rod is the sum of all the dQ values—that's what Equation 23.15 is saying. Now the critical step: The rod has some linear charge density λ. Consequently, the charge of a small segment of the rod is $dQ = \lambda\,dx$. **Densities are the link between quantities and coordinates.** Finally, "integrate over the rod" means to integrate from $x = 0$ to $x = L$. Thus the total charge on the rod is

$$Q = \int_{\text{rod}} dQ = \int_0^L \lambda\,dx \tag{23.16}$$

FIGURE 23.10 Setting up an integral to calculate the charge on a rod.

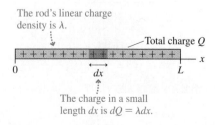

The rod's linear charge density is λ.

Total charge Q

The charge in a small length dx is $dQ = \lambda\,dx$.

Now we have an expression that can be integrated. If λ is constant, as it is for a uniformly charged rod, we can take it outside the integral to find $Q = \lambda L$. But we could also use Equation 23.16 for a nonuniformly charged rod where λ is some function of x.

This discussion reveals two key ideas that will be needed for calculating electric fields:

- Integration is the tool for summing a vast number of small pieces.
- Density is the connection between quantities and coordinates.

A Problem-Solving Strategy

Our goal is to find the electric field of a continuous distribution of charge, such as a charged rod or a charged disk. We have two basic tools to work with:

- The electric field of a point charge, and
- The principle of superposition.

We can apply these tools with a three-step strategy:

1. Divide the total charge Q into many small point-like charges ΔQ.
2. Use our knowledge of the electric field of a point charge to find the electric field of each ΔQ.
3. Calculate the net field $\vec{E}_{net}$ by summing the fields of all the ΔQ.

As you've no doubt guessed, we'll let the sum become an integral.
We will go step by step through several examples to illustrate the procedures. However, we first need to flesh out the steps of the problem-solving strategy. The aim of this problem-solving strategy is to break a difficult problem down into small steps that are individually manageable.

PROBLEM-SOLVING STRATEGY 23.2   (MP)

The electric field of a continuous distribution of charge

MODEL Model the charge distribution as a simple shape.

VISUALIZE For the pictorial representation:

- Draw a picture, establish a coordinate system, and identify the point P at which you want to calculate the electric field.
- Divide the total charge Q into small pieces of charge ΔQ, using shapes for which you *already know* how to determine $\vec{E}$. This is often, but not always, a division into point charges.
- Draw the electric field vector at P for one or two small pieces of charge. This will help you identify distances and angles that need to be calculated.

SOLVE The mathematical representation is $\vec{E}_{net} = \sum \vec{E}_i$.

- Write an algebraic expression for *each* of the three components of $\vec{E}$ (unless you are sure one or more is zero) at point P. Let the (x, y, z) coordinates of the point remain variables.
- Replace the small charge ΔQ with an equivalent expression involving a charge density and a coordinate, such as dx. **This is the critical step in making the transition from a sum to an integral** because you need a coordinate to serve as the integration variable.
- Express all angles and distances in terms of the coordinates.
- Let the sum become an integral. The integration limits for this variable must "cover" the entire charged object.

ASSESS Check that your result is consistent with any limits for which you know what the field should be.

Exercise 16

EXAMPLE 23.3 | The electric field of a line of charge

FIGURE 23.11 shows a thin, uniformly charged rod of length L with total charge Q. Find the electric field strength at radial distance r in the plane that bisects the rod.

FIGURE 23.11 A thin, uniformly charged rod.

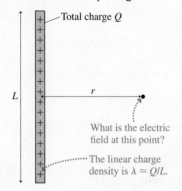

MODEL The rod is thin, so we'll assume the charge lies along a line and forms what we call a *line of charge*. The rod's linear charge density is $\lambda = Q/L$.

VISUALIZE FIGURE 23.12 illustrates the steps of the problem-solving strategy. We've chosen a coordinate system in which the rod lies along the y-axis and point P, in the bisecting plane, is on the x-axis. We've then divided the rod into N small segments of charge ΔQ, each of which is small enough to model as a point charge. For every ΔQ in the bottom half of the wire with a field that points to the right and up, there's a matching ΔQ in the top half whose field points to the right and down. The y-components of these two fields cancel, hence **the net electric field on the x-axis points straight away from the rod.** The only component we need to calculate is E_x. (This is the same reasoning on the basis of symmetry that we used in Example 23.1.)

FIGURE 23.12 Calculating the electric field of a line of charge.

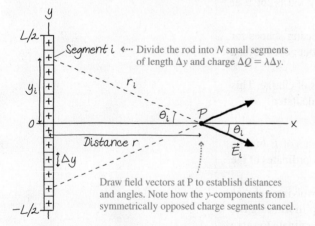

SOLVE Each of the little segments of charge can be modeled as a point charge. We know the electric field of a point charge, so we can write the x-component of $\vec{E}_i$, the electric field of segment i, as

$$(E_i)_x = E_i \cos\theta_i = \frac{1}{4\pi\epsilon_0} \frac{\Delta Q}{r_i^2} \cos\theta_i$$

where r_i is the distance from charge i to point P. You can see from the figure that $r_i = (y_i^2 + r^2)^{1/2}$ and $\cos\theta_i = r/r_i = r/(y_i^2 + r^2)^{1/2}$. With these, $(E_i)_x$ is

$$(E_i)_x = \frac{1}{4\pi\epsilon_0} \frac{\Delta Q}{y_i^2 + r^2} \frac{r}{\sqrt{y_i^2 + r^2}}$$

$$= \frac{1}{4\pi\epsilon_0} \frac{r\,\Delta Q}{(y_i^2 + r^2)^{3/2}}$$

Compare this result to the very similar calculation we did in Example 23.1. If we now sum this expression over all the charge segments, the net x-component of the electric field is

$$E_x = \sum_{i=1}^{N} (E_i)_x = \frac{1}{4\pi\epsilon_0} \sum_{i=1}^{N} \frac{r\,\Delta Q}{(y_i^2 + r^2)^{3/2}}$$

This is the same superposition we did for the $N = 3$ case in Example 23.1. The only difference is that we have now written the result as an explicit summation so that N can have any value. We want to let $N \to \infty$ and to replace the sum with an integral, but we can't integrate over Q; it's not a geometric quantity. This is where the linear charge density enters. The quantity of charge in each segment is related to its length Δy by $\Delta Q = \lambda\,\Delta y = (Q/L)\Delta y$. In terms of the linear charge density, the electric field is

$$E_x = \frac{Q/L}{4\pi\epsilon_0} \sum_{i=1}^{N} \frac{r\,\Delta y}{(y_i^2 + r^2)^{3/2}}$$

Now we're ready to let the sum become an integral. If we let $N \to \infty$, then each segment becomes an infinitesimal length $\Delta y \to dy$ while the discrete position variable y_i becomes the continuous integration variable y. The sum from $i = 1$ at the bottom end of the line of charge to $i = N$ at the top end will be replaced with an integral from $y = -L/2$ to $y = +L/2$. Thus in the limit $N \to \infty$,

$$E_x = \frac{Q/L}{4\pi\epsilon_0} \int_{-L/2}^{L/2} \frac{r\,dy}{(y^2 + r^2)^{3/2}}$$

This is a standard integral that you have learned to do in calculus and that can be found in Appendix A. Note that r is a *constant* as far as this integral is concerned. Integrating gives

$$E_x = \frac{Q/L}{4\pi\epsilon_0} \frac{y}{r\sqrt{y^2 + r^2}} \Bigg|_{-L/2}^{L/2}$$

$$= \frac{Q/L}{4\pi\epsilon_0} \left[\frac{L/2}{r\sqrt{(L/2)^2 + r^2}} - \frac{-L/2}{r\sqrt{(-L/2)^2 + r^2}} \right]$$

$$= \frac{1}{4\pi\epsilon_0} \frac{Q}{r\sqrt{r^2 + (L/2)^2}}$$

Because E_x is the *only* component of the field, the electric field strength E_{rod} at distance r from the center of a charged rod is

$$E_{\text{rod}} = \frac{1}{4\pi\epsilon_0} \frac{|Q|}{r\sqrt{r^2 + (L/2)^2}}$$

The field strength must be positive, so we added absolute value signs to Q to allow for the possibility that the charge could be negative. The only restriction is to remember that this is the electric field at a point in the plane that bisects the rod.

ASSESS Suppose we are at a point *very* far from the rod. If $r \gg L$, the length of the rod is not relevant and the rod appears to be a point charge Q in the distance. Thus in the *limiting case* $r \gg L$, we expect the rod's electric field to be that of a point charge. If $r \gg L$, the square root becomes $(r^2 + (L/2)^2)^{1/2} \approx (r^2)^{1/2} = r$ and the electric field strength at distance r becomes $E_{\text{rod}} \approx Q/4\pi\epsilon_0 r^2$, the field of a point charge. The fact that our expression of E_{rod} has the correct limiting behavior gives us confidence that we haven't made any mistakes in its derivation.

An Infinite Line of Charge

What if the rod or wire becomes very long, becoming a **line of charge,** while the linear charge density λ remains constant? To answer this question, we can rewrite the expression for E_{rod} by factoring $(L/2)^2$ out of the denominator:

$$E_{rod} = \frac{1}{4\pi\epsilon_0} \frac{|Q|}{r \cdot L/2} \frac{1}{\sqrt{1 + 4r^2/L^2}} = \frac{1}{4\pi\epsilon_0} \frac{2|\lambda|}{r} \frac{1}{\sqrt{1 + 4r^2/L^2}}$$

where $|\lambda| = |Q|/L$ is the magnitude of the linear charge density. If we now let $L \to \infty$, the last term becomes simply 1 and we're left with

$$\vec{E}_{line} = \left(\frac{1}{4\pi\epsilon_0} \frac{2|\lambda|}{r}, \begin{cases} \text{away from line if charge } + \\ \text{toward line if charge } - \end{cases} \right) \quad \text{(line of charge)} \quad (23.17)$$

where we've now included the field's direction. **FIGURE 23.13** shows the electric field vectors of an infinite line of positive charge. The vectors would point inward for a negative line of charge.

NOTE Unlike a point charge, for which the field decreases as $1/r^2$, the field of an infinitely long charged wire decreases more slowly—as only $1/r$.

The infinite line of charge is the second of our important electric field models. Although no real wire is infinitely long, the field of a realistic finite-length wire is well approximated by Equation 23.17 except at points near the end of the wire.

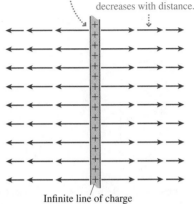

FIGURE 23.13 The electric field of an infinite line of charge.

The field points straight away from the line at all points . . .

. . . and its strength decreases with distance.

Infinite line of charge

STOP TO THINK 23.3 Which of the following actions will increase the electric field strength at the position of the dot?

a. Make the rod longer without changing the charge.
b. Make the rod shorter without changing the charge.
c. Make the rod wider without changing the charge.
d. Make the rod narrower without changing the charge.
e. Add charge to the rod.
f. Remove charge from the rod.
g. Move the dot farther from the rod.
h. Move the dot closer to the rod.

Charged rod

23.4 The Electric Fields of Rings, Disks, Planes, and Spheres

In this section we'll derive the electric fields for several important charge distributions.

EXAMPLE 23.4 | **The electric field of a ring of charge**

A thin ring of radius R is uniformly charged with total charge Q. Find the electric field at a point on the axis of the ring (perpendicular to the ring).

MODEL Because the ring is thin, we'll assume the charge lies along a circle of radius R. You can think of this as a line of charge of length $2\pi R$ wrapped into a circle. The linear charge density along the ring is $\lambda = Q/2\pi R$.

VISUALIZE **FIGURE 23.14** on the next page shows the ring and illustrates the steps of the problem-solving strategy. We've chosen a coordinate system in which the ring lies in the xy-plane and point P is on the z-axis. We've then divided the ring into N small segments of charge ΔQ, each of which can be modeled as a point charge. As you can see from the figure, the component of the field perpendicular to the axis cancels for two diametrically opposite segments. Thus we need to calculate only the z-component E_z.

Continued

FIGURE 23.14 Calculating the on-axis electric field of a ring of charge.

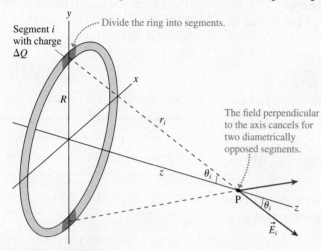

Segment i with charge ΔQ

Divide the ring into segments.

The field perpendicular to the axis cancels for two diametrically opposed segments.

SOLVE The z-component of the electric field due to segment i is

$$(E_i)_z = E_i \cos\theta_i = \frac{1}{4\pi\epsilon_0}\frac{\Delta Q}{r_i^2}\cos\theta_i$$

You can see from the figure that *every* segment of the ring, independent of i, has

$$r_i = \sqrt{z^2 + R^2}$$

$$\cos\theta_i = \frac{z}{r_i} = \frac{z}{\sqrt{z^2 + R^2}}$$

Consequently, the field of segment i is

$$(E_i)_z = \frac{1}{4\pi\epsilon_0}\frac{\Delta Q}{z^2 + R^2}\frac{z}{\sqrt{z^2 + R^2}} = \frac{1}{4\pi\epsilon_0}\frac{z}{(z^2 + R^2)^{3/2}}\Delta Q$$

The net electric field is found by summing $(E_i)_z$ due to all N segments:

$$E_z = \sum_{i=1}^{N}(E_i)_z = \frac{1}{4\pi\epsilon_0}\frac{z}{(z^2 + R^2)^{3/2}}\sum_{i=1}^{N}\Delta Q$$

We were able to bring all terms involving z to the front because z is a constant as far as the summation is concerned. Surprisingly, we don't need to convert the sum to an integral to complete this calculation. The sum of all the ΔQ around the ring is simply the ring's total charge, $\sum \Delta Q = Q$, hence the field on the axis is

$$(E_{\text{ring}})_z = \frac{1}{4\pi\epsilon_0}\frac{zQ}{(z^2 + R^2)^{3/2}}$$

This expression is valid for both positive and negative z (i.e., on either side of the ring) and for both positive and negative charge.

ASSESS It will be left as a homework problem to show that this result gives the expected limit when $z \gg R$.

FIGURE 23.15 The on-axis electric field of a ring of charge.

(a)

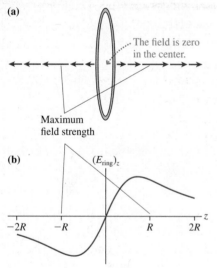

The field is zero in the center.

Maximum field strength

(b)

$(E_{\text{ring}})_z$

$-2R$ $-R$ R $2R$ z

FIGURE 23.15 shows two representations of the on-axis electric field of a positively charged ring. Figure 23.15a shows that the electric field vectors point away from the ring, increasing in length until reaching a maximum when $|z| \approx R$, then decreasing. The graph of $(E_{\text{ring}})_z$ in Figure 23.15b confirms that the field strength has a maximum on either side of the ring. Notice that the electric field at the center of the ring is zero, even though this point is surrounded by charge. You might want to spend a minute thinking about why this has to be the case.

A Disk of Charge

FIGURE 23.16 shows a disk of radius R that is uniformly charged with charge Q. This is a mathematical disk, with no thickness, and its surface charge density is

$$\eta = \frac{Q}{A} = \frac{Q}{\pi R^2} \tag{23.18}$$

We would like to calculate the on-axis electric field of this disk. Our problem-solving strategy tells us to divide a continuous charge into segments for which we already know how to find $\vec{E}$. Because we now know the on-axis electric field of a ring of charge, let's divide the disk into N very narrow rings of radius r and width Δr. One such ring, with radius r_i and charge ΔQ_i, is shown.

We need to be careful with notation. The R in Example 23.4 was the radius of the ring. Now we have many rings, and the radius of ring i is r_i. Similarly, Q was the charge on the ring. Now the charge on ring i is ΔQ_i, a small fraction of the total charge on the disk. With these changes, the electric field of ring i, with radius r_i, is

$$(E_i)_z = \frac{1}{4\pi\epsilon_0}\frac{z\,\Delta Q_i}{(z^2 + r_i^2)^{3/2}} \tag{23.19}$$

The on-axis electric field of the charged disk is the sum of the electric fields of all of the rings:

$$(E_{\text{disk}})_z = \sum_{i=1}^{N}(E_i)_z = \frac{z}{4\pi\epsilon_0}\sum_{i=1}^{N}\frac{\Delta Q_i}{(z^2 + r_i^2)^{3/2}} \tag{23.20}$$

The critical step, as always, is to relate ΔQ to a coordinate. Because we now have a surface, rather than a line, the charge in ring i is $\Delta Q = \eta \, \Delta A_i$, where ΔA_i is the area of ring i. We can find ΔA_i, as you've learned to do in calculus, by "unrolling" the ring to form a narrow rectangle of length $2\pi r_i$ and height Δr. Thus the area of ring i is $\Delta A_i = 2\pi r_i \, \Delta r$ and the charge is $\Delta Q_i = 2\pi \eta r_i \, \Delta r$. With this substitution, Equation 23.20 becomes

$$(E_{\text{disk}})_z = \frac{\eta z}{2\epsilon_0} \sum_{i=1}^{N} \frac{r_i \, \Delta r}{(z^2 + r_i^2)^{3/2}} \qquad (23.21)$$

As $N \to \infty$, $\Delta r \to dr$ and the sum becomes an integral. Adding all the rings means integrating from $r = 0$ to $r = R$; thus

$$(E_{\text{disk}})_z = \frac{\eta z}{2\epsilon_0} \int_0^R \frac{r \, dr}{(z^2 + r^2)^{3/2}} \qquad (23.22)$$

All that remains is to carry out the integration. This is straightforward if we make the variable change $u = z^2 + r^2$. Then $du = 2r \, dr$ or, equivalently, $r \, dr = \frac{1}{2} du$. At the lower integration limit $r = 0$, our new variable is $u = z^2$. At the upper limit $r = R$, the new variable is $u = z^2 + R^2$.

NOTE When changing variables in a definite integral, you *must* also change the limits of integration.

With this variable change the integral becomes

$$(E_{\text{disk}})_z = \frac{\eta z}{2\epsilon_0} \frac{1}{2} \int_{z^2}^{z^2+R^2} \frac{du}{u^{3/2}} = \frac{\eta z}{4\epsilon_0} \frac{-2}{u^{1/2}} \bigg|_{z^2}^{z^2+R^2} = \frac{\eta z}{2\epsilon_0} \left[\frac{1}{z} - \frac{1}{\sqrt{z^2 + R^2}} \right] \qquad (23.23)$$

If we multiply through by z, the on-axis electric field of a charged disk with surface charge density $\eta = Q/\pi R^2$ is

$$(E_{\text{disk}})_z = \frac{\eta}{2\epsilon_0} \left[1 - \frac{z}{\sqrt{z^2 + R^2}} \right] \qquad (23.24)$$

NOTE This expression is valid only for $z > 0$. The field for $z < 0$ has the same magnitude but points in the opposite direction.

Limiting Cases

It's a bit difficult to see what Equation 23.24 is telling us, so let's compare it to what we already know. First, you can see that the quantity in square brackets is dimensionless. The surface charge density $\eta = Q/A$ has the same units as q/r^2, so $\eta/2\epsilon_0$ has the same units as $q/4\pi\epsilon_0 r^2$. This tells us that $\eta/2\epsilon_0$ really is an electric field.

Next, let's move very far away from the disk. At distance $z \gg R$, the disk appears to be a point charge Q in the distance and the field of the disk should approach that of a point charge. If we let $z \to \infty$ in Equation 23.24, so that $z^2 + R^2 \approx z^2$, we find $(E_{\text{disk}})_z \to 0$. This is true, but not quite what we wanted. We need to let z be very large in comparison to R, but not so large as to make E_{disk} vanish. That requires a little more care in taking the limit.

We can cast Equation 23.24 into a somewhat more useful form by factoring the z^2 out of the square root to give

$$(E_{\text{disk}})_z = \frac{\eta}{2\epsilon_0} \left[1 - \frac{1}{\sqrt{1 + R^2/z^2}} \right] \qquad (23.25)$$

Now $R^2/z^2 \ll 1$ if $z \gg R$, so the second term in the square brackets is of the form $(1 + x)^{-1/2}$ where $x \ll 1$. We can then use the *binomial approximation*

$$(1 + x)^n \approx 1 + nx \quad \text{if} \quad x \ll 1 \qquad \text{(binomial approximation)}$$

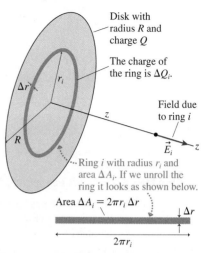

FIGURE 23.16 Calculating the on-axis field of a charged disk.

Disk with radius R and charge Q

The charge of the ring is ΔQ_i.

Field due to ring i

Ring i with radius r_i and area ΔA_i. If we unroll the ring it looks as shown below.

Area $\Delta A_i = 2\pi r_i \, \Delta r$

$2\pi r_i$

to simplify the expression in square brackets:

$$1 - \frac{1}{\sqrt{1 + R^2/z^2}} = 1 - (1 + R^2/z^2)^{-1/2} \approx 1 - \left(1 + \left(-\frac{1}{2}\right)\frac{R^2}{z^2}\right) = \frac{R^2}{2z^2} \qquad (23.26)$$

This is a good approximation when $z \gg R$. Substituting this approximation into Equation 23.25, we find that the electric field of the disk for $z \gg R$ is

$$(E_{\text{disk}})_z \approx \frac{\eta}{2\epsilon_0}\frac{R^2}{2z^2} = \frac{Q/\pi R^2}{4\epsilon_0}\frac{R^2}{z^2} = \frac{1}{4\pi\epsilon_0}\frac{Q}{z^2} \qquad \text{if } z \gg R \qquad (23.27)$$

This is, indeed, the field of a point charge Q, giving us confidence in Equation 23.24 for the on-axis electric field of a disk of charge.

EXAMPLE 23.5 | The electric field of a charged disk

A 10-cm-diameter plastic disk is charged uniformly with an extra 10^{11} electrons. What is the electric field 1.0 mm above the surface at a point near the center?

MODEL Model the plastic disk as a uniformly charged disk. We are seeking the on-axis electric field. Because the charge is negative, the field will point *toward* the disk.

SOLVE The total charge on the plastic square is $Q = N(-e) = -1.60 \times 10^{-8}$ C. The surface charge density is

$$\eta = \frac{Q}{A} = \frac{Q}{\pi R^2} = \frac{-1.60 \times 10^{-8} \text{ C}}{\pi(0.050 \text{ m})^2} = -2.04 \times 10^{-6} \text{ C/m}^2$$

The electric field at $z = 0.0010$ m, given by Equation 23.25, is

$$E_z = \frac{\eta}{2\epsilon_0}\left[1 - \frac{1}{\sqrt{1 + R^2/z^2}}\right] = -1.1 \times 10^5 \text{ N/C}$$

The minus sign indicates that the field points *toward*, rather than away from, the disk. As a vector,

$$\vec{E} = (1.1 \times 10^5 \text{ N/C, toward the disk})$$

ASSESS The total charge, -16 nC, is typical of the amount of charge produced on a small plastic object by rubbing or friction. Thus 10^5 N/C is a typical electric field strength near an object that has been charged by rubbing.

A Plane of Charge

Many electronic devices use charged, flat surfaces—disks, squares, rectangles, and so on—to steer electrons along the proper paths. These charged surfaces are called **electrodes.** Although any real electrode is finite in extent, we can often model an electrode as an infinite **plane of charge.** As long as the distance z to the electrode is small in comparison to the distance to the edges, we can reasonably treat the edges *as if* they are infinitely far away.

The electric field of a plane of charge is found from the on-axis field of a charged disk by letting the radius $R \to \infty$. That is, a disk with infinite radius is an infinite plane. From Equation 23.24, we see that the electric field of a plane of charge with surface charge density η is

$$E_{\text{plane}} = \frac{\eta}{2\epsilon_0} = \text{constant} \qquad (23.28)$$

This is a simple result, but what does it tell us? First, the field strength is directly proportional to the charge density η: more charge, bigger field. Second, and more interesting, the field strength is the same at *all* points in space, independent of the distance z. The field strength 1000 m from the plane is the same as the field strength 1 mm from the plane.

How can this be? It seems that the field should get weaker as you move away from the plane of charge. But remember that we are dealing with an *infinite* plane of charge. What does it mean to be "close to" or "far from" an infinite object? For a disk of finite radius R, whether a point at distance z is "close to" or "far from" the disk is a comparison of z to R. If $z \ll R$, the point is close to the disk. If $z \gg R$, the point is far from the disk. But as $R \to \infty$, we have no *scale* for distinguishing near and far. In essence, *every* point in space is "close to" a disk of infinite radius.

No real plane is infinite in extent, but we can interpret Equation 23.28 as saying that the field of a surface of charge, regardless of its shape, is a constant $\eta/2\epsilon_0$ for those points whose distance z to the surface is much smaller than their distance to the edge.

We do need to note that the derivation leading to Equation 23.28 considered only $z > 0$. For a positively charged plane, with $\eta > 0$, the electric field points *away from* the plane on both sides of the plane. This requires $E_z < 0$ ($\vec{E}$ pointing in the negative z-direction) on the side with $z < 0$. Thus a complete description of the electric field, valid for both sides of the plane and for either sign of η, is

$$\vec{E}_{\text{plane}} = \left(\frac{|\eta|}{2\epsilon_0}, \begin{cases} \text{away from plane if charge } + \\ \text{toward plane if charge } - \end{cases} \right) \quad \text{(plane of charge)} \quad (23.29)$$

The infinite plane of charge is the third of our important electric field models.

FIGURE 23.17 shows two views of the electric field of a positively charged plane. All the arrows would be reversed for a negatively charged plane. It would have been very difficult to anticipate this result from Coulomb's law or from the electric field of a single point charge, but step by step we have been able to use the concept of the electric field to look at increasingly complex distributions of charge.

A Sphere of Charge

The one last charge distribution for which we need to know the electric field is a **sphere of charge.** This problem is analogous to wanting to know the gravitational field of a spherical planet or star. The procedure for calculating the field of a sphere of charge is the same as we used for lines and planes, but the integrations are significantly more difficult. We will skip the details of the calculations and, for now, simply assert the result without proof. In Chapter 24 we'll use an alternative procedure to find the field of a sphere of charge.

A sphere of charge Q and radius R, be it a uniformly charged sphere or just a spherical shell, has an electric field *outside* the sphere ($r \geq R$) that is exactly the same as that of a point charge Q located at the center of the sphere:

$$\vec{E}_{\text{sphere}} = \frac{Q}{4\pi\epsilon_0 r^2} \hat{r} \quad \text{for } r \geq R \quad (23.30)$$

This assertion is analogous to our earlier assertion that the gravitational force between stars and planets can be computed as if all the mass is at the center.

FIGURE 23.18 shows the electric field of a sphere of positive charge. The field of a negative sphere would point inward.

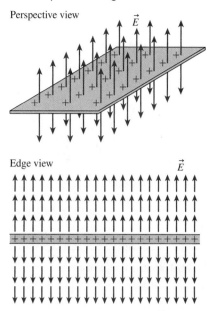

FIGURE 23.17 Two views of the electric field of a plane of charge.

Perspective view

Edge view

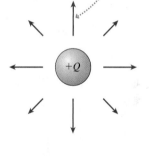

FIGURE 23.18 The electric field of a sphere of positive charge.

The electric field outside a sphere or spherical shell is the same as the field of a point charge Q at the center.

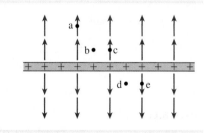

STOP TO THINK 23.4 Rank in order, from largest to smallest, the electric field strengths E_a to E_e at these five points near a plane of charge.

23.5 The Parallel-Plate Capacitor

FIGURE 23.19 shows two electrodes, one with charge $+Q$ and the other with $-Q$, placed face-to-face a distance d apart. This arrangement of two electrodes, charged equally but oppositely, is called a **parallel-plate capacitor.** Capacitors play important roles in many electric circuits. Our goal is to find the electric field both inside the capacitor (i.e., between the plates) and outside the capacitor.

NOTE The *net* charge of a capacitor is zero. Capacitors are charged by transferring electrons from one plate to the other. The plate that gains N electrons has charge $-Q = N(-e)$. The plate that loses electrons has charge $+Q$.

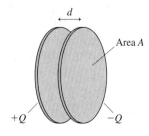

FIGURE 23.19 A parallel-plate capacitor.

Area A

$+Q$ $-Q$

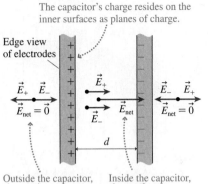

The capacitor's charge resides on the inner surfaces as planes of charge.

Edge view of electrodes

$\vec{E}_+$ $\vec{E}_-$

$\vec{E}_{net} = \vec{0}$

$\vec{E}_+$

$\vec{E}_-$ $\vec{E}_{net}$

$\vec{E}_-$ $\vec{E}_+$

$\vec{E}_{net} = \vec{0}$

d

Outside the capacitor, $\vec{E}_+$ and $\vec{E}_-$ are opposite, so the net field is zero.

Inside the capacitor, $\vec{E}_+$ and $\vec{E}_-$ are parallel, so the net field is large.

FIGURE 23.21 Ideal versus real capacitors.

(a) Ideal capacitor—edge view

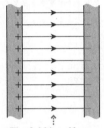

The field is uniform

(b) Real capacitor—edge view

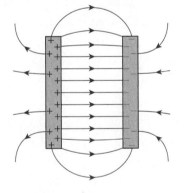

FIGURE 23.20 is an enlarged view of the capacitor plates, seen from the side. Because opposite charges attract, all of the charge is on the *inner* surfaces of the two plates. Thus the inner surfaces of the plates can be modeled as two planes of charge with equal but opposite surface charge densities. As you can see from the figure, at all points in space the electric field $\vec{E}_+$ of the positive plate points *away from* the plane of positive charges. Similarly, the field $\vec{E}_-$ of the negative plate everywhere points *toward* the plane of negative charges.

NOTE You might think the right capacitor plate would somehow "block" the electric field created by the positive plate and prevent the presence of an $\vec{E}_+$ field to the right of the capacitor. To see that it doesn't, consider an analogous situation with gravity. The strength of gravity above a table is the same as its strength below it. Just as the table doesn't block the earth's gravitational field, intervening matter or charges do not alter or block an object's electric field.

Outside the capacitor, $\vec{E}_+$ and $\vec{E}_-$ point in opposite directions and, because the field of a plane of charge is independent of the distance from the plane, have equal magnitudes. Consequently, the fields $\vec{E}_+$ and $\vec{E}_-$ add to zero outside the capacitor plates. There's *no* electric field outside an ideal parallel-plate capacitor.

Inside the capacitor, between the electrodes, field $\vec{E}_+$ points from positive to negative and has magnitude $\eta/2\epsilon_0 = Q/2\epsilon_0 A$, where A is the surface area of each electrode. Field $\vec{E}_-$ *also* points from positive to negative and *also* has magnitude $Q/2\epsilon_0 A$, so the inside field $\vec{E}_+ + \vec{E}_-$ is twice that of a plane of charge. Thus the electric field of a parallel-plate capacitor is

$$\vec{E}_{capacitor} = \begin{cases} \left(\dfrac{Q}{\epsilon_0 A}, \text{from positive to negative}\right) & \text{inside} \\ \vec{0} & \text{outside} \end{cases} \qquad (23.31)$$

FIGURE 23.21a shows the electric field—this time using field lines—of an ideal parallel-plate capacitor. Now, it's true that no real capacitor is infinite in extent, but the ideal parallel-plate capacitor is a very good approximation for all but the most precise calculations as long as the electrode separation d is much smaller than the electrodes' size. **FIGURE 23.21b** shows that the interior field of a real capacitor is virtually identical to that of an ideal capacitor but that the exterior field isn't quite zero. This weak field outside the capacitor is called the **fringe field.** We will keep things simple by always assuming the plates are very close together and using Equation 23.31 for the field inside a parallel-plate capacitor.

NOTE The shape of the electrodes—circular or square or any other shape—is not relevant as long as the electrodes are very close together.

EXAMPLE 23.6 | **The electric field inside a capacitor**

Two 1.0 cm × 2.0 cm rectangular electrodes are 1.0 mm apart. What charge must be placed on each electrode to create a uniform electric field of strength 2.0×10^6 N/C? How many electrons must be moved from one electrode to the other to accomplish this?

MODEL The electrodes can be modeled as an ideal parallel-plate capacitor because the spacing between them is much smaller than their lateral dimensions.

SOLVE The electric field strength inside the capacitor is $E = Q/\epsilon_0 A$. Thus the charge to produce a field of strength E is

$$Q = (8.85 \times 10^{-12} \text{ C}^2/\text{N m}^2)(2.0 \times 10^{-4} \text{ m}^2)(2.0 \times 10^6 \text{ N/C})$$
$$= 3.5 \times 10^{-9} \text{ C} = 3.5 \text{ nC}$$

The positive plate must be charged to +3.5 nC and the negative plate to −3.5 nC. In practice, the plates are charged by using a *battery* to move electrons from one plate to the other. The number of electrons in 3.5 nC is

$$N = \frac{Q}{e} = \frac{3.5 \times 10^{-9} \text{ C}}{1.60 \times 10^{-19} \text{ C/electron}} = 2.2 \times 10^{10} \text{ electrons}$$

Thus 2.2×10^{10} electrons are moved from one electrode to the other. Note that the capacitor *as a whole* has no net charge.

ASSESS The plate spacing does not enter the result. As long as the spacing is much smaller than the plate dimensions, as is true in this example, the field is independent of the spacing.

Uniform Electric Fields

FIGURE 23.22 shows an electric field that is the *same*—in strength and direction—at every point in a region of space. This is called a **uniform electric field.** A uniform electric field is analogous to the uniform gravitational field near the surface of the earth. Uniform fields are of great practical significance because, as you will see in the next section, computing the trajectory of a charged particle moving in a uniform electric field is a straightforward process.

The easiest way to produce a uniform electric field is with a parallel-plate capacitor, as you can see in Figure 23.21a. Indeed, our interest in capacitors is due in large measure to the fact that the electric field is uniform. Many electric field problems refer to a uniform electric field. Such problems carry an implicit assumption that the action is taking place *inside* a parallel-plate capacitor.

FIGURE 23.22 A uniform electric field.

STOP TO THINK 23.5 Rank in order, from largest to smallest, the forces F_a to F_e a proton would experience if placed at points a to e in this parallel-plate capacitor.

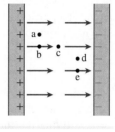

23.6 Motion of a Charged Particle in an Electric Field

Our motivation for introducing the concept of the electric field was to understand the long-range electric interaction of charges. Some charges, the *source charges,* create an electric field. Other charges then respond to that electric field. The first five sections of this chapter have focused on the electric field of the source charges. Now we turn our attention to the second half of the interaction.

FIGURE 23.23 shows a particle of charge q and mass m at a point where an electric field $\vec{E}$ has been produced by *other* charges, the source charges. The electric field exerts a force

$$\vec{F}_{on\,q} = q\vec{E}$$

on the charged particle. Notice that the force on a negatively charged particle is *opposite* in direction to the electric field vector. Signs are important!

FIGURE 23.23 The electric field exerts a force on a charged particle.

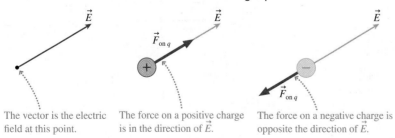

The vector is the electric field at this point.

The force on a positive charge is in the direction of $\vec{E}$.

The force on a negative charge is opposite the direction of $\vec{E}$.

If $\vec{F}_{on\,q}$ is the only force acting on q, it causes the charged particle to accelerate with

$$\vec{a} = \frac{\vec{F}_{on\,q}}{m} = \frac{q}{m}\vec{E} \qquad (23.32)$$

The technique of *gel electrophoresis* uses an electric field to measure "DNA fingerprints." DNA fragments are charged, and fragments with different charge-to-mass ratios are separated by the field.

This acceleration is the *response* of the charged particle to the source charges that created the electric field. The ratio q/m is especially important for the dynamics of charged-particle motion. It is called the **charge-to-mass ratio.** Two *equal* charges, say a proton and a Na^+ ion, will experience *equal* forces $\vec{F} = q\vec{E}$ if placed at the same point in an electric field, but their accelerations will be *different* because they have different masses and thus different charge-to-mass ratios. Two particles with different charges and masses *but* with the same charge-to-mass ratio will undergo the same acceleration and follow the same trajectory.

Motion in a Uniform Field

The motion of a charged particle in a *uniform* electric field is especially important for its basic simplicity and because of its many valuable applications. A uniform field is *constant* at all points—constant in both magnitude and direction—within the region of space where the charged particle is moving. It follows, from Equation 23.32, that **a charged particle in a uniform electric field will move with constant acceleration.** The magnitude of the acceleration is

$$a = \frac{qE}{m} = \text{constant} \tag{23.33}$$

where E is the electric field strength, and the direction of $\vec{a}$ is parallel or antiparallel to $\vec{E}$, depending on the sign of q.

Identifying the motion of a charged particle in a uniform field as being one of constant acceleration brings into play all the kinematic machinery that we developed in Chapters 2 and 4 for constant-acceleration motion. The basic trajectory of a charged particle in a uniform field is a *parabola,* analogous to the projectile motion of a mass in the near-earth uniform gravitational field. In the special case of a charged particle moving parallel to the electric field vectors, the motion is one-dimensional, analogous to the one-dimensional vertical motion of a mass tossed straight up or falling straight down.

> **NOTE** The gravitational acceleration $\vec{a}_{grav}$ always points straight down. The electric field acceleration $\vec{a}_{elec}$ can point in *any* direction. You must determine the electric field $\vec{E}$ in order to learn the direction of $\vec{a}$.

EXAMPLE 23.7 | An electron moving across a capacitor

Two 6.0-cm-diameter electrodes are spaced 5.0 mm apart. They are charged by transferring 1.0×10^{11} electrons from one electrode to the other. An electron is released from rest just above the surface of the negative electrode. How long does it take the electron to cross to the positive electrode? What is its speed as it collides with the positive electrode? Assume the space between the electrodes is a vacuum.

MODEL The electrodes form a parallel-plate capacitor. The charges *on* the electrodes cannot escape, but any additional charges *between* the capacitor plates will be accelerated by the electric field. The electric field inside a parallel-plate capacitor is a uniform field, so the electron will have constant acceleration.

VISUALIZE FIGURE 23.24 shows an edge view of the capacitor and the electron. The force on the negative electron is *opposite* the electric field, so the electron is repelled by the negative electrode as it accelerates across the gap of width d.

SOLVE The electrodes are not point charges, so we cannot use Coulomb's law to find the force on the electron. Instead, we must analyze the electron's motion in terms of the electric field inside the capacitor. The field is the agent that exerts the force on the electron,

FIGURE 23.24 An electron accelerates across a capacitor (plate separation exaggerated).

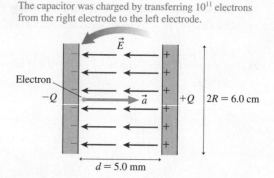

The capacitor was charged by transferring 10^{11} electrons from the right electrode to the left electrode.

causing it to accelerate. The electric field strength inside a parallel-plate capacitor with charge $Q = Ne$ is

$$E = \frac{\eta}{\epsilon_0} = \frac{Q}{\epsilon_0 A} = \frac{Ne}{\epsilon_0 \pi R^2} = 639{,}000 \text{ N/C}$$

The electron's acceleration in this field is

$$a = \frac{eE}{m} = 1.1 \times 10^{17} \text{ m/s}^2$$

where we used the electron mass $m = 9.11 \times 10^{-31}$ kg. This is an enormous acceleration compared to accelerations we're familiar with for macroscopic objects. We can use one-dimensional kinematics, with $x_i = 0$ and $v_i = 0$, to find the time required for the electron to cross the capacitor:

$$x_f = d = \tfrac{1}{2} a (\Delta t)^2$$

$$\Delta t = \sqrt{\frac{2d}{a}} = 3.0 \times 10^{-10} \text{ s} = 0.30 \text{ ns}$$

The electron's speed as it reaches the positive electrode is

$$v = a \Delta t = 3.3 \times 10^7 \text{ m/s}$$

ASSESS We used e rather than $-e$ to find the acceleration because we already knew the direction; we needed only the magnitude. The electron's speed, after traveling a mere 5 mm, is approximately 10% the speed of light.

Parallel electrodes such as those in Example 23.7 are often used to accelerate charged particles. If the positive plate has a small hole in the center, a *beam* of electrons will pass through the hole and emerge with a speed of 3.3×10^7 m/s. This is the basic idea of the *electron gun* used until quite recently in *cathode-ray tube* (CRT) devices such as televisions and computer display terminals. (A negatively charged electrode is called a *cathode,* so the physicists who first learned to produce electron beams in the late 19th century called them *cathode rays.*)

EXAMPLE 23.8 | **Deflecting an electron beam**

An electron gun creates a beam of electrons moving horizontally with a speed of 3.3×10^7 m/s. The electrons enter a 2.0-cm-long gap between two parallel electrodes where the electric field is $\vec{E} = (5.0 \times 10^4 \text{ N/C, down})$. In which direction, and by what angle, is the electron beam deflected by these electrodes?

MODEL The electric field between the electrodes is uniform. Assume that the electric field outside the electrodes is zero.

VISUALIZE FIGURE 23.25 shows an electron moving through the electric field. The electric field points down, so the force on the (negative) electrons is upward. The electrons will follow a parabolic trajectory, analogous to that of a ball thrown horizontally, except that the electrons "fall up" rather than down.

FIGURE 23.25 The deflection of an electron beam in a uniform electric field.

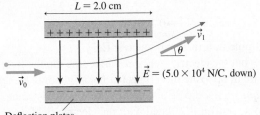

$L = 2.0$ cm

$\vec{v}_1$

θ

$\vec{E} = (5.0 \times 10^4 \text{ N/C, down})$

$\vec{v}_0$

Deflection plates

SOLVE This is a two-dimensional motion problem. The electron enters the capacitor with velocity *vector* $\vec{v}_0 = v_{0x} \hat{\imath} = 3.3 \times 10^7 \hat{\imath}$ m/s and leaves with velocity $\vec{v}_1 = v_{1x} \hat{\imath} + v_{1y} \hat{\jmath}$. The electron's angle of travel upon leaving the electric field is

$$\theta = \tan^{-1} \left(\frac{v_{1y}}{v_{1x}} \right)$$

This is the *deflection angle*. To find θ we must compute the final velocity vector $\vec{v}_1$.

There is no horizontal force on the electron, so $v_{1x} = v_{0x} = 3.3 \times 10^7$ m/s. The electron's upward acceleration has magnitude

$$a = \frac{eE}{m} = \frac{(1.60 \times 10^{-19} \text{ C})(5.0 \times 10^4 \text{ N/C})}{9.11 \times 10^{-31} \text{ kg}}$$

$$= 8.78 \times 10^{15} \text{ m/s}^2$$

We can use the fact that the horizontal velocity is constant to determine the time interval Δt needed to travel length 2.0 cm:

$$\Delta t = \frac{L}{v_{0x}} = \frac{0.020 \text{ m}}{3.3 \times 10^7 \text{ m/s}} = 6.06 \times 10^{-10} \text{ s}$$

Vertical acceleration will occur during this time interval, resulting in a final vertical velocity

$$v_{1y} = v_{0y} + a \Delta t = 5.3 \times 10^6 \text{ m/s}$$

The electron's velocity as it leaves the capacitor is thus

$$\vec{v}_1 = (3.3 \times 10^7 \, \hat{\imath} + 5.3 \times 10^6 \, \hat{\jmath}) \text{ m/s}$$

and the deflection angle θ is

$$\theta = \tan^{-1} \left(\frac{v_{1y}}{v_{1x}} \right) = 9.1°$$

ASSESS We know that the electron beam in a cathode-ray tube can be deflected enough to cover the screen, so a deflection angle of 9° seems reasonable. Our neglect of the gravitational force is seen to be justified because the acceleration of the electrons is enormous in comparison to the free-fall acceleration g.

By using two sets of deflection plates—one for vertical deflection and one for horizontal—a cathode-ray tube could steer the electrons to any point on the screen. Electrons striking a phosphor coating on the inside of the screen would then make a dot of light.

STOP TO THINK 23.6 Which electric field is responsible for the proton's trajectory?

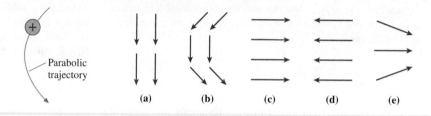

23.7 Motion of a Dipole in an Electric Field

Let us conclude this chapter by returning to one of the more striking puzzles we faced when making the observations at the beginning of Chapter 22. There you found that charged objects of *either* sign exert forces on neutral objects, such as when a comb used to brush your hair picks up pieces of paper. Our qualitative understanding of the *polarization force* was that it required two steps:

■ The charge polarizes the neutral object, creating an induced electric dipole.
■ The charge then exerts an attractive force on the near end of the dipole that is slightly stronger than the repulsive force on the far end.

We are now in a position to make that understanding more quantitative.

Dipoles in a Uniform Field

FIGURE 23.26a shows an electric dipole in a *uniform* external electric field $\vec{E}$ that has been created by source charges we do not see. That is, $\vec{E}$ is *not* the field of the dipole but, instead, is a field to which the dipole is responding. In this case, because the field is uniform, the dipole is presumably inside an unseen parallel-plate capacitor.

The net force on the dipole is the sum of the forces on the two charges forming the dipole. Because the charges $\pm q$ are equal in magnitude but opposite in sign, the two forces $\vec{F}_+ = +q\vec{E}$ and $\vec{F}_- = -q\vec{E}$ are also equal but opposite. Thus the net force on the dipole is

$$\vec{F}_{net} = \vec{F}_+ + \vec{F}_- = \vec{0} \tag{23.34}$$

There is no net force on a dipole in a uniform electric field.

There may be no net force, but the electric field *does* affect the dipole. Because the two forces in Figure 23.26a are in opposite directions but not aligned with each other, the electric field exerts a *torque* on the dipole and causes the dipole to *rotate*.

The torque rotates the dipole until it is aligned with the electric field, as shown in **FIGURE 23.26b**. In this position, the dipole experiences not only no net force but also no torque. Thus Figure 23.26b represents the *equilibrium position* for a dipole in a uniform electric field. Notice that the positive end of the dipole is in the direction in which $\vec{E}$ points.

FIGURE 23.27 shows a sample of permanent dipoles, such as water molecules, in an external electric field. All the dipoles rotate until they are aligned with the electric field. This is the mechanism by which the sample becomes *polarized*. Once the dipoles are aligned, there is an excess of positive charge at one end of the sample and an excess of negative charge at the other end. The excess charges at the ends of the sample are the basis of the polarization forces we discussed in Section 22.3.

FIGURE 23.26 A dipole in a uniform electric field.

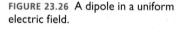

(a) The electric field exerts a torque on this dipole.

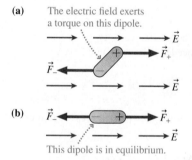

(b) This dipole is in equilibrium.

FIGURE 23.27 A sample of permanent dipoles is *polarized* in an electric field.

The dipoles align with the electric field.

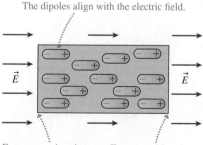

Excess negative charge on this surface   Excess positive charge on this surface

It's not hard to calculate the torque. Recall from Chapter 12 that the magnitude of a torque is the product of the force and the moment arm. **FIGURE 23.28** shows that there are two forces of the same magnitude $(F_+ = F_- = qE)$, each with the same moment arm $(d = \frac{1}{2}s\sin\theta)$. Thus the torque on the dipole is

$$\tau = 2 \times dF_+ = 2(\tfrac{1}{2}s\sin\theta)(qE) = pE\sin\theta \tag{23.35}$$

where $p = qs$ was our definition of the dipole moment. The torque is zero when the dipole is aligned with the field, making $\theta = 0$.

Also recall from Chapter 12 that the torque can be written in a compact mathematical form as the cross product between two vectors. The terms p and E in Equation 23.35 are the magnitudes of vectors, and θ is the angle between them. Thus in vector notation, the torque exerted on a dipole moment $\vec{p}$ by an electric field $\vec{E}$ is

$$\vec{\tau} = \vec{p} \times \vec{E} \tag{23.36}$$

The torque is greatest when $\vec{p}$ is perpendicular to $\vec{E}$, zero when $\vec{p}$ is aligned with or opposite to $\vec{E}$.

FIGURE 23.28 The torque on a dipole.

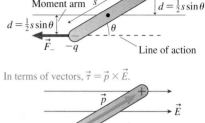

In terms of vectors, $\vec{\tau} = \vec{p} \times \vec{E}$.

EXAMPLE 23.9 | The angular acceleration of a dipole dumbbell

Two 1.0 g balls are connected by a 2.0-cm-long insulating rod of negligible mass. One ball has a charge of $+10$ nC, the other a charge of -10 nC. The rod is held in a 1.0×10^4 N/C uniform electric field at an angle of $30°$ with respect to the field, then released. What is its initial angular acceleration?

MODEL The two oppositely charged balls form an electric dipole. The electric field exerts a torque on the dipole, causing an angular acceleration.

VISUALIZE **FIGURE 23.29** shows the dipole in the electric field.

FIGURE 23.29 The dipole of Example 23.9.

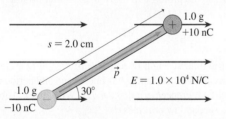

SOLVE The dipole moment is $p = qs = (1.0 \times 10^{-8}\,\text{C}) \times (0.020\,\text{m}) = 2.0 \times 10^{-10}\,\text{C m}$. The torque exerted on the dipole moment by the electric field is

$$\tau = pE\sin\theta = (2.0 \times 10^{-10}\,\text{C m})(1.0 \times 10^4\,\text{N/C})\sin 30°$$
$$= 1.0 \times 10^{-6}\,\text{N m}$$

You learned in Chapter 12 that a torque causes an angular acceleration $\alpha = \tau/I$, where I is the moment of inertia. The dipole rotates about its center of mass, which is at the center of the rod, so the moment of inertia is

$$I = m_1 r_1^2 + m_2 r_2^2 = 2m(\tfrac{1}{2}s)^2 = \tfrac{1}{2}ms^2 = 2.0 \times 10^{-7}\,\text{kg m}^2$$

Thus the rod's angular acceleration is

$$\alpha = \frac{\tau}{I} = \frac{1.0 \times 10^{-6}\,\text{N m}}{2.0 \times 10^{-7}\,\text{kg m}^2} = 5.0\,\text{rad/s}^2$$

ASSESS This value of α is the initial angular acceleration, when the rod is first released. The torque and the angular acceleration will decrease as the rod rotates toward alignment with $\vec{E}$.

Dipoles in a Nonuniform Field

Suppose that a dipole is placed in a nonuniform electric field, one in which the field strength changes with position. For example, **FIGURE 23.30** shows a dipole in the nonuniform field of a point charge. The first response of the dipole is to rotate until it is aligned with the field, with the dipole's positive end pointing in the same direction as the field. Now, however, there is a *slight difference* between the forces acting on the two ends of the dipole. This difference occurs because the electric field, which depends on the distance from the point charge, is stronger at the end of the dipole nearest the charge. This causes a net force to be exerted on the dipole.

Which way does the force point? Once the dipole is aligned, the leftward attractive force on its negative end is slightly stronger than the rightward repulsive force on its positive end. This causes a net force *toward* the point charge.

In fact, for any nonuniform electric field, **the net force on a dipole is toward the direction of the strongest field.** Because any finite-size charged object, such as a charged rod or a charged disk, has a field strength that increases as you get closer to the object, we can conclude that **a dipole will experience a net force toward any charged object.**

FIGURE 23.30 An aligned dipole is drawn toward a point charge.

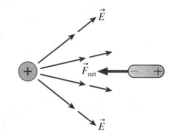

EXAMPLE 23.10 | The force on a water molecule

The water molecule H_2O has a permanent dipole moment of magnitude 6.2×10^{-30} C m. A water molecule is located 10 nm from a Na^+ ion in a saltwater solution. What force does the ion exert on the water molecule?

VISUALIZE FIGURE 23.31 shows the ion and the dipole. The forces are an action/reaction pair.

FIGURE 23.31 The interaction between an ion and a permanent dipole.

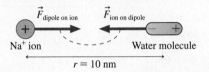

$\vec{F}_{\text{dipole on ion}}$ $\vec{F}_{\text{ion on dipole}}$

Na$^+$ ion   Water molecule

$r = 10$ nm

SOLVE A Na^+ ion has charge $q = +e$. The electric field of the ion aligns the water's dipole moment and exerts a net force on it. We could calculate the net force on the dipole as the small difference between the attractive force on its negative end and the repulsive force on its positive end. Alternatively, we know from Newton's

third law that the force $\vec{F}_{\text{dipole on ion}}$ has the same magnitude as the force $\vec{F}_{\text{ion on dipole}}$ that we are seeking. We calculated the on-axis field of a dipole in Section 23.2. An ion of charge $q = e$ will experience a force of magnitude $F = qE_{\text{dipole}} = eE_{\text{dipole}}$ when placed in that field. The dipole's electric field, which we found in Equation 23.10, is

$$E_{\text{dipole}} = \frac{1}{4\pi\epsilon_0} \frac{2p}{r^3}$$

The force on the ion at distance $r = 1.0 \times 10^{-8}$ m is

$$F_{\text{dipole on ion}} = eE_{\text{dipole}} = \frac{1}{4\pi\epsilon_0} \frac{2ep}{r^3} = 1.8 \times 10^{-14} \, \text{N}$$

Thus the force on the water molecule is $F_{\text{ion on dipole}} = 1.8 \times 10^{-14}$ N.

ASSESS While 1.8×10^{-14} N may seem like a very small force, it is $\approx 10^{11}$ times larger than the size of the earth's gravitational force on these atomic particles. Forces such as these cause water molecules to cluster around any ions that are in solution. This clustering plays an important role in the microscopic physics of solutions studied in chemistry and biochemistry.

CHALLENGE EXAMPLE 23.11 | An orbiting proton

In a vacuum chamber, a proton orbits a 1.0-cm-diameter metal ball 1.0 mm above the surface with a period of 1.0 μs. What is the charge on the ball?

MODEL Model the ball as a charged sphere. The electric field of a charged sphere is the same as that of a point charge at the center, so the radius of the ball is irrelevant. Assume that the gravitational force on the proton is extremely small compared to the electric force and can be neglected.

VISUALIZE FIGURE 23.32 shows the orbit and the force on the proton.

FIGURE 23.32 An orbiting proton.

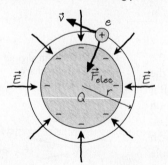

SOLVE The ball must be negative, with an inward electric field exerting an inward electric force on the positive proton. This is

exactly the necessary condition for uniform circular motion. Recall from Chapter 8 that Newton's second law for uniform circular motion is $(F_{\text{net}})_r = mv^2/r$. Here the only radial force has magnitude $F_{\text{elec}} = eE$, so the proton will move in a circular orbit if

$$eE = \frac{mv^2}{r}$$

The electric field strength of a sphere of charge Q at distance r is $E = Q/4\pi\epsilon_0 r^2$. From Chapter 4, orbital speed and period are related by $v = \text{circumference/period} = 2\pi r/T$. With these substitutions, Newton's second law becomes

$$\frac{eQ}{4\pi\epsilon_0 r^2} = \frac{4\pi^2 m}{T^2} r$$

Solving for Q, we find

$$Q = \frac{16\pi^3 \epsilon_0 m r^3}{eT^2} = 9.9 \times 10^{-12} \, \text{C}$$

where we used $r = 6.0$ mm as the radius of the proton's orbit. Q is the *magnitude* of the charge on the ball. Including the sign, we have

$$Q_{\text{ball}} = -9.9 \times 10^{-12} \, \text{C}$$

ASSESS This is not a lot of charge, but it shouldn't take much charge to affect the motion of something as light as a proton.

SUMMARY

The goal of Chapter 23 has been to learn how to calculate and use the electric field.

GENERAL PRINCIPLES

Sources of $\vec{E}$

Electric fields are created by charges.

Multiple point charges

MODEL Model objects as point charges.

VISUALIZE Establish a coordinate system and draw field vectors.

SOLVE Use superposition: $\vec{E} = \vec{E}_1 + \vec{E}_2 + \vec{E}_3 + \cdots$

Continuous distribution of charge

MODEL Model objects as simple shapes.

VISUALIZE

- Establish a coordinate system.
- Divide the charge into small segments ΔQ.
- Draw a field vector for one or two pieces of charge.

SOLVE

- Find the field of each ΔQ.
- Write $\vec{E}$ as the sum of the fields of all ΔQ. Don't forget that it's a *vector* sum; use components.
- Use the charge density (λ or η) to replace ΔQ with an integration coordinate, then integrate.

Consequences of $\vec{E}$

The electric field exerts a force on a charged particle:

$$\vec{F} = q\vec{E}$$

The force causes acceleration:

$$\vec{a} = (q/m)\vec{E}$$

Trajectories of charged particles are calculated with kinematics.

The electric field exerts a torque on a dipole:

$$\tau = pE\sin\theta$$

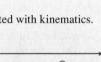

The torque tends to align the dipoles with the field.

In a nonuniform electric field, a dipole has a net force in the direction of increasing field strength.

APPLICATIONS

Four Key Electric Field Models

Point charge with charge q

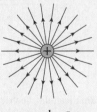

$$\vec{E} = \frac{1}{4\pi\epsilon_0}\frac{q}{r^2}\hat{r}$$

Infinite line of charge with linear charge density λ

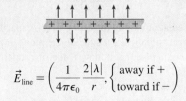

$$\vec{E}_{\text{line}} = \left(\frac{1}{4\pi\epsilon_0}\frac{2|\lambda|}{r}, \left\{\begin{array}{l}\text{away if }+\\ \text{toward if }-\end{array}\right.\right)$$

Infinite plane of charge with surface charge density η

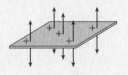

$$\vec{E}_{\text{plane}} = \left(\frac{|\eta|}{2\epsilon_0}, \left\{\begin{array}{l}\text{away if }+\\ \text{toward if }-\end{array}\right.\right)$$

Sphere of charge with total charge Q

Same as a point charge Q for $r > R$

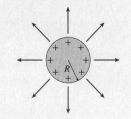

Electric dipole

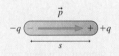

The electric dipole moment is

$$\vec{p} = (qs, \text{from negative to positive})$$

Field on axis: $\vec{E} = \dfrac{1}{4\pi\epsilon_0}\dfrac{2\vec{p}}{r^3}$

Field in bisecting plane: $\vec{E} = -\dfrac{1}{4\pi\epsilon_0}\dfrac{\vec{p}}{r^3}$

Parallel-plate capacitor

The electric field inside an ideal capacitor is a **uniform electric field**:

$$\vec{E} = \left(\frac{\eta}{\epsilon_0}, \text{from positive to negative}\right)$$

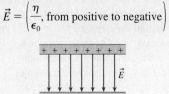

TERMS AND NOTATION

dipole moment, $\vec{p}$	uniformly charged	plane of charge	fringe field
electric field line	line of charge	sphere of charge	uniform electric field
linear charge density, λ	electrode	parallel-plate capacitor	charge-to-mass ratio, q/m
surface charge density, η			

CONCEPTUAL QUESTIONS

1. You've been assigned the task of determining the magnitude and direction of the electric field at a point in space. Give a step-by-step procedure of how you will do so. List any objects you will use, any measurements you will make, and any calculations you will need to perform. Make sure that your measurements do not disturb the charges that are creating the field.

2. Reproduce FIGURE Q23.2 on your paper. For each part, draw a dot or dots on the figure to show any position or positions (other than infinity) where $\vec{E} = \vec{0}$.

(a)

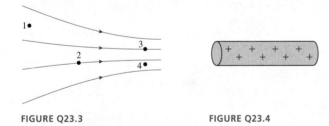

(b)

FIGURE Q23.2

3. Rank in order, from largest to smallest, the electric field strengths E_1 to E_4 at points 1 to 4 in FIGURE Q23.3. Explain.

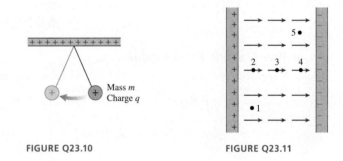

FIGURE Q23.3   FIGURE Q23.4

4. A small segment of wire in FIGURE Q23.4 contains 10 nC of charge.
   a. The segment is shrunk to one-third of its original length. What is the ratio λ_f/λ_i, where λ_i and λ_f are the initial and final linear charge densities?
   b. A proton is very far from the wire. What is the ratio F_f/F_i of the electric force on the proton after the segment is shrunk to the force before the segment was shrunk?
   c. Suppose the original segment of wire is stretched to 10 times its original length. How much charge must be *added* to the wire to keep the linear charge density unchanged?

5. An electron experiences a force of magnitude F when it is 1 cm from a very long, charged wire with linear charge density λ. If the charge density is doubled, at what distance from the wire will a proton experience a force of the same magnitude F?

6. FIGURE Q23.6 shows a hollow soda straw that has been uniformly charged with positive charge. What is the electric field at the center (inside) of the straw? Explain.

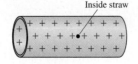

Inside straw

FIGURE Q23.6

7. The irregularly shaped area of charge in FIGURE Q23.7 has surface charge density η_i. Each dimension (x and y) of the area is reduced by a factor of 3.163.
   a. What is the ratio η_f/η_i, where η_f is the final surface charge density?
   b. An electron is very far from the area. What is the ratio F_f/F_i of the electric force on the electron after the area is reduced to the force before the area was reduced?

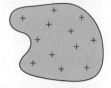

FIGURE Q23.7

8. A circular disk has surface charge density 8 nC/cm². What will the surface charge density be if the radius of the disk is doubled?

9. A sphere of radius R has charge Q. The electric field strength at distance $r > R$ is E_i. What is the ratio E_f/E_i of the final to initial electric field strengths if (a) Q is halved, (b) R is halved, and (c) r is halved (but is still $> R$)? Each part changes only one quantity; the other quantities have their initial values.

10. The ball in FIGURE Q23.10 is suspended from a large, uniformly charged positive plate. It swings with period T. If the ball is discharged, will the period increase, decrease, or stay the same? Explain.

FIGURE Q23.10   FIGURE Q23.11

11. Rank in order, from largest to smallest, the electric field strengths E_1 to E_5 at the five points in FIGURE Q23.11. Explain.

12. A parallel-plate capacitor consists of two square plates, size $L \times L$, separated by distance d. The plates are given charge $\pm Q$. What is the ratio E_f/E_i of the final to initial electric field strengths if (a) Q is doubled, (b) L is doubled, and (c) d is doubled? Each part changes only one quantity; the other quantities have their initial values.

13. A small object is released at point 3 in the center of the capacitor in FIGURE Q23.11. For each situation, does the object move to the right, to the left, or remain in place? If it moves, does it accelerate or move at constant speed?
    a. A positive object is released from rest.
    b. A neutral but polarizable object is released from rest.
    c. A negative object is released from rest.

14. A proton and an electron are released from rest in the center of a capacitor.
    a. Is the force ratio F_p/F_e greater than 1, less than 1, or equal to 1? Explain.
    b. Is the acceleration ratio a_p/a_e greater than 1, less than 1, or equal to 1? Explain.

15. Three charges are placed at the corners of the triangle in FIGURE Q23.15. The $++$ charge has twice the quantity of charge of the two $-$ charges; the net charge is zero. Is the triangle in equilibrium? If so, explain why. If not, draw the equilibrium orientation.

FIGURE Q23.15

EXERCISES AND PROBLEMS

Problems labeled ▓▓ integrate material from earlier chapters.

Exercises

Section 23.2 The Electric Field of Point Charges

1. ‖ What are the strength and direction of the electric field at the position indicated by the dot in FIGURE EX23.1? Specify the direction as an angle above or below horizontal.

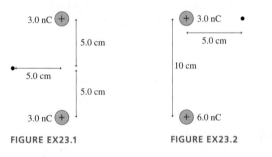

FIGURE EX23.1   FIGURE EX23.2

2. ‖ What are the strength and direction of the electric field at the position indicated by the dot in FIGURE EX23.2? Specify the direction as an angle above or below horizontal.

3. ‖ What are the strength and direction of the electric field at the position indicated by the dot in FIGURE EX23.3? Specify the direction as an angle above or below horizontal.

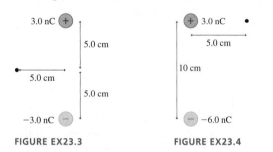

FIGURE EX23.3   FIGURE EX23.4

4. ‖ What are the strength and direction of the electric field at the position indicated by the dot in FIGURE EX23.4? Specify the direction as an angle above or below horizontal.

5. ‖ An electric dipole is formed from two charges, $\pm q$, spaced 1.0 cm apart. The dipole is at the origin, oriented along the y-axis. The electric field strength at the point $(x, y) = (0 \text{ cm}, 10 \text{ cm})$ is 360 N/C.
    a. What is the charge q? Give your answer in nC.
    b. What is the electric field strength at the point $(x, y) = (10 \text{ cm}, 0 \text{ cm})$?

6. ‖ An electric dipole is formed from ± 1.0 nC charges spaced 2.0 mm apart. The dipole is at the origin, oriented along the x-axis. What is the electric field strength at the points (a) $(x, y) = (10 \text{ cm}, 0 \text{ cm})$ and (b) $(x, y) = (0 \text{ cm}, 10 \text{ cm})$?

7. ‖ An *electret* is similar to a magnet, but rather than being permanently magnetized, it has a permanent electric dipole moment. Suppose a small electret with electric dipole moment 1.0×10^{-7} C m is 25 cm from a small ball charged to $+25$ nC, with the ball on the axis of the electric dipole. What is the magnitude of the electric force on the ball?

Section 23.3 The Electric Field of a Continuous Charge Distribution

8. ‖ The electric field strength 10.0 cm from a very long charged wire is 2000 N/C. What is the electric field strength 5.0 cm from the wire?

9. ‖ A 10-cm-long thin glass rod uniformly charged to $+10$ nC and a 10-cm-long thin plastic rod uniformly charged to -10 nC are placed side by side, 4.0 cm apart. What are the electric field strengths E_1 to E_3 at distances 1.0 cm, 2.0 cm, and 3.0 cm from the glass rod along the line connecting the midpoints of the two rods?

10. ‖ Two 10-cm-long thin glass rods uniformly charged to $+10$ nC are placed side by side, 4.0 cm apart. What are the electric field strengths E_1 to E_3 at distances 1.0 cm, 2.0 cm, and 3.0 cm to the right of the rod on the left along the line connecting the midpoints of the two rods?

11. ‖ A small glass bead charged to $+6.0$ nC is in the plane that bisects a thin, uniformly charged, 10-cm-long glass rod and is 4.0 cm from the rod's center. The bead is repelled from the rod with a force of 840 μN. What is the total charge on the rod?

12. ‖‖ The electric field 5.0 cm from a very long charged wire is (2000 N/C, toward the wire). What is the charge (in nC) on a 1.0-cm-long segment of the wire?

13. ‖‖ A 12-cm-long thin rod has the *nonuniform* charge density
CALC $\lambda(x) = (2.0 \text{ nC/cm})e^{-|x|/(6.0 \text{ cm})}$, where x is measured from the center of the rod. What is the total charge on the rod?
    Hint: This exercise requires an integration. Think about how to handle the absolute value sign.

Section 23.4 The Electric Fields of Rings, Disks, Planes, and Spheres

14. ‖ Two 10-cm-diameter charged rings face each other, 20 cm apart. The left ring is charged to -20 nC and the right ring is charged to $+20$ nC.
    a. What is the electric field $\vec{E}$, both magnitude and direction, at the midpoint between the two rings?
    b. What is the force on a proton at the midpoint?

15. ‖ Two 10-cm-diameter charged rings face each other, 20 cm apart. Both rings are charged to $+20$ nC. What is the electric field strength at (a) the midpoint between the two rings and (b) the center of the left ring?

16. ‖ Two 10-cm-diameter charged disks face each other, 20 cm apart. The left disk is charged to -50 nC and the right disk is charged to $+50$ nC.
    a. What is the electric field $\vec{E}$, both magnitude and direction, at the midpoint between the two disks?
    b. What is the force $\vec{F}$ on a -1.0 nC charge placed at the midpoint?

17. ‖ The electric field strength 2.0 cm from the surface of a 10-cm-diameter metal ball is 50,000 N/C. What is the charge (in nC) on the ball?

18. ‖ A 20 cm × 20 cm horizontal metal electrode is uniformly charged to $+80$ nC. What is the electric field strength 2.0 mm above the center of the electrode?

19. ‖ Two 2.0-cm-diameter insulating spheres have a 6.0 cm space between them. One sphere is charged to $+10$ nC, the other to -15 nC. What is the electric field strength at the midpoint between the two spheres?

20. ‖ You've hung two very large sheets of plastic facing each other with distance d between them, as shown in FIGURE EX23.20. By rubbing them with wool and silk, you've managed to give one sheet a uniform surface charge density $\eta_1 = -\eta_0$ and the other a uniform surface charge density $\eta_2 = +3\eta_0$. What are the electric field vectors at points 1, 2, and 3?

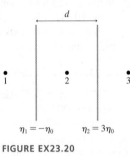

FIGURE EX23.20

21. ‖ A 2.0 m × 4.0 m flat carpet acquires a uniformly distributed charge of $-10\ \mu$C after you and your friends walk across it several times. A 2.5 μg dust particle is suspended in midair just above the center of the carpet. What is the charge on the dust particle?

Section 23.5 The Parallel-Plate Capacitor

22. ‖ Two circular disks spaced 0.50 mm apart form a parallel-plate capacitor. Transferring 3.0×10^9 electrons from one disk to the other causes the electric field strength to be 2.0×10^5 N/C. What are the diameters of the disks?

23. ‖ A parallel-plate capacitor is formed from two 6.0-cm-diameter electrodes spaced 2.0 mm apart. The electric field strength inside the capacitor is 1.0×10^6 N/C. What is the charge (in nC) on each electrode?

24. ‖‖ Air "breaks down" when the electric field strength reaches 3.0×10^6 N/C, causing a spark. A parallel-plate capacitor is made from two 4.0 cm × 4.0 cm electrodes. How many electrons must be transferred from one electrode to the other to create a spark between the electrodes?

25. ‖ Two parallel plates 1.0 cm apart are equally and oppositely charged. An electron is released from rest at the surface of the negative plate and simultaneously a proton is released from rest at the surface of the positive plate. How far from the negative plate is the point at which the electron and proton pass each other?

Section 23.6 Motion of a Charged Particle in an Electric Field

26. ‖ Two 2.0-cm-diameter disks face each other, 1.0 mm apart. They are charged to ± 10 nC.
    a. What is the electric field strength between the disks?
    b. A proton is shot from the negative disk toward the positive disk. What launch speed must the proton have to just barely reach the positive disk?

27. ‖ Honeybees acquire a charge while flying due to friction with the air. A 100 mg bee with a charge of $+23$ pC experiences an electric force in the earth's electric field, which is typically 100 N/C, directed downward.
    a. What is the ratio of the electric force on the bee to the bee's weight?
    b. What electric field strength and direction would allow the bee to hang suspended in the air?

28. ‖ An electron traveling parallel to a uniform electric field increases its speed from 2.0×10^7 m/s to 4.0×10^7 m/s over a distance of 1.2 cm. What is the electric field strength?

29. ‖ The surface charge density on an infinite charged plane is -2.0×10^{-6} C/m^2. A proton is shot straight away from the plane at 2.0×10^6 m/s. How far does the proton travel before reaching its turning point?

30. ‖ An electron in a vacuum chamber is fired with a speed of 8300 km/s toward a large, uniformly charged plate 75 cm away. The electron reaches a closest distance of 15 cm before being repelled. What is the plate's surface charge density?

31. ‖‖ A 1.0-μm-diameter oil droplet (density 900 kg/m^3) is negatively charged with the addition of 25 extra electrons. It is released from rest 2.0 mm from a very wide plane of positive charge, after which it accelerates toward the plane and collides with a speed of 3.5 m/s. What is the surface charge density of the plane?

Section 23.7 Motion of a Dipole in an Electric Field

32. ‖ The permanent electric dipole moment of the water molecule (H$_2$O) is 6.2×10^{-30} C m. What is the maximum possible torque on a water molecule in a 5.0×10^8 N/C electric field?

33. ‖ A point charge Q is distance r from a dipole consisting of charges $\pm q$ separated by distance s. The dipole is initially oriented so that Q is in the plane bisecting the dipole. Immediately after the dipole is released, what are (a) the magnitude of the force and (b) the magnitude of the torque on the dipole? You can assume $r \gg s$.

34. ‖ An ammonia molecule (NH$_3$) has a permanent electric dipole moment 5.0×10^{-30} C m. A proton is 2.0 nm from the molecule in the plane that bisects the dipole. What is the electric force of the molecule on the proton?

Problems

35. ‖ What are the strength and direction of the electric field at the position indicated by the dot in FIGURE P23.35? Give your answer (a) in component form and (b) as a magnitude and angle measured cw or ccw (specify which) from the positive x-axis.

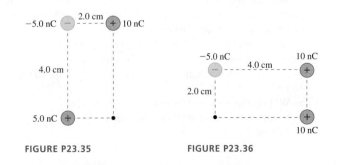

FIGURE P23.35   FIGURE P23.36

36. ‖ What are the strength and direction of the electric field at the position indicated by the dot in FIGURE P23.36? Give your answer (a) in component form and (b) as a magnitude and angle measured cw or ccw (specify which) from the positive x-axis.

37. ‖ What are the strength and direction of the electric field at the position indicated by the dot in FIGURE P23.37? Give your answer (a) in component form and (b) as a magnitude and angle measured cw or ccw (specify which) from the positive x-axis.

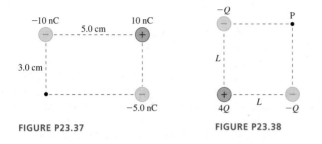

FIGURE P23.37   FIGURE P23.38

38. ‖ FIGURE P23.38 shows three charges at the corners of a square. Write the electric field at point P in component form.

39. ‖ Charges $-q$ and $+2q$ in FIGURE P23.39 are located at $x = \pm a$. Determine the electric field at points 1 to 4. Write each field in component form.

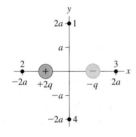

FIGURE P23.39

40. ‖ Derive Equation 23.11 for the field $\vec{E}_{dipole}$ in the plane that bisects an electric dipole.

41. ‖ FIGURE P23.41 is a cross section of two infinite lines of charge that extend out of the page. Both have linear charge density λ. Find an expression for the electric field strength E at height y above the midpoint between the lines.

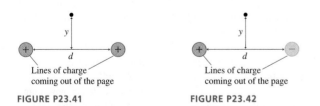

FIGURE P23.41   FIGURE P23.42

42. ‖‖ FIGURE P23.42 is a cross section of two infinite lines of charge that extend out of the page. The linear charge densities are $\pm \lambda$. Find an expression for the electric field strength E at height y above the midpoint between the lines.

43. ‖ FIGURE P23.43 shows a thin rod of length L with total charge Q.
CALC a. Find an expression for the electric field strength at point P on the axis of the rod at distance r from the center.
b. Verify that your expression has the expected behavior if $r \gg L$.
c. Evaluate E at $r = 3.0$ cm if $L = 5.0$ cm and $Q = 3.0$ nC.

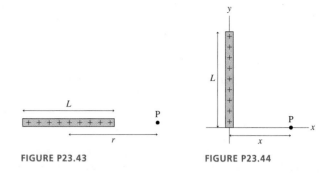

FIGURE P23.43   FIGURE P23.44

44. ‖ FIGURE P23.44 shows a thin rod of length L with total charge
CALC Q. Find an expression for the electric field $\vec{E}$ at point P. Give your answer in component form.

45. ‖ Show that the on-axis electric field of a ring of charge has the expected behavior when $z \ll R$ and when $z \gg R$.

46. ‖ A ring of radius R has total charge Q.
CALC a. At what distance along the z-axis is the electric field strength a maximum?
b. What is the electric field strength at this point?

47. ‖ Charge Q is uniformly distributed along a thin, flexible rod
CALC of length L. The rod is then bent into the semicircle shown in FIGURE P23.47.
a. Find an expression for the electric field $\vec{E}$ at the center of the semicircle.
Hint: A small piece of arc length Δs spans a small angle $\Delta \theta = \Delta s/R$, where R is the radius.
b. Evaluate the field strength if $L = 10$ cm and $Q = 30$ nC.

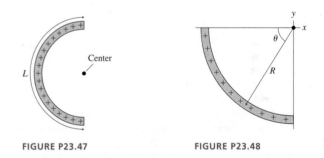

FIGURE P23.47   FIGURE P23.48

48. ‖ A plastic rod with linear charge density λ is bent into the
CALC quarter circle shown in FIGURE P23.48. We want to find the electric field at the origin.
a. Write expressions for the x- and y-components of the electric field at the origin due to a small piece of charge at angle θ.
b. Write, but do not evaluate, definite integrals for the x- and y-components of the net electric field at the origin.
c. Evaluate the integrals and write $\vec{E}_{net}$ in component form.

49. ‖ An infinite plane of charge with surface charge density $3.2 \ \mu C/m^2$ has a 20-cm-diameter circular hole cut out of it. What is the electric field strength directly over the center of the hole at a distance of 12 cm?
Hint: Can you create this charge distribution as a superposition of charge distributions for which you know the electric field?

50. ‖ A sphere of radius R and surface charge density η is positioned with its center distance $2R$ from an infinite plane with surface charge density η. At what distance from the plane, along a line toward the center of the sphere, is the electric field zero?

51. ‖ A parallel-plate capacitor has 2.0 cm × 2.0 cm electrodes with surface charge densities $\pm 1.0 \times 10^{-6}$ C/m². A proton traveling parallel to the electrodes at 1.0×10^6 m/s enters the center of the gap between them. By what distance has the proton been deflected sideways when it reaches the far edge of the capacitor? Assume the field is uniform inside the capacitor and zero outside the capacitor.

52. ‖ An electron is launched at a 45° angle and a speed of 5.0×10^6 m/s from the positive plate of the parallel-plate capacitor shown in FIGURE P23.52. The electron lands 4.0 cm away.
    a. What is the electric field strength inside the capacitor?
    b. What is the smallest possible spacing between the plates?

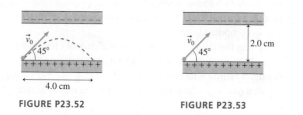

FIGURE P23.52   **FIGURE P23.53**

53. ‖‖‖ The two parallel plates in FIGURE P23.53 are 2.0 cm apart and the electric field strength between them is 1.0×10^4 N/C. An electron is launched at a 45° angle from the positive plate. What is the maximum initial speed v_0 the electron can have without hitting the negative plate?

54. ‖ A problem of practical interest is to make a beam of electrons turn a 90° corner. This can be done with the parallel-plate capacitor shown in FIGURE P23.54. An electron with kinetic energy 3.0×10^{-17} J enters through a small hole in the bottom plate of the capacitor.

FIGURE P23.54

    a. Should the bottom plate be charged positive or negative relative to the top plate if you want the electron to turn to the right? Explain.
    b. What strength electric field is needed if the electron is to emerge from an exit hole 1.0 cm away from the entrance hole, traveling at right angles to its original direction?
    Hint: The difficulty of this problem depends on how you choose your coordinate system.
    c. What minimum separation d_{min} must the capacitor plates have?

55. ‖‖‖ A *positron* is an elementary particle identical to an electron except that its charge is $+e$. An electron and a positron can rotate about their center of mass as if they were a dumbbell connected by a massless rod. What is the orbital frequency for an electron and a positron 1.0 nm apart?

56. ‖‖‖ Your physics assignment
CALC is to figure out a way to use electricity to launch a small 6.0-cm-long plastic drink stirrer. You decide that you'll charge the little plastic rod by rubbing it with fur, then hold it near a long, charged wire, as shown in FIGURE P23.56. When you let go, the electric

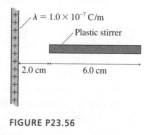

FIGURE P23.56

force of the wire on the plastic rod will shoot it away. Suppose you can uniformly charge the plastic stirrer to 10 nC and that the linear charge density of the long wire is 1.0×10^{-7} C/m. What is the net electric force on the plastic stirrer if the end closest to the wire is 2.0 cm away?
Hint: The stirrer cannot be modeled as a point charge; an integration is required.

57. ‖‖‖ The combustion of fossil fuels produces micron-sized particles of soot, one of the major components of air pollution. The terminal speeds of these particles are extremely small, so they remain suspended in air for very long periods of time. Furthermore, very small particles almost always acquire small amounts of charge from cosmic rays and various atmospheric effects, so their motion is influenced not only by gravity but also by the earth's weak electric field. Consider a small spherical particle of radius r, density ρ, and charge q. A small sphere moving with speed v experiences a drag force $F_{drag} = 6\pi\eta r v$, where η is the viscosity of the air. (This differs from the drag force you learned in Chapter 6 because there we considered macroscopic rather than microscopic objects.)
    a. A particle falling at its terminal speed v_{term} is in equilibrium with no net force. Write Newton's first law for this particle falling in the presence of a *downward* electric field of strength E, then solve to find an expression for v_{term}.
    b. Soot is primarily carbon, and carbon in the form of graphite has a density of 2200 kg/m³. In the absence of an electric field, what is the terminal speed in mm/s of a 1.0-μm-diameter graphite particle? The viscosity of air at 20°C is 1.8×10^{-5} kg/m s.
    c. The earth's electric field is typically (150 N/C, downward). In this field, what is the terminal speed in mm/s of a 1.0-μm-diameter graphite particle that has acquired 250 extra electrons?

58. ‖ A 2.0-mm-diameter glass sphere has a charge of +1.0 nC. What speed does an electron need to orbit the sphere 1.0 mm above the surface?

59. ‖ In a classical model of the hydrogen atom, the electron orbits the proton in a circular orbit of radius 0.053 nm. What is the orbital frequency? The proton is so much more massive than the electron that you can assume the proton is at rest.

60. ‖ An electric field can *induce* an electric dipole in a neutral atom or molecule by pushing the positive and negative charges in opposite directions. The dipole moment of an induced dipole is directly proportional to the electric field. That is, $\vec{p} = \alpha\vec{E}$, where α is called the *polarizability* of the molecule. A bigger field stretches the molecule farther and causes a larger dipole moment.
    a. What are the units of α?
    b. An ion with charge q is distance r from a molecule with polarizability α. Find an expression for the force $\vec{F}_{ion\,on\,dipole}$.

61. ‖ Show that an infinite line of charge with linear charge density λ exerts an attractive force on an electric dipole with magnitude $F = 2\lambda p/4\pi\epsilon_0 r^2$. Assume that r, the distance from the line, is much larger than the charge separation in the dipole.

62. ‖‖‖ The ozone molecule O_3 has a permanent dipole moment of 1.8×10^{-30} C m. Although the molecule is very slightly bent—which is why it has a dipole moment—it can be modeled as a uniform rod of length 2.5×10^{-10} m with the dipole moment perpendicular to the axis of the rod. Suppose an ozone molecule is in a 5000 N/C uniform electric field. In equilibrium, the dipole moment is aligned with the electric field. But if the molecule is rotated by a *small* angle and released, it will oscillate back and forth in simple harmonic motion. What is the frequency f of oscillation?

In Problems 63 through 66 you are given the equation(s) used to solve a problem. For each of these

   a. Write a realistic problem for which this is the correct equation(s).

   b. Finish the solution of the problem.

63. $(9.0 \times 10^9 \, \text{N m}^2/\text{C}^2) \dfrac{(2.0 \times 10^{-9} \, \text{C}) \, s}{(0.025 \, \text{m})^3} = 1150 \, \text{N/C}$

64. $(9.0 \times 10^9 \, \text{N m}^2/\text{C}^2) \dfrac{2(2.0 \times 10^{-7} \, \text{C/m})}{r} = 25{,}000 \, \text{N/C}$

65. $\dfrac{\eta}{2\epsilon_0}\left[1 - \dfrac{z}{\sqrt{z^2 + R^2}} \right] = \dfrac{1}{2} \dfrac{\eta}{2\epsilon_0}$

66. $2.0 \times 10^{12} \, \text{m/s}^2 = \dfrac{(1.60 \times 10^{-19} \, \text{C}) E}{(1.67 \times 10^{-27} \, \text{kg})}$

    $E = \dfrac{Q}{(8.85 \times 10^{-12} \, \text{C}^2/\text{N m}^2)(0.020 \, \text{m})^2}$

Challenge Problems

67. ||| A rod of length L lies along the y-axis with its center at the
CALC origin. The rod has a nonuniform linear charge density $\lambda = a|y|$, where a is a constant with the units C/m^2.

   a. Draw a graph of λ versus y over the length of the rod.

   b. Determine the constant a in terms of L and the rod's total charge Q.

   c. Find the electric field strength of the rod at distance x on the x-axis.

68. ||| a. An infinitely long *sheet* of charge of width L lies in the xy-
CALC   plane between $x = -L/2$ and $x = L/2$. The surface charge density is η. Derive an expression for the electric field $\vec{E}$ at height z above the centerline of the sheet.

   b. Verify that your expression has the expected behavior if $z \ll L$ and if $z \gg L$.

   c. Draw a graph of field strength E versus z.

69. ||| a. An infinitely long *sheet* of charge of width L lies in the xy-
CALC   plane between $x = -L/2$ and $x = L/2$. The surface charge density is η. Derive an expression for the electric field $\vec{E}$ along the x-axis for points outside the sheet $(x > L/2)$.

   b. Verify that your expression has the expected behavior if $x \gg L$.

    Hint: $\ln(1 + u) \approx u$ if $u \ll 1$.

   c. Draw a graph of field strength E versus x for $x > L/2$.

70. ||| A thin cylindrical shell of radius R and length L, like a soda
CALC straw, is uniformly charged with surface charge density η. What is the electric field strength at the center of one end of the cylinder?

71. ||| One type of ink-jet printer, called an electrostatic ink-jet printer, forms the letters by using deflecting electrodes to steer charged ink drops up and down vertically as the ink jet sweeps horizontally across the page. The ink jet forms 30-μm-diameter drops of ink, charges them by spraying 800,000 electrons on the surface, and shoots them toward the page at a speed of 20 m/s. Along the way, the drops pass through two horizontal, parallel electrodes that are 6.0 mm long, 4.0 mm wide, and spaced 1.0 mm apart. The distance from the center of the electrodes to the paper is 2.0 cm. To form the tallest letters, which have a height of 6.0 mm, the drops need to be deflected upward (or downward) by 3.0 mm. What electric field strength is needed between the electrodes to achieve this deflection? Ink, which consists of dye particles suspended in alcohol, has a density of 800 kg/m^3.

72. ||| A proton orbits a long charged wire, making 1.0×10^6 revolutions per second. The radius of the orbit is 1.0 cm. What is the wire's linear charge density?

73. ||| You have a summer intern position with a company that designs and builds nanomachines. An engineer with the company is designing a microscopic oscillator to help keep time, and you've been assigned to help him analyze the design. He wants to place a negative charge at the center of a very small, positively charged metal ring. His claim is that the negative charge will undergo simple harmonic motion at a frequency determined by the amount of charge on the ring.

   a. Consider a negative charge near the center of a positively charged ring centered on the z-axis. Show that there is a restoring force on the charge if it moves along the z-axis but stays close to the center of the ring. That is, show there's a force that tries to keep the charge at $z = 0$.

   b. Show that for *small* oscillations, with amplitude $\ll R$, a particle of mass m with charge $-q$ undergoes simple harmonic motion with frequency

$$f = \frac{1}{2\pi} \sqrt{\frac{qQ}{4\pi\epsilon_0 m R^3}}$$

    R and Q are the radius and charge of the ring.

   c. Evaluate the oscillation frequency for an electron at the center of a 2.0-μm-diameter ring charged to 1.0×10^{-13} C.

24 Gauss's Law

An electric field image of blood plasma from healthy blood. The wire in the center creates the electric field. Variations in the shape and color of the pattern can give early warning of cancer.

IN THIS CHAPTER, you will learn about and apply Gauss's law.

What is Gauss's law?

Gauss's law is a general statement about the nature of electric fields. It is more fundamental than Coulomb's law and is the first of what we will later call Maxwell's equations, the governing equations of electricity and magnetism.

Gauss's law says that the electric flux through a closed surface is proportional to the amount of charge Q_{in} enclosed within the surface. This seemingly abstract statement will be the basis of a powerful strategy for finding the electric fields of charge distributions that have a high degree of symmetry.

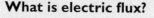

« LOOKING BACK Section 22.5 The electric field of a point charge Section 23.2 Electric field lines

What good is symmetry?

For charge distributions with a high degree of symmetry, the symmetry of the electric field must match the symmetry of the charge distribution. Important symmetries are planar symmetry, cylindrical symmetry, and spherical symmetry. The concept of symmetry plays an important role in math and science.

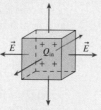

Cylindrical symmetry

What is electric flux?

The amount of electric field passing through a surface is called the electric flux. Electric flux is analogous to the amount of air or water flowing through a loop. You will learn to calculate the flux through open and closed surfaces.

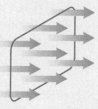

« LOOKING BACK Section 9.3 Vector dot products

How is Gauss's law used?

Gauss's law is easier to use than superposition for finding the electric field both inside and outside of charged spheres, cylinders, and planes. To use Gauss's law, you calculate the electric flux through a Gaussian surface surrounding the charge. This will turn out to be much easier than it sounds!

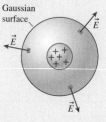

Gaussian surface

$\vec{E}$

What can we learn about conductors?

Gauss's law can be used to establish several properties of conductors in electrostatic equilibrium. In particular:

- Any excess charge is all on the surface.
- The interior electric field is zero.
- The external field is perpendicular to the surface.

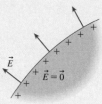

$\vec{E} = \vec{0}$

24.1 Symmetry

To continue our exploration of electric fields, suppose we knew only two things:

1. The field points away from positive charges, toward negative charges, and
2. An electric field exerts a force on a charged particle.

From this information alone, what can we deduce about the electric field of the infinitely long charged cylinder shown in **FIGURE 24.1**?

We don't know if the cylinder's diameter is large or small. We don't know if the charge density is the same at the outer edge as along the axis. All we know is that the charge is positive and the charge distribution has *cylindrical symmetry*. We say that a charge distribution is **symmetric** if there is a group of *geometric transformations* that don't cause any *physical* change.

To make this idea concrete, suppose you close your eyes while a friend transforms a charge distribution in one of the following three ways. He or she can

- *Translate* (that is, displace) the charge parallel to an axis,
- *Rotate* the charge about an axis, or
- *Reflect* the charge in a mirror.

When you open your eyes, will you be able to tell if the charge distribution has been changed? You might tell by observing a visual difference in the distribution. Or the results of an experiment with charged particles could reveal that the distribution has changed. If nothing you can see or do reveals any change, then we say that the charge distribution is symmetric under that particular transformation.

FIGURE 24.2 shows that the charge distribution of Figure 24.1 is symmetric with respect to

- Translation parallel to the cylinder axis. Shifting an infinitely long cylinder by 1 mm or 1000 m makes no noticeable or measurable change.
- Rotation by any angle about the cylinder axis. Turning a cylinder about its axis by 1° or 100° makes no detectable change.
- Reflections in any plane containing or perpendicular to the cylinder axis. Exchanging top and bottom, front and back, or left and right makes no detectable change.

A charge distribution that is symmetric under these three groups of geometric transformations is said to be *cylindrically symmetric*. Other charge distributions have other types of symmetries. Some charge distributions have no symmetry at all.

Our interest in symmetry can be summed up in a single statement:

The symmetry of the electric field must match the symmetry of the charge distribution.

If this were not true, you could use the electric field to test whether the charge distribution had undergone a transformation.

Now we're ready to see what we can learn about the electric field in Figure 24.1. Could the field look like **FIGURE 24.3a**? (Imagine this picture rotated about the axis.) That is, is this a *possible* field? This field looks the same if it's translated parallel to the

FIGURE 24.1 A charge distribution with cylindrical symmetry.

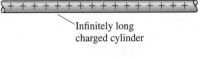

Infinitely long
charged cylinder

FIGURE 24.2 Transformations that don't change an infinite cylinder of charge.

Original
cylinder

Translation
parallel to
the axis

Rotation
about the
axis

Reflection
in plane
containing
the axis

Reflection
perpendicular
to the axis

FIGURE 24.3 Could the field of a cylindrical charge distribution look like this?

(a) Is this a possible electric field of an infinitely long charged cylinder? Suppose the charge and the field are reflected in a plane perpendicular to the axis.

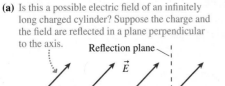
Reflection plane
$\vec{E}$
$\vec{E}$

Reflect

(b) The charge distribution is not changed by the reflection, but the field is. This field doesn't match the symmetry of the cylinder, so the cylinder's field can't look like this.

$\vec{E}$
$\vec{E}$

FIGURE 24.4 Or might the field of a cylindrical charge distribution look like this?

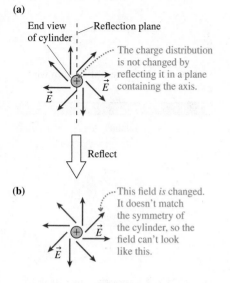

cylinder axis, if up and down are exchanged by reflecting the field in a plane coming out of the page, or if you rotate the cylinder about its axis.

But the proposed field fails one test: reflection in a plane perpendicular to the axis, a reflection that exchanges left and right. This reflection, which would *not* make any change in the charge distribution itself, produces the field shown in **FIGURE 24.3b**. This change in the field is detectable because a positively charged particle would now have a component of motion to the left instead of to the right.

The field of Figure 24.3a, which makes a distinction between left and right, is not cylindrically symmetric and thus is *not* a possible field. In general, **the electric field of a cylindrically symmetric charge distribution cannot have a component parallel to the cylinder axis.**

Well then, what about the electric field shown in **FIGURE 24.4a**? Here we're looking down the axis of the cylinder. The electric field vectors are restricted to planes perpendicular to the cylinder and thus do not have any component parallel to the cylinder axis. This field is symmetric for rotations about the axis, but it's *not* symmetric for a reflection in a plane containing the axis.

The field of **FIGURE 24.4b**, after this reflection, is easily distinguishable from the field of Figure 24.4a. Thus **the electric field of a cylindrically symmetric charge distribution cannot have a component tangent to the circular cross section.**

FIGURE 24.5 shows the only remaining possible field shape. The electric field is radial, pointing straight out from the cylinder like the bristles on a bottle brush. This is the one electric field shape matching the symmetry of the charge distribution.

FIGURE 24.5 This is the only shape for the electric field that matches the symmetry of the charge distribution.

Side view
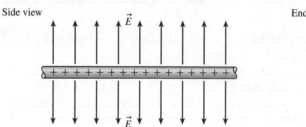

End view
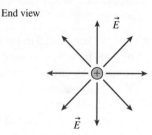

What Good Is Symmetry?

Given how little we assumed about Figure 24.1—that the charge distribution is cylindrically symmetric and that electric fields point away from positive charges—we've been able to deduce a great deal about the electric field. In particular, we've deduced the *shape* of the electric field.

Now, shape is not everything. We've learned nothing about the strength of the field or how strength changes with distance. Is E constant? Does it decrease like $1/r$ or $1/r^2$? We don't yet have a complete description of the field, but knowing what shape the field *has* to have will make finding the field strength a much easier task.

That's the good of symmetry. Symmetry arguments allow us to *rule out* many conceivable field shapes as simply being incompatible with the symmetry of the charge distribution. Knowing what doesn't happen, or can't happen, is often as useful as knowing what does happen. By the process of elimination, we're led to the one and only shape the field can possibly have. Reasoning on the basis of symmetry is a sometimes subtle but always powerful means of reasoning.

Three Fundamental Symmetries

Three fundamental symmetries appear frequently in electrostatics. The first row of **FIGURE 24.6** shows the simplest form of each symmetry. The second row shows a more complex, but more realistic, situation with the same symmetry.

FIGURE 24.6 Three fundamental symmetries.

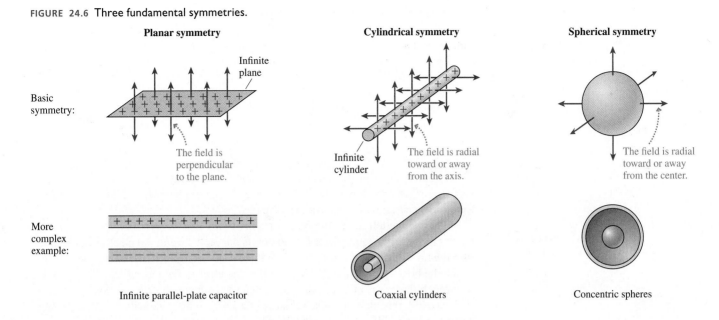

Planar symmetry

Basic symmetry:

Infinite plane

The field is perpendicular to the plane.

Cylindrical symmetry

Infinite cylinder

The field is radial toward or away from the axis.

Spherical symmetry

The field is radial toward or away from the center.

More complex example:

Infinite parallel-plate capacitor

Coaxial cylinders

Concentric spheres

NOTE Figures must be finite in extent, but the planes and cylinders in Figure 24.6 are assumed to be infinite.

Objects do exist that are extremely close to being perfect spheres, but no real cylinder or plane can be infinite in extent. Even so, the fields of infinite planes and cylinders are good models for the fields of finite planes and cylinders at points not too close to an edge or an end. The fields that we'll study in this chapter, even if idealized, have many important applications.

STOP TO THINK 24.1 A uniformly charged rod has a *finite* length L. The rod is symmetric under rotations about the axis and under reflection in any plane containing the axis. It is *not* symmetric under translations or under reflections in a plane perpendicular to the axis unless that plane bisects the rod. Which field shape or shapes match the symmetry of the rod?

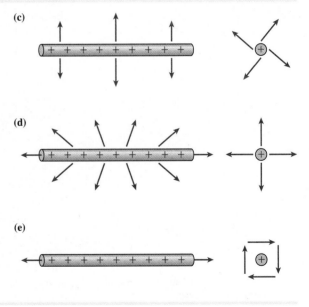

(c)

(d)

(e)

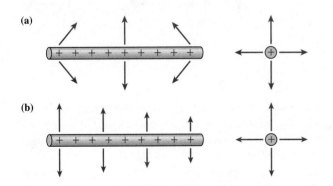

(a)

(b)

24.2 The Concept of Flux

FIGURE 24.7a on the next page shows an opaque box surrounding a region of space. We can't see what's in the box, but there's an electric field vector coming out of each face of the box. Can you figure out what's in the box?

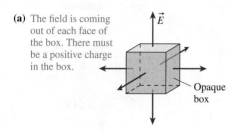

(a) The field is coming out of each face of the box. There must be a positive charge in the box.

Opaque box

FIGURE 24.7 Although we can't see into the boxes, the electric fields passing through the faces tell us something about what's in them.

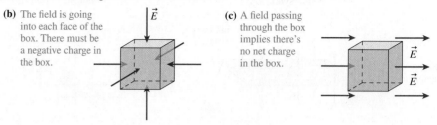

(b) The field is going into each face of the box. There must be a negative charge in the box.

(c) A field passing through the box implies there's no net charge in the box.

FIGURE 24.8 Gaussian surface surrounding a charge. A two-dimensional cross section is usually easier to draw.

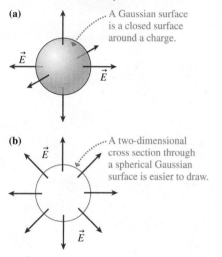

(a)

A Gaussian surface is a closed surface around a charge.

(b)

A two-dimensional cross section through a spherical Gaussian surface is easier to draw.

Of course you can. Because electric fields point away from positive charges, it seems clear that the box contains a positive charge or charges. Similarly, the box in **FIGURE 24.7b** certainly contains a negative charge.

What can we tell about the box in **FIGURE 24.7c**? The electric field points into the box on the left. An equal electric field points out on the right. An electric field passes *through* the box, but we see no evidence there's any charge (or at least any net charge) inside the box. These examples suggest that the electric field as it passes into, out of, or through the box is in some way connected to the charge within the box.

To explore this idea, suppose we surround a region of space with a *closed surface,* a surface that divides space into distinct inside and outside regions. Within the context of electrostatics, a closed surface through which an electric field passes is called a **Gaussian surface,** named after the 19th-century mathematician Karl Gauss. This is an imaginary, mathematical surface, not a physical surface, although it might coincide with a physical surface. For example, **FIGURE 24.8a** shows a spherical Gaussian surface surrounding a charge.

A closed surface must, of necessity, be a surface in three dimensions. But three-dimensional pictures are hard to draw, so we'll often look at two-dimensional cross sections through a Gaussian surface, such as the one shown in **FIGURE 24.8b**. Now we can tell from the *spherical symmetry* of the electric field vectors poking through the surface that the positive charge inside must be spherically symmetric and centered at the *center* of the sphere.

FIGURE 24.9 A Gaussian surface is most useful when it matches the shape of the field.

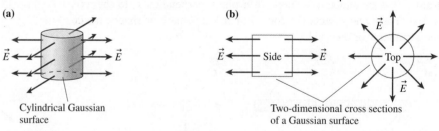

(a)

Cylindrical Gaussian surface

(b)

Two-dimensional cross sections of a Gaussian surface

A Gaussian surface is most useful when it matches the shape and symmetry of the field. For example, **FIGURE 24.9a** shows a cylindrical Gaussian surface—a *closed* cylinder—surrounding some kind of cylindrical charge distribution, such as a charged wire. **FIGURE 24.9b** simplifies the drawing by showing two-dimensional end and side views. Because the Gaussian surface matches the symmetry of the charge distribution, the electric field is everywhere *perpendicular* to the side wall and *no* field passes through the top and bottom surfaces.

For contrast, consider the spherical surface in **FIGURE 24.10**. This is also a Gaussian surface, and the protruding electric field tells us there's a positive charge inside. It might be a point charge located on the left side, but we can't really say. A Gaussian surface that doesn't match the symmetry of the charge distribution isn't terribly useful.

These examples lead us to two conclusions:

FIGURE 24.10 Not every surface is useful for learning about charge.

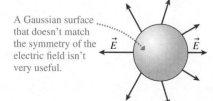

A Gaussian surface that doesn't match the symmetry of the electric field isn't very useful.

1. The electric field, in some sense, "flows" *out of* a closed surface surrounding a region of space containing a net positive charge and *into* a closed surface surrounding a net negative charge. The electric field may flow *through* a closed surface surrounding a region of space in which there is no net charge, but the *net flow* is zero.
2. The electric field pattern through the surface is particularly simple if the closed surface matches the symmetry of the charge distribution inside.

The electric field doesn't really flow like a fluid, but the metaphor is a useful one. The Latin word for flow is *flux,* and the amount of electric field passing through a surface is called the **electric flux.** Our first conclusions, stated in terms of electric flux, are

- There is an outward flux through a closed surface around a net positive charge.
- There is an inward flux through a closed surface around a net negative charge.
- There is no net flux through a closed surface around a region of space in which there is no net charge.

This chapter has been entirely qualitative thus far as we've established pictorially what we mean by symmetry, flux, and the fact that the electric flux through a closed surface has something to do with the charge inside. In the next two sections you'll learn how to calculate the electric flux through a surface and how the flux is related to the enclosed charge. That relationship, Gauss's law, will allow us to determine the electric fields of some interesting and useful charge distributions.

STOP TO THINK 24.2   This box contains

a. A positive charge.
b. A negative charge.
c. No charge.
d. A net positive charge.
e. A net negative charge.
f. No net charge.

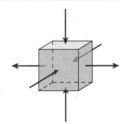

24.3 Calculating Electric Flux

Let's start with a brief overview of where this section will take us. We'll begin with a definition of flux that is easy to understand, then we'll turn that simple definition into a formidable-looking integral. We need the integral because the simple definition applies only to uniform electric fields and flat surfaces. Those are good starting points, but we'll soon need to calculate the flux of nonuniform fields through curved surfaces.

Mathematically, the flux of a nonuniform field through a curved surface is described by a special kind of integral called a *surface integral.* It's quite possible that you have not yet encountered surface integrals in your calculus course, and the "novelty factor" contributes to making this integral look worse than it really is. We will emphasize over and over the idea that an integral is just a fancy way of doing a sum, in this case the sum of the small amounts of flux through many small pieces of a surface.

The good news is that *every* surface integral we need to evaluate in this chapter, or that you will need to evaluate for the homework problems, is either zero or is so easy that you will be able to do it in your head. This seems like an astounding claim, but you will soon see it is true. The key will be to make effective use of the *symmetry* of the electric field.

The Basic Definition of Flux

Imagine holding a rectangular wire loop of area *A* in front of a fan. As **FIGURE 24.11** on the next page shows, the volume of air flowing through the loop each second depends on the angle between the loop and the direction of flow. The flow is maximum through a loop that is perpendicular to the airflow; no air goes through the same loop if it lies parallel to the flow.

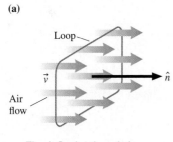
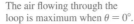
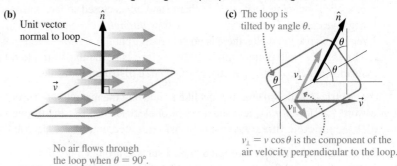

(a)

Loop

$\vec{v}$

$\hat{n}$

Air flow

The air flowing through the loop is maximum when $\theta = 0°$.

(b)

Unit vector normal to loop

$\hat{n}$

$\vec{v}$

No air flows through the loop when $\theta = 90°$.

(c) The loop is tilted by angle θ.

$\hat{n}$

θ

$v_\perp$

$v_\parallel$

$\vec{v}$

$v_\perp = v \cos\theta$ is the component of the air velocity perpendicular to the loop.

The flow direction is identified by the velocity vector $\vec{v}$. We can identify the loop's orientation by defining a unit vector $\hat{n}$ normal to the plane of the loop. Angle θ is then the angle between $\vec{v}$ and $\hat{n}$. The loop perpendicular to the flow in Figure 24.11a has $\theta = 0°$; the loop parallel to the flow in Figure 24.11b has $\theta = 90°$. You can think of θ as the angle by which a loop has been tilted away from perpendicular.

> **NOTE** A surface has two sides, so $\hat{n}$ could point either way. We'll choose the side that makes $\theta \leq 90°$.

You can see from Figure 24.11c that the velocity vector $\vec{v}$ can be decomposed into components $v_\perp = v\cos\theta$ perpendicular to the loop and $v_\parallel = v\sin\theta$ parallel to the loop. Only the perpendicular component $v_\perp$ carries air *through* the loop. Consequently, the volume of air flowing through the loop each second is

$$\text{volume of air per second (m}^3/\text{s)} = v_\perp A = vA\cos\theta \tag{24.1}$$

$\theta = 0°$ is the orientation for maximum flow through the loop, as expected, and no air flows through the loop if it is tilted $90°$.

An electric field doesn't flow in a literal sense, but we can apply the same idea to an electric field passing through a surface. **FIGURE 24.12** shows a surface of area A in a uniform electric field $\vec{E}$. Unit vector $\hat{n}$ is normal to the surface, and θ is the angle between $\hat{n}$ and $\vec{E}$. Only the component $E_\perp = E\cos\theta$ passes *through* the surface.

With this in mind, and using Equation 24.1 as an analog, let's define the *electric flux* Φ_e (uppercase Greek phi) as

$$\Phi_e = E_\perp A = EA\cos\theta \tag{24.2}$$

The electric flux measures the amount of electric field passing through a surface of area A if the normal to the surface is tilted at angle θ from the field.

Equation 24.2 looks very much like a vector dot product: $\vec{E} \cdot \vec{A} = EA\cos\theta$. For this idea to work, let's define an **area vector** $\vec{A} = A\hat{n}$ to be a vector in the direction of $\hat{n}$—that is, *perpendicular* to the surface—with a magnitude A equal to the area of the surface. Vector $\vec{A}$ has units of m^2. **FIGURE 24.13a** shows two area vectors.

FIGURE 24.12 An electric field passing through a surface.

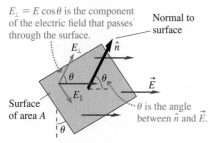

$E_\perp = E\cos\theta$ is the component of the electric field that passes through the surface.

Normal to surface

$E_\perp$

$\hat{n}$

θ

θ

$E_\parallel$

$\vec{E}$

Surface of area A

θ is the angle between $\hat{n}$ and $\vec{E}$.

θ

FIGURE 24.13 The electric flux can be defined in terms of the area vector $\vec{A}$.

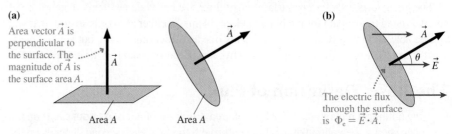

(a)

Area vector $\vec{A}$ is perpendicular to the surface. The magnitude of $\vec{A}$ is the surface area A.

$\vec{A}$

$\vec{A}$

Area A

Area A

(b)

$\vec{A}$

θ

$\vec{E}$

The electric flux through the surface is $\Phi_e = \vec{E} \cdot \vec{A}$.

FIGURE 24.13b shows an electric field passing through a surface of area A. The angle between vectors $\vec{A}$ and $\vec{E}$ is the same angle used in Equation 24.2 to define the electric flux, so Equation 24.2 really is a dot product. We can define the electric flux more concisely as

$$\Phi_e = \vec{E} \cdot \vec{A} \quad \text{(electric flux of a constant electric field)} \quad (24.3)$$

Writing the flux as a dot product helps make clear how angle θ is defined: θ is the angle between the electric field and a line *perpendicular* to the plane of the surface.

NOTE Figure 24.13b shows a circular area, but the shape of the surface is not relevant. However, Equation 24.3 is restricted to a *constant* electric field passing through a *planar* surface.

EXAMPLE 24.1 | **The electric flux inside a parallel-plate capacitor**

Two 100 cm^2 parallel electrodes are spaced 2.0 cm apart. One is charged to $+5.0$ nC, the other to -5.0 nC. A 1.0 cm $\times$ 1.0 cm surface between the electrodes is tilted to where its normal makes a $45°$ angle with the electric field. What is the electric flux through this surface?

MODEL Assume the surface is located near the center of the capacitor where the electric field is uniform. The electric flux doesn't depend on the shape of the surface.

VISUALIZE The surface is square, rather than circular, but otherwise the situation looks like Figure 24.13b.

SOLVE In Chapter 23, we found the electric field inside a parallel-plate capacitor to be

$$E = \frac{Q}{\epsilon_0 A_{\text{plates}}} = \frac{5.0 \times 10^{-9} \text{ C}}{(8.85 \times 10^{-12} \text{ C}^2/\text{N m}^2)(1.0 \times 10^{-2} \text{ m}^2)}$$

$$= 5.65 \times 10^4 \text{ N/C}$$

A 1.0 cm $\times$ 1.0 cm surface has $A = 1.0 \times 10^{-4}$ m^2. The electric flux through this surface is

$$\Phi_e = \vec{E} \cdot \vec{A} = EA \cos\theta$$

$$= (5.65 \times 10^4 \text{ N/C})(1.0 \times 10^{-4} \text{ m}^2) \cos 45°$$

$$= 4.0 \text{ N m}^2/\text{C}$$

ASSESS The units of electric flux are the product of electric field and area units: N m^2/C.

The Electric Flux of a Nonuniform Electric Field

Our initial definition of the electric flux assumed that the electric field $\vec{E}$ was constant over the surface. How should we calculate the electric flux if $\vec{E}$ varies from point to point on the surface? We can answer this question by returning to the analogy of air flowing through a loop. Suppose the airflow varies from point to point. We can still find the total volume of air passing through the loop each second by dividing the loop into many small areas, finding the flow through each small area, then adding them. Similarly, **the electric flux through a surface can be calculated as the sum of the fluxes through smaller pieces of the surface.** Because flux is a scalar, adding fluxes is easier than adding electric fields.

FIGURE 24.14 shows a surface in a nonuniform electric field. Imagine dividing the surface into many small pieces of area δA. Each little area has an area vector $\delta\vec{A}$ perpendicular to the surface. Two of the little pieces are shown in the figure. The electric fluxes through these two pieces differ because the electric fields are different.

Consider the small piece i where the electric field is $\vec{E}_i$. The small electric flux $\delta\Phi_i$ through area $(\delta\vec{A})_i$ is

$$\delta\Phi_i = \vec{E}_i \cdot (\delta\vec{A})_i \quad (24.4)$$

The flux through every other little piece of the surface is found the same way. The total electric flux through the entire surface is then the sum of the fluxes through each of the small areas:

$$\Phi_e = \sum_i \delta\Phi_i = \sum_i \vec{E}_i \cdot (\delta\vec{A})_i \quad (24.5)$$

Now let's go to the limit $\delta\vec{A} \to d\vec{A}$. That is, the little areas become infinitesimally small, and there are infinitely many of them. Then the sum becomes an integral, and the electric flux through the surface is

$$\Phi_e = \int_{\text{surface}} \vec{E} \cdot d\vec{A} \quad (24.6)$$

The integral in Equation 24.6 is called a **surface integral.**

FIGURE 24.14 A surface in a nonuniform electric field.

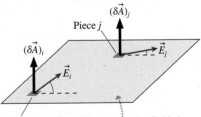

Piece i The total area A can be divided into many small pieces of area δA. $\vec{E}$ may be different at each piece.

Equation 24.6 may look rather frightening if you haven't seen surface integrals before. Despite its appearance, a surface integral is no more complicated than integrals you know from calculus. After all, what does $\int f(x)\,dx$ really mean? This expression is a shorthand way to say "Divide the x-axis into many little segments of length δx, evaluate the function $f(x)$ in each of them, then add up $f(x)\,\delta x$ for all the segments along the line." The integral in Equation 24.6 differs only in that we're dividing a surface into little pieces instead of a line into little segments. In particular, we're summing the fluxes through a vast number of very tiny pieces.

You may be thinking, "OK, I understand the idea, but I don't know what to *do*. In calculus, I learned formulas for evaluating integrals such as $\int x^2\,dx$. How do I evaluate a surface integral?" This is a good question. We'll deal with evaluation shortly, and it will turn out that the surface integrals in electrostatics are quite easy to evaluate. But don't confuse *evaluating* the integral with understanding what the integral *means*. The surface integral in Equation 24.6 is simply a shorthand notation for the summation of the electric fluxes through a vast number of very tiny pieces of a surface.

The electric field might be different at every point on the surface, but suppose it isn't. That is, suppose a flat surface is in a uniform electric field $\vec{E}$. A field that is the same at every single point on a surface is a constant as far as the integration of Equation 24.6 is concerned, so we can take it outside the integral. In that case,

$$\Phi_e = \int_{\text{surface}} \vec{E} \cdot d\vec{A} = \int_{\text{surface}} E\cos\theta\,dA = E\cos\theta \int_{\text{surface}} dA \tag{24.7}$$

The integral that remains in Equation 24.7 tells us to add up all the little areas into which the full surface was subdivided. But the sum of all the little areas is simply the area of the surface:

$$\int_{\text{surface}} dA = A \tag{24.8}$$

This idea—that the surface integral of dA is the area of the surface—is one we'll use to evaluate most of the surface integrals of electrostatics. If we substitute Equation 24.8 into Equation 24.7, we find that the electric flux in a uniform electric field is $\Phi_e = EA\cos\theta$. We already knew this, from Equation 24.2, but it was important to see that the surface integral of Equation 24.6 gives the correct result for the case of a uniform electric field.

The Flux Through a Curved Surface

Most of the Gaussian surfaces we considered in the last section were curved surfaces. FIGURE 24.15 shows an electric field passing through a curved surface. How do we find the electric flux through this surface? Just as we did for a flat surface!

Divide the surface into many small pieces of area δA. For each, define the area vector $\delta\vec{A}$ perpendicular to the surface *at that point*. Compared to Figure 24.14, the only difference that the curvature of the surface makes is that the $\delta\vec{A}$ are no longer parallel to each other. Find the small electric flux $\delta\Phi_i = \vec{E}_i \cdot (\delta\vec{A})_i$ through each little area, then add them all up. The result, once again, is

$$\Phi_e = \int_{\text{surface}} \vec{E} \cdot d\vec{A} \tag{24.9}$$

We *assumed*, in deriving this expression the first time, that the surface was flat and that all the $\delta\vec{A}$ were parallel to each other. But that assumption wasn't necessary. The *meaning* of Equation 24.9—a summation of the fluxes through a vast number of very tiny pieces—is unchanged if the pieces lie on a curved surface.

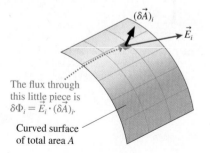

FIGURE 24.15 A curved surface in an electric field.

$(\delta\vec{A})_i$

$\vec{E}_i$

The flux through this little piece is $\delta\Phi_i = \vec{E}_i \cdot (\delta\vec{A})_i$.

Curved surface of total area A

We seem to be getting more and more complex, using surface integrals first for nonuniform fields and now for curved surfaces. But consider the two situations shown in **FIGURE 24.16**. The electric field $\vec{E}$ in Figure 24.16a is everywhere tangent, or parallel, to the curved surface. We don't need to know the magnitude of $\vec{E}$ to recognize that $\vec{E} \cdot d\vec{A}$ is *zero at every point* on the surface because $\vec{E}$ is perpendicular to $d\vec{A}$ at every point. Thus $\Phi_e = 0$. A tangent electric field never pokes through the surface, so it has no flux through the surface.

The electric field in Figure 24.16b is everywhere perpendicular to the surface *and* has the same magnitude E at every point. $\vec{E}$ differs in direction at different points on a curved surface, but at any particular point $\vec{E}$ is parallel to $d\vec{A}$ and $\vec{E} \cdot d\vec{A}$ is simply $E\,dA$. In this case,

$$\Phi_e = \int_{\text{surface}} \vec{E} \cdot d\vec{A} = \int_{\text{surface}} E\,dA = E \int_{\text{surface}} dA = EA \qquad (24.10)$$

As we evaluated the integral, the fact that E has the same magnitude at every point on the surface allowed us to bring the constant value outside the integral. We then used the fact that the integral of dA over the surface is the surface area A.

We can summarize these two situations with a Tactics Box.

TACTICS BOX 24.1   (MP)

Evaluating surface integrals

❶ If the electric field is everywhere tangent to a surface, the electric flux through the surface is $\Phi_e = 0$.

❷ If the electric field is everywhere perpendicular to a surface *and* has the same magnitude E at every point, the electric flux through the surface is $\Phi_e = EA$.

These two results will be of immeasurable value for using Gauss's law because *every* flux we'll need to calculate will be one of these situations. This is the basis for our earlier claim that the evaluation of surface integrals is not going to be difficult.

The Electric Flux Through a Closed Surface

Our final step, to calculate the electric flux through a closed surface such as a box, a cylinder, or a sphere, requires nothing new. We've already learned how to calculate the electric flux through flat and curved surfaces, and a closed surface is nothing more than a surface that happens to be closed.

However, the mathematical notation for the surface integral over a closed surface differs slightly from what we've been using. It is customary to use a little circle on the integral sign to indicate that the surface integral is to be performed over a closed surface. With this notation, the electric flux through a closed surface is

$$\Phi_e = \oint \vec{E} \cdot d\vec{A} \qquad (24.11)$$

Only the notation has changed. The electric flux is still the summation of the fluxes through a vast number of tiny pieces, pieces that now cover a closed surface.

NOTE A closed surface has a distinct inside and outside. The area vector $d\vec{A}$ is defined to always point *toward the outside*. This removes an ambiguity that was present for a single surface, where $d\vec{A}$ could point to either side.

Now we're ready to calculate the flux through a closed surface.

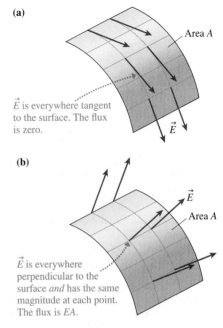

FIGURE 24.16 Electric fields that are everywhere tangent to or everywhere perpendicular to a curved surface.

(a)

Area A

$\vec{E}$ is everywhere tangent to the surface. The flux is zero.

$\vec{E}$

(b)

$\vec{E}$

Area A

$\vec{E}$ is everywhere perpendicular to the surface *and* has the same magnitude at each point. The flux is EA.

TACTICS BOX 24.2

Finding the flux through a closed surface

❶ Choose a Gaussian surface made up of pieces that are everywhere tangent to the electric field or everywhere perpendicular to the electric field.

❷ Use Tactics Box 24.1 to evaluate the surface integrals over these surfaces, then add the results.

Exercise 10

EXAMPLE 24.2 | Calculating the electric flux through a closed cylinder

A charge distribution with cylindrical symmetry has created the electric field $\vec{E} = E_0(r^2/r_0^2)\hat{r}$, where E_0 and r_0 are constants and where unit vector $\hat{r}$ lies in the xy-plane. Calculate the electric flux through a closed cylinder of length L and radius R that is centered along the z-axis.

MODEL The electric field extends radially outward from the z-axis with cylindrical symmetry. The z-component is $E_z = 0$. The cylinder is a Gaussian surface.

VISUALIZE FIGURE 24.17a is a view of the electric field looking along the z-axis. The field strength increases with increasing radial distance, and it's symmetric about the z-axis. FIGURE 24.17b is the closed Gaussian surface for which we need to calculate the electric flux. We can place the cylinder anywhere along the z-axis because the electric field extends forever in that direction.

SOLVE To calculate the flux, we divide the closed cylinder into three surfaces: the top, the bottom, and the cylindrical wall. The electric field is tangent to the surface at every point on the top and bottom surfaces. Hence, according to step 1 in Tactics Box 24.1, the flux through those two surfaces is zero. For the cylindrical wall, the electric field is perpendicular to the surface at every point *and* has the constant magnitude $E = E_0(R^2/r_0^2)$ at every point on the surface. Thus, from step 2 in Tactics Box 24.1,

$$\Phi_{\text{wall}} = EA_{\text{wall}}$$

If we add the three pieces, the net flux through the closed surface is

$$\Phi_e = \oint \vec{E} \cdot d\vec{A} = \Phi_{\text{top}} + \Phi_{\text{bottom}} + \Phi_{\text{wall}} = 0 + 0 + EA_{\text{wall}}$$
$$= EA_{\text{wall}}$$

We've evaluated the surface integral, using the two steps in Tactics Box 24.1, and there was nothing to it! To finish, all we need to recall is that the surface area of a cylindrical wall is circumference × height, or $A_{\text{wall}} = 2\pi RL$. Thus

$$\Phi_e = \left(E_0 \frac{R^2}{r_0^2}\right)(2\pi RL) = \frac{2\pi LR^3}{r_0^2} E_0$$

ASSESS LR^3/r_0^2 has units of m², an area, so this expression for Φ_e has units of N m²/C. These are the correct units for electric flux, giving us confidence in our answer. Notice the important role played by symmetry. The electric field was perpendicular to the wall and of constant value at every point on the wall *because* the Gaussian surface had the same symmetry as the charge distribution. We would not have been able to evaluate the surface integral in such an easy way for a surface of any other shape. Symmetry is the key.

FIGURE 24.17 The electric field and the closed surface through which we will calculate the electric flux.

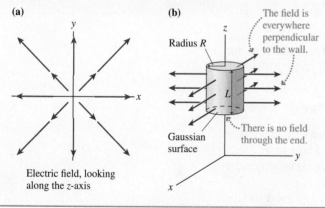

(a) Electric field, looking along the z-axis

(b) Radius R — The field is everywhere perpendicular to the wall. — Gaussian surface — There is no field through the end.

STOP TO THINK 24.3 The total electric flux through this box is

a. 0 N m²/C
b. 1 N m²/C
c. 2 N m²/C
d. 4 N m²/C
e. 6 N m²/C
f. 8 N m²/C

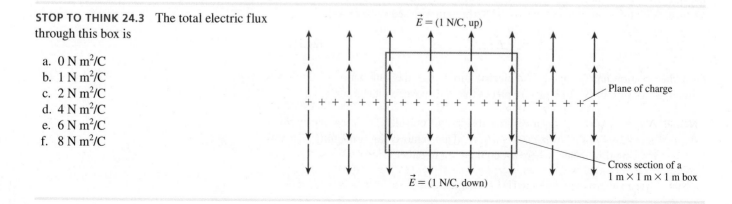

$\vec{E} = (1 \text{ N/C, up})$

$\vec{E} = (1 \text{ N/C, down})$

Plane of charge

Cross section of a 1 m × 1 m × 1 m box

24.4 Gauss's Law

The last section was long, but knowing how to calculate the electric flux through a closed surface is essential for the main topic of this chapter: Gauss's law. Gauss's law is equivalent to Coulomb's law for static charges, although Gauss's law will look very different.

The purpose of learning Gauss's law is twofold:

- Gauss's law allows the electric fields of some continuous distributions of charge to be found much more easily than does Coulomb's law.
- Gauss's law is valid for *moving* charges, but Coulomb's law is not (although it's a very good approximation for velocities that are much less than the speed of light). Thus Gauss's law is ultimately a more fundamental statement about electric fields than is Coulomb's law.

Let's start with Coulomb's law for the electric field of a point charge. **FIGURE 24.18** shows a spherical Gaussian surface of radius r centered on a positive charge q. Keep in mind that this is an imaginary, mathematical surface, not a physical surface. There is a net flux through this surface because the electric field points outward at every point on the surface. To evaluate the flux, given formally by the surface integral of Equation 24.11, notice that the electric field is perpendicular to the surface at every point on the surface *and*, from Coulomb's law, it has the same magnitude $E = q/4\pi\epsilon_0 r^2$ at every point on the surface. This simple situation arises because **the Gaussian surface has the same symmetry as the electric field.**

Thus we know, without having to do any hard work, that the flux integral is

$$\Phi_e = \oint \vec{E} \cdot d\vec{A} = EA_{\text{sphere}} \tag{24.12}$$

The surface area of a sphere of radius r is $A_{\text{sphere}} = 4\pi r^2$. If we use A_{sphere} and the Coulomb-law expression for E in Equation 24.12, we find that the electric flux through the spherical surface is

$$\Phi_e = \frac{q}{4\pi\epsilon_0 r^2}\, 4\pi r^2 = \frac{q}{\epsilon_0} \tag{24.13}$$

You should examine the logic of this calculation closely. We really did evaluate the surface integral of Equation 24.11, although it may appear, at first, as if we didn't do much. The integral was easily evaluated, we reiterate for emphasis, because the closed surface on which we performed the integration matched the *symmetry* of the charge distribution.

NOTE We found Equation 24.13 for a positive charge, but it applies equally to negative charges. According to Equation 24.13, Φ_e is negative if q is negative. And that's what we would expect from the basic definition of flux, $\vec{E} \cdot \vec{A}$. The electric field of a negative charge points inward, while the area vector of a closed surface points outward, making the dot product negative.

Electric Flux Is Independent of Surface Shape and Radius

Notice something interesting about Equation 24.13. The electric flux depends on the amount of charge but *not* on the radius of the sphere. Although this may seem a bit surprising, it's really a direct consequence of what we *mean* by flux. Think of the fluid analogy with which we introduced the term "flux." If fluid flows outward from a central point, all the fluid crossing a small-radius spherical surface will, at some later time, cross a large-radius spherical surface. No fluid is lost along the way, and no new fluid is created. Similarly, the point charge in **FIGURE 24.19** is the only source of electric field. Every electric field line passing through a small-radius spherical surface also passes through a large-radius spherical surface. Hence the electric flux is independent of r.

FIGURE 24.18 A spherical Gaussian surface surrounding a point charge.

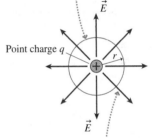

Cross section of a Gaussian sphere of radius r. This is a mathematical surface, not a physical surface.

Point charge q

The electric field is everywhere perpendicular to the surface *and* has the same magnitude at every point.

FIGURE 24.19 The electric flux is the same through *every* sphere centered on a point charge.

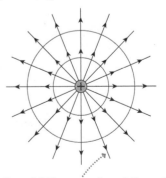

Every field line passes through the smaller *and* the larger sphere. The flux through the two spheres is the same.

NOTE This argument hinges on the fact that Coulomb's law is an inverse-square force law. The electric field strength, which is proportional to $1/r^2$, decreases with distance. But the surface area, which increases in proportion to r^2, exactly compensates for this decrease. Consequently, the electric flux of a point charge through a spherical surface is independent of the radius of the sphere.

This conclusion about the flux has an extremely important generalization. **FIGURE 24.20a** shows a point charge and a closed Gaussian surface of arbitrary shape and dimensions. All we know is that the charge is *inside* the surface. What is the electric flux through this arbitrary surface?

One way to answer the question is to approximate the surface as a patchwork of spherical and radial pieces. The spherical pieces are centered on the charge and the radial pieces lie along lines extending outward from the charge. (Figure 24.20 is a two-dimensional drawing so you need to imagine these arcs as actually being pieces of a spherical shell.) The figure, of necessity, shows fairly large pieces that don't match the actual surface all that well. However, we can make this approximation as good as we want by letting the pieces become sufficiently small.

The electric field is everywhere tangent to the radial pieces. Hence the electric flux through the radial pieces is zero. The spherical pieces, although at varying distances from the charge, form a *complete sphere*. That is, any line drawn radially outward from the charge will pass through exactly one spherical piece, and no radial lines can avoid passing through a spherical piece. You could even imagine, as **FIGURE 24.20b** shows, sliding the spherical pieces in and out *without changing the angle they subtend* until they come together to form a complete sphere.

Consequently, the electric flux through these spherical pieces that, when assembled, form a complete sphere must be exactly the same as the flux q/ϵ_0 through a spherical Gaussian surface. In other words, **the flux through *any* closed surface surrounding a point charge q is**

$$\Phi_e = \oint \vec{E} \cdot d\vec{A} = \frac{q}{\epsilon_0} \tag{24.14}$$

This surprisingly simple result is a consequence of the fact that Coulomb's law is an inverse-square force law. Even so, the reasoning that got us to Equation 24.14 is rather subtle and well worth reviewing.

Charge Outside the Surface

The closed surface shown in **FIGURE 24.21a** has a point charge q outside the surface but no charges inside. Now what can we say about the flux? By approximating this surface with spherical and radial pieces *centered on the charge,* as we did in Figure 24.20, we can reassemble the surface into the equivalent surface of **FIGURE 24.21b**. This closed

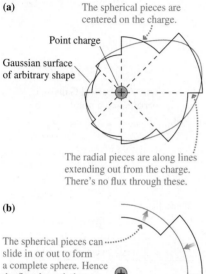

FIGURE 24.20 An arbitrary Gaussian surface can be approximated with spherical and radial pieces.

(a)

The spherical pieces are centered on the charge.

Point charge

Gaussian surface of arbitrary shape

The radial pieces are along lines extending out from the charge. There's no flux through these.

(b)

The spherical pieces can slide in or out to form a complete sphere. Hence the flux through the pieces is the same as the flux through a sphere.

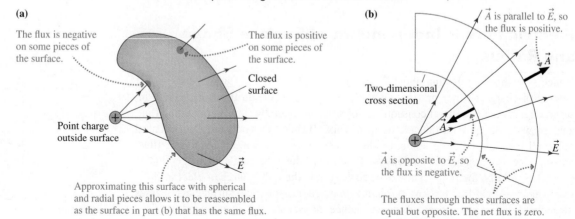

FIGURE 24.21 A point charge outside a Gaussian surface.

(a)

The flux is negative on some pieces of the surface.

The flux is positive on some pieces of the surface.

Closed surface

Point charge outside surface

$\vec{E}$

Approximating this surface with spherical and radial pieces allows it to be reassembled as the surface in part (b) that has the same flux.

(b)

$\vec{A}$ is parallel to $\vec{E}$, so the flux is positive.

$\vec{A}$

Two-dimensional cross section

$\vec{A}$

$\vec{E}$

$\vec{A}$ is opposite to $\vec{E}$, so the flux is negative.

The fluxes through these surfaces are equal but opposite. The net flux is zero.

surface consists of sections of two spherical shells, and it is equivalent in the sense that the electric flux through this surface is the same as the electric flux through the original surface of Figure 24.21a.

If the electric field were a fluid flowing outward from the charge, all the fluid *entering* the closed region through the first spherical surface would later *exit* the region through the second spherical surface. There is no *net* flow into or out of the closed region. Similarly, every electric field line entering this closed volume through one spherical surface exits through the other spherical surface.

Mathematically, the electric fluxes through the two spherical surfaces have the same magnitude because Φ_e is independent of r. But they have *opposite signs* because the outward-pointing area vector $\vec{A}$ is parallel to $\vec{E}$ on one surface but opposite to $\vec{E}$ on the other. The sum of the fluxes through the two surfaces is zero, and we are led to the conclusion that **the net electric flux is zero through a closed surface that does not contain any net charge.** Charges outside the surface do not produce a net flux through the surface.

This isn't to say that the flux through a small piece of the surface is zero. In fact, as Figure 24.21a shows, nearly every piece of the surface has an electric field either entering or leaving and thus has a nonzero flux. But some of these are positive and some are negative. When summed over the *entire* surface, the positive and negative contributions exactly cancel to give no *net* flux.

Multiple Charges

Finally, consider an arbitrary Gaussian surface and a group of charges $q_1, q_2, q_3, \ldots$ such as those shown in **FIGURE 24.22**. What is the net electric flux through the surface?

By definition, the net flux is

$$\Phi_e = \oint \vec{E} \cdot d\vec{A}$$

From the principle of superposition, the electric field is $\vec{E} = \vec{E}_1 + \vec{E}_2 + \vec{E}_3 + \cdots$, where $\vec{E}_1, \vec{E}_2, \vec{E}_3, \ldots$ are the fields of the individual charges. Thus the flux is

$$\Phi_e = \oint \vec{E}_1 \cdot d\vec{A} + \oint \vec{E}_2 \cdot d\vec{A} + \oint \vec{E}_3 \cdot d\vec{A} + \cdots \tag{24.15}$$
$$= \Phi_1 + \Phi_2 + \Phi_3 + \cdots$$

where $\Phi_1, \Phi_2, \Phi_3, \ldots$ are the fluxes through the Gaussian surface due to the individual charges. That is, the net flux is the sum of the fluxes due to individual charges. But we know what those are: q/ϵ_0 for the charges inside the surface and zero for the charges outside. Thus

$$\Phi_e = \left(\frac{q_1}{\epsilon_0} + \frac{q_2}{\epsilon_0} + \cdots + \frac{q_i}{\epsilon_0} \text{ for all charges inside the surface} \right) \tag{24.16}$$
$$+ (0 + 0 + \cdots + 0 \text{ for all charges outside the surface})$$

We define

$$Q_{\text{in}} = q_1 + q_2 + \cdots + q_i \text{ for all charges inside the surface} \tag{24.17}$$

as the total charge enclosed *within* the surface. With this definition, we can write our result for the net electric flux in a very neat and compact fashion. For any *closed* surface enclosing total charge Q_{in}, the net electric flux through the surface is

$$\Phi_e = \oint \vec{E} \cdot d\vec{A} = \frac{Q_{\text{in}}}{\epsilon_0} \tag{24.18}$$

This result for the electric flux is known as **Gauss's law.**

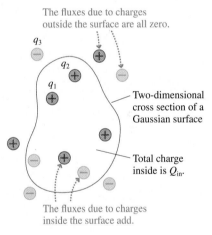

FIGURE 24.22 Charges both inside and outside a Gaussian surface.

The fluxes due to charges outside the surface are all zero.

q_3

q_2

q_1

Two-dimensional cross section of a Gaussian surface

Total charge inside is Q_{in}.

The fluxes due to charges inside the surface add.

What Does Gauss's Law Tell Us?

In one sense, Gauss's law doesn't say anything new or anything that we didn't already know from Coulomb's law. After all, we derived Gauss's law from Coulomb's law. But in another sense, Gauss's law is more important than Coulomb's law. Gauss's law states a very general property of electric fields—namely, that charges create electric fields in just such a way that the net flux of the field is the same through *any* surface surrounding the charges, no matter what its size and shape may be. This fact may have been implied by Coulomb's law, but it was by no means obvious. And Gauss's law will turn out to be particularly useful later when we combine it with other electric and magnetic field equations.

Gauss's law is the mathematical statement of our observations in Section 24.2. There we noticed a net "flow" of electric field out of a closed surface containing charges. Gauss's law quantifies this idea by making a specific connection between the "flow," now measured as electric flux, and the amount of charge.

But is it useful? Although to some extent Gauss's law is a formal statement about electric fields, not a tool for solving practical problems, there are exceptions: Gauss's law will allow us to find the electric fields of some very important and very practical charge distributions much more easily than if we had to rely on Coulomb's law. We'll consider some examples in the next section.

STOP TO THINK 24.4 These are two-dimensional cross sections through three-dimensional closed spheres and a cube. Rank in order, from largest to smallest, the electric fluxes Φ_a to Φ_e through surfaces a to e.

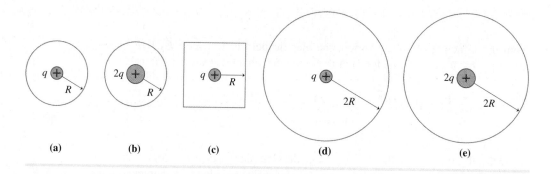

(a) (b) (c) (d) (e)

24.5 Using Gauss's Law

In this section, we'll use Gauss's law to determine the electric fields of several important charge distributions. Some of these you already know, from Chapter 23; others will be new. Three important observations can be made about using Gauss's law:

1. Gauss's law applies only to a *closed* surface, called a Gaussian surface.
2. A Gaussian surface is not a physical surface. It need not coincide with the boundary of any physical object (although it could if we wished). It is an imaginary, mathematical surface in the space surrounding one or more charges.
3. We can't find the electric field from Gauss's law alone. We need to apply Gauss's law in situations where, from symmetry and superposition, we already can guess the *shape* of the field.

These observations and our previous discussion of symmetry and flux lead to the following strategy for solving electric field problems with Gauss's law.

Gauss's law

MODEL Model the charge distribution as a distribution with symmetry.

VISUALIZE Draw a picture of the charge distribution.
- Determine the symmetry of its electric field.
- Choose and draw a Gaussian surface with the *same symmetry*.
- You need not enclose all the charge within the Gaussian surface.
- Be sure every part of the Gaussian surface is either tangent to or perpendicular to the electric field.

SOLVE The mathematical representation is based on Gauss's law

$$\Phi_e = \oint \vec{E} \cdot d\vec{A} = \frac{Q_{in}}{\epsilon_0}$$

- Use Tactics Boxes 24.1 and 24.2 to evaluate the surface integral.

ASSESS Check that your result has correct units and significant figures, is reasonable, and answers the question.

Exercise 19

EXAMPLE 24.3 | Outside a sphere of charge

In Chapter 23 we asserted, without proof, that the electric field outside a sphere of total charge Q is the same as the field of a point charge Q at the center. Use Gauss's law to prove this result.

MODEL The charge distribution within the sphere need not be uniform (i.e., the charge density might increase or decrease with r), but it must have spherical symmetry in order for us to use Gauss's law. We will assume that it does.

VISUALIZE FIGURE 24.23 shows a sphere of charge Q and radius R. We want to find $\vec{E}$ outside this sphere, for distances $r > R$. The spherical symmetry of the charge distribution tells us that the electric field must point *radially outward* from the sphere. Although Gauss's law is true for any surface surrounding the charged sphere, it is useful only if we choose a Gaussian surface to match the spherical symmetry of the charge distribution and the field. Thus a spherical surface of radius $r > R$ *concentric with* the charged sphere will be our Gaussian

FIGURE 24.23 A spherical Gaussian surface surrounding a sphere of charge.

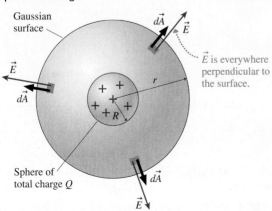

surface. Because this surface surrounds the entire sphere of charge, the enclosed charge is simply $Q_{in} = Q$.

SOLVE Gauss's law is

$$\Phi_e = \oint \vec{E} \cdot d\vec{A} = \frac{Q_{in}}{\epsilon_0} = \frac{Q}{\epsilon_0}$$

To calculate the flux, notice that the electric field is everywhere perpendicular to the spherical surface. And although we don't know the electric field magnitude E, spherical symmetry dictates that E must have the same value at all points equally distant from the center of the sphere. Thus we have the simple result that the net flux through the Gaussian surface is

$$\Phi_e = EA_{sphere} = 4\pi r^2 E$$

where we used the fact that the surface area of a sphere is $A_{sphere} = 4\pi r^2$. With this result for the flux, Gauss's law is

$$4\pi r^2 E = \frac{Q}{\epsilon_0}$$

Thus the electric field at distance r outside a sphere of charge is

$$E_{outside} = \frac{1}{4\pi\epsilon_0}\frac{Q}{r^2}$$

Or in vector form, making use of the fact that $\vec{E}$ is radially outward,

$$\vec{E}_{outside} = \frac{1}{4\pi\epsilon_0}\frac{Q}{r^2}\hat{r}$$

where $\hat{r}$ is a radial unit vector.

ASSESS The field is exactly that of a point charge Q, which is what we wanted to show.

The derivation of the electric field of a sphere of charge depended crucially on a proper choice of the Gaussian surface. We would not have been able to evaluate the flux integral so simply for any other choice of surface. It's worth noting that the result of Example 24.3 can also be proven by the superposition of point-charge fields, but it requires a difficult three-dimensional integral and about a page of algebra. We obtained the answer using Gauss's law in just a few lines. Where Gauss's law works, it works *extremely* well! However, it works only in situations, such as this, with a very high degree of symmetry.

EXAMPLE 24.4 | Inside a sphere of charge

What is the electric field *inside* a uniformly charged sphere?

MODEL We haven't considered a situation like this before. To begin, we don't know if the field strength is increasing or decreasing as we move outward from the center of the sphere. But the field inside must have spherical symmetry. That is, the field must point radially inward or outward, and the field strength can depend only on r. This is sufficient information to solve the problem because it allows us to choose a Gaussian surface.

VISUALIZE FIGURE 24.24 shows a spherical Gaussian surface with radius $r \leq R$ *inside,* and *concentric with,* the sphere of charge. This surface matches the symmetry of the charge distribution, hence $\vec{E}$ is perpendicular to this surface and the field strength E has the same value at all points on the surface.

FIGURE 24.24 A spherical Gaussian surface inside a uniform sphere of charge.

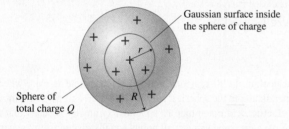

Gaussian surface inside the sphere of charge

Sphere of total charge Q

SOLVE The flux integral is identical to that of Example 24.3:

$$\Phi_e = EA_{\text{sphere}} = 4\pi r^2 E$$

Consequently, Gauss's law is

$$\Phi_e = 4\pi r^2 E = \frac{Q_{\text{in}}}{\epsilon_0}$$

The difference between this example and Example 24.3 is that Q_{in} is no longer the total charge of the sphere. Instead, Q_{in} is the amount of charge *inside* the Gaussian sphere of radius r. Because the charge distribution is *uniform,* the volume charge density is

$$\rho = \frac{Q}{V_R} = \frac{Q}{\frac{4}{3}\pi R^3}$$

The charge enclosed in a sphere of radius r is thus

$$Q_{\text{in}} = \rho V_r = \left(\frac{Q}{\frac{4}{3}\pi R^3}\right)\left(\frac{4}{3}\pi r^3\right) = \frac{r^3}{R^3}Q$$

The amount of enclosed charge increases with the cube of the distance r from the center and, as expected, $Q_{\text{in}} = Q$ if $r = R$. With this expression for Q_{in}, Gauss's law is

$$4\pi r^2 E = \frac{(r^3/R^3)Q}{\epsilon_0}$$

Thus the electric field at radius r inside a uniformly charged sphere is

$$E_{\text{inside}} = \frac{1}{4\pi\epsilon_0}\frac{Q}{R^3}r$$

The electric field strength inside the sphere increases *linearly* with the distance r from the center.

ASSESS The field inside and the field outside a sphere of charge match at the boundary of the sphere, $r = R$, where both give $E = Q/4\pi\epsilon_0 R^2$. In other words, the field strength is *continuous* as we cross the boundary of the sphere. These results are shown graphically in FIGURE 24.25.

FIGURE 24.25 The electric field strength of a uniform sphere of charge of radius R.

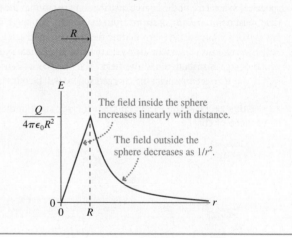

The field inside the sphere increases linearly with distance.

The field outside the sphere decreases as $1/r^2$.

EXAMPLE 24.5 | The electric field of a long, charged wire

In Chapter 23, we used superposition to find the electric field of a long, charged wire with linear charge density λ (C/m). It was not an easy derivation. Find the electric field using Gauss's law.

MODEL A long, charged wire can be modeled as an infinitely long line of charge.

VISUALIZE FIGURE 24.26 shows an infinitely long line of charge. We can use the symmetry of the situation to see that the only

FIGURE 24.26 A Gaussian surface around a charged wire.

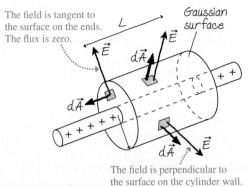

The field is tangent to the surface on the ends. The flux is zero.

The field is perpendicular to the surface on the cylinder wall.

possible shape of the electric field is to point straight into or out from the wire, rather like the bristles on a bottle brush. The shape of the field suggests that we choose our Gaussian surface to be a cylinder of radius r and length L, centered on the wire. Because Gauss's law refers to *closed* surfaces, we must include the ends of the cylinder as part of the surface.

SOLVE Gauss's law is

$$\Phi_e = \oint \vec{E} \cdot d\vec{A} = \frac{Q_{in}}{\epsilon_0}$$

where Q_{in} is the charge *inside* the closed cylinder. We have two tasks: to evaluate the flux integral, and to determine how much charge is inside the closed surface. The wire has linear charge density λ, so the amount of charge inside a cylinder of length L is simply

$$Q_{in} = \lambda L$$

Finding the net flux is just as straightforward. We can divide the flux through the entire closed surface into the flux through each end plus the flux through the cylindrical wall. The electric field $\vec{E}$, pointing straight out from the wire, is tangent to the end surfaces at every point. Thus the flux through these two surfaces is zero. On the wall, $\vec{E}$ is perpendicular to the surface and has the same strength E at every point. Thus

$$\Phi_e = \Phi_{top} + \Phi_{bottom} + \Phi_{wall} = 0 + 0 + EA_{cyl} = 2\pi rLE$$

where we used $A_{cyl} = 2\pi rL$ as the surface area of a cylindrical wall of radius r and length L. Once again, the proper choice of the Gaussian surface reduces the flux integral merely to finding a surface area. With these expressions for Q_{in} and Φ_e, Gauss's law is

$$\Phi_e = 2\pi rLE = \frac{Q_{in}}{\epsilon_0} = \frac{\lambda L}{\epsilon_0}$$

Thus the electric field at distance r from a long, charged wire is

$$E_{wire} = \frac{\lambda}{2\pi\epsilon_0 r}$$

ASSESS This agrees exactly with the result of the more complex derivation in Chapter 23. Notice that the result does not depend on our choice of L. A Gaussian surface is an imaginary device, not a physical object. We needed a finite-length cylinder to do the flux calculation, but the electric field of an *infinitely* long wire can't depend on the length of an imaginary cylinder.

Example 24.5, for the electric field of a long, charged wire, contains a subtle but important idea, one that often occurs when using Gauss's law. The Gaussian cylinder of length L encloses only some of the wire's charge. The pieces of the charged wire outside the cylinder are not enclosed by the Gaussian surface and consequently do not contribute anything to the net flux. Even so, *they are essential* to the use of Gauss's law because it takes the *entire* charged wire to produce an electric field with cylindrical symmetry. In other words, the charge outside the cylinder may not contribute to the flux, but it affects the *shape* of the electric field. Our ability to write $\Phi_e = EA_{cyl}$ depended on knowing that E is the same at every point on the wall of the cylinder. That would not be true for a charged wire of finite length, so we cannot use Gauss's law to find the electric field of a finite-length charged wire.

EXAMPLE 24.6 | **The electric field of a plane of charge**

Use Gauss's law to find the electric field of an infinite plane of charge with surface charge density η (C/m^2).

MODEL A uniformly charged flat electrode can be modeled as an infinite plane of charge.

VISUALIZE FIGURE 24.27 on the next page shows a uniformly charged plane with surface charge density η. We will assume that the plane extends infinitely far in all directions, although we obviously have to show "edges" in our drawing. The planar

symmetry allows the electric field to point only straight toward or away from the plane. With this in mind, choose as a Gaussian surface a cylinder with length L and faces of area A centered on the plane of charge. Although we've drawn them as circular, the shape of the faces is not relevant.

SOLVE The electric field is perpendicular to both faces of the cylinder, so the total flux through both faces is $\Phi_{faces} = 2EA$. (The fluxes add rather than cancel because the area vector $\vec{A}$ points

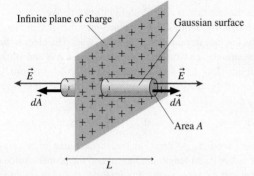

FIGURE 24.27 The Gaussian surface extends to both sides of a plane of charge.

outward on each face.) There's *no* flux through the wall of the cylinder because the field vectors are tangent to the wall. Thus the net flux is simply

$$\Phi_e = 2EA$$

The charge inside the cylinder is the charge contained in area A of the plane. This is

$$Q_{in} = \eta A$$

With these expressions for Q_{in} and Φ_e, Gauss's law is

$$\Phi_e = 2EA = \frac{Q_{in}}{\epsilon_0} = \frac{\eta A}{\epsilon_0}$$

Thus the electric field of an infinite charged plane is

$$E_{plane} = \frac{\eta}{2\epsilon_0}$$

This agrees with the result in Chapter 23.

ASSESS This is another example of a Gaussian surface enclosing only some of the charge. Most of the plane's charge is outside the Gaussian surface and does not contribute to the flux, but it does affect the shape of the field. We wouldn't have planar symmetry, with the electric field exactly perpendicular to the plane, without all the rest of the charge on the plane.

The plane of charge is an especially good example of how powerful Gauss's law can be. Finding the electric field of a plane of charge via superposition was a difficult and tedious derivation. With Gauss's law, once you see how to apply it, the problem is simple enough to solve in your head!

You might wonder, then, why we bothered with superposition at all. The reason is that Gauss's law, powerful though it may be, is effective only in a limited number of situations where the field is highly symmetric. Superposition always works, even if the derivation is messy, because superposition goes directly back to the fields of individual point charges. It's good to use Gauss's law when you can, but superposition is often the only way to attack real-world charge distributions.

STOP TO THINK 24.5 Which Gaussian surface would allow you to use Gauss's law to calculate the electric field outside a uniformly charged cube?

a. A sphere whose center coincides with the center of the charged cube
b. A cube whose center coincides with the center of the charged cube and that has parallel faces
c. Either a or b
d. Neither a nor b

24.6 Conductors in Electrostatic Equilibrium

Consider a charged conductor, such as a charged metal electrode, in electrostatic equilibrium. That is, there is no current through the conductor and the charges are all stationary. One very important conclusion is that **the electric field is zero at all points within a conductor in electrostatic equilibrium.** That is, $\vec{E}_{in} = \vec{0}$. If this weren't true, the electric field would cause the charge carriers to move and thus violate the assumption that all the charges are at rest. Let's use Gauss's law to see what else we can learn.

At the Surface of a Conductor

FIGURE 24.28 shows a Gaussian surface just barely inside the physical surface of a conductor that's in electrostatic equilibrium. The electric field is zero at all points within the conductor, hence the electric flux Φ_e through this Gaussian surface must be zero. But if $\Phi_e = 0$, Gauss's law tells us that $Q_{in} = 0$. That is, there's no net charge within this surface. There are charges—electrons and positive ions—but no *net* charge.

If there's no net charge in the interior of a conductor in electrostatic equilibrium, then **all the excess charge on a charged conductor resides on the exterior surface of the conductor.** Any charges added to a conductor quickly spread across the surface until reaching positions of electrostatic equilibrium, but there is no net charge *within* the conductor.

There may be no electric field within a charged conductor, but the presence of net charge requires an exterior electric field in the space outside the conductor. **FIGURE 24.29** shows that **the electric field right at the surface of the conductor has to be perpendicular to the surface.** To see that this is so, suppose $\vec{E}_{surface}$ had a component tangent to the surface. This component of $\vec{E}_{surface}$ would exert a force on the surface charges and cause a surface current, thus violating the assumption that all charges are at rest. The only exterior electric field consistent with electrostatic equilibrium is one that is perpendicular to the surface.

We can use Gauss's law to relate the field strength at the surface to the charge density on the surface. **FIGURE 24.30** shows a small Gaussian cylinder with faces very slightly above and below the surface of a charged conductor. The charge inside this Gaussian cylinder is ηA, where η is the surface charge density at this point on the conductor. There's a flux $\Phi = AE_{surface}$ through the outside face of this cylinder but, unlike Example 24.6 for the plane of charge, *no* flux through the inside face because $\vec{E}_{in} = \vec{0}$ within the conductor. Furthermore, there's no flux through the wall of the cylinder because $\vec{E}_{surface}$ is perpendicular to the surface. Thus the net flux is $\Phi_e = AE_{surface}$. Gauss's law is

$$\Phi_e = AE_{surface} = \frac{Q_{in}}{\epsilon_0} = \frac{\eta A}{\epsilon_0} \qquad (24.19)$$

from which we can conclude that the electric field at the surface of a charged conductor is

$$\vec{E}_{surface} = \left(\frac{\eta}{\epsilon_0}, \text{ perpendicular to surface} \right) \qquad (24.20)$$

In general, the surface charge density η is *not* constant on the surface of a conductor but depends on the shape of the conductor. If we can determine η, by either calculating it or measuring it, then Equation 24.20 tells us the electric field at that point on the surface. Alternatively, we can use Equation 24.20 to deduce the charge density on the conductor's surface if we know the electric field just outside the conductor.

Charges and Fields Within a Conductor

FIGURE 24.31 shows a charged conductor with a hole inside. Can there be charge on this interior surface? To find out, we place a Gaussian surface around the hole, infinitesimally close but entirely within the conductor. The electric flux Φ_e through this Gaussian surface is zero because the electric field is zero everywhere inside the conductor. Thus we must conclude that $Q_{in} = 0$. There's no net charge inside this Gaussian surface and thus no charge on the surface of the hole. Any excess charge resides on the *exterior* surface of the conductor, not on any interior surfaces.

Furthermore, because there's no electric field inside the conductor and no charge inside the hole, the electric field inside the hole must also be zero. This conclusion has an important practical application. For example, suppose we need to exclude the electric field from the region in **FIGURE 24.32a** on the next page enclosed within dashed lines. We can do so by surrounding this region with the neutral conducting box of **FIGURE 24.32b**.

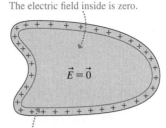

FIGURE 24.28 A Gaussian surface just inside a conductor's surface.

The electric field inside is zero.

$\vec{E} = \vec{0}$

The flux through the Gaussian surface is zero. Hence all the excess charge must be on the surface.

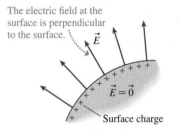

FIGURE 24.29 The electric field at the surface of a charged conductor.

The electric field at the surface is perpendicular to the surface.

$\vec{E}$

$\vec{E} = \vec{0}$

Surface charge

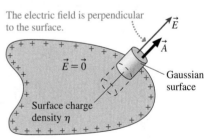

FIGURE 24.30 A Gaussian surface extending through the surface has a flux only through the outer face.

The electric field is perpendicular to the surface.

$\vec{E}$

$\vec{A}$

$\vec{E} = \vec{0}$

Gaussian surface

Surface charge density η

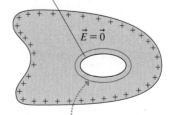

FIGURE 24.31 A Gaussian surface surrounding a hole inside a conductor.

A hollow completely enclosed by the conductor

$\vec{E} = \vec{0}$

The flux through the Gaussian surface is zero. There's no net charge inside, hence no charge on this interior surface.

FIGURE 24.32 The electric field can be excluded from a region of space by surrounding it with a conducting box.

(a) Parallel-plate capacitor

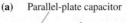

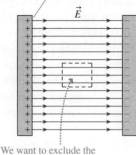

We want to exclude the electric field from this region.

(b) The conducting box has been polarized and has induced surface charges.

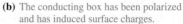

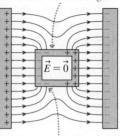

The electric field is perpendicular to all conducting surfaces.

FIGURE 24.33 A charge in the hole causes a net charge on the interior and exterior surfaces.

The flux through the Gaussian surface is zero, hence there's no *net* charge inside this surface. There must be charge $-q$ on the inside surface to balance point charge q.

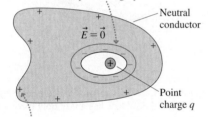

Neutral conductor

$\vec{E} = \vec{0}$

Point charge q

The outer surface must have charge $+q$ so that the conductor remains neutral.

This region of space is now a hole inside a conductor, thus the interior electric field is zero. The use of a conducting box to exclude electric fields from a region of space is called **screening.** Solid metal walls are ideal, but in practice wire screen or wire mesh—sometimes called a *Faraday cage*—provides sufficient screening for all but the most sensitive applications. The price we pay is that the exterior field is now very complicated.

Finally, **FIGURE 24.33** shows a charge q inside a hole within a neutral conductor. The electric field *within* the conductor is still zero, hence the electric flux through the Gaussian surface is zero. But $\Phi_e = 0$ requires $Q_{in} = 0$. Consequently, the charge inside the hole attracts an equal charge of opposite sign, and charge $-q$ now lines the inner surface of the hole.

The conductor as a whole is neutral, so moving $-q$ to the surface of the hole must leave $+q$ behind somewhere else. Where is it? It can't be in the interior of the conductor, as we've seen, and that leaves only the exterior surface. In essence, an internal charge polarizes the conductor just as an external charge would. Net charge $-q$ moves to the inner surface and net charge $+q$ is left behind on the exterior surface.

In summary, conductors in electrostatic equilibrium have the properties described in Tactics Box 24.3.

TACTICS BOX 24.3 (MP)

Finding the electric field of a conductor in electrostatic equilibrium

❶ The electric field is zero at all points within the volume of the conductor.
❷ Any excess charge resides entirely on the *exterior* surface.
❸ The external electric field at the surface of a charged conductor is perpendicular to the surface and of magnitude η/ϵ_0, where η is the surface charge density at that point.
❹ The electric field is zero inside any hole within a conductor unless there is a charge in the hole.

Exercises 20–24 🖉

EXAMPLE 24.7 | **The electric field at the surface of a charged metal sphere**

A 2.0-cm-diameter brass sphere has been given a charge of 2.0 nC. What is the electric field strength at the surface?

MODEL Brass is a conductor. The excess charge resides on the surface.

VISUALIZE The charge distribution has spherical symmetry. The electric field points radially outward from the surface.

SOLVE We can solve this problem in two ways. One uses the fact that a sphere, because of its complete symmetry, is the one shape for which any excess charge will spread out to a *uniform* surface charge density. Thus

$$\eta = \frac{q}{A_{\text{sphere}}} = \frac{q}{4\pi R^2} = \frac{2.0 \times 10^{-9}\,\text{C}}{4\pi (0.010\,\text{m})^2} = 1.59 \times 10^{-6}\,\text{C/m}^2$$

From Equation 24.20, we know the electric field at the surface has strength

$$E_{\text{surface}} = \frac{\eta}{\epsilon_0} = \frac{1.59 \times 10^{-6}\,\text{C/m}^2}{8.85 \times 10^{-12}\,\text{C}^2/\text{N}\,\text{m}^2} = 1.8 \times 10^5\,\text{N/C}$$

Alternatively, we could have used the result, obtained earlier in the chapter, that the electric field strength outside a sphere of charge Q is $E_{\text{outside}} = Q_{\text{in}}/(4\pi\epsilon_0 r^2)$. But $Q_{\text{in}} = q$ and, at the surface, $r = R$. Thus

$$E_{\text{surface}} = \frac{1}{4\pi\epsilon_0}\frac{q}{R^2} = (9.0 \times 10^9\,\text{N}\,\text{m}^2/\text{C}^2)\frac{2.0 \times 10^{-9}\,\text{C}}{(0.010\,\text{m})^2}$$

$$= 1.8 \times 10^5\,\text{N/C}$$

As we can see, both methods lead to the same result.

CHALLENGE **EXAMPLE 24.8** | The electric field of a slab of charge

An infinite slab of charge of thickness $2a$ is centered in the xy-plane. The charge density is $\rho = \rho_0(1 - |z|/a)$. Find the electric field strengths inside and outside this slab of charge.

MODEL The charge density is not uniform. Starting at ρ_0 in the xy-plane, it decreases linearly with distance above and below the xy-plane until reaching zero at $z = \pm a$, the edges of the slab.

VISUALIZE FIGURE 24.34 shows an edge view of the slab of charge and, as Gaussian surfaces, side viwews of two cylinders with cross-section area A. By symmetry, the electric field must point away from the xy-plane; the field cannot have an x- or y-component.

FIGURE 24.34 Two cylindrical Gaussian surfaces for an infinite slab of charge.

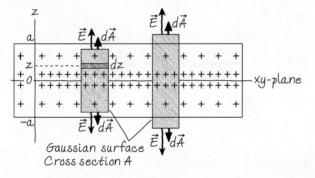

SOLVE Gauss's law is

$$\Phi_e = \oint \vec{E} \cdot d\vec{A} = \frac{Q_{\text{in}}}{\epsilon_0}$$

With symmetry, finding the net flux is straightforward. The electric field is perpendicular to the faces of the cylinders and pointing outward, so the total flux through the faces is $\Phi_{\text{faces}} = 2EA$, where E may depend on distance z. The field is parallel to the walls of the cylinders, so $\Phi_{\text{wall}} = 0$. Thus the net flux is simply

$$\Phi_e = 2EA$$

Because the charge density is not uniform, we need to integrate to find Q_{in}, the charge *inside* the cylinder. We can slice the cylinder into small slabs of infinitesimal thickness dz and volume $dV = A\,dz$. Figure 24.34 shows one such little slab at distance z from the xy-plane. The charge in this little slab is

$$dq = \rho\,dV = \rho_0\left(1 - \frac{z}{a}\right)A\,dz$$

where we assumed that z is positive. Because the charge is symmetric about $z = 0$, we can avoid difficulties with the absolute value sign in the charge density by integrating from 0 and multiplying by 2. For the Gaussian cylinder that ends inside the slab of charge, at distance z, the total charge inside is

$$Q_{\text{in}} = \int dq = 2\int_0^z \rho_0\left(1 - \frac{z}{a}\right)A\,dz$$

$$= 2\rho_0 A\left[z\Big|_0^z - \frac{1}{2a}z^2\Big|_0^z\right]$$

$$= 2\rho_0 A z\left(1 - \frac{z}{2a}\right)$$

Gauss's law inside the slab is then

$$\Phi_e = 2E_{\text{inside}}A = \frac{Q_{\text{in}}}{\epsilon_0} = \frac{2\rho_0 A z}{\epsilon_0}\left(1 - \frac{z}{2a}\right)$$

The area A cancels, as it must because it was an arbitrary choice, leaving

$$E_{\text{inside}} = \frac{\rho_0 z}{\epsilon_0}\left(1 - \frac{z}{2a}\right)$$

The field strength is zero at $z = 0$, then increases as z increases. This expression is valid only above the xy-plane, for $z > 0$, but the field strength is symmetric on the other side.

For the Gaussian cylinder that extends outside the slab of charge, the integral for Q has to end at $z = a$. Thus

$$Q_{\text{in}} = 2\rho_0 A a\left(1 - \frac{a}{2a}\right) = \rho_0 A a$$

independent of distance z. With this, Gauss's law gives

$$E_{\text{outside}} = \frac{Q_{\text{in}}}{2\epsilon_0 A} = \frac{\rho_0 a}{2\epsilon_0}$$

This matches E_{inside} at the surface, $z = a$, so the field is continuous as it crosses the boundary.

ASSESS Outside a sphere of charge, the field is the same as that of a point charge at the center. Similarly, the field outside an infinite slab of charge should be the same as that of an infinite charged plane. We found, by integration, that the total charge in an area A of the slab is $Q = \rho_0 A a$. If we squished this charge into a plane, the surface charge density would be $\eta = Q/A = \rho_0 a$. Thus our expression for E_{outside} could be written $\eta/2\epsilon_0$, which matches the field we found in Example 24.6 for a plane of charge.

SUMMARY

The goal of Chapter 24 has been to learn about and apply Gauss's law.

GENERAL PRINCIPLES

Gauss's Law

For any *closed* surface enclosing net charge Q_{in}, the net electric flux through the surface is

$$\Phi_e = \oint \vec{E} \cdot d\vec{A} = \frac{Q_{in}}{\epsilon_0}$$

The electric flux Φ_e is the same for *any* closed surface enclosing charge Q_{in}.

To solve electric field problems with Gauss's law:

MODEL Model the charge distribution as one with symmetry.

VISUALIZE Draw a picture of the charge distribution. Draw a Gaussian surface with the same symmetry as the electric field, every part of which is either tangent to or perpendicular to the electric field.

SOLVE Apply Gauss's law and Tactics Boxes 24.1 and 24.2 to evaluate the surface integral.

ASSESS Is the result reasonable?

Symmetry

The symmetry of the electric field must match the symmetry of the charge distribution.

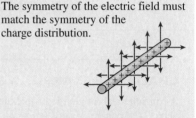

In practice, Φ_e is computable only if the symmetry of the Gaussian surface matches the symmetry of the charge distribution.

IMPORTANT CONCEPTS

Charge creates the electric field that is responsible for the electric flux.

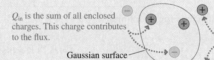

Q_{in} is the sum of all enclosed charges. This charge contributes to the flux.

Gaussian surface

Charges outside the surface contribute to the electric field, but they don't contribute to the flux.

Flux is the amount of electric field passing through a surface of area A:

$$\Phi_e = \vec{E} \cdot \vec{A}$$

where $\vec{A}$ is the **area vector.**

For closed surfaces:
A net flux in or out indicates that the surface encloses a net charge.

Field lines through but with no *net* flux mean that the surface encloses no *net* charge.

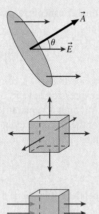

Surface integrals calculate the flux by summing the fluxes through many small pieces of the surface:

$$\Phi_e = \sum \vec{E} \cdot \delta\vec{A}$$
$$\rightarrow \int \vec{E} \cdot d\vec{A}$$

Two important situations:

If the electric field is everywhere tangent to the surface, then

$$\Phi_e = 0$$

If the electric field is everywhere perpendicular to the surface *and* has the same strength E at all points, then

$$\Phi_e = EA$$

APPLICATIONS

Conductors in electrostatic equilibrium

- The electric field is zero at all points within the conductor.
- Any excess charge resides entirely on the exterior surface.
- The external electric field is perpendicular to the surface and of magnitude η/ϵ_0, where η is the surface charge density.
- The electric field is zero inside any hole within a conductor unless there is a charge in the hole.

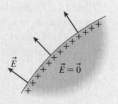

TERMS AND NOTATION

symmetric	electric flux, Φ_e	surface integral	screening
Gaussian surface	area vector, $\vec{A}$	Gauss's law	

CONCEPTUAL QUESTIONS

1. Suppose you have the uniformly charged cube in **FIGURE Q24.1**. Can you use symmetry alone to deduce the *shape* of the cube's electric field? If so, sketch and describe the field shape. If not, why not?

FIGURE Q24.1

2. **FIGURE Q24.2** shows cross sections of three-dimensional closed surfaces. They have a flat top and bottom surface above and below the plane of the page. However, the electric field is everywhere parallel to the page, so there is no flux through the top or bottom surface. The electric field is uniform over each face of the surface. For each, does the surface enclose a net positive charge, a net negative charge, or no net charge? Explain.

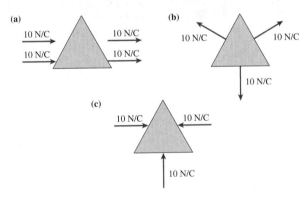

FIGURE Q24.2

3. The square and circle in **FIGURE Q24.3** are in the same uniform field. The diameter of the circle equals the edge length of the square. Is Φ_{square} larger than, smaller than, or equal to Φ_{circle}? Explain.

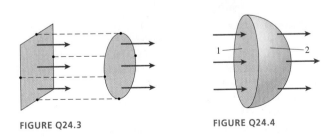

FIGURE Q24.3   **FIGURE Q24.4**

4. In **FIGURE Q24.4**, where the field is uniform, is the magnitude of Φ_1 larger than, smaller than, or equal to the magnitude of Φ_2? Explain.

5. What is the electric flux through each of the surfaces in **FIGURE Q24.5**? Give each answer as a multiple of q/ϵ_0.

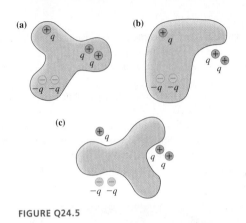

FIGURE Q24.5

6. What is the electric flux through each of the surfaces A to E in **FIGURE Q24.6**? Give each answer as a multiple of q/ϵ_0.

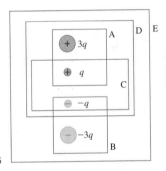

FIGURE Q24.6

7. The charged balloon in **FIGURE Q24.7** expands as it is blown up, increasing in size from the initial to final diameters shown. Do the electric field strengths at points 1, 2, and 3 increase, decrease, or stay the same? Explain your reasoning for each.

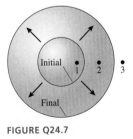

FIGURE Q24.7

8. The two spheres in **FIGURE Q24.8** on the next page surround equal charges. Three students are discussing the situation.
Student 1: The fluxes through spheres A and B are equal because they enclose equal charges.
Student 2: But the electric field on sphere B is weaker than the electric field on sphere A. The flux depends on the electric field strength, so the flux through A is larger than the flux through B.

Student 3: I thought we learned that flux was about surface area. Sphere B is larger than sphere A, so I think the flux through B is larger than the flux through A.

Which of these students, if any, do you agree with? Explain.

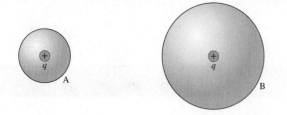

FIGURE Q24.8

9. The sphere and ellipsoid in **FIGURE Q24.9** surround equal charges. Four students are discussing the situation.

Student 1: The fluxes through A and B are equal because the average radius is the same.

Student 2: I agree that the fluxes are equal, but that's because they enclose equal charges.

Student 3: The electric field is not perpendicular to the surface for B, and that makes the flux through B less than the flux through A.

Student 4: I don't think that Gauss's law even applies to a situation like B, so we can't compare the fluxes through A and B.

Which of these students, if any, do you agree with? Explain.

 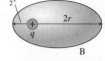

FIGURE Q24.9

10. A small, metal sphere hangs by an insulating thread within the larger, hollow conducting sphere of **FIGURE Q24.10**. A conducting wire extends from the small sphere through, but not touching, a small hole in the hollow sphere. A charged rod is used to transfer positive charge to the protruding wire. After the charged rod has touched the wire and been removed, are the following surfaces positive, negative, or not charged? Explain.
a. The small sphere.
b. The inner surface of the hollow sphere.
c. The outer surface of the hollow sphere.

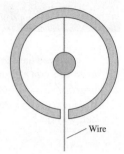

FIGURE Q24.10

EXERCISES AND PROBLEMS

Problems labeled [ ] integrate material from earlier chapters.

Exercises

Section 24.1 Symmetry

1. | **FIGURE EX24.1** shows two cross sections of two infinitely long coaxial cylinders. The inner cylinder has a positive charge, the outer cylinder has an equal negative charge. Draw this figure on your paper, then draw electric field vectors showing the shape of the electric field.

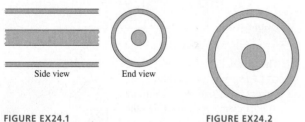

Side view   End view

FIGURE EX24.1   **FIGURE EX24.2**

2. | **FIGURE EX24.2** shows a cross section of two concentric spheres. The inner sphere has a negative charge. The outer sphere has a positive charge larger in magnitude than the charge on the inner sphere. Draw this figure on your paper, then draw electric field vectors showing the shape of the electric field.

3. | **FIGURE EX24.3** shows a cross section of two infinite parallel planes of charge. Draw this figure on your paper, then draw electric field vectors showing the shape of the electric field.

++++++++++++++++++

FIGURE EX24.3 ++++++++++++++++++

Section 24.2 The Concept of Flux

4. | The electric field is constant over each face of the cube shown in **FIGURE EX24.4**. Does the box contain positive charge, negative charge, or no charge? Explain.

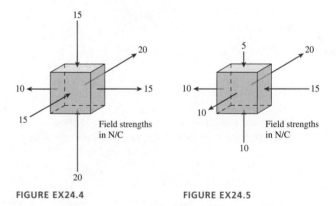

Field strengths in N/C   Field strengths in N/C

FIGURE EX24.4   **FIGURE EX24.5**

5. | The electric field is constant over each face of the cube shown in **FIGURE EX24.5**. Does the box contain positive charge, negative charge, or no charge? Explain.

6. | The cube in **FIGURE EX24.6** contains negative charge. The electric field is constant over each face of the cube. Does the missing electric field vector on the front face point in or out? What strength must this field exceed?

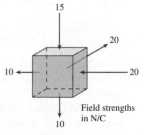

Field strengths in N/C

FIGURE EX24.6

7. | The cube in **FIGURE EX24.7** contains negative charge. The electric field is constant over each face of the cube. Does the missing electric field vector on the front face point in or out? What strength must this field exceed?

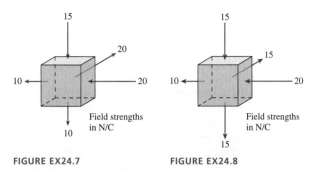

FIGURE EX24.7   **FIGURE EX24.8**

8. | The cube in **FIGURE EX24.8** contains no net charge. The electric field is constant over each face of the cube. Does the missing electric field vector on the front face point in or out? What is the field strength?

Section 24.3 Calculating Electric Flux

9. ‖   What is the electric flux through the surface shown in **FIGURE EX24.9**?

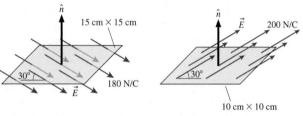

FIGURE EX24.9   **FIGURE EX24.10**

10. ‖   What is the electric flux through the surface shown in **FIGURE EX24.10**?

11. ‖   The electric flux through the surface shown in **FIGURE EX24.11** is 25 N m²/C. What is the electric field strength?

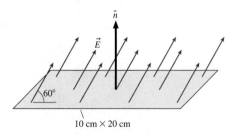

FIGURE EX24.11

12. ‖   A 2.0 cm × 3.0 cm rectangle lies in the xy-plane. What is the magnitude of the electric flux through the rectangle if
a. $\vec{E} = (100\hat{\imath} - 200\hat{k})$ N/C?
b. $\vec{E} = (100\hat{\imath} - 200\hat{\jmath})$ N/C?

13. ‖   A 2.0 cm × 3.0 cm rectangle lies in the xz-plane. What is the magnitude of the electric flux through the rectangle if
a. $\vec{E} = (100\hat{\imath} - 200\hat{k})$ N/C?
b. $\vec{E} = (100\hat{\imath} - 200\hat{\jmath})$ N/C?

14. ‖   A 3.0-cm-diameter circle lies in the xz-plane in a region where the electric field is $\vec{E} = (1500\hat{\imath} + 1500\hat{\jmath} - 1500\hat{k})$ N/C. What is the electric flux through the circle?

15. ‖   A 1.0 cm × 1.0 cm × 1.0 cm box with its edges aligned with the xyz-axes is in the electric field $\vec{E} = (350x + 150)\hat{\imath}$ N/C, where x is in meters. What is the net electric flux through the box?

16. | What is the net electric flux through the two cylinders shown in **FIGURE EX24.16**? Give your answer in terms of R and E.

(a)   **(b)**

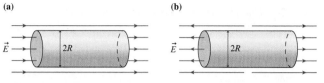

FIGURE EX24.16

Section 24.4 Gauss's Law

Section 24.5 Using Gauss's Law

17. | **FIGURE EX24.17** shows three charges. Draw these charges on your paper four times. Then draw two-dimensional cross sections of three-dimensional closed surfaces through which the electric flux is (a) $2q/\epsilon_0$, (b) q/ϵ_0, (c) 0, and (d) $5q/\epsilon_0$.

FIGURE EX24.17   **FIGURE EX24.18**

18. | **FIGURE EX24.18** shows three charges. Draw these charges on your paper four times. Then draw two-dimensional cross sections of three-dimensional closed surfaces through which the electric flux is (a) $-q/\epsilon_0$, (b) q/ϵ_0, (c) $3q/\epsilon_0$, and (d) $4q/\epsilon_0$.

19. | **FIGURE EX24.19** shows three Gaussian surfaces and the electric flux through each. What are the three charges q_1, q_2, and q_3?

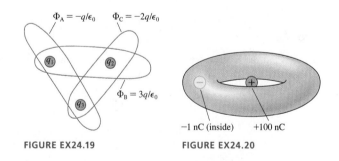

FIGURE EX24.19   **FIGURE EX24.20**

20. ‖   What is the net electric flux through the torus (i.e., doughnut shape) of **FIGURE EX24.20**?

21. ‖   What is the net electric flux through the cylinder of **FIGURE EX24.21**?

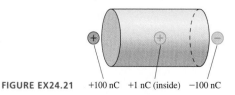

FIGURE EX24.21   +100 nC +1 nC (inside) −100 nC

22. ‖ The net electric flux through an octahedron is -1000 N m^2/C. How much charge is enclosed within the octahedron?

23. ‖ 55.3 million excess electrons are inside a closed surface. What is the net electric flux through the surface?

Section 24.6 Conductors in Electrostatic Equilibrium

24. | A spark occurs at the tip of a metal needle if the electric field strength exceeds 3.0×10^6 N/C, the field strength at which air breaks down. What is the minimum surface charge density for producing a spark?

25. ‖ The electric field strength just above one face of a copper penny is 2000 N/C. What is the surface charge density on this face of the penny?

26. | The conducting box in **FIGURE EX24.26** has been given an excess negative charge. The surface density of excess electrons at the center of the top surface is 5.0×10^{10} electrons/m^2. What are the electric field strengths E_1 to E_3 at points 1 to 3?

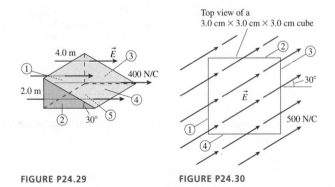

FIGURE EX24.26   FIGURE EX24.27

27. ‖ **FIGURE EX24.27** shows a hollow cavity within a neutral conductor. A point charge Q is inside the cavity. What is the net electric flux through the closed surface that surrounds the conductor?

28. | A thin, horizontal, 10-cm-diameter copper plate is charged to 3.5 nC. If the electrons are uniformly distributed on the surface, what are the strength and direction of the electric field
   a. 0.1 mm above the center of the top surface of the plate?
   b. at the plate's center of mass?
   c. 0.1 mm below the center of the bottom surface of the plate?

Problems

29. ‖ Find the electric fluxes Φ_1 to Φ_5 through surfaces 1 to 5 in **FIGURE P24.29**.

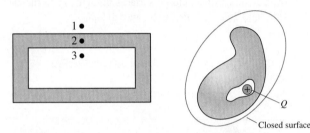

FIGURE P24.29   FIGURE P24.30

30. | **FIGURE P24.30** shows four sides of a 3.0 cm $\times$ 3.0 cm $\times$ 3.0 cm cube.
   a. What are the electric fluxes Φ_1 to Φ_4 through sides 1 to 4?
   b. What is the net flux through these four sides?

31. ‖ A tetrahedron has an equilateral triangle base with 20-cm-long edges and three equilateral triangle sides. The base is parallel to the ground, and a vertical uniform electric field of strength 200 N/C passes upward through the tetrahedron.
   a. What is the electric flux through the base?
   b. What is the electric flux through each of the three sides?

32. | Charges $q_1 = -4Q$ and $q_2 = +2Q$ are located at $x = -a$ and $x = +a$, respectively. What is the net electric flux through a sphere of radius $2a$ centered (a) at the origin and (b) at $x = 2a$?

33. ‖ A 10 nC charge is at the center of a 2.0 m $\times$ 2.0 m $\times$ 2.0 m cube. What is the electric flux through the top surface of the cube?

34. ‖ A spherically symmetric charge distribution produces the electric field $\vec{E} = (5000r^2)\hat{r}$ N/C, where r is in m.
   a. What is the electric field strength at $r = 20$ cm?
   b. What is the electric flux through a 40-cm-diameter spherical surface that is concentric with the charge distribution?
   c. How much charge is inside this 40-cm-diameter spherical surface?

35. ‖ A neutral conductor contains a hollow cavity in which there is a $+100$ nC point charge. A charged rod then transfers -50 nC to the conductor. Afterward, what is the charge (a) on the inner wall of the cavity, and (b) on the exterior surface of the conductor?

36. ‖ A hollow metal sphere has inner radius a and outer radius b. The hollow sphere has charge $+2Q$. A point charge $+Q$ sits at the center of the hollow sphere.
   a. Determine the electric fields in the three regions $r \le a$, $a < r < b$, and $r \ge b$.
   b. How much charge is on the inside surface of the hollow sphere? On the exterior surface?

37. ‖‖ A 20-cm-radius ball is uniformly charged to 80 nC.
   a. What is the ball's volume charge density (C/m^3)?
   b. How much charge is enclosed by spheres of radii 5, 10, and 20 cm?
   c. What is the electric field strength at points 5, 10, and 20 cm from the center?

38. ‖‖ **FIGURE P24.38** shows a solid metal sphere at the center of a hollow metal sphere. What is the total charge on (a) the exterior of the inner sphere, (b) the inside surface of the hollow sphere, and (c) the exterior surface of the hollow sphere?

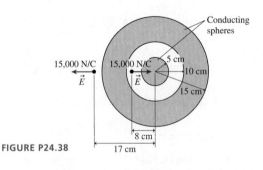

FIGURE P24.38

39. ‖ The earth has a vertical electric field at the surface, pointing down, that averages 100 N/C. This field is maintained by various atmospheric processes, including lightning. What is the excess charge on the surface of the earth?

40. ‖ Figure 24.32b showed a conducting box inside a parallel-plate capacitor. The electric field inside the box is $\vec{E} = \vec{0}$. Suppose the surface charge on the exterior of the box could be frozen. Draw a picture of the electric field inside the box after the box, with its frozen charge, is removed from the capacitor.
   Hint: Superposition.

41. ‖ A hollow metal sphere has 6 cm and 10 cm inner and outer radii, respectively. The surface charge density on the inside surface is -100 nC/m². The surface charge density on the exterior surface is $+100$ nC/m². What are the strength and direction of the electric field at points 4, 8, and 12 cm from the center?

42. ‖ A positive point charge q sits at the center of a hollow spherical shell. The shell, with radius R and negligible thickness, has net charge $-2q$. Find an expression for the electric field strength (a) inside the sphere, $r < R$, and (b) outside the sphere, $r > R$. In what direction does the electric field point in each case?

43. ‖ Find the electric field inside and outside a hollow plastic ball of radius R that has charge Q uniformly distributed on its outer surface.

44. ‖ A uniformly charged ball of radius a and charge $-Q$ is at the center of a hollow metal shell with inner radius b and outer radius c. The hollow sphere has net charge $+2Q$. Determine the electric field strength in the four regions $r \le a$, $a < r < b$, $b \le r \le c$, and $r > c$.

45. ‖ The three parallel planes of charge shown in FIGURE P24.45 have surface charge densities $-\frac{1}{2}\eta$, η, and $-\frac{1}{2}\eta$. Find the electric fields $\vec{E}_1$ to $\vec{E}_4$ in regions 1 to 4.

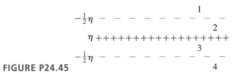

FIGURE P24.45

46. ‖ An infinite slab of charge of thickness $2z_0$ lies in the xy-plane between $z = -z_0$ and $z = +z_0$. The volume charge density ρ (C/m³) is a constant.
   a. Use Gauss's law to find an expression for the electric field strength inside the slab $(-z_0 \le z \le z_0)$.
   b. Find an expression for the electric field strength above the slab $(z \ge z_0)$.
   c. Draw a graph of E from $z = 0$ to $z = 3z_0$.

47. ‖ FIGURE P24.47 shows an infinitely wide conductor parallel to and distance d from an infinitely wide plane of charge with surface charge density η. What are the electric fields $\vec{E}_1$ to $\vec{E}_4$ in regions 1 to 4?

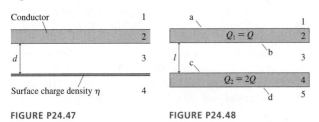

FIGURE P24.47   FIGURE P24.48

48. ‖ FIGURE P24.48 shows two very large slabs of metal that are parallel and distance l apart. The top and bottom of each slab has surface area A. The thickness of each slab is so small in comparison to its lateral dimensions that the surface area around the sides is negligible. Metal 1 has total charge $Q_1 = Q$ and metal 2 has total charge $Q_2 = 2Q$. Assume Q is positive. In terms of Q and A, determine
   a. The electric field strengths E_1 to E_5 in regions 1 to 5.
   b. The surface charge densities η_a to η_d on the four surfaces a to d.

49. ‖ A long, thin straight wire with linear charge density λ runs down the center of a thin, hollow metal cylinder of radius R. The cylinder has a net linear charge density 2λ. Assume λ is positive. Find expressions for the electric field strength (a) inside the cylinder, $r < R$, and (b) outside the cylinder, $r > R$. In what direction does the electric field point in each of the cases?

50. ‖ A very long, uniformly charged cylinder has radius R and linear charge density λ. Find the cylinder's electric field strength (a) outside the cylinder, $r \ge R$, and (b) inside the cylinder, $r \le R$. (c) Show that your answers to parts a and b match at the boundary, $r = R$.

51. ‖ The electric field must be zero inside a *conductor* in electrostatic equilibrium, but not inside an insulator. It turns out that we can still apply Gauss's law to a Gaussian surface that is entirely within an insulator by replacing the right-hand side of Gauss's law, Q_{in}/ϵ_0, with Q_{in}/ϵ, where ϵ is the *permittivity* of the material. (Technically, ϵ_0 is called the *vacuum permittivity*.) Suppose that a 50 nC point charge is surrounded by a thin, 32-cm-diameter spherical rubber shell and that the electric field strength inside the rubber shell is 2500 N/C. What is the permittivity of rubber?

52. ‖ The electric field must be zero inside a *conductor* in electrostatic equilibrium, but not inside an insulator. It turns out that we can still apply Gauss's law to a Gaussian surface that is entirely within an insulator by replacing the right-hand side of Gauss's law, Q_{in}/ϵ_0, with Q_{in}/ϵ, where ϵ is the *permittivity* of the material. (Technically, ϵ_0 is called the *vacuum permittivity*.) Suppose a long, straight wire with linear charge density 250 nC/m is covered with insulation whose permittivity is $2.5\epsilon_0$. What is the electric field strength at a point inside the insulation that is 1.5 mm from the axis of the wire?

53. ‖‖ A long cylinder with radius b and volume charge density ρ has a spherical hole with radius $a < b$ centered on the axis of the cylinder. What is the electric field strength inside the hole at radial distance $r < a$ in a plane that is perpendicular to the cylinder through the center of the hole?
   Hint: Can you create this charge distribution as a superposition of charge distributions for which you can use Gauss's law to find the electric field?

54. ‖ A spherical shell has inner radius R_{in} and outer radius R_{out}. The shell contains total charge Q, uniformly distributed. The interior of the shell is empty of charge and matter.
   a. Find the electric field strength outside the shell, $r \ge R_{out}$.
   b. Find the electric field strength in the interior of the shell, $r \le R_{in}$.
   c. Find the electric field strength within the shell, $R_{in} \le r \le R_{out}$.
   d. Show that your solutions match at both the inner and outer boundaries.

55. ‖‖ An early model of the atom, proposed by Rutherford after his discovery of the atomic nucleus, had a positive point charge $+Ze$ (the nucleus) at the center of a sphere of radius R with uniformly distributed negative charge $-Ze$. Z is the atomic number, the number of protons in the nucleus and the number of electrons in the negative sphere.
   a. Show that the electric field strength inside this atom is

   $$E_{in} = \frac{Ze}{4\pi\epsilon_0}\left(\frac{1}{r^2} - \frac{r}{R^3}\right)$$

   b. What is E at the surface of the atom? Is this the expected value? Explain.
   c. A uranium atom has $Z = 92$ and $R = 0.10$ nm. What is the electric field strength at $r = \frac{1}{2}R$?

56. ‖ Newton's law of gravity and Coulomb's law are both inverse-square laws. Consequently, there should be a "Gauss's law for gravity."
   CALC
   a. The electric field was defined as $\vec{E} = \vec{F}_{on\ q}/q$, and we used this to find the electric field of a point charge. Using analogous reasoning, what is the gravitational field $\vec{g}$ of a point mass?

Write your answer using the unit vector $\hat{r}$, but be careful with signs; the gravitational force between two "like masses" is attractive, not repulsive.

b. What is Gauss's law for gravity, the gravitational equivalent of Equation 24.18? Use Φ_G for the gravitational flux, $\vec{g}$ for the gravitational field, and M_{in} for the enclosed mass.

c. A spherical planet is discovered with mass M, radius R, and a mass density that varies with radius as $\rho = \rho_0(1 - r/2R)$, where ρ_0 is the density at the center. Determine ρ_0 in terms of M and R.

Hint: Divide the planet into infinitesimal shells of thickness dr, then sum (i.e., integrate) their masses.

d. Find an expression for the gravitational field strength inside the planet at distance $r < R$.

Challenge Problems

57. ‖‖ All examples of Gauss's law have used highly symmetric
CALC surfaces where the flux integral is either zero or EA. Yet we've claimed that the net $\Phi_e = Q_{in}/\epsilon_0$ is independent of the surface. This is worth checking. **FIGURE CP24.57** shows a cube of edge length L centered on a long thin wire with linear charge density λ. The flux through one face of the cube is *not* simply EA because, in this case, the electric field varies in both strength and direction. But you can calculate the flux by actually doing the flux integral.

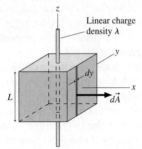

FIGURE CP24.57

a. Consider the face parallel to the yz-plane. Define area $d\vec{A}$ as a strip of width dy and height L with the vector pointing in the x-direction. One such strip is located at position y. Use the known electric field of a wire to calculate the electric flux $d\Phi$ through this little area. Your expression should be written in terms of y, which is a variable, and various constants. It should not explicitly contain any angles.

b. Now integrate $d\Phi$ to find the total flux through this face.

c. Finally, show that the net flux through the cube is $\Phi_e = Q_{in}/\epsilon_0$.

58. ‖‖ An infinite cylinder of radius R has a linear charge density λ.
CALC The volume charge density (C/m^3) within the cylinder ($r \le R$) is $\rho(r) = r\rho_0/R$, where ρ_0 is a constant to be determined.

a. Draw a graph of ρ versus x for an x-axis that crosses the cylinder perpendicular to the cylinder axis. Let x range from $-2R$ to $2R$.

b. The charge within a small volume dV is $dq = \rho\,dV$. The integral of $\rho\,dV$ over a cylinder of length L is the total charge $Q = \lambda L$ within the cylinder. Use this fact to show that $\rho_0 = 3\lambda/2\pi R^2$.

Hint: Let dV be a cylindrical shell of length L, radius r, and thickness dr. What is the volume of such a shell?

c. Use Gauss's law to find an expression for the electric field strength E inside the cylinder, $r \le R$, in terms of λ and R.

d. Does your expression have the expected value at the surface, $r = R$? Explain.

59. ‖‖ A sphere of radius R has total charge Q. The volume charge
CALC density (C/m^3) within the sphere is $\rho(r) = C/r^2$, where C is a constant to be determined.

a. The charge within a small volume dV is $dq = \rho\,dV$. The integral of $\rho\,dV$ over the entire volume of the sphere is the total charge Q. Use this fact to determine the constant C in terms of Q and R.

Hint: Let dV be a spherical shell of radius r and thickness dr. What is the volume of such a shell?

b. Use Gauss's law to find an expression for the electric field strength E inside the sphere, $r \le R$, in terms of Q and R.

c. Does your expression have the expected value at the surface, $r = R$? Explain.

60. ‖‖ A sphere of radius R has total charge Q. The volume charge
CALC density (C/m^3) within the sphere is

$$\rho = \rho_0\left(1 - \frac{r}{R}\right)$$

This charge density decreases linearly from ρ_0 at the center to zero at the edge of the sphere.

a. Show that $\rho_0 = 3Q/\pi R^3$.

b. Show that the electric field inside the sphere points radially outward with magnitude

$$E = \frac{Qr}{4\pi\epsilon_0 R^3}\left(4 - 3\frac{r}{R}\right)$$

c. Show that your result of part b has the expected value at $r = R$.

61. ‖‖ A spherical ball of charge has radius R and total charge Q. The
CALC electric field strength inside the ball ($r \le R$) is $E(r) = r^4 E_{max}/R^4$.

a. What is E_{max} in terms of Q and R?

b. Find an expression for the volume charge density $\rho(r)$ inside the ball as a function of r.

c. Verify that your charge density gives the total charge Q when integrated over the volume of the ball.

25 The Electric Potential

City lights seen from space show where millions of lightbulbs are transforming electric energy into light and thermal energy.

IN THIS CHAPTER, you will learn to use the electric potential and electric potential energy.

What is electric potential energy?

Recall that potential energy is an interaction energy. Charged particles that interact via the electric force have electric potential energy U. You'll learn that there's a close analogy with gravitational potential energy.

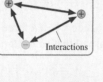

« LOOKING BACK Section 10.1 Potential energy
« LOOKING BACK Section 10.5 Energy diagrams

What is the electric potential?

You've seen that source charges create an electric field. Source charges also create an electric potential. The electric potential V

- Exists everywhere in space.
- Is a scalar.
- Causes charges to have potential energy.
- Is measured in volts.

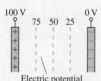

What potentials are especially important?

We'll calculate the electric potential of four important charge distributions: a point charge, a charged sphere, a ring of charge, and a parallel-plate capacitor. Finding the potential of a continuous charge distribution is similar to calculating electric fields, but easier because potential is a scalar.

Potential of a point charge

« LOOKING BACK Section 23.3 The electric field

How is potential represented?

Electric potential is a fairly abstract idea, so it will be important to visualize how the electric potential varies in space. One way of doing so is with equipotential surfaces. These are mathematical surfaces, not physical surfaces, with the same value of the potential V at every point.

How is electric potential used?

A charged particle q in an electric potential V has electric potential energy $U = qV$.

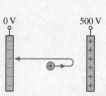

- Charged particles accelerate as they move through a potential difference.
- Mechanical energy is conserved:
$$K_f + qV_f = K_i + qV_i$$

« LOOKING BACK Section 10.4 Energy conservation

Why is energy important in electricity?

Energy allows things to happen. You want your lights to light, your computer to compute, and your music to play. All these require energy—*electric* energy. This is the first of two chapters that explore electric energy and its connection to electric forces and fields. You'll then be prepared to understand electric circuits—which are all about how energy is transformed and transferred from sources, such as batteries, to devices that utilize and dissipate the energy.

25.1 Electric Potential Energy

We started our study of electricity with electric forces and fields. But in electricity, just as in mechanics, *energy* is also a powerful idea. This chapter and the next will explore how energy is used in electricity, introduce the important concept of *electric potential*, and lay the groundwork for our upcoming study of electric circuits.

It's been many chapters since we dealt much with work and energy, but these ideas will now be *essential* to our story. Consequently, the Looking Back recommendations in the chapter preview are especially important. You will recall that a system's mechanical energy $E_{mech} = K + U$ is conserved for particles that interact with each other via *conservative forces*, where K and U are the kinetic and potential energy. That is,

$$\Delta E_{mech} = \Delta K + \Delta U = 0 \tag{25.1}$$

We need to be careful with notation because we are now using E to represent the electric field strength. To avoid confusion, we will represent mechanical energy either as the sum $K + U$ or as E_{mech}, with an explicit subscript.

> **NOTE** Recall that for any X, the *change* in X is $\Delta X = X_{final} - X_{initial}$.

A key idea of Chapters 9 and 10 was that energy is the energy *of a system*, and clearly defining the system is crucial. Kinetic energy $K = \frac{1}{2}mv^2$ is a system's *energy of motion*. For a multiparticle system, K is the sum of the kinetic energies of each particle in the system.

Potential energy U is the *interaction energy* of the system. Suppose the particles of the system move from some initial set of positions i to final positions f. As the particles move, the action/reaction pairs of forces between the particles—the interaction forces—do work and the system's potential energy changes. In « Section 10.1 we defined the *change* in potential energy to be

$$\Delta U = -W_{interaction}(i \rightarrow f) \tag{25.2}$$

where the notation means the work done by the interaction forces as the configuration changes from i to f. This rather abstract definition will make more sense when we see specific applications.

Recall that *work* is done when a force acts on a particle as it is being displaced. In « Section 9.3 you learned that a *constant* force $\vec{F}$ does work

$$W = \vec{F} \cdot \Delta \vec{r} = F \Delta r \cos\theta \tag{25.3}$$

as the particle undergoes displacement $\Delta\vec{r}$, where θ is the angle between the two vectors. FIGURE 25.1 reminds you of the three special cases $\theta = 0°$, $90°$, and $180°$.

> **NOTE** Work is *not* the oft-remembered "force times distance." Work is force times distance only in the one very special case in which the force is both constant *and* parallel to the displacement.

If the force is *not* constant, we can calculate the work by dividing the path into many small segments of length dx, finding the work done in each segment, and then summing (i.e., integrating) from the start of the path to the end. The work done in one such segment is $dW = F(x)\cos\theta \, dx$, where $F(x)$ indicates that the force is a function of position x. Thus the work done on the particle as it moves from x_i to x_f is

$$W = \int_{x_i}^{x_f} F(x) \cos\theta \, dx \tag{25.4}$$

Finally, recall that a *conservative force* is one for which the work done on a particle as it moves from position i to position f is *independent of the path followed*. We'll assert for now, and prove later, that **the electric force is a conservative force,** and thus we can define an electric potential energy.

FIGURE 25.1 The work done by a constant force.

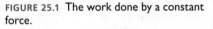
The particle undergoes displacement $\Delta\vec{r}$.

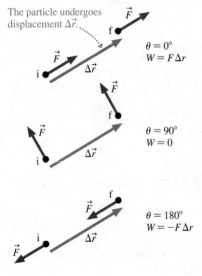

$\theta = 0°$
$W = F\Delta r$

$\theta = 90°$
$W = 0$

$\theta = 180°$
$W = -F\Delta r$

A Gravitational Analogy

Gravity, like electricity, is a long-range force. Much as we defined the electric field $\vec{E} = \vec{F}_{\text{on } q}/q$, we can also define a *gravitational field*—the agent that exerts gravitational forces on masses—as $\vec{F}_{\text{on } m}/m$. But $\vec{F}_{\text{on } m} = m\vec{g}$ near the earth's surface; thus the familiar $\vec{g} = (9.80 \text{ N/kg, down})$ is really the gravitational field! Notice how we've written the units of $\vec{g}$ as N/kg, but you can easily show that N/kg = m/s^2. The gravitational field near the earth's surface is a *uniform* field in the downward direction.

FIGURE 25.2 shows a particle of mass m falling in the gravitational field. The gravitational force is in the same direction as the particle's displacement, so the gravitational field does a *positive* amount of work on the particle. The gravitational force is constant, hence the work done by the gravitational field is

$$W_{\text{G}} = F_{\text{G}} \Delta r \cos 0° = mg|y_{\text{f}} - y_{\text{i}}| = mgy_{\text{i}} - mgy_{\text{f}} \qquad (25.5)$$

We have to be careful with signs because Δr, the magnitude of the displacement vector, must be a positive number.

Now we can see how the definition of ΔU in Equation 25.2 makes sense. The *change* in gravitational potential energy is

$$\Delta U_{\text{G}} = U_{\text{f}} - U_{\text{i}} = -W_{\text{G}}(\text{i} \rightarrow \text{f}) = mgy_{\text{f}} - mgy_{\text{i}} \qquad (25.6)$$

Comparing the initial and final terms on the two sides of the equation, we see that the gravitational potential energy near the earth is the familiar quantity

$$U_{\text{G}} = U_0 + mgy \qquad (25.7)$$

where U_0 is the value of U_{G} at $y = 0$. We usually choose $U_0 = 0$, in which case $U_{\text{G}} = mgy$, but such a choice is not necessary.

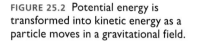

FIGURE 25.2 Potential energy is transformed into kinetic energy as a particle moves in a gravitational field.

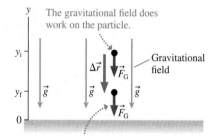

The net force on the particle is down. It gains kinetic energy (i.e., speeds up) as it loses potential energy.

A Uniform Electric Field

FIGURE 25.3 shows a charged particle inside a parallel-plate capacitor with electrode spacing d. This is a uniform electric field, and the situation looks very much like Figure 25.2 for a mass in a uniform gravitational field. The one difference is that $\vec{g}$ always points down whereas the positive-to-negative electric field can point in any direction. To deal with this, let's define a coordinate axis s that points *from* the negative plate, which we define to be $s = 0$, *toward* the positive plate. The electric field $\vec{E}$ then points in the negative s-direction, just as the gravitational field $\vec{g}$ points in the negative y-direction. This s-axis, which is valid no matter how the capacitor is oriented, is analogous to the y-axis used for gravitational potential energy.

A positive charge q inside the capacitor speeds up and gains kinetic energy as it "falls" toward the negative plate. Is the charge losing potential energy as it gains kinetic energy? Indeed it is, and the calculation of the potential energy is just like the calculation of gravitational potential energy. The electric field exerts a *constant* force $F = qE$ on the charge in the direction of motion; thus the work done on the charge by the electric field is

$$W_{\text{elec}} = F \Delta r \cos 0° = qE|s_{\text{f}} - s_{\text{i}}| = qEs_{\text{i}} - qEs_{\text{f}} \qquad (25.8)$$

where we again have to be careful with the signs because $s_{\text{f}} < s_{\text{i}}$.

The work done by the electric field causes the *electric* potential energy to change by

$$\Delta U_{\text{elec}} = U_{\text{f}} - U_{\text{i}} = -W_{\text{elec}}(\text{i} \rightarrow \text{f}) = qEs_{\text{f}} - qEs_{\text{i}} \qquad (25.9)$$

Comparing the initial and final terms on the two sides of the equation, we see that the **electric potential energy** of charge q in a uniform electric field is

$$U_{\text{elec}} = U_0 + qEs \qquad (25.10)$$

where s is measured from the negative plate and U_0 is the potential energy at the negative plate ($s = 0$). It will often be convenient to choose $U_0 = 0$, but the choice has no physical consequences because it doesn't affect ΔU_{elec}. Equation 25.10 was derived with the assumption that q is positive, but it is valid for either sign of q.

FIGURE 25.3 The electric field does work on the charged particle.

The electric field does work on the particle.

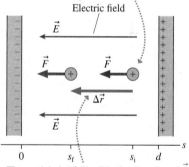

The particle is "falling" in the direction of $\vec{E}$.

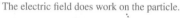

NOTE Although Equation 25.10 is sometimes called "the potential energy of charge q," it is really the potential energy of the charge + capacitor system.

FIGURE 25.4 shows positive and negative charged particles moving inside a parallel-plate capacitor. For a positive charge, U_{elec} decreases and K increases as the charge moves toward the negative plate (decreasing s). Thus a positive charge is going "downhill" if it moves in the direction of the electric field. A positive charge moving opposite the field direction is going "uphill," slowing as it transforms kinetic energy into electric potential energy.

FIGURE 25.4 A charged particle exchanges kinetic and potential energy as it moves in an electric field.

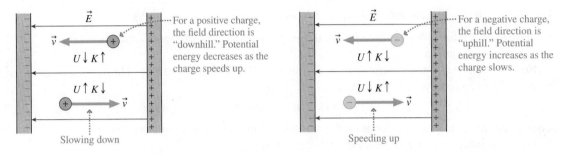

If we choose $U_0 = 0$, so that potential energy is zero at the negative plate, then a negative charged particle has *negative* potential energy. You learned in Chapter 10 that there's nothing wrong with negative potential energy—it's simply less than the potential energy at some arbitrarily chosen reference location. The more important point, from Equation 25.10, is that the potential energy *increases* (becomes less negative) as a negative charge moves toward the negative plate. A negative charge moving in the field direction is going "uphill," transforming kinetic energy into electric potential energy as it slows.

FIGURE 25.5 is an *energy diagram* for a positively charged particle in an electric field. Recall that an energy diagram is a graphical representation of how kinetic and potential energies are transformed as a particle moves. For positive q, the electric potential energy given by Equation 25.10 increases linearly from 0 at the negative plate (with $U_0 = 0$) to qEd at the positive plate. The total mechanical energy—which is under your control—is constant. If $E_{mech} < qEd$, as shown here, a positively charged particle projected from the negative plate will gradually slow (transforming kinetic energy into potential energy) until it reaches a *turning point* where $U_{elec} = E_{mech}$. But if you project the particle with greater speed, such that $E_{mech} > qEd$, it will be able to cross the gap to collide with the positive plate.

FIGURE 25.5 The energy diagram for a positively charged particle in a uniform electric field.

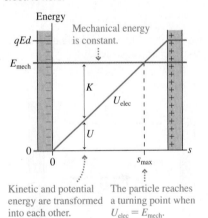

Kinetic and potential energy are transformed into each other.

The particle reaches a turning point when $U_{elec} = E_{mech}$.

EXAMPLE 25.1 | Conservation of energy

A 2.0 cm × 2.0 cm parallel-plate capacitor with a 2.0 mm spacing is charged to ±1.0 nC. First a proton and then an electron are released from rest at the midpoint of the capacitor.

a. What is each particle's energy?

b. What is each particle's speed as it reaches the plate?

MODEL The mechanical energy is conserved. A parallel-plate capacitor has a uniform electric field.

VISUALIZE FIGURE 25.6 is a before-and-after pictorial representation, as you learned to draw in Part II. Each particle is released from rest ($K = 0$) and moves "downhill" toward lower potential energy. Thus the proton moves toward the negative plate, the electron toward the positive plate.

FIGURE 25.6 A proton and an electron in a capacitor.

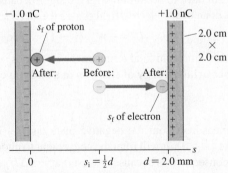

SOLVE a. The s-axis was defined to point from the negative toward the positive plate of the capacitor. Both charged particles have $s_i = \frac{1}{2}d$, where $d = 2.0$ mm is the plate separation. If we let $U_0 = 0$, defining the negative plate as our zero-energy reference point, then the proton $(q = e)$ has energy

$$E_{\text{mech p}} = K_i + U_i = 0 + eE\left(\tfrac{1}{2}d\right)$$

while the electron $(q = -e)$ has

$$E_{\text{mech e}} = K_i + U_i = 0 - eE\left(\tfrac{1}{2}d\right)$$

The electric field inside the parallel-plate capacitor, from Chapter 23, is

$$E = \frac{Q}{\epsilon_0 A} = 2.82 \times 10^5 \text{ N/C}$$

Thus the particles' energies can be calculated to be

$$E_{\text{mech p}} = 4.5 \times 10^{-17} \text{ J and } E_{\text{mech e}} = -4.5 \times 10^{-17} \text{ J}$$

Notice that the electron's mechanical energy is negative.

b. Conservation of mechanical energy requires $K_f + U_f = K_i + U_i = E_{\text{mech}}$. The proton collides with the negative plate, so $U_f = 0$, and the final kinetic energy is $K_f = \frac{1}{2}m_p v_f^2 = E_{\text{mech p}}$. Thus the proton's impact speed is

$$(v_f)_p = \sqrt{\frac{2E_{\text{mech p}}}{m_p}} = 2.3 \times 10^5 \text{ m/s}$$

Similarly, the electron collides with the positive plate, where $U_f = qEd = -eEd = 2E_{\text{mech e}}$. Thus energy conservation for the electron is

$$K_f = \tfrac{1}{2}m_e v_f^2 = E_{\text{mech e}} - U_f = E_{\text{mech e}} - 2E_{\text{mech e}} = -E_{\text{mech e}}$$

We found the electron's mechanical energy to be negative, so K_f is positive. The electron reaches the positive plate with speed

$$(v_f)_e = \sqrt{\frac{-2E_{\text{mech e}}}{m_e}} = 1.0 \times 10^7 \text{ m/s}$$

ASSESS Even though both particles have mechanical energy with the same magnitude, the electron has a much greater final speed due to its much smaller mass.

STOP TO THINK 25.1 A glass rod is positively charged. The figure shows an end view of the rod. A negatively charged particle moves in a circular arc around the glass rod. Is the work done on the charged particle by the rod's electric field positive, negative, or zero?

Motion of negatively charged particle

End view of charged rod

25.2 The Potential Energy of Point Charges

FIGURE 25.7a shows two charges q_1 and q_2, which we will assume to be like charges. These two charges interact, and the energy of their interaction can be found by calculating the work done by the electric field of q_1 on q_2 as q_2 moves from position x_i to position x_f. We'll assume that q_1 has been glued down and is unable to move, as shown in **FIGURE 25.7b**.

The force is entirely in the direction of motion, so $\cos\theta = 1$. Thus

$$W_{\text{elec}} = \int_{x_i}^{x_f} F_{1 \text{ on } 2}\, dx = \int_{x_i}^{x_f} \frac{Kq_1 q_2}{x^2}\, dx = Kq_1 q_2 \left.\frac{-1}{x}\right|_{x_i}^{x_f} = -\frac{Kq_1 q_2}{x_f} + \frac{Kq_1 q_2}{x_i} \quad (25.11)$$

The potential energy of the two charges is related to the work done by

$$\Delta U_{\text{elec}} = U_f - U_i = -W_{\text{elec}}(i \rightarrow f) = \frac{Kq_1 q_2}{x_f} - \frac{Kq_1 q_2}{x_i} \quad (25.12)$$

By comparing the left and right sides of the equation we see that the potential energy of the two-point-charge system is

$$U_{\text{elec}} = \frac{Kq_1 q_2}{x} \quad (25.13)$$

FIGURE 25.7 The interaction between two point charges.

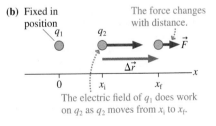

(a) $\vec{F}_{2 \text{ on } 1}$ q_1 q_2 $\vec{F}_{1 \text{ on } 2}$

Like charges exert repulsive forces.

(b) Fixed in position q_1 q_2 The force changes with distance.

$\vec{F}$

$\Delta\vec{r}$

The electric field of q_1 does work on q_2 as q_2 moves from x_i to x_f.

We could include a constant U_0, as we did in Equation 25.10, for the potential energy of a charge in a uniform electric field, but it is customary to set $U_0 = 0$.

We chose to integrate along the *x*-axis for convenience, but all that matters is the *distance* between the charges. Thus a general expression for the electric potential energy is

$$U_{elec} = \frac{Kq_1q_2}{r} = \frac{1}{4\pi\epsilon_0}\frac{q_1q_2}{r} \quad \text{(two point charges)} \quad (25.14)$$

This is explicitly the energy *of the system*, not the energy of just q_1 or q_2.

> **NOTE** The electric potential energy of two point charges looks *almost* the same as the force between the charges. The difference is the *r* in the denominator of the potential energy compared to the r^2 in Coulomb's law.

Three important points need to be noted:

- The choice $U_0 = 0$ is equivalent to saying that the potential energy of two charged particles is zero only when they are infinitely far apart. This makes sense because two charged particles cease interacting only when they are infinitely far apart.
- We derived Equation 25.14 for two like charges, but it is equally valid for two opposite charges. The potential energy of two like charges is *positive* and of two opposite charges is *negative*.
- Because the electric field outside a *sphere of charge* is the same as that of a point charge at the center, Equation 25.14 is also the electric potential energy of two charged spheres. Distance *r* is the distance between their centers.

Charged-Particle Interactions

FIGURE 25.8a shows the potential-energy curve—a hyperbola—for two like charges as a function of the distance *r* between them. Also shown is the total energy line for two charged particles shot toward each other with equal but opposite momenta.

You can see that the total energy line crosses the potential-energy curve at r_{min}. This is a turning point. The two charges gradually slow down, because of the repulsive force between them, until the distance separating them is r_{min}. At this point, the kinetic energy is zero and both particles are instantaneously at rest. Both then reverse direction and move apart, speeding up as they go. r_{min} is the *distance of closest approach*.

Two opposite charges are a little trickier because of the negative energies. Negative total energies seem troubling at first, but they characterize *bound systems*. **FIGURE 25.8b** shows two oppositely charged particles shot apart from each other with equal but opposite momenta. If $E_{mech} < 0$, as shown, then their total energy line crosses the potential-energy curve at r_{max}. That is, the particles slow down, lose kinetic energy, reverse directions at *maximum separation* r_{max}, and then "fall" back together. They cannot escape from each other. Although moving in three dimensions rather than one, the electron and proton of a hydrogen atom are a realistic example of a bound system, and their mechanical energy is negative.

Two oppositely charged particles *can* escape from each other if $E_{mech} > 0$. They'll slow down, but eventually the potential energy vanishes and the particles still have kinetic energy. The threshold condition for escape is $E_{mech} = 0$, which will allow the particles to reach infinite separation $(U \rightarrow 0)$ at infinitesimally slow speed $(K \rightarrow 0)$. The initial speed that gives $E_{mech} = 0$ is called the *escape speed*.

> **NOTE** Real particles can't be infinitely far apart, but because U_{elec} decreases with distance, there comes a point when $U_{elec} = 0$ is an excellent approximation. Two charged particles for which $U_{elec} \approx 0$ are sometimes described as "far apart" or "far away."

FIGURE 25.8 The potential-energy diagrams for two like charges and two opposite charges.

(a) Like charges

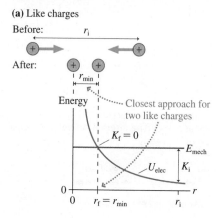

(b) Opposite charges

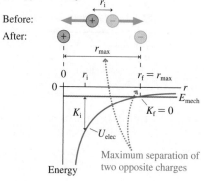

The Electric Force Is a Conservative Force

Potential energy can be defined only if the force is *conservative,* meaning that the work done on the particle as it moves from position i to position f is independent of the path followed between i and f. **FIGURE 25.9** demonstrates that electric force is indeed conservative.

FIGURE 25.9 The work done on q_2 is independent of the path from i to f.

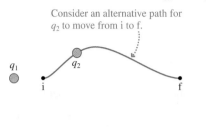

Consider an alternative path for q_2 to move from i to f.

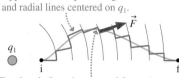

Approximate the path using circular arcs and radial lines centered on q_1.

The electric force is a *central force.* As a result, zero work is done as q_2 moves along a circular arc because the force is perpendicular to the displacement.

All the work is done along the radial line segments, which are equivalent to a straight line from i to f. This is the work that was calculated in Equation 25.11.

EXAMPLE 25.2 | Approaching a charged sphere

A proton is fired from far away at a 1.0-mm-diameter glass sphere that has been charged to $+100$ nC. What initial speed must the proton have to just reach the surface of the glass?

MODEL Energy is conserved. The glass sphere can be modeled as a charged particle, so the potential energy is that of two point charges. The glass is so much more massive than the proton that it remains at rest as the proton moves. The proton starts "far away," which we interpret as sufficiently far to make $U_i \approx 0$.

VISUALIZE **FIGURE 25.10** shows the before-and-after pictorial representation. To "just reach" the glass sphere means that the proton comes to rest, $v_f = 0$, as it reaches $r_f = 0.50$ mm, the *radius* of the sphere.

SOLVE Conservation of energy $K_f + U_f = K_i + U_i$ is

$$0 + \frac{Kq_p q_{sphere}}{r_f} = \tfrac{1}{2}mv_i^2 + 0$$

FIGURE 25.10 A proton approaching a glass sphere.

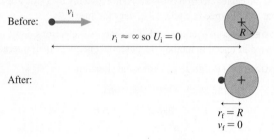

The proton charge is $q_p = e$. With this, we can solve for the proton's initial speed:

$$v_i = \sqrt{\frac{2Keq_{sphere}}{mr_f}} = 1.86 \times 10^7 \text{ m/s}$$

EXAMPLE 25.3 | Escape speed

An interaction between two elementary particles causes an electron and a positron (a positive electron) to be shot out back to back with equal speeds. What minimum speed must each have when they are 100 fm apart in order to escape each other?

MODEL Energy is conserved. The particles end "far apart," which we interpret as sufficiently far to make $U_f \approx 0$.

VISUALIZE **FIGURE 25.11** shows the before-and-after pictorial representation. The minimum speed to escape is the speed that allows the particles to reach $r_f = \infty$ with $v_f = 0$.

FIGURE 25.11 An electron and a positron flying apart.

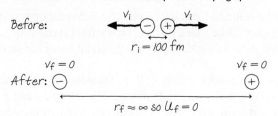

SOLVE U_{elec} is the potential energy of the electron + positron system. Similarly, K is the *total* kinetic energy of the system. The electron and the positron, with equal masses and equal speeds, have equal kinetic energies. Conservation of energy $K_f + U_f = K_i + U_i$ is

$$0 + 0 + 0 = \tfrac{1}{2}mv_i^2 + \tfrac{1}{2}mv_i^2 + \frac{Kq_e q_p}{r_i} = mv_i^2 - \frac{Ke^2}{r_i}$$

Using $r_i = 100$ fm $= 1.0 \times 10^{-13}$ m, we can calculate the minimum initial speed to be

$$v_i = \sqrt{\frac{Ke^2}{mr_i}} = 5.0 \times 10^7 \text{ m/s}$$

ASSESS v_i is a little more than 10% the speed of light, just about the limit of what a "classical" calculation can predict. We would need to use the theory of relativity if v_i were any larger.

Multiple Point Charges

If more than two charges are present, their potential energy is the sum of the potential energies due to all pairs of charges:

$$U_{\text{elec}} = \sum_{i<j} \frac{Kq_iq_j}{r_{ij}} \tag{25.15}$$

where r_{ij} is the distance between q_i and q_j. The summation contains the $i < j$ restriction to ensure that each pair of charges is counted only once.

> **NOTE** For energy conservation problems, it's necessary to calculate only the potential energy for those pairs of charges for which the distance r_{ij} changes. The potential energy of any pair that doesn't move is an additive constant with no physical consequences.

EXAMPLE 25.4 | **Launching an electron**

Three electrons are spaced 1.0 mm apart along a vertical line. The outer two electrons are fixed in position.

a. Is the center electron at a point of stable or unstable equilibrium?

b. If the center electron is displaced horizontally by a small distance, what will its speed be when it is very far away?

MODEL Energy is conserved. The outer two electrons don't move, so we don't need to include the potential energy of their interaction.

VISUALIZE FIGURE 25.12 shows the situation.

FIGURE 25.12 Three electrons.

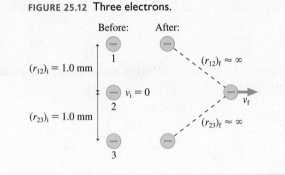

SOLVE a. The center electron is in equilibrium *exactly* in the center because the two electric forces on it balance. But if it moves a little to the right or left, no matter how little, then the horizontal components of the forces from both outer electrons will push the center electron farther away. This is an unstable equilibrium for horizontal displacements, like being on the top of a hill.

b. A small displacement will cause the electron to move away. If the displacement is only infinitesimal, the initial conditions are $(r_{12})_i = (r_{23})_i = 1.0$ mm and $v_i = 0$. "Far away" is interpreted as $r_f \rightarrow \infty$, where $U_f \approx 0$. There are now *two* terms in the potential energy, so conservation of energy $K_f + U_f = K_i + U_i$ gives

$$\frac{1}{2}mv_f^2 + 0 + 0 = 0 + \left[\frac{Kq_1q_2}{(r_{12})_i} + \frac{Kq_2q_3}{(r_{23})_i} \right]$$

$$= \left[\frac{Ke^2}{(r_{12})_i} + \frac{Ke^2}{(r_{23})_i} \right]$$

This is easily solved to give

$$v_f = \sqrt{\frac{2}{m}\left[\frac{Ke^2}{(r_{12})_i} + \frac{Ke^2}{(r_{23})_i} \right]} = 1000 \text{ m/s}$$

STOP TO THINK 25.2 Rank in order, from largest to smallest, the potential energies U_a to U_d of these four pairs of charges. Each + symbol represents the same amount of charge.

(a) (b) (c) (d)

FIGURE 25.13 The electric field does work as a dipole rotates.

The electric forces exert a torque on the dipole.

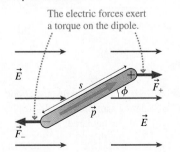

25.3 The Potential Energy of a Dipole

The electric dipole has been our model for understanding how charged objects interact with neutral objects. In Chapter 23 we found that an electric field exerts a *torque* on a dipole. We can complete the picture by calculating the potential energy of an electric dipole in a uniform electric field.

FIGURE 25.13 shows a dipole in an electric field $\vec{E}$. Recall that the dipole moment $\vec{p}$ is a vector that points from $-q$ to q with magnitude $p = qs$. The forces $\vec{F}_+$ and $\vec{F}_-$ exert a torque on the dipole, but now we're interested in calculating the *work* done by these forces as the dipole rotates from angle ϕ_i to angle ϕ_f.

When a force component F_s acts through a small displacement ds, the force does work $dW = F_s \, ds$. If we exploit the rotational-linear motion analogy from Chapter 12, where torque τ is the analog of force and angular displacement $\Delta\phi$ is the analog of linear displacement, then a torque acting through a small angular displacement $d\phi$ does work $dW = \tau \, d\phi$. From Chapter 23, the torque on the dipole in Figure 25.13 is $\tau = -pE \sin\phi$, where the minus sign is due to the torque trying to cause a clockwise rotation. Thus the work done by the electric field on the dipole as it rotates through the small angle $d\phi$ is

$$dW_{elec} = -pE \sin\phi \, d\phi \qquad (25.16)$$

The total work done by the electric field as the dipole turns from ϕ_i to ϕ_f is

$$W_{elec} = -pE \int_{\phi_i}^{\phi_f} \sin\phi \, d\phi = pE \cos\phi_f - pE \cos\phi_i \qquad (25.17)$$

The potential energy associated with the work done on the dipole is

$$\Delta U_{dipole} = U_f - U_i = -W_{elec}(i \to f) = -pE \cos\phi_f + pE \cos\phi_i \qquad (25.18)$$

By comparing the left and right sides of Equation 25.18, we see that the potential energy of an electric dipole $\vec{p}$ in a uniform electric field $\vec{E}$ is

$$U_{dipole} = -pE \cos\phi = -\vec{p} \cdot \vec{E} \qquad (25.19)$$

FIGURE 25.14 shows the energy diagram of a dipole. The potential energy is minimum at $\phi = 0°$ where the dipole is aligned with the electric field. This is a point of stable equilibrium. A dipole exactly opposite $\vec{E}$, at $\phi = \pm 180°$, is at a point of unstable equilibrium. Any disturbance will cause it to flip around. A frictionless dipole with mechanical energy E_{mech} will oscillate back and forth between turning points on either side of $\phi = 0°$.

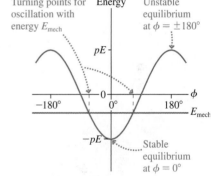

FIGURE 25.14 The energy of a dipole in an electric field.

EXAMPLE 25.5 | **Rotating a molecule**

The water molecule is a permanent electric dipole with dipole moment 6.2×10^{-30} C m. A water molecule is aligned in an electric field with field strength 1.0×10^7 N/C. How much energy is needed to rotate the molecule 90°?

MODEL The molecule is at the point of minimum energy. It won't spontaneously rotate 90°. However, an external force that supplies energy, such as a collision with another molecule, can cause the water molecule to rotate.

SOLVE The molecule starts at $\phi_i = 0°$ and ends at $\phi_f = 90°$. The increase in potential energy is

$$\Delta U_{dipole} = U_f - U_i = -pE \cos 90° - (-pE \cos 0°)$$
$$= pE = 6.2 \times 10^{-23} \text{ J}$$

This is the energy needed to rotate the molecule 90°.

ASSESS ΔU_{dipole} is significantly less than $k_B T$ at room temperature. Thus collisions with other molecules can easily supply the energy to rotate the water molecules and keep them from staying aligned with the electric field.

25.4 The Electric Potential

We introduced the concept of the *electric field* in Chapter 22 because action at a distance raised concerns and difficulties. The field provides an intermediary through which two charges exert forces on each other. Charge q_1 somehow alters the space around it by creating an electric field $\vec{E}_1$. Charge q_2 then responds to the field, experiencing force $\vec{F} = q_2\vec{E}_1$.

In defining the electric field, we separated the charges that are the *source* of the field from the charge *in* the field. The force on charge q is related to the electric field of the source charges by

force on q by sources = [charge q] $\times$ [alteration of space by the source charges]

FIGURE 25.15 Source charges alter the space around them by creating an electric potential.

The potential at this point is V.

The source charges alter the space around them by creating an electric potential.

Source charges

If charge q is in the potential, the electric potential energy is $U_{q + \text{sources}} = qV$.

This battery is labeled 1.5 volts. As we'll soon see, a battery is a source of electric potential.

Let's try a similar procedure for the potential energy. The electric potential energy is due to the interaction of charge q with other charges, so let's write

potential energy of q + sources

= [charge q] × [*potential* for interaction with the source charges]

FIGURE 25.15 shows this idea schematically.

In analogy with the electric field, we will define the **electric potential** V (or, for brevity, just *the potential*) as

$$V \equiv \frac{U_{q + \text{sources}}}{q} \qquad (25.20)$$

Charge q is used as a probe to determine the electric potential, but the value of V is *independent of q*. **The electric potential, like the electric field, is a property of the source charges.** And, like the electric field, the electric potential fills the space around the source charges. It is there whether or not another charge is there to experience it.

In practice, we're usually more interested in knowing the potential energy if a charge q happens to be at a point in space where the electric potential of the source charges is V. Turning Equation 25.20 around, we see that the electric potential energy is

$$U_{q + \text{sources}} = qV \qquad (25.21)$$

Once the potential has been determined, it's very easy to find the potential energy.

The unit of electric potential is the joule per coulomb, which is called the **volt** V:

$$1 \text{ volt} = 1 \text{ V} \equiv 1 \text{ J/C}$$

This unit is named for Alessandro Volta, who invented the electric battery in the year 1800. Microvolts (μV), millivolts (mV), and kilovolts (kV) are commonly used units.

NOTE Once again, commonly used symbols are in conflict. The symbol V is widely used to represent *volume,* and now we're introducing the same symbol to represent *potential*. To make matters more confusing, V is the abbreviation for *volts*. In printed text, V for potential is italicized and V for volts is not, but you can't make such a distinction in handwritten work. This is not a pleasant state of affairs, but these are the commonly accepted symbols. It's incumbent upon you to be especially alert to the *context* in which a symbol is used.

Using the Electric Potential

The electric potential is an abstract idea, and it will take some practice to see just what it means and how it is useful. We'll use multiple representations—words, pictures, graphs, and analogies—to explain and describe the electric potential.

NOTE It is unfortunate that the terms *potential* and *potential energy* are so much alike. Despite the similar names, they are very different concepts and are not interchangeable. **TABLE 25.1** will help you to distinguish between the two.

TABLE 25.1 Distinguishing electric potential and potential energy

The *electric potential* is a property of the source charges and, as you'll soon see, is related to the electric field. The electric potential is present whether or not a charged particle is there to experience it. Potential is measured in J/C, or V.

The *electric potential energy* is the interaction energy of a charged particle with the source charges. Potential energy is measured in J.

Basically, knowing the electric potential in a region of space allows us to determine whether a charged particle speeds up or slows down as it moves through that region. **FIGURE 25.16** illustrates this idea. Here a group of source charges, which remains hidden offstage, has created an electric potential V that increases from left to right. A charged particle q, which for now we'll assume to be positive, has electric potential energy $U = qV$. If the particle moves to the right, its potential energy increases and so, by energy conservation, its kinetic energy must decrease. **A positive charge slows down as it moves into a region of higher electric potential.**

FIGURE 25.16 A charged particle speeds up or slows down as it moves through a potential difference.

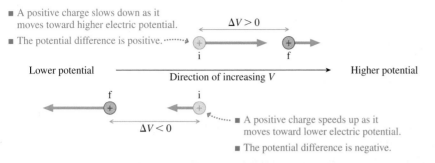

It is customary to say that the particle moves through a **potential difference** $\Delta V = V_f - V_i$. The potential difference between two points is often called the **voltage.** The particle moving to the right moves through a positive potential difference ($\Delta V > 0$ because $V_f > V_i$), so we can say that a positively charged particle slows down as it moves through a positive potential difference.

The particle moving to the left in Figure 25.16 travels in the direction of decreasing electric potential—through a negative potential difference—and is losing potential energy. It speeds up as it transforms potential energy into kinetic energy. A negatively charged particle would slow down because its potential energy qV would increase as V decreases. **TABLE 25.2** summarizes these ideas.

If a particle moves through a potential difference ΔV, its electric potential energy changes by $\Delta U = q\Delta V$. We can write the conservation of energy equation in terms of the electric potential as $\Delta K + \Delta U = \Delta K + q\Delta V = 0$ or, as is often more practical,

$$K_f + qV_f = K_i + qV_i \tag{25.22}$$

Conservation of energy is the basis of a powerful problem-solving strategy.

TABLE 25.2 Charged particles moving in an electric potential

	Electric potential	
	Increasing ($\Delta V > 0$)	Decreasing ($\Delta V < 0$)
+ charge	Slows down	Speeds up
− charge	Speeds up	Slows down

PROBLEM-SOLVING STRATEGY 25.1   (MP)

Conservation of energy in charge interactions

MODEL Define the system. If possible, model it as an isolated system for which mechanical energy is conserved.

VISUALIZE Draw a before-and-after pictorial representation. Define symbols, list known values, and identify what you're trying to find.

SOLVE The mathematical representation is based on the law of conservation of mechanical energy:

$$K_f + qV_f = K_i + qV_i$$

- Is the electric potential given in the problem statement? If not, you'll need to use a known potential, such as that of a point charge, or calculate the potential using the procedure given later, in Problem-Solving Strategy 25.2.
- K_i and K_f are the sums of the kinetic energies of all moving particles.
- Some problems may need additional conservation laws, such as conservation of charge or conservation of momentum.

ASSESS Check that your result has correct units and significant figures, is reasonable, and answers the question.

Exercise 22   ✎

EXAMPLE 25.6 | **Moving through a potential difference**

A proton with a speed of 2.0×10^5 m/s enters a region of space in which there is an electric potential. What is the proton's speed after it moves through a potential difference of 100 V? What will be the final speed if the proton is replaced by an electron?

MODEL The system is the charge plus the unseen source charges creating the potential. This is an isolated system, so mechanical energy is conserved.

VISUALIZE FIGURE 25.17 is a before-and-after pictorial representation of a charged particle moving through a potential difference. A positive charge *slows down* as it moves into a region of higher potential ($K \rightarrow U$). A negative charge *speeds up* ($U \rightarrow K$).

FIGURE 25.17 A charged particle moving through a potential difference.

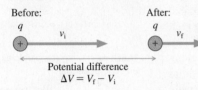

SOLVE The potential energy of charge q is $U = qV$. Conservation of energy, now expressed in terms of the electric potential V, is $K_f + qV_f = K_i + qV_i$, or

$$K_f = K_i - q\,\Delta V$$

where $\Delta V = V_f - V_i$ is the potential difference through which the particle moves. In terms of the speeds, energy conservation is

$$\tfrac{1}{2}mv_f^2 = \tfrac{1}{2}mv_i^2 - q\,\Delta V$$

We can solve this for the final speed:

$$v_f = \sqrt{v_i^2 - \frac{2q}{m}\Delta V}$$

For a proton, with $q = e$, the final speed is

$$(v_f)_p = \sqrt{(2.0 \times 10^5 \text{ m/s})^2 - \frac{2(1.60 \times 10^{-19} \text{ C})(100 \text{ V})}{1.67 \times 10^{-27} \text{ kg}}}$$

$$= 1.4 \times 10^5 \text{ m/s}$$

An electron, though, with $q = -e$ and a different mass, reaches speed $(v_f)_e = 5.9 \times 10^6$ m/s.

ASSESS The proton slowed down and the electron sped up, as we expected. Note that the electric potential *already existed* in space due to other charges that are not explicitly seen in the problem. The electron and proton have nothing to do with creating the potential. Instead, they *respond* to the potential by having potential energy $U = qV$.

STOP TO THINK 25.3 A proton is released from rest at point B, where the potential is 0 V. Afterward, the proton

a. Remains at rest at B.
b. Moves toward A with a steady speed.
c. Moves toward A with an increasing speed.
d. Moves toward C with a steady speed.
e. Moves toward C with an increasing speed.

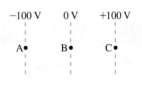

25.5 The Electric Potential Inside a Parallel-Plate Capacitor

FIGURE 25.18 A parallel-plate capacitor.

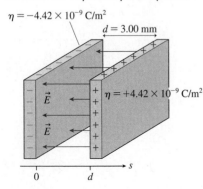

We began this chapter with the potential energy of a charge inside a parallel-plate capacitor. Now let's investigate the electric potential. FIGURE 25.18 shows two parallel electrodes, separated by distance d, with surface charge density $\pm\eta$. As a specific example, we'll let $d = 3.00$ mm and $\eta = 4.42 \times 10^{-9}$ C/m². The electric field inside the capacitor, as you learned in Chapter 23, is

$$\vec{E} = \left(\frac{\eta}{\epsilon_0}, \text{ from positive toward negative}\right)$$

$$= (500 \text{ N/C, from right to left})$$

(25.23)

This electric field is due to the *source charges* on the capacitor plates.

Graphical representations of the electric potential inside a capacitor

A graph of potential versus *s*. You can see the potential increasing from 0.0 V at the negative plate to 1.5 V at the positive plate.

A three-dimensional view showing **equipotential surfaces.** These are mathematical surfaces, not physical surfaces, with the same value of *V* at every point. The equipotential surfaces of a capacitor are planes parallel to the capacitor plates. The capacitor plates are also equipotential surfaces.

A two-dimensional **contour map.** The capacitor plates and the equipotential surfaces are seen edge-on, so you need to imagine them extending above and below the plane of the page.

A three-dimensional **elevation graph.** The potential is graphed vertically versus the *s*-coordinate on one axis and a generalized "*yz*-coordinate" on the other axis. Viewing the right face of the elevation graph gives you the potential graph.

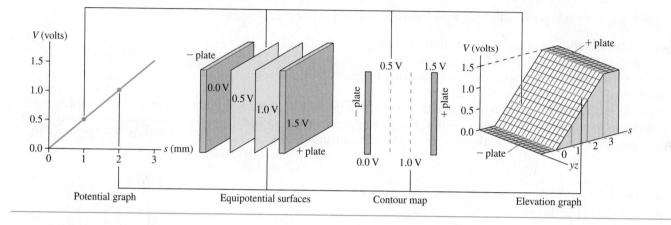

Potential graph   Equipotential surfaces   Contour map   Elevation graph

FIGURE 25.20 Equipotentials and electric field vectors inside a parallel-plate capacitor.

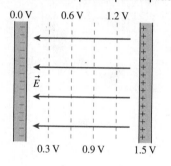

FIGURE 25.21 Using a battery to charge a capacitor to a precise value of ΔV_C.

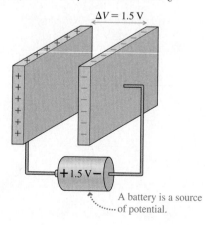

A battery is a source of potential.

These four graphical representations show the same information from different perspectives, and the connecting lines help you see how they are related. If you think of the elevation graph as a "mountain," then the contour lines on the contour map are like the lines of a topographic map.

The potential graph and the contour map are the two representations most widely used in practice because they are easy to draw. Their limitation is that they are trying to convey three-dimensional information in a two-dimensional presentation. When you see graphs or contour maps, you need to imagine the three-dimensional equipotential surfaces or the three-dimensional elevation graph.

There's nothing special about showing equipotential surfaces or contour lines every 0.5 V. We chose these intervals because they were convenient. As an alternative, **FIGURE 25.20** shows how the contour map looks if the contour lines are spaced every 0.3 V. Contour lines and equipotential surfaces are *imaginary* lines and surfaces drawn to help us visualize how the potential changes in space. Drawing the map more than one way reinforces the idea that there is an electric potential at *every* point inside the capacitor, not just at the points where we happened to draw a contour line or an equipotential surface.

Figure 25.20 also shows the electric field vectors. Notice that

- The electric field vectors are perpendicular to the equipotential surfaces.
- The electric field points in the direction of decreasing potential. In other words, the electric field points "downhill" on a graph or map of the electric potential.

Chapter 26 will present a more in-depth exploration of the connection between the electric field and the electric potential. There you will find that these observations are always true. They are not unique to the parallel-plate capacitor.

Finally, you might wonder how we can arrange a capacitor to have a surface charge density of precisely 4.42×10^{-9} C/m². Simple! As **FIGURE 25.21** shows, we use wires to attach 3.00-mm-spaced capacitor plates to a 1.5 V battery. This is another topic that we'll explore in Chapter 26, but it's worth noting now that **a battery is a source of potential.** That's why batteries are labeled in volts, and it's a major reason we need to thoroughly understand the concept of potential.

EXAMPLE 25.7 | Measuring the speed of a proton

You've been assigned the task of measuring the speed of protons as they emerge from a small accelerator. To do so, you decide to measure how much voltage is needed across a parallel-plate capacitor to stop the protons. The capacitor you choose has a 2.0 mm plate separation and a small hole in one plate that you shoot the protons through. By filling the space between the plates with a low-density gas, you can see (with a microscope) a slight glow from the region where the protons collide with and excite the gas molecules. The width of the glow tells you how far the protons travel before being stopped and reversing direction. Varying the voltage across the capacitor gives the following data:

Voltage (V)	Glow width (mm)
1000	1.7
1250	1.3
1500	1.1
1750	1.0
2000	0.8

What value will you report for the speed of the protons?

MODEL The system is the proton plus the capacitor charges. This is an isolated system, so mechanical energy is conserved.

VISUALIZE FIGURE 25.22 is a before-and-after pictorial representation of the proton entering the capacitor with speed v_i and later reaching a turning point with $v_f = 0$ m/s after traveling distance s_f = glow width. For the protons to slow, the hole through which they pass has to be in the negative plate. The s-axis has $s = 0$ at this point.

FIGURE 25.22 A proton stopping in a capacitor.

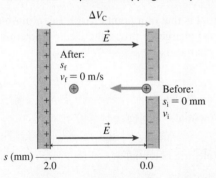

SOLVE The conservation of energy equation, with the proton having charge $q = e$, is $K_f + eV_f = K_i + eV_i$. The initial potential

energy is zero, because the capacitor's electric potential is zero at $s_i = 0$, and the final kinetic energy is zero. Equation 25.28 for the potential inside the capacitor gives

$$eV_f = e\left(\frac{s_f}{d}\Delta V_C\right) = K_i = \tfrac{1}{2}mv_i^2$$

Solving for the distance traveled, you find

$$s_f = \frac{dmv_i^2}{2e}\frac{1}{\Delta V_C}$$

Thus a graph of the distance traveled versus the *inverse* of the capacitor voltage should be a straight line with zero y-intercept and slope $dmv_i^2/2e$. You can use the experimentally determined slope to find the proton speed.

FIGURE 25.23 is a graph of s_f versus $1/\Delta V_C$. It has the expected shape, and the slope of the best-fit line is seen to be 1.72 V m. The units are those of the rise-over-run. Using the slope, you calculate the proton speed:

$$v_i = \sqrt{\frac{2e}{dm} \times \text{slope}} = \sqrt{\frac{2(1.60 \times 10^{-19}\,\text{C})(1.72\,\text{V m})}{(0.0020\,\text{m})(1.67 \times 10^{-27}\,\text{kg})}}$$

$$= 4.1 \times 10^5\,\text{m/s}$$

FIGURE 25.23 A graph of the data.

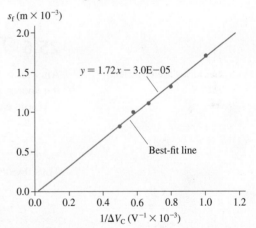

ASSESS This would be a very high speed for a macroscopic object but is quite typical of the speeds of charged particles.

In writing the electric potential inside a parallel-plate capacitor, we made the choice that $V_- = 0$ V at the negative plate. But that is not the only possible choice. FIGURE 25.24 on the next page shows three parallel-plate capacitors, each having the same capacitor voltage $\Delta V_C = V_+ - V_- = 100$ V, but each with a different choice for the location of the zero point of the electric potential. Notice the *terminal symbols* (lines with small circles at the end) showing how the potential, from a battery or a power supply, is applied to each plate; these symbols are common in electronics.

FIGURE 25.24 These three choices for $V = 0$ represent the same physical situation. These are contour maps, showing the edges of the equipotential surfaces.

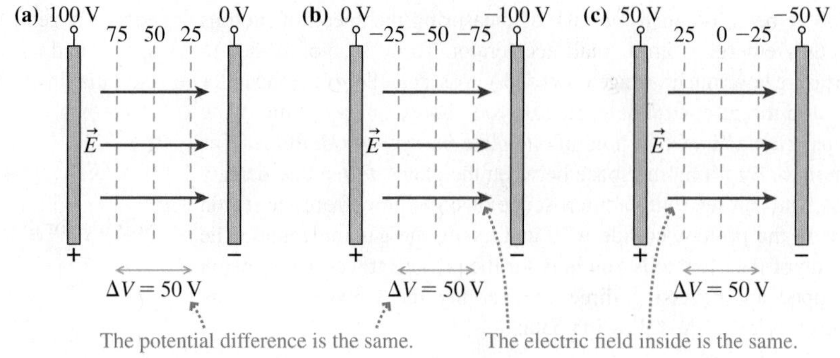

(a) 100 V   75 50 25   0 V
(b) 0 V −25 −50 −75 −100 V
(c) 50 V   25 0 −25   −50 V

$\vec{E}$

$\Delta V = 50$ V

$\Delta V = 50$ V

$\Delta V = 50$ V

The potential difference is the same.   The electric field inside is the same.

The important thing to notice is that the three contour maps in Figure 25.24 represent the *same physical situation*. The potential difference between any two points is the same in all three maps. The electric field is the same in all three. We may *prefer* one of these figures over the others, but there is no measurable physical difference between them.

STOP TO THINK 25.4 Rank in order, from largest to smallest, the potentials V_a to V_e at the points a to e.

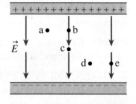

25.6 The Electric Potential of a Point Charge

FIGURE 25.25 Measuring the electric potential of charge q.

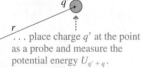

To determine the potential of q at this point . . .

q

q'

r

. . . place charge q' at the point as a probe and measure the potential energy $U_{q'+q}$.

q

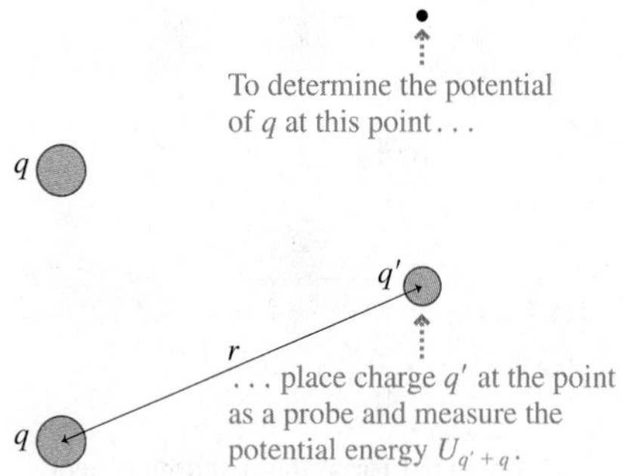

Another important electric potential is that of a point charge. Let q in **FIGURE 25.25** be the source charge, and let a second charge q' probe the electric potential of q. The potential energy of the two point charges is

$$U_{q'+q} = \frac{1}{4\pi\epsilon_0} \frac{qq'}{r} \tag{25.29}$$

Thus, by definition, the electric potential of charge q is

$$V = \frac{U_{q'+q}}{q'} = \frac{1}{4\pi\epsilon_0} \frac{q}{r} \quad \text{(electric potential of a point charge)} \tag{25.30}$$

The potential of Equation 25.30 extends through all of space, showing the influence of charge q, but it weakens with distance as $1/r$. This expression for V assumes that we have chosen $V = 0$ V to be at $r = \infty$. This is the most logical choice for a point charge because the influence of charge q ends at infinity.

The expression for the electric potential of charge q is similar to that for the electric field of charge q. The difference most quickly seen is that V depends on $1/r$ whereas $\vec{E}$ depends on $1/r^2$. But it is also important to notice that **the potential is a scalar** whereas the field is a vector. Thus the mathematics of using the potential are much easier than the vector mathematics using the electric field requires.

For example, the electric potential 1.0 cm from a $+1.0$ nC charge is

$$V_{1\,\text{cm}} = \frac{1}{4\pi\epsilon_0} \frac{q}{r} = (9.0 \times 10^9 \,\text{N m}^2/\text{C}^2) \frac{1.0 \times 10^{-9}\,\text{C}}{0.010\,\text{m}} = 900\,\text{V}$$

1 nC is typical of the electrostatic charge produced by rubbing, and you can see that such a charge creates a fairly large potential nearby. Why are we not shocked and injured

when working with the "high voltages" of such charges? The sensation of being shocked is a result of current, not potential. Some high-potential sources simply do not have the ability to generate much current. We will look at this issue in Chapter 28.

Visualizing the Potential of a Point Charge

FIGURE 25.26 shows four graphical representations of the electric potential of a point charge. These match the four representations of the electric potential inside a capacitor, and a comparison of the two is worthwhile. This figure assumes that q is positive; you may want to think about how the representations would change if q were negative.

FIGURE 25.26 Four graphical representations of the electric potential of a point charge.

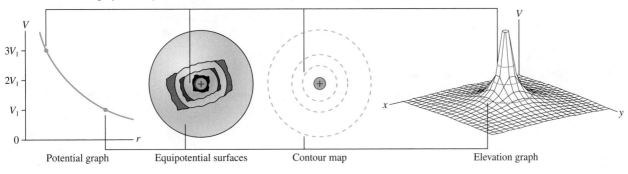

Potential graph   Equipotential surfaces   Contour map   Elevation graph

STOP TO THINK 25.5   Rank in order, from largest to smallest, the potential differences ΔV_{ab}, ΔV_{ac}, and ΔV_{bc} between points a and b, points a and c, and points b and c.

The Electric Potential of a Charged Sphere

In practice, you are more likely to work with a charged sphere, of radius R and total charge Q, than with a point charge. Outside a uniformly charged sphere, the electric potential is identical to that of a point charge Q at the center. That is,

$$V = \frac{1}{4\pi\epsilon_0}\frac{Q}{r} \qquad \text{(sphere of charge, } r \geq R) \qquad (25.31)$$

We can cast this result in a more useful form. It is customary to speak of charging an electrode, such as a sphere, "to" a certain potential, as in "Bob charged the sphere to a potential of 3000 volts." This potential, which we will call V_0, is the potential right on the surface of the sphere. We can see from Equation 25.31 that

$$V_0 = V(\text{at } r = R) = \frac{Q}{4\pi\epsilon_0 R} \qquad (25.32)$$

Consequently, a sphere of radius R that is charged to potential V_0 has total charge

$$Q = 4\pi\epsilon_0 R V_0 \qquad (25.33)$$

If we substitute this expression for Q into Equation 25.31, we can write the potential outside a sphere that is charged to potential V_0 as

$$V = \frac{R}{r}V_0 \qquad \text{(sphere charged to potential } V_0) \qquad (25.34)$$

Equation 25.34 tells us that the potential of a sphere is V_0 on the surface and decreases inversely with the distance. The potential at $r = 3R$ is $\frac{1}{3}V_0$.

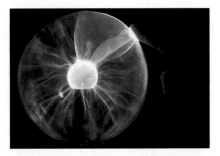

A *plasma ball* consists of a small metal ball charged to a potential of about 2000 V inside a hollow glass sphere. The electric field of the high-voltage ball is sufficient to cause a gas breakdown at this pressure, creating "lightning bolts" between the ball and the glass sphere.

EXAMPLE 25.8 | **A proton and a charged sphere**

A proton is released from rest at the surface of a 1.0-cm-diameter sphere that has been charged to +1000 V.

a. What is the charge of the sphere?

b. What is the proton's speed at 1.0 cm from the sphere?

MODEL Energy is conserved. The potential outside the charged sphere is the same as the potential of a point charge at the center.

VISUALIZE FIGURE 25.27 shows the situation.

FIGURE 25.27 A sphere and a proton.

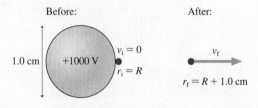

SOLVE a. The charge of the sphere is

$$Q = 4\pi\epsilon_0 R V_0 = 0.56 \times 10^{-9} \text{ C} = 0.56 \text{ nC}$$

b. A sphere charged to $V_0 = +1000$ V is positively charged. The proton will be repelled by this charge and move away from the sphere. The conservation of energy equation $K_f + eV_f = K_i + eV_i$, with Equation 25.34 for the potential of a sphere, is

$$\tfrac{1}{2}mv_f^2 + \frac{eR}{r_f}V_0 = \tfrac{1}{2}mv_i^2 + \frac{eR}{r_i}V_0$$

The proton starts from the surface of the sphere, $r_i = R$, with $v_i = 0$. When the proton is 1.0 cm from the *surface* of the sphere, it has $r_f = 1.0$ cm $+ R = 1.5$ cm. Using these, we can solve for v_f:

$$v_f = \sqrt{\frac{2eV_0}{m}\left(1 - \frac{R}{r_f}\right)} = 3.6 \times 10^5 \text{ m/s}$$

ASSESS This example illustrates how the ideas of electric potential and potential energy work together, yet they are *not* the same thing.

25.7 The Electric Potential of Many Charges

Suppose there are many source charges $q_1, q_2, \ldots$. The electric potential V at a point in space is the sum of the potentials due to each charge:

$$V = \sum_i \frac{1}{4\pi\epsilon_0}\frac{q_i}{r_i} \tag{25.35}$$

where r_i is the distance from charge q_i to the point in space where the potential is being calculated. In other words, **the electric potential, like the electric field, obeys the principle of superposition.**

As an example, the contour map and elevation graph in FIGURE 25.28 show that the potential of an electric dipole is the sum of the potentials of the positive and negative charges. Potentials such as these have many practical applications. For example, electrical activity within the body can be monitored by measuring equipotential lines on the skin. Figure 25.28c shows that the equipotentials near the heart are a slightly distorted but recognizable electric dipole.

FIGURE 25.28 The electric potential of an electric dipole.

(a) Contour map

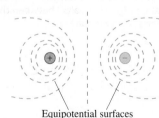

Equipotential surfaces

(b) Elevation graph

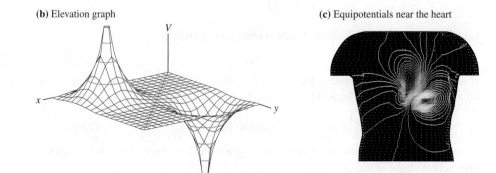

(c) Equipotentials near the heart

EXAMPLE 25.9 | **The potential of two charges**

What is the electric potential at the point indicated in FIGURE 25.29?

▶ **FIGURE 25.29** Finding the potential of two charges.

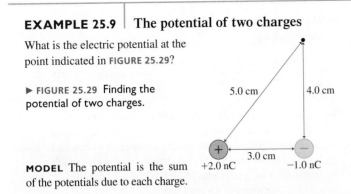

MODEL The potential is the sum of the potentials due to each charge.

SOLVE The potential at the indicated point is

$$V = \frac{1}{4\pi\epsilon_0}\frac{q_1}{r_1} + \frac{1}{4\pi\epsilon_0}\frac{q_2}{r_2}$$

$$= (9.0 \times 10^9 \text{ N m}^2/\text{C}^2)\left(\frac{2.0 \times 10^{-9}\text{ C}}{0.050\text{ m}} + \frac{-1.0 \times 10^{-9}\text{ C}}{0.040\text{ m}}\right)$$

$$= 135 \text{ V}$$

ASSESS The potential is a *scalar*, so we found the net potential by adding two numbers. We don't need any angles or components to calculate the potential.

A Continuous Distribution of Charge

Equation 25.35 is the basis for determining the potential of a continuous distribution of charge, such as a charged rod or a charged disk. The procedure is much like the one you learned in Chapter 23 for calculating the electric field of a continuous distribution of charge, but *easier* because the potential is a scalar. We will continue to assume that the object is *uniformly charged,* meaning that the charges are evenly spaced over the object.

PROBLEM-SOLVING STRATEGY 25.2   (MP)

The electric potential of a continuous distribution of charge

MODEL Model the charge distribution as a simple shape.

VISUALIZE For the pictorial representation:

■ Draw a picture, establish a coordinate system, and identify the point P at which you want to calculate the electric potential.

■ Divide the total charge Q into small pieces of charge ΔQ, using shapes for which you *already know* how to determine V. This division is often, but not always, into point charges.

■ Identify distances that need to be calculated.

SOLVE The mathematical representation is $V = \sum V_i$.

■ Use superposition to form an algebraic expression for the potential at P. Let the (x, y, z) coordinates of the point remain as variables.

■ Replace the small charge ΔQ with an equivalent expression involving a *charge density* and a *coordinate,* such as dx. **This is the critical step in making the transition from a sum to an integral** because you need a coordinate to serve as the integration variable.

■ All distances must be expressed in terms of the coordinates.

■ Let the sum become an integral. The integration limits will depend on the coordinate system you have chosen.

ASSESS Check that your result is consistent with any limits for which you know what the potential should be.

Exercise 29 ✐

EXAMPLE 25.10 | **The potential of a ring of charge**

A thin, uniformly charged ring of radius R has total charge Q. Find the potential at distance z on the axis of the ring.

MODEL Because the ring is thin, we'll assume the charge lies along a circle of radius R.

VISUALIZE FIGURE 25.30 on the next page illustrates the problem-

solving strategy. We've chosen a coordinate system in which the ring lies in the xy-plane and point P is on the z-axis. We've then divided the ring into N small segments of charge ΔQ, each of which can be modeled as a point charge. The distance r_i between segment i and point P is

Continued

$$r_i = \sqrt{R^2 + z^2}$$

Note that r_i is a constant distance, the same for every charge segment.

FIGURE 25.30 Finding the potential of a ring of charge.

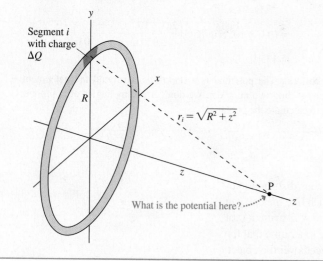

Segment i with charge ΔQ

R

$r_i = \sqrt{R^2 + z^2}$

z

P

What is the potential here? ·······

SOLVE The potential V at P is the sum of the potentials due to each segment of charge:

$$V = \sum_{i=1}^{N} V_i = \sum_{i=1}^{N} \frac{1}{4\pi\epsilon_0} \frac{\Delta Q}{r_i} = \frac{1}{4\pi\epsilon_0} \frac{1}{\sqrt{R^2 + z^2}} \sum_{i=1}^{N} \Delta Q$$

We were able to bring all terms involving z to the front because z is a constant as far as the summation is concerned. Surprisingly, we don't need to convert the sum to an integral to complete this calculation. The sum of all the ΔQ charge segments around the ring is simply the ring's total charge, $\sum(\Delta Q) = Q$; hence the electric potential on the axis of a charged ring is

$$V_{\text{ring on axis}} = \frac{1}{4\pi\epsilon_0} \frac{Q}{\sqrt{R^2 + z^2}}$$

ASSESS From far away, the ring appears as a point charge Q in the distance. Thus we expect the potential of the ring to be that of a point charge when $z \gg R$. You can see that $V_{\text{ring}} \approx Q/4\pi\epsilon_0 z$ when $z \gg R$, which is, indeed, the potential of a point charge Q.

CHALLENGE EXAMPLE 25.11 | The potential of a charged disk

A thin plastic disk of radius R is uniformly coated with charge until it receives total charge Q.

a. What is the potential at distance z along the axis of the disk?

b. What is the potential energy if an electron is 1.00 cm from a 35.0-cm-diameter disk that has been charged to $+5.00$ nC?

MODEL Model the disk as a uniformly charged disk of zero thickness, radius R, and charge Q. The disk has uniform surface charge density $\eta = Q/A = Q/\pi R^2$. We can take advantage of now knowing the on-axis potential of a ring of charge.

VISUALIZE Orient the disk in the xy-plane, as shown in **FIGURE 25.31**, with point P at distance z. Then divide the disk into *rings* of equal width Δr. Ring i has radius r_i and charge ΔQ_i.

FIGURE 25.31 Finding the potential of a disk of charge.

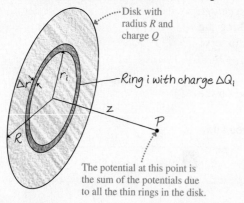

Disk with radius R and charge Q

Δr r_i

Ring i with charge ΔQ_i

z

P

R

The potential at this point is the sum of the potentials due to all the thin rings in the disk.

SOLVE a. We can use the result of Example 25.10 to write the potential at distance z of ring i as

$$V_i = \frac{1}{4\pi\epsilon_0} \frac{\Delta Q_i}{\sqrt{r_i^2 + z^2}}$$

The potential at P due to all the rings is the sum

$$V = \sum_i V_i = \frac{1}{4\pi\epsilon_0} \sum_{i=1}^{N} \frac{\Delta Q_i}{\sqrt{r_i^2 + z^2}}$$

The critical step is to relate ΔQ_i to a coordinate. Because we now have a surface, rather than a line, the charge in ring i is $\Delta Q_i = \eta \Delta A_i$, where ΔA_i is the area of ring i. We can find ΔA_i, as you've learned to do in calculus, by "unrolling" the ring to form a narrow rectangle of length $2\pi r_i$ and height Δr. Thus the area of ring i is $\Delta A_i = 2\pi r_i \Delta r$ and the charge is

$$\Delta Q_i = \eta \Delta A_i = \frac{Q}{\pi R^2} 2\pi r_i \Delta r = \frac{2Q}{R^2} r_i \Delta r$$

With this substitution, the potential at P is

$$V = \frac{1}{4\pi\epsilon_0} \sum_{i=1}^{N} \frac{2Q}{R^2} \frac{r_i \Delta r_i}{\sqrt{r_i^2 + z^2}} \rightarrow \frac{Q}{2\pi\epsilon_0 R^2} \int_0^R \frac{r\,dr}{\sqrt{r^2 + z^2}}$$

where, in the last step, we let $N \rightarrow \infty$ and the sum become an integral. This integral can be found in Appendix A, but it's not hard to evaluate with a change of variables. Let $u = r^2 + z^2$, in which case $r\,dr = \frac{1}{2}\,du$. Changing variables requires that we also change the integration limits. You can see that $u = z^2$ when $r = 0$, and $u = R^2 + z^2$ when $r = R$. With these changes, the on-axis potential of a charged disk is

$$V_{\text{disk on axis}} = \frac{Q}{2\pi\epsilon_0 R^2} \int_{z^2}^{R^2+z^2} \frac{\frac{1}{2}\,du}{u^{1/2}} = \frac{Q}{2\pi\epsilon_0 R^2} u^{1/2} \Big|_{z^2}^{R^2+z^2}$$

$$= \frac{Q}{2\pi\epsilon_0 R^2} \left(\sqrt{R^2 + z^2} - z \right)$$

b. To calculate the potential energy, we first need to determine the potential of the disk at $z = 1.00$ cm. Using $R = 0.0175$ m and $Q = 5.00$ nC, you can calculate $V = 3870$ V. The electron's charge is $q = -e = -1.60 \times 10^{-19}$ C, so the potential energy with an electron at $z = 1.00$ cm is $U = qV = -6.19 \times 10^{-16}$ J.

ASSESS Although we had to go through a number of steps, this procedure is easier than evaluating the electric field because we do not have to worry about vector components.

SUMMARY

The goals of Chapter 25 have been to use the electric potential and electric potential energy.

GENERAL PRINCIPLES

Sources of Potential

The **electric potential** V, like the electric field, is created by source charges. Two major tools for calculating the potential are:

- The potential of a point charge, $V = \dfrac{1}{4\pi\epsilon_0}\dfrac{q}{r}$
- The principle of superposition

For multiple point charges

Use superposition: $V = V_1 + V_2 + V_3 + \cdots$

For a continuous distribution of charge

MODEL Model as a simple charge distribution.

VISUALIZE Draw a pictorial representation.

- Establish a coordinate system.
- Identify where the potential will be calculated.

SOLVE Set up a sum.

- Divide the charge into point-like ΔQ.
- Find the potential due to each ΔQ.
- Use the charge density (λ or η) to replace ΔQ with an integration coordinate, then sum by integrating.

V is easier to calculate than $\vec{E}$ because potential is a scalar.

Electric Potential Energy

If charge q is placed in an electric potential V, the system's **electric potential energy** (interaction energy) is

$$U = qV$$

Point charges and dipoles

The electric potential energy of two **point charges** is

$$U_{q_1 + q_2} = \frac{Kq_1q_2}{r} = \frac{1}{4\pi\epsilon_0}\frac{q_1q_2}{r}$$

The potential energy of two opposite charges is negative.
The potential energy in an electric field of an **electric dipole** with dipole moment $\vec{p}$ is

$$U_{\text{dipole}} = -pE\cos\theta = -\vec{p}\cdot\vec{E}$$

Solving conservation of energy problems

MODEL Model as an isolated system.

VISUALIZE Draw a before-and-after representation.

SOLVE Mechanical energy is conserved.

- Mathematically $K_f + qV_f = K_i + qV_i$.
- K is the sum of the kinetic energies of all particles.
- V is the potential due to the source charges.

APPLICATIONS

Graphical representations of the potential:

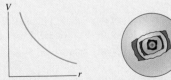

Potential graph   **Equipotential surfaces**

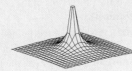

Contour map   **Elevation graph**

Sphere of charge Q

Same as a point charge
if $r \geq R$

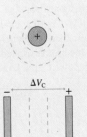

Parallel-plate capacitor

$V = Es$, where s is measured from the negative plate. The electric field inside is

$$E = \frac{\Delta V_C}{d}$$

Units

Electric potential: $1\ \text{V} = 1\ \text{J/C}$

Electric field: $1\ \text{V/m} = 1\ \text{N/C}$

TERMS AND NOTATION

electric potential energy, U volt, V   voltage, ΔV contour map
electric potential, V potential difference, ΔV equipotential surface   elevation graph

CONCEPTUAL QUESTIONS

1. a. Charge q_1 is distance r from a positive point charge Q. Charge $q_2 = q_1/3$ is distance $2r$ from Q. What is the ratio U_1/U_2 of their potential energies due to their interactions with Q?
   b. Charge q_1 is distance s from the negative plate of a parallel-plate capacitor. Charge $q_2 = q_1/3$ is distance $2s$ from the negative plate. What is the ratio U_1/U_2 of their potential energies?

2. **FIGURE Q25.2** shows the potential energy of a proton ($q = +e$) and a lead nucleus ($q = +82e$). The horizontal scale is in units of *femtometers,* where 1 fm = 10^{-15} m.
   a. A proton is fired toward a lead nucleus from very far away. How much initial kinetic energy does the proton need to reach a turning point 10 fm from the nucleus? Explain.
   b. How much kinetic energy does the proton of part a have when it is 20 fm from the nucleus and moving toward it, before the collision?

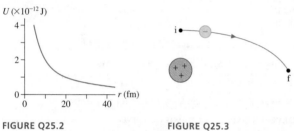

FIGURE Q25.2   **FIGURE Q25.3**

3. An electron moves along the trajectory of **FIGURE Q25.3** from i to f.
   a. Does the electric potential energy increase, decrease, or stay the same? Explain.
   b. Is the electron's speed at f greater than, less than, or equal to its speed at i? Explain.

4. Two protons are launched with the same speed from point 1 inside the parallel-plate capacitor of **FIGURE Q25.4**. Points 2 and 3 are the same distance from the negative plate.
   a. Is $\Delta U_{1\rightarrow 2}$, the change in potential energy along the path $1 \rightarrow 2$, larger than, smaller than, or equal to $\Delta U_{1\rightarrow 3}$?
   b. Is the proton's speed v_2 at point 2 larger than, smaller than, or equal to v_3? Explain.

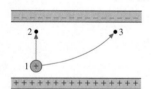

FIGURE Q25.4

5. Rank in order, from most positive to most negative, the potential energies U_a to U_f of the six electric dipoles in the uniform electric field of **FIGURE Q25.5**. Explain.

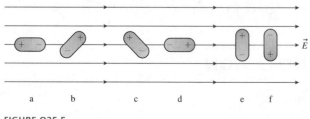

FIGURE Q25.5

6. **FIGURE Q25.6** shows the electric potential along the x-axis.
   a. Draw a graph of the potential energy of a 0.1 C charged particle. Provide a numerical scale for both axes.
   b. If the charged particle is shot toward the right from $x = 1$ m with 1.0 J of kinetic energy, where is its turning point? Use your graph to explain.

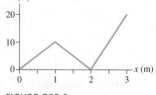

FIGURE Q25.6

7. A capacitor with plates separated by distance d is charged to a potential difference ΔV_C. All wires and batteries are disconnected, then the two plates are pulled apart (with insulated handles) to a new separation of distance $2d$.
   a. Does the capacitor charge Q change as the separation increases? If so, by what factor? If not, why not?
   b. Does the electric field strength E change as the separation increases? If so, by what factor? If not, why not?
   c. Does the potential difference ΔV_C change as the separation increases? If so, by what factor? If not, why not?

8. Rank in order, from largest to smallest, the electric potentials V_a to V_e at points a to e in **FIGURE Q25.8**. Explain.

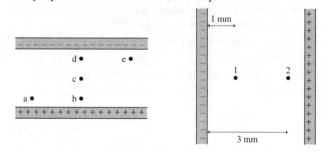

FIGURE Q25.8   **FIGURE Q25.9**

9. **FIGURE Q25.9** shows two points inside a capacitor. Let $V = 0$ V at the negative plate.
   a. What is the ratio V_2/V_1 of the electric potentials? Explain.
   b. What is the ratio E_2/E_1 of the electric field strengths?

10. **FIGURE Q25.10** shows two points near a positive point charge.
    a. What is the ratio V_2/V_1 of the electric potentials? Explain.
    b. What is the ratio E_2/E_1 of the electric field strengths?

FIGUREQ25.10

11. **FIGURE Q25.11** shows three points near two point charges. The charges have equal magnitudes. For each part, rank in order, from most positive to most negative, the potentials V_a to V_c.

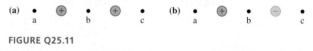

FIGURE Q25.11

12. Reproduce **FIGURE Q25.12** on your paper. Then draw a dot (or dots) on the figure to show the position (or positions) at which the electric potential is zero.

FIGURE Q25.12

In Section 25.1, we found that the electric potential energy of a charge q in the uniform electric field of a parallel-plate capacitor is

$$U_{elec} = U_{q + sources} = qEs \qquad (25.24)$$

We've set the constant term U_0 to zero. U_{elec} is the energy of q interacting with the source charges on the capacitor plates.

Our new view of the interaction is to separate the role of charge q from the role of the source charges by defining the electric potential $V = U_{q + sources}/q$. Thus the electric potential inside a parallel-plate capacitor is

$$V = Es \qquad \text{(electric potential inside a parallel-plate capacitor)} \qquad (25.25)$$

where s **is the distance from the *negative* electrode.** The electric potential, like the electric field, exists at *all points* inside the capacitor. The electric potential is created by the source charges on the capacitor plates and exists whether or not charge q is inside the capacitor.

FIGURE 25.19 illustrates the important point that the electric potential increases linearly from the negative plate, where $V_- = 0$, to the positive plate, where $V_+ = Ed$. Let's define the *potential difference* ΔV_C between the two capacitor plates to be

$$\Delta V_C = V_+ - V_- = Ed \qquad (25.26)$$

In our specific example, $\Delta V_C = (500 \text{ N/C})(0.0030 \text{ m}) = 1.5 \text{ V}$. The units work out because $1.5 \text{ (N m)/C} = 1.5 \text{ J/C} = 1.5 \text{ V}$.

> **NOTE** People who work with circuits would call ΔV_C "the voltage across the capacitor" or simply "the capacitor voltage."

Equation 25.26 has an interesting implication. Thus far, we've determined the electric field inside a capacitor by specifying the surface charge density η on the plates. Alternatively, we could specify the capacitor voltage ΔV_C (i.e., the potential difference between the capacitor plates) and then determine the electric field strength as

$$E = \frac{\Delta V_C}{d} \qquad (25.27)$$

In fact, this is how E is determined in practical applications because it's easy to measure ΔV_C with a voltmeter but difficult, in practice, to know the value of η.

Equation 25.27 implies that the units of electric field are volts per meter, or V/m. We have been using electric field units of newtons per coulomb. In fact, as you can show as a homework problem, these units are equivalent to each other. That is,

$$1 \text{ N/C} = 1 \text{ V/m}$$

> **NOTE** Volts per meter are the electric field units used by scientists and engineers in practice. We will now adopt them as our standard electric field units.

Returning to the electric potential, we can substitute Equation 25.27 for E into Equation 25.25 for V. Thus the electric potential inside the capacitor is

$$V = Es = \frac{s}{d}\Delta V_C \qquad (25.28)$$

The potential increases linearly from $V_- = 0 \text{ V}$ at the negative plate $(s = 0)$ to $V_+ = \Delta V_C$ at the positive plate $(s = d)$.

Visualizing Electric Potential

Let's explore the electric potential inside the capacitor by looking at several different, but related, ways that the potential can be represented graphically.

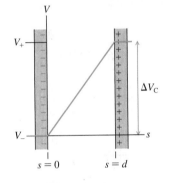

FIGURE 25.19 The electric potential of a parallel-plate capacitor increases linearly from the negative to the positive plate.

Section 25.5 The Electric Potential Inside a Parallel-Plate Capacitor

21. | Show that 1 V/m = 1 N/C.

22. ‖ a. What is the potential of an ordinary AA or AAA battery? (If you're not sure, find one and look at the label.)

b. An AA battery is connected to a parallel-plate capacitor having 4.0 cm × 4.0 cm plates spaced 1.0 mm apart. How much charge does the battery supply to each plate?

23. ‖ A 3.0-cm-diameter parallel-plate capacitor has a 2.0 mm spacing. The electric field strength inside the capacitor is 1.0×10^5 V/m.

a. What is the potential difference across the capacitor?

b. How much charge is on each plate?

24. ‖ Two 2.00 cm × 2.00 cm plates that form a parallel-plate capacitor are charged to ±0.708 nC. What are the electric field strength inside and the potential difference across the capacitor if the spacing between the plates is (a) 1.00 mm and (b) 2.00 mm?

25. ‖ Two 2.0-cm-diameter disks spaced 2.0 mm apart form a parallel-plate capacitor. The electric field between the disks is 5.0×10^5 V/m.

a. What is the voltage across the capacitor?

b. An electron is launched from the negative plate. It strikes the positive plate at a speed of 2.0×10^7 m/s. What was the electron's speed as it left the negative plate?

26. ‖ In FIGURE EX25.26, a proton is fired with a speed of 200,000 m/s from the midpoint of the capacitor toward the positive plate.

a. Show that this is insufficient speed to reach the positive plate.

b. What is the proton's speed as it collides with the negative plate?

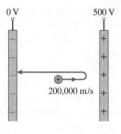

FIGURE EX25.26

Section 25.6 The Electric Potential of a Point Charge

27. | a. What is the electric potential at points A, B, and C in FIGURE EX25.27?

b. What are the potential differences $\Delta V_{AB} = V_B - V_A$ and $\Delta V_{CB} = V_B - V_C$?

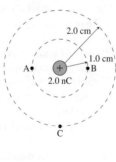

FIGURE EX25.27

28. ‖ A 1.0-mm-diameter ball bearing has 2.0×10^9 excess electrons. What is the ball bearing's potential?

29. | In a semiclassical model of the hydrogen atom, the electron orbits the proton at a distance of 0.053 nm.

a. What is the electric potential of the proton at the position of the electron?

b. What is the electron's potential energy?

Section 25.7 The Electric Potential of Many Charges

30. | What is the electric potential at the point indicated with the dot in FIGURE EX25.30?

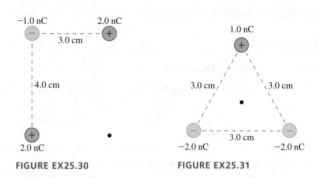

FIGURE EX25.30   FIGURE EX25.31

31. | What is the electric potential at the point indicated with the dot in FIGURE EX25.31?

32. ‖ The electric potential at the dot in FIGURE EX25.32 is 3140 V. What is charge q?

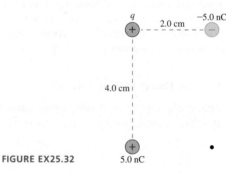

FIGURE EX25.32

33. ‖ A −2.0 nC charge and a +2.0 nC charge are located on the x-axis at $x = -1.0$ cm and $x = +1.0$ cm, respectively.

a. Other than at infinity, is there a position or positions on the x-axis where the electric field is zero? If so, where?

b. Other than at infinity, at what position or positions on the x-axis is the electric potential zero?

c. Sketch graphs of the electric field strength and the electric potential along the x-axis.

34. ‖ Two point charges q_a and q_b are located on the x-axis at $x = a$ and $x = b$. FIGURE EX25.34 is a graph of V, the electric potential.

a. What are the signs of q_a and q_b?

b. What is the ratio $|q_a/q_b|$?

c. Draw a graph of E_x, the x-component of the electric field, as a function of x.

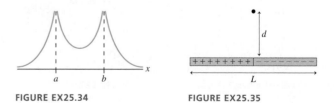

FIGURE EX25.34   FIGURE EX25.35

35. | The two halves of the rod in FIGURE EX25.35 are uniformly charged to ±Q. What is the electric potential at the point indicated by the dot?

36. ‖ A 5.0-cm-diameter metal ball has a surface charge density of $10 \ \mu C/m^2$. How much work is required to remove one electron from this ball?

Problems

37. ‖‖ Two point charges 2.0 cm apart have an electric potential energy $-180 \ \mu J$. The total charge is 30 nC. What are the two charges?

38. ‖‖ A -10.0 nC point charge and a $+20.0$ nC point charge are 15.0 cm apart on the x-axis.
    a. What is the electric potential at the point on the x-axis where the electric field is zero?
    b. What is the magnitude of the electric field at the point on the x-axis, between the charges, where the electric potential is zero?

39. ‖‖ A $+3.0$ nC charge is at $x = 0$ cm and a -1.0 nC charge is at $x = 4$ cm. At what point or points on the x-axis is the electric potential zero?

40. ‖ A -3.0 nC charge is on the x-axis at $x = -9$ cm and a $+4.0$ nC charge is on the x-axis at $x = 16$ cm. At what point or points on the y-axis is the electric potential zero?

41. ‖ Two small metal cubes with masses 2.0 g and 4.0 g are tied together by a 5.0-cm-long massless string and are at rest on a frictionless surface. Each is charged to $+2.0 \ \mu C$.
    a. What is the energy of this system?
    b. What is the tension in the string?
    c. The string is cut. What is the speed of each cube when they are far apart?
    Hint: There are *two* conserved quantities. Make use of both.

42. ‖ The four 1.0 g spheres shown in FIGURE P25.42 are released simultaneously and allowed to move away from each other. What is the speed of each sphere when they are very far apart?

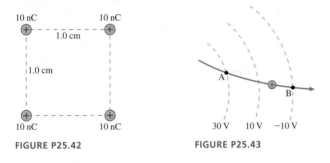

FIGURE P25.42   FIGURE P25.43

43. ‖ A proton's speed as it passes point A is 50,000 m/s. It follows the trajectory shown in FIGURE P25.43. What is the proton's speed at point B?

BIO ‖ Living cells "pump" singly ionized sodium ions, Na^+, from the inside of the cell to the outside to maintain a membrane potential $\Delta V_{membrane} = V_{in} - V_{out} = -70$ mV. It is called *pumping* because work must be done to move a positive ion from the negative inside of the cell to the positive outside, and it must go on continuously because sodium ions "leak" back through the cell wall by diffusion.
    a. How much work must be done to move one sodium ion from the inside of the cell to the outside?
    b. At rest, the human body uses energy at the rate of approximately 100 W to maintain basic metabolic functions. It has been estimated that 20% of this energy is used to operate the sodium pumps of the body. Estimate—to one significant figure—the number of sodium ions pumped per second.

45. ‖‖ An arrangement of source charges produces the electric potential $V = 5000x^2$ along the x-axis, where V is in volts and x is in meters. What is the maximum speed of a 1.0 g, 10 nC charged particle that moves in this potential with turning points at ± 8.0 cm?

46. ‖ A proton moves along the x-axis where some arrangement of charges has produced the potential $V(x) = V_0 \sin(2\pi x/\lambda)$, where $V_0 = 5000$ V and $\lambda = 1.0$ mm.
    a. What minimum speed must the proton have at $x = 0$ to move down the axis without being reflected?
    b. What is the maximum speed reached by a proton that at $x = 0$ has the speed you calculated in part a?

47. ‖ The electron gun in an old TV picture tube accelerates electrons between two parallel plates 1.2 cm apart with a 25 kV potential difference between them. The electrons enter through a small hole in the negative plate, accelerate, then exit through a small hole in the positive plate. Assume that the holes are small enough not to affect the electric field or potential.
    a. What is the electric field strength between the plates?
    b. With what speed does an electron exit the electron gun if its entry speed is close to zero?
    NOTE The exit speed is so fast that we really need to use the theory of relativity to compute an accurate value. Your answer to part b is in the right range but a little too big.

48. ‖ A room with 3.0-m-high ceilings has a metal plate on the floor with $V = 0$ V and a separate metal plate on the ceiling. A 1.0 g glass ball charged to $+4.9$ nC is shot straight up at 5.0 m/s. How high does the ball go if the ceiling voltage is (a) $+3.0 \times 10^6$ V and (b) -3.0×10^6 V?

49. ‖ A group of science and engineering students embarks on a quest to make an electrostatic projectile launcher. For their first trial, a horizontal, frictionless surface is positioned next to the 12-cm-diameter sphere of a Van de Graaff generator, and a small, 5.0 g plastic cube is placed on the surface with its center 2.0 cm from the edge of the sphere. The cube is given a positive charge, and then the Van de Graaff generator is turned on, charging the sphere to a potential of 200,000 V in a negligible amount of time. How much charge does the plastic cube need to achieve a final speed of a mere 3.0 m/s? Does this seem like a practical projectile launcher?

50. ‖ Two 2.0 g plastic buttons each with $+50$ nC of charge are placed on a frictionless surface 2.0 cm (measured between centers) on either side of a 5.0 g button charged to $+250$ nC. All three are released simultaneously.
    a. How many interactions are there that have a potential energy?
    b. What is the final speed of each button?

51. ‖ What is the escape speed of an electron launched from the surface of a 1.0-cm-diameter glass sphere that has been charged to 10 nC?

52. ‖‖ An electric dipole has dipole moment p. If $r \gg s$, where s is the separation between the charges, show that the electric potential of the dipole can be written

$$V = \frac{1}{4\pi\epsilon_0} \frac{p \cos\theta}{r^2}$$

where r is the distance from the center of the dipole and θ is the angle from the dipole axis.

53. ‖‖ Three electrons form an equilateral triangle 1.0 nm on each side. A proton is at the center of the triangle. What is the potential energy of this group of charges?

54. ||| A 2.0-mm-diameter glass bead is positively charged. The potential difference between a point 2.0 mm from the bead and a point 4.0 mm from the bead is 500 V. What is the charge on the bead?

55. || Your lab assignment for the week is to measure the amount of charge on the 6.0-cm-diameter metal sphere of a Van de Graaff generator. To do so, you're going to use a spring with spring constant 0.65 N/m to launch a small, 1.5 g bead horizontally toward the sphere. You can reliably charge the bead to 2.5 nC, and your plan is to use a video camera to measure the bead's closest approach to the edge of the sphere as you change the compression of the spring. Your data are as follows:

Compression (cm)	Closest approach (cm)
1.6	5.5
1.9	2.6
2.2	1.6
2.5	0.4

Use an appropriate graph of the data to determine the sphere's charge in nC. You can assume that the bead's motion is entirely horizontal, that the spring is so far away that the bead has no interaction with the sphere as it's launched, and that the approaching bead does not alter the charge distribution on the sphere.

56. || A proton is fired from far away toward the nucleus of an iron atom. Iron is element number 26, and the diameter of the nucleus is 9.0 fm. What initial speed does the proton need to just reach the surface of the nucleus? Assume the nucleus remains at rest.

57. ||| A proton is fired from far away toward the nucleus of a mercury atom. Mercury is element number 80, and the diameter of the nucleus is 14.0 fm. If the proton is fired at a speed of 4.0×10^7 m/s, what is its closest approach to the surface of the nucleus? Assume the nucleus remains at rest.

58. || In the form of radioactive decay known as *alpha decay,* an unstable nucleus emits a helium-atom nucleus, which is called an *alpha particle.* An alpha particle contains two protons and two neutrons, thus having mass $m = 4$ u and charge $q = 2e$. Suppose a uranium nucleus with 92 protons decays into thorium, with 90 protons, and an alpha particle. The alpha particle is initially at rest at the surface of the thorium nucleus, which is 15 fm in diameter. What is the speed of the alpha particle when it is detected in the laboratory? Assume the thorium nucleus remains at rest.

59. || One form of nuclear radiation, *beta decay,* occurs when a neutron changes into a proton, an electron, and a neutral particle called a *neutrino:* $n \rightarrow p^+ + e^- + \nu$ where ν is the symbol for a neutrino. When this change happens to a neutron within the nucleus of an atom, the proton remains behind in the nucleus while the electron and neutrino are ejected from the nucleus. The ejected electron is called a *beta particle.* One nucleus that exhibits beta decay is the isotope of hydrogen ^{3}H, called *tritium,* whose nucleus consists of one proton (making it hydrogen) and two neutrons (giving tritium an atomic mass $m = 3$ u). Tritium is radioactive, and it decays to helium: ^{3}H $\rightarrow$ ^{3}He $+ e^- + \nu$.
  a. Is charge conserved in the beta decay process? Explain.
  b. Why is the final product a helium atom? Explain.
  c. The nuclei of both ^{3}H and ^{3}He have radii of 1.5×10^{-15} m. With what minimum speed must the electron be ejected if it is to escape from the nucleus and not fall back?

60. || Two 10-cm-diameter electrodes 0.50 cm apart form a parallel-plate capacitor. The electrodes are attached by metal wires to the terminals of a 15 V battery. After a long time, the capacitor is disconnected from the battery but is not discharged. What are the charge on each electrode, the electric field strength inside the capacitor, and the potential difference between the electrodes
  a. Right after the battery is disconnected?
  b. After insulating handles are used to pull the electrodes away from each other until they are 1.0 cm apart?
  c. After the original electrodes (not the modified electrodes of part b) are expanded until they are 20 cm in diameter?

61. || Two 10-cm-diameter electrodes 0.50 cm apart form a parallel-plate capacitor. The electrodes are attached by metal wires to the terminals of a 15 V battery. What are the charge on each electrode, the electric field strength inside the capacitor, and the potential difference between the electrodes
  a. While the capacitor is attached to the battery?
  b. After insulating handles are used to pull the electrodes away from each other until they are 1.0 cm apart? The electrodes remain connected to the battery during this process.
  c. After the original electrodes (not the modified electrodes of part b) are expanded until they are 20 cm in diameter while remaining connected to the battery?

62. || Electrodes of area A are spaced distance d apart to form a CALC parallel-plate capacitor. The electrodes are charged to $\pm q$.
  a. What is the infinitesimal increase in electric potential energy dU if an infinitesimal amount of charge dq is moved from the negative electrode to the positive electrode?
  b. An uncharged capacitor can be charged to $\pm Q$ by transferring charge dq over and over and over. Use your answer to part a to show that the potential energy of a capacitor charged to $\pm Q$ is $U_{cap} = \frac{1}{2} Q \, \Delta V_C$.

63. || a. Find an algebraic expression for the electric field strength E_0 at the surface of a charged sphere in terms of the sphere's potential V_0 and radius R.
  b. What is the electric field strength at the surface of a 1.0-cm-diameter marble charged to 500 V?

64. || Two spherical drops of mercury each have a charge of 0.10 nC and a potential of 300 V at the surface. The two drops merge to form a single drop. What is the potential at the surface of the new drop?

65. || A Van de Graaff generator is a device for generating a large electric potential by building up charge on a hollow metal sphere. A typical classroom-demonstration model has a diameter of 30 cm.
  a. How much charge is needed on the sphere for its potential to be 500,000 V?
  b. What is the electric field strength just outside the surface of the sphere when it is charged to 500,000 V?

66. || FIGURE P25.66 shows two uniformly charged spheres. What is the potential difference between points a and b? Which point is at the higher potential?
  Hint: The potential at any point is the superposition of the potentials due to *all* charges.

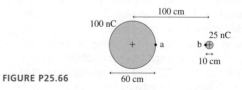

FIGURE P25.66

67. || Two positive point charges q are located on the y-axis at $y = \pm \frac{1}{2} s$.

a. Find an expression for the potential along the x-axis.

b. Draw a graph of V versus x for $-\infty < x < \infty$. For comparison, use a dotted line to show the potential of a point charge $2q$ located at the origin.

68. ‖ The arrangement of charges shown in FIGURE P25.68 is called a *linear electric quadrupole*. The positive charges are located at $y = \pm s$. Notice that the net charge is zero. Find an expression for the electric potential on the y-axis at distances $y \gg s$. Give your answer in terms of the *quadrupole moment, $Q = 2qs^2$*.

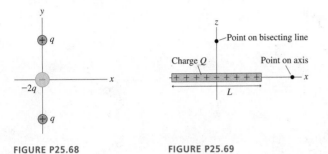

FIGURE P25.68   FIGURE P25.69

69. ‖ FIGURE P25.69 shows a thin rod of length L and charge Q. Find
CALC an expression for the electric potential a distance x away from the center of the rod on the axis of the rod.

70. ‖‖ FIGURE P25.69 shows a thin rod of length L and charge Q. Find
CALC an expression for the electric potential a distance z away from the center of rod on the line that bisects the rod.

71. ‖ FIGURE P25.71 shows a thin rod with charge Q that has been bent into a semicircle of radius R. Find an expression for the electric potential at the center.

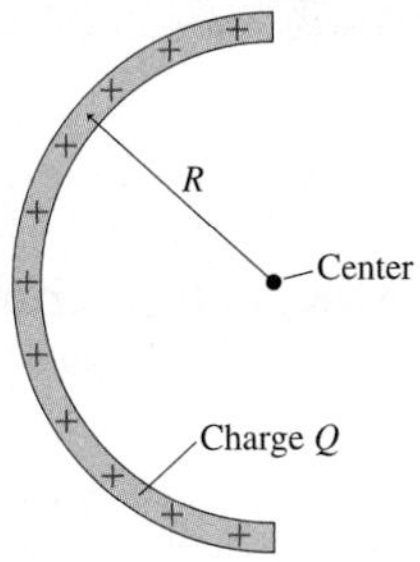

FIGURE P25.71

72. ‖ A disk with a hole has inner radius R_{in} and outer radius R_{out}.
CALC The disk is uniformly charged with total charge Q. Find an expression for the on-axis electric potential at distance z from the center of the disk. Verify that your expression has the correct behavior when $R_{in} \rightarrow 0$.

73. ‖ The wire in FIGURE P25.73 has linear charge density λ. What is
CALC the electric potential at the center of the semicircle?

FIGURE P25.73

In Problems 74 through 76 you are given the equation(s) used to solve a problem. For each of these,

a. Write a realistic problem for which this is the correct equation(s).

b. Finish the solution of the problem.

74. $\dfrac{(9.0 \times 10^9 \text{ N m}^2/\text{C}^2)q_1q_2}{0.030 \text{ m}} = 90 \times 10^{-6} \text{ J}$

$q_1 + q_2 = 40 \text{ nC}$

75. $\frac{1}{2}(1.67 \times 10^{-27} \text{ kg})(2.5 \times 10^6 \text{ m/s})^2 + 0 =$

$\frac{1}{2}(1.67 \times 10^{-27} \text{ kg})v_i^2 +$

$\dfrac{(9.0 \times 10^9 \text{ N m}^2/\text{C}^2)(2.0 \times 10^{-9} \text{ C})(1.60 \times 10^{-19} \text{ C})}{0.0010 \text{ m}}$

76. $\dfrac{(9.0 \times 10^9 \text{ N m}^2/\text{C}^2)(3.0 \times 10^{-9} \text{ C})}{0.030 \text{ m}} +$

$\dfrac{(9.0 \times 10^9 \text{ N m}^2/\text{C}^2)(3.0 \times 10^{-9} \text{ C})}{(0.030 \text{ m}) + d} = 1200 \text{ V}$

Challenge Problems

77. ‖‖‖ An electric dipole consists of 1.0 g spheres charged to ± 2.0 nC at the ends of a 10-cm-long massless rod. The dipole rotates on a frictionless pivot at its center. The dipole is held perpendicular to a uniform electric field with field strength 1000 V/m, then released. What is the dipole's angular velocity at the instant it is aligned with the electric field?

78. ‖‖‖ A proton and an alpha particle ($q = +2e$, $m = 4$ u) are fired directly toward each other from far away, each with an initial speed of $0.010c$. What is their distance of closest approach, as measured between their centers?

79. ‖‖‖ Bead A has a mass of 15 g and a charge of -5.0 nC. Bead B has a mass of 25 g and a charge of -10.0 nC. The beads are held 12 cm apart (measured between their centers) and released. What maximum speed is achieved by each bead?

80. ‖‖‖ Two 2.0-mm-diameter beads, C and D, are 10 mm apart, measured between their centers. Bead C has mass 1.0 g and charge 2.0 nC. Bead D has mass 2.0 g and charge -1.0 nC. If the beads are released from rest, what are the speeds v_C and v_D at the instant the beads collide?

81. ‖‖‖ A thin rod of length L and total charge Q has the nonuniform
CALC linear charge distribution $\lambda(x) = \lambda_0 x/L$, where x is measured from the rod's left end.

a. What is λ_0 in terms of Q and L?

b. What is the electric potential on the axis at distance d left of the rod's left end?

82. ‖‖‖ A hollow cylindrical shell of length L and radius R has charge
CALC Q uniformly distributed along its length. What is the electric potential at the center of the cylinder?

26 Potential and Field

These solar cells are *photovoltaic* cells, meaning that light creates a voltage— a potential difference.

IN THIS CHAPTER, you will learn how the electric potential is related to the electric field.

How are electric potential and field related?

The electric field and the electric potential are intimately connected. In fact, they are simply two different perspectives on how source charges alter the space around them.

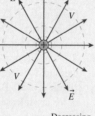

- The electric potential can be found if you know the electric field.
- The electric field can be found if you know the electric potential.
- Electric field lines are always perpendicular to equipotential surfaces.
- The electric field points "downhill" in the direction of decreasing potential.
- The electric field is stronger where equipotentials are closer together.

≪ LOOKING BACK Sections 25.4– 25.6 The electric potential and its graphical representations

What are the properties of conductors?

You'll learn about the properties of conductors in electrostatic equilibrium, finding the same results as using Gauss's law:

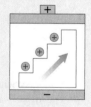

- Any excess charge is on the surface.
- The interior electric field is zero.
- The exterior electric field is perpendicular to the surface.
- The entire conductor is an equipotential.

What are sources of electric potential?

A potential difference—voltage—is created by separating positive and negative charges.

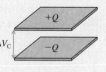

- Work must be done to separate charges. The work done per charge is called the emf of a device. Emf is measured in volts.
- We'll use a charge escalator model of a battery in which chemical reactions "lift" charges from one terminal to the other.

What is a capacitor?

Any two electrodes with equal and opposite charges form a capacitor. Their capacitance indicates their capacity for storing charge. The energy stored in a capacitor will lead us to recognize that electric energy is stored in the electric field.

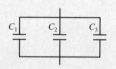

≪ LOOKING BACK Section 23.5 Parallel-plate capacitors

How are capacitors used?

Capacitors are important circuit elements that store charge and energy.

- You'll learn to work with combinations of capacitors arranged in series and parallel.
- You'll learn that an insulator—called a dielectric—between the capacitor plates alters the capacitor in useful ways.

26.1 Connecting Potential and Field

FIGURE 26.1 shows the four key ideas of force, field, potential energy, and potential. The electric field and the electric potential were based on force and potential energy. We know, from Chapters 9 and 10, that force and potential energy are closely related. The focus of this chapter is to establish a similar relationship between the electric field and the electric potential. **The electric potential and electric field are not two distinct entities but, instead, two different perspectives or two different mathematical representations of how source charges alter the space around them.**

If this is true, we should be able to find the electric potential from the electric field. Chapter 25 introduced all the pieces we need to do so. We used the potential energy of charge q and the source charges to define the electric potential as

$$V \equiv \frac{U_{q + \text{sources}}}{q} \tag{26.1}$$

Potential energy is defined in terms of the work done by force $\vec{F}$ on charge q as it moves from position i to position f:

$$\Delta U = -W(\text{i} \rightarrow \text{f}) = -\int_{s_i}^{s_f} F_s \, ds = -\int_i^f \vec{F} \cdot d\vec{s} \tag{26.2}$$

But the force exerted on charge q by the electric field is $\vec{F} = q\vec{E}$. Putting these three pieces together, you can see that the charge q cancels out and the potential difference between two points in space is

$$\Delta V = V_f - V_i = -\int_{s_i}^{s_f} E_s \, ds = -\int_i^f \vec{E} \cdot d\vec{s} \tag{26.3}$$

where s is the position along a line from point i to point f. That is, **we can find the potential difference between two points if we know the electric field.**

NOTE The minus sign tells us that the potential *decreases* along the field direction.

A graphical interpretation of Equation 26.3 is

$$V_f = V_i - (\text{area under the } E_s\text{-versus-}s \text{ curve between } s_i \text{ and } s_f) \tag{26.4}$$

Notice, because of the minus sign in Equation 26.3, that the area is *subtracted* from V_i.

FIGURE 26.1 The four key ideas.

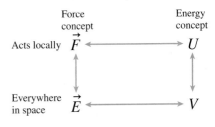

EXAMPLE 26.1 | Finding the potential

FIGURE 26.2 is a graph of E_x, the x-component of the electric field, versus position along the x-axis. Find and graph $V(x)$. Assume $V = 0$ V at $x = 0$ m.

FIGURE 26.2 Graph of E_x versus x.

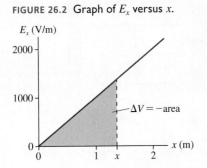

MODEL The potential difference is the *negative* of the area under the curve.

VISUALIZE E_x is positive throughout this region of space, meaning that $\vec{E}$ points in the positive x-direction.

SOLVE We can see that $E_x = 1000x$ V/m, where x is in m. Thus

$$V_f = V(x) = 0 - (\text{area under the } E_x \text{ curve})$$
$$= -\tfrac{1}{2} \times \text{base} \times \text{height} = -\tfrac{1}{2}(x)(1000x) = -500x^2 \text{ V}$$

FIGURE 26.3 shows that the electric potential in this region of space is parabolic, decreasing from 0 V at $x = 0$ m to -2000 V at $x = 2$ m.

FIGURE 26.3 Graph of V versus x.

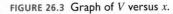
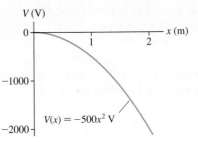

ASSESS The electric field points in the direction in which V is *decreasing*. We'll soon see that this is a general rule.

Finding the potential from the electric field

❶ Draw a picture and identify the point at which you wish to find the potential. Call this position f.
❷ Choose the zero point of the potential, often at infinity. Call this position i.
❸ Establish a coordinate axis from i to f along which you already know or can easily determine the electric field component E_s.
❹ Carry out the integration of Equation 26.3 to find the potential.

Exercise 1

FIGURE 26.4 Finding the potential of a point charge.

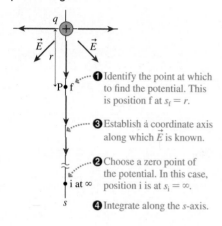

❶ Identify the point at which to find the potential. This is position f at $s_f = r$.

❸ Establish a coordinate axis along which $\vec{E}$ is known.

❷ Choose a zero point of the potential. In this case, position i is at $s_i = \infty$.

❹ Integrate along the s-axis.

To see how this works, let's use the electric field of a point charge to find its electric potential. FIGURE 26.4 identifies a point P at $s_f = r$ at which we want to know the potential and calls this position f. We've chosen position i to be at $s_i = \infty$ and identified that as the zero point of the potential. The integration of Equation 26.3 is straight inward along the radial line from i to f:

$$\Delta V = V(r) - V(\infty) = -\int_{\infty}^{r} E_s \, ds = \int_{r}^{\infty} E_s \, ds \qquad (26.5)$$

The electric field is radially outward. Its s-component is

$$E_s = \frac{1}{4\pi\epsilon_0} \frac{q}{s^2}$$

Thus the potential at distance r from a point charge q is

$$V(r) = V(\infty) + \frac{q}{4\pi\epsilon_0} \int_{r}^{\infty} \frac{ds}{s^2} = V(\infty) + \frac{q}{4\pi\epsilon_0} \frac{-1}{s} \Big|_{r}^{\infty} = 0 + \frac{1}{4\pi\epsilon_0} \frac{q}{r} \qquad (26.6)$$

We've rediscovered the potential of a point charge that you learned in Chapter 25:

$$V_{\text{point charge}} = \frac{1}{4\pi\epsilon_0} \frac{q}{r} \qquad (26.7)$$

EXAMPLE 26.2 | **The potential of a parallel-plate capacitor**

In Chapter 23, the electric field inside a capacitor was found to be

$$\vec{E} = \left(\frac{Q}{\epsilon_0 A}, \text{ from positive to negative} \right)$$

Find the electric potential inside the capacitor. Let $V = 0$ V at the negative plate.

MODEL The electric field inside a capacitor is a uniform field.

VISUALIZE FIGURE 26.5 shows the capacitor and establishes a point P where we want to find the potential. We've chosen an s-axis measured from the negative plate, which is the zero point of the potential.

SOLVE We'll integrate along the s-axis from $s_i = 0$ (where $V_i = 0$ V) to $s_f = s$. Notice that $\vec{E}$ points in the negative s-direction, so $E_s = -Q/\epsilon_0 A$. $Q/\epsilon_0 A$ is a constant, so

$$V(s) = V_f = V_i - \int_{0}^{s} E_s \, ds = -\left(-\frac{Q}{\epsilon_0 A} \right) \int_{0}^{s} ds = \frac{Q}{\epsilon_0 A} s = Es$$

ASSESS $V = Es$ is the capacitor potential we deduced in Chapter 25 by working directly with the potential energy. The potential

FIGURE 26.5 Finding the potential inside a capacitor.

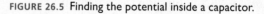

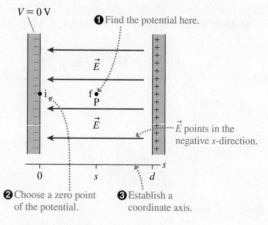

$V = 0$ V

❶ Find the potential here.

$\vec{E}$

$\vec{E}$

$\vec{E}$ points in the negative s-direction.

❷ Choose a zero point of the potential.

❸ Establish a coordinate axis.

increases linearly from $V = 0$ at the negative plate to $V = Ed$ at the positive plate. Here we found the potential by explicitly recognizing the connection between the potential and the field.

26.2 Finding the Electric Field from the Potential

FIGURE 26.6 shows two points i and f separated by a very small distance Δs, so small that the electric field is essentially constant over this very short distance. The work done by the electric field as a charge q moves through this small distance is $W = F_s \Delta s = qE_s \Delta s$. Consequently, the potential difference between these two points is

$$\Delta V = \frac{\Delta U_{q + \text{sources}}}{q} = \frac{-W}{q} = -E_s \Delta s \qquad (26.8)$$

In terms of the potential, the component of the electric field in the s-direction is $E_s = -\Delta V/\Delta s$. In the limit $\Delta s \to 0$,

$$E_s = -\frac{dV}{ds} \qquad (26.9)$$

Now we have reversed Equation 26.3 and can find the electric field from the potential. We'll begin with examples where the field is parallel to a coordinate axis, then we'll look at what Equation 26.9 tells us about the geometry of the field and the potential.

Field Parallel to a Coordinate Axis

The derivative in Equation 26.9 gives E_s, the component of the electric field parallel to the displacement $\Delta \vec{s}$. It doesn't tell us about the electric field component perpendicular to $\Delta \vec{s}$. Thus Equation 26.9 is most useful if we can use symmetry to select a coordinate axis that is parallel to $\vec{E}$ and along which the perpendicular component of $\vec{E}$ is known to be zero.

For example, suppose we knew the potential of a point charge to be $V = q/4\pi\epsilon_0 r$ but didn't remember the electric field. Symmetry requires that the field point straight outward from the charge, with only a radial component E_r. If we choose the s-axis to be in the radial direction, parallel to $\vec{E}$, we can use Equation 26.9 to find

$$E_r = -\frac{dV}{dr} = -\frac{d}{dr}\left(\frac{q}{4\pi\epsilon_0 r}\right) = \frac{1}{4\pi\epsilon_0}\frac{q}{r^2} \qquad (26.10)$$

This is, indeed, the well-known electric field of a point charge.

Equation 26.9 is especially useful for a continuous distribution of charge because calculating V, which is a scalar, is usually much easier than calculating the vector $\vec{E}$ directly from the charge. Once V is known, $\vec{E}$ is found simply by taking a derivative.

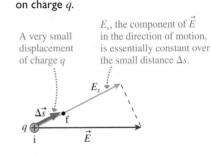

FIGURE 26.6 The electric field does work on charge q.

A very small displacement of charge q

E_s, the component of $\vec{E}$ in the direction of motion, is essentially constant over the small distance Δs.

EXAMPLE 26.3 | **The electric field of a ring of charge**

In Chapter 25, we found the on-axis potential of a ring of radius R and charge Q to be

$$V_{\text{ring}} = \frac{1}{4\pi\epsilon_0}\frac{Q}{\sqrt{z^2 + R^2}}$$

Find the on-axis electric field of a ring of charge.

SOLVE Symmetry requires the electric field along the axis to point straight outward from the ring with only a z-component E_z. The electric field at position z is

$$E_z = -\frac{dV}{dz} = -\frac{d}{dz}\left(\frac{1}{4\pi\epsilon_0}\frac{Q}{\sqrt{z^2 + R^2}}\right)$$

$$= \frac{1}{4\pi\epsilon_0}\frac{zQ}{(z^2 + R^2)^{3/2}}$$

ASSESS This result is in perfect agreement with the electric field we found in Chapter 23, but this calculation was easier because we didn't have to deal with angles.

A geometric interpretation of Equation 26.9 is that the electric field is the negative of the *slope* of the V-versus-s graph. This interpretation should be familiar. You learned in Chapter 10 that the force on a particle is the negative of the slope of the potential-energy graph: $F = -dU/ds$. In fact, Equation 26.9 is simply $F = -dU/ds$ with both sides divided by q to yield E and V. This geometric interpretation is an important step in developing an understanding of potential.

EXAMPLE 26.4 | **Finding E from the slope of V**

FIGURE 26.7 is a graph of the electric potential in a region of space where $\vec{E}$ is parallel to the x-axis. Draw a graph of E_x versus x.

FIGURE 26.7 Graph of V versus position x.

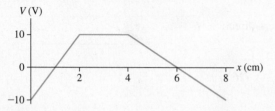

MODEL The electric field is the *negative* of the slope of the potential graph.

SOLVE There are three regions of different slope:

$$0 < x < 2 \text{ cm} \begin{cases} \Delta V/\Delta x = (20 \text{ V})/(0.020 \text{ m}) = 1000 \text{ V/m} \\ E_x = -1000 \text{ V/m} \end{cases}$$

$$2 < x < 4 \text{ cm} \begin{cases} \Delta V/\Delta x = 0 \text{ V/m} \\ E_x = 0 \text{ V/m} \end{cases}$$

$$4 < x < 8 \text{ cm} \begin{cases} \Delta V/\Delta x = (-20 \text{ V})/(0.040 \text{ m}) = -500 \text{ V/m} \\ E_x = 500 \text{ V/m} \end{cases}$$

The results are shown in **FIGURE 26.8**.

FIGURE 26.8 Graph of E_x versus position x.

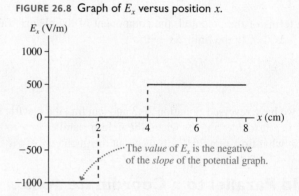

The *value* of E_x is the negative of the *slope* of the potential graph.

ASSESS The electric field $\vec{E}$ points to the left (E_x is negative) for $0 < x < 2$ cm and to the right (E_x is positive) for $4 < x < 8$ cm. Notice that **the electric field is zero in a region of space where the potential is not changing.**

STOP TO THINK 26.1 Which potential graph describes the electric field at the left?

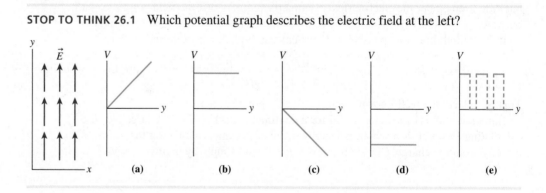

The Geometry of Potential and Field

FIGURE 26.9 The electric field at P is related to the shape of the equipotential surfaces.

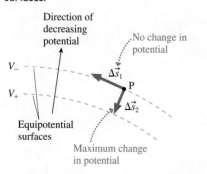

Equations 26.3 for V in terms of E_s and 26.9 for E_s in terms of V have profound implications for the geometry of the potential and the field. **FIGURE 26.9** shows two equipotential surfaces, with V_+ positive relative to V_-. To learn about the electric field $\vec{E}$ at point P, allow a charge to move through the two displacements $\Delta\vec{s}_1$ and $\Delta\vec{s}_2$. Displacement $\Delta\vec{s}_1$ is *tangent* to the equipotential surface, hence a charge moving in this direction experiences *no* potential difference. According to Equation 26.9, the electric field component along a direction of *constant* potential is $E_s = -dV/ds = 0$. In other words, **the electric field component tangent to the equipotential is $E_\parallel = 0$.**

Displacement $\Delta\vec{s}_2$ is *perpendicular* to the equipotential surface. There is a potential difference along $\Delta\vec{s}_2$, hence the electric field component is

$$E_\perp = -\frac{dV}{ds} \approx -\frac{\Delta V}{\Delta s} = -\frac{V_+ - V_-}{\Delta s_2}$$

EXERCISES AND PROBLEMS

Problems labeled ▨ integrate material from earlier chapters.

Exercises

Section 25.1 Electric Potential Energy

1. ‖ The electric field strength is 20,000 N/C inside a parallel-plate capacitor with a 1.0 mm spacing. An electron is released from rest at the negative plate. What is the electron's speed when it reaches the positive plate?

2. ‖ The electric field strength is 50,000 N/C inside a parallel-plate capacitor with a 2.0 mm spacing. A proton is released from rest at the positive plate. What is the proton's speed when it reaches the negative plate?

3. ‖ A proton is released from rest at the positive plate of a parallel-plate capacitor. It crosses the capacitor and reaches the negative plate with a speed of 50,000 m/s. What will be the final speed of an electron released from rest at the negative plate?

4. ‖ A proton is released from rest at the positive plate of a parallel-plate capacitor. It crosses the capacitor and reaches the negative plate with a speed of 50,000 m/s. The experiment is repeated with a He$^+$ ion (charge e, mass 4 u). What is the ion's speed at the negative plate?

Section 25.2 The Potential Energy of Point Charges

5. ‖ What is the potential energy of the electron-proton interactions in FIGURE EX25.5? The electrons are fixed and cannot move.

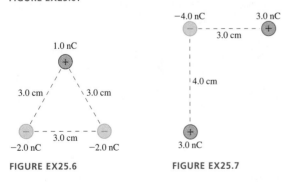

FIGURE EX25.5

6. ‖ What is the electric potential energy of the group of charges in FIGURE EX25.6?

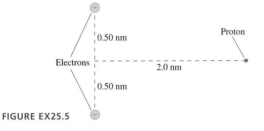

FIGURE EX25.6

FIGURE EX25.7

7. ‖ What is the electric potential energy of the group of charges in FIGURE EX25.7?

8. ‖ Two positive point charges are 5.0 cm apart. If the electric potential energy is 72 μJ, what is the magnitude of the force between the two charges?

Section 25.3 The Potential Energy of a Dipole

9. ‖ A water molecule perpendicular to an electric field has 1.0×10^{-21} J more potential energy than a water molecule aligned with the field. The dipole moment of a water molecule is 6.2×10^{-30} C m. What is the strength of the electric field?

10. ‖ FIGURE EX25.10 shows the potential energy of an electric dipole. Consider a dipole that oscillates between $\pm 60°$.
    a. What is the dipole's mechanical energy?
    b. What is the dipole's kinetic energy when it is aligned with the electric field?

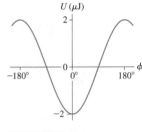

FIGURE EX25.10

Section 25.4 The Electric Potential

11. ‖ What is the speed of a proton that has been accelerated from rest through a potential difference of -1000 V?

12. ‖ What is the speed of an electron that has been accelerated from rest through a potential difference of 1000 V?

13. ‖ What potential difference is needed to accelerate an electron from rest to a speed of 2.0×10^6 m/s?

14. ‖ What potential difference is needed to accelerate a He$^+$ ion (charge $+e$, mass 4 u) from rest to a speed of 2.0×10^6 m/s?

15. ‖ A proton with an initial speed of 800,000 m/s is brought to rest by an electric field.
    a. Did the proton move into a region of higher potential or lower potential?
    b. What was the potential difference that stopped the proton?

16. ‖ An electron with an initial speed of 500,000 m/s is brought to rest by an electric field.
    a. Did the electron move into a region of higher potential or lower potential?
    b. What was the potential difference that stopped the electron?

17. ‖ Through what potential difference must a proton be accelerated to reach the speed it would have by falling 100 m in vacuum?

18. ‖ In *proton-beam therapy*, a high-energy beam of protons is BIO fired at a tumor. As the protons stop in the tumor, their kinetic energy breaks apart the tumor's DNA, thus killing the tumor cells. For one patient, it is desired to deposit 0.10 J of proton energy in the tumor. To create the proton beam, protons are accelerated from rest through a 10,000 kV potential difference. What is the total charge of the protons that must be fired at the tumor?

19. ‖ A student wants to make a very small particle accelerator using a 9.0 V battery. What speed will (a) a proton and (b) an electron have after being accelerated from rest through the 9.0 V potential difference?

20. ‖ Physicists often use a different unit of energy, the *electron volt*, when dealing with energies at the atomic level. One electron volt, abbreviated eV, is defined as the amount of kinetic energy gained by an electron upon accelerating through a 1.0 V potential difference.
    a. What is 1.0 electron volt in joules?
    b. What is the speed of a proton with 5000 eV of kinetic energy?

You can see that the electric field is inversely proportional to Δs_2, the spacing between the equipotential surfaces. Furthermore, because $(V_+ - V_-) > 0$, the minus sign tells us that the electric field is *opposite* in direction to $\Delta \vec{s}_2$. In other words, $\vec{E}$ **is perpendicular to the equipotential surfaces and points "downhill" in the direction of** *decreasing* **potential.**

These important ideas are summarized in FIGURE 26.10.

FIGURE 26.10 The geometry of the potential and the field.

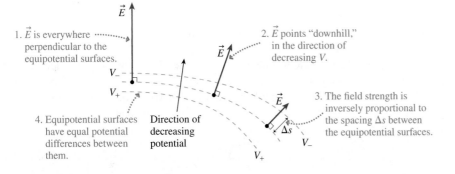

1. $\vec{E}$ is everywhere perpendicular to the equipotential surfaces.

2. $\vec{E}$ points "downhill," in the direction of decreasing V.

3. The field strength is inversely proportional to the spacing Δs between the equipotential surfaces.

4. Equipotential surfaces have equal potential differences between them.

Direction of decreasing potential

Mathematically, we can calculate the individual components of $\vec{E}$ at any point by extending Equation 26.9 to three dimensions:

$$\vec{E} = E_x \hat{\imath} + E_y \hat{\jmath} + E_z \hat{k} = -\left(\frac{\partial V}{\partial x}\hat{\imath} + \frac{\partial V}{\partial y}\hat{\jmath} + \frac{\partial V}{\partial z}\hat{k}\right) \qquad (26.11)$$

where $\partial V/\partial x$ is the partial derivative of V with respect to x while y and z are held constant. You may recognize from calculus that the expression in parentheses is the *gradient* of V, written ∇V. Thus, $\vec{E} = -\nabla V$. More advanced treatments of the electric field make extensive use of this mathematical relationship, but for the most part we'll limit our investigations to those we can analyze graphically.

EXAMPLE 26.5 | **Finding the electric field from the equipotential surfaces**

In FIGURE 26.11 a 1 cm × 1 cm grid is superimposed on a contour map of the potential. Estimate the strength and direction of the electric field at points 1, 2, and 3. Show your results graphically by drawing the electric field vectors on the contour map.

FIGURE 26.11 Equipotential lines.

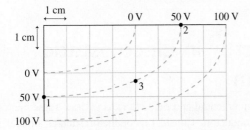

MODEL The electric field is perpendicular to the equipotential lines, points "downhill," and depends on the slope of the potential hill.

VISUALIZE The potential is highest on the bottom and the right. An elevation graph of the potential would look like the lower-right quarter of a bowl or a football stadium.

SOLVE Some distant but unseen source charges have created an electric field and potential. We do not need to see the source charges to relate the field to the potential. Because $E \approx -\Delta V/\Delta s$, the electric field is stronger where the equipotential lines are closer

together and weaker where they are farther apart. If Figure 26.11 were a topographic map, you would interpret the closely spaced contour lines at the bottom of the figure as a steep slope.

FIGURE 26.12 shows how measurements of Δs from the grid are combined with values of ΔV to determine $\vec{E}$. Point 3 requires an estimate of the spacing between the 0 V and the 100 V surfaces. Notice that we're using the 0 V and 100 V equipotential surfaces to determine $\vec{E}$ at a point on the 50 V equipotential.

FIGURE 26.12 The electric field at points 1 to 3.

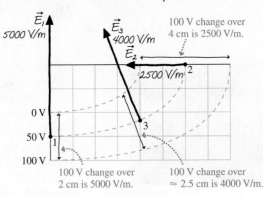

ASSESS The *directions* of $\vec{E}$ are found by drawing downhill vectors perpendicular to the equipotentials. The distances between the equipotential surfaces are needed to determine the field strengths.

Kirchhoff's Loop Law

FIGURE 26.13 The potential difference between points 1 and 2 is the same along either path.

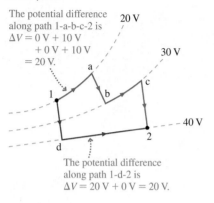

The potential difference along path 1-a-b-c-2 is
$\Delta V = 0\,V + 10\,V$
$\quad\quad + 0\,V + 10\,V$
$\quad = 20\,V.$

The potential difference along path 1-d-2 is
$\Delta V = 20\,V + 0\,V = 20\,V.$

FIGURE 26.13 shows two points, 1 and 2, in a region of electric field and potential. You learned in Chapter 25 that the work done in moving a charge between points 1 and 2 is *independent of the path*. Consequently, the potential difference between points 1 and 2 along any two paths that join them is $\Delta V = 20$ V. This must be true in order for the idea of an equipotential surface to make sense.

Now consider the path 1–a–b–c–2–d–1 that ends where it started. What is the potential difference "around" this closed path? The potential increases by 20 V in moving from 1 to 2, but then decreases by 20 V in moving from 2 back to 1. Thus $\Delta V = 0$ V around the closed path.

The numbers are specific to this example, but the idea applies to any loop (i.e., a closed path) through an electric field. The situation is analogous to hiking on the side of a mountain. You may walk uphill during parts of your hike and downhill during other parts, but if you return to your starting point your *net* change of elevation is zero. So for any path that starts and ends at the same point, we can conclude that

$$\Delta V_{\text{loop}} = \sum_i (\Delta V)_i = 0 \qquad (26.12)$$

Stated in words, **the sum of all the potential differences encountered while moving around a loop or closed path is zero.** This statement is known as **Kirchhoff's loop law.**

Kirchhoff's loop law is a statement of energy conservation because a charge that moves around a loop and returns to its starting point has $\Delta U = q\,\Delta V = 0$. Kirchhoff's loop law and a second Kirchhoff's law you'll meet in the next chapter will turn out to be the two fundamental principles of circuit analysis.

STOP TO THINK 26.2 Which set of equipotential surfaces matches this electric field?

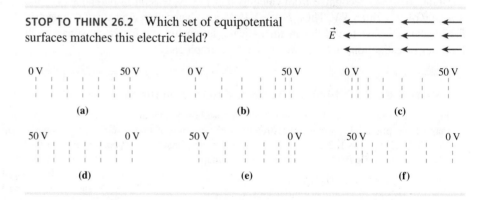

26.3 A Conductor in Electrostatic Equilibrium

The basic relationships between potential and field allow us to draw some interesting and important conclusions about conductors. Consider a conductor, such as a metal, that is in electrostatic equilibrium. The conductor may be charged, but all the charges are at rest.

You learned in Chapter 22 that any excess charges on a conductor in electrostatic equilibrium are always located on the *surface* of the conductor. Using similar reasoning, we can conclude that **the electric field is zero at any interior point of a conductor in electrostatic equilibrium**. Why? If the field were other than zero, then there would be a force $\vec{F} = q\vec{E}$ on the charge carriers and they would move, creating a current. But there are no currents in a conductor in electrostatic equilibrium, so it must be that $\vec{E} = \vec{0}$ at all interior points.

The two points inside the conductor in FIGURE 26.14 are connected by a line that remains entirely inside the conductor. We can find the potential difference $\Delta V = V_2 - V_1$

A *corona discharge* occurs at pointed metal tips where the electric field can be very strong.

between these points by using Equation 26.3 to integrate E_s along the line from 1 to 2. But $E_s = 0$ at all points along the line, because $\vec{E} = \vec{0}$; thus the value of the integral is zero and $\Delta V = 0$. In other words, **any two points inside a conductor in electrostatic equilibrium are at the same potential.**

When a conductor is in electrostatic equilibrium, the *entire conductor* is at the same potential. If we charge a metal sphere, then the entire sphere is at a single potential. Similarly, a charged metal rod or wire is at a single potential *if* it is in electrostatic equilibrium.

If $\vec{E} = \vec{0}$ inside a charged conductor but $\vec{E} \neq \vec{0}$ outside, what happens right at the surface? If the entire conductor is at the same potential, then the surface is an equipotential surface. You have seen that the electric field is always perpendicular to an equipotential surface, hence **the exterior electric field $\vec{E}$ of a charged conductor is perpendicular to the surface.**

We can also conclude that the electric field, and thus the surface charge density, is largest at sharp points. This follows from our earlier discovery that the field at the surface of a sphere of radius R can be written $E = V_0/R$. If we approximate the rounded corners of a conductor with sections of spheres, all of which are at the same potential V_0, the field strength will be largest at the corners with the smallest radii of curvature—the sharpest points.

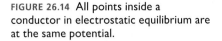

FIGURE 26.14 All points inside a conductor in electrostatic equilibrium are at the same potential.

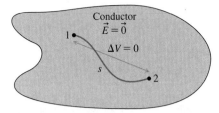

FIGURE 26.15 Properties of a conductor in electrostatic equilibrium.

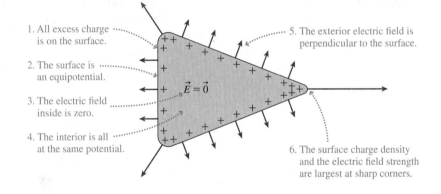

1. All excess charge is on the surface.

2. The surface is an equipotential.

3. The electric field inside is zero.

4. The interior is all at the same potential.

$\vec{E} = \vec{0}$

5. The exterior electric field is perpendicular to the surface.

6. The surface charge density and the electric field strength are largest at sharp corners.

FIGURE 26.15 summarizes what we know about conductors in electrostatic equilibrium. These are important and practical conclusions because conductors are the primary components of electrical devices.

We can use similar reasoning to estimate the electric field and potential between two charged conductors. As an example, **FIGURE 26.16** shows a negatively charged metal sphere near a flat metal plate. The surfaces of the sphere and the flat plate are equipotentials, hence the electric field must be perpendicular to both. Close to a surface, the electric field is still *nearly* perpendicular to the surface. Consequently, **an equipotential surface close to an electrode must roughly match the shape of the electrode.**

In between, the equipotential surfaces *gradually* change as they "morph" from one electrode shape to the other. It's not hard to sketch a contour map showing a plausible set of equipotential surfaces. You can then draw electric field lines (field lines are easier to draw than field vectors) that are perpendicular to the equipotentials, point "downhill," and are closer together where the contour line spacing is smaller.

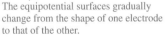

FIGURE 26.16 Estimating the field and potential between two charged conductors.

The field lines are perpendicular to the equipotential surfaces.

0 V   50 V

10 V   40 V

20 V   30 V

The equipotential surfaces gradually change from the shape of one electrode to that of the other.

STOP TO THINK 26.3 Three charged metal spheres of different radii are connected by a thin metal wire. The potential and electric field at the surface of each sphere are V and E. Which of the following is true?

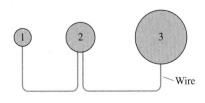

a. $V_1 = V_2 = V_3$ and $E_1 = E_2 = E_3$

b. $V_1 = V_2 = V_3$ and $E_1 > E_2 > E_3$

c. $V_1 > V_2 > V_3$ and $E_1 = E_2 = E_3$

d. $V_1 > V_2 > V_3$ and $E_1 > E_2 > E_3$

e. $V_3 > V_2 > V_1$ and $E_3 = E_2 = E_1$

f. $V_3 > V_2 > V_1$ and $E_3 > E_2 > E_1$

26.4 Sources of Electric Potential

We've now studied many properties of the electric potential and seen how potential and field are connected, but we've not said much about how an electric potential is created. Simply put, **an electric potential difference is created by separating positive and negative charge.** Shuffling your feet on the carpet transfers electrons from the carpet to you, creating a potential difference between you and a doorknob that causes a spark and a shock as you touch it. Charging a capacitor by moving electrons from one plate to the other creates a potential difference across the capacitor.

As FIGURE 26.17 shows, moving charge from one electrode to another creates an electric field $\vec{E}$ pointing from the positive toward the negative electrode. As a consequence, there is a potential difference between the electrodes that is given by

$$\Delta V = V_{\text{pos}} - V_{\text{neg}} = -\int_{\text{neg}}^{\text{pos}} E_s \, ds$$

where the integral runs from any point on the negative electrode to any point on the positive. The *net* charge is zero, but pulling the positive and negative charge apart creates a potential difference.

Now electric forces try to bring positive and negative charges together, so **a nonelectrical process is needed to separate charge.** As an example, the **Van de Graaff generator** shown in FIGURE 26.18a separates charges mechanically. A moving plastic or leather belt is charged, then the charge is mechanically transported via the conveyor belt to the spherical electrode at the top of the insulating column. The charging of the belt could be done by friction, but in practice a *corona discharge* created by the strong electric field at the tip of a needle is more efficient and reliable.

FIGURE 26.17 A charge separation creates a potential difference.

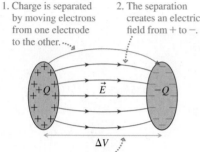

1. Charge is separated by moving electrons from one electrode to the other.
2. The separation creates an electric field from + to −.
3. Because of the electric field, there's a potential difference between the electrodes.

FIGURE 26.18 A Van de Graaff generator.

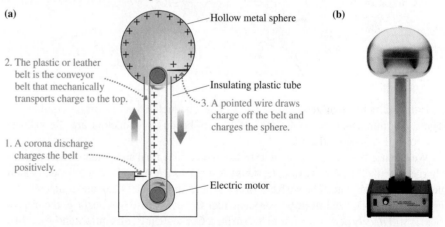

(a)

Hollow metal sphere

2. The plastic or leather belt is the conveyor belt that mechanically transports charge to the top.

Insulating plastic tube

3. A pointed wire draws charge off the belt and charges the sphere.

1. A corona discharge charges the belt positively.

Electric motor

(b)

A Van de Graaff generator has two noteworthy features:

- Charge is *mechanically* transported from the negative side to the positive side. This charge separation creates a potential difference between the spherical electrode and its surroundings.
- The electric field of the spherical electrode exerts a downward force on the positive charges moving up the belt. Consequently, *work must be done* to "lift" the positive charges. The work is done by the electric motor that runs the belt.

A classroom-demonstration Van de Graaff generator like the one shown in FIGURE 26.18b creates a potential difference of several hundred thousand volts between the upper sphere and its surroundings. The maximum potential is reached when the electric field near the sphere becomes large enough to cause a breakdown of the air. This produces a spark and temporarily discharges the sphere. A large Van de Graaff generator surrounded by vacuum can reach a potential of 20 MV or more. These generators are used to accelerate protons for nuclear physics experiments.

Batteries and emf

The most common source of electric potential is a **battery,** which uses *chemical reactions* to separate charge. A battery consists of chemicals, called *electrolytes,* sandwiched between two electrodes made of different metals. Chemical reactions in the electrolytes transport ions (i.e., charges) from one electrode to the other. This chemical process pulls positive and negative charges apart, creating a potential difference between the terminals of the battery. When the chemicals are used up, the reactions cease and the battery is dead.

We can sidestep the chemistry details by introducing the **charge escalator model** of a battery.

MODEL 26.1

Charge escalator model of a battery

A battery uses chemical reactions to separate charge.

- The charge escalator "lifts" positive charges from the negative terminal to the positive terminal. This requires *work*, with the energy being supplied by the chemical reactions.
- The work done *per charge* is called the **emf** of the battery: $\mathcal{E} = W_{chem}/q$.
- The charge separation creates a potential difference ΔV_{bat} between the terminals. An *ideal battery* has $\Delta V_{bat} = \mathcal{E}$.
- Limitations: $\Delta V_{bat} < \mathcal{E}$ if current flows through the battery. In most cases, the difference is small and a battery can be considered ideal.

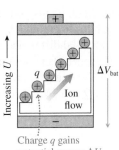

Charge q gains potential energy ΔU.

Emf is pronounced as the sequence of letters e-m-f. The symbol for emf is $\mathcal{E}$, a script E, and the units of emf are joules per coulomb, or volts. The rating of a battery, such as 1.5 V or 9 V, is the battery's emf.

The key idea is that **emf is work,** specifically the work done *per charge* to pull positive and negative charges apart. This work can be done by mechanical forces, chemical reactions, or—as you'll see later—magnetic forces. *Because* work is done, charges gain potential energy and their separation creates a potential difference ΔV_{bat} between the positive and negative terminals of the battery. This is called the **terminal voltage.**

In an **ideal battery,** which has no internal energy losses, the work W_{chem} done to move charge q from the negative to the positive terminal goes entirely to increasing the potential energy of the charge, and so $\Delta V_{bat} = \mathcal{E}$. In practice, the terminal voltage is slightly less than the emf when current flows through a battery—we'll discuss this in Chapter 28—but the difference usually small and in most cases we can model batteries as being ideal.

Batteries in Series

Many consumer goods, from flashlights to digital cameras, use more than one battery. Why? A particular type of battery, such as an AA or AAA battery, produces a fixed emf determined by the chemical reactions inside. The emf of one battery, often 1.5 V, is not sufficient to light a lightbulb or power a camera. But just as you can reach the third floor of a building by taking three escalators in succession, we can produce a larger potential difference by placing two or more batteries *in series.*

FIGURE 26.19 shows two batteries with the positive terminal of one literally touching the negative terminal of the next. Flashlight batteries usually are arranged like this.

FIGURE 26.19 Batteries in series.

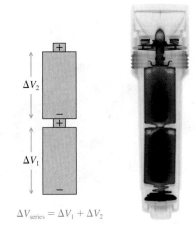

$\Delta V_{series} = \Delta V_1 + \Delta V_2$

Other devices, such as cameras, achieve the same effect by using conducting metal wires between one battery and the next. Either way, the total potential difference of batteries in series is simply the sum of their individual terminal voltages:

$$\Delta V_{\text{series}} = \Delta V_1 + \Delta V_2 + \cdots \qquad \text{(batteries in series)} \qquad (26.13)$$

STOP TO THINK 26.4 What total potential difference is created by these three batteries?

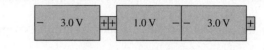

26.5 Capacitance and Capacitors

FIGURE 26.20 shows two electrodes that have been charged to $\pm Q$. Their net charge is zero, but something has separated positive and negative charges. Consequently, there is a potential difference ΔV between the electrodes.

It seems plausible that ΔV is directly proportional to Q. That is, doubling the amount of charge on the electrodes will double the potential difference. We can write this as $Q = C\Delta V$, where the proportionality constant

$$C = \frac{Q}{\Delta V_C} \qquad (26.14)$$

is called the **capacitance** of the two electrodes. The two electrodes themselves form a **capacitor,** so we've written a subscript C on ΔV_C to indicate that this is the **capacitor voltage,** the potential difference between the positive and negative electrodes.

> **NOTE** You've already met the parallel-plate capacitor, but a capacitor can be formed from any two electrodes. The electrodes of a capacitor always have *equal but opposite charges* (zero net charge), and the Q appearing in equations is the magnitude (always positive) of this amount of charge.

The SI unit of capacitance is the **farad,** named in honor of Michael Faraday. One farad is defined as

$$1 \text{ farad} = 1 \text{ F} = 1 \text{ C/V}$$

One farad turns out to be an enormous amount of capacitance. Practical capacitors are usually measured in units of microfarads (μF) or picofarads ($1 \text{ pF} = 10^{-12} \text{ F}$).

Turning Equation 26.14 around, we see that the amount of charge on a capacitor that has been charged to ΔV_C is

$$Q = C\Delta V_C \qquad \text{(charge on a capacitor)} \qquad (26.15)$$

The amount of charge is determined jointly by the potential difference *and* by a property of the electrodes called capacitance. As we'll see, **capacitance depends only on the geometry of the electrodes.**

The Parallel-Plate Capacitor

A parallel-plate capacitor consists of two flat electrodes (the plates) facing each other with a plate separation d that is small compared to the sizes of the plates. You learned in Chapter 25 that the potential difference across a parallel-plate capacitor is related to the electric field inside by $\Delta V_C = Ed$. And you know from Chapter 23 that the electric field inside a parallel-plate capacitor is

$$E = \frac{Q}{\epsilon_0 A} \qquad (26.16)$$

FIGURE 26.20 Two equally but oppositely charged electrodes form a capacitor.

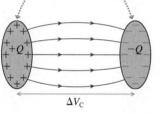

The separated charge has created a potential difference even though the net charge is zero.

Capacitors are important elements in electric circuits.

where A is the surface area of the plates. Combining these gives

$$Q = \frac{\epsilon_0 A}{d} \Delta V_C \qquad (26.17)$$

You can see that the charge is proportional to the potential difference, as expected. So from the definition of capacitance, Equation 26.14, we find that the capacitance of a parallel-plate capacitor is

$$C = \frac{Q}{\Delta V_C} = \frac{\epsilon_0 A}{d} \qquad \text{(parallel-plate capacitor)} \qquad (26.18)$$

The capacitance is a purely *geometric* property of the electrodes, depending only on their surface area and spacing. Capacitors of other shapes will have different formulas for their capacitance, but all will depend entirely on geometry. A cylindrical capacitor is the topic of Challenge Example 26.11, and a homework problem will let you analyze a spherical capacitor.

EXAMPLE 26.6 | **Charging a capacitor**

The spacing between the plates of a 1.0 μF capacitor is 0.050 mm.

a. What is the surface area of the plates?

b. How much charge is on the plates if this capacitor is charged to 1.5 V?

MODEL Assume the capacitor is a parallel-plate capacitor.

SOLVE a. From the definition of capacitance,

$$A = \frac{dC}{\epsilon_0} = 5.65 \text{ m}^2$$

b. The charge is $Q = C\, \Delta V_C = 1.5 \times 10^{-6} \text{ C} = 1.5\ \mu\text{C}$.

ASSESS The surface area needed to construct a 1.0 μF capacitor (a fairly typical value) is enormous. We'll see in Section 26.7 how the area can be reduced by inserting an insulator between the capacitor plates.

Charging a Capacitor

All well and good, but *how* does a capacitor get charged? By connecting it to a battery! FIGURE 26.21a shows the two plates of a capacitor shortly after two conducting wires have connected them to the two terminals of a battery. At this instant, the battery's charge escalator is moving charge from one capacitor plate to the other, and it is this work done by the battery that charges the capacitor. (The connecting wires are conductors, and you learned in Chapter 22 that charges can move through conductors as a *current*.) The capacitor voltage ΔV_C steadily increases as the charge separation continues.

FIGURE 26.21 A parallel-plate capacitor is charged by a battery.

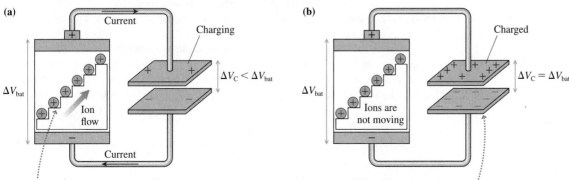

(a) Current Charging $\Delta V_C < \Delta V_{bat}$ ΔV_{bat} Ion flow Current

The charge escalator moves charge from one plate to the other. ΔV_C increases as the charge separation increases.

(b) Charged $\Delta V_C = \Delta V_{bat}$ ΔV_{bat} Ions are not moving

When $\Delta V_C = \Delta V_{bat}$, the current stops and the capacitor is fully charged.

The keys on computer keyboards are capacitor switches. Pressing the key pushes two capacitor plates closer together, increasing their capacitance. A larger capacitor can hold more charge, so a momentary current carries charge from the battery (or power supply) to the capacitor. This current is sensed, and the keystroke is then recorded.

But this process cannot continue forever. The growing positive charge on the upper capacitor plate exerts a repulsive force on new charges coming up the escalator, and eventually the capacitor charge gets so large that no new charges can arrive. The capacitor in FIGURE 26.21b is now *fully charged*. In Chapter 28 we'll analyze how long the charging process takes, but it is typically less than a nanosecond for a capacitor connected directly to a battery with copper wires.

Once the capacitor is fully charged, with charges no longer in motion, the positive capacitor plate, the upper wire, and the positive terminal of the battery form a single conductor in electrostatic equilibrium. This is an important idea, and it wasn't true while the capacitor was charging. As you just learned, any two points in a conductor in electrostatic equilibrium are at the same potential. **Thus the positive plate of a fully charged capacitor is at the same potential as the positive terminal of the battery.**

Similarly, the negative plate of a fully charged capacitor is at the same potential as the negative terminal of the battery. Consequently, the potential difference ΔV_C between the capacitor plates exactly matches the potential difference ΔV_{bat} between the battery terminals. **A capacitor attached to a battery charges until $\Delta V_C = \Delta V_{bat}$.** Once the capacitor is charged, you can disconnect it from the battery; it will maintain this charge and potential difference until and unless something—a current—allows positive charge to move back to the negative plate. An ideal capacitor in vacuum would stay charged forever.

Combinations of Capacitors

Two or more capacitors are often joined together. FIGURE 26.22 illustrates two basic combinations: **parallel capacitors** and **series capacitors.** Notice that a capacitor, no matter what its actual geometric shape, is represented in *circuit diagrams* by two parallel lines.

FIGURE 26.22 Parallel and series capacitors.

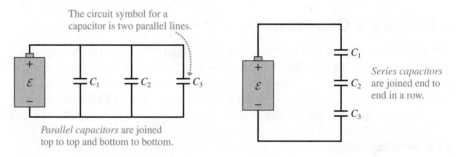

FIGURE 26.23 Replacing two parallel capacitors with an equivalent capacitor.

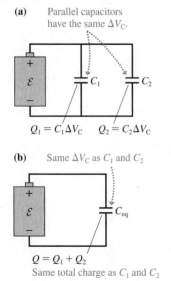

(a) Parallel capacitors have the same ΔV_C.

$Q_1 = C_1 \Delta V_C$ $Q_2 = C_2 \Delta V_C$

(b) Same ΔV_C as C_1 and C_2

$Q = Q_1 + Q_2$
Same total charge as C_1 and C_2

NOTE The terms "parallel capacitors" and "parallel-plate capacitor" do not describe the same thing. The former term describes how two or more capacitors are connected to each other, the latter describes how a particular capacitor is constructed.

As we'll show, parallel or series capacitors (or, as is sometimes said, capacitors "in parallel" or "in series") can be represented by a single **equivalent capacitance.** We'll demonstrate this first with the two parallel capacitors C_1 and C_2 of FIGURE 26.23a. Because the two top electrodes are connected by a conducting wire, they form a single conductor in electrostatic equilibrium. Thus the two top electrodes are at the same potential. Similarly, the two connected bottom electrodes are at the same potential. Consequently, two (or more) capacitors in parallel each have the *same* potential difference ΔV_C between the two electrodes.

The charges on the two capacitors are $Q_1 = C_1 \Delta V_C$ and $Q_2 = C_2 \Delta V_C$. Altogether, the battery's charge escalator moved total charge $Q = Q_1 + Q_2$ from the negative electrodes to the positive electrodes. Suppose, as in FIGURE 26.23b, we replaced the two capacitors with a single capacitor having charge $Q = Q_1 + Q_2$ and potential difference ΔV_C. This capacitor is equivalent to the original two in the sense that the battery can't tell the difference. In either case, the battery has to establish the same potential difference and move the same amount of charge.

By definition, the capacitance of this equivalent capacitor is

$$C_{eq} = \frac{Q}{\Delta V_C} = \frac{Q_1 + Q_2}{\Delta V_C} = \frac{Q_1}{\Delta V_C} + \frac{Q_2}{\Delta V_C} = C_1 + C_2 \qquad (26.19)$$

This analysis hinges on the fact that **parallel capacitors each have the same potential difference ΔV_C.** We could easily extend this analysis to more than two capacitors. If capacitors $C_1, C_2, C_3, \ldots$ are in parallel, their equivalent capacitance is

$$C_{eq} = C_1 + C_2 + C_3 + \cdots \qquad \text{(parallel capacitors)} \qquad (26.20)$$

Neither the battery nor any other part of a circuit can tell if the parallel capacitors are replaced by a single capacitor having capacitance C_{eq}.

Now consider the two series capacitors in FIGURE 26.24a. The center section, consisting of the bottom plate of C_1, the top plate of C_2, and the connecting wire, is electrically isolated. The battery cannot remove charge from or add charge to this section. If it starts out with no net charge, it must end up with no net charge. As a consequence, the two capacitors in series have equal charges $\pm Q$. The battery transfers Q from the bottom of C_2 to the top of C_1. This transfer polarizes the center section, as shown, but it still has $Q_{net} = 0$.

The potential differences across the two capacitors are $\Delta V_1 = Q/C_1$ and $\Delta V_2 = Q/C_2$. The total potential difference across both capacitors is $\Delta V_C = \Delta V_1 + \Delta V_2$. Suppose, as in FIGURE 26.24b, we replaced the two capacitors with a single capacitor having charge Q and potential difference $\Delta V_C = \Delta V_1 + \Delta V_2$. This capacitor is equivalent to the original two because the battery has to establish the same potential difference and move the same amount of charge in either case.

By definition, the capacitance of this equivalent capacitor is $C_{eq} = Q/\Delta V_C$. The inverse of the equivalent capacitance is thus

$$\frac{1}{C_{eq}} = \frac{\Delta V_C}{Q} = \frac{\Delta V_1 + \Delta V_2}{Q} = \frac{\Delta V_1}{Q} + \frac{\Delta V_2}{Q} = \frac{1}{C_1} + \frac{1}{C_2} \qquad (26.21)$$

This analysis hinges on the fact that **series capacitors each have the same charge Q.** We could easily extend this analysis to more than two capacitors. If capacitors $C_1, C_2, C_3, \ldots$ are in series, their equivalent capacitance is

$$C_{eq} = \left(\frac{1}{C_1} + \frac{1}{C_2} + \frac{1}{C_3} + \cdots \right)^{-1} \qquad \text{(series capacitors)} \qquad (26.22)$$

NOTE Be careful to avoid the common error of adding the inverses but forgetting to invert the sum.

Let's summarize the key facts before looking at a numerical example:

- Parallel capacitors all have the same potential difference ΔV_C. Series capacitors all have the same amount of charge $\pm Q$.
- The equivalent capacitance of a parallel combination of capacitors is *larger* than any single capacitor in the group. The equivalent capacitance of a series combination of capacitors is *smaller* than any single capacitor in the group.

FIGURE 26.24 Replacing two series capacitors with an equivalent capacitor.

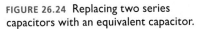

(a) Series capacitors have the same Q.

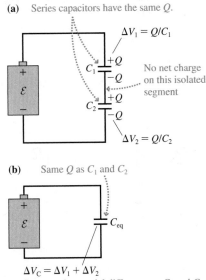

$\Delta V_1 = Q/C_1$

No net charge on this isolated segment

$\Delta V_2 = Q/C_2$

(b) Same Q as C_1 and C_2

$\Delta V_C = \Delta V_1 + \Delta V_2$
Same total potential difference as C_1 and C_2

EXAMPLE 26.7 | A capacitor circuit

Find the charge on and the potential difference across each of the three capacitors in FIGURE 26.25.

FIGURE 26.25 A capacitor circuit.

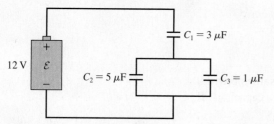

MODEL Assume the battery is ideal, with $\Delta V_{bat} = \mathcal{E} = 12$ V. Use the results for parallel and series capacitors.

SOLVE The three capacitors are neither in parallel nor in series, but we can break them down into smaller groups that are. A useful method of *circuit analysis* is first to combine elements until reaching a single equivalent element, then to reverse the process and calculate values for each element. FIGURE 26.26 shows the analysis of this circuit. Notice that we redraw the circuit after every step. The equivalent capacitance of the 3 μF and 6 μF capacitors in series is found from

$$C_{eq} = \left(\frac{1}{3\ \mu F} + \frac{1}{6\ \mu F}\right)^{-1} = \left(\frac{2}{6} + \frac{1}{6}\right)^{-1} \mu F = 2\ \mu F$$

Once we get to the single equivalent capacitance, we find that $\Delta V_C = \Delta V_{bat} = 12$ V and $Q = C\Delta V_C = 24$ μC. Now we can reverse direction. Capacitors in series all have the same charge, so the charge on C_1 and on C_{2+3} is ± 24 μC. This is enough to determine that $\Delta V_1 = 8$ V and $\Delta V_{2+3} = 4$ V. Capacitors in parallel all have the same potential difference, so $\Delta V_2 = \Delta V_3 = 4$ V. This is enough to find that $Q_2 = 20$ μC and $Q_3 = 4$ μC. The charge on and the potential difference across each of the three capacitors are shown in the final step of Figure 26.26.

ASSESS Notice that we had two important checks of internal consistency. $\Delta V_1 + \Delta V_{2+3} = 8$ V + 4 V add up to the 12 V we had found for the 2 μF equivalent capacitor. Then $Q_2 + Q_3 = 20$ μC + 4 μC add up to the 24 μC we had found for the 6 μF equivalent capacitor. We'll do much more circuit analysis of this type in Chapter 28, but it's worth noting now that circuit analysis becomes nearly foolproof *if* you make use of these checks of internal consistency.

FIGURE 26.26 Analyzing the capacitor circuit.

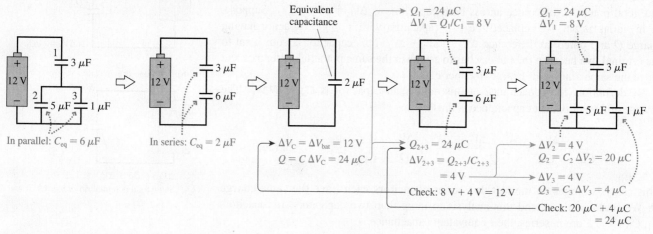

STOP TO THINK 26.5 Rank in order, from largest to smallest, the equivalent capacitance $(C_{eq})_a$ to $(C_{eq})_d$ of circuits a to d.

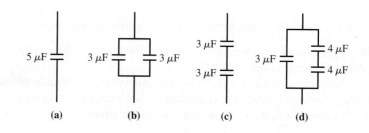

26.6 The Energy Stored in a Capacitor

Capacitors are important elements in electric circuits because of their ability to store energy. FIGURE 26.27 shows a capacitor being charged. The instantaneous value of the charge on the two plates is $\pm q$, and this charge separation has established a potential difference $\Delta V = q/C$ between the two electrodes.

An additional charge dq is in the process of being transferred from the negative to the positive electrode. The battery's charge escalator must do work to lift charge dq "uphill" to a higher potential. Consequently, the potential energy of dq + capacitor increases by

$$dU = dq\,\Delta V = \frac{q\,dq}{C} \qquad (26.23)$$

NOTE Energy must be conserved. This increase in the capacitor's potential energy is provided by the battery.

The total energy transferred from the battery to the capacitor is found by integrating Equation 26.23 from the start of charging, when $q = 0$, until the end, when $q = Q$. Thus we find that the energy stored in a charged capacitor is

$$U_C = \frac{1}{C}\int_0^Q q\,dq = \frac{Q^2}{2C} \qquad (26.24)$$

In practice, it is often easier to write the stored energy in terms of the capacitor's potential difference $\Delta V_C = Q/C$. This is

$$U_C = \frac{Q^2}{2C} = \tfrac{1}{2}C(\Delta V_C)^2 \qquad (26.25)$$

The potential energy stored in a capacitor depends on the *square* of the potential difference across it. This result is reminiscent of the potential energy $U = \tfrac{1}{2}k(\Delta x)^2$ stored in a spring, and a charged capacitor really is analogous to a stretched spring. A stretched spring holds the energy until we release it, then that potential energy is transformed into kinetic energy. Likewise, a charged capacitor holds energy until we discharge it. Then the potential energy is transformed into the kinetic energy of moving charges (the current).

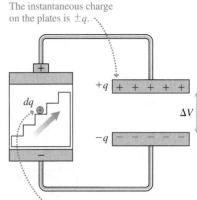

FIGURE 26.27 The charge escalator does work on charge dq as the capacitor is being charged.

The instantaneous charge on the plates is $\pm q$.

$+q$

$-q$

ΔV

dq

The charge escalator does work $dq\,\Delta V$ to move charge dq from the negative plate to the positive plate.

EXAMPLE 26.8 | Storing energy in a capacitor

How much energy is stored in a 220 μF camera-flash capacitor that has been charged to 330 V? What is the average power dissipation if this capacitor is discharged in 1.0 ms?

SOLVE The energy stored in the charged capacitor is

$$U_C = \tfrac{1}{2}C(\Delta V_C)^2 = \tfrac{1}{2}(220 \times 10^{-6}\,\text{F})(330\,\text{V})^2 = 12\,\text{J}$$

If this energy is released in 1.0 ms, the average power dissipation is

$$P = \frac{\Delta E}{\Delta t} = \frac{12\,\text{J}}{1.0 \times 10^{-3}\,\text{s}} = 12{,}000\,\text{W}$$

ASSESS The stored energy is equivalent to raising a 1 kg mass 1.2 m. This is a rather large amount of energy, which you can see by imagining the damage a 1 kg mass could do after falling 1.2 m. When this energy is released very quickly, which is possible in an electric circuit, it provides an *enormous* amount of power.

The usefulness of a capacitor stems from the fact that it can be charged slowly (the charging rate is usually limited by the battery's ability to transfer charge) but then can release the energy very quickly. A mechanical analogy would be using a crank to slowly stretch the spring of a catapult, then quickly releasing the energy to launch a massive rock.

The capacitor described in Example 26.8 is typical of the capacitors used in the flash units of cameras. The camera batteries charge a capacitor, then the energy stored in the capacitor is quickly discharged into a *flashlamp*. The charging process in a camera takes several seconds, which is why you can't fire a camera flash twice in quick succession.

An important medical application of capacitors is the *defibrillator*. A heart attack or a serious injury can cause the heart to enter a state known as *fibrillation* in which the heart muscles twitch randomly and cannot pump blood. A strong electric shock through the chest completely stops the heart, giving the cells that control the heart's rhythm a chance to restore the proper heartbeat. A defibrillator has a large capacitor that can store

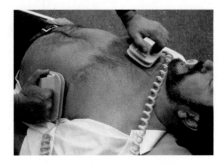

A defibrillator, which can restore a normal heartbeat, discharges a capacitor through the patient's chest.

up to 360 J of energy. This energy is released in about 2 ms through two "paddles" pressed against the patient's chest. It takes several seconds to charge the capacitor, which is why, on television medical shows, you hear an emergency room doctor or nurse shout "Charging!"

The Energy in the Electric Field

We can "see" the potential energy of a stretched spring in the tension of the coils. If a charged capacitor is analogous to a stretched spring, where is the stored energy? It's in the electric field!

FIGURE 26.28 shows a parallel-plate capacitor in which the plates have area A and are separated by distance d. The potential difference across the capacitor is related to the electric field inside the capacitor by $\Delta V_C = Ed$. The capacitance, which we found in Equation 26.18, is $C = \epsilon_0 A/d$. Substituting these into Equation 26.25, we find that the energy stored in the capacitor is

$$U_C = \tfrac{1}{2}C(\Delta V_C)^2 = \frac{\epsilon_0 A}{2d}(Ed)^2 = \frac{\epsilon_0}{2}(Ad)E^2 \tag{26.26}$$

The quantity Ad is the volume *inside* the capacitor, the region in which the capacitor's electric field exists. (Recall that an ideal capacitor has $\vec{E} = \vec{0}$ everywhere except between the plates.) Although we talk about "the energy stored in the capacitor," Equation 26.26 suggests that, strictly speaking, **the energy is stored in the capacitor's electric field.**

Because Ad is the volume in which the energy is stored, we can define an **energy density** u_E of the electric field:

$$u_E = \frac{\text{energy stored}}{\text{volume in which it is stored}} = \frac{U_C}{Ad} = \frac{\epsilon_0}{2}E^2 \tag{26.27}$$

The energy density has units J/m^3. We've derived Equation 26.27 for a parallel-plate capacitor, but it turns out to be the correct expression for any electric field.

From this perspective, charging a capacitor stores energy in the capacitor's electric field as the field grows in strength. Later, when the capacitor is discharged, the energy is released as the field collapses.

We first introduced the electric field as a way to visualize how a long-range force operates. But if the field can store energy, the field must be real, not merely a pictorial device. We'll explore this idea further in Chapter 31, where we'll find that the energy transported by a light wave—the very real energy of warm sunshine—is the energy of electric and magnetic fields.

26.7 Dielectrics

FIGURE 26.29a shows a parallel-plate capacitor with the plates separated by vacuum, the perfect insulator. Suppose the capacitor is charged to voltage $(\Delta V_C)_0$, then disconnected from the battery. The charge on the plates will be $\pm Q_0$, where $Q_0 = C_0(\Delta V_C)_0$. We'll use a subscript 0 in this section to refer to a vacuum-insulated capacitor.

FIGURE 26.29 Vacuum-insulated and dielectric-filled capacitors.

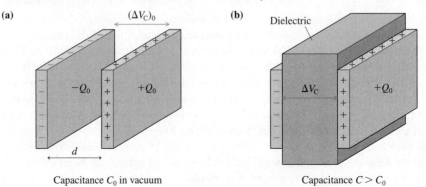

(a) Capacitance C_0 in vacuum

(b) Capacitance $C > C_0$

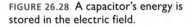

FIGURE 26.28 A capacitor's energy is stored in the electric field.

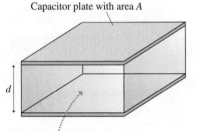

Capacitor plate with area A

d

The capacitor's energy is stored in the electric field in volume Ad between the plates.

Now suppose, as in FIGURE 26.29b, an insulating material, such as oil or glass or plastic, is slipped between the capacitor plates. We'll assume for now that the insulator is of thickness d and completely fills the space. An insulator in an electric field is called a **dielectric,** for reasons that will soon become clear, so we call this a *dielectric-filled capacitor.* How does a dielectric-filled capacitor differ from the vacuum-insulated capacitor?

The charge on the capacitor plates does not change. The insulator doesn't allow charge to move through it, and the capacitor has been disconnected from the battery, so no charge can be added to or removed from either plate. That is, $Q = Q_0$. Nonetheless, measurements of the capacitor voltage with a voltmeter would find that the voltage has decreased: $\Delta V_C < (\Delta V_C)_0$. Consequently, based on the definition of capacitance, the capacitance has increased:

$$C = \frac{Q}{\Delta V_C} > \frac{Q_0}{(\Delta V_C)_0} = C_0$$

Example 26.6 found that the plate size needed to make a 1 μF capacitor is unreasonably large. It appears that we can get more capacitance *with the same plates* by filling the capacitor with an insulator.

We can utilize two tools you learned in Chapter 23, superposition and polarization, to understand the properties of dielectric-filled capacitors. Figure 23.27 showed how an insulating material becomes *polarized* in an external electric field. FIGURE 26.30a reproduces the basic ideas from that earlier figure. The electric dipoles in Figure 26.30a could be permanent dipoles, such as water molecules, or simply induced dipoles due to a slight charge separation in the atoms. However the dipoles originate, their alignment in the electric field—the *polarization* of the material—produces an excess positive charge on one surface, an excess negative charge on the other. The insulator as a whole is still neutral, but the external electric field separates positive and negative charge.

FIGURE 26.30b represents the polarized insulator as simply two sheets of charge with surface charge densities $\pm \eta_{induced}$. The size of $\eta_{induced}$ depends both on the strength of the electric field and on the properties of the insulator. These two sheets of charge create an electric field—a situation we analyzed in Chapter 23. In essence, the two sheets of induced charge act just like the two charged plates of a parallel-plate capacitor. The **induced electric field** (keep in mind that this field is due to the insulator responding to the external electric field) is

$$\vec{E}_{induced} = \begin{cases} \left(\dfrac{\eta_{induced}}{\epsilon_0}, \text{from positive to negative} \right) & \text{inside the insulator} \\ \vec{0} & \text{outside the insulator} \end{cases} \quad (26.28)$$

It is because an insulator in an electric field has *two* sheets of induced *electric* charge that we call it a *dielectric*, with the prefix *di*, meaning *two,* the same as in "diatomic" and "dipole."

Inserting a Dielectric into a Capacitor

FIGURE 26.31 on the next page shows what happens when you insert a dielectric into a capacitor. The capacitor plates have their own surface charge density $\eta_0 = Q_0/A$. This creates the electric field $\vec{E}_0 = (\eta_0/\epsilon_0,$ from positive to negative) into which the dielectric is placed. The dielectric responds with induced surface charge density $\eta_{induced}$ and the induced electric field $\vec{E}_{induced}$. Notice that $\vec{E}_{induced}$ points *opposite* to $\vec{E}_0$. By the principle of superposition, another important lesson from Chapter 23, the net electric field between the capacitor plates is the *vector* sum of these two fields:

$$\vec{E} = \vec{E}_0 + \vec{E}_{induced} = (E_0 - E_{induced}, \text{from positive to negative}) \quad (26.29)$$

The presence of the dielectric weakens the electric field, from E_0 to $E_0 - E_{induced}$, but the field still points from the positive capacitor plate to the negative capacitor plate. The field is weakened because the induced surface charge in the dielectric acts to counter the electric field of the capacitor plates.

FIGURE 26.30 An insulator in an external electric field.

(a) The insulator is polarized.

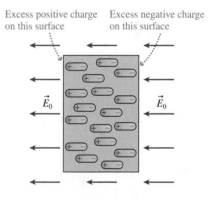

Excess positive charge on this surface   Excess negative charge on this surface

(b) The polarized insulator—a dielectric—can be represented as two sheets of surface charge. This surface charge creates an electric field inside the insulator.

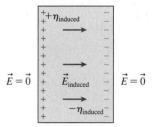

FIGURE 26.31 The consequences of filling a capacitor with a dielectric.

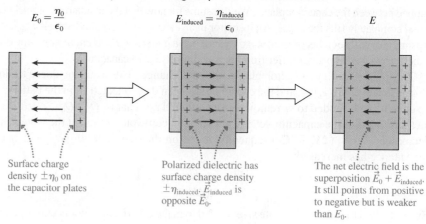

Surface charge density $\pm\eta_0$ on the capacitor plates

Polarized dielectric has surface charge density $\pm\eta_{induced}.$ $\vec{E}_{induced}$ is opposite $\vec{E}_0.$

The net electric field is the superposition $\vec{E}_0 + \vec{E}_{induced}.$ It still points from positive to negative but is weaker than $E_0.$

Let's define the **dielectric constant** κ (Greek *kappa*) as

$$\kappa \equiv \frac{E_0}{E} \qquad (26.30)$$

Equivalently, the field strength inside a dielectric in an external field is $E = E_0/\kappa$. The dielectric constant is the factor by which a dielectric *weakens* an electric field, so $\kappa \geq 1$. You can see from the definition that κ is a pure number with no units.

The dielectric constant, like density or specific heat, is a property of a material. Easily polarized materials have larger dielectric constants than materials not easily polarized. Vacuum has $\kappa = 1$ exactly, and low-pressure gases have $\kappa \approx 1$. (Air has $\kappa_{air} = 1.00$ to three significant figures, so we won't worry about the very slight effect air has on capacitors.) TABLE 26.1 lists the dielectric constants for different materials.

The electric field inside the capacitor, although weakened, is still uniform. Consequently, the potential difference across the capacitor is

$$\Delta V_C = Ed = \frac{E_0}{\kappa}d = \frac{(\Delta V_C)_0}{\kappa} \qquad (26.31)$$

where $(\Delta V_C)_0 = E_0 d$ was the voltage of the vacuum-insulated capacitor. The presence of a dielectric reduces the capacitor voltage, the observation with which we started this section. Now we see why; it is due to the polarization of the material. Further, the new capacitance is

$$C = \frac{Q}{\Delta V_C} = \frac{Q_0}{(\Delta V_C)_0/\kappa} = \kappa\frac{Q_0}{(\Delta V_C)_0} = \kappa C_0 \qquad (26.32)$$

Filling a capacitor with a dielectric increases the capacitance by a factor equal to the dielectric constant. This ranges from virtually no increase for an air-filled capacitor to a capacitance 300 times larger if the capacitor is filled with strontium titanate.

We'll leave it as a homework problem to show that the induced surface charge density is

$$\eta_{induced} = \eta_0\left(1 - \frac{1}{\kappa}\right) \qquad (26.33)$$

$\eta_{induced}$ ranges from nearly zero when $\kappa \approx 1$ to $\approx \eta_0$ when $\kappa \gg 1$.

> **NOTE** We assumed that the capacitor was disconnected from the battery after being charged, so Q couldn't change. If you insert a dielectric while a capacitor is attached to a battery, then it will be ΔV_C, fixed at the battery voltage, that can't change. In this case, more charge will flow from the battery until $Q = \kappa Q_0$. In both cases, the capacitance increases to $C = \kappa C_0$.

TABLE 26.1 Properties of dielectrics

Material	Dielectric constant κ	Dielectric strength $E_{max}\,(10^6$ V/m$)$
Vacuum	1	—
Air (1 atm)	1.0006	3
Teflon	2.1	60
Polystyrene plastic	2.6	24
Mylar	3.1	7
Paper	3.7	16
Pyrex glass	4.7	14
Pure water (20°C)	80	—
Titanium dioxide	110	6
Strontium titanate	300	8

EXAMPLE 26.9 | A water-filled capacitor

A 5.0 nF parallel-plate capacitor is charged to 160 V. It is then disconnected from the battery and immersed in distilled water. What are (a) the capacitance and voltage of the water-filled capacitor and (b) the energy stored in the capacitor before and after its immersion?

MODEL Pure distilled water is a good insulator. (The conductivity of tap water is due to dissolved ions.) Thus the immersed capacitor has a dielectric between the electrodes.

SOLVE a. From Table 26.1, the dielectric constant of water is $\kappa = 80$. The presence of the dielectric increases the capacitance to

$$C = \kappa C_0 = 80 \times 5.0 \text{ nF} = 400 \text{ nF}$$

At the same time, the voltage decreases to

$$\Delta V_C = \frac{(\Delta V_C)_0}{\kappa} = \frac{160 \text{ V}}{80} = 2.0 \text{ V}$$

b. The presence of a dielectric does not alter the derivation leading to Equation 26.25 for the energy stored in a capacitor. Right after being disconnected from the battery, the stored energy was

$$(U_C)_0 = \tfrac{1}{2} C_0 (\Delta V_C)_0^2 = \tfrac{1}{2} (5.0 \times 10^{-9} \text{ F})(160 \text{ V})^2 = 6.4 \times 10^{-5} \text{ J}$$

After being immersed, the stored energy is

$$U_C = \tfrac{1}{2} C (\Delta V_C)^2 = \tfrac{1}{2} (400 \times 10^{-9} \text{ F})(2.0 \text{ V})^2 = 8.0 \times 10^{-7} \text{ J}$$

ASSESS Water, with its large dielectric constant, has a *big* effect on the capacitor. But where did the energy go? We learned in Chapter 23 that a dipole is drawn into a region of stronger electric field. The electric field inside the capacitor is much stronger than just outside the capacitor, so the polarized dielectric is actually *pulled* into the capacitor. The "lost" energy is the work the capacitor's electric field did pulling in the dielectric.

EXAMPLE 26.10 | Energy density of a defibrillator

A defibrillator unit contains a 150 μF capacitor that is charged to 2100 V. The capacitor plates are separated by a 0.050-mm-thick insulator with dielectric constant 120.

a. What is the area of the capacitor plates?

b. What are the stored energy and the energy density in the electric field when the capacitor is charged?

MODEL Model the defibrillator as a parallel-plate capacitor with a dielectric.

SOLVE a. The capacitance of a parallel-plate capacitor in a vacuum is $C_0 = \epsilon_0 A/d$. A dielectric increases the capacitance by the factor κ, to $C = \kappa C_0$, so the area of the capacitor plates is

$$A = \frac{Cd}{\kappa \epsilon_0} = \frac{(150 \times 10^{-6} \text{ F})(5.0 \times 10^{-5} \text{ m})}{120(8.85 \times 10^{-12} \text{ C}^2/\text{N m}^2)} = 7.1 \text{ m}^2$$

Although the surface area is very large, Figure 26.32 shows how very large sheets of very thin metal can be rolled up into capacitors that you hold in your hand.

b. The energy stored in the capacitor is

$$U_C = \tfrac{1}{2} C (\Delta V_C)^2 = \tfrac{1}{2} (150 \times 10^{-6} \text{ F})(2100 \text{ V})^2 = 330 \text{ J}$$

Because the dielectric has increased C by a factor of κ, the energy density of Equation 26.27 is increased by a factor of κ to $u_E = \tfrac{1}{2} \kappa \epsilon_0 E^2$. The electric field strength in the capacitor is

$$E = \frac{\Delta V_C}{d} = \frac{2100 \text{ V}}{5.0 \times 10^{-5} \text{ m}} = 4.2 \times 10^7 \text{ V/m}$$

Consequently, the energy density is

$$u_E = \tfrac{1}{2} (120)(8.85 \times 10^{-12} \text{ C}^2/\text{N m}^2)(4.2 \times 10^7 \text{ V/m})^2$$
$$= 9.4 \times 10^5 \text{ J/m}^3$$

ASSESS 330 J is a substantial amount of energy—equivalent to that of a 1 kg mass traveling at 25 m/s. And it can be delivered very quickly as the capacitor is discharged through the patient's chest.

Solid or liquid dielectrics allow a set of electrodes to have more capacitance than they would if filled with air. Not surprisingly, as **FIGURE 26.32** shows, this is important in the production of practical capacitors. In addition, dielectrics allow capacitors to be charged to higher voltages. All materials have a maximum electric field they can sustain without *breakdown*—the production of a spark. The breakdown electric field of air, as we've noted previously, is about 3×10^6 V/m. In general, a material's maximum sustainable electric field is called its **dielectric strength.** Table 26.1 includes dielectric strengths for air and the solid dielectrics. (The breakdown of water is extremely sensitive to ions and impurities in the water, so water doesn't have a well-defined dielectric strength.)

Many materials have dielectric strengths much larger than air. Teflon, for example, has a dielectric strength 20 times that of air. Consequently, a Teflon-filled capacitor can be safely charged to a voltage 20 times larger than an air-filled capacitor with the same plate separation. An air-filled capacitor with a plate separation of 0.2 mm can be charged only to 600 V, but a capacitor with a 0.2-mm-thick Teflon sheet could be charged to 12,000 V.

FIGURE 26.32 A practical capacitor.

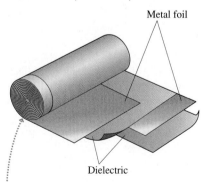

Metal foil

Dielectric

Many real capacitors are a rolled-up sandwich of metal foils and thin, insulating dielectrics.

CHALLENGE **EXAMPLE** 26.11 | The Geiger counter: A cylindrical capacitor

The radiation detector known as a *Geiger counter* consists of a 25-mm-diameter cylindrical metal tube, sealed at the ends, with a 1.0-mm-diameter wire along its axis. The wire and cylinder are separated by a low-pressure gas whose dielectric strength is 1.0×10^6 V/m.

a. What is the capacitance per unit length?

b. What is the maximum potential difference between the wire and the tube?

MODEL Model the Geiger counter as two infinitely long, concentric, conducting cylinders. Applying a potential difference between the cylinders charges them like a capacitor; indeed, they *are* a cylindrical capacitor. The charge on an infinitely long conductor is infinite, but the linear charge density λ is the charge per unit length (C/m), which is finite. So for a cylindrical capacitor we compute the capacitance per unit length $C = \lambda/\Delta V$, in F/m, rather than an absolute capacitance. To avoid breakdown of the gas, the field strength at the surface of the wire—the point of maximum field strength—must not exceed the dielectric strength.

VISUALIZE FIGURE 26.33 shows a cross section of the Geiger counter tube. We've chosen to let the outer cylinder be positive, with an inward-pointing electric field, but a negative outer cylinder would lead to the same answer since it's only the field strength that we're interested in.

FIGURE 26.33 Cross section of a Geiger counter tube.

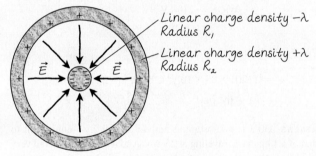

Linear charge density $-\lambda$
Radius R_1

Linear charge density $+\lambda$
Radius R_2

SOLVE a. Gauss's law tells us that the electric field between the cylinders is due only to the charge on the inner cylinder. Thus $\vec{E}$ is the field of a long, charged wire—a field we found in Chapter 23 using superposition and again in Chapter 24 using Gauss's law. It is

$$\vec{E} = \left(\frac{\lambda}{2\pi\epsilon_0 r}, \text{ inward}\right)$$

where λ is the magnitude of the linear charge density. We need to connect this field to the potential difference between the wire and the outer cylinder. For that, we need to use Equation 26.3:

$$\Delta V = V_f - V_i = -\int_{s_i}^{s_f} E_s \, ds$$

We'll integrate along a radial line from $s_i = R_1$ on the surface of the inner cylinder to $s_f = R_2$ at the outer cylinder. The field component E_s is negative because the field points inward. Thus the potential difference is

$$\Delta V = -\int_{R_1}^{R_2} \left(-\frac{\lambda}{2\pi\epsilon_0 s}\right) ds = \frac{\lambda}{2\pi\epsilon_0} \int_{R_1}^{R_2} \frac{ds}{s}$$

$$= \frac{\lambda}{2\pi\epsilon_0} \ln s \Big|_{R_1}^{R_2} = \frac{\lambda}{2\pi\epsilon_0} \ln\left(\frac{R_2}{R_1}\right)$$

We see that the applied potential difference and the linear charge density are related by

$$\frac{\lambda}{2\pi\epsilon_0} = \frac{\Delta V}{\ln(R_2/R_1)}$$

Thus the capacitance per unit length is

$$C = \frac{\lambda}{\Delta V_C} = \frac{2\pi\epsilon_0}{\ln(R_2/R_1)}$$

$$= \frac{2\pi(8.85 \times 10^{-12} \text{ C}^2/\text{N m}^2)}{\ln(25)} = 17 \text{ pF/m}$$

You should convince yourself that the units of ϵ_0 are equivalent to F/m.

b. Using the above in the expression for $\vec{E}$, we find the electric field strength at distance r is

$$E = \frac{\Delta V}{r \ln(R_2/R_1)}$$

The field strength is a maximum at the surface of the wire, where it reaches

$$E_{max} = \frac{\Delta V}{R_1 \ln(R_2/R_1)}$$

The maximum applied voltage will bring E_{max} to the dielectric strength, $E_{max} = 1.0 \times 10^6$ V/m. Thus the maximum potential difference between the wire and the tube is

$$\Delta V_{max} = R_1 E_{max} \ln\left(\frac{R_2}{R_1}\right)$$

$$= (5.0 \times 10^{-4} \text{ m})(1.0 \times 10^6 \text{ V/m}) \ln(25)$$

$$= 1600 \text{ V}$$

ASSESS This is the *maximum* possible voltage, but it's not practical to operate right at the maximum. Real Geiger counters operate with typically a 1000 V potential difference to avoid an accidental breakdown of the gas. If a high-speed charged particle from a radioactive decay then happens to pass through the tube, it will collide with and ionize a number of the gas atoms. Because the tube is already very close to breakdown, the addition of these extra ions and electrons is enough to push it over the edge: A breakdown of the gas occurs, with a spark jumping across the tube. The "clicking" sounds of a Geiger counter are made by amplifying the current pulses associated with the sparks.

SUMMARY

The goal of Chapter 26 has been to learn how the electric potential is related to the electric field.

GENERAL PRINCIPLES

Connecting V and $\vec{E}$

The electric potential and the electric field are two different perspectives of how source charges alter the space around them. V and $\vec{E}$ are related by

$$\Delta V = V_f - V_i = -\int_{s_i}^{s_f} E_s \, ds$$

where s is measured from point i to point f and E_s is the component of $\vec{E}$ parallel to the line of integration.

Graphically

$\Delta V =$ the negative of the area under the E_s graph

$$E_s = -\frac{dV}{ds}$$
$\quad=$ the negative of the slope of the potential graph

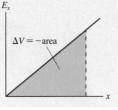

The Geometry of Potential and Field

The electric field

- Is perpendicular to the equipotential surfaces.
- Points "downhill" in the direction of decreasing V.
- Is inversely proportional to the spacing Δs between the equipotential surfaces.

Conservation of Energy

The sum of all potential differences around a closed path is zero.

$$\sum (\Delta V)_i = 0$$

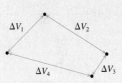

IMPORTANT CONCEPTS

A **battery** is a **source of potential**. The charge escalator in a battery uses chemical reactions to move charges from the negative terminal to the positive terminal:

$$\Delta V_{bat} = \mathcal{E}$$

where the emf $\mathcal{E}$ is the work per charge done by the charge escalator.

For a **conductor in electrostatic equilibrium**

- The interior electric field is zero.
- The exterior electric field is perpendicular to the surface.
- The surface is an equipotential.
- The interior is at the same potential as the surface.

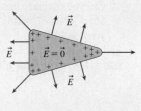

APPLICATIONS

Capacitors

The **capacitance** of two conductors charged to $\pm Q$ is

$$C = \frac{Q}{\Delta V_C}$$

A parallel-plate capacitor has

$$C = \frac{\epsilon_0 A}{d}$$

Filling the space between the plates with a **dielectric** of dielectric constant κ increases the capacitance to $C = \kappa C_0$.

The energy stored in a capacitor is $u_C = \frac{1}{2} C (\Delta V_C)^2$.

This energy is stored in the electric field at density $u_E = \frac{1}{2} \kappa \epsilon_0 E^2$.

Combinations of capacitors

Series capacitors

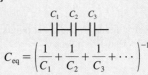

$$C_{eq} = \left(\frac{1}{C_1} + \frac{1}{C_2} + \frac{1}{C_3} + \cdots \right)^{-1}$$

Parallel capacitors

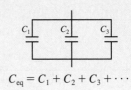

$$C_{eq} = C_1 + C_2 + C_3 + \cdots$$

TERMS AND NOTATION

Kirchhoff's loop law	terminal voltage, ΔV_{bat}	farad, F	dielectric
Van de Graaff generator	ideal battery	parallel capacitors	induced electric field
battery	capacitance, C	series capacitors	dielectric constant, κ
charge escalator model	capacitor	equivalent capacitance, C_{eq}	dielectric strength
emf, $\mathcal{E}$	capacitor voltage, ΔV_C	energy density, u_E	

CONCEPTUAL QUESTIONS

1. FIGURE Q26.1 shows the x-component of $\vec{E}$ as a function of x. Draw a graph of V versus x in this same region of space. Let $V = 0$ V at $x = 0$ m and include an appropriate vertical scale.

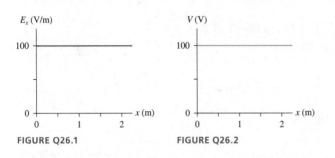

FIGURE Q26.1

FIGURE Q26.2

2. FIGURE Q26.2 shows the electric potential as a function of x. Draw a graph of E_x versus x in this same region of space.
3. a. Suppose that $\vec{E} = \vec{0}$ V/m throughout some region of space. Can you conclude that $V = 0$ V in this region? Explain.
   b. Suppose that $V = 0$ V throughout some region of space. Can you conclude that $\vec{E} = \vec{0}$ V/m in this region? Explain.
4. Estimate the electric fields $\vec{E}_1$ and $\vec{E}_2$ at points 1 and 2 in FIGURE Q26.4. Don't forget that $\vec{E}$ is a vector.

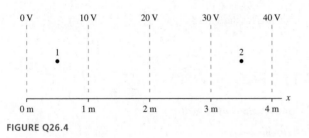

FIGURE Q26.4

5. Estimate the electric fields $\vec{E}_1$ and $\vec{E}_2$ at points 1 and 2 in FIGURE Q26.5. Don't forget that $\vec{E}$ is a vector.

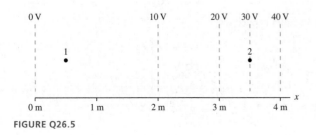

FIGURE Q26.5

6. An electron is released from rest at $x = 2$ m in the potential shown in FIGURE Q26.6. Does it move? If so, to the left or to the right? Explain.

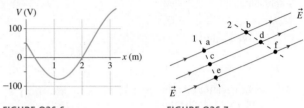

FIGURE Q26.6

FIGURE Q26.7

7. FIGURE Q26.7 shows an electric field diagram. Dashed lines 1 and 2 are two surfaces in space, not physical objects.
   a. Is the electric potential at point a higher than, lower than, or equal to the electric potential at point b? Explain.
   b. Rank in order, from largest to smallest, the magnitudes of the potential differences ΔV_{ab}, ΔV_{cd}, and ΔV_{ef}.
   c. Is surface 1 an equipotential surface? What about surface 2? Explain why or why not.
8. FIGURE Q26.8 shows a negatively charged electroscope. The gold leaf stands away from the rigid metal post. Is the electric potential of the leaf higher than, lower than, or equal to the potential of the post? Explain.

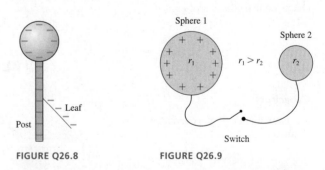

FIGURE Q26.8

FIGURE Q26.9

9. The two metal spheres in FIGURE Q26.9 are connected by a metal wire with a switch in the middle. Initially the switch is open. Sphere 1, with the larger radius, is given a positive charge. Sphere 2, with the smaller radius, is neutral. Then the switch is closed. Afterward, sphere 1 has charge Q_1, is at potential V_1, and the electric field strength at its surface is E_1. The values for sphere 2 are Q_2, V_2, and E_2.
   a. Is V_1 larger than, smaller than, or equal to V_2? Explain.
   b. Is Q_1 larger than, smaller than, or equal to Q_2? Explain.
   c. Is E_1 larger than, smaller than, or equal to E_2? Explain.

10. **FIGURE Q26.10** shows a 3 V battery with metal wires attached to each end. What are the potential differences $\Delta V_{12} = V_2 - V_1$, $\Delta V_{23} = V_3 - V_2$, $\Delta V_{34} = V_4 - V_3$, and $\Delta V_{41} = V_1 - V_4$?

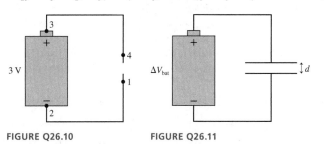

FIGURE Q26.10   **FIGURE Q26.11**

11. The parallel-plate capacitor in **FIGURE Q26.11** is connected to a battery having potential difference ΔV_{bat}. Without breaking any of the connections, insulating handles are used to increase the plate separation to $2d$.

a. Does the potential difference ΔV_C change as the separation increases? If so, by what factor? If not, why not?
b. Does the capacitance change? If so, by what factor? If not, why not?
c. Does the capacitor charge Q change? If so, by what factor? If not, why not?

12. Rank in order, from largest to smallest, the potential differences $(\Delta V_C)_1$ to $(\Delta V_C)_4$ of the four capacitors in **FIGURE Q26.12**. Explain.

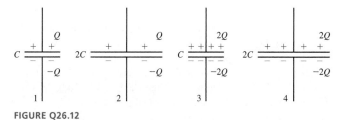

FIGURE Q26.12

EXERCISES AND PROBLEMS

Problems labeled [ ] integrate material from earlier chapters.

Exercises

Section 26.1 Connecting Potential and Field

1. ‖ What is the potential difference between $x_i = 10$ cm and $x_f = 30$ cm in the uniform electric field $E_x = 1000$ V/m?
2. ‖ What is the potential difference between $y_i = -5$ cm and $y_f = 5$ cm in the uniform electric field $\vec{E} = (20{,}000\hat{\imath} - 50{,}000\hat{\jmath})$ V/m?
3. ‖ **FIGURE EX26.3** is a graph of E_x. What is the potential difference between $x_i = 1.0$ m and $x_f = 3.0$ m?

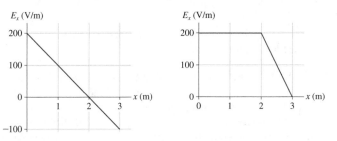

FIGURE EX26.3   **FIGURE EX26.4**

4. ‖ **FIGURE EX26.4** is a graph of E_x. The potential at the origin is -50 V. What is the potential at $x = 3.0$ m?
5. ‖ a. Which point in **FIGURE EX26.5**, A or B, has a larger electric potential?
   b. What is the potential difference between A and B?

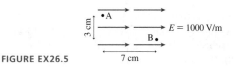

FIGURE EX26.5

Section 26.2 Finding the Electric Field from the Potential

6. ‖ Two flat, parallel electrodes 2.5 cm apart are kept at potentials of 20 V and 35 V. Estimate the electric field strength between them.

7. ‖ What are the magnitude and direction of the electric field at the dot in **FIGURE EX26.7**?

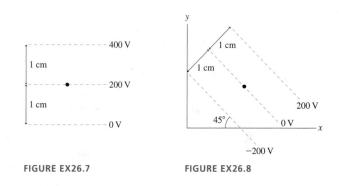

FIGURE EX26.7   **FIGURE EX26.8**

8. ‖ What are the magnitude and direction of the electric field at the dot in **FIGURE EX26.8**?
9. ‖ **FIGURE EX26.9** shows a graph of V versus x in a region of space. The potential is independent of y and z. What is E_x at (a) $x = -2$ cm, (b) $x = 0$ cm, and (c) $x = 2$ cm?

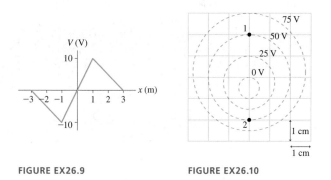

FIGURE EX26.9   **FIGURE EX26.10**

10. ‖ Determine the magnitude and direction of the electric field at points 1 and 2 in **FIGURE EX26.10**.

11. ‖ **FIGURE EX26.11** is a graph of V versus x. Draw the corresponding graph of E_x versus x.

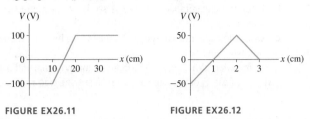

FIGURE EX26.11   **FIGURE EX26.12**

12. ‖ **FIGURE EX26.12** is a graph of V versus x. Draw the corresponding graph of E_x versus x.

13. ‖ The electric potential in a region of uniform electric field is -1000 V at $x = -1.0$ m and $+1000$ V at $x = +1.0$ m. What is E_x?

14. ‖ **CALC** The electric potential along the x-axis is $V = 100x^2$ V, where x is in meters. What is E_x at (a) $x = 0$ m and (b) $x = 1$ m?

15. ‖ **CALC** The electric potential along the x-axis is $V = 100e^{-2x}$ V, where x is in meters. What is E_x at (a) $x = 1.0$ m and (b) $x = 2.0$ m?

16. ∣ What is the potential difference ΔV_{34} in **FIGURE EX26.16**?

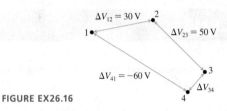

FIGURE EX26.16

Section 26.4 Sources of Electric Potential

17. ∣ How much work does the charge escalator do to move $1.0~\mu$C of charge from the negative terminal to the positive terminal of a 1.5 V battery?

18. ∣ How much charge does a 9.0 V battery transfer from the negative to the positive terminal while doing 27 J of work?

19. ∣ How much work does the electric motor of a Van de Graaff generator do to lift a positive ion ($q = e$) if the potential of the spherical electrode is 1.0 MV?

20. ∣ Light from the sun allows a solar cell to move electrons from the positive to the negative terminal, doing 2.4×10^{-19} J of work per electron. What is the emf of this solar cell?

Section 26.5 Capacitance and Capacitors

21. ‖ Two 3.0-cm-diameter aluminum electrodes are spaced 0.50 mm apart. The electrodes are connected to a 100 V battery.
    a. What is the capacitance?
    b. What is the magnitude of the charge on each electrode?

22. ‖ What is the capacitance of the two metal spheres shown in **FIGURE EX26.22**?

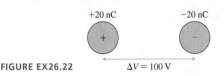

FIGURE EX26.22 $\Delta V = 100$ V

23. ∣ You need to construct a 100 pF capacitor for a science project. You plan to cut two $L \times L$ metal squares and insert small spacers between their corners. The thinnest spacers you have are 0.20 mm thick. What is the proper value of L?

24. ∣ A switch that connects a battery to a $10~\mu$F capacitor is closed. Several seconds later you find that the capacitor plates are charged to $\pm 30~\mu$C. What is the emf of the battery?

25. ∣ A $6~\mu$F capacitor, a $10~\mu$F capacitor, and a $16~\mu$F capacitor are connected in series. What is their equivalent capacitance?

26. ∣ A $6~\mu$F capacitor, a $10~\mu$F capacitor, and a $16~\mu$F capacitor are connected in parallel. What is their equivalent capacitance?

27. ∣ What is the equivalent capacitance of the three capacitors in **FIGURE EX26.27**?

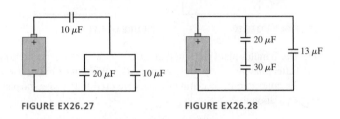

FIGURE EX26.27   **FIGURE EX26.28**

28. ∣ What is the equivalent capacitance of the three capacitors in **FIGURE EX26.28**?

29. ∣ You need a capacitance of $50~\mu$F, but you don't happen to have a $50~\mu$F capacitor. You do have a $30~\mu$F capacitor. What additional capacitor do you need to produce a total capacitance of $50~\mu$F? Should you join the two capacitors in parallel or in series?

30. ∣ You need a capacitance of $50~\mu$F, but you don't happen to have a $50~\mu$F capacitor. You do have a $75~\mu$F capacitor. What additional capacitor do you need to produce a total capacitance of $50~\mu$F? Should you join the two capacitors in parallel or in series?

Section 26.6 The Energy Stored in a Capacitor

31. ‖ To what potential should you charge a $1.0~\mu$F capacitor to store 1.0 J of energy?

32. ‖ 50 pJ of energy is stored in a $2.0~\text{cm} \times 2.0~\text{cm} \times 2.0~\text{cm}$ region of uniform electric field. What is the electric field strength?

33. ‖ A 2.0-cm-diameter parallel-plate capacitor with a spacing of 0.50 mm is charged to 200 V. What are (a) the total energy stored in the electric field and (b) the energy density?

34. ‖ **BIO** The $90~\mu$F capacitor in a defibrillator unit supplies an average of 6500 W of power to the chest of the patient during a discharge lasting 5.0 ms. To what voltage is the capacitor charged?

Section 26.7 Dielectrics

35. ‖ Two $4.0~\text{cm} \times 4.0~\text{cm}$ metal plates are separated by a 0.20-mm-thick piece of Teflon.
    a. What is the capacitance?
    b. What is the maximum potential difference between the plates?

36. ‖ Two $5.0~\text{mm} \times 5.0~\text{mm}$ electrodes are held 0.10 mm apart and are attached to a 9.0 V battery. Without disconnecting the battery, a 0.10-mm-thick sheet of Mylar is inserted between the electrodes. What are the capacitor's potential difference, electric field, and charge (a) before and (b) after the Mylar is inserted?
    Hint: Section 26.7 considered a capacitor with *isolated* plates. What changes, and what doesn't, when the plates stay connected to the battery?

37. ‖ A typical cell has a layer of negative charge on the inner
BIO surface of the cell wall and a layer of positive charge on the
outside surface, thus making the cell wall a capacitor. What is
the capacitance of a 50-μm-diameter cell with a 7.0-nm-thick
cell wall whose dielectric constant is 9.0? Because the
cell's diameter is much larger than the wall thickness, it is reasonable to ignore the curvature of the cell and think of it as a
parallel-plate capacitor.

Problems

38. ‖ The electric field in a region of space is $E_x = 5000x$ V/m,
CALC where x is in meters.
   a. Graph E_x versus x over the region -1 m $\leq x \leq 1$ m.
   b. Find an expression for the potential V at position x. As a
   reference, let $V = 0$ V at the origin.
   c. Graph V versus x over the region -1 m $\leq x \leq 1$ m.
39. ‖‖ The electric field in a region of space is $E_x = -1000x$ V/m,
CALC where x is in meters.
   a. Graph E_x versus x over the region -1 m $\leq x \leq 1$ m.
   b. What is the potential difference between $x_i = -20$ cm and
   $x_f = 30$ cm?
40. ‖ An infinitely long cylinder of radius R has linear charge den-
CALC sity λ. The potential on the surface of the cylinder is V_0, and
the electric field outside the cylinder is $E_r = \lambda/2\pi\epsilon_0 r$. Find the
potential relative to the surface at a point that is distance r from
the axis, assuming $r > R$.
41. ‖ FIGURE P26.41 is an edge view of three charged metal
electrodes. Let the left electrode be the zero point of the electric
potential. What are V and $\vec{E}$ at (a) $x = 0.5$ cm, (b) $x = 1.5$ cm,
and (c) $x = 2.5$ cm?

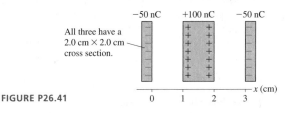

All three have a
2.0 cm × 2.0 cm
cross section.

-50 nC $+100$ nC -50 nC

FIGURE P26.41

42. ‖ Use the on-axis potential of a charged disk from Chapter 25
to find the on-axis electric field of a charged disk.
43. ‖ a. Use the methods of
   Chapter 25 to find the
   potential at distance
   x on the axis of the
   charged rod shown in
   FIGURE P26.43.
   b. Use the result of part a
   to find the electric field
   at distance x on the axis of a rod.

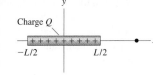

Charge Q

$-L/2$ $L/2$

FIGURE P26.43

44. ‖ It is postulated that the radial electric field of a group of
charges falls off as $E_r = C/r^n$, where C is a constant, r is the
distance from the center of the group, and n is an unknown
exponent. To test this hypothesis, you make a *field probe*
consisting of two needle tips spaced 1.00 mm apart. You
orient the needles so that a line between the tips points to the
center of the charges, then use a voltmeter to read the potential
difference between the tips. After you take measurements
at several distances from the center of the group, your data are
as follows:

Distance (cm)	Potential difference (mV)
2.0	34.7
4.0	6.6
6.0	2.1
8.0	1.2
10.0	0.6

Use an appropriate graph of the data to determine the constants
C and n.

45. ‖ Engineers discover that the electric potential between two
CALC electrodes can be modeled as $V(x) = V_0 \ln(1 + x/d)$, where V_0 is
a constant, x is the distance from the first electrode in the direction
of the second, and d is the distance between the electrodes. What
is the electric field strength midway between the electrodes?
46. ‖ The electric potential in a region of space is $V = (150x^2 -$
CALC $200y^2)$ V, where x and y are in meters. What are the strength and
direction of the electric field at $(x, y) = (2.0$ m, 2.0 m$)$? Give the
direction as an angle cw or ccw (specify which) from the positive
x-axis.
47. ‖ The electric potential in a region of space is $V = 200/\sqrt{x^2 + y^2}$,
CALC where x and y are in meters. What are the strength and direction of
the electric field at $(x, y) = (2.0$ m, 1.0 m$)$? Give the direction as
an angle cw or ccw (specify which) from the positive x-axis.
48. ‖ Consider a large, thin, electrically neutral conducting plate in
CALC the xy-plane at $z = 0$ and a point charge q on the z-axis at distance
$z_{charge} = d$. What is the electric field on and above the plate?
Although the plate is neutral, electric forces from the point charge
polarize the conducting plate and cause it to have some complex
distribution of surface charge. The electric field and potential are then
a superposition of fields and potentials due to the point charge *and*
the plate's surface charge. That's complicated! However, it is shown
in more advanced classes that the field and potential outside the plate
$(z \geq 0)$ are exactly the same as the field and potential of the original
charge q plus a "mirror image" charge $-q$ located at $z_{image} = -d$.
   a. Find an expression for the electric potential in the yz-plane for
   $z \geq 0$ (i.e., in the space above the plate).
   b. We know that electric fields are perpendicular to conductors
   in electrostatic equilibrium, so the field at the surface of the
   plate has only a z-component. Find an expression for the field
   E_z on the surface of the plate $(z = 0)$ as a function of distance
   y away from the z-axis.
   c. A $+10$ nC point charge is 2.0 cm above a large conducting
   plate. What is the electric field strength at the surface of the
   plate (i) directly beneath the point charge and (ii) 2.0 cm away
   from being directly beneath the point charge?
49. ‖ Metal sphere 1 has a positive charge of 6.0 nC. Metal sphere
2, which is twice the diameter of sphere 1, is initially uncharged.
The spheres are then connected together by a long, thin metal
wire. What are the final charges on each sphere?
50. ‖ The metal spheres in FIGURE P26.50 are charged to ± 300 V.
Draw this figure on your paper, then draw a plausible contour
map of the potential, showing and labeling the -300 V, -200 V,
-100 V, . . . , 300 V equipotential surfaces.

-300 V $+300$ V

FIGURE P26.50

51. ‖ The potential at the center of a 4.0-cm-diameter copper
sphere is 500 V, relative to $V = 0$ V at infinity. How much excess
charge is on the sphere?

52. ‖ The electric potential is 40 V at point A near a uniformly charged sphere. At point B, 2.0 μm farther away from the sphere, the potential has decreased by 0.16 mV. How far is point A from the center of the sphere?

53. ‖ Two 2.0 cm × 2.0 cm metal electrodes are spaced 1.0 mm apart and connected by wires to the terminals of a 9.0 V battery.
   a. What are the charge on each electrode and the potential difference between them?
   The wires are disconnected, and insulated handles are used to pull the plates apart to a new spacing of 2.0 mm.
   b. What are the charge on each electrode and the potential difference between them?

54. ‖ Two 2.0 cm × 2.0 cm metal electrodes are spaced 1.0 mm apart and connected by wires to the terminals of a 9.0 V battery.
   a. What are the charge on each electrode and the potential difference between them?
   While the plates are still connected to the battery, insulated handles are used to pull them apart to a new spacing of 2.0 mm.
   b. What are the charge on each electrode and the potential difference between them?

55. ‖ Find expressions for the equivalent capacitance of (a) N identical capacitors C in parallel and (b) N identical capacitors C in series.

56. ‖ What are the charge on and the potential difference across each capacitor in FIGURE P26.56?

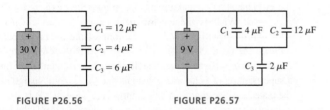

FIGURE P26.56   FIGURE P26.57

57. ‖ What are the charge on and the potential difference across each capacitor in FIGURE P26.57?

58. ‖ What are the charge on and the potential difference across each capacitor in FIGURE P26.58?

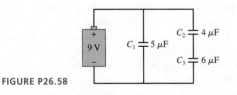

FIGURE P26.58

59. ‖ You have three 12 μF capacitors. Draw diagrams showing how you could arrange all three so that their equivalent capacitance is (a) 4.0 μF, (b) 8.0 μF, (c) 18 μF, and (d) 36 μF.

60. ‖ Six identical capacitors with capacitance C are connected as shown in FIGURE P26.60.
   a. What is the equivalent capacitance of these six capacitors?
   b. What is the potential difference between points a and b?

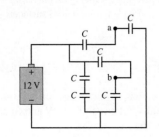

FIGURE P26.60

61. ‖ Initially, the switch in FIGURE P26.61 is in position A and capacitors C_2 and C_3 are uncharged. Then the switch is flipped to position B. Afterward, what are the charge on and the potential difference across each capacitor?

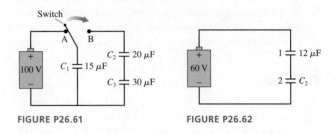

FIGURE P26.61   FIGURE P26.62

62. ‖ A battery with an emf of 60 V is connected to the two capacitors shown in FIGURE P26.62. Afterward, the charge on capacitor 2 is 450 μC. What is the capacitance of capacitor 2?

63. ‖ Capacitors $C_1 = 10$ μF and $C_2 = 20$ μF are each charged to 10 V, then disconnected from the battery without changing the charge on the capacitor plates. The two capacitors are then connected in parallel, with the positive plate of C_1 connected to the negative plate of C_2 and vice versa. Afterward, what are the charge on and the potential difference across each capacitor?

64. ‖ An isolated 5.0 μF parallel-plate capacitor has 4.0 mC of charge. An external force changes the distance between the electrodes until the capacitance is 2.0 μF. How much work is done by the external force?

65. ‖ An ideal parallel-plate capacitor has a uniform electric field between the plates, zero field outside. By superposition, half the field strength is due to one plate and half due to the other.
   a. The plates of a parallel-plate capacitor are oppositely charged and attract each other. Find an expression in terms of C, ΔV_C, and the plate separation d for the force one plate exerts on the other.
   b. What is the attractive force on each plate of a 100 pF capacitor with a 1.0 mm plate spacing when charged to 1000 V?

66. ‖ High-frequency signals are often
CALC transmitted along a *coaxial cable*, such as the one shown in FIGURE P26.66. For example, the cable TV hookup coming into your home is a coaxial cable. The signal is carried on a wire of radius R_1 while the outer conductor of radius R_2 is grounded (i.e., at $V = 0$ V). An insulating material fills the space between them, and an insulating plastic coating goes around the outside.

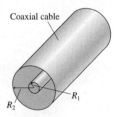

Coaxial cable

FIGURE P26.66

   a. Find an expression for the capacitance per meter of a coaxial cable. Assume that the insulating material between the cylinders is air.
   b. Evaluate the capacitance per meter of a cable having $R_1 = 0.50$ mm and $R_2 = 3.0$ mm.

67. ‖ The flash unit in a camera uses a 3.0 V battery to charge a capacitor. The capacitor is then discharged through a flashlamp. The discharge takes 10 μs, and the average power dissipated in the flashlamp is 10 W. What is the capacitance of the capacitor?

68. ‖ The label rubbed off one of the capacitors you are using to build a circuit. To find out its capacitance, you place it in series with a 10 μF capacitor and connect them to a 9.0 V battery. Using your voltmeter, you measure 6.0 V across the unknown capacitor. What is the unknown capacitor's capacitance?

69. ‖ A capacitor being charged has a *current* carrying charge to and
CALC away from the plates. In the next chapter we will define current to
be dQ/dt, the rate of charge flow. What is the current to a 10 μF
capacitor whose voltage is increasing at the rate of 2.0 V/s?

70. ‖ The current that charges a capacitor transfers energy that is
CALC stored in the capacitor's electric field. Consider a 2.0 μF capacitor,
initially uncharged, that is storing energy at a constant 200 W rate.
What is the capacitor voltage 2.0 μs after charging begins?

71. ‖ A typical cell has a membrane potential of -70 mV, meaning
BIO that the potential inside the cell is 70 mV less than the potential
outside due to a layer of negative charge on the inner surface of
the cell wall and a layer of positive charge on the outer surface.
This effectively makes the cell wall a charged capacitor. Because
a cell's diameter is much larger than the wall thickness, it is
reasonable to ignore the curvature of the cell and think of it as a
parallel-plate capacitor. How much energy is stored in the electric
field of a 50-μm-diameter cell with a 7.0-nm-thick cell wall
whose dielectric constant is 9.0?

72. ‖ A nerve cell in its resting state has a membrane potential of
BIO -70 mV, meaning that the potential inside the cell is 70 mV less
than the potential outside due to a layer of negative charge on the
inner surface of the cell wall and a layer of positive charge on the
outer surface. This effectively makes the cell wall a charged
capacitor. When the nerve cell fires, sodium ions, Na$^+$, flood
through the cell wall to briefly switch the membrane potential to
$+40$ mV. Model the central body of a nerve cell—the *soma*—as
a 50-μm-diameter sphere with a 7.0-nm-thick cell wall whose
dielectric constant is 9.0. Because a cell's diameter is much larger
than the wall thickness, it is reasonable to ignore the curvature of
the cell and think of it as a parallel-plate capacitor. How many
sodium ions enter the cell as it fires?

73. ‖ Derive Equation 26.33 for the induced surface charge density
on the dielectric in a capacitor.

74. ‖ A vacuum-insulated parallel-plate capacitor with plate
separation d has capacitance C_0. What is the capacitance if an
insulator with dielectric constant κ and thickness $d/2$ is slipped
between the electrodes without changing the plate separation?

In Problems 75 through 77 you are given the equation(s) used to solve
a problem. For each of these, you are to
  a. Write a realistic problem for which this is the correct equation(s).
  b. Finish the solution of the problem.

75. $2az$ V/m $= -\dfrac{dV}{dz}$, where a is a constant with units of V/m^2
$V(z = 0) = 10$ V

76. 400 nC $= (100$ V$)C$

$C = \dfrac{(8.85 \times 10^{-12} \text{ C}^2/\text{N m}^2)(0.10 \text{ m} \times 0.10 \text{ m})}{d}$

77. $\left(\dfrac{1}{3 \mu F} + \dfrac{1}{6 \mu F}\right)^{-1} + C = 4 \mu F$

Challenge Problems

78. ‖‖ Two 5.0-cm-diameter metal disks separated by a 0.50-mm-
thick piece of Pyrex glass are charged to a potential difference of
1000 V. What are (a) the surface charge density on the disks and
(b) the surface charge density on the glass?

79. ‖‖ An electric dipole at the origin consists of two charges $\pm q$
CALC spaced distance s apart along the y-axis.
  a. Find an expression for the potential $V(x, y)$ at an arbitrary
     point in the xy-plane. Your answer will be in terms of q, s,
     x, and y.
  b. Use the binomial approximation to simplify your result of
     part a when $s \ll x$ and $s \ll y$.
  c. Assuming $s \ll x$ and y, find expressions for E_x and E_y, the
     components of $\vec{E}$ for a dipole.
  d. What is the on-axis field $\vec{E}$? Does your result agree with
     Equation 23.10?
  e. What is the field $\vec{E}$ on the bisecting axis? Does your result
     agree with Equation 23.11?

80. ‖‖ Charge is uniformly distributed with charge density ρ inside
CALC a very long cylinder of radius R. Find the potential difference
between the surface and the axis of the cylinder.

81. ‖‖ Consider a uniformly charged sphere of radius R and total
CALC charge Q. The electric field E_{out} *outside* the sphere $(r \geq R)$ is
simply that of a point charge Q. In Chapter 24, we used Gauss's
law to find that the electric field E_{in} *inside* the sphere $(r \leq R)$ is
radially outward with field strength

$$E_{in} = \frac{1}{4\pi\epsilon_0}\frac{Q}{R^3}r$$

  a. The electric potential V_{out} *outside* the sphere is that of a point
     charge Q. Find an expression for the electric potential V_{in} at
     position r *inside* the sphere. As a reference, let $V_{in} = V_{out}$ at
     the surface of the sphere.
  b. What is the ratio $V_{center}/V_{surface}$?
  c. Graph V versus r for $0 \leq r \leq 3R$.

82. ‖‖ a. Find an expression for the capacitance of a *spherical
CALC   capacitor*, consisting of concentric spherical shells of radii
         R_1 (inner shell) and R_2 (outer shell).
     b. A spherical capacitor with a 1.0 mm gap between the
        shells has a capacitance of 100 pF. What are the diameters
        of the two spheres?

83. ‖‖ Each capacitor in **FIGURE CP26.83** has capacitance C. What is
the equivalent capacitance between points a and b?

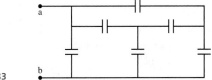

FIGURE CP26.83

27 Current and Resistance

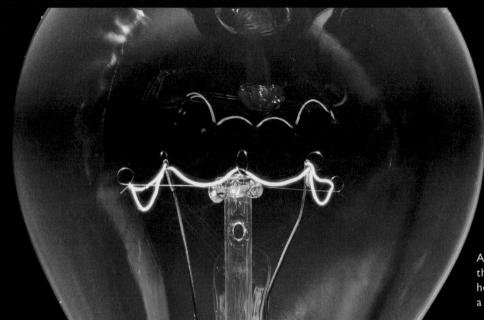

A lightbulb filament is a very thin tungsten wire that is heated until it glows by passing a current through it.

IN THIS CHAPTER, you will learn how and why charge moves through a wire as a current.

What is current?

Current is the flow of charge through a conductor. We can't see charge moving, but two indicators of current are:

- A nearby compass needle is deflected.
- A wire with a current gets warm.

Current I is measured in amperes, a charge flow rate of one coulomb per second. You know this informally as "amps."

« LOOKING BACK Section 23.6 The motion of charges in electric fields

How does current flow?

We'll develop a model of conduction:

- Connecting a wire to a battery causes a nonuniform surface charge distribution.
- The surface charges create an electric field inside the wire.
- The electric field pushes the sea of electrons through the metal.
- Electrons are the charge carriers in metals, but it is customary to treat current as the motion of *positive* charges.

Current is I. Current density $J = I/A$ is the amount of current per square meter.

What law governs current?

Current is governed by Kirchhoff's junction law.

- The current is the same everywhere in a circuit with no junctions.
- The sum of currents entering a junction equals the sum leaving.

What are resistivity and resistance?

Collisions of electrons with atoms cause a conductor to *resist* the motion of charges.

- Resistivity is an electrical property of a material, such as copper.
- Resistance is a property of a specific wire or circuit element based on the material of which it is made *and* its size and shape.

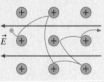

What is Ohm's law?

Ohm's law says that the current flowing through a wire or circuit element depends on both the potential difference across it and the element's resistance:

$$I = \Delta V/R$$

« LOOKING BACK Section 26.4 Sources of potential

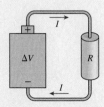
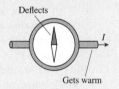

27.1 The Electron Current

We've focused thus far on situations in which charges are in static equilibrium. Now it's time to explore the *controlled motion* of charges—currents. Let's begin with a simple question: How does a capacitor get discharged? **FIGURE 27.1a** shows a charged capacitor. If, as in **FIGURE 27.1b**, we connect the two capacitor plates with a metal wire, a conductor, the plates quickly become neutral; that is, the capacitor has been *discharged*. Charge has somehow moved from one plate to the other.

FIGURE 27.1 A capacitor is discharged by a metal wire.

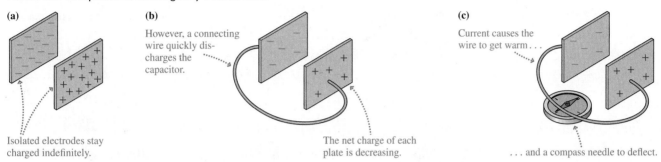

(a)

Isolated electrodes stay charged indefinitely.

(b)

However, a connecting wire quickly discharges the capacitor.

The net charge of each plate is decreasing.

(c)

Current causes the wire to get warm . . .

. . . and a compass needle to deflect.

In Chapter 22, we defined **current** as the motion of charges. It would seem that the capacitor is discharged by a current in the connecting wire. Let's see what else we can observe. **FIGURE 27.1c** shows that the connecting wire gets warm. If the wire is very thin in places, such as the thin filament in a lightbulb, the wire gets hot enough to glow. The current-carrying wire also deflects a compass needle, an observation we'll explore further in Chapter 29. For now, we will use "makes the wire warm" and "deflects a compass needle" as *indicators* that a current is present in a wire.

Charge Carriers

The charges that move in a conductor are called the *charge carriers*. **FIGURE 27.2** reminds you of the microscopic model of a metallic conductor that we introduced in Chapter 22. The outer electrons of metal atoms—the valence electrons—are only weakly bound to the nuclei. When the atoms come together to form a solid, the outer electrons become detached from their parent nuclei to form a fluid-like *sea of electrons* that can move through the solid. That is, **electrons are the charge carriers in metals.** Notice that the metal as a whole remains electrically neutral. This is not a perfect model because it overlooks some quantum effects, but it provides a reasonably good description of current in a metal.

> **NOTE** Electrons are the charge carriers in *metals*. Other conductors, such as ionic solutions or semiconductors, have different charge carriers. We will focus on metals because of their importance to circuits, but don't think that electrons are *always* the charge carrier.

The conduction electrons in a metal, like molecules in a gas, undergo random thermal motions, but there is no *net* motion. We can change that by pushing on the sea of electrons with an electric field, causing the entire sea of electrons to move in one direction like a gas or liquid flowing through a pipe. This net motion, which takes place at what we'll call the **drift speed** v_d, is superimposed on top of the random thermal motions of the individual electrons. The drift speed is quite small. As we'll establish later, 10^{-4} m/s is a fairly typical value for v_d.

As **FIGURE 27.3** shows, the entire sea of electrons moves from left to right at the drift speed. Suppose an observer could count the electrons as they pass through this cross section of the wire. Let's define the **electron current** i_e to be the number of electrons *per second* that pass through a cross section of a wire or other conductor. The

FIGURE 27.2 The sea of electrons is a model of electrons in a metal.

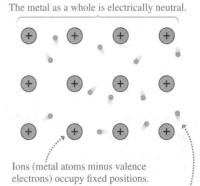

The metal as a whole is electrically neutral.

Ions (metal atoms minus valence electrons) occupy fixed positions.

The conduction electrons are bound to the solid as a whole, not any particular atom. They are free to move around.

FIGURE 27.3 The electron current.

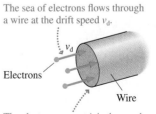

The sea of electrons flows through a wire at the drift speed v_d.

v_d

Electrons

Wire

The electron current i_e is the number of electrons passing through this cross section of the wire per second.

units of electron current are s^{-1}. Stated another way, the number N_e of electrons that pass through the cross section during the time interval Δt is

$$N_e = i_e \Delta t \qquad (27.1)$$

Not surprisingly, the electron current depends on the electrons' drift speed. To see how, **FIGURE 27.4** shows the sea of electrons moving through a wire at the drift speed v_d. The electrons passing through a particular cross section of the wire during the interval Δt are shaded. How many of them are there?

FIGURE 27.4 The sea of electrons moves to the right with drift speed v_d.

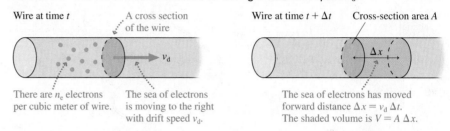

The electrons travel distance $\Delta x = v_d \Delta t$ to the right during the interval Δt, forming a cylinder of charge with volume $V = A \Delta x$. If the *number density* of conduction electrons is n_e electrons per cubic meter, then the total number of electrons in the cylinder is

$$N_e = n_e V = n_e A \Delta x = n_e A v_d \Delta t \qquad (27.2)$$

Comparing Equations 27.1 and 27.2, you can see that the electron current in the wire is

$$i_e = n_e A v_d \qquad (27.3)$$

You can increase the electron current—the number of electrons per second moving through the wire—by making them move faster, by having more of them per cubic meter, or by increasing the size of the pipe they're flowing through. That all makes sense.

In most metals, each atom contributes one valence electron to the sea of electrons. Thus the number of conduction electrons per cubic meter is the same as the number of atoms per cubic meter, a quantity that can be determined from the metal's mass density. **TABLE 27.1** gives values of the conduction-electron density n_e for several metals.

TABLE 27.1 Conduction-electron density in metals

Metal	Electron density (m^{-3})
Aluminum	6.0×10^{28}
Copper	8.5×10^{28}
Iron	8.5×10^{28}
Gold	5.9×10^{28}
Silver	5.8×10^{28}

EXAMPLE 27.1 | **The size of the electron current**

What is the electron current in a 2.0-mm-diameter copper wire if the electron drift speed is 1.0×10^{-4} m/s?

SOLVE This is a straightforward calculation. The wire's cross-section area is $A = \pi r^2 = 3.14 \times 10^{-6}$ m². Table 27.1 gives the electron density for copper as 8.5×10^{28} m^{-3}. Thus we find

$$i_e = n_e A v_d = 2.7 \times 10^{19} \text{ s}^{-1}$$

ASSESS This is an incredible number of electrons to pass through a section of the wire every second. The number is high not because the sea of electrons moves fast—in fact, it moves at literally a snail's pace—but because the density of electrons is so enormous. This is a fairly typical electron current.

STOP TO THINK 27.1 These four wires are made of the same metal. Rank in order, from largest to smallest, the electron currents i_a to i_d.

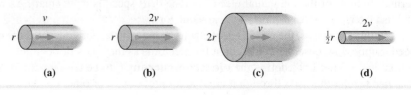

Discharging a Capacitor

FIGURE 27.5 shows a capacitor charged to ± 16 nC as it is being discharged by a 2.0-mm-diameter, 20-cm-long copper wire. *How long does it take* to discharge the capacitor? We've noted that a fairly typical drift speed of the electron current through a wire is 10^{-4} m/s. At this rate, it would take 2000 s, or about a half hour, for an electron to travel 20 cm.

But this isn't what happens. As far as our senses are concerned, the discharge of a capacitor is instantaneous. So what's wrong with our simple calculation?

The important point we overlooked is that the wire is *already full* of electrons. As an analogy, think of water in a hose. If the hose is already full of water, adding a drop to one end immediately (or very nearly so) pushes a drop out the other end. Likewise with the wire. As soon as the excess electrons move from the negative capacitor plate into the wire, they immediately (or very nearly so) push an equal number of electrons out the other end of the wire and onto the positive plate, thus neutralizing it. We don't have to wait for electrons to move all the way through the wire from one plate to the other. Instead, we just need to slightly rearrange the charges on the plates *and* in the wire.

Let's do a rough estimate of how much rearrangement is needed and how long the discharge takes. Using the conduction-electron density of copper in Table 27.1, we can calculate that there are 5×10^{22} conduction electrons in the wire. The negative plate in **FIGURE 27.6**, with $Q = -16$ nC, has 10^{11} excess electrons, far fewer than in the wire. In fact, the length of copper wire needed to hold 10^{11} electrons is a mere 4×10^{-13} m.

The instant the wire joins the capacitor plates together, the repulsive forces between the excess 10^{11} electrons on the negative plate cause them to push their way into the wire. As they do, 10^{11} electrons are squeezed out of the final 4×10^{-13} m of the wire and onto the positive plate. If the electrons all move together, and if they move at the typical drift speed of 10^{-4} m/s—both less than perfect assumptions but fine for making an estimate—it takes 4×10^{-9} s, or 4 ns, to move 4×10^{-13} m and discharge the capacitor. And, indeed, this is the right order of magnitude for how long the electrons take to rearrange themselves so that the capacitor plates are neutral.

STOP TO THINK 27.2 Why does the light in a room come on instantly when you flip a switch several meters away?

FIGURE 27.5 How long does it take to discharge this capacitor?

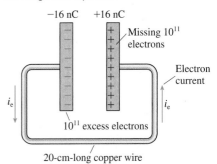

FIGURE 27.6 The sea of electrons needs only a minuscule rearrangement.

1. The 10^{11} excess electrons on the negative plate move into the wire.

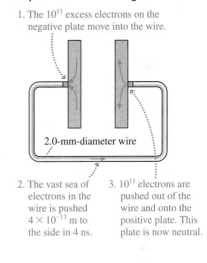

2.0-mm-diameter wire

2. The vast sea of electrons in the wire is pushed 4×10^{-13} m to the side in 4 ns.

3. 10^{11} electrons are pushed out of the wire and onto the positive plate. This plate is now neutral.

27.2 Creating a Current

Suppose you want to slide a book across the table to your friend. You give it a quick push to start it moving, but it begins slowing down because of friction as soon as you take your hand off. The book's kinetic energy is transformed into thermal energy, leaving the book and the table slightly warmer. The only way to keep the book moving at a constant speed is to *continue pushing it*.

As **FIGURE 27.7** shows, the sea of electrons is similar to the book. If you push the sea of electrons, you create a current of electrons moving through the conductor. But the electrons aren't moving in a vacuum. Collisions between the electrons and the atoms of the metal transform the electrons' kinetic energy into the thermal energy of the metal, making the metal warmer. (Recall that "makes the wire warm" is one of our indicators of a current.) Consequently, the sea of electrons will quickly slow down and stop *unless you continue pushing*. How do you push on electrons? With an electric field!

One of the important conclusions of Chapter 24 was that $\vec{E} = \vec{0}$ inside a conductor in electrostatic equilibrium. But a conductor with electrons moving through it is *not* in electrostatic equilibrium. **An electron current is a nonequilibrium motion of charges sustained by an internal electric field.**

Thus the quick answer to "What creates a current?" is "An electric field." But why is there an electric field in a current-carrying wire?

FIGURE 27.7 Sustaining the electron current with an electric field.

Because of friction, a steady push is needed to move the book at steady speed.

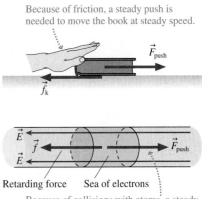

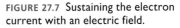

Retarding force   Sea of electrons

Because of collisions with atoms, a steady push is needed to move the sea of electrons at steady speed.

Establishing the Electric Field in a Wire

FIGURE 27.8a shows two metal wires attached to the plates of a charged capacitor. The wires are conductors, so some of the charges on the capacitor plates become spread out along the wires as a surface charge. (Remember that all excess charge on a conductor is located on the surface.)

This is an electrostatic situation, with no current and no charges in motion. Consequently—because this is always true in electrostatic equilibrium—the electric field inside the wire is zero. Symmetry requires there to be equal amounts of charge to either side of each point to make $\vec{E} = \vec{0}$ at that point; hence the surface charge density must be uniform along each wire except near the ends (where the details need not concern us). We implied this uniform density in Figure 27.8a by drawing equally spaced $+$ and $-$ symbols along the wire. Remember that a positively charged surface is a surface that is *missing* electrons.

FIGURE 27.8 The surface charge on the wires before and after they are connected.

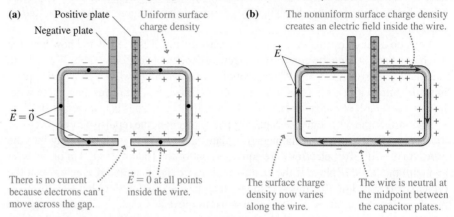

Now we connect the ends of the wires together. What happens? The excess electrons on the negative wire suddenly have an opportunity to move onto the positive wire that is missing electrons. Within a *very* brief interval of time ($\approx 10^{-9}$ s), the sea of electrons shifts slightly and the surface charge is rearranged into a *nonuniform* distribution like that shown in **FIGURE 27.8b**. The surface charge near the positive and negative plates remains strongly positive and negative because of the large amount of charge on the capacitor plates, but the midpoint of the wire, halfway between the positive and negative plates, is now electrically neutral. The new surface charge density on the wire varies from positive at the positive capacitor plate through zero at the midpoint to negative at the negative plate.

This nonuniform distribution of surface charge has an *extremely* important consequence. **FIGURE 27.9** shows a section from a wire on which the surface charge density becomes more positive toward the left and more negative toward the right. Calculating the exact electric field is complicated, but we can understand the basic idea if we *model* this section of wire with four circular rings of charge.

FIGURE 27.9 A varying surface charge distribution creates an internal electric field inside the wire.

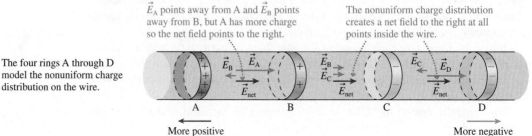

In Chapter 23, we found that the on-axis field of a ring of charge

- Points away from a positive ring, toward a negative ring;
- Is proportional to the amount of charge on the ring; and
- Decreases with distance away from the ring.

The field midway between rings A and B is well approximated as $\vec{E}_{net} \approx \vec{E}_A + \vec{E}_B$. Ring A has more charge than ring B, so $\vec{E}_{net}$ points away from A.

The analysis of Figure 27.9 leads to a very important conclusion:

> A *nonuniform* distribution of surface charges along a wire creates a net electric field *inside* the wire that points from the more positive end of the wire toward the more negative end of the wire. This is the internal electric field $\vec{E}$ that pushes the electron current through the wire.

Note that the surface charges are *not* the moving charges of the current. Further, the current—the moving charges—is *inside* the wire, not on the surface. In fact, as the next example shows, the electric field inside a current-carrying wire can be established with an extremely small amount of surface charge.

EXAMPLE 27.2 | **The surface charge on a current-carrying wire**

Table 23.1 in Chapter 23 gave a typical electric field strength in a current-carrying wire as 0.01 N/C or, as we would now say, 0.01 V/m. (We'll verify this value later in this chapter.) Two 2.0-mm-diameter rings are 2.0 mm apart. They are charged to $\pm Q$. What value of Q causes the electric field at the midpoint to be 0.010 V/m?

MODEL Use the on-axis electric field of a ring of charge from Chapter 23.

VISUALIZE FIGURE 27.10 shows the two rings. Both contribute equally to the field strength, so the electric field strength of the

positive ring is $E_+ = 0.0050$ V/m. The distance $z = 1.0$ mm is half the ring spacing.

SOLVE Chapter 23 found the on-axis electric field of a ring of charge Q to be

$$E_+ = \frac{1}{4\pi\epsilon_0} \frac{zQ}{(z^2 + R^2)^{3/2}}$$

Thus the charge needed to produce the desired field is

$$Q = \frac{4\pi\epsilon_0(z^2 + R^2)^{3/2}}{z} E_+$$

$$= \frac{\left((0.0010 \text{ m})^2 + (0.0010 \text{ m})^2\right)^{3/2}}{(9.0 \times 10^9 \text{ N m}^2/\text{C}^2)(0.0010 \text{ m})}(0.0050 \text{ V/m})$$

$$= 1.6 \times 10^{-18} \text{ C}$$

ASSESS The electric field of a ring of charge is largest at $z \approx R$, so these two rings are a simple but reasonable model for estimating the electric field inside a 2.0-mm-diameter wire. We find that the surface charge needed to establish the electric field is *very small*. A mere 10 electrons have to be moved from one ring to the other to charge them to $\pm 1.6 \times 10^{-18}$ C. The resulting electric field is sufficient to drive a sizable electron current through the wire.

FIGURE 27.10 The electric field of two charged rings.

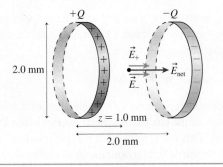

STOP TO THINK 27.3 The two charged rings are a model of the surface charge distribution along a wire. Rank in order, from largest to smallest, the electron currents E_a to E_e at the midpoint between the rings.

(a) (b) (c) (d) (e)

A Model of Conduction

Electrons don't just magically move through a wire as a current. They move because an electric field inside the wire—a field created by a nonuniform surface charge density on the wire—pushes on the sea of electrons to create the electron current. The field has to *keep* pushing because the electrons continuously lose energy in collisions with the positive ions that form the structure of the solid. These collisions provide a drag force, much like friction.

We will model the conduction electrons—those electrons that make up the sea of electrons—as free particles moving through the lattice of the metal. In the absence of an electric field, the electrons, like the molecules in a gas, move randomly in all directions with a distribution of speeds. If we assume that the average thermal energy of the electrons is given by the same $\frac{3}{2}k_BT$ that applies to an ideal gas, we can calculate that the average electron speed at room temperature is $\approx 10^5$ m/s. This estimate turns out, for quantum physics reasons, to be not quite right, but it correctly indicates that the conduction electrons are moving very fast.

However, an individual electron does not travel far before colliding with an ion and being scattered to a new direction. **FIGURE 27.11a** shows that an electron bounces back and forth between collisions, but its *average* velocity is zero, and it undergoes no *net* displacement. This is similar to molecules in a container of gas.

Suppose we now turn on an electric field. **FIGURE 27.11b** shows that the steady electric force causes the electrons to move along *parabolic trajectories* between collisions. Because of the curvature of the trajectories, the negatively charged electrons begin to drift slowly in the direction opposite the electric field. The motion is similar to a ball moving in a pinball machine with a slight downward tilt. An individual electron ricochets back and forth between the ions at a high rate of speed, but now there is a slow *net* motion in the "downhill" direction. Even so, this net displacement is a *very* small effect superimposed on top of the much larger thermal motion. Figure 27.11b has greatly exaggerated the rate at which the drift would occur.

Suppose an electron just had a collision with an ion and has rebounded with velocity $\vec{v}_0$. The acceleration of the electron between collisions is

$$a_x = \frac{F}{m} = \frac{eE}{m} \tag{27.4}$$

where E is the electric field strength inside the wire and m is the mass of the electron. (We'll assume that $\vec{E}$ points in the negative x-direction.) The field causes the x-component of the electron's velocity to increase linearly with time:

$$v_x = v_{0x} + a_x\,\Delta t = v_{0x} + \frac{eE}{m}\,\Delta t \tag{27.5}$$

The electron speeds up, with increasing kinetic energy, until its next collision with an ion. The collision transfers much of the electron's kinetic energy to the ion and thus to the thermal energy of the metal. **This energy transfer is the "friction" that raises the temperature of the wire.** The electron then rebounds, in a random direction, with a new initial velocity $\vec{v}_0$, and starts the process all over.

FIGURE 27.12a shows how the velocity abruptly changes due to a collision. Notice that the acceleration (the slope of the line) is the same before and after the collision. **FIGURE 27.12b** follows an electron through a series of collisions. You can see that each collision "resets" the velocity. The primary observation we can make from Figure 27.12b is that this repeated process of speeding up and colliding gives the electron a nonzero *average* velocity. **The magnitude of the electron's average velocity, due to the electric field, is the *drift speed* v_d of the electron.**

If we observe all the electrons in the metal at one instant of time, their average velocity is

$$v_d = \overline{v_x} = \overline{v_{0x}} + \frac{eE}{m}\overline{\Delta t} \tag{27.6}$$

FIGURE 27.11 A microscopic view of a conduction electron moving through a metal.

(a) No electric field   Ions in the lattice of the metal

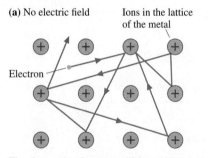

The electron has frequent collisions with ions, but it undergoes no net displacement.

(b) With an electric field   Parabolic trajectories in the electric field

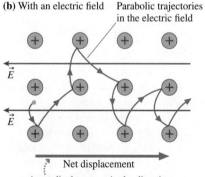

Net displacement

A net displacement in the direction opposite to $\vec{E}$ is superimposed on the random thermal motion.

Because the coulomb is the unit of charge, and because currents are charges in motion, it seemed quite natural in the 19th century to define current as the *rate,* in coulombs per second, at which charge moves through a wire. If Q is the total amount of charge that has moved past a point in the wire, we define the current I in the wire to be the rate of charge flow:

$$I \equiv \frac{dQ}{dt} \qquad (27.9)$$

For a *steady current,* which will be our primary focus, the amount of charge delivered by current I during the time interval Δt is

$$Q = I\,\Delta t \qquad (27.10)$$

The SI unit for current is the coulomb per second, which is called the **ampere** A:

$$1 \text{ ampere} = 1 \text{ A} \equiv 1 \text{ coulomb per second} = 1 \text{ C/s}$$

The current unit is named after the French scientist André Marie Ampère, who made major contributions to the study of electricity and magnetism in the early 19th century. The *amp* is an informal abbreviation of ampere. Household currents are typically ≈ 1 A. For example, the current through a 100 watt lightbulb is 0.85 A, meaning that 0.85 C of charge flow through the bulb every second. Currents in consumer electronics, such as stereos and computers, are much less. They are typically measured in milliamps $(1 \text{ mA} = 10^{-3} \text{ A})$ or microamps $(1 \text{ }\mu\text{A} = 10^{-6} \text{ A})$.

Equation 27.10 is closely related to Equation 27.1, which said that the number of electrons delivered during a time interval Δt is $N_e = i_e\,\Delta t$. Each electron has charge of magnitude e; hence the total charge of N_e electrons is $Q = eN_e$. Consequently, the conventional current I and the electron current i_e are related by

$$I = \frac{Q}{\Delta t} = \frac{eN_e}{\Delta t} = ei_e \qquad (27.11)$$

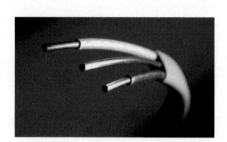

The electron current in the copper wire of Examples 27.1 and 27.3 was 2.7×10^{19} electrons/s. To find the conventional current, multiply by e to get 4.3 C/s, or 4.3 A.

Because electrons are the charge carriers, the rate at which charge moves is e times the rate at which the electrons move.

In one sense, the current I and the electron current i_e differ by only a scale factor. The electron current i_e, the rate at which electrons move through a wire, is more *fundamental* because it looks directly at the charge carriers. The current I, the rate at which the charge of the electrons moves through the wire, is more *practical* because we can measure charge more easily than we can count electrons.

Despite the close connection between i_e and I, there's one extremely important distinction. Because currents were known and studied before it was known what the charge carriers are, **the direction of current is *defined* to be the direction in which positive charges *seem* to move.** Thus the direction of the current I is the same as that of the internal electric field $\vec{E}$. But because the charge carriers turned out to be negative, at least for a metal, **the direction of the current I in a metal is opposite the direction of motion of the electrons.**

The situation shown in **FIGURE 27.13** may seem disturbing, but it makes no real difference. A capacitor is discharged regardless of whether positive charges move toward the negative plate or negative charges move toward the positive plate. The primary application of current is the analysis of circuits, and in a circuit—a macroscopic device—we simply can't tell what is moving through the wires. All of our calculations will be correct and all of our circuits will work perfectly well if we choose to think of current as the flow of positive charge. The distinction is important only at the microscopic level.

FIGURE 27.13 The current I is opposite the direction of motion of the electrons in a metal.

The current I is in the direction that positive charges would move. It is in the direction of $\vec{E}$.

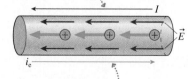

The electron current i_e is the motion of actual charge carriers. It is opposite to $\vec{E}$ and I.

The Current Density in a Wire

We found the electron current in a wire of cross-section area A to be $i_e = n_e A v_d$. Thus the current I is

$$I = ei_e = n_e ev_d A \qquad (27.12)$$

The quantity $n_e ev_d$ depends on the charge carriers and on the internal electric field that determines the drift speed, whereas A is simply a physical dimension of the wire. It will be useful to separate these quantities by defining the **current density** J in a wire as the current per square meter of cross section:

$$J = \text{current density} \equiv \frac{I}{A} = n_e ev_d \qquad (27.13)$$

The current density has units of A/m^2. A specific piece of metal, shaped into a wire with cross-section area A, carries current $I = JA$.

EXAMPLE 27.4 | **Finding the electron drift speed**

A 1.0 A current passes through a 1.0-mm-diameter aluminum wire. What are the current density and the drift speed of the electrons in the wire?

SOLVE We can find the drift speed from the current density. The current density is

$$J = \frac{I}{A} = \frac{I}{\pi r^2} = \frac{1.0 \text{ A}}{\pi (0.00050 \text{ m})^2} = 1.3 \times 10^6 \text{ A/m}^2$$

The electron drift speed is thus

$$v_d = \frac{J}{n_e e} = 1.3 \times 10^{-4} \text{ m/s} = 0.13 \text{ mm/s}$$

where the conduction-electron density for aluminum was taken from Table 27.1.

ASSESS We earlier used 1.0×10^{-4} m/s as a typical electron drift speed. This example shows where that value comes from.

Charge Conservation and Current

FIGURE 27.14 shows two identical lightbulbs in the wire connecting two charged capacitor plates. Both bulbs glow as the capacitor is discharged. How do you think the brightness of bulb A compares to that of bulb B? Is one brighter than the other? Or are they equally bright? Think about this before going on.

You might have predicted that B is brighter than A because the current I, which carries positive charges from plus to minus, reaches B first. In order to be glowing, B must use up some of the current, leaving less for A. Or perhaps you realized that the actual charge carriers are electrons, moving from minus to plus. The conventional current I may be mathematically equivalent, but physically it's the negative electrons rather than positive charge that actually move. Because the electron current gets to A first, you might have predicted that A is brighter than B.

In fact, both bulbs are equally bright. This is an important observation, one that demands an explanation. After all, "something" gets used up to make the bulb glow, so why don't we observe a decrease in the current? Current is the amount of charge moving through the wire per second. There are only two ways to decrease I: either decrease the amount of charge, or decrease the charge's drift speed through the wire. Electrons, the charge carriers, are charged particles. The lightbulb can't destroy electrons without violating both the law of conservation of mass and the law of conservation of charge. Thus the amount of charge (i.e., the *number* of electrons) cannot be changed by a lightbulb.

Do charges slow down after passing through the bulb? This is a little trickier, so consider the fluid analogy shown in **FIGURE 27.15**. Suppose the water flows into one end at a rate of 2.0 kg/s. Is it possible that the water, after turning a paddle wheel, flows out the other end at a rate of only 1.5 kg/s? That is, does turning the paddle wheel cause the water current to decrease?

We can't destroy water molecules any more than we can destroy electrons, we can't increase the density of water by pushing the molecules closer together, and there's nowhere to store extra water inside the pipe. Each drop of water entering the left end pushes a drop out the right end; hence water flows out at exactly the same rate it flows in.

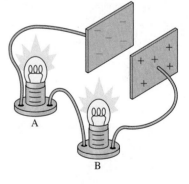

FIGURE 27.14 How does the brightness of bulb A compare to that of bulb B?

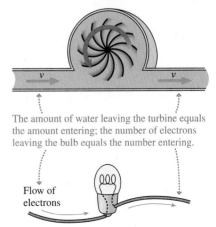

FIGURE 27.15 A current dissipates energy, but the flow is unchanged.

The amount of water leaving the turbine equals the amount entering; the number of electrons leaving the bulb equals the number entering.

Flow of electrons

The same is true for electrons in a wire. **The rate of electrons leaving a lightbulb (or any other device) is exactly the same as the rate of electrons entering the lightbulb. The current does not change.** A lightbulb doesn't "use up" current, but it *does*—like the paddlewheel in the fluid analogy—use energy. The kinetic energy of the electrons is dissipated by their collisions with the ions in the lattice of the metal (the atomic-level friction) as the electrons move through the atoms, making the wire hotter until, in the case of the lightbulb filament, it glows. The lightbulb affects the amount of current *everywhere* in the wire, a process we'll examine later in the chapter, but the current doesn't change as it passes through the bulb.

There are many issues that we'll need to look at before we can say that we understand how currents work, and we'll take them one at a time. For now, we draw a first important conclusion: **Due to conservation of charge, the current must be the same at all points in a current-carrying wire.**

FIGURE 27.16 The sum of the currents into a junction must equal the sum of the currents leaving the junction.

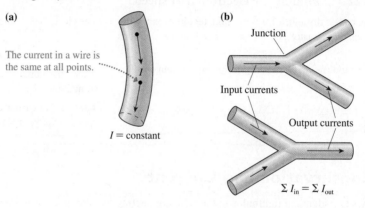

(a)

The current in a wire is the same at all points.

I

$I = $ constant

(b)

Junction

Input currents

Output currents

$\Sigma I_{in} = \Sigma I_{out}$

FIGURE 27.16a summarizes the situation in a single wire. But what about FIGURE 27.16b, where two wires merge into one and another wire splits into two? A point where a wire branches is called a **junction.** The presence of a junction doesn't change our basic reasoning. We cannot create or destroy electrons in the wire, and neither can we store them in the junction. The rate at which electrons flow into one *or many* wires must be exactly balanced by the rate at which they flow out of others. For a *junction*, the law of conservation of charge requires that

$$\sum I_{in} = \sum I_{out} \qquad (27.14)$$

where, as usual, the Σ symbol means summation.

This basic conservation statement—that the sum of the currents into a junction equals the sum of the currents leaving—is called **Kirchhoff's junction law.** The junction law, together with *Kirchhoff's loop law* that you met in Chapter 26, will play an important role in circuit analysis in the next chapter.

STOP TO THINK 27.4 What are the magnitude and the direction of the current in the fifth wire?

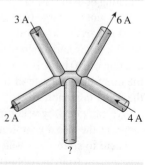

3 A

6 A

2 A

4 A

?

27.4 Conductivity and Resistivity

The current density $J = n_e e v_d$ is directly proportional to the electron drift speed v_d. We earlier used the microscopic model of conduction to find that the drift speed is $v_d = e\tau E/m$, where τ is the mean time between collisions and m is the mass of an electron. Combining these, we find the current density is

$$J = n_e e v_d = n_e e \left(\frac{e\tau E}{m} \right) = \frac{n_e e^2 \tau}{m} E \qquad (27.15)$$

The quantity $n_e e^2 \tau/m$ depends *only* on the conducting material. According to Equation 27.15, a given electric field strength will generate a larger current density in a material with a larger electron density n_e or longer times τ between collisions than in materials with smaller values. In other words, such a material is a *better conductor* of current.

It makes sense, then, to define the **conductivity** σ of a material as

$$\sigma = \text{conductivity} = \frac{n_e e^2 \tau}{m} \qquad (27.16)$$

Conductivity, like density, characterizes a material as a whole. All pieces of copper (at the same temperature) have the same value of σ, but the conductivity of copper is different from that of aluminum. Notice that the mean time between collisions τ can be inferred from measured values of the conductivity.

With this definition of conductivity, Equation 27.15 becomes

$$J = \sigma E \qquad (27.17)$$

This is a result of fundamental importance. Equation 27.17 tells us three things:

1. Current is caused by an electric field exerting forces on the charge carriers.
2. The current density, and hence the current $I = JA$, depends linearly on the strength of the electric field. To double the current, you must double the strength of the electric field that pushes the charges along.
3. The current density also depends on the *conductivity* of the material. Different conducting materials have different conductivities because they have different values of the electron density and, especially, different values of the mean time between electron collisions with the lattice of atoms.

The value of the conductivity is affected by the structure of a metal, by any impurities, and by the temperature. As the temperature increases, so do the thermal vibrations of the lattice atoms. This makes them "bigger targets" and causes collisions to be more frequent, thus lowering τ and decreasing the conductivity. Metals conduct better at low temperatures than at high temperatures.

For many practical applications of current it will be convenient to use the inverse of the conductivity, called the **resistivity**:

$$\rho = \text{resistivity} = \frac{1}{\sigma} = \frac{m}{n_e e^2 \tau} \qquad (27.18)$$

The resistivity of a material tells us how reluctantly the electrons move in response to an electric field. **TABLE 27.2** gives measured values of the resistivity and conductivity for several metals and for carbon. You can see that they vary quite a bit, with copper and silver being the best two conductors.

The units of conductivity, from Equation 27.17, are those of J/E, namely $A\,C/N\,m^2$. These are clearly awkward. In the next section we will introduce a new unit called the *ohm*, symbolized by Ω (uppercase Greek omega). It will then turn out that resistivity has units of $\Omega\,m$ and conductivity has units of $\Omega^{-1}m^{-1}$.

This woman is measuring her percentage body fat by gripping a device that sends a small electric current through her body. Because muscle and fat have different resistivities, the amount of current allows the fat-to-muscle ratio to be determined.

TABLE 27.2 Resistivity and conductivity of conducting materials

Material	Resistivity $(\Omega\,m)$	Conductivity $(\Omega^{-1}m^{-1})$
Aluminum	2.8×10^{-8}	3.5×10^{7}
Copper	1.7×10^{-8}	6.0×10^{7}
Gold	2.4×10^{-8}	4.1×10^{7}
Iron	9.7×10^{-8}	1.0×10^{7}
Silver	1.6×10^{-8}	6.2×10^{7}
Tungsten	5.6×10^{-8}	1.8×10^{7}
Nichrome*	1.5×10^{-6}	6.7×10^{5}
Carbon	3.5×10^{-5}	2.9×10^{4}

*Nickel-chromium alloy used for heating wires.

EXAMPLE 27.5 | The electric field in a wire

A 2.0-mm-diameter aluminum wire carries a current of 800 mA. What is the electric field strength inside the wire?

SOLVE The electric field strength is

$$E = \frac{J}{\sigma} = \frac{I}{\sigma \pi r^2} = \frac{0.80 \text{ A}}{(3.5 \times 10^7 \ \Omega^{-1} \text{ m}^{-1})\pi(0.0010 \text{ m})^2} = 0.0073 \text{ V/m}$$

where the conductivity of aluminum was taken from Table 27.2.

ASSESS This is a *very* small field in comparison with those we calculated in Chapters 22 and 23. This calculation justifies the claim in Table 23.1 that a typical electric field strength inside a current-carrying wire is ≈ 0.01 V/m. It takes *very few* surface charges on a wire to create the weak electric field necessary to push a considerable current through the wire. The reason, once again, is the enormous value of the charge-carrier density n_e. Even though the electric field is very tiny and the drift speed is agonizingly slow, a wire can carry a substantial current due to the vast number of charge carriers able to move.

Superconductors have unusual magnetic properties. Here a small permanent magnet levitates above a disk of the high-temperature superconductor $YBa_2Cu_3O_7$ that has been cooled to liquid-nitrogen temperature.

Superconductivity

In 1911, the Dutch physicist Heike Kamerlingh Onnes was studying the conductivity of metals at very low temperatures. Scientists had just recently discovered how to liquefy helium, and this opened a whole new field of *low-temperature physics*. As we noted above, metals become better conductors (i.e., they have higher conductivity and lower resistivity) at lower temperatures. But the effect is gradual. Onnes, however, found that mercury suddenly and dramatically loses *all* resistance to current when cooled below a temperature of 4.2 K. This complete loss of resistance at low temperatures is called **superconductivity.**

Later experiments established that the resistivity of a superconducting metal is not just small, it is truly zero. The electrons are moving in a frictionless environment, and charge will continue to move through a superconductor *without an electric field*. Superconductivity was not understood until the 1950s, when it was explained as being a specific quantum effect.

Superconducting wires can carry enormous currents because the wires are not heated by electrons colliding with the atoms. Very strong magnetic fields can be created with superconducting electromagnets, but applications remained limited for many decades because all known superconductors required temperatures less than 20 K. This situation changed dramatically in 1986 with the discovery of *high-temperature superconductors*. These ceramic-like materials are superconductors at temperatures as "high" as 125 K. Although $-150°C$ may not seem like a high temperature to you, the technology for producing such temperatures is simple and inexpensive. Thus many new superconductor applications are likely to appear in coming years.

STOP TO THINK 27.5 Rank in order, from largest to smallest, the current densities J_a to J_d in these four wires.

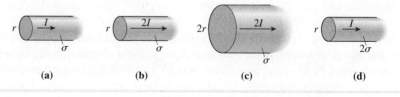

The one major difference between a capacitor and a battery is the duration of the current. The current discharging a capacitor is transient, ceasing as soon as the excess charge on the capacitor plates is removed. In contrast, the current supplied by a battery is *sustained*.

We can use the charge escalator model of a battery to understand why. FIGURE 27.19 shows the charge escalator creating a potential difference ΔV_{bat} by lifting positive charge from the negative terminal to the positive terminal. Once at the positive terminal, positive charges can move *through the wire* as current I. In essence, the charges are "falling downhill" through the wire, losing the energy they gained on the escalator. This energy transfer to the wire warms the wire.

Eventually the charges find themselves back at the negative terminal of the battery, where they can ride the escalator back up and repeat the journey. A battery, unlike a charged capacitor, has an internal source of energy (the chemical reactions) that keeps the charge escalator running. It is the charge escalator that *sustains* the current in the wire by providing a continually renewed supply of charge at the battery terminals.

An important consequence of the charge escalator model, one you learned in the previous chapter, is that **a battery is a source of potential difference.** It is true that charges flow through a wire connecting the battery terminals, but current is a *consequence* of the battery's potential difference. The battery's emf is the *cause;* current, heat, light, sound, and so on are all *effects* that happen when the battery is used in certain ways.

Distinguishing cause and effect will be vitally important for understanding how a battery functions in a circuit. The reasoning is as follows:

1. A battery is a source of potential difference ΔV_{bat}. An ideal battery has $\Delta V_{bat} = \mathcal{E}$.
2. The battery creates a potential difference $\Delta V_{wire} = \Delta V_{bat}$ between the ends of a wire.
3. The potential difference ΔV_{wire} causes an electric field $E = \Delta V_{wire}/L$ in the wire.
4. The electric field establishes a current $I = JA = \sigma AE$ in the wire.
5. The magnitude of the current is determined *jointly* by the battery and the wire's resistance R to be $I = \Delta V_{wire}/R$.

Resistors and Ohmic Materials

Circuit textbooks often write Ohm's law as $V = IR$ rather than $I = \Delta V/R$. This can be misleading until you have sufficient experience with circuit analysis. First, Ohm's law relates the current to the potential *difference* between the ends of the conductor. Engineers and circuit designers *mean* "potential difference" when they use the symbol V, but the symbol is easily misinterpreted as simply "the potential." Second, $V = IR$ or even $\Delta V = IR$ suggests that a current I causes a potential difference ΔV. As you have seen, current is a *consequence* of a potential difference; hence $I = \Delta V/R$ is a better description of cause and effect.

Despite its name, Ohm's law is *not* a law of nature. It is limited to those materials whose resistance R remains constant—or very nearly so—during use. The materials to which Ohm's law applies are called *ohmic*. FIGURE 27.20a shows that the current through an ohmic material is directly proportional to the potential difference. Doubling the potential difference doubles the current. Metal and other conductors are ohmic devices.

Because the resistance of metals is small, a circuit made exclusively of metal wires would have enormous currents and would quickly deplete the battery. It is useful to limit the current in a circuit with ohmic devices, called **resistors,** whose resistance is significantly larger than the metal wires. Resistors are made with poorly conducting materials, such as carbon, or by depositing very thin metal films on an insulating substrate.

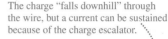

FIGURE 27.19 A battery's charge escalator causes a sustained current in a wire.

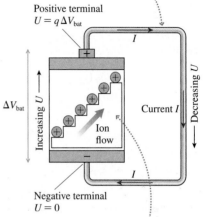

The charge "falls downhill" through the wire, but a current can be sustained because of the charge escalator.

The charge escalator "lifts" charge from the negative side to the positive side. Charge q gains energy $\Delta U = q\Delta V_{bat}$.

FIGURE 27.20 Current-versus-potential-difference graphs for ohmic and nonohmic materials.

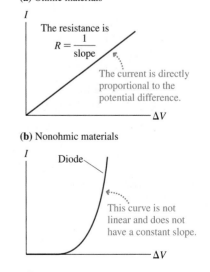

(a) Ohmic materials

The resistance is $R = \dfrac{1}{slope}$

The current is directly proportional to the potential difference.

(b) Nonohmic materials

Diode

This curve is not linear and does not have a constant slope.

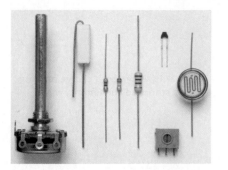

The resistors used in circuits range from a few ohms to millions of ohms of resistance.

Some materials and devices are *nonohmic*, meaning that the current through the device is *not* directly proportional to the potential difference. For example, FIGURE 27.20b shows the *I*-versus-ΔV graph of a commonly used semiconductor device called a *diode*. Diodes do not have a well-defined resistance. Batteries, where $\Delta V = \mathcal{E}$ is determined by chemical reactions, and capacitors, where the relationship between *I* and ΔV differs from that of a resistor, are important nonohmic devices.

We can identify three important classes of ohmic circuit materials:

1. *Wires* are metals with very small resistivities ρ and thus very small resistances ($R \ll 1\ \Omega$). An **ideal wire** has $R = 0\ \Omega$; hence the potential difference between the ends of an ideal wire is $\Delta V = 0$ V *even if there is a current in it.* We will usually adopt the *ideal-wire model* of assuming that any connecting wires in a circuit are ideal.

2. *Resistors* are poor conductors with resistances usually in the range 10^1 to $10^6\ \Omega$. They are used to control the current in a circuit. Most resistors in a circuit have a specified value of *R*, such as 500 Ω. The filament in a lightbulb (a tungsten wire with a high resistance due to an extremely small cross-section area *A*) functions as a resistor as long as it is glowing, but the filament is slightly nonohmic because the value of its resistance when hot is larger than its room-temperature value.

3. *Insulators* are materials such as glass, plastic, or air. An **ideal insulator** has $R = \infty\ \Omega$; hence there is no current in an insulator even if there is a potential difference across it ($I = \Delta V/R = 0$ A). This is why insulators can be used to hold apart two conductors at different potentials. All practical insulators have $R \gg 10^9\ \Omega$ and can be treated, for our purposes, as ideal.

NOTE Ohm's law will be an important part of circuit analysis in the next chapter because resistors are essential components of almost any circuit. However, it is important that you apply Ohm's law *only* to the resistors and not to anything else.

FIGURE 27.21 **The potential along a wire-resistor-wire combination.**

(a) The current is constant along the wire-resistor-wire combination.

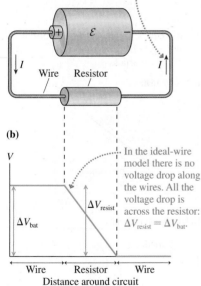

(b)

In the ideal-wire model there is no voltage drop along the wires. All the voltage drop is across the resistor: $\Delta V_{\text{resist}} = \Delta V_{\text{bat}}$.

FIGURE 27.21a shows a resistor connected to a battery with current-carrying wires. There are no junctions; hence the current *I* through the resistor is the same as the current in each wire. Because the wire's resistance is *much* less than that of the resistor, $R_{\text{wire}} \ll R_{\text{resist}}$, the potential difference $\Delta V_{\text{wire}} = IR_{\text{wire}}$ between the ends of each wire is *much* less than the potential difference $\Delta V_{\text{resist}} = IR_{\text{resist}}$ across the resistor.

If we assume ideal wires with $R_{\text{wire}} = 0\ \Omega$, then $\Delta V_{\text{wire}} = 0$ V and *all* the voltage drop occurs across the resistor. In this **ideal-wire model,** shown in FIGURE 27.21b, the wires are equipotentials, and the segments of the voltage graph corresponding to the wires are horizontal. As we begin circuit analysis in the next chapter, we will assume that all wires are ideal unless stated otherwise. Thus our analysis will be focused on the resistors.

EXAMPLE 27.7 | **A battery and a resistor**

What resistor would have a 15 mA current if connected across the terminals of a 9.0 V battery?

MODEL Assume the resistor is connected to the battery with ideal wires.

SOLVE Connecting the resistor to the battery with ideal wires makes $\Delta V_{\text{resist}} = \Delta V_{\text{bat}} = 9.0$ V. From Ohm's law, the resistance giving a 15 mA current is

$$R = \frac{\Delta V_{\text{resist}}}{I} = \frac{9.0\ \text{V}}{0.015\ \text{A}} = 600\ \Omega$$

ASSESS Currents of a few mA and resistances of a few hundred ohms are quite typical of real circuits.

STOP TO THINK 27.6 A wire connects the positive and negative terminals of a battery. Two identical wires connect the positive and negative terminals of an identical battery. Rank in order, from largest to smallest, the currents I_a to I_d at points a to d.

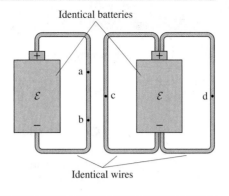

Identical batteries

Identical wires

CHALLENGE EXAMPLE 27.8 | Measuring body composition

The woman in the photo on page 753 is gripping a device that measures body fat. To illustrate how this works, **FIGURE 27.22** models an upper arm as part muscle and part fat, showing the resistivities of each. Nonconductive elements, such as skin and bone, have been ignored. This is obviously not a picture of the actual structure, but gathering all the fat tissue together and all the muscle tissue together is a model that predicts the arm's electrical character quite well.

A 0.87 mA current is recorded when a 0.60 V potential difference is applied across an upper arm having the dimensions shown in the figure. What are the percentages of muscle and fat in this person's upper arm?

FIGURE 27.22 A simple model for the resistance of an arm.

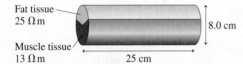

Fat tissue
25 Ω m

Muscle tissue
13 Ω m

8.0 cm

25 cm

MODEL Model the muscle and the fat as separate resistors connected to a 0.60 V battery. Assume the connecting wires to be ideal, with no "loss" of potential along the wires.

VISUALIZE **FIGURE 27.23** shows the circuit, with the side-by-side muscle and fat resistors connected to the two terminals of the battery.

FIGURE 27.23 Circuit for passing current through the upper arm.

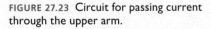
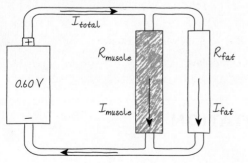

I_{total}

R_{muscle} R_{fat}

0.60 V

I_{muscle} I_{fat}

SOLVE The measured current of 0.87 mA is I_{total}, the current traveling from the battery to the arm and later back to the battery. This current splits at the junction between the two resistors. Kirchhoff's junction law, for the conservation of charge, requires

$$I_{total} = I_{muscle} + I_{fat}$$

The current through each resistor can be found from Ohm's law: $I = \Delta V/R$. Each resistor has $\Delta V = 0.60$ V because each is connected to the battery terminals by lossless, ideal wires, but they have different resistances.

Let the fraction of muscle tissue be x; the fraction of fat is then $1 - x$. If the cross-section area of the upper arm is $A = \pi r^2$, then the muscle resistor has $A_{muscle} = xA$ while the fat resistor has $A_{fat} = (1 - x)A$. The resistances are related to the resistivities and the geometry by

$$R_{muscle} = \frac{\rho_{muscle} L}{A_{muscle}} = \frac{\rho_{muscle} L}{x\pi r^2}$$

$$R_{fat} = \frac{\rho_{fat} L}{A_{fat}} = \frac{\rho_{fat} L}{(1 - x)\pi r^2}$$

The currents are thus

$$I_{muscle} = \frac{\Delta V}{R_{muscle}} = \frac{x\pi r^2 \Delta V}{\rho_{muscle} L} = 0.93x \text{ mA}$$

$$I_{fat} = \frac{\Delta V}{R_{fat}} = \frac{(1 - x)\pi r^2 \Delta V}{\rho_{fat} L} = 0.48(1 - x) \text{ mA}$$

The sum of these is the total current:

$$I_{total} = 0.87 \text{ mA} = 0.93x \text{ mA} + 0.48(1 - x) \text{ mA}$$
$$= (0.48 + 0.45x) \text{ mA}$$

Solving, we find $x = 0.87$. This subject's upper arm is 87% muscle tissue, 13% fat tissue.

ASSESS The percentages seem reasonable for a healthy adult. A real measurement of body fat requires a more detailed model of the human body, because the current passes through both arms and across the chest, but the principles are the same.

SUMMARY

The goal of Chapter 27 has been to learn how and why charge moves through a wire as a current.

GENERAL PRINCIPLES

Current is a nonequilibrium motion of charges sustained by an electric field. Nonuniform surface charge density creates an electric field in a wire. The electric field pushes the electron current i_e in a direction opposite to $\vec{E}$. The conventional current I is in the direction in which positive charge *seems* to move.

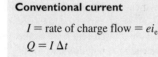

Conservation of Charge

The current is the same at any two points in a wire.
At a junction,

$$\sum I_{in} = \sum I_{out}$$

This is **Kirchhoff's junction law.**

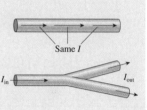

Electron current

i_e = rate of electron flow

$N_e = i_e \, \Delta t$

Conventional current

I = rate of charge flow = $e i_e$

$Q = I \, \Delta t$

Current density

$J = I/A$

IMPORTANT CONCEPTS

Sea of electrons

Conduction electrons move freely around the positive ions that form the atomic lattice.

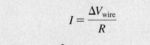

Conduction

An electric field causes a slow drift at speed v_d to be superimposed on the rapid but random thermal motions of the electrons.

Collisions of electrons with the ions transfer energy to the atoms. This makes the wire warm and lightbulbs glow. More collisions mean a higher resistivity ρ and a lower conductivity σ.

The **drift speed** is $v_d = \dfrac{e\tau}{m} E$, where τ is the mean time between collisions.

The electron current is related to the drift speed by

$$i_e = n_e A v_d$$

where n_e is the electron density.

An electric field E in a conductor causes a current density $J = n_e e v_d = \sigma E$, where the **conductivity** is

$$\sigma = \frac{n_e e^2 \tau}{m}$$

The **resistivity** is $\rho = 1/\sigma$.

APPLICATIONS

Resistors

A potential difference ΔV_{wire} between the ends of a wire creates an electric field inside the wire:

$$E_{wire} = \frac{\Delta V_{wire}}{L}$$

The electric field causes a current in the direction of decreasing potential.

The size of the current is

$$I = \frac{\Delta V_{wire}}{R}$$

where $R = \dfrac{\rho L}{A}$ is the wire's **resistance.**
This is **Ohm's law.**

TERMS AND NOTATION

current, I	ampere, A	resistivity, ρ	resistor
drift speed, v_d	current density, J	superconductivity	ideal wire
electron current, i_e	junction	resistance, R	ideal insulator
mean time between	Kirchhoff's junction law	ohm, Ω	ideal-wire model
collisions, τ	conductivity, σ	Ohm's law	

FIGURE 27.12 The electron velocity as a function of time.

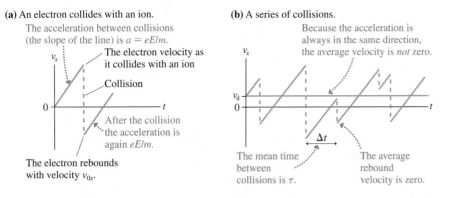

(a) An electron collides with an ion.

The acceleration between collisions
(the slope of the line) is $a = eE/m$.

The electron velocity as
it collides with an ion

v_x

Collision

0

t

After the collision
the acceleration is
again eE/m.

The electron rebounds
with velocity v_{0x}.

(b) A series of collisions.

Because the acceleration is
always in the same direction,
the average velocity is *not* zero.

v_x

v_d

0

t

Δt

The mean time
between
collisions is τ.

The average
rebound
velocity is zero.

where a bar over a quantity indicates an average value. The average value of v_{0x}, the velocity with which an electron rebounds after a collision, is zero. We know this because, in the absence of an electric field, the sea of electrons moves neither right nor left.

The quantity Δt is the time between collisions, so the average value of Δt is the **mean time between collisions,** which we designate τ. The mean time between collisions, analogous to the mean free path between collisions in the kinetic theory of gases, depends on the metal's temperature but can be considered a constant in the equations below.

Thus the average speed at which the electrons are pushed along by the electric field is

$$v_d = \frac{e\tau}{m} E \qquad (27.7)$$

We can complete our model of conduction by using Equation 27.7 for v_d in the electron-current equation $i_e = n_e A v_d$. Upon doing so, we find that an electric field strength E in a wire of cross-section area A causes an electron current

$$i_e = \frac{n_e e \tau A}{m} E \qquad (27.8)$$

The electron density n_e and the mean time between collisions τ are properties of the metal.

Equation 27.8 is the main result of this model of conduction. We've found that **the electron current is directly proportional to the electric field strength.** A stronger electric field pushes the electrons faster and thus increases the electron current.

EXAMPLE 27.3 | **Collisions in a copper wire**

Example 27.1 found the electron current to be 2.7×10^{19} s^{-1} for a 2.0-mm-diameter copper wire in which the electron drift speed is 1.0×10^{-4} m/s. If an internal electric field of 0.020 V/m is needed to sustain this current, a typical value, how many collisions per second, on average, do electrons in copper undergo?

MODEL Use the model of conduction.

SOLVE From Equation 27.7, the mean time between collisions is

$$\tau = \frac{mv_d}{eE} = 2.8 \times 10^{-14} \text{ s}$$

The average number of collisions per second is the inverse:

$$\text{Collision rate} = \frac{1}{\tau} = 3.5 \times 10^{13} \text{ s}^{-1}$$

ASSESS This was another straightforward calculation simply to illustrate the incredibly large collision rate of conduction electrons.

27.3 Current and Current Density

We have developed the idea of a current as the motion of electrons through metals. But the properties of currents were known and used for a century before the discovery that electrons are the charge carriers in metals. We need to connect our ideas about the electron current to the conventional definition of current.

CONCEPTUAL QUESTIONS

1. Suppose a time machine has just brought you forward from 1750 (post-Newton but pre-electricity) and you've been shown the lightbulb demonstration of FIGURE Q27.1. Do observations or *simple* measurements you might make—measurements that must make sense to you with your 1700s knowledge—prove that something is *flowing* through the wires? Or might you advance an alternative hypothesis for why the bulb is glowing? If your answer to the first question is yes, state what observations and/or measurements are relevant and the reasoning from which you can infer that something must be flowing. If not, can you offer an alternative hypothesis about why the bulb glows that could be tested?

FIGURE Q27.1

2. Consider a lightbulb circuit such as the one in FIGURE Q27.1.
   a. From the simple observations and measurements you can make on this circuit, can you distinguish a current composed of positive charge carriers from a current composed of negative charge carriers? If so, describe how you can tell which it is. If not, why not?
   b. One model of current is the motion of discrete charged particles. Another model is that current is the flow of a continuous charged fluid. Do simple observations and measurements on this circuit provide evidence in favor of either one of these models? If so, describe how. If not, why not?
3. The electron drift speed in a wire is exceedingly slow—typically only a fraction of a millimeter per second. Yet when you turn on a flashlight switch, the light comes on almost instantly. Resolve this apparent paradox.
4. Is FIGURE Q27.4 a possible surface charge distribution for a current-carrying wire? If so, in which direction is the current? If not, why not?

FIGURE Q27.4

5. What is the difference between current and current density?
6. All the wires in FIGURE Q27.6 are made of the same material and have the same diameter. Rank in order, from largest to smallest, the currents I_a to I_d. Explain.

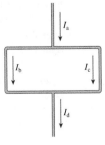

FIGURE Q27.6

7. Both batteries in FIGURE Q27.7 are ideal and identical, and all lightbulbs are the same. Rank in order, from brightest to least bright, the brightness of bulbs a to c. Explain.

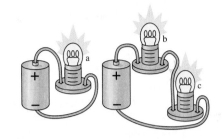

FIGURE Q27.7

8. Both batteries in FIGURE Q27.8 are ideal and identical, and all lightbulbs are the same. Rank in order, from brightest to least bright, the brightness of bulbs a to c. Explain.

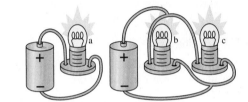

FIGURE Q27.8

9. The wire in FIGURE Q27.9 consists of two segments of different diameters but made from the same metal. The current in segment 1 is I_1.

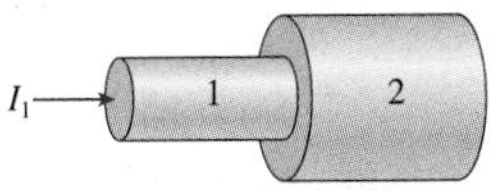

FIGURE Q27.9

   a. Compare the currents in the two segments. That is, is I_2 greater than, less than, or equal to I_1? Explain.
   b. Compare the current densities J_1 and J_2 in the two segments.
   c. Compare the electric field strengths E_1 and E_2 in the two segments.
   d. Compare the drift speeds $(v_d)_1$ and $(v_d)_2$ in the two segments.
10. The current in a wire is doubled. What happens to (a) the current density, (b) the conduction-electron density, (c) the mean time between collisions, and (d) the electron drift speed? Are each of these doubled, halved, or unchanged? Explain.
11. The wires in FIGURE Q27.11 are all made of the same material. Rank in order, from largest to smallest, the resistances R_a to R_e of these wires. Explain.

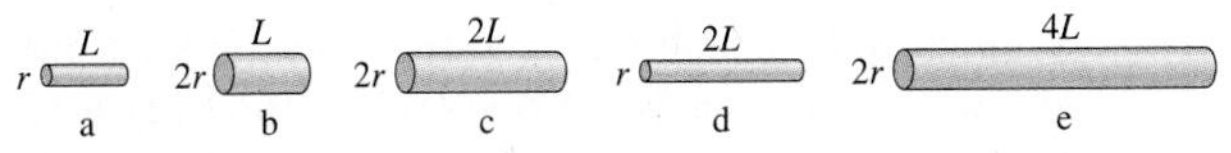

r ⬤ L a $2r$ ⬤ L b $2r$ ⬤ $2L$ c r ⬤ $2L$ d $2r$ ⬤ $4L$ e

FIGURE Q27.11

12. Which, if any, of these statements are true? (More than one may be true.) Explain. Assume the batteries are ideal.
   a. A battery supplies the energy to a circuit.
   b. A battery is a source of potential difference; the potential difference between the terminals of the battery is always the same.
   c. A battery is a source of current; the current leaving the battery is always the same.

EXERCISES AND PROBLEMS

Problems labeled [   ] integrate material from earlier chapters.

Exercises

Section 27.1 The Electron Current

1. ‖ 1.0×10^{20} electrons flow through a cross section of a 2.0-mm-diameter iron wire in 5.0 s. What is the electron drift speed?

2. ‖ The electron drift speed in a 1.0-mm-diameter gold wire is 5.0×10^{-5} m/s. How long does it take 1 mole of electrons to flow through a cross section of the wire?

3. ‖ 1.0×10^{16} electrons flow through a cross section of a silver wire in 320 μs with a drift speed of 8.0×10^{-4} m/s. What is the diameter of the wire?

4. ‖ Electrons flow through a 1.6-mm-diameter aluminum wire at 2.0×10^{-4} m/s. How many electrons move through a cross section of the wire each day?

Section 27.2 Creating a Current

5. ∣ The electron drift speed is 2.0×10^{-4} m/s in a metal with a mean time between collisions of 5.0×10^{-14} s. What is the electric field strength?

6. ‖ a. How many conduction electrons are there in a 1.0-mm-diameter gold wire that is 10 cm long?
   b. How far must the sea of electrons in the wire move to deliver -32 nC of charge to an electrode?

7. ‖ A 2.0×10^{-3} V/m electric field creates a 3.5×10^{17} electrons/s current in a 1.0-mm-diameter aluminum wire. What are (a) the drift speed and (b) the mean time between collisions for electrons in this wire?

8. ‖ The mean time between collisions in iron is 4.2×10^{-15} s. What electron current is driven through a 1.8-mm-diameter iron wire by a 0.065 V/m electric field?

Section 27.3 Current and Current Density

9. ∣ The wires leading to and from a 0.12-mm-diameter lightbulb filament are 1.5 mm in diameter. The wire to the filament carries a current with a current density of 4.5×10^5 A/m². What are (a) the current and (b) the current density in the filament?

10. ‖ The current in a 100 watt lightbulb is 0.85 A. The filament inside the bulb is 0.25 mm in diameter.
    a. What is the current density in the filament?
    b. What is the electron current in the filament?

11. ‖ In an integrated circuit, the current density in a 2.5-μm-thick $\times$ 75-μm-wide gold film is 7.5×10^5 A/m². How much charge flows through the film in 15 min?

12. ∣ The current in an electric hair dryer is 10.0 A. How many electrons flow through the hair dryer in 5.0 min?

13. ∣ When a nerve cell fires, charge is transferred across the cell
BIO membrane to change the cell's potential from negative to positive. For a typical nerve cell, 9.0 pC of charge flows in a time of 0.50 ms. What is the average current through the cell membrane?

14. ‖ The current in a 2.0 $\times$ 2.0 mm square aluminum wire is 2.5 A. What are (a) the current density and (b) the electron drift speed?

15. ‖ A hollow copper wire with an inner diameter of 1.0 mm and an outer diameter of 2.0 mm carries a current of 10 A. What is the current density in the wire?

16. ∣ A car battery is rated at 90 A h, meaning that it can supply a 90 A current for 1 h before being completely discharged. If you leave your headlights on until the battery is completely dead, how much charge leaves the battery?

Section 27.4 Conductivity and Resistivity

17. ∣ What is the mean time between collisions for electrons in (a) an aluminum wire and (b) an iron wire?

18. ‖ The electric field in a 2.0 mm $\times$ 2.0 mm square aluminum wire is 0.012 V/m. What is the current in the wire?

19. ‖ A 15-cm-long nichrome wire is connected across the terminals of a 1.5 V battery.
    a. What is the electric field inside the wire?
    b. What is the current density inside the wire?
    c. If the current in the wire is 2.0 A, what is the wire's diameter?

20. ∣ What electric field strength is needed to create a 5.0 A current in a 2.0-mm-diameter iron wire?

21. ∣ A 0.0075 V/m electric field creates a 3.9 mA current in a 1.0-mm-diameter wire. What material is the wire made of?

22. ‖ A 3.0-mm-diameter wire carries a 12 A current when the electric field is 0.085 V/m. What is the wire's resistivity?

23. ‖ A 0.50-mm-diameter silver wire carries a 20 mA current. What are (a) the electric field and (b) the electron drift speed in the wire?

24. ∣ The two segments of the wire in FIGURE EX27.24 have equal diameters but different conductivities σ_1 and σ_2. Current I passes through this wire. If the conductivities have the ratio $\sigma_2/\sigma_1 = 2$, what is the ratio E_2/E_1 of the electric field strengths in the two segments of the wire?

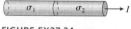

FIGURE EX27.24

Section 27.5 Resistance and Ohm's Law

25. ∣ A 1.5 V battery provides 0.50 A of current.
    a. At what rate (C/s) is charge lifted by the charge escalator?
    b. How much work does the charge escalator do to lift 1.0 C of charge?
    c. What is the power output of the charge escalator?

26. ‖ Wires 1 and 2 are made of the same metal. Wire 2 has twice the length and twice the diameter of wire 1. What are the ratios (a) ρ_2/ρ_1 of the resistivities and (b) R_2/R_1 of the resistances of the two wires?

27. ∣ What is the resistance of
    a. A 2.0-m-long gold wire that is 0.20 mm in diameter?
    b. A 10-cm-long piece of carbon with a 1.0 mm $\times$ 1.0 mm square cross section?

28. ‖ An engineer cuts a 1.0-m-long, 0.33-mm-diameter piece of wire, connects it across a 1.5 V battery, and finds that the current in the wire is 8.0 A. Of what material is the wire made?

29. ∣ The electric field inside a 30-cm-long copper wire is 5.0 mV/m. What is the potential difference between the ends of the wire?

30. ‖ a. How long must a 0.60-mm-diameter aluminum wire be to have a 0.50 A current when connected to the terminals of a 1.5 V flashlight battery?
    b. What is the current if the wire is half this length?

31. ‖ The terminals of a 0.70 V watch battery are connected by a 100-m-long gold wire with a diameter of 0.10 mm. What is the current in the wire?

32. ‖ BIO The femoral artery is the large artery that carries blood to the leg. What is the resistance of a 20-cm-long column of blood in a 1.0-cm-diameter femoral artery? The conductivity of blood is $0.63 \ \Omega^{-1} \, m^{-1}$.

33. ‖ Pencil "lead" is actually carbon. What is the current if a 9.0 V potential difference is applied between the ends of a 0.70-mm-diameter, 6.0-cm-long lead from a mechanical pencil?

34. ‖‖ The resistance of a very fine aluminum wire with a $10 \ \mu m \times 10 \ \mu m$ square cross section is 1000 Ω. A 1000 Ω resistor is made by wrapping this wire in a spiral around a 3.0-mm-diameter glass core. How many turns of wire are needed?

35. ‖ **FIGURE EX27.35** is a current-versus-potential-difference graph for a material. What is the material's resistance?

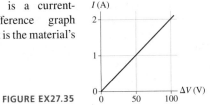

FIGURE EX27.35

36. ‖ A circuit calls for a 0.50-mm-diameter copper wire to be stretched between two points. You don't have any copper wire, but you do have aluminum wire in a wide variety of diameters. What diameter aluminum wire will provide the same resistance?

37. ‖ Household wiring often uses 2.0-mm-diameter copper wires. The wires can get rather long as they snake through the walls from the fuse box to the farthest corners of your house. What is the potential difference across a 20-m-long, 2.0-mm-diameter copper wire carrying an 8.0 A current?

38. ‖ An ideal battery would produce an extraordinarily large current if "shorted" by connecting the positive and negative terminals with a short wire of very low resistance. Real batteries do not. The current of a real battery is limited by the fact that the battery itself has resistance. What is the resistance of a 9.0 V battery that produces a 21 A current when shorted by a wire of negligible resistance?

Problems

39. ‖ For what electric field strength would the current in a 2.0-mm-diameter nichrome wire be the same as the current in a 1.0-mm-diameter aluminum wire in which the electric field strength is 0.0080 V/m?

40. ‖ The electron beam inside an old television picture tube is 0.40 mm in diameter and carries a current of $50 \ \mu A$. This electron beam impinges on the inside of the picture tube screen.
    a. How many electrons strike the screen each second?
    b. What is the current density in the electron beam?
    c. The electrons move with a velocity of 4.0×10^7 m/s. What electric field strength is needed to accelerate electrons from rest to this velocity in a distance of 5.0 mm?
    d. Each electron transfers its kinetic energy to the picture tube screen upon impact. What is the *power* delivered to the screen by the electron beam?

41. ‖ Energetic particles, such as protons, can be detected with a *silicon detector*. When a particle strikes a thin piece of silicon, it creates a large number of free electrons by ionizing silicon atoms. The electrons flow to an electrode on the surface of the detector, and this current is then amplified and detected. In one experiment, each incident proton creates, on average, 35,000 electrons; the electron current is amplified by a factor of 100; and the experimenters record an amplified current of $3.5 \ \mu A$. How many protons are striking the detector per second?

42. ‖ **FIGURE P27.42** shows a 4.0-cm-wide plastic film being wrapped onto a 2.0-cm-diameter roller that turns at 90 rpm. The plastic has a uniform surface charge density of $-2.0 \ nC/cm^2$.

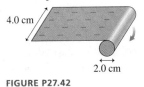

FIGURE P27.42

    a. What is the current of the moving film?
    b. How long does it take the roller to accumulate a charge of $-10 \ \mu C$?

43. ‖ A sculptor has asked you to help electroplate gold onto a brass statue. You know that the charge carriers in the ionic solution are singly charged gold ions, and you've calculated that you must deposit 0.50 g of gold to reach the necessary thickness. How much current do you need, in mA, to plate the statue in 3.0 hours?

44. ‖ In a classic model of the hydrogen atom, the electron moves around the proton in a circular orbit of radius 0.053 nm.
    a. What is the electron's orbital frequency?
    b. What is the effective current of the electron?

45. ‖ You've been asked to determine whether a new material your company has made is ohmic and, if so, to measure its electrical conductivity. Taking a 0.50 mm × 1.0 mm × 45 mm sample, you wire the ends of the long axis to a power supply and then measure the current for several different potential differences. Your data are as follows:

Voltage (V)	Current (A)
0.200	0.47
0.400	1.06
0.600	1.53
0.800	1.97

Use an appropriate graph of the data to determine whether the material is ohmic and, if so, its conductivity.

46. ‖ BIO The biochemistry that takes place inside cells depends on various elements, such as sodium, potassium, and calcium, that are dissolved in water as ions. These ions enter cells through narrow pores in the cell membrane known as *ion channels*. Each ion channel, which is formed from a specialized protein molecule, is selective for one type of ion. Measurements with microelectrodes have shown that a 0.30-nm-diameter potassium ion (K^+) channel carries a current of 1.8 pA.
    a. How many potassium ions pass through if the ion channel opens for 1.0 ms?
    b. What is the current density in the ion channel?

47. ‖ The starter motor of a car engine draws a current of 150 A from the battery. The copper wire to the motor is 5.0 mm in diameter and 1.2 m long. The starter motor runs for 0.80 s until the car engine starts.
    a. How much charge passes through the starter motor?
    b. How far does an electron travel along the wire while the starter motor is on?

48. ‖ A 1.5-m-long wire is made of a metal with the same electron density as copper. The wire is connected across the terminals of a 9.0 V battery. What conductivity would the metal need for the drift velocity of electrons in the wire to be 60 mph? By what factor is this larger than the conductivity of copper?

49. ‖ The resistivity of a metal increases slightly with increased temperature. This can be expressed as $\rho = \rho_0[1 + \alpha(T - T_0)]$, where T_0 is a reference temperature, usually 20°C, and α is the *temperature coefficient of resistivity*. For copper, $\alpha = 3.9 \times 10^{-3} \, °C^{-1}$. Suppose a long, thin copper wire has a resistance of 0.25 Ω at 20°C. At what temperature, in °C, will its resistance be 0.30 Ω?

50. ‖ Variations in the resistivity of blood can give valuable clues
BIO about changes in various properties of the blood. Suppose a medical device attaches two electrodes into a 1.5-mm-diameter vein at positions 5.0 cm apart. What is the blood resistivity if a 9.0 V potential difference causes a 230 μA current through the blood in the vein?

51. ‖ The conducting path between the right hand and the left hand
BIO can be modeled as a 10-cm-diameter, 160-cm-long cylinder. The average resistivity of the interior of the human body is 5.0 Ω m. Dry skin has a much higher resistivity, but skin resistance can be made negligible by soaking the hands in salt water. If skin resistance is neglected, what potential difference between the hands is needed for a lethal shock of 100 mA across the chest? Your result shows that even small potential differences can produce dangerous currents when the skin is wet.

52. ‖‖ The conductive tissues of the upper leg can be modeled as a
BIO 40-cm-long, 12-cm-diameter cylinder of muscle and fat. The resistivities of muscle and fat are 13 Ω m and 25 Ω m, respectively. One person's upper leg is 82% muscle, 18% fat. What current is measured if a 1.5 V potential difference is applied between the person's hip and knee?

53. ‖ The resistivity of a metal increases slightly with increased
CALC temperature. This can be expressed as $\rho = \rho_0[1 + \alpha(T - T_0)]$, where T_0 is a reference temperature, usually 20°C, and α is the *temperature coefficient of resistivity*.
    a. First find an expression for the current I through a wire of length L, cross-section area A, and temperature T when connected across the terminals of an ideal battery with terminal voltage ΔV. Then, because the *change* in resistance is small, use the *binomial approximation* to simplify your expression. Your final expression should have the temperature coefficient α in the numerator.
    b. For copper, $\alpha = 3.9 \times 10^{-3} \, °C^{-1}$. Suppose a 2.5-m-long, 0.40-mm-diameter copper wire is connected across the terminals of a 1.5 V ideal battery. What is the current in the wire at 20°C?
    c. What is the rate, in A/°C, at which the current changes with temperature as the wire heats up?

54. ‖ Electrical engineers sometimes use a wire's *conductance*, $G = \sigma A/L$, instead of its resistance.
    a. Write Ohm's law in terms of conductance, starting with "$\Delta V =$".
    b. What is the conductance of a 5.4-cm-long, 0.15-mm-diameter tungsten wire?
    c. A 1.5 A current flows through the wire of part b. What is the potential difference between the ends of the wire?

55. ‖ You need to design a 1.0 A fuse that "blows" if the current exceeds 1.0 A. The fuse material in your stockroom melts at a current density of 500 A/cm². What diameter wire of this material will do the job?

56. ‖ A hollow metal cylinder has inner radius a, outer radius b, length L, and conductivity σ. The current I is *radially* outward from the inner surface to the outer surface.
    a. Find an expression for the electric field strength inside the metal as a function of the radius r from the cylinder's axis.
    b. Evaluate the electric field strength at the inner and outer surfaces of an iron cylinder if $a = 1.0$ cm, $b = 2.5$ cm, $L = 10$ cm, and $I = 25$ A.

57. ‖ A hollow metal sphere has inner radius a, outer radius b, and conductivity σ. The current I is *radially* outward from the inner surface to the outer surface.
    a. Find an expression for the electric field strength inside the metal as a function of the radius r from the center.
    b. Evaluate the electric field strength at the inner and outer surfaces of a copper sphere if $a = 1.0$ cm, $b = 2.5$ cm, and $I = 25$ A.

58. ‖ The total amount of charge in coulombs that has entered a
CALC wire at time t is given by the expression $Q = 4t - t^2$, where t is in seconds and $t \geq 0$.
    a. Find an expression for the current in the wire at time t.
    b. Graph I versus t for the interval $0 \leq t \leq 4$ s.

59. ‖ The total amount of charge that has entered a wire at time t is
CALC given by the expression $Q = (20 \, C)(1 - e^{-t/(2.0 \, s)})$, where t is in seconds and $t \geq 0$.
    a. Find an expression for the current in the wire at time t.
    b. What is the maximum value of the current?
    c. Graph I versus t for the interval $0 \leq t \leq 10$ s.

60. ‖ The current in a wire at time t is given by the expression
CALC $I = (2.0 \, A)e^{-t/(2.0 \, \mu s)}$, where t is in microseconds and $t \geq 0$.
    a. Find an expression for the total amount of charge (in coulombs) that has entered the wire at time t. The initial conditions are $Q = 0$ C at $t = 0 \, \mu s$.
    b. Graph Q versus t for the interval $0 \leq t \leq 10 \, \mu s$.

61. ‖ The current supplied by a battery slowly decreases as the bat-
CALC tery runs down. Suppose that the current as a function of time is $I = (0.75 \, A)e^{-t/(6 \, h)}$. What is the total number of electrons transported from the positive electrode to the negative electrode by the charge escalator from the time the battery is first used until it is completely dead?

62. ‖ The two wires in FIGURE P27.62 are made of the same material. What are the current and the electron drift speed in the 2.0-mm-diameter segment of the wire?

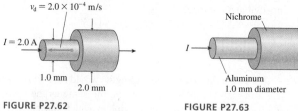

FIGURE P27.62   FIGURE P27.63

63. ‖ What diameter should the nichrome wire in FIGURE P27.63 be in order for the electric field strength to be the same in both wires?

64. ‖ An aluminum wire consists of the three segments shown in FIGURE P27.64. The current in the top segment is 10 A. For each of these three segments, find the
    a. Current I.
    b. Current density J.
    c. Electric field E.
    d. Drift velocity v_d.
    e. Electron current i.
    Place your results in a table for easy viewing.

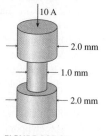

FIGURE P27.64

65. || A wire of radius R has a current density that increases linearly
CALC with distance from the center of the wire: $J(r) = kr$, where k is
a constant. Find an expression for k in terms of R and the total
current I carried by the wire.

66. || A 0.60-mm-diameter wire made from an alloy (a combina-
CALC tion of different metals) has a conductivity that decreases linearly
with distance from the center of the wire: $\sigma(r) = \sigma_0 - cr$, with
$\sigma_0 = 5.0 \times 10^7 \ \Omega^{-1} m^{-1}$ and $c = 1.2 \times 10^{11} \ \Omega^{-1} m^{-2}$. What is the
resistance of a 4.0 m length of this wire?

67. || A 20-cm-long hollow nichrome tube of inner diameter
2.8 mm, outer diameter 3.0 mm is connected, at its ends, to a 3.0 V
battery. What is the current in the tube?

68. || The batteries in **FIGURE P27.68** are identical. Both resistors
have equal currents. What is the resistance of the resistor on the
right?

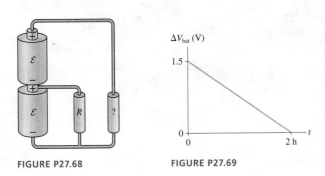

FIGURE P27.68	**FIGURE P27.69**

69. || A 1.5 V flashlight battery is connected to a wire with a
resistance of 3.0 Ω. **FIGURE P27.69** shows the battery's potential
difference as a function of time. What is the total charge lifted by
the charge escalator?

70. || Two 10-cm-diameter metal plates 1.0 cm apart are charged to
± 12.5 nC. They are suddenly connected together by a 0.224-mm-
diameter copper wire stretched taut from the center of one plate to
the center of the other.
   a. What is the maximum current in the wire?
   b. Does the current increase with time, decrease with time, or
      remain steady? Explain.
   c. What is the total amount of energy dissipated in the wire?

71. || A long, round wire has resistance R. What will the wire's re-
sistance be if you stretch it to twice its initial length?

72. || You've decided to protect your house by placing a 5.0-m-tall
iron lightning rod next to the house. The top is sharpened to a
point and the bottom is in good contact with the ground. From
your research, you've learned that lightning bolts can carry up to
50 kA of current and last up to 50 μs.

a. How much charge is delivered by a lightning bolt with these
   parameters?
b. You don't want the potential difference between the top and
   bottom of the lightning rod to exceed 100 V. What minimum
   diameter must the rod have?

Challenge Problems

73. ||| A 5.0-mm-diameter proton beam carries a total current of
CALC 1.5 mA. The current density in the proton beam, which increases
with distance from the center, is given by $J = J_{edge}(r/R)$, where R
is the radius of the beam and J_{edge} is the current density at the edge.
   a. How many protons per second are delivered by this proton beam?
   b. Determine the value of J_{edge}.

74. ||| **FIGURE CP27.74** shows a wire that is made of two equal-
diameter segments with conductivities σ_1 and σ_2. When current
I passes through the wire, a thin layer of charge appears at the
boundary between the segments.
   a. Find an expression for the surface charge density η on the
      boundary. Give your result in terms of I, σ_1, σ_2, and the wire's
      cross-section area A.
   b. A 1.0-mm-diameter wire made of copper and iron segments
      carries a 5.0 A current. How much charge accumulates at the
      boundary between the segments?

FIGURE CP27.74   Surface charge density η

75. ||| A 300 μF capacitor is charged to 9.0 V, then connected in
CALC parallel with a 5000 Ω resistor. The capacitor will discharge
because the resistor provides a conducting pathway between the
capacitor plates, but much more slowly than if the plates were
connected by a wire. Let $t = 0$ s be the instant the fully charged
capacitor is first connected to the resistor. At what time has the
capacitor voltage decreased by half, to 4.5 V?
   Hint: The current through the resistor is related to the rate at
which charge is *leaving* the capacitor. Consequently, you'll need
a minus sign that you might not have expected.

76. ||| A thin metal cylinder of length L and radius R_1 is coaxial
CALC with a thin metal cylinder of length L and a larger radius R_2.
The space between the two coaxial cylinders is filled with a
material that has resistivity ρ. The two cylinders are connected
to the terminals of a battery with potential difference ΔV, causing
current I to flow *radially* from the inner cylinder to the outer
cylinder. Find an expression for the resistance of this device.

28 Fundamentals of Circuits

A microprocessor, the heart of a powerful computer, is a complex device. Even so, a microprocessor operates on the basis of just a few fundamental physical principles.

IN THIS CHAPTER, you will learn the fundamental physical principles that govern electric circuits.

What is a circuit?

Circuits—from flashlights to computers—are the controlled motion of charges through conductors and resistors.

- This chapter focuses on DC circuits, meaning *direct current*, in which potentials and currents are constant.
- You'll learn to draw circuit diagrams.

« LOOKING BACK Section 26.4 Sources of potential

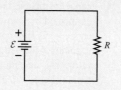

How are circuits analyzed?

Any circuit, no matter how complex, can be analyzed with Kirchhoff's two laws:

- The junction law (charge conservation) relates the currents at a junction.
- The loop law (energy conservation) relates the voltages around a closed loop.
- We'll also use Ohm's law for resistors.

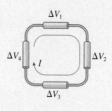

How do circuits use energy?

Circuits use energy to do things, such as lighting a bulb or turning a motor. You'll learn to calculate power, the rate at which the battery supplies energy to a circuit and the rate at which a resistor dissipates energy. Many circuit elements are *rated* by their power consumption in watts.

How are resistors combined?

Resistors, like capacitors, often occur in series or in parallel. These combinations of resistors can be simplified by replacing them with a single resistor with equivalent resistance.

« LOOKING BACK Sections 27.3–27.5 Current, resistance, and Ohm's law

What is an *RC* circuit?

Capacitors are charged and discharged by current flowing through a resistor. These important circuits are called *RC* circuits. Their uses range from heart defibrillators to digital electronics. You'll learn that capacitors charge and discharge exponentially with time constant $\tau = RC$.

« LOOKING BACK Section 26.5 Capacitors

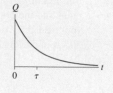

Why are circuits important?

We live in an electronic era, and electric circuits surround you: your household wiring, the ignition system in your car, your music and communication devices, and your tablets and computers. Electric circuits are one of the most important applications of physics, and in this chapter you will see how the seemingly abstract ideas of electric charge, field, and potential are the foundation for many of the things we take for granted in the 21st century.

28.1 Circuit Elements and Diagrams

The last several chapters have focused on the physics of electric forces, fields, and potentials. Now we'll put those ideas to use by looking at one of the most important applications of electricity: the controlled motion of charges in *electric circuits*. This chapter is not about circuit design—you will see that in more advanced courses—but about understanding the fundamental ideas that underlie all circuits.

FIGURE 28.1 shows an electric circuit in which a resistor and a capacitor are connected by wires to a battery. To understand the functioning of this circuit, we do not need to know whether the wires are bent or straight, or whether the battery is to the right or to the left of the resistor. The literal picture of Figure 28.1 provides many irrelevant details. It is customary when describing or analyzing circuits to use a more abstract picture called a **circuit diagram.** A circuit diagram is a *logical* picture of what is connected to what.

A circuit diagram also replaces pictures of the circuit elements with symbols. FIGURE 28.2 shows the basic symbols that we will need. The longer line at one end of the battery symbol represents the positive terminal of the battery. Notice that a lightbulb, like a wire or a resistor, has two "ends," and current passes *through* the bulb. It is often useful to think of a lightbulb as a resistor that gives off light when a current is present. A lightbulb filament is not a perfectly ohmic material, but the resistance of a *glowing* lightbulb remains reasonably constant if you don't change ΔV by much.

FIGURE 28.1 An electric circuit.

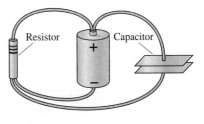

FIGURE 28.2 A library of basic symbols used for electric circuit drawings.

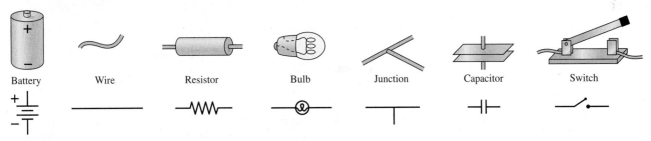

FIGURE 28.3 is a circuit diagram of the circuit shown in Figure 28.1. Notice how the circuit elements are labeled. The battery's emf $\mathcal{E}$ is shown beside the battery, and + and − symbols, even though somewhat redundant, are shown beside the terminals. We would use numerical values for $\mathcal{E}$, R, and C if we knew them. The wires, which in practice may bend and curve, are shown as straight-line connections between the circuit elements.

FIGURE 28.3 A circuit diagram for the circuit of Figure 28.1.

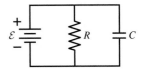

STOP TO THINK 28.1 Which of these diagrams represent the same circuit?

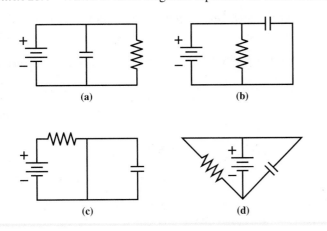

FIGURE 28.4 Kirchhoff's junction law.

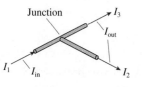
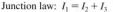

Junction law: $I_1 = I_2 + I_3$

FIGURE 28.5 Kirchhoff's loop law.

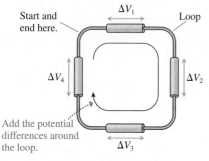

Loop law: $\Delta V_1 + \Delta V_2 + \Delta V_3 + \Delta V_4 = 0$

28.2 Kirchhoff's Laws and the Basic Circuit

We are now ready to begin analyzing circuits. To analyze a circuit means to find:

1. The potential difference across each circuit component.
2. The current in each circuit component.

Because charge is conserved, the total current into the junction of FIGURE 28.4 must equal the total current leaving the junction. That is,

$$\sum I_{in} = \sum I_{out} \tag{28.1}$$

This statement, which you met in Chapter 27, is **Kirchhoff's junction law.**

Because energy is conserved, a charge that moves around a closed path has $\Delta U = 0$. We apply this idea to the circuit of FIGURE 28.5 by adding all of the potential differences *around* the loop formed by the circuit. Doing so gives

$$\Delta V_{loop} = \sum (\Delta V)_i = 0 \tag{28.2}$$

where $(\Delta V)_i$ is the potential difference of the ith component in the loop. This statement, introduced in Chapter 26, is **Kirchhoff's loop law.**

Kirchhoff's loop law can be true only if at least one of the $(\Delta V)_i$ is negative. To apply the loop law, we need to explicitly identify which potential differences are positive and which are negative.

TACTICS BOX 28.1 (MP)

Using Kirchhoff's loop law

❶ **Draw a circuit diagram.** Label all known and unknown quantities.

❷ **Assign a direction to the current.** Draw and label a current arrow I to show your choice.

■ If you know the actual current direction, choose that direction.

■ If you don't know the actual current direction, make an arbitrary choice. All that will happen if you choose wrong is that your value for I will end up negative.

❸ **"Travel" around the loop.** Start at any point in the circuit, then go all the way around the loop in the direction you assigned to the current in step 2. As you go through each circuit element, ΔV is interpreted to mean $\Delta V = V_{downstream} - V_{upstream}$.

■ For an ideal battery in the negative-to-positive direction:

$$\Delta V_{bat} = +\mathcal{E}$$

→ Travel

Potential increases

■ For an ideal battery in the positive-to-negative direction:

$$\Delta V_{bat} = -\mathcal{E}$$

→ Travel

Potential decreases

■ For a resistor: $\Delta V_{res} = -\Delta V_R = -IR$

 → I

Potential decreases

❹ **Apply the loop law:** $\sum (\Delta V)_i = 0$

Exercises 4–7

NOTE Ohm's law gives us only the *magnitude* $\Delta V_R = IR$ of the potential difference across a resistor. For using Kirchhoff's law, $\Delta V_{res} = V_{downstream} - V_{upstream} = -\Delta V_R$.

The Basic Circuit

The most basic electric circuit is a single resistor connected to the two terminals of a battery. FIGURE 28.6a shows a literal picture of the circuit elements and the connecting wires; FIGURE 28.6b is the circuit diagram. Notice that this is a **complete circuit,** forming a continuous path between the battery terminals.

The resistor might be a known resistor, such as "a 10 Ω resistor," or it might be some other resistive device, such as a lightbulb. Regardless of what the resistor is, it is called the **load.** The battery is called the **source.**

FIGURE 28.7 shows the use of Kirchhoff's loop law to analyze this circuit. Two things are worth noting:

1. This circuit has no junctions, so the current I is the same in all four sides of the circuit. Kirchhoff's junction law is not needed.
2. We're assuming the ideal-wire model, in which there are *no* potential differences along the connecting wires.

Kirchhoff's loop law, with two circuit elements, is

$$\Delta V_{\text{loop}} = \sum (\Delta V)_i = \Delta V_{\text{bat}} + \Delta V_{\text{res}} \qquad (28.3)$$
$$= 0$$

Let's look at each of the two voltages in Equation 28.3:

1. The potential *increases* as we travel through the battery on our clockwise journey around the loop. We enter the negative terminal and, farther downstream, exit the positive terminal after having gained potential $\mathcal{E}$. Thus

$$\Delta V_{\text{bat}} = +\mathcal{E}$$

2. The potential of a conductor *decreases* in the direction of the current, which we've indicated with the + and − signs in Figure 28.7. Thus

$$\Delta V_{\text{res}} = V_{\text{downstream}} - V_{\text{upstream}} = -IR$$

NOTE Determining which potential differences are positive and which are negative is perhaps *the* most important step in circuit analysis.

With this information, the loop equation becomes

$$\mathcal{E} - IR = 0 \qquad (28.4)$$

We can solve the loop equation to find that the current in the circuit is

$$I = \frac{\mathcal{E}}{R} \qquad (28.5)$$

We can then use the current to find that the magnitude of the resistor's potential difference is

$$\Delta V_R = IR = \mathcal{E} \qquad (28.6)$$

This result should come as no surprise. The potential energy that the charges gain in the battery is subsequently lost as they "fall" through the resistor.

NOTE The current that the battery delivers depends jointly on the emf of the battery and the resistance of the load.

FIGURE 28.6 The basic circuit of a resistor connected to a battery.

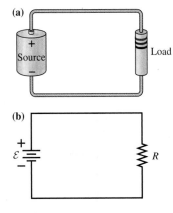

FIGURE 28.7 Analysis of the basic circuit using Kirchhoff's loop law.

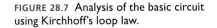

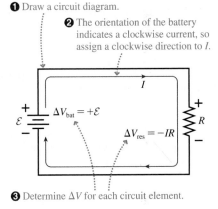

EXAMPLE 28.1 | Two resistors and two batteries

Analyze the circuit shown in **FIGURE 28.8**.

a. Find the current in and the potential difference across each resistor.

b. Draw a graph showing how the potential changes around the circuit, starting from $V = 0$ V at the negative terminal of the 6 V battery.

FIGURE 28.8 Circuit for Example 28.1.

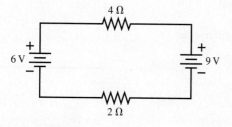

MODEL Assume ideal connecting wires and ideal batteries, for which $\Delta V_{bat} = \mathcal{E}$.

VISUALIZE In **FIGURE 28.9**, we've redrawn the circuit and defined $\mathcal{E}_1, \mathcal{E}_2, R_1$, and R_2. Because there are no junctions, the current is the same through *each* component in the circuit. With some thought, we might deduce whether the current is cw or ccw, but we do not need to know in advance of our analysis. We will choose a clockwise direction and solve for the value of I. If our solution is positive, then the current really is cw. If the solution should turn out to be negative, we will know that the current is ccw.

FIGURE 28.9 Analyzing the circuit.

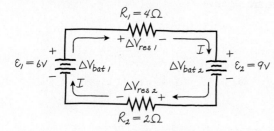

SOLVE a. How do we deal with *two* batteries? Can charge flow "backward" through a battery, from positive to negative? Consider the charge escalator analogy. Left to itself, a charge escalator lifts charge from lower to higher potential. But it *is* possible to run down an up escalator, as many of you have probably done. If two escalators are placed "head to head," whichever is stronger will, indeed, force the charge to run down the up escalator of the other battery. The current in a battery *can* be from positive to negative if driven in that direction by a larger emf from a second battery. Indeed, this is how rechargeable batteries are recharged.

Kirchhoff's loop law, going clockwise from the negative terminal of battery 1, is

$$\Delta V_{loop} = \sum (\Delta V)_i = \Delta V_{bat\,1} + \Delta V_{res\,1}$$
$$+ \Delta V_{bat\,2} + \Delta V_{res\,2} = 0$$

All the signs are $+$ because this is a formal statement of *adding* potential differences around the loop. Next we can evaluate each ΔV. As we go cw, the charges *gain* potential in battery 1 but *lose* potential in battery 2. Thus $\Delta V_{bat\,1} = +\mathcal{E}_1$ and $\Delta V_{bat\,2} = -\mathcal{E}_2$. There is a *loss* of potential in traveling through each resistor, because we're traversing them in the direction we assigned to the current, so $\Delta V_{res\,1} = -IR_1$ and $\Delta V_{res\,2} = -IR_2$. Thus Kirchhoff's loop law becomes

$$\sum (\Delta V)_i = \mathcal{E}_1 - IR_1 - \mathcal{E}_2 - IR_2$$
$$= \mathcal{E}_1 - \mathcal{E}_2 - I(R_1 + R_2) = 0$$

We can solve this equation to find the current in the loop:

$$I = \frac{\mathcal{E}_1 - \mathcal{E}_2}{R_1 + R_2} = \frac{6\text{ V} - 9\text{ V}}{4\ \Omega + 2\ \Omega} = -0.50\text{ A}$$

The value of I is negative; hence the actual current in this circuit is 0.50 A *counterclockwise*. You perhaps anticipated this from the orientation of the 9 V battery with its larger emf.

The potential difference across the 4 Ω resistor is

$$\Delta V_{res\,1} = -IR_1 = -(-0.50\text{ A})(4\ \Omega) = +2.0\text{ V}$$

Because the current is actually ccw, the resistor's potential *increases* in the cw direction of our travel around the loop. Similarly, $\Delta V_{res\,2} = 1.0$ V.

b. **FIGURE 28.10** shows the potential experienced by charges flowing around the circuit. The distance s is measured from the 6 V battery's negative terminal, and we have chosen to let $V = 0$ V at that point. The potential ends at the value from which it started.

FIGURE 28.10 A graphical presentation of how the potential changes around the loop.

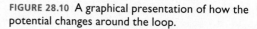
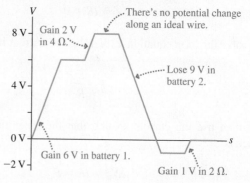

ASSESS Notice how the potential *drops* 9 V upon passing through battery 2 in the cw direction. It then gains 1 V upon passing through R_2 to end at the starting potential.

STOP TO THINK 28.2 What is ΔV across the unspecified circuit element? Does the potential increase or decrease when traveling through this element in the direction assigned to I?

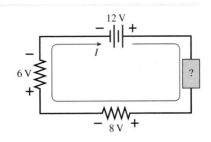

28.3 Energy and Power

The circuit of **FIGURE 28.11** has two identical lightbulbs, A and B. Which is brighter? Or are they equally bright? Think about this before going on.

You might have been tempted to say that A is brighter. After all, the current gets to A first, so A might "use up" some of the current and leave less for B. But this would violate the law of conservation of charge. There are no junctions between A and B, so the current through the two bulbs must be the same. Hence the bulbs are equally bright.

What a bulb uses is not current but *energy*. The battery's charge escalator is an energy-transfer process, transferring chemical energy E_{chem} stored in the battery to the potential energy U of the charges. That energy is, in turn, transformed into the *thermal energy* of the resistors, increasing their temperature until—in the case of lightbulb filaments—they glow.

A charge gains potential energy $\Delta U = q\,\Delta V_{bat}$ as it moves up the charge escalator in the battery. For an ideal battery, with $\Delta V_{bat} = \mathcal{E}$, the battery supplies energy $\Delta U = q\mathcal{E}$ as it lifts charge q from the negative to the positive terminal.

It is useful to know the *rate* at which the battery supplies energy to the charges. Recall from Chapter 9 that the rate at which energy is transferred is *power,* measured in joules per second or *watts*. If energy $\Delta U = q\mathcal{E}$ is transferred to charge q, then the *rate* at which energy is transferred from the battery to the moving charges is

$$P_{bat} = \text{rate of energy transfer} = \frac{dU}{dt} = \frac{dq}{dt}\mathcal{E} \qquad (28.7)$$

But dq/dt, the rate at which charge moves through the battery, is the current I. Hence the power supplied by a battery, or the rate at which the battery (or any other source of emf) transfers energy to the charges passing through it, is

$$P_{bat} = I\mathcal{E} \qquad \text{(power delivered by an emf)} \qquad (28.8)$$

$I\mathcal{E}$ has units of J/s, or W. For example, a 120 V battery that generates 2 A of current is delivering 240 W of power to the circuit.

Energy Dissipation in Resistors

P_{bat} is the energy transferred per second from the battery's store of chemicals to the moving charges that make up the current. But what happens to this energy? Where does it end up? **FIGURE 28.12**, a section of a current-carrying resistor, reminds you of our microscopic model of conduction. The electrons accelerate in the electric field, transforming potential energy into kinetic, then collide with atoms in the lattice. The collisions transfer the electron's kinetic energy to the *thermal* energy of the lattice. The potential energy was acquired in the battery, so the entire energy-transfer process looks like

$$E_{chem} \rightarrow U \rightarrow K \rightarrow E_{th}$$

The net result is that **the battery's chemical energy is transferred to the thermal energy of the resistors,** raising their temperature.

Consider a charge q that moves all the way through a resistor with a potential difference ΔV_R between its two ends. The charge *loses potential energy* $\Delta U = -q\,\Delta V_R$,

FIGURE 28.11 Which lightbulb is brighter?

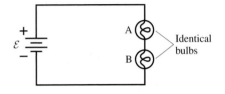

FIGURE 28.12 A current-carrying resistor dissipates energy.

The electric field causes electrons to speed up. The energy transformation is $U \rightarrow K$.

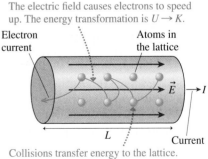

Collisions transfer energy to the lattice. The energy transformation is $K \rightarrow E_{th}$.

and, after a vast number of collisions, all that energy is transformed into thermal energy. Thus the resistor's *increase in thermal energy* due to this one charge is

$$\Delta E_{th} = q \, \Delta V_R \tag{28.9}$$

The *rate* at which energy is transferred from the current to the resistor is then

$$P_R = \frac{dE_{th}}{dt} = \frac{dq}{dt} \Delta V_R = I \Delta V_R \tag{28.10}$$

Power—so many joules per second—is the rate at which energy is *dissipated* by the resistor as charge flows through it. The resistor, in turn, transfers this energy to the air and to the circuit board on which it is mounted, causing the circuit and all its surroundings to heat up.

From our analysis of the basic circuit, in which a single resistor is connected to a battery, we learned that $\Delta V_R = \mathcal{E}$. That is, the potential difference across the resistor is exactly the emf supplied by the battery. But then Equations 28.8 and 28.10, for P_{bat} and P_R, are numerically equal, and we find that

$$P_R = P_{bat} \tag{28.11}$$

The answer to the question "What happens to the energy supplied by the battery?" is "The battery's chemical energy is transformed into the thermal energy of the resistor." The *rate* at which the battery supplies energy is exactly equal to the *rate* at which the resistor dissipates energy. This is, of course, exactly what we would have expected from energy conservation.

EXAMPLE 28.2 | The power of light

How much current is "drawn" by a 100 W lightbulb connected to a 120 V outlet?

MODEL Most household appliances, such as a 100 W lightbulb or a 1500 W hair dryer, have a power rating. The rating does *not* mean that these appliances *always* dissipate that much power. These appliances are intended for use at a standard household voltage of 120 V, and their rating is the power they will dissipate *if* operated with a potential difference of 120 V. Their power consumption will differ from the rating if they are operated at any other potential difference.

SOLVE Because the lightbulb is operating as intended, it will dissipate 100 W of power. Thus

$$I = \frac{P_R}{\Delta V_R} = \frac{100 \text{ W}}{120 \text{ V}} = 0.833 \text{ A}$$

ASSESS A current of 0.833 A in this lightbulb transfers 100 J/s to the thermal energy of the filament, which, in turn, dissipates 100 J/s as heat and light to its surroundings.

A resistor obeys Ohm's law, $\Delta V_R = IR$. (Remember that Ohm's law gives only the *magnitude* of ΔV_R.) This gives us two alternative ways of writing the power dissipated by a resistor. We can either substitute IR for ΔV_R or substitute $\Delta V_R/R$ for I. Thus

$$P_R = I \Delta V_R = I^2 R = \frac{(\Delta V_R)^2}{R} \qquad \text{(power dissipated by a resistor)} \tag{28.12}$$

If the same current I passes through several resistors in series, then $P_R = I^2 R$ tells us that most of the power will be dissipated by the largest resistance. This is why a lightbulb filament glows but the connecting wires do not. Essentially *all* of the power supplied by the battery is dissipated by the high-resistance lightbulb filament and essentially no power is dissipated by the low-resistance wires. The filament gets very hot, but the wires do not.

EXAMPLE 28.3 | The power of sound

Most loudspeakers are designed to have a resistance of 8 Ω. If an 8 Ω loudspeaker is connected to a stereo amplifier with a rating of 100 W, what is the maximum possible current to the loudspeaker?

MODEL The rating of an amplifier is the *maximum* power it can deliver. Most of the time it delivers far less, but the maximum might be reached for brief, intense sounds like cymbal crashes.

SOLVE The loudspeaker is a resistive load. The maximum current to the loudspeaker occurs when the amplifier delivers maximum power $P_{max} = (I_{max})^2 R$. Thus

$$I_{max} = \sqrt{\frac{P_{max}}{R}} = \sqrt{\frac{100 \text{ W}}{8 \text{ Ω}}} = 3.5 \text{ A}$$

Kilowatt Hours

The energy dissipated (i.e., transformed into thermal energy) by a resistor during time Δt is $E_{th} = P_R \Delta t$. The product of watts and seconds is joules, the SI unit of energy. However, your local electric company prefers to use a different unit, the *kilowatt hour,* to measure the energy you use each month.

A load that consumes P_R kW of electricity for Δt hours has used $P_R \Delta t$ **kilowatt hours** of energy, abbreviated kWh. For example, a 4000 W electric water heater uses 40 kWh of energy in 10 hours. A 1500 W hair dryer uses 0.25 kWh of energy in 10 minutes. Despite the rather unusual name, a kilowatt hour is a unit of energy. A homework problem will let you find the conversion factor from kilowatt hours to joules.

Your monthly electric bill specifies the number of kilowatt hours you used last month. This is the amount of energy that the electric company delivered to you, via an electric current, and that you transformed into light and thermal energy inside your home. The cost of electricity varies throughout the country, but the average cost of electricity in the United States is approximately 10¢ per kWh ($0.10/kWh). Thus it costs about $4.00 to run your water heater for 10 hours, about 2.5¢ to dry your hair.

The electric meter on the side of your house or apartment records the kilowatt hours of electric energy that you use.

STOP TO THINK 28.3 Rank in order, from largest to smallest, the powers P_a to P_d dissipated in resistors a to d.

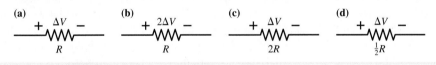

(a) $+\ \Delta V\ -$ R

(b) $+\ 2\Delta V\ -$ R

(c) $+\ \Delta V\ -$ $2R$

(d) $+\ \Delta V\ -$ $\frac{1}{2}R$

28.4 Series Resistors

Consider the three lightbulbs in **FIGURE 28.13**. The batteries are identical and the bulbs are identical. You learned in the previous section that B and C are equally bright, because the current is the same through both, but how does the brightness of B compare to that of A? Think about this before going on.

FIGURE 28.14a shows two resistors placed end to end between points a and b. Resistors that are aligned end to end, *with no junctions between them,* are called **series resistors** or, sometimes, resistors "in series." Because there are no junctions, the current I must be the same through each of these resistors. That is, the current out of the last resistor in a series is equal to the current into the first resistor.

The potential differences across the two resistors are $\Delta V_1 = IR_1$ and $\Delta V_2 = IR_2$. The total potential difference ΔV_{ab} between points a and b is the sum of the individual potential differences:

$$\Delta V_{ab} = \Delta V_1 + \Delta V_2 = IR_1 + IR_2 = I(R_1 + R_2) \tag{28.13}$$

FIGURE 28.13 How does the brightness of bulb B compare to that of A?

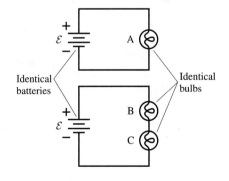

FIGURE 28.14 Replacing two series resistors with an equivalent resistor.

(a) Two resistors in series

(b) An equivalent resistor

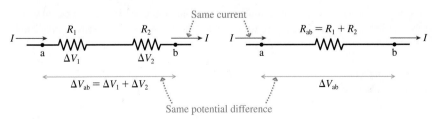

Suppose, as in **FIGURE 28.14b**, we replaced the two resistors with a single resistor having current I and potential difference $\Delta V_{ab} = \Delta V_1 + \Delta V_2$. We can then use Ohm's law to find that the resistance R_{ab} between points a and b is

$$R_{ab} = \frac{\Delta V_{ab}}{I} = \frac{I(R_1 + R_2)}{I} = R_1 + R_2 \qquad (28.14)$$

Because the battery has to establish the same potential difference across the load and provide the same current in both cases, the two resistors R_1 and R_2 act exactly the same as a *single* resistor of value $R_1 + R_2$. We can say that the single resistor R_{ab} is *equivalent* to the two resistors in series.

There was nothing special about having only two resistors. If we have N resistors in series, their **equivalent resistance** is

$$R_{eq} = R_1 + R_2 + \cdots + R_N \qquad \text{(series resistors)} \qquad (28.15)$$

The current and the power output of the battery will be unchanged if the N series resistors are replaced by the single resistor R_{eq}. The key idea in this analysis is that **resistors in series all have the same current.**

NOTE Compare this idea to what you learned in Chapter 26 about capacitors in series. The end-to-end connections are the same, but the equivalent capacitance is *not* the sum of the individual capacitances.

Now we can answer the lightbulb question posed at the beginning of this section. Suppose the resistance of each lightbulb is R. The battery drives current $I_A = \mathcal{E}/R$ through bulb A. Bulbs B and C are in series, with an equivalent resistance $R_{eq} = 2R$, but the battery has the same emf $\mathcal{E}$. Thus the current through bulbs B and C is $I_{B+C} = \mathcal{E}/R_{eq} = \mathcal{E}/2R = \frac{1}{2}I_A$. Bulb B has only half the current of bulb A, so B is dimmer.

Many people predict that A and B should be equally bright. It's the same battery, so shouldn't it provide the same current to both circuits? No! A battery is a source of emf, *not* a source of current. In other words, the battery's emf is the same no matter how the battery is used. When you buy a 1.5 V battery you're buying a device that provides a specified amount of potential difference, not a specified amount of current. The battery does provide the current to the circuit, but the *amount* of current depends on the resistance of the load. Your 1.5 V battery causes 1 A to pass through a 1.5 Ω load but only 0.1 A to pass through a 15 Ω load. As an analogy, think about a water faucet. The pressure in the water main underneath the street is a fixed and unvarying quantity set by the water company, but the amount of water coming out of a faucet depends on how far you open it. A faucet opened slightly has a "high resistance," so only a little water flows. A wide-open faucet has a "low resistance," and the water flow is large.

In summary, **a battery provides a fixed and unvarying emf (potential difference). It does *not* provide a fixed and unvarying current. The amount of current depends jointly on the battery's emf *and* the resistance of the circuit attached to the battery.**

EXAMPLE 28.4 | A series resistor circuit

a. What is the current in the circuit of **FIGURE 28.15a**?

b. Draw a graph of potential versus position in the circuit, going cw from $V = 0$ V at the battery's negative terminal.

MODEL The three resistors are end to end, with no junctions between them, and thus are in series. Assume ideal connecting wires and an ideal battery.

SOLVE a. The battery "acts" the same—it provides the same current at the same potential difference—if we replace the three series

resistors by their equivalent resistance

$$R_{eq} = 15 \ \Omega + 4 \ \Omega + 8 \ \Omega = 27 \ \Omega$$

This is shown as an equivalent circuit in **FIGURE 28.15b**. Now we have a circuit with a single battery and a single resistor, for which we know the current to be

$$I = \frac{\mathcal{E}}{R_{eq}} = \frac{9 \text{ V}}{27 \ \Omega} = 0.333 \text{ A}$$

FIGURE 28.15 Analyzing a circuit with series resistors.

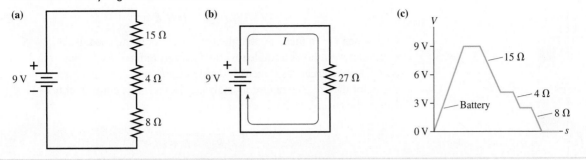

(a)

(b)

(c)

b. $I = 0.333$ A is the current in each of the three resistors in the original circuit. Thus the potential differences across the resistors are $\Delta V_{\text{res }1} = -IR_1 = -5.0$ V, $\Delta V_{\text{res }2} = -IR_2 = -1.3$ V, and $\Delta V_{\text{res }3} = -IR_3 = -2.7$ V for the 15 Ω, the 4 Ω, and the 8 Ω resistors. **FIGURE 28.15c** shows that the potential increases by 9 V due to the battery, then decreases by 9 V in three steps.

Ammeters

A device that measures the current in a circuit element is called an **ammeter.** Because charge flows *through* circuit elements, an ammeter must be placed *in series* with the circuit element whose current is to be measured.

FIGURE 28.16a shows a simple one-resistor circuit with an unknown emf $\mathcal{E}$. We can measure the current in the circuit by inserting the ammeter as shown in **FIGURE 28.16b**. Notice that we have to *break the connection* between the battery and the resistor in order to insert the ammeter. Now the current in the resistor has to first pass through the ammeter.

Because the ammeter is now in series with the resistor, the total resistance seen by the battery is $R_{\text{eq}} = 6\ \Omega + R_{\text{ammeter}}$. In order that the ammeter measure the current without changing the current, the ammeter's resistance must, in this case, be $\ll 6\ \Omega$. Indeed, an ideal ammeter has $R_{\text{ammeter}} = 0\ \Omega$ and thus has no effect on the current. Real ammeters come very close to this ideal.

The ammeter in Figure 28.16b reads 0.50 A, meaning that the current through the 6 Ω resistor is $I = 0.50$ A. Thus the resistor's potential difference is $\Delta V_R = IR = 3.0$ V. If the ammeter is ideal, with no resistance and thus no potential difference across it, then, from Kirchhoff's loop law, the battery's emf is $\mathcal{E} = \Delta V_R = 3.0$ V.

FIGURE 28.16 An ammeter measures the current in a circuit element.

(a)

(b)

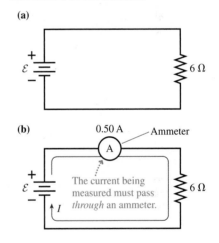

0.50 A   Ammeter

The current being measured must pass *through* an ammeter.

STOP TO THINK 28.4 What are the current and the potential at points a to e?

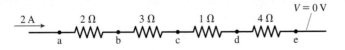

$V = 0$ V

28.5 Real Batteries

Real batteries, like ideal batteries, use chemical reactions to separate charge, create a potential difference, and provide energy to the circuit. However, real batteries also provide a slight resistance to the charges on the charge escalator. They have what is called an **internal resistance,** which is symbolized by r. **FIGURE 28.17** on the next page shows both an ideal and a real battery.

From our vantage point outside a battery, we cannot see $\mathcal{E}$ and r separately. To the user, the battery provides a potential difference ΔV_{bat} called the **terminal voltage.** $\Delta V_{\text{bat}} = \mathcal{E}$ for an ideal battery, but the presence of the internal resistance affects ΔV_{bat}. Suppose the current in the battery is I. As charges travel from the negative to the

FIGURE 28.17 An ideal battery and a real battery.

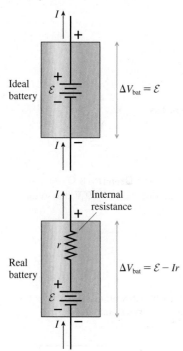

positive terminal, the potential increases by $\mathcal{E}$ but *decreases* by $\Delta V_{int} = -Ir$ due to the internal resistance. Thus the terminal voltage of the battery is

$$\Delta V_{bat} = \mathcal{E} - Ir \leq \mathcal{E} \qquad (28.16)$$

Only when $I = 0$, meaning that the battery is not being used, is $\Delta V_{bat} = \mathcal{E}$.

FIGURE 28.18 shows a single resistor R connected to the terminals of a battery having emf $\mathcal{E}$ and internal resistance r. Resistances R and r are in series, so we can replace them, for the purpose of circuit analysis, with a single equivalent resistor $R_{eq} = R + r$. Hence the current in the circuit is

$$I = \frac{\mathcal{E}}{R_{eq}} = \frac{\mathcal{E}}{R + r} \qquad (28.17)$$

If $r \ll R$, so that the internal resistance of the battery is negligible, then $I \approx \mathcal{E}/R$, exactly the result we found before. But the current decreases significantly as r increases.

FIGURE 28.18 A single resistor connected to a real battery is in series with the battery's internal resistance, giving $R_{eq} = R + r$.

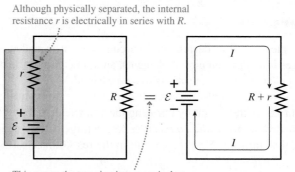

Although physically separated, the internal resistance r is electrically in series with R.

This means the two circuits are equivalent.

We can use Ohm's law to find that the potential difference across the load resistor R is

$$\Delta V_R = IR = \frac{R}{R + r}\mathcal{E} \qquad (28.18)$$

Similarly, the potential difference across the terminals of the battery is

$$\Delta V_{bat} = \mathcal{E} - Ir = \mathcal{E} - \frac{r}{R + r}\mathcal{E} = \frac{R}{R + r}\mathcal{E} \qquad (28.19)$$

The potential difference across the resistor is equal to the potential difference between the *terminals* of the battery, where the resistor is attached, *not* equal to the battery's emf. Notice that $\Delta V_{bat} = \mathcal{E}$ only if $r = 0$ (an ideal battery with no internal resistance).

EXAMPLE 28.5 | **Lighting up a flashlight**

A 6 Ω flashlight bulb is powered by a 3 V battery with an internal resistance of 1 Ω. What are the power dissipation of the bulb and the terminal voltage of the battery?

MODEL Assume ideal connecting wires but not an ideal battery.

VISUALIZE The circuit diagram looks like Figure 28.18. R is the resistance of the bulb's filament.

SOLVE Equation 28.17 gives us the current:

$$I = \frac{\mathcal{E}}{R + r} = \frac{3\text{ V}}{6\ \Omega + 1\ \Omega} = 0.43\text{ A}$$

This is 15% less than the 0.5 A an ideal battery would supply. The potential difference across the resistor is $\Delta V_R = IR = 2.6$ V, thus the power dissipation is

$$P_R = I\Delta V_R = 1.1\text{ W}$$

The battery's terminal voltage is

$$\Delta V_{bat} = \frac{R}{R + r}\mathcal{E} = \frac{6\ \Omega}{6\ \Omega + 1\ \Omega}\,3\text{ V} = 2.6\text{ V}$$

ASSESS 1 Ω is a typical internal resistance for a flashlight battery. The internal resistance causes the battery's terminal voltage to be 0.4 V less than its emf in this circuit.

A Short Circuit

In **FIGURE 28.19** we've replaced the resistor with an ideal wire having $R_{wire} = 0\ \Omega$. When a connection of very low or zero resistance is made between two points in a circuit that are normally separated by a higher resistance, we have what is called a **short circuit.** The wire in Figure 28.17 is *shorting out* the battery.

If the battery were ideal, shorting it with an ideal wire $(R = 0\ \Omega)$ would cause the current to be $I = \mathcal{E}/0 = \infty$. The current, of course, cannot really become infinite. Instead, the battery's internal resistance r becomes the only resistance in the circuit. If we use $R = 0\ \Omega$ in Equation 28.17, we find that the *short-circuit current* is

$$I_{short} = \frac{\mathcal{E}}{r} \qquad (28.20)$$

A 3 V battery with 1 Ω internal resistance generates a short circuit current of 3 A. This is the *maximum possible current* that this battery can produce. Adding any external resistance R will decrease the current to a value less than 3 A.

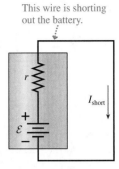

FIGURE 28.19 The short-circuit current of a battery.

This wire is shorting out the battery.

EXAMPLE 28.6 | A short-circuited battery

What is the short-circuit current of a 12 V car battery with an internal resistance of 0.020 Ω? What happens to the power supplied by the battery?

SOLVE The short-circuit current is

$$I_{short} = \frac{\mathcal{E}}{r} = \frac{12\ \text{V}}{0.02\ \Omega} = 600\ \text{A}$$

Power is generated by chemical reactions in the battery and dissipated by the load resistance. But with a short-circuited battery, the load resistance is *inside* the battery! The "shorted" battery has to dissipate power $P = I^2r = 7200\ \text{W}$ *internally*.

ASSESS This value is realistic. Car batteries are designed to drive the starter motor, which has a very small resistance and can draw a current of a few hundred amps. That is why the battery cables are so thick. A shorted car battery can produce an *enormous* amount of current. The normal response of a shorted car battery is to explode; it simply cannot dissipate this much power. Shorting a flashlight battery can make it rather hot, but your life is not in danger. Although the voltage of a car battery is relatively small, a car battery can be dangerous and should be treated with great respect.

Most of the time a battery is used under conditions in which $r \ll R$ and the internal resistance is negligible. The ideal battery model is fully justified in that case. Thus we will assume that batteries are ideal *unless stated otherwise*. But keep in mind that batteries (and other sources of emf) do have an internal resistance, and this internal resistance limits the current of the battery.

28.6 Parallel Resistors

FIGURE 28.20 is another lightbulb puzzle. Initially the switch is open. The current is the same through bulbs A and B and they are equally bright. Bulb C is not glowing. What happens to the brightness of A and B when the switch is closed? And how does the brightness of C then compare to that of A and B? Think about this before going on.

FIGURE 28.21a on the next page shows two resistors aligned side by side with their ends connected at c and d. Resistors connected *at both ends* are called **parallel resistors** or, sometimes, resistors "in parallel." The left ends of both resistors are at the same potential V_c. Likewise, the right ends are at the same potential V_d. Thus the potential *differences* ΔV_1 and ΔV_2 are the *same* and are simply ΔV_{cd}.

Kirchhoff's junction law applies at the junctions. The input current I splits into currents I_1 and I_2 at the left junction. On the right, the two currents are recombined into current I. According to the junction law,

$$I = I_1 + I_2 \qquad (28.21)$$

We can apply Ohm's law to each resistor, along with $\Delta V_1 = \Delta V_2 = \Delta V_{cd}$, to find that the current is

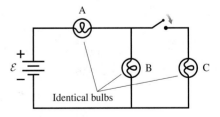

FIGURE 28.20 What happens to the brightness of the bulbs when the switch is closed?

Identical bulbs

FIGURE 28.21 Replacing two parallel resistors with an equivalent resistor.

(a) Two resistors in parallel

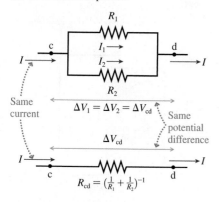

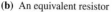

(b) An equivalent resistor

Two identical resistors***

In series	$R_{eq} = 2R$
In parallel	$R_{eq} = \dfrac{R}{2}$

*$R_1 = R_2 = R$

$$I = \frac{\Delta V_1}{R_1} + \frac{\Delta V_2}{R_2} = \frac{\Delta V_{cd}}{R_1} + \frac{\Delta V_{cd}}{R_2} = \Delta V_{cd}\left(\frac{1}{R_1} + \frac{1}{R_2}\right) \qquad (28.22)$$

Suppose, as in **FIGURE 28.21b**, we replaced the two resistors with a single resistor having current I and potential difference ΔV_{cd}. This resistor is equivalent to the original two because the battery has to establish the same potential difference and provide the same current in either case. A second application of Ohm's law shows that the resistance between points c and d is

$$R_{cd} = \frac{\Delta V_{cd}}{I} = \left(\frac{1}{R_1} + \frac{1}{R_2}\right)^{-1} \qquad (28.23)$$

The single resistor R_{cd} draws the same current as resistors R_1 and R_2, so, as far as the battery is concerned, resistor R_{cd} is *equivalent* to the two resistors in parallel.

There is nothing special about having chosen two resistors to be in parallel. If we have N resistors in parallel, the *equivalent resistance* is

$$R_{eq} = \left(\frac{1}{R_1} + \frac{1}{R_2} + \cdots + \frac{1}{R_N}\right)^{-1} \qquad \text{(parallel resistors)} \qquad (28.24)$$

The behavior of the circuit will be unchanged if the N parallel resistors are replaced by the single resistor R_{eq}. The key idea of this analysis is that **resistors in parallel all have the same potential difference.**

NOTE Don't forget to take the inverse—the -1 exponent in Equation 28.24—after adding the inverses of all the resistances.

EXAMPLE 28.7 | **A parallel resistor circuit**

The three resistors of **FIGURE 28.22** are connected to a 9 V battery. Find the potential difference across and the current through each resistor.

FIGURE 28.22 Parallel resistor circuit of Example 28.7.

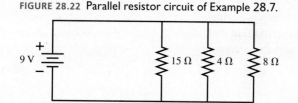

MODEL The resistors are in parallel. Assume an ideal battery and ideal connecting wires.

SOLVE The three parallel resistors can be replaced by a single equivalent resistor

$$R_{eq} = \left(\frac{1}{15\,\Omega} + \frac{1}{4\,\Omega} + \frac{1}{8\,\Omega}\right)^{-1} = (0.4417\,\Omega^{-1})^{-1} = 2.26\,\Omega$$

The equivalent circuit is shown in **FIGURE 28.23a**, from which we find the current to be

$$I = \frac{\mathcal{E}}{R_{eq}} = \frac{9\,\text{V}}{2.26\,\Omega} = 3.98\,\text{A}$$

The potential difference across R_{eq} is $\Delta V_{eq} = \mathcal{E} = 9.0$ V. Now we have to be careful. Current I divides at the junction into the smaller currents I_1, I_2, and I_3 shown in **FIGURE 28.23b**. However, the division is *not* into three equal currents. According to Ohm's law, resistor i has current $I_i = \Delta V_i / R_i$. Because the three resistors are each connected to the battery by ideal wires, as is the equivalent resistor, their potential differences are equal:

$$\Delta V_1 = \Delta V_2 = \Delta V_3 = \Delta V_{eq} = 9.0\,\text{V}$$

Thus the currents are

$$I_1 = \frac{9\,\text{V}}{15\,\Omega} = 0.60\,\text{A} \qquad I_2 = \frac{9\,\text{V}}{4\,\Omega} = 2.25\,\text{A} \qquad I_3 = \frac{9\,\text{V}}{8\,\Omega} = 1.13\,\text{A}$$

ASSESS The *sum* of the three currents is 3.98 A, as required by Kirchhoff's junction law.

FIGURE 28.23 The parallel resistors can be replaced by a single equivalent resistor.

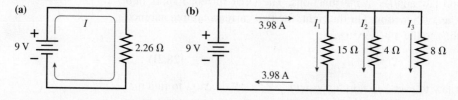

The result of Example 28.7 seems surprising. The equivalent of a parallel combination of 15 Ω, 4 Ω, and 8 Ω was found to be 2.26 Ω. How can the equivalent of a group of resistors be *less* than any single resistance in the group? Shouldn't more resistors imply more resistance? The answer is yes for resistors in series but not for resistors in parallel. Even though a resistor is an obstacle to the flow of charge, parallel resistors provide more pathways for charge to get through. Consequently, the equivalent of several resistors in parallel is always *less* than any single resistor in the group.

Complex combinations of resistors can often be reduced to a single equivalent resistance through a step-by-step application of the series and parallel rules. The final example in this section illustrates this idea.

Summary of series and parallel resistors

	I	ΔV
Series	Same	Add
Parallel	Add	Same

EXAMPLE 28.8 | A combination of resistors

What is the equivalent resistance of the group of resistors shown in FIGURE 28.24?

FIGURE 28.24 A combination of resistors.

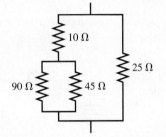

MODEL This circuit contains both series and parallel resistors.

SOLVE Reduction to a single equivalent resistance is best done in a series of steps, with the circuit being redrawn after each step. The procedure is shown in FIGURE 28.25. Note that the 10 Ω and 25 Ω resistors are *not* in parallel. They are connected at their top ends

but not at their bottom ends. Resistors must be connected to each other at *both* ends to be in parallel. Similarly, the 10 Ω and 45 Ω resistors are *not* in series because of the junction between them. If the original group of four resistors occurred within a larger circuit, they could be replaced with a single 15.4 Ω resistor without having any effect on the rest of the circuit.

FIGURE 28.25 The combination is reduced to a single equivalent resistor.

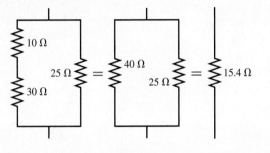

To return to the lightbulb question that began this section, FIGURE 28.26 has redrawn the circuit with each bulb shown as a resistance R. Initially, before the switch is closed, bulbs A and B are in series with equivalent resistance $2R$. The current from the battery is

$$I_{before} = \frac{\mathcal{E}}{2R} = \frac{1}{2}\frac{\mathcal{E}}{R}$$

This is the current in both bulbs.

Closing the switch places bulbs B and C in parallel. The equivalent resistance of two identical resistors in parallel is $R_{eq} = \frac{1}{2}R$. This equivalent resistance of B and C is in series with bulb A; hence the total resistance of the circuit is $\frac{3}{2}R$ and the current leaving the battery is

$$I_{after} = \frac{\mathcal{E}}{3R/2} = \frac{2}{3}\frac{\mathcal{E}}{R} > I_{before}$$

Closing the switch *decreases* the circuit resistance and thus *increases* the current leaving the battery.

All the charge flows through A, so A *increases* in brightness when the switch is closed. The current I_{after} then splits at the junction. Bulbs B and C have equal resistance, so the current splits equally. The current in B is $\frac{1}{3}(\mathcal{E}/R)$, which is *less* than I_{before}. Thus B *decreases* in brightness when the switch is closed. Bulb C has the same brightness as bulb B.

FIGURE 28.26 The lightbulbs of Figure 28.20 with the switch open and closed.

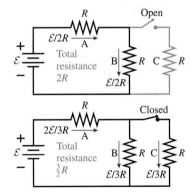

Voltmeters

A device that measures the potential difference across a circuit element is called a **voltmeter.** Because potential difference is measured *across* a circuit element, from

FIGURE 28.27 A voltmeter measures the potential difference across an element.

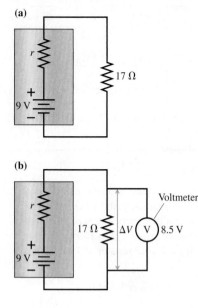

one side to the other, a voltmeter is placed in *parallel* with the circuit element whose potential difference is to be measured.

FIGURE 28.27a shows a simple circuit in which a 17 Ω resistor is connected across a 9 V battery with an unknown internal resistance. By connecting a voltmeter across the resistor, as shown in FIGURE 28.27b, we can measure the potential difference across the resistor. Unlike an ammeter, using a voltmeter does *not* require us to break the connections.

Because the voltmeter is now in parallel with the resistor, the total resistance seen by the battery is $R_{eq} = (1/17\ \Omega + 1/R_{voltmeter})^{-1}$. In order that the voltmeter measure the voltage without changing the voltage, the voltmeter's resistance must, in this case, be $\gg 17\ \Omega$. Indeed, an *ideal voltmeter* has $R_{voltmeter} = \infty\ \Omega$, and thus has no effect on the voltage. Real voltmeters come very close to this ideal, and we will always assume them to be so.

The voltmeter in Figure 28.27b reads 8.5 V. This is less than $\mathcal{E}$ because of the battery's internal resistance. Equation 28.18 found an expression for the resistor's potential difference ΔV_R. That equation is easily solved for the internal resistance r:

$$r = \frac{\mathcal{E} - \Delta V_R}{\Delta V_R} R = \frac{0.5\ \text{V}}{8.5\ \text{V}}\, 17\ \Omega = 1.0\ \Omega$$

Here a voltmeter reading was the one piece of experimental data we needed in order to determine the battery's internal resistance.

STOP TO THINK 28.5 Rank in order, from brightest to dimmest, the identical bulbs A to D.

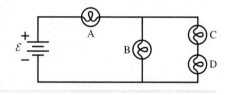

28.7 Resistor Circuits

We can use the information in this chapter to analyze a variety of more complex but more realistic circuits.

PROBLEM-SOLVING STRATEGY 28.1 (MP)

Resistor circuits

MODEL Model wires as ideal and, where appropriate, batteries as ideal.

VISUALIZE Draw a circuit diagram. Label all known and unknown quantities.

SOLVE Base your mathematical analysis on Kirchhoff's laws and on the rules for series and parallel resistors.

- Step by step, reduce the circuit to the smallest possible number of equivalent resistors.
- Write Kirchhoff's loop law for each independent loop in the circuit.
- Determine the current through and the potential difference across the equivalent resistors.
- Rebuild the circuit, using the facts that the current is the same through all resistors in series and the potential difference is the same for all parallel resistors.

ASSESS Use two important checks as you rebuild the circuit.

- Verify that the sum of the potential differences across series resistors matches ΔV for the equivalent resistor.
- Verify that the sum of the currents through parallel resistors matches I for the equivalent resistor.

Exercise 26

EXAMPLE 28.9 | **Analyzing a complex circuit**

Find the current through and the potential difference across each of the four resistors in the circuit shown in FIGURE 28.28.

FIGURE 28.28 A complex resistor circuit.

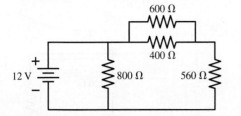

MODEL Assume an ideal battery, with no internal resistance, and ideal connecting wires.

VISUALIZE Figure 28.28 shows the circuit diagram. We'll keep redrawing the diagram as we analyze the circuit.

SOLVE First, we break the circuit down, step by step, into one with a single resistor. FIGURE 28.29a shows this done in three steps. The final battery-and-resistor circuit is our basic circuit, with current

$$I = \frac{\mathcal{E}}{R} = \frac{12 \text{ V}}{400 \text{ } \Omega} = 0.030 \text{ A} = 30 \text{ mA}$$

The potential difference across the 400 Ω resistor is $\Delta V_{400} = \Delta V_{bat} = \mathcal{E} = 12$ V.

Second, we rebuild the circuit, step by step, finding the currents and potential differences at each step. FIGURE 28.29b repeats the steps of Figure 28.29a exactly, but in reverse order. The 400 Ω resistor came from two 800 Ω resistors in parallel. Because $\Delta V_{400} = 12$ V, it must be true that each $\Delta V_{800} = 12$ V. The current through each 800 Ω is then $I = \Delta V/R = 15$ mA. The checkpoint is to note that 15 mA + 15 mA = 30 mA.

The right 800 Ω resistor was formed by 240 Ω and 560 Ω in series. Because $I_{800} = 15$ mA, it must be true that $I_{240} = I_{560} = 15$ mA. The potential difference across each is $\Delta V = IR$, so $\Delta V_{240} = 3.6$ V and $\Delta V_{560} = 8.4$ V. Here the checkpoint is to note that 3.6 V + 8.4 V = 12 V = ΔV_{800}, so the potential differences add as they should.

Finally, the 240 Ω resistor came from 600 Ω and 400 Ω in parallel, so they each have the same 3.6 V potential difference as their 240 Ω equivalent. The currents are $I_{600} = 6$ mA and $I_{400} = 9$ mA. Note that 6 mA + 9 mA = 15 mA, which is our third checkpoint. We now know all currents and potential differences.

ASSESS We *checked our work* at each step of the rebuilding process by verifying that currents summed properly at junctions and that potential differences summed properly along a series of resistances. This "check as you go" procedure is extremely important. It provides you, the problem solver, with a built-in error finder that will immediately inform you if a mistake has been made.

FIGURE 28.29 The step-by-step circuit analysis.

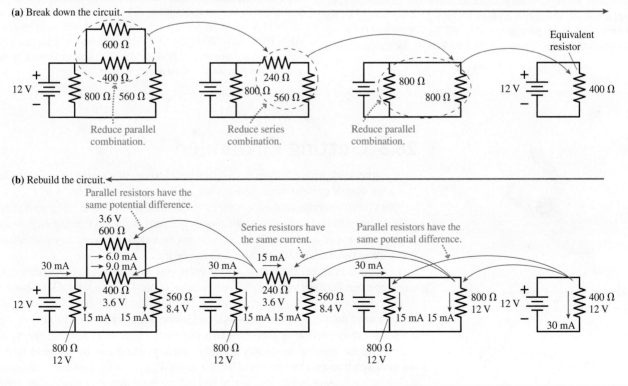

EXAMPLE 28.10 | Analyzing a two-loop circuit

Find the current through and the potential difference across the 100 Ω resistor in the circuit of FIGURE 28.30.

FIGURE 28.30 A two-loop circuit.

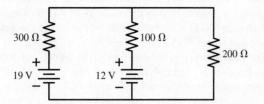

MODEL Assume ideal batteries and ideal connecting wires.

VISUALIZE Figure 28.30 shows the circuit diagram. None of the resistors are connected in series or in parallel, so this circuit cannot be reduced to a simpler circuit.

SOLVE Kirchhoff's loop law applies to *any* loop. To analyze a multiloop problem, we need to write a loop-law equation for each loop. **FIGURE 28.31** redraws the circuit and defines clockwise currents I_1 in the left loop and I_2 in the right loop. But what about the middle branch? Let's assign a downward current I_3 to the middle branch. If we apply Kirchhoff's junction law $\sum I_{in} = \sum I_{out}$ to the junction above the 100 Ω resistor, as shown in the blow-up of Figure 28.31, we see that $I_1 = I_2 + I_3$ and thus $I_3 = I_1 - I_2$. If I_3 ends up being a positive number, then the current in the middle branch really is downward. A negative I_3 will signify an upward current.

Kirchhoff's loop law for the left loop, going clockwise from the lower-left corner, is

$$\sum (\Delta V)_i = 19\ \text{V} - (300\ \Omega)I_1 - (100\ \Omega)I_3 - 12\ \text{V} = 0$$

We're traveling through the 100 Ω resistor in the direction of I_3, the "downhill" direction, so the potential decreases. The 12 V battery is traversed positive to negative, so there we have $\Delta V = -\mathcal{E} = -12$ V. For the right loop, we're going to travel "uphill" through the 100 Ω resistor, opposite to I_3, and gain potential. Thus the loop law for the right loop is

$$\sum (\Delta V)_i = 12\ \text{V} + (100\ \Omega)I_3 - (200\ \Omega)I_2 = 0$$

FIGURE 28.31 Applying Kirchhoff's laws.

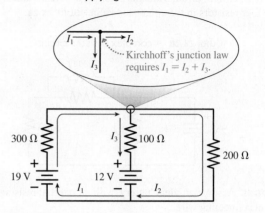

If we substitute $I_3 = I_1 - I_2$ and then rearrange the terms, we find that the two independent loops have given us two simultaneous equations in the two unknowns I_1 and I_2:

$$400I_1 - 100I_2 = 7$$
$$-100I_1 + 300I_2 = 12$$

We can eliminate I_2 by multiplying through the first equation by 3 and then adding the two equations. This gives $1100I_1 = 33$, from which $I_1 = 0.030$ A = 30 mA. Using this value in either of the two loop equations gives $I_2 = 0.050$ A = 50 mA. Because $I_2 > I_1$, the current through the 100 Ω resistor is $I_3 = I_1 - I_2 = -20$ mA, or, because of the minus sign, 20 mA upward. The potential difference across the 100 Ω resistor is $\Delta V_{100\ \Omega} = I_3 R = 2.0$ V, with the bottom end more positive.

ASSESS The three "legs" of the circuit are in parallel, so they must have the same potential difference across them. The left leg has $\Delta V = 19$ V $- (0.030$ A$)(300\ \Omega) = 10$ V, the middle leg has $\Delta V = 12$ V $- (0.020$ A$)(100\ \Omega) = 10$ V, and the right leg has $\Delta V = (0.050$ A$)(200\ \Omega) = 10$ V. Consistency checks such as these are very important. Had we made a numerical error in our circuit analysis, we would have caught it at this point.

28.8 Getting Grounded

The circular prong of a three-prong plug is a connection to ground.

People who work with electronics are often heard to say that something is "grounded." It always sounds quite serious, perhaps somewhat mysterious. What is it?

The circuit analysis procedures we have discussed so far deal only with potential *differences*. Although we are free to choose the zero point of potential anywhere that is convenient, our analysis of circuits has not revealed any need to establish a zero point. Potential differences are all we have needed.

Difficulties can begin to arise, however, if you want to connect two *different* circuits together. Perhaps you would like to connect your DVD to your television or your computer monitor to the computer itself. Incompatibilities can arise unless all the circuits to be connected have a *common* reference point for the potential.

You learned previously that the earth itself is a conductor. Suppose we have two circuits. If we connect *one* point of each circuit to the earth by an ideal wire, and we also agree to call the potential of the earth $V_{earth} = 0$ V, then both circuits have a common reference point. But notice something very important: *one* wire connects the circuit to the earth, but there is not a second wire returning to the circuit. That is, the wire connecting the circuit to the earth is not part of a complete circuit, so there is *no current* in this wire! Because the wire is an equipotential, it gives one point in the circuit the same potential as the earth, but it does *not* in any way change how the circuit functions. A circuit connected to the earth in this way is said to be **grounded,** and the wire is called the *ground wire.*

FIGURE 28.32a shows a fairly simple circuit with a 10 V battery and two resistors in series. The symbol beneath the circuit is the *ground symbol.* It indicates that a wire has been connected between the negative battery terminal and the earth, but the presence of the ground wire does not affect the circuit's behavior. The total resistance is 8 Ω + 12 Ω = 20 Ω, so the current in the loop is $I = (10 \text{ V})/(20 \text{ }\Omega) = 0.50$ A. The potential differences across the two resistors are found, using Ohm's law, to be $\Delta V_8 = 4$ V and $\Delta V_{12} = 6$ V. These are the same values that we would find if the ground wire were *not* present. So what has grounding the circuit accomplished?

FIGURE 28.32b shows the actual potential at several points in the circuit. By definition, $V_{\text{earth}} = 0$ V. The negative battery terminal and the bottom of the 12 Ω resistor are connected by ideal wires to the earth, so *the* potential at these two points must also be zero. The positive terminal of the battery is 10 V more positive than the negative terminal, so $V_{\text{neg}} = 0$ V implies $V_{\text{pos}} = +10$ V. Similarly, the fact that the potential *decreases* by 6 V as charge flows through the 12 Ω resistor now implies that *the* potential at the junction of the resistors must be +6 V. The potential difference across the 8 Ω resistor is 4 V, so the top has to be at +10 V. This agrees with the potential at the positive battery terminal, as it must because these two points are connected by an ideal wire.

All that grounding the circuit does is allow us to have *specific values* for the potential at each point in the circuit. Now we can say "The voltage at the resistor junction is 6 V," whereas before all we could say was "There is a 6 V potential difference across the 12 Ω resistor."

There is one important lesson from this: **Being grounded does not affect the circuit's behavior under normal conditions.** You cannot use "because it is grounded" to *explain* anything about a circuit's behavior.

We added "under normal conditions" because there is one exception. Most circuits are enclosed in a case of some sort that is held away from the circuit with insulators. Sometimes a circuit breaks or malfunctions in such a way that the case comes into electrical contact with the circuit. If the circuit uses high voltage, or even ordinary 120 V household voltage, anyone touching the case could be injured or killed by electrocution. To prevent this, many appliances or electrical instruments have the case itself grounded. Grounding ensures that the potential of the case will always remain at 0 V and be safe. If a malfunction occurs that connects the case to the circuit, a large current will pass through the ground wire to the earth and cause a fuse to blow. This is the *only* time the ground wire would ever have a current, and it is *not* a normal operation of the circuit.

FIGURE 28.32 A circuit that is grounded at one point.

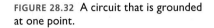
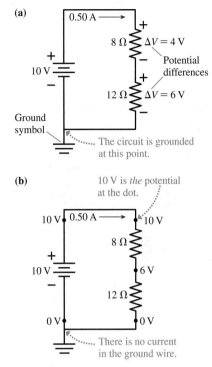

(a)

(b)

EXAMPLE 28.11 | A grounded circuit

Suppose the circuit of Figure 28.32 were grounded at the junction between the two resistors instead of at the bottom. Find the potential at each corner of the circuit.

VISUALIZE FIGURE 28.33 shows the new circuit. (It is customary to draw the ground symbol so that its "point" is always down.)

SOLVE Changing the ground point does not affect the circuit's behavior. The current is still 0.50 A, and the potential differences across the two resistors are still 4 V and 6 V. All that has happened is that we have moved the $V = 0$ V reference point. Because the earth has $V_{\text{earth}} = 0$ V, the junction itself now has a potential of 0 V. The potential decreases by 4 V as charge flows through the 8 Ω resistor.

FIGURE 28.33 Circuit of Figure 28.32 grounded at the point between the resistors.

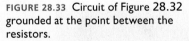
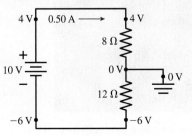

Continued

Because it *ends* at 0 V, the potential at the top of the 8 Ω resistor must be +4 V. Similarly, the potential decreases by 6 V through the 12 Ω resistor. Because it *starts* at 0 V, the bottom of the 12 Ω resistor must be at −6 V. The negative battery terminal is at the same potential as the bottom of the 12 Ω resistor, because they are connected by a wire, so $V_{neg} = -6$ V. Finally, the potential increases by 10 V as the charge flows through the battery, so $V_{pos} = +4$ V, in agreement, as it should be, with the potential at the top of the 8 Ω resistor.

ASSESS A negative voltage means only that the potential at that point is less than the potential at some other point that we chose to call $V = 0$ V. Only potential *differences* are physically meaningful, and only potential differences enter into Ohm's law: $I = \Delta V/R$. The potential difference across the 12 Ω resistor in this example is 6 V, decreasing from top to bottom, regardless of which point we choose to call $V = 0$ V.

28.9 RC Circuits

A resistor circuit has a steady current. By adding a capacitor and a switch, we can make a circuit in which the current varies with time as the capacitor charges and discharges. Circuits with resistors and capacitors are called **RC circuits.** *RC* circuits are at the heart of timekeeping circuits in applications ranging from the intermittent windshield wipers on your car to computers and other digital electronics.

FIGURE 28.34a shows a charged capacitor, a switch, and a resistor. The capacitor has charge Q_0 and potential difference $\Delta V_0 = Q_0/C$. There is no current, so the potential difference across the resistor is zero. Then, at $t = 0$, the switch closes and the capacitor begins to discharge through the resistor.

How long does the capacitor take to discharge? How does the current through the resistor vary as a function of time? To answer these questions, **FIGURE 28.34b** shows the circuit at some point in time after the switch was closed.

Kirchhoff's loop law is valid for any circuit, not just circuits with batteries. The loop law applied to the circuit of Figure 28.34b, going around the loop cw, is

$$\Delta V_{cap} + \Delta V_{res} = \frac{Q}{C} - IR = 0 \qquad (28.25)$$

Q and I in this equation are the *instantaneous* values of the capacitor charge and the resistor current.

The current I is the rate at which charge flows through the resistor: $I = dq/dt$. But the charge flowing through the resistor is charge that was *removed* from the capacitor. That is, an infinitesimal charge dq flows through the resistor when the capacitor charge *decreases* by dQ. Thus $dq = -dQ$, and the resistor current is related to the instantaneous capacitor charge by

$$I = -\frac{dQ}{dt} \qquad (28.26)$$

Now I is positive when Q is decreasing, as we would expect. The reasoning that has led to Equation 28.26 is rather subtle but very important. You'll see the same reasoning later in other contexts.

If we substitute Equation 28.26 into Equation 28.25 and then divide by R, the loop law for the *RC* circuit becomes

$$\frac{dQ}{dt} + \frac{Q}{RC} = 0 \qquad (28.27)$$

Equation 28.27 is a first-order differential equation for the capacitor charge Q, but one that we can solve by direct integration. First, we rearrange Equation 28.27 to get all the charge terms on one side of the equation:

$$\frac{dQ}{Q} = -\frac{1}{RC} dt$$

The product RC is a constant for any particular circuit.

FIGURE 28.34 Discharging a capacitor.

(a) Before the switch closes

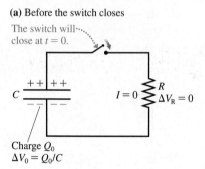

The switch will close at $t = 0$.

Charge Q_0
$\Delta V_0 = Q_0/C$

(b) After the switch closes

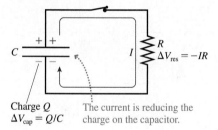

Charge Q
$\Delta V_{cap} = Q/C$

The current is reducing the charge on the capacitor.

The rear flasher on a bike helmet flashes on and off. The timing is controlled by an *RC* circuit.

The capacitor charge was Q_0 at $t = 0$ when the switch was closed. We want to integrate from these starting conditions to charge Q at a later time t. That is,

$$\int_{Q_0}^{Q} \frac{dQ}{Q} = -\frac{1}{RC} \int_0^t dt \qquad (28.28)$$

Both are well-known integrals, giving

$$\ln Q \Big|_{Q_0}^{Q} = \ln Q - \ln Q_0 = \ln\left(\frac{Q}{Q_0}\right) = -\frac{t}{RC}$$

We can solve for the capacitor charge Q by taking the exponential of both sides, then multiplying by Q_0. Doing so gives

$$Q = Q_0 e^{-t/RC} \qquad (28.29)$$

Notice that $Q = Q_0$ at $t = 0$, as expected.

The argument of an exponential function must be dimensionless, so the quantity RC must have dimensions of time. It is useful to define the **time constant** τ to be

$$\tau = RC \qquad (28.30)$$

We can then write Equation 28.29 as

$$Q = Q_0 e^{-t/\tau} \quad \text{(capacitor discharging)} \qquad (28.31)$$

The capacitor voltage, directly proportional to the charge, also decays exponentially as

$$\Delta V_C = \Delta V_0 e^{-t/\tau} \qquad (28.32)$$

The meaning of Equation 28.31 is easier to understand if we portray it graphically. **FIGURE 28.35a** shows the capacitor charge as a function of time. The charge decays exponentially, starting from Q_0 at $t = 0$ and asymptotically approaching zero as $t \rightarrow \infty$. The time constant τ is the time at which the charge has decreased to e^{-1} (about 37%) of its initial value. At time $t = 2\tau$, the charge has decreased to e^{-2} (about 13%) of its initial value. A voltage graph would have the same shape.

NOTE The *shape* of the graph of Q is always the same, regardless of the specific value of the time constant τ.

We find the resistor current by using Equation 28.26:

$$I = -\frac{dQ}{dt} = \frac{Q_0}{\tau} e^{-t/\tau} = \frac{Q_0}{RC} e^{-t/\tau} = \frac{\Delta V_0}{R} e^{-t/\tau} = I_0 e^{-t/\tau} \qquad (28.33)$$

where $I_0 = \Delta V_0 / R$ is the initial current, immediately after the switch closes. **FIGURE 28.35b** is a graph of the resistor current versus t. You can see that the current undergoes the same exponential decay, with the same time constant, as the capacitor charge.

NOTE There's no specific time at which the capacitor has been discharged, because Q approaches zero asymptotically, but the charge and current have dropped to less than 1% of their initial values at $t = 5\tau$. Thus 5τ is a reasonable answer to the question "How long does it take to discharge a capacitor?"

FIGURE 28.35 The decay curves of the capacitor charge and the resistor current.

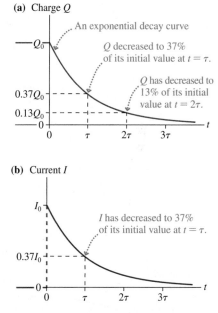

(a) Charge Q

An exponential decay curve

Q decreased to 37% of its initial value at $t = \tau$.

Q has decreased to 13% of its initial value at $t = 2\tau$.

(b) Current I

I has decreased to 37% of its initial value at $t = \tau$.

EXAMPLE 28.12 | Measuring capacitance

To determine the capacitance of an unmarked capacitor, you set up the circuit shown in **FIGURE 28.36**. After holding the switch in position a for several seconds, you suddenly flip it—at $t = 0$ s—to position b while monitoring the resistor voltage with a voltmeter. Your measurements are shown in the table. What is the capacitance? And what was the resistor current 5.0 s after the switch changed position?

Time (s)	Voltage (V)
0.0	9.0
2.0	5.4
4.0	2.7
6.0	1.6
8.0	1.0

FIGURE 28.36 An *RC* circuit for measuring capacitance.

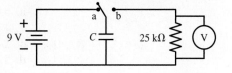

Continued

MODEL The battery charges the capacitor to 9.0 V. Then, when the switch is flipped to position b, the capacitor discharges through the 25,000 Ω resistor with time constant $\tau = RC$.

SOLVE With the switch in position b, the resistor is in parallel with the capacitor and both have the same potential difference $\Delta V_R = \Delta V_C = Q/C$ at all times. The capacitor charge decays exponentially as

$$Q = Q_0 e^{-t/\tau}$$

Consequently, the resistor (and capacitor) voltage also decays exponentially:

$$\Delta V_R = \frac{Q_0}{C} e^{-t/\tau} = \Delta V_0 e^{-t/\tau}$$

where $\Delta V_0 = 9.0$ V is the potential difference at the instant the switch closes. To analyze exponential decays, we take the natural logarithm of both sides. This gives

$$\ln(\Delta V_R) = \ln(\Delta V_0) + \ln(e^{-t/\tau}) = \ln(\Delta V_0) - \frac{1}{\tau} t$$

This result tells us that a graph of $\ln(\Delta V_R)$ versus t—a *semi-log graph*—should be linear with y-intercept $\ln(\Delta V_0)$ and slope $-1/\tau$. If this turns out to be true, we can determine τ and hence C from an experimental measurement of the slope.

FIGURE 28.37 is a graph of $\ln(\Delta V_R)$ versus t. It is, indeed, linear with a negative slope. From the y-intercept of the best-fit line, we find $\Delta V_0 = e^{2.20} = 9.0$ V, as expected. This gives us confidence in our analysis. Using the slope, we find

$$\tau = -\frac{1}{\text{slope}} = -\frac{1}{-0.28 \text{ s}^{-1}} = 3.6 \text{ s}$$

With this, we can calculate

$$C = \frac{\tau}{R} = \frac{3.6 \text{ s}}{25,000 \ \Omega} = 1.4 \times 10^{-4} \text{ F} = 140 \ \mu\text{F}$$

The initial current is $I_0 = (9.0 \text{ V})/(25,000 \ \Omega) = 360 \ \mu\text{A}$. Current also decays exponentially with the same time constant, so the current after 5.0 s is

$$I = I_0 e^{-t/\tau} = (360 \ \mu\text{A}) e^{-(5.0 \text{ s})/(3.6 \text{ s})} = 90 \ \mu\text{A}$$

FIGURE 28.37 A semi-log graph of the data.

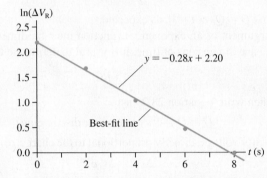

ASSESS The time constant of an exponential decay can be estimated as the time required to decay to one-third of the initial value. Looking at the data, we see that the voltage drops to one-third of the initial 9.0 V in just under 4 s. This is consistent with the more precise $\tau = 3.6$ s, so we have confidence in our results.

Charging a Capacitor

FIGURE 28.38 Charging a capacitor.

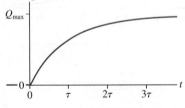

(b) Charge Q

FIGURE 28.38a shows a circuit that charges a capacitor. After the switch is closed, the battery's charge escalator moves charge from the bottom electrode of the capacitor to the top electrode. The resistor, by limiting the current, slows the process but doesn't stop it. The capacitor charges until $\Delta V_C = \mathcal{E}$; then the charging current ceases. The full charge of the capacitor is $Q_0 = C(\Delta V_C)_{\max} = C\mathcal{E}$.

The analysis is much like that of discharging a capacitor. As a homework problem, you can show that the capacitor charge and the circuit current at time t are

$$Q = Q_0(1 - e^{-t/\tau})$$
$$I = I_0 e^{-t/\tau} \qquad \text{(capacitor charging)} \qquad (28.34)$$

where $I_0 = \mathcal{E}/R$ and, again, $\tau = RC$. The capacitor charge's "upside-down decay" to Q_0 is shown graphically in **FIGURE 28.38b**.

STOP TO THINK 28.6 The time constant for the discharge of this capacitor is

a. 5 s.
b. 4 s.
c. 2 s.
d. 1 s.
e. The capacitor doesn't discharge because the resistors cancel each other.

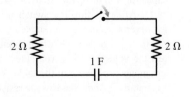

CHALLENGE **EXAMPLE** 28.13 | Energy dissipated during a capacitor discharge

The switch in FIGURE 28.39 has been in position a for a long time. It is suddenly switched to position b for 1.0 s, then back to a. How much energy is dissipated by the 5500 Ω resistor?

FIGURE 28.39 Circuit of a switched capacitor.

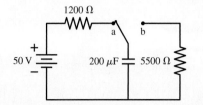

MODEL With the switch in position a, the capacitor charges through the 1200 Ω resistor with time constant $\tau_{charge} = (1200\ \Omega)(2.0 \times 10^{-4}\ F) = 0.24$ s. Because the switch has been in position a for a "long time," which we interpret as being much longer than 0.24 s, we will assume that the capacitor is fully charged to 50 V when the switch is changed to position b. The capacitor then discharges through the 5500 Ω resistor until the switch is returned to position a. Assume ideal wires.

SOLVE Let $t = 0$ s be the time when the switch is moved from a to b, initiating the discharge. The battery and 1200 Ω resistor are irrelevant during the discharge, so the circuit looks like that of Figure 28.34b. The time constant is $\tau = (5500\ \Omega)(2.0 \times 10^{-4}\ F) = 1.1$ s, so the capacitor voltage decreases from 50 V at $t = 0$ s to

$$\Delta V_C = (50\ V)e^{-(1.0\ s)/(1.1\ s)} = 20\ V$$

at $t = 1.0$ s.

There are two ways to determine the energy dissipated in the resistor. We learned in Section 28.3 that a resistor dissipates energy at the rate $dE/dt = P_R = I^2 R$. The current decays exponentially as $I = I_0 \exp(-t/\tau)$, with $I_0 = \Delta V_0/R = 9.09$ mA. We can find the energy dissipated during a time T by integrating:

$$\Delta E = \int_0^T I^2 R\ dt = I_0^2 R \int_0^T e^{-2t/\tau}\ dt = -\tfrac{1}{2}\tau I_0^2 R e^{-2t/\tau}\Big|_0^T$$

$$= \tfrac{1}{2}\tau I_0^2 R(1 - e^{-2T/\tau})$$

The 2 in the exponent appears because we squared the expression for I. Evaluating for $T = 1.0$ s, we find

$$\Delta E = \tfrac{1}{2}(1.1\ s)(0.00909\ A)^2(5500\ \Omega)(1 - e^{-(2.0\ s)/(1.1\ s)}) = 0.21\ J$$

Alternatively, we can use the known capacitor voltages at $t = 0$ s and $t = 1.0$ s and $U_C = \tfrac{1}{2}C(\Delta V_C)^2$ to calculate the energy stored in the capacitor at these times:

$$U_C(t = 0.0\ s) = \tfrac{1}{2}(2.0 \times 10^{-4}\ F)(50\ V)^2 = 0.25\ J$$

$$U_C(t = 1.0\ s) = \tfrac{1}{2}(2.0 \times 10^{-4}\ F)(20\ V)^2 = 0.04\ J$$

The capacitor has lost $\Delta E = 0.21$ J of energy, and this energy was dissipated by the current through the resistor.

ASSESS Not every problem can be solved in two ways, but doing so when it's possible gives us great confidence in our result.

SUMMARY

The goal of Chapter 28 has been to learn the fundamental physical principles that govern electric circuits.

GENERAL STRATEGY

Solving Circuit Problems

MODEL Assume that wires and, where appropriate, batteries are ideal.

VISUALIZE Draw a circuit diagram. Label all quantities.

SOLVE Base the solution on Kirchhoff's laws.

- Reduce the circuit to the smallest possible number of equivalent resistors.
- Write one loop equation for each independent loop.
- Find the current and the potential difference.
- Rebuild the circuit to find I and ΔV for each resistor.

ASSESS Verify that

- The sum of potential differences across series resistors matches ΔV for the equivalent resistor.
- The sum of the currents through parallel resistors matches I for the equivalent resistor.

Kirchhoff's loop law

For a closed loop:

- Assign a direction to the current I.
- $\sum (\Delta V)_i = 0$

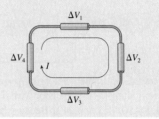

Kirchhoff's junction law

For a junction:

- $\sum I_{in} = \sum I_{out}$

IMPORTANT CONCEPTS

Ohm's Law

A potential difference ΔV between the ends of a conductor with resistance R creates a current

$$I = \frac{\Delta V}{R}$$

Signs of ΔV for Kirchhoff's loop law

Travel →   Travel → I →

$$\Delta V_{bat} = +\mathcal{E} \qquad \Delta V_{bat} = -\mathcal{E} \qquad \Delta V_{res} = -IR$$

The **energy used by a circuit** is supplied by the emf $\mathcal{E}$ of the battery through the energy transformations

$$E_{chem} \rightarrow U \rightarrow K \rightarrow E_{th}$$

The battery *supplies* energy at the rate

$$P_{bat} = I\mathcal{E}$$

The resistors *dissipate* energy at the rate

$$P_R = I\Delta V_R = I^2 R = \frac{(\Delta V_R)^2}{R}$$

APPLICATIONS

Equivalent resistance

Groups of resistors can often be reduced to one equivalent resistor.

Series resistors

$$R_{eq} = R_1 + R_2 + R_3 + \cdots$$

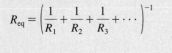

Parallel resistors

$$R_{eq} = \left(\frac{1}{R_1} + \frac{1}{R_2} + \frac{1}{R_3} + \cdots \right)^{-1}$$

RC circuits

The charge on and current through a discharging capacitor are

$$Q = Q_0 e^{-t/\tau}$$

$$I = -\frac{dQ}{dt} = \frac{Q_0}{\tau} e^{-t/\tau} = I_0 e^{-t/\tau}$$

where $\tau = RC$ is the **time constant**.

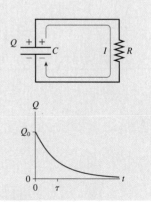

TERMS AND NOTATION

circuit diagram	source	internal resistance, r	grounded
Kirchhoff's junction law	kilowatt hour, kWh	terminal voltage, ΔV_{bat}	RC circuit
Kirchhoff's loop law	series resistors	short circuit	time constant, τ
complete circuit	equivalent resistance, R_{eq}	parallel resistors	
load	ammeter	voltmeter	

CONCEPTUAL QUESTIONS

1. Rank in order, from largest to smallest, the currents I_a to I_d through the four resistors in FIGURE Q28.1.

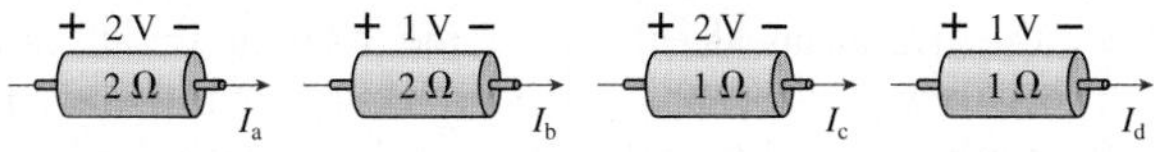

FIGURE Q28.1

2. The tip of a flashlight bulb is touching the top of the 3 V battery in FIGURE Q28.2. Does the bulb light? Why or why not?

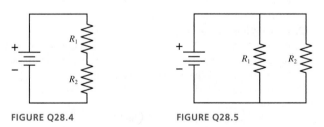

FIGURE Q28.2 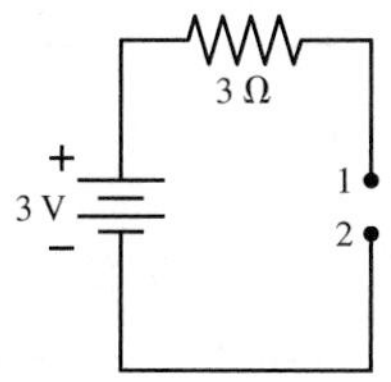   FIGURE Q28.3

3. The wire is broken on the right side of the circuit in FIGURE Q28.3. What is the potential difference $V_1 - V_2$ between points 1 and 2? Explain.

4. The circuit of FIGURE Q28.4 has two resistors, with $R_1 > R_2$. Which of the two resistors dissipates the larger amount of power? Explain.

FIGURE Q28.4   FIGURE Q28.5

5. The circuit of FIGURE Q28.5 has two resistors, with $R_1 > R_2$. Which of the two resistors dissipates the larger amount of power? Explain.

6. Rank in order, from largest to smallest, the powers P_a to P_d dissipated by the four resistors in FIGURE Q28.6.

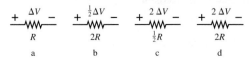

FIGURE Q28.6

7. Are the two resistors in FIGURE Q28.7 in series or in parallel? Explain.

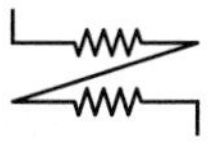

FIGURE Q28.7

8. A battery with internal resistance r is connected to a load resistance R. If R is increased, does the terminal voltage of the battery increase, decrease, or stay the same? Explain.

9. Initially bulbs A and B in FIGURE Q28.9 are glowing. What happens to each bulb if the switch is closed? Does it get brighter, stay the same, get dimmer, or go out? Explain.

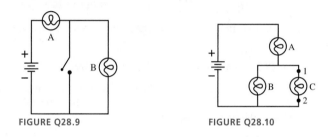

FIGURE Q28.9   FIGURE Q28.10

10. Bulbs A, B, and C in FIGURE Q28.10 are identical, and all are glowing.
    a. Rank in order, from most to least, the brightnesses of the three bulbs. Explain.
    b. Suppose a wire is connected between points 1 and 2. What happens to each bulb? Does it get brighter, stay the same, get dimmer, or go out? Explain.

11. Bulbs A and B in FIGURE Q28.11 are identical, and both are glowing. Bulb B is removed from its socket. Does the potential difference ΔV_{12} between points 1 and 2 increase, stay the same, decrease, or become zero? Explain.

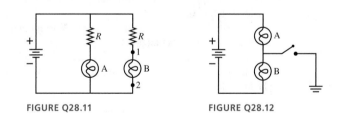

FIGURE Q28.11   FIGURE Q28.12

12. Bulbs A and B in FIGURE Q28.12 are identical, and both are glowing. What happens to each bulb when the switch is closed? Does its brightness increase, stay the same, decrease, or go out? Explain.

13. FIGURE Q28.13 shows the voltage as a function of time of a capacitor as it is discharged (separately) through three different resistors. Rank in order, from largest to smallest, the values of the resistances R_1 to R_3.

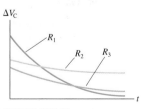

FIGURE Q28.13

EXERCISES AND PROBLEMS

Problems labeled ▨ integrate material from earlier chapters.

Exercises

Section 28.1 Circuit Elements and Diagrams

1. | Draw a circuit diagram for the circuit of FIGURE EX28.1.

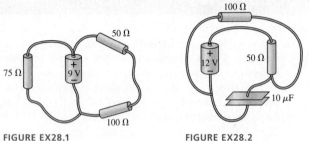

FIGURE EX28.1

FIGURE EX28.2

2. | Draw a circuit diagram for the circuit of FIGURE EX28.2.

Section 28.2 Kirchhoff's Laws and the Basic Circuit

3. ‖ In FIGURE EX28.3, what is the magnitude of the current in the wire to the right of the junction? Does the charge in this wire flow to the right or to the left?

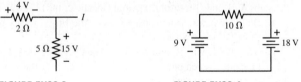

FIGURE EX28.3

FIGURE EX28.4

4. | a. What are the magnitude and direction of the current in the 10 Ω resistor in FIGURE EX28.4?
    b. Draw a graph of the potential as a function of the distance traveled through the circuit, traveling cw from $V = 0$ V at the lower left corner.

5. | a. What are the magnitude and direction of the current in the 18 Ω resistor in FIGURE EX28.5?
    b. Draw a graph of the potential as a function of the distance traveled through the circuit, traveling cw from $V = 0$ V at the lower left corner.

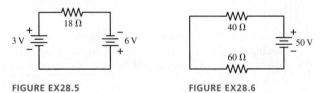

FIGURE EX28.5

FIGURE EX28.6

6. | What is the magnitude of the potential difference across each resistor in FIGURE EX28.6?

Section 28.3 Energy and Power

7. | What is the resistance of a 1500 W (120 V) hair dryer? What is the current in the hair dryer when it is used?

8. | How much power is dissipated by each resistor in FIGURE EX28.8?

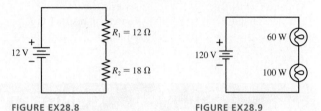

FIGURE EX28.8

FIGURE EX28.9

9. ‖ A 60 W lightbulb and a 100 W lightbulb are placed in the circuit shown in FIGURE EX28.9. Both bulbs are glowing.
   a. Which bulb is brighter? Or are they equally bright?
   b. Calculate the power dissipated by each bulb.

10. ‖ The five identical bulbs in FIGURE EX28.10 are all glowing. The battery is ideal. What is the order of brightness of the bulbs, from brightest to dimmest? Some may be equal.
    A. $P = S > Q = R = T$
    B. $P = S = T > Q = R$
    C. $P > S = T > Q = R$
    D. $P > Q = R > S = T$

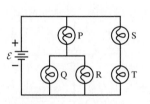

FIGURE EX28.10

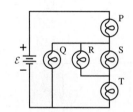

FIGURE EX28.11

11. ‖ The five identical bulbs in FIGURE EX28.11 are all glowing. The battery is ideal. What is the order of brightness of the bulbs, from brightest to dimmest? Some may be equal.
    A. $P = T > Q = R = S$
    B. $P > Q = R = S > T$
    C. $P = T > Q > R = S$
    D. $P > Q > T > R = S$

12. | 1 kWh is how many joules?

13. ‖ A standard 100 W (120 V) lightbulb contains a 7.0-cm-long tungsten filament. The high-temperature resistivity of tungsten is 9.0×10^{-7} Ω m. What is the diameter of the filament?

14. | A typical American family uses 1000 kWh of electricity a month.
    a. What is the average current in the 120 V power line to the house?
    b. On average, what is the resistance of a household?

15. | A waterbed heater uses 450 W of power. It is on 25% of the time, off 75%. What is the annual cost of electricity at a billing rate of $0.12/kWh?

Section 28.4 Series Resistors

Section 28.5 Real Batteries

16. | What is the value of resistor R in FIGURE EX28.16?

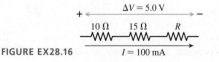

FIGURE EX28.16

17. | The battery in FIGURE EX28.17 is short-circuited by an ideal ammeter having zero resistance.
  a. What is the battery's internal resistance?
  b. How much power is dissipated inside the battery?

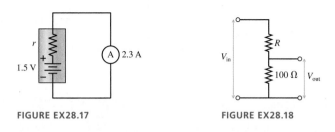

FIGURE EX28.17

FIGURE EX28.18

18. ‖ The circuit in FIGURE EX28.18 is called a *voltage divider*. What value of R will make $V_{out} = V_{in}/10$?

19. ‖ The voltage across the terminals of a 9.0 V battery is 8.5 V when the battery is connected to a 20 Ω load. What is the battery's internal resistance?

20. ‖ Compared to an ideal battery, by what percentage does the battery's internal resistance reduce the potential difference across the 20 Ω resistor in FIGURE EX28.20?

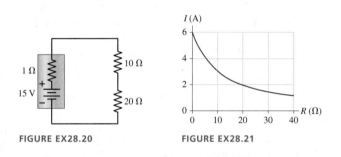

FIGURE EX28.20

FIGURE EX28.21

21. | A variable resistor R is connected across the terminals of a battery. FIGURE EX28.21 shows the current in the circuit as R is varied. What are the emf and internal resistance of the battery?

Section 28.6 Parallel Resistors

22. | Two of the three resistors in FIGURE EX28.22 are unknown but equal. The total resistance between points a and b is 75 Ω. What is the value of R?

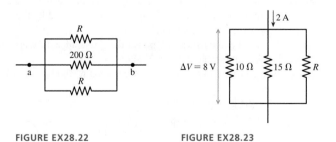

FIGURE EX28.22

FIGURE EX28.23

23. ‖ What is the value of resistor R in FIGURE EX28.23?

24. | A metal wire of resistance R is cut into two pieces of equal length. The two pieces are connected together side by side. What is the resistance of the two connected wires?

25. | What is the equivalent resistance between points a and b in FIGURE EX28.25?

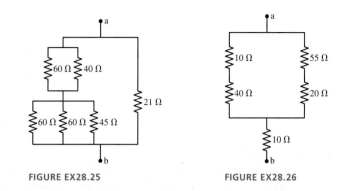

FIGURE EX28.25

FIGURE EX28.26

26. | What is the equivalent resistance between points a and b in FIGURE EX28.26?

27. | What is the equivalent resistance between points a and b in FIGURE EX28.27?

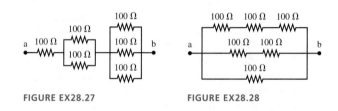

FIGURE EX28.27

FIGURE EX28.28

28. | What is the equivalent resistance between points a and b in FIGURE EX28.28?

29. ‖ The 10 Ω resistor in FIGURE EX28.29 is dissipating 40 W of power. How much power are the other two resistors dissipating?

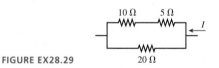

FIGURE EX28.29

Section 28.8 Getting Grounded

30. ‖ In FIGURE EX28.30, what is the value of the potential at points a and b?

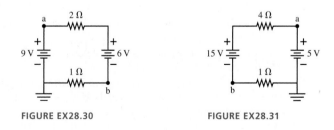

FIGURE EX28.30

FIGURE EX28.31

31. ‖‖ In FIGURE EX28.31, what is the value of the potential at points a and b?

Section 28.9 *RC* Circuits

32. | Show that the product RC has units of s.

33. | What is the time constant for the discharge of the capacitors in **FIGURE EX28.33**?

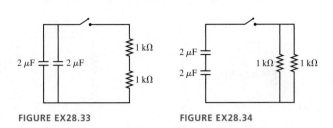

FIGURE EX28.33　　　　**FIGURE EX28.34**

34. | What is the time constant for the discharge of the capacitors in **FIGURE EX28.34**?

35. ‖ A 10 μF capacitor initially charged to 20 μC is discharged through a 1.0 kΩ resistor. How long does it take to reduce the capacitor's charge to 10 μC?

36. | The switch in **FIGURE EX28.36** has been in position a for a long time. It is changed to position b at $t = 0$ s. What are the charge Q on the capacitor and the current I through the resistor (a) immediately after the switch is closed? (b) at $t = 50$ μs? (c) at $t = 200$ μs?

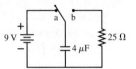

FIGURE EX28.36

37. ‖ What value resistor will discharge a 1.0 μF capacitor to 10% of its initial charge in 2.0 ms?

38. ‖ A capacitor is discharged through a 100 Ω resistor. The discharge current decreases to 25% of its initial value in 2.5 ms. What is the value of the capacitor?

Problems

39. ‖ It seems hard to justify spending $4.00 for a compact fluorescent lightbulb when an ordinary incandescent bulb costs 50¢. To see if this makes sense, compare a 60 W incandescent bulb lasting 1000 hours to a 15 W compact fluorescent bulb having a lifetime of 10,000 hours. Both bulbs produce the same amount of visible light and are interchangeable. If electricity costs $0.10/kWh, what is the total cost—purchase plus energy—to obtain 10,000 hours of light from each type of bulb? This is called the *life-cycle cost*.

40. ‖ A refrigerator has a 1000 W compressor, but the compressor runs only 20% of the time.
    a. If electricity costs $0.10/kWh, what is the monthly (30 day) cost of running the refrigerator?
    b. A more energy-efficient refrigerator with an 800 W compressor costs $100 more. If you buy the more expensive refrigerator, how many months will it take to recover your additional cost?

41. ‖ Two 75 W (120 V) lightbulbs are wired in series, then the combination is connected to a 120 V supply. How much power is dissipated by each bulb?

42. ‖ An electric eel develops a 450 V potential difference between
BIO its head and tail. The eel can stun a fish or other prey by using this potential difference to drive a 0.80 A current pulse for 1.0 ms. What are (a) the energy delivered by this pulse and (b) the total charge that flows?

43. ‖ You have a 2.0 Ω resistor, a 3.0 Ω resistor, a 6.0 Ω resistor, and a 6.0 V battery. Draw a diagram of a circuit in which all three resistors are used and the battery delivers 9.0 W of power.

44. ‖ A 2.0-m-long, 1.0-mm-diameter wire has a variable resistivity
CALC given by

$$\rho(x) = (2.5 \times 10^{-6})\left[1 + \left(\frac{x}{1.0 \text{ m}}\right)^2\right] \Omega \text{ m}$$

where x is measured from one end of the wire. What is the current if this wire is connected to the terminals of a 9.0 V battery?

45. ‖ What is the equivalent resistance between points a and b in **FIGURE P28.45**?

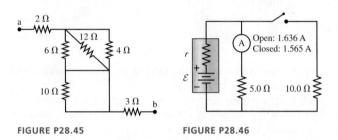

FIGURE P28.45　　　　**FIGURE P28.46**

46. ‖ What are the emf and internal resistance of the battery in **FIGURE P28.46**?

47. ‖ A string of holiday lights can be wired in series, but all the bulbs go out if one burns out because that breaks the circuit. Most lights today are wired in series, but each bulb has a special fuse that short-circuits the bulb—making a connection around it—if it burns out, thus keeping the other lights on. Suppose a string of 50 lights is connected in this way and plugged into a 120 V outlet. By what factor does the power dissipated by each remaining bulb increase when the first bulb burns out?

48. ‖ The circuit shown in **FIGURE P28.48** is inside a 15-cm-diameter balloon filled with helium that is kept at a constant pressure of 1.2 atm. How long will it take the balloon's diameter to increase to 16 cm?

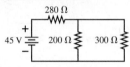

FIGURE P28.48

49. ‖ Suppose you have resistors 2.5 kΩ, 3.5 kΩ, and 4.5 kΩ and a 100 V power supply. What is the ratio of the total power delivered to the resistors if they are connected in parallel to the total power delivered if they are connected in series?

50. ‖ A lightbulb is in series with a 2.0 Ω resistor. The lightbulb dissipates 10 W when this series circuit is connected to a 9.0 V battery. What is the current through the lightbulb? There are two possible answers; give both of them.

51. ‖ a. Load resistor R is attached to a battery of emf $\mathcal{E}$ and
CALC internal resistance r. For what value of the resistance R, in terms of $\mathcal{E}$ and r, will the power dissipated by the load resistor be a maximum?
    b. What is the maximum power that the load can dissipate if the battery has $\mathcal{E} = 9.0$ V and $r = 1.0$ Ω?

52. ‖ The ammeter in **FIGURE P28.52** reads 3.0 A. Find I_1, I_2, and $\mathcal{E}$.

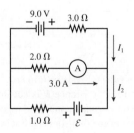

FIGURE P28.52

53. ‖ What is the current in the 2 Ω resistor in FIGURE P28.53?

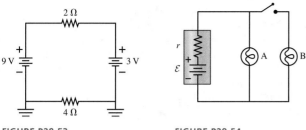

FIGURE P28.53

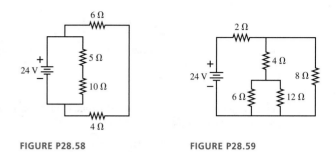

FIGURE P28.54

54. ‖ For an ideal battery $(r = 0\ \Omega)$, closing the switch in FIGURE P28.54 does not affect the brightness of bulb A. In practice, bulb A dims *just a little* when the switch closes. To see why, assume that the 1.50 V battery has an internal resistance $r = 0.50\ \Omega$ and that the resistance of a glowing bulb is $R = 6.00\ \Omega$.
   a. What is the current through bulb A when the switch is open?
   b. What is the current through bulb A after the switch has closed?
   c. By what percentage does the current through A change when the switch is closed?

55. ‖ What are the battery current I_{bat} and the potential difference $V_a - V_b$ between points a and b when the switch in FIGURE P28.55 is (a) open and (b) closed?

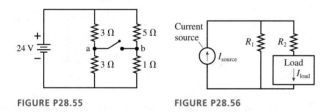

FIGURE P28.55   FIGURE P28.56

56. ‖ A battery is a voltage source, always providing the same potential difference regardless of the current. It is possible to make a *current source* that always provides the same current regardless of the potential difference. The circuit in FIGURE P28.56 is called a *current divider*. It sends a fraction of the source current to the load. Find an expression for I_{load} in terms of R_1, R_2, and I_{source}. You can assume that the load's resistance is much less than R_2.

57. ‖ A circuit you're building needs an ammeter that goes from 0 mA to a full-scale reading of 50 mA. Unfortunately, the only ammeter in the storeroom goes from 0 μA to a full-scale reading of only 500 μA. Fortunately, you've just finished a physics class, and you realize that you can make this ammeter work by putting a resistor in parallel with it, as shown in FIGURE P28.57. You've measured that the resistance of the ammeter is 50.0 Ω, not the 0 Ω of an ideal ammeter.
   a. What value of R must you use so that the meter will go to full scale when the current I is 50 mA?
   b. What is the effective resistance of your ammeter?

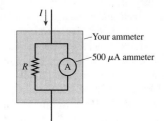

FIGURE P28.57

58. ‖ For the circuit shown in FIGURE P28.58, find the current through and the potential difference across each resistor. Place your results in a table for ease of reading.

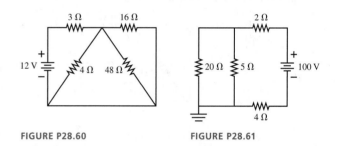

FIGURE P28.58   FIGURE P28.59

59. ‖ For the circuit shown in FIGURE P28.59, find the current through and the potential difference across each resistor. Place your results in a table for ease of reading.

60. ‖ For the circuit shown in FIGURE P28.60, find the current through and the potential difference across each resistor. Place your results in a table for ease of reading.

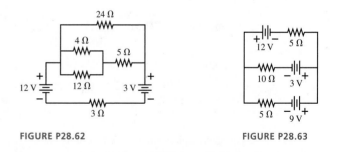

FIGURE P28.60   FIGURE P28.61

61. ‖ What is the current through the 20 Ω resistor in FIGURE P28.61?

62. ‖ For the circuit shown in FIGURE P28.62, find the current through and the potential difference across each resistor. Place your results in a table for ease of reading.

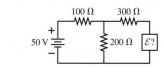

FIGURE P28.62   FIGURE P28.63

63. ‖ What is the current through the 10 Ω resistor in FIGURE P28.63? Is the current from left to right or right to left?

64. ‖ For what emf $\mathcal{E}$ does the 200 Ω resistor in FIGURE P28.64 dissipate no power? Should the emf be oriented with its positive terminal at the top or at the bottom?

FIGURE P28.64

65. ‖ A 12 V car battery dies not so much because its voltage drops but because chemical reactions increase its internal resistance. A good battery connected with jumper cables can both start the engine and recharge the dead battery. Consider the automotive circuit of FIGURE P28.65.

a. How much current could the good battery alone drive through the starter motor?

b. How much current is the dead battery alone able to drive through the starter motor?

c. With the jumper cables attached, how much current passes through the starter motor?

d. With the jumper cables attached, how much current passes through the dead battery, and in which direction?

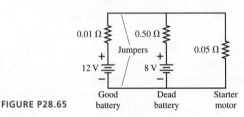

FIGURE P28.65

66. ‖ How much current flows through the bottom wire in FIGURE P28.66, and in which direction?

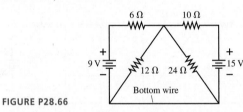

FIGURE P28.66

67. ‖ The capacitor in an RC circuit is discharged with a time constant of 10 ms. At what time after the discharge begins are (a) the charge on the capacitor reduced to half its initial value and (b) the energy stored in the capacitor reduced to half its initial value?

68. ‖ A circuit you're using discharges a 20 μF capacitor through an unknown resistor. After charging the capacitor, you close a switch at $t = 0$ s and then monitor the resistor current with an ammeter. Your data are as follows:

Time (s)	Current (μA)
0.5	890
1.0	640
1.5	440
2.0	270
2.5	200

Use an appropriate graph of the data to determine (a) the resistance and (b) the initial capacitor voltage.

69. ‖ A 150 μF defibrillator capacitor is charged to 1500 V. When
BIO fired through a patient's chest, it loses 95% of its charge in 40 ms. What is the resistance of the patient's chest?

70. ‖ A 50 μF capacitor that had been charged to 30 V is discharged through a resistor. FIGURE P28.70 shows the capacitor voltage as a function of time. What is the value of the resistance?

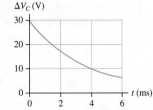

FIGURE P28.70

71. ‖ A 0.25 μF capacitor is charged to 50 V. It is then connected in series with a 25 Ω resistor and a 100 Ω resistor and allowed to discharge completely. How much energy is dissipated by the 25 Ω resistor?

72. | A 70 μF capacitor is discharged through two parallel resistors, 15 kΩ and 25 kΩ. By what factor will the time constant of this circuit increase if the resistors are instead placed in series with each other?

73. ‖ The capacitor in FIGURE P28.73
CALC begins to charge after the switch closes at $t = 0$ s.

a. What is ΔV_C a very long time after the switch has closed?

b. What is Q_{max} in terms of $\mathcal{E}$, R, and C?

c. In this circuit, does $I = +dQ/dt$ or $-dQ/dt$? Explain.

d. Find an expression for the current I at time t. Graph I from $t = 0$ to $t = 5\tau$.

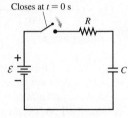

FIGURE P28.73

74. ‖ The capacitors in FIGURE P28.74 are charged and the switch closes at $t = 0$ s. At what time has the current in the 8 Ω resistor decayed to half the value it had immediately after the switch was closed?

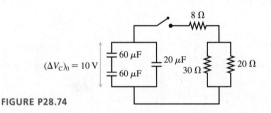

FIGURE P28.74

75. ‖ The flash on a compact camera stores energy in a 120 μF capacitor that is charged to 220 V. When the flash is fired, the capacitor is quickly discharged through a lightbulb with 5.0 Ω of resistance.

a. Light from the flash is essentially finished after two time constants have elapsed. For how long does this flash illuminate the scene?

b. At what rate is the lightbulb dissipating energy 250 μs after the flash is fired?

c. What total energy is dissipated by the lightbulb?

76. ‖ Large capacitors can hold a potentially dangerous charge long
BIO after a circuit has been turned off, so it is important to make sure they are discharged before you touch them. Suppose a 120 μF capacitor from a camera flash unit retains a voltage of 150 V when an unwary student removes it from the camera. If the student accidentally touches the two terminals with his hands, and if the resistance of his body between his hands is 1.8 kΩ, for how long will the current across his chest exceed the danger level of 50 mA?

77. ‖ Digital circuits require actions to take place at precise times, so they are controlled by a *clock* that generates a steady sequence of rectangular voltage pulses. One of the most widely used integrated circuits for creating clock pulses is called a 555 timer. FIGURE P28.77 shows how the timer's output pulses, oscillating between 0 V and 5 V, are controlled with two resistors and a capacitor. The circuit manufacturer tells users that T_H, the time the clock output spends in the high (5 V) state, is $T_H = (R_1 + R_2)C \times \ln 2$. Similarly, the time spent in the low (0 V) state is $T_L = R_2C \times \ln 2$. You need to design a clock that

will oscillate at 10 MHz and will spend 75% of each cycle in the high state. You will be using a 500 pF capacitor. What values do you need to specify for R_1 and R_2?

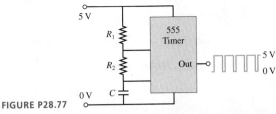

FIGURE P28.77

Challenge Problems

78. ‖ What power is dissipated by the $2\,\Omega$ resistor in FIGURE CP28.78?

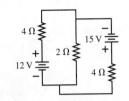

FIGURE CP28.78

79. ‖ You've made the finals of the Science Olympics! As one of your tasks, you're given 1.0 g of aluminum and asked to make a wire, using all the aluminum, that will dissipate 7.5 W when connected to a 1.5 V battery. What length and diameter will you choose for your wire?

80. ‖ The switch in FIGURE CP28.80 has been closed for a very long time.
   a. What is the charge on the capacitor?
   b. The switch is opened at $t = 0$ s. At what time has the charge on the capacitor decreased to 10% of its initial value?

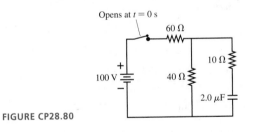

FIGURE CP28.80

81. ‖ The capacitor in Figure 28.38a begins to charge after the
CALC switch closes at $t = 0$ s. Analyze this circuit and show that $Q = Q_{max}(1 - e^{-t/\tau})$, where $Q_{max} = C\mathcal{E}$.

82. ‖ The switch in Figure 28.38a closes at $t = 0$ s and, after a very
CALC long time, the capacitor is fully charged. Find expressions for (a) the total energy supplied by the battery as the capacitor is being charged, (b) total energy dissipated by the resistor as the capacitor is being charged, and (c) the energy stored in the capacitor when it is fully charged. Your expressions will be in terms of $\mathcal{E}$, R, and C. (d) Do your results for parts a to c show that energy is conserved? Explain.

83. ‖ An *oscillator circuit* is important to many applications. A simple oscillator circuit can be built by adding a neon gas tube to an *RC* circuit, as shown in FIGURE CP28.83. Gas is normally a good insulator, and the resistance of the gas tube is essentially infinite when the light is off. This allows the capacitor to charge. When the capacitor voltage reaches a value V_{on}, the electric field inside the tube becomes strong enough to ionize the neon gas. Visually, the tube lights with an orange glow. Electrically, the ionization of the gas provides a very-low-resistance path through the tube. The capacitor very rapidly (we can think of it as instantaneously) discharges through the tube and the capacitor voltage drops. When the capacitor voltage has dropped to a value V_{off}, the electric field inside the tube becomes too weak to sustain the ionization and the neon light turns off. The capacitor then starts to charge again. The capacitor voltage oscillates between V_{off}, when it starts charging, and V_{on}, when the light comes on to discharge it.
   a. Show that the oscillation period is

$$T = RC \ln\left(\frac{\mathcal{E} - V_{off}}{\mathcal{E} - V_{on}}\right)$$

   b. A neon gas tube has $V_{on} = 80$ V and $V_{off} = 20$ V. What resistor value should you choose to go with a 10 μF capacitor and a 90 V battery to make a 10 Hz oscillator?

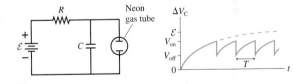

FIGURE CP28.83

29 The Magnetic Field

The aurora occurs when high-energy charged particles from the sun are steered into the upper atmosphere by the earth's magnetic field.

IN THIS CHAPTER, you will learn about magnetism and the magnetic field.

What is magnetism?
Magnetism is an interaction between *moving* charges.

- Magnetic forces, similar to electric forces, are due to the action of magnetic fields.
- A magnetic field $\vec{B}$ is created by a moving charge.
- Magnetic interactions are understood in terms of magnetic poles: north and south.
- Magnetic poles never occur in isolation. All magnets are dipoles, with two poles.
- Practical magnetic fields are created by currents—collections of moving charges.
- Magnetic materials, such as iron, occur because electrons have an inherent magnetic dipole called electron spin.

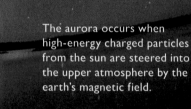

Magnetic field lines

What fields are especially important?
We will develop and use three important magnetic field models.

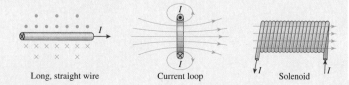

Long, straight wire   Current loop   Solenoid

How do charges respond to magnetic fields?
A charged particle *moving* in a magnetic field experiences a force perpendicular to both $\vec{B}$ and $\vec{v}$. The perpendicular force causes charged particles to move in circular orbits in a uniform magnetic field. This cyclotron motion has many important applications.

« LOOKING BACK Sections 8.2–8.3 Circular motion
« LOOKING BACK Section 12.10 The cross product

How do currents respond to magnetic fields?
Currents are moving charged particles, so:

- There's a force on a current-carrying wire in a magnetic field.
- Two parallel current-carrying wires attract or repel each other.
- There's a torque on a current loop in a magnetic field. This is how motors work.

Why is magnetism important?
Magnetism is much more important than a way to hold a shopping list on the refrigerator door. Motors and generators are based on magnetic forces. Many forms of data storage, from hard disks to the stripe on your credit card, are magnetic. Magnetic resonance imaging (MRI) is essential to modern medicine. Magnetic levitation trains are being built around the world. And the earth's magnetic field keeps the solar wind from sterilizing the surface. There would be no life and no modern technology without magnetism.

29.1 Magnetism

We began our investigation of electricity in Chapter 22 by looking at the results of simple experiments with charged rods. We'll do the same with magnetism.

Discovering magnetism

Experiment 1

If a bar magnet is taped to a piece of cork and allowed to float in a dish of water, it always turns to align itself in an approximate north-south direction. The end of a magnet that points north is called the *north-seeking pole,* or simply the **north pole.** The other end is the **south pole.**

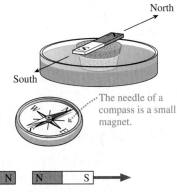

North

South

The needle of a compass is a small magnet.

Experiment 2

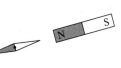

If the north pole of one magnet is brought near the north pole of another magnet, they repel each other. Two south poles also repel each other, but the north pole of one magnet exerts an attractive force on the south pole of another magnet.

Experiment 3

The north pole of a bar magnet attracts one end of a compass needle and repels the other. Apparently the compass needle itself is a little bar magnet with a north pole and a south pole.

Experiment 4

Cutting a bar magnet in half produces two weaker but still complete magnets, each with a north pole and a south pole. No matter how small the magnets are cut, even down to microscopic sizes, each piece remains a complete magnet with two poles.

Experiment 5

Magnets can pick up some objects, such as paper clips, but not all. If an object is attracted to one end of a magnet, it is also attracted to the other end. Most materials, including copper (a penny), aluminum, glass, and plastic, experience no force from a magnet.

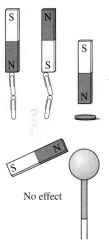

Experiment 6

A magnet does not affect an electroscope. A charged rod exerts a weak *attractive* force on *both* ends of a magnet. However, the force is the same as the force on a metal bar that isn't a magnet, so it is simply a polarization force like the ones we studied in Chapter 22. Other than polarization forces, charges have *no effects* on magnets.

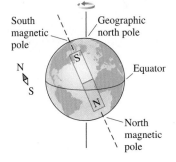

No effect

What do these experiments tell us?

- First, magnetism is not the same as electricity. **Magnetic poles and electric charges share some similar behavior, but they are not the same.**
- Magnetism is a long-range force. Paper clips leap upward to a magnet.
- Magnets have two poles, called north and south poles, and thus are **magnetic dipoles.** Two like poles exert repulsive forces on each other; two opposite poles attract. The behavior is *analogous* to electric charges, but, as noted, magnetic poles and electric charges are *not* the same. Unlike charges, isolated north or south poles do not exist.
- The poles of a bar magnet can be identified by using it as a compass. The poles of other magnets, such as flat refrigerator magnets, can be identified by testing them against a bar magnet. A pole that attracts a known north pole and repels a known south pole must be a south magnetic pole.
- Materials that are attracted to a magnet are called **magnetic materials.** The most common magnetic material is iron. Magnetic materials are attracted to *both* poles of a magnet. This attraction is analogous to how neutral objects are attracted to both positively and negatively charged rods by the polarization force. The difference is that *all* neutral objects are attracted to a charged rod whereas only a few materials are attracted to a magnet.

Our goal is to develop a theory of magnetism to explain these observations.

Compasses and Geomagnetism

The north pole of a compass needle is attracted toward the geographic north pole of the earth. Apparently the earth itself is a large magnet, as shown in **FIGURE 29.1.** The

FIGURE 29.1 The earth is a large magnet.

South magnetic pole

Geographic north pole

N
S

Equator

North magnetic pole

reasons for the earth's magnetism are complex, but geophysicists think that the earth's magnetic poles arise from currents in its molten iron core. Two interesting facts about the earth's magnetic field are (1) that the magnetic poles are offset slightly from the geographic poles of the earth's rotation axis, and (2) that the geographic north pole is actually a *south* magnetic pole! You should be able to use what you have learned thus far to convince yourself that this is the case.

STOP TO THINK 29.1 Does the compass needle rotate clockwise (cw), counterclockwise (ccw), or not at all?

Positive rod

Pivot

29.2 The Discovery of the Magnetic Field

As electricity began to be seriously studied in the 18th century, some scientists speculated that there might be a connection between electricity and magnetism. Interestingly, the link between electricity and magnetism was discovered *in the midst of a classroom lecture demonstration* in 1819 by the Danish scientist Hans Christian Oersted. Oersted was using a battery—a fairly recent invention—to produce a large current in a wire. By chance, a compass was sitting next to the wire, and Oersted noticed that the current caused the compass needle to turn. In other words, the compass responded as if a magnet had been brought near.

Oersted had long been interested in a possible connection between electricity and magnetism, so the significance of this serendipitous observation was immediately apparent to him. Oersted's discovery that **magnetism is caused by an electric current** is illustrated in FIGURE 29.2. Part c of the figure demonstrates an important **right-hand rule** that relates the orientation of the compass needles to the direction of the current.

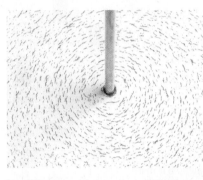
Iron filings reveal the magnetic field around a current-carrying wire.

FIGURE 29.2 Response of compass needles to a current in a straight wire.

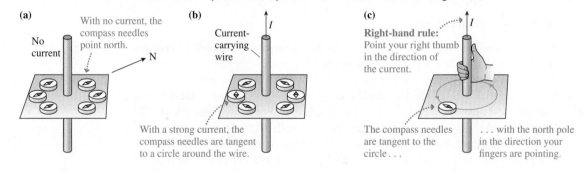

(a)
No current

With no current, the compass needles point north.

N

(b)
Current-carrying wire

I

With a strong current, the compass needles are tangent to a circle around the wire.

(c)

I

Right-hand rule: Point your right thumb in the direction of the current.

The compass needles are tangent to the circle . . .

. . . with the north pole in the direction your fingers are pointing.

FIGURE 29.3 The notation for vectors and currents perpendicular to the page.

× × × ×   • • • •

× × × ×   • • • •

Vectors into page   Vectors out of page

⊗   ⊙

Current into page   Current out of page

Magnetism is more demanding than electricity in requiring a three-dimensional perspective of the sort shown in Figure 29.2. But since two-dimensional figures are easier to draw, we will make as much use of them as we can. Consequently, we will often need to indicate field vectors or currents that are perpendicular to the page. FIGURE 29.3 shows the notation we will use. FIGURE 29.4 demonstrates this notation by showing the compasses around a current that is directed into the page. To use the right-hand rule, point your right thumb in the direction of the current (into the page). Your fingers will curl cw, and that is the direction in which the north poles of the compass needles point.

The Magnetic Field

We introduced the idea of a *field* as a way to understand the long-range electric force. Although this idea appeared rather far-fetched, it turned out to be very useful.

We need a similar idea to understand the long-range force exerted by a current on a compass needle.

Let us define the **magnetic field** $\vec{B}$ as having the following properties:

1. A magnetic field is created at *all* points in space surrounding a current-carrying wire.
2. The magnetic field at each point is a vector. It has both a magnitude, which we call the *magnetic field strength B,* and a direction.
3. The magnetic field exerts forces on magnetic poles. The force on a north pole is parallel to $\vec{B}$; the force on a south pole is opposite $\vec{B}$.

FIGURE 29.5 shows a compass needle in a magnetic field. The field vectors are shown at several points, but keep in mind that the field is present at *all* points in space. A magnetic force is exerted on each of the two poles of the compass, parallel to $\vec{B}$ for the north pole and opposite $\vec{B}$ for the south pole. This pair of opposite forces exerts a torque on the needle, rotating the needle until it is parallel to the magnetic field at that point.

Notice that the north pole of the compass needle, when it reaches the equilibrium position, is in the direction of the magnetic field. Thus a compass needle can be used as a probe of the magnetic field, just as a charge was a probe of the electric field. **Magnetic forces cause a compass needle to become aligned parallel to a magnetic field, with the north pole of the compass showing the direction of the magnetic field at that point.**

Look back at the compass alignments around the current-carrying wire in Figure 29.4. Because compass needles align with the magnetic field, the magnetic field at each point must be tangent to a circle around the wire. **FIGURE 29.6a** shows the magnetic field by drawing field vectors. Notice that the field is weaker (shorter vectors) at greater distances from the wire.

Another way to picture the field is with the use of **magnetic field lines.** These are imaginary lines drawn through a region of space so that

■ A tangent to a field line is in the direction of the magnetic field, and
■ The field lines are closer together where the magnetic field strength is larger.

FIGURE 29.6b shows the magnetic field lines around a current-carrying wire. Notice that magnetic field lines form loops, with no beginning or ending point. This is in contrast to electric field lines, which stop and start on charges.

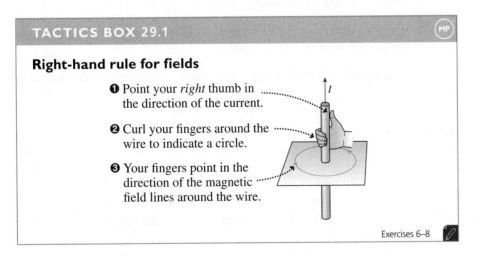

TACTICS BOX 29.1   (MP)

Right-hand rule for fields

❶ Point your *right* thumb in the direction of the current.

❷ Curl your fingers around the wire to indicate a circle.

❸ Your fingers point in the direction of the magnetic field lines around the wire.

Exercises 6–8

NOTE The magnetic field of a current-carrying wire is very different from the electric field of a charged wire. The electric field of a charged wire points radially outward (positive wire) or inward (negative wire).

Two Kinds of Magnetism?

You might be concerned that we have introduced two kinds of magnetism. We opened this chapter discussing permanent magnets and their forces. Then, without warning, we switched to the magnetic forces caused by a current. It is not at all obvious that these

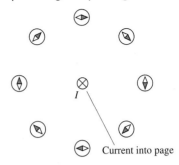

FIGURE 29.4 The orientation of the compasses is given by the right-hand rule.

Current into page

FIGURE 29.5 The magnetic field exerts forces on the poles of a compass.

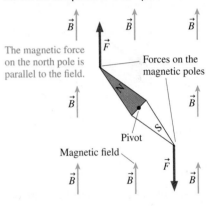

The magnetic force on the north pole is parallel to the field.

Forces on the magnetic poles

Pivot

Magnetic field

FIGURE 29.6 The magnetic field around a current-carrying wire.

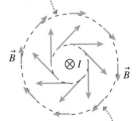

(a) The magnetic field *vectors* are tangent to circles around the wire, pointing in the direction given by the right-hand rule.

The field is weaker farther from the wire.

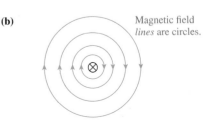

(b)

Magnetic field *lines* are circles.

forces are the same kind of magnetism as that exhibited by stationary chunks of metal called "magnets." Perhaps there are two different types of magnetic forces, one having to do with currents and the other being responsible for permanent magnets. One of the major goals for our study of magnetism is to see that these two quite different ways of producing magnetic effects are really just two different aspects of a *single* magnetic force.

STOP TO THINK 29.2 The magnetic field at position P points

a. Up.
b. Down.
c. Into the page.
d. Out of the page.

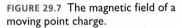

FIGURE 29.7 The magnetic field of a moving point charge.

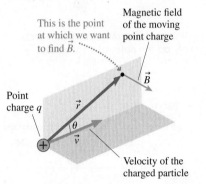

This is the point at which we want to find $\vec{B}$.

Magnetic field of the moving point charge

$\vec{B}$

Point charge q

$\vec{r}$

θ

$\vec{v}$

Velocity of the charged particle

TABLE 29.1 Typical magnetic field strengths

Field source	Field strength (T)
Surface of the earth	5×10^{-5}
Refrigerator magnet	5×10^{-3}
Laboratory magnet	0.1 to 1
Superconducting magnet	10

FIGURE 29.8 Two views of the magnetic field of a moving positive charge.

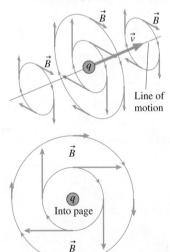

$\vec{B}$

$\vec{B}$

$\vec{v}$

$\vec{B}$

q

Line of motion

$\vec{B}$

$\vec{B}$

q

Into page

$\vec{B}$

29.3 The Source of the Magnetic Field: Moving Charges

Figure 29.6 is a qualitative picture of the wire's magnetic field. Our first task is to turn that picture into a quantitative description. Because current in a wire generates a magnetic field, and a current is a collection of moving charges, our starting point is the idea that **moving charges are the source of the magnetic field.** FIGURE 29.7 shows a charged particle q moving with velocity $\vec{v}$. The magnetic field of this moving charge is found to be

$$\vec{B}_{\text{point charge}} = \left(\frac{\mu_0}{4\pi} \frac{qv \sin\theta}{r^2}, \text{ direction given by the right-hand rule} \right) \quad (29.1)$$

where r is the distance from the charge and θ is the angle between $\vec{v}$ and $\vec{r}$.

Equation 29.1 is called the **Biot-Savart law** for a point charge (rhymes with *Leo* and *bazaar*), named for two French scientists whose investigations were motivated by Oersted's observations. It is analogous to Coulomb's law for the electric field of a point charge. Notice that the Biot-Savart law, like Coulomb's law, is an inverse-square law. However, the Biot-Savart law is somewhat more complex than Coulomb's law because the magnetic field depends on the angle θ between the charge's velocity and the line to the point where the field is evaluated.

> **NOTE** A moving charge has both a magnetic field *and* an electric field. What you know about electric fields has not changed.

The SI unit of magnetic field strength is the **tesla,** abbreviated as T. The tesla is defined as

$$1 \text{ tesla} = 1 \text{ T} \equiv 1 \text{ N/A m}$$

You will see later in the chapter that this definition is based on the magnetic force on a current-carrying wire. One tesla is quite a large field; most magnetic fields are a small fraction of a tesla. TABLE 29.1 lists a few magnetic field strengths.

The constant μ_0 in Equation 29.1 is called the **permeability constant.** Its value is

$$\mu_0 = 4\pi \times 10^{-7} \text{ T m/A} = 1.257 \times 10^{-6} \text{ T m/A}$$

This constant plays a role in magnetism similar to that of the permittivity constant ϵ_0 in electricity.

The right-hand rule for finding the direction of $\vec{B}$ is similar to the rule used for a current-carrying wire: Point your right thumb in the direction of $\vec{v}$. The magnetic field vector $\vec{B}$ is perpendicular to the plane of $\vec{r}$ and $\vec{v}$, pointing in the direction in which your fingers curl. In other words, the $\vec{B}$ vectors are tangent to circles drawn about the charge's line of motion. FIGURE 29.8 shows a more complete view of the magnetic field of a moving positive charge. Notice that $\vec{B}$ is zero along the line of motion, where $\theta = 0°$ or 180°, due to the $\sin\theta$ term in Equation 29.1.

> **NOTE** The vector arrows in Figure 29.8 would have the same lengths but be reversed in direction for a negative charge.

The requirement that a charge be moving to generate a magnetic field is explicit in Equation 29.1. If the speed v of the particle is zero, the magnetic field (but not the electric field!) is zero. This helps to emphasize a fundamental distinction between electric and magnetic fields: **All charges create electric fields, but only *moving* charges create magnetic fields.**

EXAMPLE 29.1 | **The magnetic field of a proton**

A proton moves with velocity $\vec{v} = 1.0 \times 10^7 \,\hat{\imath}$ m/s. As it passes the origin, what is the magnetic field at the (x, y, z) positions (1 mm, 0 mm, 0 mm), (0 mm, 1 mm, 0 mm), and (1 mm, 1 mm, 0 mm)?

MODEL The magnetic field is that of a moving charged particle.

VISUALIZE FIGURE 29.9 shows the geometry. The first point is on the x-axis, directly in front of the proton, with $\theta_1 = 0°$. The second point is on the y-axis, with $\theta_2 = 90°$, and the third is in the xy-plane.

FIGURE 29.9 The magnetic field of Example 29.1.

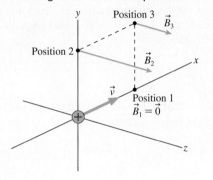

SOLVE Position 1, which is along the line of motion, has $\theta_1 = 0°$. Thus $\vec{B}_1 = \vec{0}$. Position 2 (at 0 mm, 1 mm, 0 mm) is at distance $r_2 = 1$ mm $= 0.001$ m. Equation 29.1, the Biot-Savart law, gives us the magnetic field strength at this point as

$$B = \frac{\mu_0}{4\pi} \frac{qv\sin\theta_2}{r_2^2}$$

$$= \frac{4\pi \times 10^{-7}\,\text{T m/A}}{4\pi} \frac{(1.60 \times 10^{-19}\text{C})(1.0 \times 10^7\,\text{m/s})\sin 90°}{(0.0010\,\text{m})^2}$$

$$= 1.60 \times 10^{-13}\,\text{T}$$

According to the right-hand rule, the field points in the positive z-direction. Thus

$$\vec{B}_2 = 1.60 \times 10^{-13}\,\hat{k}\,\text{T}$$

where $\hat{k}$ is the unit vector in the positive z-direction. The field at position 3, at (1 mm, 1 mm, 0 mm), also points in the z-direction, but it is weaker than at position 2 both because r is larger *and* because θ is smaller. From geometry we know $r_3 = \sqrt{2}$ mm $= 0.00141$ m and $\theta_3 = 45°$. Another calculation using Equation 29.1 gives

$$\vec{B}_3 = 0.57 \times 10^{-13}\,\hat{k}\,\text{T}$$

ASSESS The magnetic field of a single moving charge is *very* small.

Superposition

The Biot-Savart law is the starting point for generating all magnetic fields, just as our earlier expression for the electric field of a point charge was the starting point for generating all electric fields. Magnetic fields, like electric fields, have been found experimentally to obey the principle of superposition. If there are n moving point charges, the net magnetic field is given by the vector sum

$$\vec{B}_{\text{total}} = \vec{B}_1 + \vec{B}_2 + \cdots + \vec{B}_n \tag{29.2}$$

where each individual $\vec{B}$ is calculated with Equation 29.1. The principle of superposition will be the basis for calculating the magnetic fields of several important current distributions.

The Vector Cross Product

In « Section 22.5, we found that the electric field of a point charge can be written

$$\vec{E} = \frac{1}{4\pi\epsilon_0} \frac{q}{r^2} \hat{r}$$

where $\hat{r}$ is a *unit vector* that points from the charge to the point at which we wish to calculate the field. Unit vector $\hat{r}$ expresses the idea "away from q."

The unit vector $\hat{r}$ also allows us to write the Biot-Savart law more concisely, but we'll need to use the form of vector multiplication called the *cross product*. To remind you, FIGURE 29.10 shows two vectors, $\vec{C}$ and $\vec{D}$, with angle α between them. The **cross product** of $\vec{C}$ and $\vec{D}$ is defined to be the vector

$$\vec{C} \times \vec{D} = (CD\sin\alpha, \text{ direction given by the right-hand rule}) \tag{29.3}$$

The symbol $\times$ between the vectors is *required* to indicate a cross product.

FIGURE 29.10 The cross product $\vec{C} \times \vec{D}$ is a vector perpendicular to the plane of vectors $\vec{C}$ and $\vec{D}$.

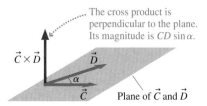

The cross product is perpendicular to the plane. Its magnitude is $CD\sin\alpha$.

Plane of $\vec{C}$ and $\vec{D}$

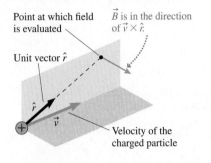

FIGURE 29.11 Unit vector $\hat{r}$ defines the direction from the moving charge to the point at which we evaluate the field.

Point at which field is evaluated

$\vec{B}$ is in the direction of $\vec{v} \times \hat{r}$.

Unit vector $\hat{r}$

$\hat{r}$

$\vec{v}$

Velocity of the charged particle

NOTE The cross product of two vectors and the right-hand rule used to determine the direction of the cross product were introduced in « Section 12.10 to describe torque and angular momentum. A review is worthwhile.

The Biot-Savart law, Equation 29.1, can be written in terms of the cross product as

$$\vec{B}_{\text{point charge}} = \frac{\mu_0}{4\pi} \frac{q\vec{v} \times \hat{r}}{r^2} \quad \text{(magnetic field of a point charge)} \quad (29.4)$$

where unit vector $\hat{r}$, shown in FIGURE 29.11, points from charge q to the point at which we want to evaluate the field. This expression for the magnetic field $\vec{B}$ has magnitude $(\mu_0/4\pi)qv \sin\theta/r^2$ (because the magnitude of $\hat{r}$ is 1) and points in the correct direction (given by the right-hand rule), so it agrees completely with Equation 29.1.

EXAMPLE 29.2 | **The magnetic field direction of a moving electron**

The electron in FIGURE 29.12 is moving to the right. What is the direction of the electron's magnetic field at the dot?

$\vec{v}$

$\hat{r}$

▶ FIGURE 29.12 A moving electron.

VISUALIZE Because the charge is negative, the magnetic field points *opposite* the direction of $\vec{v} \times \hat{r}$. Unit vector $\hat{r}$ points from the charge toward the dot. We can use the right-hand rule to find that $\vec{v} \times \hat{r}$ points *into* the page. Thus the electron's magnetic field at the dot points *out* of the page.

STOP TO THINK 29.3 The positive charge is moving straight out of the page. What is the direction of the magnetic field at the dot?

$\vec{v}$ out of page

a. Up   b. Down   c. Left   d. Right

29.4 The Magnetic Field of a Current

Moving charges are the source of magnetic fields, but the magnetic fields of current-carrying wires—immense numbers of charges moving together—are much more important than the feeble magnetic fields of individual charges. Real current-carrying wires, with their twists and turns, have very complex fields. However, we can once again focus on the physics by using simplified models. It turns out that three common magnetic field models are the basis for understanding a wide variety of magnetic phenomena. We present them here together as a reference; the next few sections of this chapter will be devoted to justifying and explaining these results.

MODEL 29.1

Three key magnetic fields

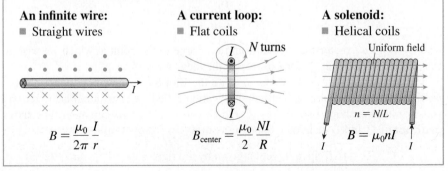

An infinite wire:
- Straight wires

$$B = \frac{\mu_0}{2\pi} \frac{I}{r}$$

A current loop:
- Flat coils

N turns

$$B_{\text{center}} = \frac{\mu_0}{2} \frac{NI}{R}$$

A solenoid:
- Helical coils

Uniform field

$n = N/L$

$$B = \mu_0 nI$$

To begin, we need to write the Biot-Savart law in terms of current. **FIGURE 29.13a** shows a current-carrying wire. The wire as a whole is electrically neutral, but current I represents the motion of positive charge carriers through the wire. Suppose the small amount of moving charge ΔQ spans the small length Δs. The charge has velocity $\vec{v} = \Delta \vec{s}/\Delta t$, where the vector $\Delta \vec{s}$, which is parallel to $\vec{v}$, is the charge's displacement vector. If ΔQ is small enough to treat as a point charge, the magnetic field it creates at a point in space is proportional to $(\Delta Q)\vec{v}$. We can write $(\Delta Q)\vec{v}$ in terms of the wire's current I as

$$(\Delta Q)\vec{v} = \Delta Q \frac{\Delta \vec{s}}{\Delta t} = \frac{\Delta Q}{\Delta t}\Delta \vec{s} = I\Delta \vec{s} \qquad (29.5)$$

where we used the definition of current, $I = \Delta Q/\Delta t$.

If we replace $q\vec{v}$ in the Biot-Savart law with $I\Delta \vec{s}$, we find that the magnetic field of a very short segment of wire carrying current I is

$$\vec{B}_{\text{current segment}} = \frac{\mu_0}{4\pi}\frac{I\Delta \vec{s} \times \hat{r}}{r^2} \qquad (29.6)$$

(magnetic field of a very short segment of current)

Equation 29.6 is still the Biot-Savart law, only now written in terms of current rather than the motion of an individual charge. **FIGURE 29.13b** shows the direction of the current segment's magnetic field as determined by using the right-hand rule.

Equation 29.6 is the basis of a strategy for calculating the magnetic field of a current-carrying wire. You will recognize that it is the same basic strategy you learned for calculating the electric field of a continuous distribution of charge. The goal is to break a problem down into small steps that are individually manageable.

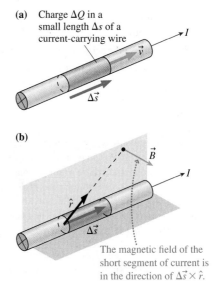

FIGURE 29.13 Relating the charge velocity $\vec{v}$ to the current I.

(a) Charge ΔQ in a small length Δs of a current-carrying wire

(b)

The magnetic field of the short segment of current is in the direction of $\Delta \vec{s} \times \hat{r}$.

PROBLEM-SOLVING STRATEGY 29.1   (MP)

The magnetic field of a current

MODEL Model the wire as a simple shape.

VISUALIZE For the pictorial representation:
- Draw a picture, establish a coordinate system, and identify the point P at which you want to calculate the magnetic field.
- Divide the current-carrying wire into small segments for which you *already know* how to determine $\vec{B}$. This is usually, though not always, a division into very short segments of length Δs.
- Draw the magnetic field vector for one or two segments. This will help you identify distances and angles that need to be calculated.

SOLVE The mathematical representation is $\vec{B}_{\text{net}} = \Sigma \vec{B}_i$.
- Write an algebraic expression for *each* of the three components of $\vec{B}$ (unless you are sure one or more is zero) at point P. Let the (x, y, z)-coordinates of the point remain as variables.
- Express all angles and distances in terms of the coordinates.
- Let $\Delta s \rightarrow ds$ and the sum become an integral. Think carefully about the integration limits for this variable; they will depend on the boundaries of the wire and on the coordinate system you have chosen to use.

ASSESS Check that your result is consistent with any limits for which you know what the field should be.

The key idea here, as it was in Chapter 23, is that **integration is summation.** We need to add up the magnetic field contributions of a vast number of current segments, and we'll do that by letting the sum become an integral.

EXAMPLE 29.3 | The magnetic field of a long, straight wire

A long, straight wire carries current I in the positive x-direction. Find the magnetic field at distance r from the wire.

MODEL Model the wire as being infinitely long.

VISUALIZE FIGURE 29.14 illustrates the steps in the problem-solving strategy. We've chosen a coordinate system with point P on the y-axis. We've then divided the wire into small segments, labeled with index i, each containing a small amount ΔQ of *moving charge*. Unit vector $\hat{r}$ and angle θ_i are shown for segment i. You should use the right-hand rule to convince yourself that $\vec{B}_i$ points *out of the page,* in the positive z-direction. This is the direction no matter where segment i happens to be along the x-axis. Consequently, B_x (the component of $\vec{B}$ parallel to the wire) and B_y (the component of $\vec{B}$ straight away from the wire) are zero. The only component of $\vec{B}$ we need to evaluate is B_z, the component tangent to a circle around the wire.

FIGURE 29.14 Calculating the magnetic field of a long, straight wire carrying current I.

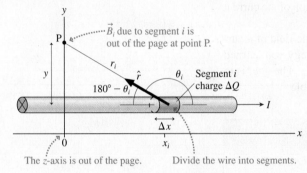

SOLVE We can use the Biot-Savart law to find the field $(B_i)_z$ of segment i. The cross product $\Delta\vec{s}_i \times \hat{r}$ has magnitude $(\Delta x)(1)\sin\theta_i$, hence

$$(B_i)_z = \frac{\mu_0}{4\pi}\frac{I\Delta x\sin\theta_i}{r_i^2} = \frac{\mu_0}{4\pi}\frac{I\sin\theta_i}{r_i^2}\Delta x = \frac{\mu_0}{4\pi}\frac{I\sin\theta_i}{x_i^2+y^2}\Delta x$$

where we wrote the distance r_i in terms of x_i and y. We also need to express θ_i in terms of x_i and y. Because $\sin(180°-\theta)=\sin\theta$, this is

$$\sin\theta_i = \sin(180°-\theta_i) = \frac{y}{r_i} = \frac{y}{\sqrt{x_i^2+y^2}}$$

With this expression for $\sin\theta_i$, the magnetic field of segment i is

$$(B_i)_z = \frac{\mu_0}{4\pi}\frac{Iy}{(x_i^2+y^2)^{3/2}}\Delta x$$

Now we're ready to sum the magnetic fields of all the segments. The superposition is a vector sum, but in this case only the z-components are nonzero. Summing all the $(B_i)_z$ gives

$$B_{\text{wire}} = \frac{\mu_0 Iy}{4\pi}\sum_i\frac{\Delta x}{(x_i^2+y^2)^{3/2}} \rightarrow \frac{\mu_0 Iy}{4\pi}\int_{-\infty}^{\infty}\frac{dx}{(x^2+y^2)^{3/2}}$$

Only at the very last step did we convert the sum to an integral. Then our model of the wire as being infinitely long sets the integration limits at $\pm\infty$. This is a standard integral that can be found in Appendix A or with integration software. Evaluation gives

$$B_{\text{wire}} = \frac{\mu_0 Iy}{4\pi}\frac{x}{y^2(x^2+y^2)^{1/2}}\bigg|_{-\infty}^{\infty} = \frac{\mu_0}{2\pi}\frac{I}{y}$$

This is the magnitude of the field. The field direction is determined by using the right-hand rule.

The coordinate system was our choice, and there's nothing special about the y-axis. The symbol y is simply the distance from the wire, which is better represented by r. With this change, the magnetic field is

$$\vec{B}_{\text{wire}} = \left(\frac{\mu_0}{2\pi}\frac{I}{r}, \text{ tangent to a circle around the wire}\atop \text{in the right-hand direction}\right)$$

ASSESS FIGURE 29.15 shows the magnetic field of a current-carrying wire. Compare this to Figure 29.2 and convince yourself that the direction shown agrees with the right-hand rule.

FIGURE 29.15 The magnetic field of a long, straight wire carrying current I.

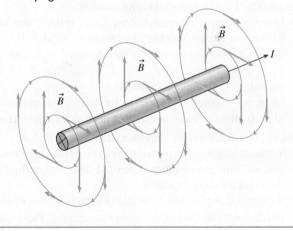

The magnetic field of an infinite, straight wire was the first of our key magnetic field models. Example 29.3 has shown that the field has magnitude

$$B_{\text{wire}} = \frac{\mu_0}{2\pi}\frac{I}{r} \quad \text{(long, straight wire)} \tag{29.7}$$

The field direction, encircling the wire, is given by the right-hand rule.

EXAMPLE 29.4 | **The magnetic field strength near a heater wire**

A 1.0-m-long, 1.0-mm-diameter nichrome heater wire is connected to a 12 V battery. What is the magnetic field strength 1.0 cm away from the wire?

MODEL 1 cm is much less than the 1 m length of the wire, so model the wire as infinitely long.

SOLVE The current through the wire is $I = \Delta V_{bat}/R$, where the wire's resistance R is

$$R = \frac{\rho L}{A} = \frac{\rho L}{\pi r^2} = 1.91 \ \Omega$$

The nichrome resistivity $\rho = 1.50 \times 10^{-6} \ \Omega \, m$ was taken from Table 27.2. Thus the current is $I = (12 \ V)/(1.91 \ \Omega) = 6.28 \ A$. The magnetic field strength at distance $d = 1.0 \ cm = 0.010 \ m$ from the wire is

$$B_{wire} = \frac{\mu_0}{2\pi} \frac{I}{d} = (2.0 \times 10^{-7} \, T \, m/A) \frac{6.28 \ A}{0.010 \ m}$$
$$= 1.3 \times 10^{-4} \ T$$

ASSESS The magnetic field of the wire is slightly more than twice the strength of the earth's magnetic field.

Motors, loudspeakers, metal detectors, and many other devices generate magnetic fields with *coils* of wire. The simplest coil is a single-turn circular loop of wire. A circular loop of wire with a circulating current is called a **current loop.**

EXAMPLE 29.5 | **The magnetic field of a current loop**

FIGURE 29.16a shows a current loop, a circular loop of wire with radius R that carries current I. Find the magnetic field of the current loop at distance z on the axis of the loop.

FIGURE 29.16 A current loop.

(a) A practical current loop   **(b)** An ideal current loop

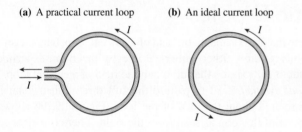

MODEL Real coils need wires to bring the current in and out, but we'll model the coil as a current moving around the full circle shown in FIGURE 29.16b.

VISUALIZE FIGURE 29.17 shows a loop for which we've assumed that the current is circulating ccw. We've chosen a coordinate system in which the loop lies at $z = 0$ in the xy-plane. Let segment i be the segment at the top of the loop. Vector $\Delta \vec{s}_i$ is parallel to the x-axis and unit vector $\hat{r}$ is in the yz-plane, thus angle θ_i, the angle between $\Delta \vec{s}_i$ and $\hat{r}$, is 90°.

FIGURE 29.17 Calculating the magnetic field of a current loop.

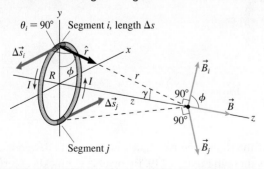

The direction of $\vec{B}_i$, the magnetic field due to the current in segment i, is given by the cross product $\Delta \vec{s}_i \times \hat{r}$. $\vec{B}_i$ must be perpendicular to $\Delta \vec{s}_i$ *and* perpendicular to $\hat{r}$. You should convince yourself that $\vec{B}_i$ in Figure 29.17 points in the correct direction. Notice that the y-component of $\vec{B}_i$ is canceled by the y-component of magnetic field $\vec{B}_j$ due to the current segment at the bottom of the loop, 180° away. In fact, *every* current segment on the loop can be paired with a segment 180° away, on the opposite side of the loop, such that the x- and y-components of $\vec{B}$ cancel and the components of $\vec{B}$ parallel to the z-axis add. In other words, the symmetry of the loop requires the on-axis magnetic field to point along the z-axis. Knowing that we need to sum only the z-components will simplify our calculation.

SOLVE We can use the Biot-Savart law to find the z-component $(B_i)_z = B_i \cos \phi$ of the magnetic field of segment i. The cross product $\Delta \vec{s}_i \times \hat{r}$ has magnitude $(\Delta s)(1) \sin 90° = \Delta s$, thus

$$(B_i)_z = \frac{\mu_0}{4\pi} \frac{I \Delta s}{r^2} \cos \phi = \frac{\mu_0 I \cos \phi}{4\pi(z^2 + R^2)} \Delta s$$

where we used $r = (z^2 + R^2)^{1/2}$. You can see, because $\phi + \gamma = 90°$, that angle ϕ is also the angle between $\hat{r}$ and the radius of the loop. Hence $\cos \phi = R/r$, and

$$(B_i)_z = \frac{\mu_0 I R}{4\pi(z^2 + R^2)^{3/2}} \Delta s$$

The final step is to sum the magnetic fields due to all the segments:

$$B_{loop} = \sum_i (B_i)_z = \frac{\mu_0 I R}{4\pi(z^2 + R^2)^{3/2}} \sum_i \Delta s$$

In this case, unlike the straight wire, none of the terms multiplying Δs depends on the position of segment i, so all these terms can be factored out of the summation. We're left with a summation that adds up the lengths of all the small segments. But this is just the total length of the wire, which is the circumference $2\pi R$. Thus the on-axis magnetic field of a current loop is

$$B_{loop} = \frac{\mu_0 I R}{4\pi(z^2 + R^2)^{3/2}} 2\pi R = \frac{\mu_0}{2} \frac{I R^2}{(z^2 + R^2)^{3/2}}$$

In practice, current often passes through a *coil* consisting of *N turns* of wire. If the turns are all very close together, so that the magnetic field of each is essentially the same, then the magnetic field of a coil is *N* times the magnetic field of a current loop. The magnetic field at the center ($z = 0$) of an *N*-turn coil, or *N*-turn current loop, is

$$B_{\text{coil center}} = \frac{\mu_0}{2}\frac{NI}{R} \qquad (N\text{-turn current loop}) \qquad (29.8)$$

This is the second of our key magnetic field models.

EXAMPLE 29.6 | Matching the earth's magnetic field

What current is needed in a 5-turn, 10-cm-diameter coil to cancel the earth's magnetic field at the center of the coil?

MODEL One way to create a zero-field region of space is to generate a magnetic field equal to the earth's field but pointing in the opposite direction. The vector sum of the two fields is zero.

VISUALIZE FIGURE 29.18 shows a five-turn coil of wire. The magnetic field is five times that of a single current loop.

SOLVE The earth's magnetic field, from Table 29.1, is 5×10^{-5} T. We can use Equation 29.8 to find that the current needed to generate a 5×10^{-5} T field is

$$I = \frac{2RB}{\mu_0 N} = \frac{2(0.050\ \text{m})(5.0 \times 10^{-5}\ \text{T})}{5(4\pi \times 10^{-7}\ \text{T m/A})} = 0.80\ \text{A}$$

FIGURE 29.18 A coil of wire.

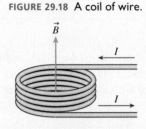

ASSESS A 0.80 A current is easily produced. Although there are better ways to cancel the earth's field than using a simple coil, this illustrates the idea.

29.5 Magnetic Dipoles

We were able to calculate the on-axis magnetic field of a current loop, but determining the field at off-axis points requires either numerical integrations or an experimental mapping of the field. FIGURE 29.19 shows the full magnetic field of a current loop. This is a field with *rotational symmetry,* so to picture the full three-dimensional field, imagine Figure 29.19a rotated about the axis of the loop. Figure 29.19b shows the magnetic field in the plane of the loop as seen from the right. There is a clear sense, seen in the photo of Figure 29.19c, that the magnetic field leaves the loop on one side, "flows" around the outside, then returns to the loop.

FIGURE 29.19 The magnetic field of a current loop.

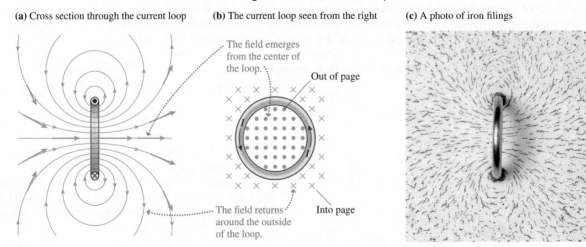

(a) Cross section through the current loop (b) The current loop seen from the right (c) A photo of iron filings

The field emerges from the center of the loop.

Out of page

Into page

The field returns around the outside of the loop.

There are two versions of the right-hand rule that you can use to determine which way a loop's field points. Try these in Figure 29.19. Being able to quickly ascertain the field direction of a current loop is an important skill.

> ### TACTICS BOX 29.2
>
> **Finding the magnetic field direction of a current loop**
>
> Use either of the following methods to find the magnetic field direction:
>
> ❶ Point your right thumb in the direction of the current at any point on the loop and let your fingers curl through the center of the loop. Your fingers are then pointing in the direction in which $\vec{B}$ leaves the loop.
>
> ❷ Curl the fingers of your right hand around the loop in the direction of the current. Your thumb is then pointing in the direction in which $\vec{B}$ leaves the loop.
>
> Exercises 18–20

A Current Loop Is a Magnetic Dipole

A current loop has two distinct sides. Bar magnets and flat refrigerator magnets also have two distinct sides or ends, so you might wonder if current loops are related to these permanent magnets. Consider the following experiments with a current loop. Notice that we're showing the magnetic field only in the plane of the loop.

Investigating current loops

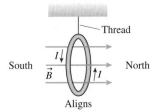

Aligns

A current loop hung by a thread aligns itself with the magnetic field pointing north.

Repels

The north pole of a permanent magnet repels the side of a current loop from which the magnetic field is emerging.

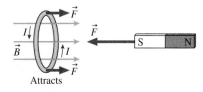

Attracts

The south pole of a permanent magnet attracts the side of a current loop from which the magnetic field is emerging.

These investigations show that **a current loop is a magnet,** just like a permanent magnet. A magnet created by a current in a coil of wire is called an **electromagnet.** An electromagnet picks up small pieces of iron, influences a compass needle, and acts in every way like a permanent magnet.

In fact, FIGURE 29.20 shows that a flat permanent magnet and a current loop generate the same magnetic field—the field of a magnetic dipole. For both, **you can identify the north pole as the face or end** *from which* **the magnetic field emerges.** The magnetic fields of both point *into* the south pole.

FIGURE 29.20 A current loop has magnetic poles and generates the same magnetic field as a flat permanent magnet.

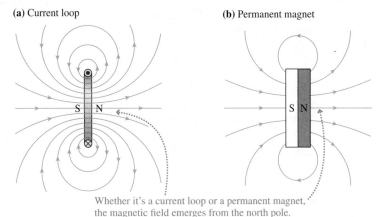

(a) Current loop   **(b)** Permanent magnet

Whether it's a current loop or a permanent magnet, the magnetic field emerges from the north pole.

One of the goals of this chapter is to show that magnetic forces exerted by currents and magnetic forces exerted by permanent magnets are just two different aspects of a single magnetism. This connection between permanent magnets and current loops will turn out to be a big piece of the puzzle.

The Magnetic Dipole Moment

The expression for the electric field of an electric dipole was considerably simplified when we considered the field at distances significantly larger than the size of the charge separation s. The on-axis field of an electric dipole when $z \gg s$ is

$$\vec{E}_{\text{dipole}} = \frac{1}{4\pi\epsilon_0} \frac{2\vec{p}}{z^3}$$

where the electric dipole moment $\vec{p} = (qs$, from negative to positive charge).

The on-axis magnetic field of a current loop is

$$B_{\text{loop}} = \frac{\mu_0}{2} \frac{IR^2}{(z^2 + R^2)^{3/2}}$$

If z is much larger than the diameter of the current loop, $z \gg R$, we can make the approximation $(z^2 + R^2)^{3/2} \rightarrow z^3$. Then the loop's field is

$$B_{\text{loop}} \approx \frac{\mu_0}{2} \frac{IR^2}{z^3} = \frac{\mu_0}{4\pi} \frac{2(\pi R^2)I}{z^3} = \frac{\mu_0}{4\pi} \frac{2AI}{z^3} \tag{29.9}$$

where $A = \pi R^2$ is the area of the loop.

A more advanced treatment of current loops shows that, if z is much larger than the size of the loop, Equation 29.9 is the on-axis magnetic field of a current loop of *any* shape, not just a circular loop. The shape of the loop affects the nearby field, but the distant field depends only on the current I and the area A enclosed within the loop. With this in mind, let's define the **magnetic dipole moment** $\vec{\mu}$ of a current loop enclosing area A to be

$$\vec{\mu} = (AI, \text{ from the south pole to the north pole})$$

The SI units of the magnetic dipole moment are $A\,m^2$.

> **NOTE** Don't confuse the magnetic dipole moment $\vec{\mu}$ with the constant μ_0 in the Biot-Savart law.

The magnetic dipole moment, like the electric dipole moment, is a vector. It has the same direction as the on-axis magnetic field. Thus the right-hand rule for determining the direction of $\vec{B}$ also shows the direction of $\vec{\mu}$. **FIGURE 29.21** shows the magnetic dipole moment of a circular current loop.

Because the on-axis magnetic field of a current loop points in the same direction as $\vec{\mu}$, we can combine Equation 29.9 and the definition of $\vec{\mu}$ to write the on-axis field of a magnetic dipole as

$$\vec{B}_{\text{dipole}} = \frac{\mu_0}{4\pi} \frac{2\vec{\mu}}{z^3} \quad \text{(on the axis of a magnetic dipole)} \tag{29.10}$$

If you compare $\vec{B}_{\text{dipole}}$ to $\vec{E}_{\text{dipole}}$, you can see that the magnetic field of a magnetic dipole is very similar to the electric field of an electric dipole.

A permanent magnet also has a magnetic dipole moment and its on-axis magnetic field is given by Equation 29.10 when z is much larger than the size of the magnet. Equation 29.10 and laboratory measurements of the on-axis magnetic field can be used to determine a permanent magnet's dipole moment.

FIGURE 29.21 The magnetic dipole moment of a circular current loop.

The magnetic dipole moment is perpendicular to the loop, in the direction of the right-hand rule. The magnitude of $\vec{\mu}$ is AI.

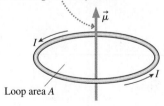

Loop area A

EXAMPLE 29.7 | **Measuring current in a superconducting loop**

You'll learn in Chapter 30 that a current can be *induced* in a closed loop of wire. If the loop happens to be made of a superconducting material, with zero resistance, the induced current will—in principle—persist forever. The current cannot be measured with an ammeter because any real ammeter has resistance that will quickly stop the current. Instead, physicists measure the persistent current in a superconducting loop by measuring its magnetic field. What is the current in a 3.0-mm-diameter superconducting loop if the axial magnetic field is 9.0 μT at a distance of 2.5 cm?

MODEL The measurements are made far enough from the loop in comparison to its radius ($z \gg R$) that we can model the loop as a magnetic dipole rather than using the exact expression for the on-axis field of a current loop.

SOLVE The axial magnetic field strength of a dipole is

$$B = \frac{\mu_0}{4\pi}\frac{2\mu}{z^3} = \frac{\mu_0}{4\pi}\frac{2\pi R^2 I}{z^3} = \frac{\mu_0 R^2 I}{2}\frac{1}{z^3}$$

where we used $\mu = AI = \pi R^2 I$ for the magnetic dipole moment of a circular loop of radius R. Thus the current is

$$I = \frac{2z^3 B}{\mu_0 R^2} = \frac{2(0.025\ \text{m})^3(9.0\times 10^{-6}\ \text{T})}{(1.26\times 10^{-6}\ \text{T m/A})(0.0015\ \text{m})^2}$$
$$= 99\ \text{A}$$

ASSESS This would be a very large current for ordinary wire. An important property of superconducting wires is their ability to carry current that would melt an ordinary wire.

STOP TO THINK 29.4 What is the current direction in this loop? And which side of the loop is the north pole?

a. Current cw; north pole on top
b. Current cw; north pole on bottom
c. Current ccw; north pole on top
d. Current ccw; north pole on bottom

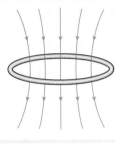

29.6 Ampère's Law and Solenoids

In principle, the Biot-Savart law can be used to calculate the magnetic field of any current distribution. In practice, the integrals are difficult to evaluate for anything other than very simple situations. We faced a similar situation for calculating electric fields, but we discovered an alternative method—Gauss's law—for calculating the electric field of charge distributions with a high degree of symmetry.

Likewise, there's an alternative method, called *Ampère's law,* for calculating the magnetic fields of current distributions with a high degree of symmetry. Whereas Gauss's law is written in terms of a surface integral, Ampère's law is based on the mathematical procedure called a *line integral.*

Line Integrals

We've flirted with the idea of a line integral ever since introducing the concept of work in Chapter 9, but now we need to take a more serious look at what a line integral represents and how it is used. **FIGURE 29.22a** shows a curved line that goes from an initial point i to a final point f.

FIGURE 29.22 Integrating along a line from i to f.

(a)

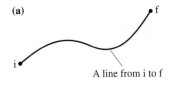

A line from i to f

(b)

The line can be divided into many small segments. The sum of all the Δs's is the length l of the line.

Suppose, as shown in FIGURE 29.22b, we divide the line into many small segments of length Δs. The first segment is Δs_1, the second is Δs_2, and so on. The sum of all the Δs's is the length l of the line between i and f. We can write this mathematically as

$$l = \sum_k \Delta s_k \rightarrow \int_i^f ds \qquad (29.11)$$

where, in the last step, we let $\Delta s \rightarrow ds$ and the sum become an integral.

This integral is called a **line integral.** All we've done is to subdivide a line into infinitely many infinitesimal pieces, then add them up. This is exactly what you do in calculus when you evaluate an integral such as $\int x \, dx$. In fact, an integration along the x-axis *is* a line integral, one that happens to be along a straight line. Figure 29.22 differs only in that the line is curved. **The underlying idea in both cases is that an integral is just a fancy way of doing a sum.**

The line integral of Equation 29.11 is not terribly exciting. FIGURE 29.23a makes things more interesting by allowing the line to pass through a magnetic field. FIGURE 29.23b again divides the line into small segments, but this time $\Delta \vec{s}_k$ is the displacement vector of segment k. The magnetic field at this point in space is $\vec{B}_k$.

Suppose we were to evaluate the dot product $\vec{B}_k \cdot \Delta \vec{s}_k$ at each segment, then add the values of $\vec{B}_k \cdot \Delta \vec{s}_k$ due to every segment. Doing so, and again letting the sum become an integral, we have

$$\sum_k \vec{B}_k \cdot \Delta \vec{s}_k \rightarrow \int_i^f \vec{B} \cdot d\vec{s} = \text{the line integral of } \vec{B} \text{ from i to f}$$

Once again, the integral is just a shorthand way to say: Divide the line into lots of little pieces, evaluate $\vec{B}_k \cdot \Delta \vec{s}_k$ for each piece, then add them up.

Although this process of evaluating the integral could be difficult, the only line integrals we'll need to deal with fall into two simple cases. If the magnetic field is *everywhere perpendicular* to the line, then $\vec{B} \cdot d\vec{s} = 0$ at every point along the line and the integral is zero. If the magnetic field is *everywhere tangent* to the line *and* has the same magnitude B at every point, then $\vec{B} \cdot d\vec{s} = B \, ds$ at every point and

$$\int_i^f \vec{B} \cdot d\vec{s} = \int_i^f B \, ds = B \int_i^f ds = Bl \qquad (29.12)$$

We used Equation 29.11 in the last step to integrate ds along the line.

Tactics Box 29.3 summarizes these two situations.

FIGURE 29.23 Integrating $\vec{B}$ along a line from i to f.

(a)

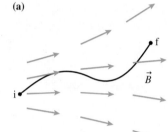

The line passes through a magnetic field.

(b)

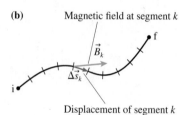

Magnetic field at segment k

$\vec{B}_k$

$\Delta \vec{s}_k$

Displacement of segment k

TACTICS BOX 29.3 (MP)

Evaluating line integrals

❶ If $\vec{B}$ is everywhere perpendicular to a line, the line integral of $\vec{B}$ is

$$\int_i^f \vec{B} \cdot d\vec{s} = 0$$

❷ If $\vec{B}$ is everywhere tangent to a line of length l *and* has the same magnitude B at every point, then

$$\int_i^f \vec{B} \cdot d\vec{s} = Bl$$

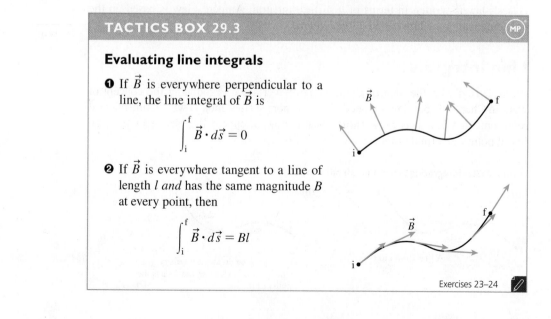

Exercises 23–24

Ampère's Law

FIGURE 29.24 shows a wire carrying current I into the page and the magnetic field at distance r. The magnetic field of a current-carrying wire is everywhere tangent to a circle around the wire *and* has the same magnitude $\mu_0 I/2\pi r$ at all points on the circle. According to Tactics Box 29.3, these conditions allow us to easily evaluate the line integral of $\vec{B}$ along a circular path around the wire. Suppose we were to integrate the magnetic field *all the way around* the circle. That is, the initial point i of the integration path and the final point f will be the same point. This would be a line integral around a *closed curve,* which is denoted

$$\oint \vec{B} \cdot d\vec{s}$$

The little circle on the integral sign indicates that the integration is performed around a closed curve. The notation has changed, but the meaning has not.

Because $\vec{B}$ is tangent to the circle *and* of constant magnitude at every point on the circle, we can use Option 2 from Tactics Box 29.3 to write

$$\oint \vec{B} \cdot d\vec{s} = Bl = B(2\pi r) \tag{29.13}$$

where, in this case, the path length l is the circumference $2\pi r$ of the circle. The magnetic field strength of a current-carrying wire is $B = \mu_0 I/2\pi r$, thus

$$\oint \vec{B} \cdot d\vec{s} = \mu_0 I \tag{29.14}$$

The interesting result is that the line integral of $\vec{B}$ around the current-carrying wire is independent of the radius of the circle. Any circle, from one touching the wire to one far away, would give the same result. The integral depends only on the amount of current passing *through* the circle that we integrated around.

This is reminiscent of Gauss's law. In our investigation of Gauss's law, we started with the observation that the electric flux Φ_e through a sphere surrounding a point charge depends only on the amount of charge inside, not on the radius of the sphere. After examining several cases, we concluded that the shape of the surface wasn't relevant. The electric flux through *any* closed surface enclosing total charge Q_{in} turned out to be $\Phi_e = Q_{in}/\epsilon_0$.

Although we'll skip the details, the same type of reasoning that we used to prove Gauss's law shows that the result of Equation 29.14

- Is independent of the shape of the curve around the current.
- Is independent of where the current passes through the curve.
- Depends only on the total amount of current through the area enclosed by the integration path.

Thus whenever total current $I_{through}$ passes through an area bounded by a *closed curve,* the line integral of the magnetic field around the curve is

$$\oint \vec{B} \cdot d\vec{s} = \mu_0 I_{through} \tag{29.15}$$

This result for the magnetic field is known as **Ampère's law.**

To make practical use of Ampère's law, we need to determine which currents are positive and which are negative. The right-hand rule is once again the proper tool. If you curl your right fingers around the closed path in the direction in which you are going to integrate, then any current passing though the bounded area in the direction of your thumb is a positive current. Any current in the opposite direction is a negative current. In **FIGURE 29.25,** for example, currents I_2 and I_4 are positive, I_3 is negative. Thus $I_{through} = I_2 - I_3 + I_4$.

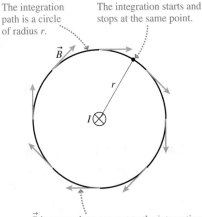

FIGURE 29.24 Integrating the magnetic field around a wire.

The integration path is a circle of radius r.

The integration starts and stops at the same point.

$\vec{B}$ is everywhere tangent to the integration path and has constant magnitude.

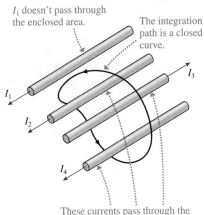

FIGURE 29.25 Using Ampère's law.

I_1 doesn't pass through the enclosed area.

The integration path is a closed curve.

These currents pass through the bounded area.

NOTE The integration path of Ampère's law is a mathematical curve through space. It does not have to match a physical surface or boundary, although it could if we want it to.

In one sense, Ampère's law doesn't tell us anything new. After all, we derived Ampère's law from the Biot-Savart law. But in another sense, Ampère's law is more important than the Biot-Savart law because it states a very general property about magnetic fields. We will use Ampère's law to find the magnetic fields of some important current distributions that have a high degree of symmetry.

EXAMPLE 29.8 | The magnetic field inside a current-carrying wire

A wire of radius R carries current I. Find the magnetic field *inside* the wire at distance $r < R$ from the axis.

MODEL Assume the current density is uniform over the cross section of the wire.

VISUALIZE FIGURE 29.26 shows a cross section through the wire. The wire has cylindrical symmetry, with all the charges moving parallel to the wire, so the magnetic field *must* be tangent to circles that are concentric with the wire. We don't know how the strength of the magnetic field depends on the distance from the center—that's what we're going to find—but the symmetry of the situation dictates the *shape* of the magnetic field.

FIGURE 29.26 Using Ampère's law inside a current-carrying wire.

By symmetry, the magnetic field must be tangent to the circle.

Current-carrying wire of radius R

Closed integration path

$\vec{B}$

r R

I $\vec{B}$

I_{through} is the current inside radius r.

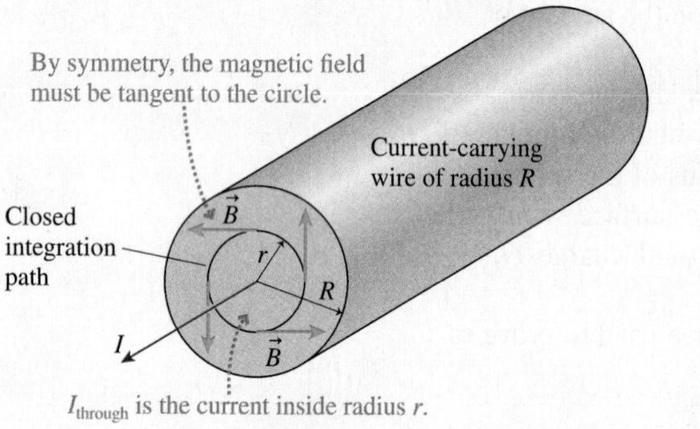

SOLVE To find the field strength at radius r, we draw a circle of radius r. The amount of current passing through this circle is

$$I_{\text{through}} = JA_{\text{circle}} = \pi r^2 J$$

where J is the current density. Our assumption of a uniform current density allows us to use the full current I passing through a wire of radius R to find that

$$J = \frac{I}{A} = \frac{I}{\pi R^2}$$

Thus the current through the circle of radius r is

$$I_{\text{through}} = \frac{r^2}{R^2} I$$

Let's integrate $\vec{B}$ around the circumference of this circle. According to Ampère's law,

$$\oint \vec{B} \cdot d\vec{s} = \mu_0 I_{\text{through}} = \frac{\mu_0 r^2}{R^2} I$$

We know from the symmetry of the wire that $\vec{B}$ is everywhere tangent to the circle *and* has the same magnitude at all points on the circle. Consequently, the line integral of $\vec{B}$ around the circle can be evaluated using Option 2 of Tactics Box 29.3:

$$\oint \vec{B} \cdot d\vec{s} = Bl = 2\pi r B$$

where $l = 2\pi r$ is the path length. If we substitute this expression into Ampère's law, we find that

$$2\pi r B = \frac{\mu_0 r^2}{R^2} I$$

Solving for B, we find that the magnetic field strength at radius r *inside* a current-carrying wire is

$$B = \frac{\mu_0 I}{2\pi R^2} r$$

ASSESS The magnetic field *inside* a wire increases linearly with distance from the center until, at the surface of the wire, $B = \mu_0 I / 2\pi R$ matches our earlier solution for the magnetic field *outside* a current-carrying wire. This agreement at $r = R$ gives us confidence in our result. The magnetic field strength both inside and outside the wire is shown graphically in FIGURE 29.27.

FIGURE 29.27 Graphical representation of the magnetic field of a current-carrying wire.

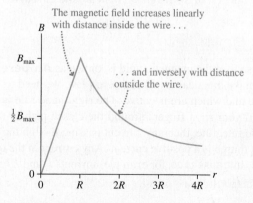

The magnetic field increases linearly with distance inside the wire . . .

B

B_{max}

. . . and inversely with distance outside the wire.

$\frac{1}{2} B_{\text{max}}$

0

0 R $2R$ $3R$ $4R$ r

The Magnetic Field of a Solenoid

In our study of electricity, we made extensive use of the idea of a uniform electric field: a field that is the same at every point in space. We found that two closely spaced, parallel charged plates generate a uniform electric field between them, and this was one reason we focused so much attention on the parallel-plate capacitor.

Similarly, there are many applications of magnetism for which we would like to generate a **uniform magnetic field,** a field having the same magnitude and the same direction at every point within some region of space. None of the sources we have looked at thus far produces a uniform magnetic field.

In practice, a uniform magnetic field is generated with a **solenoid.** A solenoid, shown in **FIGURE 29.28**, is a helical coil of wire with the same current I passing through each loop in the coil. Solenoids may have hundreds or thousands of coils, often called *turns,* sometimes wrapped in several layers.

FIGURE 29.28 A solenoid.

FIGURE 29.29 Using superposition to find the magnetic field of a stack of current loops.

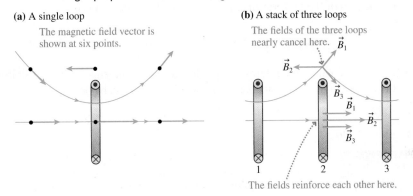

(a) A single loop

The magnetic field vector is shown at six points.

(b) A stack of three loops

The fields of the three loops nearly cancel here. $\vec{B}_1$

$\vec{B}_2$

$\vec{B}_3$ $\vec{B}_1$

$\vec{B}_2$

$\vec{B}_3$

1   2   3

The fields reinforce each other here.

We can understand a solenoid by thinking of it as a stack of current loops. **FIGURE 29.29a** shows the magnetic field of a single current loop at three points on the axis and three points equally distant from the axis. The field directly above the loop is opposite in direction to the field inside the loop. **FIGURE 29.29b** then shows three parallel loops. We can use information from Figure 29.29b to draw the magnetic fields of each loop at the center of loop 2 and at a point above loop 2.

The superposition of the three fields at the center of loop 2 produces a *stronger* field than that of loop 2 alone. But the superposition at the point above loop 2 produces a net magnetic field that is very much weaker than the field at the center of the loop. We've used only three current loops to illustrate the idea, but these tendencies are reinforced by including more loops. With many current loops along the same axis, **the field in the center is strong and roughly parallel to the axis, whereas the field outside the loops is very close to zero.**

FIGURE 29.30a is a photo of the magnetic field of a short solenoid. You can see that the magnetic field inside the coils is nearly uniform (i.e., the field lines are nearly parallel) and the field outside is much weaker. Our goal of producing a uniform magnetic field can be achieved by increasing the number of coils until we have an *ideal solenoid* that is infinitely long and in which the coils are as close together as possible. As **FIGURE 29.30b** **shows, the magnetic field inside an ideal solenoid is uniform and parallel to the axis; the magnetic field outside is zero.** No real solenoid is ideal, but a very uniform magnetic field can be produced near the center of a tightly wound solenoid whose length is much larger than its diameter.

We can use Ampère's law to calculate the field of an ideal solenoid. **FIGURE 29.31** on the next page shows a cross section through an infinitely long solenoid. The integration path that we'll use is a rectangle of width l, enclosing N turns of the solenoid coil. Because this is a mathematical curve, not a physical boundary, there's no difficulty with letting it protrude through the wall of the solenoid wherever we wish. The solenoid's magnetic field direction, given by the right-hand rule, is left to right, so we'll integrate around this path in the ccw direction.

FIGURE 29.30 The magnetic field of a solenoid.

(a) A short solenoid

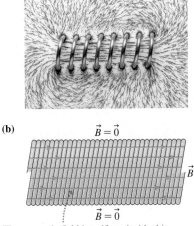

(b)

$\vec{B} = \vec{0}$

$\vec{B}$

$\vec{B} = \vec{0}$

The magnetic field is uniform inside this section of an ideal, infinitely long solenoid. The magnetic field outside the solenoid is zero.

FIGURE 29.31 A closed path inside and outside an ideal solenoid.

This is the integration path for Ampère's law. There are N turns inside.

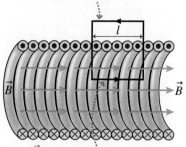

$\vec{B}$ is tangent to the integration path along the bottom edge.

Each of the N wires enclosed by the integration path carries current I, so the total current passing through the rectangle is $I_{\text{through}} = NI$. Ampère's law is thus

$$\oint \vec{B} \cdot d\vec{s} = \mu_0 I_{\text{through}} = \mu_0 NI \tag{29.16}$$

The line integral around this path is the sum of the line integrals along each side. Along the bottom, where $\vec{B}$ is parallel to $d\vec{s}$ and of constant value B, the integral is simply Bl. The integral along the top is zero because the magnetic field outside an ideal solenoid is zero.

The left and right sides sample the magnetic field both inside and outside the solenoid. The magnetic field outside is zero, and the interior magnetic field is everywhere *perpendicular* to the line of integration. Consequently, as we recognized in Option 1 of Tactics Box 29.3, the line integral is zero.

Only the integral along the bottom path is nonzero, leading to

$$\oint \vec{B} \cdot d\vec{s} = Bl = \mu_0 NI$$

Thus the strength of the uniform magnetic field inside a solenoid is

$$B_{\text{solenoid}} = \frac{\mu_0 NI}{l} = \mu_0 nI \quad \text{(solenoid)} \tag{29.17}$$

where $n = N/l$ is the number of turns per unit length. Measurements that need a uniform magnetic field are often conducted inside a solenoid, which can be built quite large.

EXAMPLE 29.9 | **Generating an MRI magnetic field**

A 1.0-m-long MRI solenoid generates a 1.2 T magnetic field. To produce such a large field, the solenoid is wrapped with superconducting wire that can carry a 100 A current. How many turns of wire does the solenoid need?

MODEL Assume that the solenoid is ideal.

SOLVE Generating a magnetic field with a solenoid is a trade-off between current and turns of wire. A larger current requires fewer turns, but the resistance of ordinary wires causes them to overheat if the current is too large. For a superconducting wire that can carry 100 A with no resistance, we can use Equation 29.17 to find the required number of turns:

$$N = \frac{lB}{\mu_0 I} = \frac{(1.0 \text{ m})(1.2 \text{ T})}{(4\pi \times 10^{-7} \text{ T m/A})(100 \text{ A})} = 9500 \text{ turns}$$

ASSESS The solenoid coil requires a large number of turns, but that's not surprising for generating a very strong field. If the wires are 1 mm in diameter, there would be 10 layers with approximately 1000 turns per layer.

The magnetic field of a finite-length solenoid is approximately uniform *inside* the solenoid and weak, but not zero, outside. As **FIGURE 29.32** shows, the magnetic field outside the solenoid looks like that of a bar magnet. Thus **a solenoid is an electromagnet,** and you can use the right-hand rule to identify the north-pole end. A solenoid with many turns and a large current can be a very powerful magnet.

FIGURE 29.32 The magnetic fields of a finite-length solenoid and of a bar magnet.

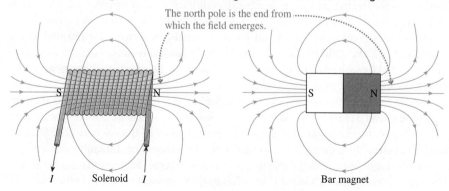

The north pole is the end from which the field emerges.

Solenoid

Bar magnet

29.7 The Magnetic Force on a Moving Charge

It's time to switch our attention from how magnetic fields are generated to how magnetic fields exert forces and torques. Oersted discovered that a current passing through a wire causes a magnetic torque to be exerted on a nearby compass needle. Upon hearing of Oersted's discovery, André-Marie Ampère, for whom the SI unit of current is named, reasoned that the current was acting like a magnet and, if this were true, that two current-carrying wires should exert magnetic forces on each other.

To find out, Ampère set up two parallel wires that could carry large currents either in the same direction or in opposite (or "antiparallel") directions. **FIGURE 29.33** shows the outcome of his experiment. Notice that, for currents, "likes" attract and "opposites" repel. This is the opposite of what would have happened had the wires been charged and thus exerting electric forces on each other. Ampère's experiment showed that **a magnetic field exerts a force on a current.**

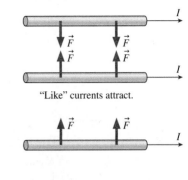

FIGURE 29.33 The forces between parallel current-carrying wires.

"Like" currents attract.

"Opposite" currents repel.

Magnetic Force

Because a current consists of moving charges, Ampère's experiment implies that **a magnetic field exerts a force on a *moving* charge.** It turns out that the magnetic force is somewhat more complex than the electric force, depending not only on the charge's velocity but also on how the velocity vector is oriented relative to the magnetic field. Consider the following experiments:

Investigating the magnetic force on a charged particle

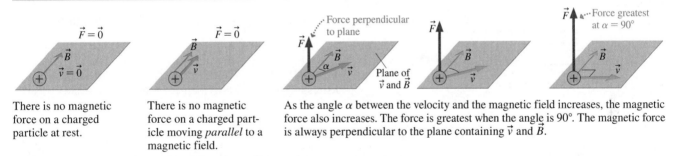

There is no magnetic force on a charged particle at rest.

There is no magnetic force on a charged particle moving *parallel* to a magnetic field.

As the angle α between the velocity and the magnetic field increases, the magnetic force also increases. The force is greatest when the angle is 90°. The magnetic force is always perpendicular to the plane containing $\vec{v}$ and $\vec{B}$.

Notice that the relationship among $\vec{v}$, $\vec{B}$, and $\vec{F}$ is exactly the same as the geometric relationship among vectors $\vec{C}$, $\vec{D}$, and $\vec{C} \times \vec{D}$. The magnetic force on a charge q as it moves through a magnetic field $\vec{B}$ with velocity $\vec{v}$ can be written

$$\vec{F}_{\text{on } q} = q\vec{v} \times \vec{B} = (qvB \sin \alpha, \text{ direction of right-hand rule}) \qquad (29.18)$$

where α is the angle between $\vec{v}$ and $\vec{B}$.

The right-hand rule is that of the cross product, shown in **FIGURE 29.34**. Notice that **the magnetic force on a moving charged particle is perpendicular to both $\vec{v}$ and $\vec{B}$.**

The magnetic force has several important properties:

- Only a *moving* charge experiences a magnetic force. There is no magnetic force on a charge at rest ($\vec{v} = \vec{0}$) in a magnetic field.
- There is no force on a charge moving parallel ($\alpha = 0°$) or antiparallel ($\alpha = 180°$) to a magnetic field.
- When there is a force, the force is perpendicular to *both* $\vec{v}$ and $\vec{B}$.
- The force on a negative charge is in the direction *opposite* to $\vec{v} \times \vec{B}$.
- For a charge moving perpendicular to $\vec{B}$ ($\alpha = 90°$), the magnitude of the magnetic force is $F = |q|vB$.

FIGURE 29.34 The right-hand rule for magnetic forces.

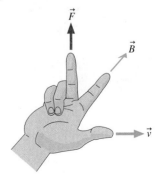

FIGURE 29.35 shows the relationship among $\vec{v}$, $\vec{B}$, and $\vec{F}$ for four moving charges. (The *source* of the magnetic field isn't shown, only the field itself.) You can see the inherent three-dimensionality of magnetism, with the force perpendicular to both $\vec{v}$ and $\vec{B}$. The magnetic force is very different from the electric force, which is parallel to the electric field.

FIGURE 29.35 Magnetic forces on moving charges.

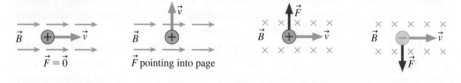

EXAMPLE 29.10 | The magnetic force on an electron

A long wire carries a 10 A current from left to right. An electron 1.0 cm above the wire is traveling to the right at a speed of 1.0×10^7 m/s. What are the magnitude and the direction of the magnetic force on the electron?

MODEL The magnetic field is that of a long, straight wire.

VISUALIZE FIGURE 29.36 shows the current and an electron moving to the right. The right-hand rule tells us that the wire's magnetic

FIGURE 29.36 An electron moving parallel to a current-carrying wire.

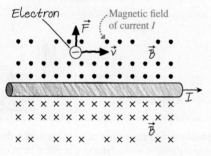
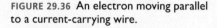

field above the wire is out of the page, so the electron is moving perpendicular to the field.

SOLVE The electron charge is negative, thus the direction of the force is opposite the direction of $\vec{v} \times \vec{B}$. The right-hand rule shows that $\vec{v} \times \vec{B}$ points down, toward the wire, so $\vec{F}$ points up, away from the wire. The magnitude of the force is $|q|vB = evB$. The field is that of a long, straight wire:

$$B = \frac{\mu_0 I}{2\pi r} = 2.0 \times 10^{-4} \text{ T}$$

Thus the magnitude of the force on the electron is

$$F = evB = (1.60 \times 10^{-19} \text{ C})(1.0 \times 10^7 \text{ m/s})(2.0 \times 10^{-4} \text{ T})$$
$$= 3.2 \times 10^{-16} \text{ N}$$

The force on the electron is $\vec{F} = (3.2 \times 10^{-16} \text{ N, up})$.

ASSESS This force will cause the electron to curve away from the wire.

We can draw an interesting and important conclusion at this point. You have seen that the magnetic field is *created by* moving charges. Now you also see that magnetic forces are *exerted on* moving charges. Thus it appears that **magnetism is an interaction between moving charges.** Any two charges, whether moving or stationary, interact with each other through the electric field. In addition, two *moving* charges interact with each other through the magnetic field.

Cyclotron Motion

Many important applications of magnetism involve the motion of charged particles in a magnetic field. Older television picture tubes use magnetic fields to steer electrons through a vacuum from the electron gun to the screen. Microwave generators, which are used in applications ranging from ovens to radar, use a device called a *magnetron* in which electrons oscillate rapidly in a magnetic field.

You've just seen that there is no force on a charge that has velocity $\vec{v}$ parallel or antiparallel to a magnetic field. Consequently, **a magnetic field has no effect on a charge moving parallel or antiparallel to the field.** To understand the motion of charged particles in magnetic fields, we need to consider only motion *perpendicular* to the field.

FIGURE 29.37 shows a positive charge q moving with a velocity $\vec{v}$ in a plane that is perpendicular to a *uniform* magnetic field $\vec{B}$. According to the right-hand rule, the

FIGURE 29.37 Cyclotron motion of a charged particle moving in a uniform magnetic field.

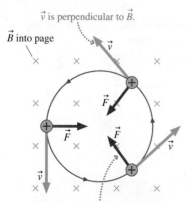

The magnetic force is always perpendicular to $\vec{v}$, causing the particle to move in a circle.

magnetic force on this particle is *perpendicular* to the velocity $\vec{v}$. A force that is always perpendicular to $\vec{v}$ changes the *direction* of motion, by deflecting the particle sideways, but it cannot change the particle's speed. Thus **a particle moving perpendicular to a uniform magnetic field undergoes uniform circular motion at constant speed.** This motion is called the **cyclotron motion** of a charged particle in a magnetic field.

> **NOTE** A negative charge will orbit in the opposite direction from that shown in Figure 29.37 for a positive charge.

You've seen many analogies to cyclotron motion earlier in this text. For a mass moving in a circle at the end of a string, the tension force is always perpendicular to $\vec{v}$. For a satellite moving in a circular orbit, the gravitational force is always perpendicular to $\vec{v}$. Now, for a charged particle moving in a magnetic field, it is the magnetic force of strength $F = qvB$ that points toward the center of the circle and causes the particle to have a centripetal acceleration.

Newton's second law for circular motion, which you learned in Chapter 8, is

$$F = qvB = ma_r = \frac{mv^2}{r} \qquad (29.19)$$

Thus the radius of the cyclotron orbit is

$$r_{cyc} = \frac{mv}{qB} \qquad (29.20)$$

The inverse dependence on B indicates that the size of the orbit can be decreased by increasing the magnetic field strength.

We can also determine the frequency of the cyclotron motion. Recall from your earlier study of circular motion that the frequency of revolution f is related to the speed and radius by $f = v/2\pi r$. A rearrangement of Equation 29.20 gives the **cyclotron frequency:**

$$f_{cyc} = \frac{qB}{2\pi m} \qquad (29.21)$$

where the ratio q/m is the particle's *charge-to-mass ratio*. Notice that the cyclotron frequency depends on the charge-to-mass ratio and the magnetic field strength but *not* on the charge's speed.

Electrons undergoing circular motion in a magnetic field. You can see the electrons' path because they collide with a low-density gas that then emits light.

EXAMPLE 29.11 | The radius of cyclotron motion

In **FIGURE 29.38**, an electron is accelerated from rest through a potential difference of 500 V, then injected into a uniform magnetic field. Once in the magnetic field, it completes half a revolution in 2.0 ns. What is the radius of its orbit?

MODEL Energy is conserved as the electron is accelerated by the potential difference. The electron then undergoes cyclotron motion in the magnetic field, although it completes only half a revolution before hitting the back of the acceleration electrode.

SOLVE The electron accelerates from rest $(v_i = 0 \text{ m/s})$ at $V_i = 0$ V to speed v_f at $V_f = 500$ V. We can use conservation of energy $K_f + qV_f = K_i + qV_i$ to find the speed v_f with which it enters the magnetic field:

$$\tfrac{1}{2}mv_f^2 + (-e)V_f = 0 + 0$$

$$v_f = \sqrt{\frac{2eV_f}{m}} = \sqrt{\frac{2(1.60 \times 10^{-19}\,\text{C})(500\,\text{V})}{9.11 \times 10^{-31}\,\text{kg}}}$$

$$= 1.33 \times 10^7 \text{ m/s}$$

The cyclotron radius in the magnetic field is $r_{cyc} = mv/eB$, but we first need to determine the field strength. Were it not for the electrode,

FIGURE 29.38 An electron is accelerated, then injected into a magnetic field.

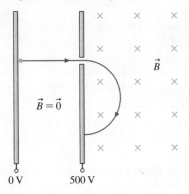

$\vec{B} = \vec{0}$

$\vec{B}$

0 V 500 V

Continued

the electron would undergo circular motion with period $T = 4.0$ ns. Hence the cyclotron frequency is $f = 1/T = 2.5 \times 10^8$ Hz. We can use the cyclotron frequency to determine that the magnetic field strength is

$$B = \frac{2\pi m f_{cyc}}{e} = \frac{2\pi (9.11 \times 10^{-31} \text{ kg})(2.50 \times 10^8 \text{ Hz})}{1.60 \times 10^{-19} \text{ C}}$$

$$= 8.94 \times 10^{-3} \text{ T}$$

Thus the radius of the electron's orbit is

$$r_{cyc} = \frac{mv}{qB} = 8.5 \times 10^{-3} \text{ m} = 8.5 \text{ mm}$$

ASSESS A 17-mm-diameter orbit is similar to what is seen in the photo just before this example, so this seems to be a typical size for electrons moving in modest magnetic fields.

FIGURE 29.39a shows a more general situation in which the charged particle's velocity $\vec{v}$ is neither parallel nor perpendicular to $\vec{B}$. The component of $\vec{v}$ parallel to $\vec{B}$ is not affected by the field, so the charged particle spirals around the magnetic field lines in a helical trajectory. The radius of the helix is determined by $\vec{v}_\perp$, the component of $\vec{v}$ perpendicular to $\vec{B}$.

FIGURE 29.39 In general, charged particles spiral along helical trajectories around the magnetic field lines. This motion is responsible for the earth's aurora.

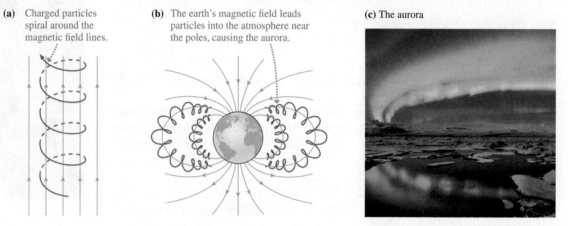

(a) Charged particles spiral around the magnetic field lines.

(b) The earth's magnetic field leads particles into the atmosphere near the poles, causing the aurora.

(c) The aurora

The motion of charged particles in a magnetic field is responsible for the earth's aurora. High-energy particles and radiation streaming out from the sun, called the *solar wind,* create ions and electrons as they strike molecules high in the atmosphere. Some of these charged particles become trapped in the earth's magnetic field, creating what is known as the *Van Allen radiation belt.*

As FIGURE 29.39b shows, the electrons spiral along the magnetic field lines until the field leads them into the atmosphere. The shape of the earth's magnetic field is such that most electrons enter the atmosphere near the magnetic poles. There they collide with oxygen and nitrogen atoms, exciting the atoms and causing them to emit auroral light, as seen in FIGURE 29.39c.

STOP TO THINK 29.5 An electron moves perpendicular to a magnetic field. What is the direction of $\vec{B}$?

a. Left
d. Right

b. Up
e. Down

c. Into the page
f. Out of the page

The Cyclotron

Physicists studying the structure of the atomic nucleus and of elementary particles usually use a device called a *particle accelerator.* The first practical particle accelerator, invented in the 1930s, was the **cyclotron.** Cyclotrons remain important for many applications of nuclear physics, such as the creation of radioisotopes for medicine.

A cyclotron, shown in FIGURE 29.40, consists of an evacuated chamber within a large, uniform magnetic field. Inside the chamber are two hollow conductors shaped like the letter D and hence called "dees." The dees are made of copper, which doesn't affect the magnetic field; are open along the straight sides; and are separated by a small gap. A charged particle, typically a proton, is injected into the magnetic field from a source near the center of the cyclotron, and it begins to move in and out of the dees in a circular cyclotron orbit.

The cyclotron operates by taking advantage of the fact that the cyclotron frequency f_{cyc} of a charged particle is independent of the particle's speed. An *oscillating* potential difference ΔV is connected across the dees and adjusted until its frequency is exactly the cyclotron frequency. There is almost no electric field inside the dees (you learned in Chapter 24 that the electric field inside a hollow conductor is zero), but a strong electric field points from the positive to the negative dee in the gap between them.

Suppose the proton emerges into the gap from the positive dee. The electric field in the gap *accelerates* the proton across the gap into the negative dee, and it gains kinetic energy $e \Delta V$. A half cycle later, when it next emerges into the gap, the potential of the dees (whose potential difference is oscillating at f_{cyc}) will have changed sign. The proton will *again* be emerging from the positive dee and will *again* accelerate across the gap and gain kinetic energy $e \Delta V$.

This pattern will continue orbit after orbit. The proton's kinetic energy increases by $2e \Delta V$ every orbit, so after N orbits its kinetic energy is $K = 2Ne \Delta V$ (assuming that its initial kinetic energy was near zero). The radius of its orbit increases as it speeds up; hence the proton follows the *spiral* path shown in Figure 29.40 until it finally reaches the outer edge of the dee. It is then directed out of the cyclotron and aimed at a target. Although ΔV is modest, usually a few hundred volts, the fact that the proton can undergo many thousands of orbits before reaching the outer edge allows it to acquire a very large kinetic energy.

FIGURE 29.40 A cyclotron.

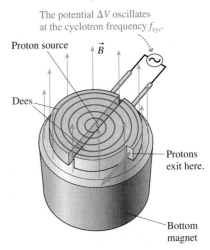

The potential ΔV oscillates at the cyclotron frequency f_{cyc}.

Proton source · $\vec{B}$

Dees

Protons exit here.

Bottom magnet

The Hall Effect

A charged particle moving through a vacuum is deflected sideways, perpendicular to $\vec{v}$, by a magnetic field. In 1879, a graduate student named Edwin Hall showed that the same is true for the charges moving through a conductor as part of a current. This phenomenon—now called the **Hall effect**—is used to gain information about the charge carriers in a conductor. It is also the basis of a widely used technique for measuring magnetic field strengths.

FIGURE 29.41a shows a magnetic field perpendicular to a flat, current-carrying conductor. You learned in Chapter 27 that the charge carriers move through a conductor at the drift speed v_d. Their motion is perpendicular to $\vec{B}$, so each charge carrier experiences a magnetic force $F_B = ev_dB$ perpendicular to both $\vec{B}$ and the current I. However, for the first time we have a situation in which it *does* matter whether the charge carriers are positive or negative.

FIGURE 29.41b, with the field out of the page, shows that positive charge carriers moving in the direction of I are pushed toward the bottom surface of the conductor. This creates an excess positive charge on the bottom surface and leaves an excess negative charge on the top. FIGURE 29.41c, where the electrons in an electron current i move opposite the direction of I, shows that electrons would be pushed toward the bottom surface. (Be sure to use the right-hand rule and the sign of the electron charge to confirm the deflections shown in these figures.) Thus the sign of the excess charge on the bottom surface is the same as the sign of the charge carriers. Experimentally, the bottom surface is negative when the conductor is a metal, and this is one more piece of evidence that the charge carriers in metals are electrons.

Electrons are deflected toward the bottom surface once the current starts flowing, but the process can't continue indefinitely. As excess charge accumulates on the top and bottom surfaces, it acts like the charge on the plates of a capacitor, creating a potential difference ΔV between the two surfaces and an electric field $E = \Delta V/w$ *inside* the conductor of width w. This electric field increases until the upward electric force $\vec{F}_E$

FIGURE 29.41 In a magnetic field, the charge carriers of a current are deflected to one side.

(a)

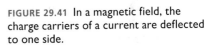

$\vec{B}$

The charge carriers are deflected to the side.

w · $\vec{F}$ · $\vec{v}$

t

Area $A = wt$

I

(b) Top surface is negative.

Conventional current of positive charge carriers

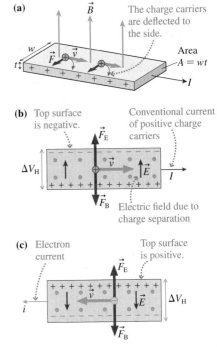

$\vec{F}_E$

ΔV_H

$\vec{v}$ · $\vec{E}$

I

$\vec{F}_B$

Electric field due to charge separation

(c) Electron current

Top surface is positive.

$\vec{F}_E$

$\vec{v}$ · $\vec{E}$

ΔV_H

i

$\vec{F}_B$

on the charge carriers exactly balances the downward magnetic force $\vec{F}_B$. Once the forces are balanced, a steady state is reached in which the charge carriers move in the direction of the current and no additional charge is deflected to the surface.

The steady-state condition, in which $F_B = F_E$, is

$$F_B = ev_dB = F_E = eE = e\frac{\Delta V}{w} \tag{29.22}$$

Thus the steady-state potential difference between the two surfaces of the conductor, which is called the **Hall voltage** ΔV_H, is

$$\Delta V_H = wv_dB \tag{29.23}$$

You learned in Chapter 27 that the drift speed is related to the current density J by $J = nev_d$, where n is the charge-carrier density (charge carriers per m^3). Thus

$$v_d = \frac{J}{ne} = \frac{I/A}{ne} = \frac{I}{wtne} \tag{29.24}$$

where $A = wt$ is the cross-section area of the conductor. If we use this expression for v_d in Equation 29.23, we find that the Hall voltage is

$$\Delta V_H = \frac{IB}{tne} \tag{29.25}$$

The Hall voltage is very small for metals in laboratory-sized magnetic fields, typically in the microvolt range. Even so, measurements of the Hall voltage in a known magnetic field are used to determine the charge-carrier density n. Interestingly, the Hall voltage is larger for *poor* conductors that have smaller charge-carrier densities. A laboratory probe for measuring magnetic field strengths, called a *Hall probe,* measures ΔV_H for a poor conductor whose charge-carrier density is known. The magnetic field is then determined from Equation 29.25.

EXAMPLE 29.12 | **Measuring the magnetic field**

A Hall probe consists of a strip of the metal bismuth that is 0.15 mm thick and 5.0 mm wide. Bismuth is a poor conductor with charge-carrier density 1.35×10^{25} m^{-3}. The Hall voltage on the probe is 2.5 mV when the current through it is 1.5 A. What is the strength of the magnetic field, and what is the electric field strength inside the bismuth?

VISUALIZE The bismuth strip looks like Figure 29.41a. The thickness is $t = 1.5 \times 10^{-4}$ m and the width is $w = 5.0 \times 10^{-3}$ m.

SOLVE Equation 29.25 gives the Hall voltage. We can rearrange the equation to find that the magnetic field is

$$B = \frac{tne}{I}\Delta V_H$$

$$= \frac{(1.5 \times 10^{-4}\text{ m})(1.35 \times 10^{25}\text{ m}^{-3})(1.60 \times 10^{-19}\text{ C})}{1.5\text{ A}}0.0025\text{ V}$$

$$= 0.54\text{ T}$$

The electric field created inside the bismuth by the excess charge on the surface is

$$E = \frac{\Delta V_H}{w} = \frac{0.0025\text{ V}}{5.0 \times 10^{-3}\text{ m}} = 0.50\text{ V/m}$$

ASSESS 0.54 T is a fairly typical strength for a laboratory magnet.

29.8 Magnetic Forces on Current-Carrying Wires

Ampère's observation of magnetic forces between current-carrying wires motivated us to look at the magnetic forces on moving charges. We're now ready to apply that knowledge to Ampère's experiment. As a first step, let us find the force exerted by a uniform magnetic field on a long, straight wire carrying current I through the field. As FIGURE 29.42a shows, there's *no* force on a current-carrying wire *parallel* to a magnetic field. This shouldn't be surprising; it follows from the fact that there is no force on a charged particle moving parallel to $\vec{B}$.

FIGURE 29.45 Magnetic forces between parallel current-carrying wires.

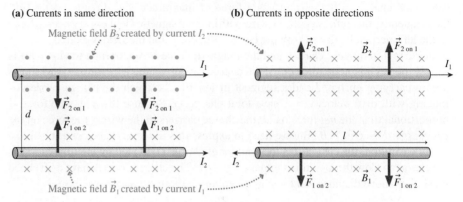

(a) Currents in same direction

(b) Currents in opposite directions

As Figure 29.45a shows, the current I_2 in the lower wire creates a magnetic field $\vec{B}_2$ at the position of the upper wire. $\vec{B}_2$ points out of the page, perpendicular to current I_1. **It is field $\vec{B}_2$, due to the lower wire, that exerts a magnetic force on the upper wire.** Using the right-hand rule, you can see that the force on the upper wire is downward, thus attracting it toward the lower wire. The field of the lower current is not a uniform field, but it is the *same* at all points along the upper wire because the two wires are parallel. Consequently, we can use the field of a long, straight wire, with $r = d$, to determine the magnetic force exerted by the lower wire on the upper wire:

$$F_{\text{parallel wires}} = I_1 l B_2 = I_1 l \frac{\mu_0 I_2}{2\pi d} = \frac{\mu_0 l I_1 I_2}{2\pi d} \qquad (29.27)$$

As an exercise, you should convince yourself that the current in the upper wire exerts an upward-directed magnetic force on the lower wire with exactly the same magnitude. You should also convince yourself, using the right-hand rule, that the forces are repulsive and tend to push the wires apart if the two currents are in opposite directions.

Thus two parallel wires exert equal but opposite forces on each other, as required by Newton's third law. **Parallel wires carrying currents in the same direction attract each other; parallel wires carrying currents in opposite directions repel each other.**

EXAMPLE 29.14 | **A current balance**

Two stiff, 50-cm-long, parallel wires are connected at the ends by metal springs. Each spring has an unstretched length of 5.0 cm and a spring constant of 0.025 N/m. The wires push each other apart when a current travels around the loop. How much current is required to stretch the springs to lengths of 6.0 cm?

MODEL Two parallel wires carrying currents in opposite directions exert repulsive magnetic forces on each other.

VISUALIZE FIGURE 29.46 shows the "circuit." The springs are conductors, allowing a current to travel around the loop. In equilibrium, the repulsive magnetic forces between the wires are balanced by the restoring forces $F_{\text{Sp}} = k \, \Delta y$ of the springs.

SOLVE Figure 29.46 shows the forces on the lower wire. The net force is zero, hence the magnetic force is $F_B = 2F_{\text{Sp}}$. The force between the wires is given by Equation 29.27 with $I_1 = I_2 = I$:

$$F_B = \frac{\mu_0 l I^2}{2\pi d} = 2F_{\text{Sp}} = 2k \, \Delta y$$

FIGURE 29.46 The current-carrying wires of Example 29.14.

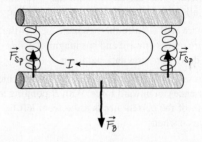

where k is the spring constant and $\Delta y = 1.0$ cm is the amount by which each spring stretches. Solving for the current, we find

$$I = \sqrt{\frac{4\pi k d \, \Delta y}{\mu_0 l}} = 17 \text{ A}$$

ASSESS Devices in which a magnetic force balances a mechanical force are called *current balances*. They can be used to make very accurate current measurements.

29.9 Forces and Torques on Current Loops

You have seen that a current loop is a magnetic dipole, much like a permanent magnet. We will now look at some important features of how current loops behave in magnetic fields. This discussion will be largely qualitative, but it will highlight some of the important properties of magnets and magnetic fields. We will use these ideas in the next section to make the connection between electromagnets and permanent magnets.

FIGURE 29.47a shows two current loops. Using what we just learned about the forces between parallel and antiparallel currents, you can see that **parallel current loops exert attractive magnetic forces on each other if the currents circulate in the same direction; they repel each other when the currents circulate in opposite directions.**

We can think of these forces in terms of magnetic poles. Recall that the north pole of a current loop is the side from which the magnetic field emerges, which you can determine with the right-hand rule. FIGURE 29.47b shows the north and south magnetic poles of the current loops. When the currents circulate in the same direction, a north and a south pole face each other and exert attractive forces on each other. When the currents circulate in opposite directions, the two like poles repel each other.

Here, at last, we have a real connection to the behavior of magnets that opened our discussion of magnetism—namely, that like poles repel and opposite poles attract. Now we have an *explanation* for this behavior, at least for electromagnets. **Magnetic poles attract or repel because the moving charges in one current exert attractive or repulsive magnetic forces on the moving charges in the other current.** Our tour through interacting moving charges is finally starting to show some practical results!

Now let's consider what happens to a current loop in a magnetic field. FIGURE 29.48 shows a square current loop in a uniform magnetic field along the z-axis. As we've learned, the field exerts magnetic forces on the currents in each of the four sides of the loop. Their directions are given by the right-hand rule. Forces $\vec{F}_{\text{front}}$ and $\vec{F}_{\text{back}}$ are opposite to each other and cancel. Forces $\vec{F}_{\text{top}}$ and $\vec{F}_{\text{bottom}}$ also add to give no net force, but because $\vec{F}_{\text{top}}$ and $\vec{F}_{\text{bottom}}$ don't act along the same line they will *rotate* the loop by exerting a torque on it.

Recall that torque is the magnitude of the force F multiplied by the moment arm d, the distance between the pivot point and the line of action. Both forces have the same moment arm $d = \frac{1}{2} l \sin\theta$, hence the torque on the loop—a torque exerted by the magnetic field—is

$$\tau = 2Fd = 2(IlB)\left(\tfrac{1}{2} l \sin\theta\right) = (Il^2)B\sin\theta = \mu B \sin\theta \qquad (29.28)$$

where $\mu = Il^2 = IA$ is the loop's magnetic dipole moment.

Although we derived Equation 29.28 for a square loop, the result is valid for a current loop of any shape. Notice that Equation 29.28 looks like another example of a cross product. We earlier defined the magnetic dipole moment vector $\vec{\mu}$ to be a vector perpendicular to the current loop in a direction given by the right-hand rule. Figure 29.48 shows that θ is the angle between $\vec{B}$ and $\vec{\mu}$, hence the torque on a magnetic dipole is

$$\vec{\tau} = \vec{\mu} \times \vec{B} \qquad (29.29)$$

The torque is zero when the magnetic dipole moment $\vec{\mu}$ is aligned parallel or antiparallel to the magnetic field, and is maximum when $\vec{\mu}$ is perpendicular to the field. It is this magnetic torque that causes a compass needle—a magnetic moment—to rotate until it is aligned with the magnetic field.

An Electric Motor

The torque on a current loop in a magnetic field is the basis for how an electric motor works. As FIGURE 29.49 on the next page shows, the *armature* of a motor is a coil of wire wound on an axle. When a current passes through the coil, the magnetic field exerts a torque on the armature and causes it to rotate. If the current were steady, the armature

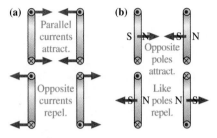

FIGURE 29.47 Two alternative but equivalent ways to view magnetic forces.

FIGURE 29.48 A uniform magnetic field exerts a torque on a current loop.

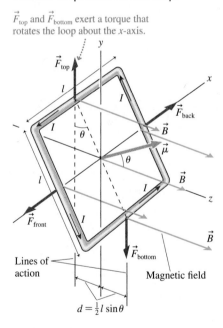

would oscillate back and forth around the equilibrium position until (assuming there's some friction or damping) it stopped with the plane of the coil perpendicular to the field. To keep the motor turning, a device called a *commutator* reverses the current direction in the coils every 180°. (Notice that the commutator is split, so the positive terminal of the battery sends current into whichever wire touches the right half of the commutator.) The current reversal prevents the armature from ever reaching an equilibrium position, so the magnetic torque keeps the motor spinning as long as there is a current.

FIGURE 29.49 A simple electric motor.

STOP TO THINK 29.6   What is the current direction in the loop?

a. Out of the page at the top of the loop, into the page at the bottom
b. Out of the page at the bottom of the loop, into the page at the top

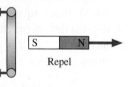

29.10 Magnetic Properties of Matter

Our theory has focused mostly on the magnetic properties of currents, yet our everyday experience is mostly with permanent magnets. We have seen that current loops and solenoids have magnetic poles and exhibit behaviors like those of permanent magnets, but we still lack a specific connection between electromagnets and permanent magnets. The goal of this section is to complete our understanding by developing an atomic-level view of the magnetic properties of matter.

Atomic Magnets

A plausible explanation for the magnetic properties of materials is the orbital motion of the atomic electrons. **FIGURE 29.50** shows a simple, classical model of an atom in which a negative electron orbits a positive nucleus. In this picture of the atom, the electron's motion is that of a current loop! It is a microscopic current loop, to be sure, but a current loop nonetheless. Consequently, an orbiting electron acts as a tiny magnetic dipole, with a north pole and a south pole. You can think of the magnetic dipole as an atomic-size magnet.

FIGURE 29.50 A classical orbiting electron is a tiny magnetic dipole.

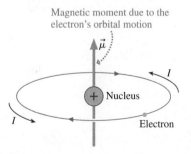

However, the atoms of most elements contain many electrons. Unlike the solar system, where all of the planets orbit in the same direction, electron orbits are arranged to oppose each other: one electron moves counterclockwise for every electron that moves clockwise. Thus the magnetic moments of individual orbits tend to cancel each other and the *net* magnetic moment is either zero or very small.

The cancellation continues as the atoms are joined into molecules and the molecules into solids. When all is said and done, the net magnetic moment of any bulk matter due to the orbiting electrons is so small as to be negligible. There are various subtle magnetic effects that can be observed under laboratory conditions, but orbiting electrons cannot explain the very strong magnetic effects of a piece of iron.

The Electron Spin

The key to understanding atomic magnetism was the 1922 discovery that electrons have an *inherent magnetic moment*. Perhaps this shouldn't be surprising. An electron has a *mass,* which allows it to interact with gravitational fields, and a *charge,* which allows it to interact with electric fields. There's no reason an electron shouldn't also interact with magnetic fields, and to do so it comes with a magnetic moment.

An electron's inherent magnetic moment, shown in FIGURE 29.51, is often called the electron *spin* because, in a classical picture, a spinning ball of charge would have a magnetic moment. This classical picture is not a realistic portrayal of how the electron really behaves, but its inherent magnetic moment makes it seem *as if* the electron were spinning. While it may not be spinning in a literal sense, an electron really is a microscopic magnet.

We must appeal to the results of quantum physics to find out what happens in an atom with many electrons. The spin magnetic moments, like the orbital magnetic moments, tend to oppose each other as the electrons are placed into their shells, causing the net magnetic moment of a *filled* shell to be zero. However, atoms containing an odd number of electrons must have at least one valence electron with an unpaired spin. These atoms have a net magnetic moment due to the electron's spin.

But atoms with magnetic moments don't necessarily form a solid with magnetic properties. For most elements, the magnetic moments of the atoms are randomly arranged when the atoms join together to form a solid. As FIGURE 29.52 shows, this random arrangement produces a solid whose net magnetic moment is very close to zero. This agrees with our common experience that most materials are not magnetic.

Ferromagnetism

It happens that in iron, and a few other substances, the spins interact with each other in such a way that atomic magnetic moments tend to all line up in the *same* direction, as shown in FIGURE 29.53. Materials that behave in this fashion are called **ferromagnetic,** with the prefix *ferro* meaning "iron-like."

In ferromagnetic materials, the individual magnetic moments add together to create a *macroscopic* magnetic dipole. The material has a north and a south magnetic pole, generates a magnetic field, and aligns parallel to an external magnetic field. In other words, it is a magnet!

Although iron is a magnetic material, a typical piece of iron is not a strong permanent magnet. You need not worry that a steel nail, which is mostly iron and is easily lifted with a magnet, will leap from your hands and pin itself against the hammer because of its own magnetism. It turns out, as shown in FIGURE 29.54 on the next page, that a piece of iron is divided into small regions, typically less than 100 μm in size, called **magnetic domains.** The magnetic moments of all the iron atoms within each domain are perfectly aligned, so each individual domain, like Figure 29.53, is a strong magnet.

FIGURE 29.51 Magnetic moment of the electron.

The arrow represents the inherent magnetic moment of the electron.

FIGURE 29.52 The random magnetic moments of the atoms in a typical solid.

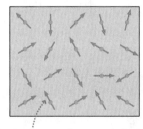

The atomic magnetic moments due to unpaired electrons point in random directions. The sample has no net magnetic moment.

FIGURE 29.53 In a ferromagnetic material, the atomic magnetic moments are aligned.

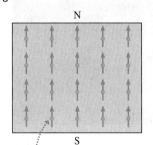

The atomic magnetic moments are aligned. The sample has north and south magnetic poles.

FIGURE 29.54 Magnetic domains in a ferromagnetic material. The net magnetic dipole is nearly zero.

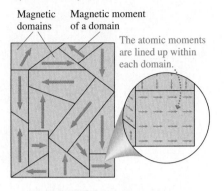

However, the various magnetic domains that form a larger solid, such as you might hold in your hand, are randomly arranged. Their magnetic dipoles largely cancel, much like the cancellation that occurs on the atomic scale for nonferromagnetic substances, so the solid as a whole has only a small magnetic moment. That is why the nail is not a strong permanent magnet.

Induced Magnetic Dipoles

If a ferromagnetic substance is subjected to an *external* magnetic field, the external field exerts a torque on the magnetic dipole of each domain. The torque causes many of the domains to rotate and become aligned with the external field, just as a compass needle aligns with a magnetic field, although internal forces between the domains generally prevent the alignment from being perfect. In addition, atomic-level forces between the spins can cause the *domain boundaries* to move. Domains that are aligned along the external field become larger at the expense of domains that are opposed to the field. These changes in the size and orientation of the domains cause the material to develop a *net magnetic dipole* that is aligned with the external field. This magnetic dipole has been *induced* by the external field, so it is called an **induced magnetic dipole.**

FIGURE 29.55 The magnetic field of the solenoid creates an induced magnetic dipole in the iron.

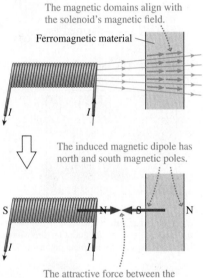

> **NOTE** The induced magnetic dipole is analogous to the polarization forces and induced electric dipoles that you studied in Chapter 23.

FIGURE 29.55 shows a ferromagnetic material near the end of a solenoid. The magnetic moments of the domains align with the solenoid's field, creating an induced magnetic dipole whose south pole faces the solenoid's north pole. Consequently, the magnetic force between the poles pulls the ferromagnetic object to the electromagnet.

The fact that a magnet attracts and picks up ferromagnetic objects was one of the basic observations about magnetism with which we started the chapter. Now we have an *explanation* of how it works, based on three ideas:

1. Electrons are microscopic magnets due to their spin.
2. A ferromagnetic material in which the spins are aligned is organized into magnetic domains.
3. The individual domains align with an external magnetic field to produce an induced magnetic dipole moment for the entire object.

The object's magnetic dipole may not return to zero when the external field is removed because some domains remain "frozen" in the alignment they had in the external field. Thus a ferromagnetic object that has been in an external field may be left with a net magnetic dipole moment after the field is removed. In other words, the object has become a **permanent magnet.** A permanent magnet is simply a ferromagnetic material in which a majority of the magnetic domains are aligned with each other to produce a net magnetic dipole moment.

Whether or not a ferromagnetic material can be made into a permanent magnet depends on the internal crystalline structure of the material. *Steel* is an alloy of iron with other elements. An alloy of mostly iron with the right percentages of chromium and nickel produces *stainless steel,* which has virtually no magnetic properties at all because its particular crystalline structure is not conducive to the formation of domains. A very different steel alloy called Alnico V is made with 51% iron, 24% cobalt, 14% nickel, 8% aluminum, and 3% copper. It has extremely prominent magnetic properties and is used to make high-quality permanent magnets.

So we've come full circle. One of our initial observations about magnetism was that a permanent magnet can exert forces on some materials but not others. The *theory* of magnetism that we then proceeded to develop was about the interactions between moving charges. What moving charges had to do with permanent magnets was not obvious. But finally, by considering magnetic effects at the atomic level, we found that properties of permanent magnets and magnetic materials can be traced to the interactions of vast numbers of electron spins.

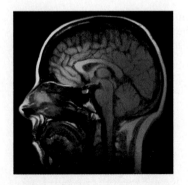

Magnetic resonance imaging, or MRI, uses the magnetic properties of atoms as a noninvasive probe of the human body.

STOP TO THINK 29.7 Which magnet or magnets induced this magnetic dipole?

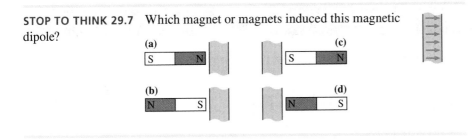

(a) S | N
(b) N | S
(c) S | N
(d) N | S

CHALLENGE EXAMPLE 29.15 | Designing a loudspeaker

A loudspeaker consists of a paper cone wrapped at the bottom with several turns of fine wire. As **FIGURE 29.56** shows, this coil sits in a narrow gap between the poles of a circular magnet. To produce sound, the amplifier drives a current through the coil. The magnetic field then exerts a force on this current, pushing the cone and thus pushing the air to create a sound wave. An ideal speaker would experience only forces from the magnetic field, thus responding only to the current from the amplifier. Real speakers are balanced so as to come close to this ideal unless driven very hard.

Consider a 5.5 g loudspeaker cone with a 5.0-cm-diameter, 20-turn coil having a resistance of 8.0 Ω. There is a 0.18 T field in the gap between the poles. These values are typical of the loudspeakers found in car stereo systems. What is the oscillation amplitude of this speaker if driven by a 100 Hz oscillatory voltage from the amplifier with a peak value of 12 V?

FIGURE 29.56 The coil and magnet of a loudspeaker.

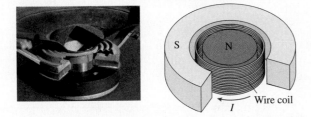

Wire coil

MODEL Model the loudspeaker as ideal, responding only to magnetic forces. These forces cause the cone to accelerate. We'll use kinematics to relate the acceleration to the displacement.

VISUALIZE **FIGURE 29.57** shows the coil in the gap between the magnet poles. Magnetic fields go from north to south poles, so the field is radially outward. Consequently, the field at all points is perpendicular to the circular current. According to the right-hand rule, the magnetic force on the current is into or out of the

FIGURE 29.57 The magnetic field in the gap, from north to south, is perpendicular to the current.

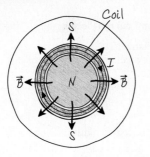

page, depending on whether the current is counterclockwise or clockwise, respectively.

SOLVE We can write the output voltage of the amplifier as $\Delta V = V_0 \cos \omega t$, where $V_0 = 12$ V is the peak voltage and $\omega = 2\pi f = 628$ rad/s is the angular frequency at 100 Hz. The voltage drives current

$$I = \frac{\Delta V}{R} = \frac{V_0 \cos \omega t}{R}$$

through the coil, where R is the coil's resistance. This causes the oscillating in-and-out force that drives the speaker cone back and forth. Even though the coil isn't a straight wire, the fact that the magnetic field is everywhere perpendicular to the current means that we can calculate the magnetic force as $F = IlB$ where l is the total length of the wire in the coil. The circumference of the coil is $\pi(0.050 \text{ m}) = 0.157$ m so 20 turns gives $l = 3.1$ m. The cone responds to the force by accelerating with $a = F/m$. Combining these pieces, we find the cone's acceleration is

$$a = \frac{IlB}{m} = \frac{V_0 lB \cos \omega t}{mR} = a_{max} \cos \omega t$$

It is straightforward to evaluate $a_{max} = 152$ m/s^2.

From kinematics, $a = dv/dt$ and $v = dx/dt$. We need to integrate twice to find the displacement. First,

$$v = \int a \, dt = a_{max} \int \cos \omega t \, dt = \frac{a_{max}}{\omega} \sin \omega t$$

The integration constant is zero because we know, from simple harmonic motion, that the average velocity is zero. Integrating again, we get

$$x = \int v \, dt = \frac{a_{max}}{\omega} \int \sin \omega t \, dt = -\frac{a_{max}}{\omega^2} \cos \omega t$$

where the integration constant is again zero if we assume the oscillation takes place around the origin. The minus sign tells us that the displacement and acceleration are out of phase. The *amplitude* of the oscillation, which we seek, is

$$A = \frac{a_{max}}{\omega^2} = \frac{152 \text{ m/s}^2}{(628 \text{ rad/s})^2} = 3.8 \times 10^{-4} \text{ m} = 0.38 \text{ mm}$$

ASSESS If you've ever placed your hand on a loudspeaker cone, you know that you can feel a slight vibration. An amplitude of 0.38 mm is consistent with this observation. The fact that the amplitude increases with the inverse square of the frequency explains why you can sometimes *see* the cone vibrating with an amplitude of several millimeters for low-frequency bass notes.

SUMMARY

The goal of Chapter 29 has been to learn about magnetism and the magnetic field.

GENERAL PRINCIPLES

At its most fundamental level, **magnetism** is an interaction between moving charges. The magnetic field of one moving charge exerts a force on another moving charge.

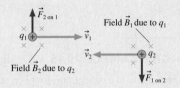

Magnetic Fields

The **Biot-Savart law** for a moving point charge

$$\vec{B} = \frac{\mu_0}{4\pi} \frac{q\vec{v} \times \hat{r}}{r^2}$$

Magnetic field of a current

MODEL Model wires as simple shapes.

VISUALIZE Divide the wire into short segments.

SOLVE Use superposition:

- Find the field of each segment Δs.
- Find $\vec{B}$ by summing the fields of all Δs, usually as an integral.

An alternative method for fields with a high degree of symmetry is **Ampère's law**:

$$\oint \vec{B} \cdot d\vec{s} = \mu_0 I_{through}$$

where $I_{through}$ is the current through the area bounded by the integration path.

Magnetic Forces

The magnetic force on a moving charge is

$$\vec{F} = q\vec{v} \times \vec{B}$$

The force is perpendicular to $\vec{v}$ and $\vec{B}$.

The magnetic force on a current-carrying wire is

$$\vec{F} = I\vec{l} \times \vec{B}$$

$\vec{F} = \vec{0}$ for a charge or current moving parallel to $\vec{B}$.

The magnetic torque on a magnetic dipole is

$$\vec{\tau} = \vec{\mu} \times \vec{B}$$

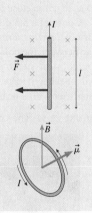

APPLICATIONS

Wire

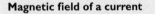

$$B = \frac{\mu_0}{2\pi} \frac{I}{r}$$

Solenoid

Uniform field

$$B = \frac{\mu_0 NI}{l}$$

Loop

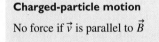

$$B_{center} = \frac{\mu_0}{2} \frac{NI}{R}$$

Flat magnet

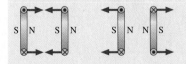

Right-hand rule

Point your right thumb in the direction of I. Your fingers curl in the direction of $\vec{B}$. For a dipole, $\vec{B}$ emerges from the side that is the north pole.

Charged-particle motion

No force if $\vec{v}$ is parallel to $\vec{B}$

Circular motion at the cyclotron frequency $f_{cyc} = qB/2\pi m$ if $\vec{v}$ is perpendicular to $\vec{B}$

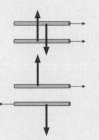

Parallel wires and current loops

Parallel currents attract.
Opposite currents repel.

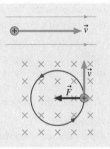

TERMS AND NOTATION

north pole	Biot-Savart law	line integral	Hall effect
south pole	tesla, T	Ampère's law	Hall voltage, ΔV_H
magnetic dipole	permeability constant, μ_0	uniform magnetic field	ferromagnetic
magnetic material	cross product	solenoid	magnetic domain
right-hand rule	current loop	cyclotron motion	induced magnetic dipole
magnetic field, $\vec{B}$	electromagnet	cyclotron frequency, f_{cyc}	permanent magnet
magnetic field lines	magnetic dipole moment, $\vec{\mu}$	cyclotron	

CONCEPTUAL QUESTIONS

1. The lightweight glass sphere in **FIGURE Q29.1** hangs by a thread. The north pole of a bar magnet is brought near the sphere.
   a. Suppose the sphere is electrically neutral. Is it attracted to, repelled by, or not affected by the magnet? Explain.
   b. Answer the same question if the sphere is positively charged.

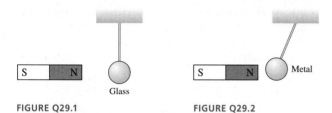

FIGURE Q29.1   **FIGURE Q29.2**

2. The metal sphere in **FIGURE Q29.2** hangs by a thread. When the north pole of a magnet is brought near, the sphere is strongly attracted to the magnet. Then the magnet is reversed and its south pole is brought near the sphere. How does the sphere respond? Explain.

3. You have two electrically neutral metal cylinders that exert strong attractive forces on each other. You have no other metal objects. Can you determine if *both* of the cylinders are magnets, or if one is a magnet and the other is just a piece of iron? If so, how? If not, why not?

4. What is the current direction in the wire of **FIGURE Q29.4**? Explain.

FIGURE Q29.4   **FIGURE Q29.5**

5. What is the current direction in the wire of **FIGURE Q29.5**? Explain.

6. What is the *initial* direction of deflection for the charged particles entering the magnetic fields shown in **FIGURE Q29.6**?

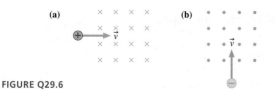

FIGURE Q29.6

7. What is the *initial* direction of deflection for the charged particles entering the magnetic fields shown in **FIGURE Q29.7**?

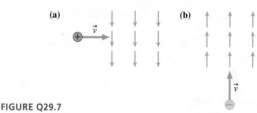

FIGURE Q29.7

8. Determine the magnetic field direction that causes the charged particles shown in **FIGURE Q29.8** to experience the indicated magnetic force.

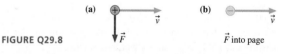

FIGURE Q29.8

9. Determine the magnetic field direction that causes the charged particles shown in **FIGURE Q29.9** to experience the indicated magnetic force.

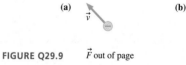

FIGURE Q29.9 $\vec{F}$ out of page $\vec{v}$ into page

10. You have a horizontal cathode-ray tube (CRT) for which the controls have been adjusted such that the electron beam *should* make a single spot of light exactly in the center of the screen. You observe, however, that the spot is deflected to the right. It is possible that the CRT is broken. But as a clever scientist, you realize that your laboratory might be in either an electric or a magnetic field. Assuming that you do not have a compass, any magnets, or any charged rods, how can you use the CRT itself to determine whether the CRT is broken, is in an electric field, or is in a magnetic field? You cannot remove the CRT from the room.

11. The south pole of a bar magnet is brought toward the current loop of **FIGURE Q29.11**. Does the bar magnet attract, repel, or have no effect on the loop? Explain.

FIGURE Q29.11

12. Give a step-by-step explanation, using both words and pictures, of how a permanent magnet can pick up a piece of nonmagnetized iron.

EXERCISES AND PROBLEMS

Problems labeled   integrate material from earlier chapters.

Exercises

Section 29.3 The Source of the Magnetic Field: Moving Charges

1. | What is the magnetic field strength at points 2 to 4 in **FIGURE EX29.1**? Assume that the wires overlap closely at 2 and 3, that each point is the same distance from nearby wires, and that all other wires are too far away to contribute to the field.

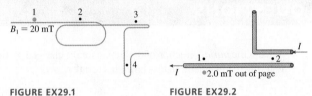

FIGURE EX29.1   **FIGURE EX29.2**

2. | Points 1 and 2 in **FIGURE EX29.2** are the same distance from the wires as the point where $B = 2.0$ mT. What are the strength and direction of $\vec{B}$ at points 1 and 2?

3. | A proton moves along the x-axis with $v_x = 1.0 \times 10^7$ m/s. As it passes the origin, what are the strength and direction of the magnetic field at the (x, y, z) positions (a) (1 cm, 0 cm, 0 cm), (b) (0 cm, 1 cm, 0 cm), and (c) (0 cm, -2 cm, 0 cm)?

4. || An electron moves along the z-axis with $v_z = 2.0 \times 10^7$ m/s. As it passes the origin, what are the strength and direction of the magnetic field at the (x, y, z) positions (a) (1 cm, 0 cm, 0 cm), (b) (0 cm, 0 cm, 1 cm), and (c) (0 cm, 1 cm, 1 cm)?

5. || What is the magnetic field at the position of the dot in **FIGURE EX29.5**? Give your answer as a vector.

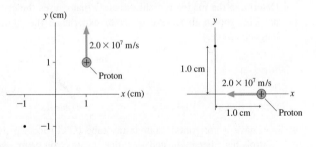

FIGURE EX29.5   **FIGURE EX29.6**

6. || What is the magnetic field at the position of the dot in **FIGURE EX29.6**? Give your answer as a vector.

Section 29.4 The Magnetic Field of a Current

7. | What currents are needed to generate the magnetic field strengths of Table 29.1 at a point 1.0 cm from a long, straight wire?

8. || The element niobium, which is a metal, is a superconductor (i.e., no electrical resistance) at temperatures below 9 K. However, the superconductivity is destroyed if the magnetic field at the surface of the metal reaches or exceeds 0.10 T. What is the maximum current in a straight, 3.0-mm-diameter superconducting niobium wire?

9. | Although the evidence is weak, there has been concern in recent years over possible health effects from the magnetic fields generated by electric transmission lines. A typical high-voltage transmission line is 20 m above the ground and carries a 200 A current at a potential of 110 kV.
   a. What is the magnetic field strength on the ground directly under such a transmission line?
   b. What percentage is this of the earth's magnetic field of 50 μT?

10. | A biophysics experiment uses a very sensitive magnetic field probe to determine the current associated with a nerve impulse traveling along an axon. If the peak field strength 1.0 mm from an axon is 8.0 pT, what is the peak current carried by the axon?

11. || The magnetic field at the center of a 1.0-cm-diameter loop is 2.5 mT.
   a. What is the current in the loop?
   b. A long straight wire carries the same current you found in part a. At what distance from the wire is the magnetic field 2.5 mT?

12. || What are the magnetic fields at points a to c in **FIGURE EX29.12**? Give your answers as vectors.

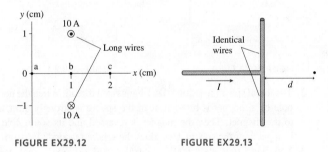

FIGURE EX29.12   **FIGURE EX29.13**

13. | A wire carries current I into the junction shown in **FIGURE EX29.13**. What is the magnetic field at the dot?

14. || What are the magnetic field strength and direction at points a to c in **FIGURE EX29.14**?

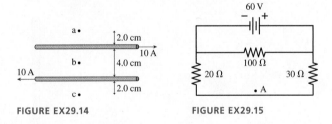

FIGURE EX29.14   **FIGURE EX29.15**

15. || Point A in **FIGURE EX29.15** is 2.0 mm from the wire. What is the magnetic field strength at point A? You can assume that the wire is very long and that all other wires are too far away to contribute to the magnetic field.

Section 29.5 Magnetic Dipoles

16. || The on-axis magnetic field strength 10 cm from a small bar magnet is 5.0 μT.
   a. What is the bar magnet's magnetic dipole moment?
   b. What is the on-axis field strength 15 cm from the magnet?

17. ‖   A 100 A current circulates around a 2.0-mm-diameter super-conducting ring.
   a.   What is the ring's magnetic dipole moment?
   b.   What is the on-axis magnetic field strength 5.0 cm from the ring?

18. ‖‖   A small, square loop carries a 25 A current. The on-axis magnetic field strength 50 cm from the loop is 7.5 nT. What is the edge length of the square?

19. ‖   The earth's magnetic dipole moment is 8.0×10^{22} A m².
   a.   What is the magnetic field strength on the surface of the earth at the earth's north magnetic pole? How does this compare to the value in Table 29.1? You can assume that the current loop is deep inside the earth.
   b.   Astronauts discover an earth-size planet without a magnetic field. To create a magnetic field at the north pole with the same strength as earth's, they propose running a current through a wire around the equator. What size current would be needed?

Section 29.6 Ampère's Law and Solenoids

20. ‖   What is the line integral of $\vec{B}$ between points i and f in **FIGURE EX29.20**?

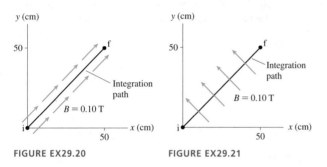

FIGURE EX29.20   **FIGURE EX29.21**

21. ‖   What is the line integral of $\vec{B}$ between points i and f in **FIGURE EX29.21**?

22. ‖   The value of the line integral of $\vec{B}$ around the closed path in **FIGURE EX29.22** is 3.77×10^{-6} T m. What is I_3?

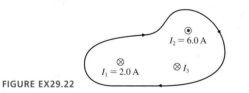

$I_2 = 6.0$ A
$I_1 = 2.0$ A $\otimes I_3$

FIGURE EX29.22

23. ‖   The value of the line integral of $\vec{B}$ around the closed path in **FIGURE EX29.23** is 1.38×10^{-5} T m. What are the direction (in or out of the page) and magnitude of I_3?

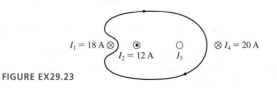

$I_1 = 18$ A $\otimes$ $I_2 = 12$ A I_3 $\otimes I_4 = 20$ A

FIGURE EX29.23

24. ‖   What is the line integral of $\vec{B}$ between points i and f in **FIGURE EX29.24**?

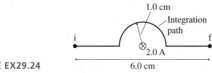

1.0 cm
Integration path
i   f
$\otimes$ 2.0 A
6.0 cm

FIGURE EX29.24

25. ‖   Magnetic resonance imaging needs a magnetic field strength
BIO of 1.5 T. The solenoid is 1.8 m long and 75 cm in diameter. It is tightly wound with a single layer of 2.0-mm-diameter superconducting wire. What size current is needed?

Section 29.7 The Magnetic Force on a Moving Charge

26. ‖   A proton moves in the magnetic field $\vec{B} = 0.50\,\hat{\imath}$ T with a speed of 1.0×10^7 m/s in the directions shown in **FIGURE EX29.26**. For each, what is magnetic force $\vec{F}$ on the proton? Give your answers in component form.

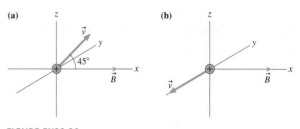

FIGURE EX29.26

27. ‖   An electron moves in the magnetic field $\vec{B} = 0.50\,\hat{\imath}$ T with a speed of 1.0×10^7 m/s in the directions shown in **FIGURE EX29.27**. For each, what is magnetic force $\vec{F}$ on the electron? Give your answers in component form.

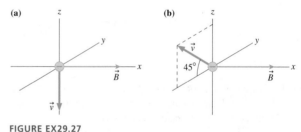

FIGURE EX29.27

28. | Radio astronomers detect electromagnetic radiation at 45 MHz from an interstellar gas cloud. They suspect this radiation is emitted by electrons spiraling in a magnetic field. What is the magnetic field strength inside the gas cloud?

29. ‖   To five significant figures, what are the cyclotron frequencies in a 3.0000 T magnetic field of the ions (a) O_2^+, (b) N_2^+, and (c) CO^+? The atomic masses are shown in the table; the mass of the missing electron is less than 0.001 u and is not relevant at this level of precision. Although N_2^+ and CO^+ both have a nominal molecular mass of 28, they are easily distinguished by virtue of their slightly different cyclotron frequencies. Use the following constants: 1 u = 1.6605×10^{-27} kg, $e = 1.6022 \times 10^{-19}$ C.

Atomic masses	
^{12}C	12.000
^{14}N	14.003
^{16}O	15.995

30. ‖   For your senior project, you would like to build a cyclotron that will accelerate protons to 10% of the speed of light. The largest vacuum chamber you can find is 50 cm in diameter. What magnetic field strength will you need?

31. ‖ The microwaves in a microwave oven are produced in a special tube called a *magnetron*. The electrons orbit the magnetic field at 2.4 GHz, and as they do so they emit 2.4 GHz electromagnetic waves.
   a. What is the magnetic field strength?
   b. If the maximum diameter of the electron orbit before the electron hits the wall of the tube is 2.5 cm, what is the maximum electron kinetic energy?

32. ‖ The Hall voltage across a conductor in a 55 mT magnetic field is 1.9 μV. When used with the same current in a different magnetic field, the voltage across the conductor is 2.8 μV. What is the strength of the second field?

Section 29.8 Magnetic Forces on Current-Carrying Wires

33. ‖ What magnetic field strength and direction will levitate the 2.0 g wire in FIGURE EX29.33?

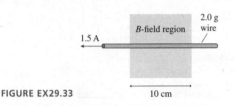

FIGURE EX29.33

34. ‖ The two 10-cm-long parallel wires in FIGURE EX29.34 are separated by 5.0 mm. For what value of the resistor R will the force between the two wires be 5.4×10^{-5} N?

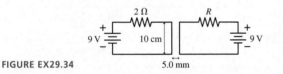

FIGURE EX29.34

35. ‖ The right edge of the circuit in FIGURE EX29.35 extends into a 50 mT uniform magnetic field. What are the magnitude and direction of the net force on the circuit?

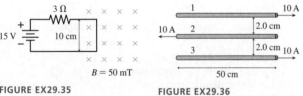

FIGURE EX29.35   **FIGURE EX29.36**

36. ‖ What is the net force (magnitude and direction) on each wire in FIGURE EX29.36?

37. ‖ FIGURE EX29.37 is a cross section through three long wires with linear mass density 50 g/m. They each carry equal currents in the directions shown. The lower two wires are 4.0 cm apart and are attached to a table. What current I will allow the upper wire to "float" so as to form an equilateral triangle with the lower wires?

FIGURE EX29.37

Section 29.9 Forces and Torques on Current Loops

38. ‖ A square current loop 5.0 cm on each side carries a 500 mA current. The loop is in a 1.2 T uniform magnetic field. The axis of the loop, perpendicular to the plane of the loop, is 30° away from the field direction. What is the magnitude of the torque on the current loop?

39. ‖ A small bar magnet experiences a 0.020 N m torque when the axis of the magnet is at 45° to a 0.10 T magnetic field. What is the magnitude of its magnetic dipole moment?

40. ‖ a. What is the magnitude of the torque on the current loop in FIGURE EX29.40?
   b. What is the loop's equilibrium orientation?

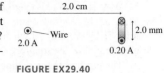

FIGURE EX29.40

Problems

41. ‖ A long wire carrying a 5.0 A current perpendicular to the xy-plane intersects the x-axis at $x = -2.0$ cm. A second, parallel wire carrying a 3.0 A current intersects the x-axis at $x = +2.0$ cm. At what point or points on the x-axis is the magnetic field zero if (a) the two currents are in the same direction and (b) the two currents are in opposite directions?

42. ‖ The two insulated wires in FIGURE P29.42 cross at a 30° angle but do not make electrical contact. Each wire carries a 5.0 A current. Points 1 and 2 are each 4.0 cm from the intersection and equally distant from both wires. What are the magnitude and direction of the magnetic fields at points 1 and 2?

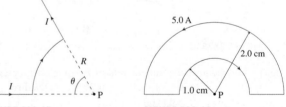

FIGURE P29.42   **FIGURE P29.43**

43. ‖ What are the strength and direction of the magnetic field at the center of the loop in FIGURE P29.43?

44. ‖ At what distance on the axis of a current loop is the magnetic field half the strength of the field at the center of the loop? Give your answer as a multiple of R.

45. ‖ Find an expression for the magnetic field strength at the center (point P) of the circular arc in FIGURE P29.45.

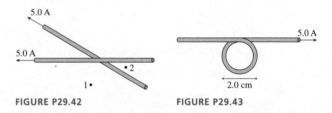

FIGURE P29.45   **FIGURE P29.46**

46. ‖ What are the strength and direction of the magnetic field at point P in FIGURE P29.46?

47. ‖ A scientist measuring the resistivity of a new metal alloy left her ammeter in another lab, but she does have a magnetic field probe. So she creates a 6.5-m-long, 2.0-mm-diameter wire of the material, connects it to a 1.5 V battery, and measures a 3.0 mT magnetic field 1.0 mm from the surface of the wire. What is the material's resistivity?

48. ‖ A 2.5-m-long, 2.0-mm-diameter aluminum wire has a potential difference of 1.5 V between its ends. Consider an electron halfway between the center of the wire and the surface that is moving parallel to the wire at the drift speed. What is the ratio of the electric force on the electron to the magnetic force on the electron?

49. ‖ Your employer asks you to build a 20-cm-long solenoid with an interior field of 5.0 mT. The specifications call for a single layer of wire, wound with the coils as close together as possible. You have two spools of wire available. Wire with a #18 gauge has a diameter of 1.02 mm and has a maximum current rating of 6 A. Wire with a #26 gauge is 0.41 mm in diameter and can carry up to 1 A. Which wire should you use, and what current will you need?

50. ‖ The magnetic field strength at the north pole of a 2.0-cm-diameter, 8-cm-long Alnico magnet is 0.10 T. To produce the same field with a solenoid of the same size, carrying a current of 2.0 A, how many turns of wire would you need?

51. ‖ The earth's magnetic field, with a magnetic dipole moment of 8.0×10^{22} A m^2, is generated by currents within the molten iron of the earth's outer core. Suppose we model the core current as a 3000-km-diameter current loop made from a 1000-km-diameter "wire." The loop diameter is measured from the centers of this very fat wire.
   a. What is the current in the current loop?
   b. What is the current density J in the current loop?
   c. To decide whether this is a large or a small current density, compare it to the current density of a 1.0 A current in a 1.0-mm-diameter wire.

52. ‖‖ Weak magnetic fields can be measured at the surface of the brain.
BIO Although the currents causing these fields are quite complicated, we can estimate their size by modeling them as a current loop around the equator of a 16-cm-diameter (the width of a typical head) sphere. What current is needed to produce a 3.0 pT field—the strength measured for one subject—at the pole of this sphere?

53. ‖ The heart produces a weak magnetic field that can be used
BIO to diagnose certain heart problems. It is a dipole field produced by a current loop in the outer layers of the heart.
   a. It is estimated that the field at the center of the heart is 90 pT. What current must circulate around an 8.0-cm-diameter loop, about the size of a human heart, to produce this field?
   b. What is the magnitude of the heart's magnetic dipole moment?

54. ‖ What is the magnetic field strength at the center of the semicircle in FIGURE P29.54?

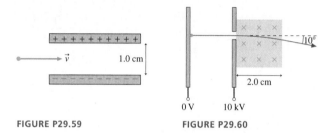

FIGURE P29.54

55. ‖ The *toroid* of FIGURE P29.55 is a coil of wire wrapped around a doughnut-shaped ring (a *torus*). Toroidal magnetic fields are used to confine fusion plasmas.
   a. From symmetry, what must be the *shape* of the magnetic field in this toroid? Explain.
   b. Consider a toroid with N closely spaced turns carrying current I. Use Ampère's law to find an expression for the magnetic field strength at a point inside the torus at distance r from the axis.
   c. Is a toroidal magnetic field a uniform field? Explain.

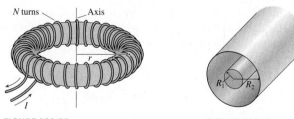

FIGURE P29.55   FIGURE P29.56

56. ‖ The coaxial cable shown in FIGURE P29.56 consists of a solid inner conductor of radius R_1 surrounded by a hollow, very thin outer conductor of radius R_2. The two carry equal currents I, but in *opposite* directions. The current density is uniformly distributed over each conductor.
   a. Find expressions for three magnetic fields: within the inner conductor, in the space between the conductors, and outside the outer conductor.
   b. Draw a graph of B versus r from $r = 0$ to $r = 2R_2$ if $R_1 = \frac{1}{3}R_2$.

57. ‖ A long, hollow wire has inner radius R_1 and outer radius R_2. The wire carries current I uniformly distributed across the area of the wire. Use Ampère's law to find an expression for the magnetic field strength in the three regions $0 < r < R_1$, $R_1 < r < R_2$, and $R_2 < r$.

58. ‖ A proton moving in a uniform magnetic field with $\vec{v}_1 = 1.00 \times 10^6 \hat{\imath}$ m/s experiences force $\vec{F}_1 = 1.20 \times 10^{-16} \hat{k}$ N. A second proton with $\vec{v}_2 = 2.00 \times 10^6 \hat{\jmath}$ m/s experiences $\vec{F}_2 = -4.16 \times 10^{-16} \hat{k}$ N in the same field. What is $\vec{B}$? Give your answer as a magnitude and an angle measured ccw from the $+x$-axis.

59. ‖ An electron travels with speed 1.0×10^7 m/s between the two parallel charged plates shown in FIGURE P29.59. The plates are separated by 1.0 cm and are charged by a 200 V battery. What magnetic field strength and direction will allow the electron to pass between the plates without being deflected?

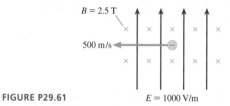

FIGURE P29.59   FIGURE P29.60

60. ‖ An electron in a cathode-ray tube is accelerated through a potential difference of 10 kV, then passes through the 2.0-cm-wide region of uniform magnetic field in FIGURE P29.60. What field strength will deflect the electron by 10°?

61. ‖ An antiproton (same properties as a proton except that $q = -e$) is moving in the combined electric and magnetic fields of FIGURE P29.61. What are the magnitude and direction of the antiproton's acceleration at this instant?

62. ‖ a. A 65-cm-diameter cyclotron uses a 500 V oscillating potential difference between the dees. What is the maximum kinetic energy of a proton if the magnetic field strength is 0.75 T?
   b. How many revolutions does the proton make before leaving the cyclotron?

63. ‖ An antiproton is identical to a proton except it has the opposite charge, $-e$. To study antiprotons, they must be confined in an ultrahigh vacuum because they will annihilate—producing gamma rays—if they come into contact with the protons of ordinary matter. One way of confining antiprotons is to keep them in a magnetic field. Suppose that antiprotons are created with a speed of 1.5×10^4 m/s and then trapped in a 2.0 mT magnetic field. What minimum diameter must the vacuum chamber have to allow these antiprotons to circulate without touching the walls?

64. ‖ **FIGURE P29.64** shows a *mass spectrometer,* an analytical instrument used to identify the various molecules in a sample by measuring their charge-to-mass ratio q/m. The sample is ionized, the positive ions are accelerated (starting from rest) through a potential difference ΔV, and they then enter a region of uniform magnetic field. The field bends the ions into circular trajectories, but after just half a circle they either strike the wall or pass through a small opening to a detector. As the accelerating voltage is slowly increased, different ions reach the detector and are measured. Consider a mass spectrometer with a 200.00 mT magnetic field and an 8.0000 cm spacing between the entrance and exit holes. To five significant figures, what accelerating potential differences ΔV are required to detect the ions (a) O_2^+, (b) N_2^+, and (c) CO^+? See Exercise 29 for atomic masses; the mass of the missing electron is less than 0.001 u and is not relevant at this level of precision. Although N_2^+ and CO^+ both have a nominal molecular mass of 28, they are easily distinguished by virtue of their slightly different accelerating voltages. Use the following constants: $1\ u = 1.6605 \times 10^{-27}$ kg, $e = 1.6022 \times 10^{-19}$ C.

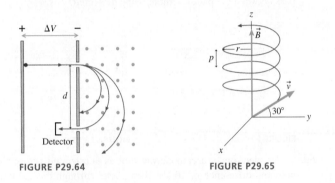

FIGURE P29.64 **FIGURE P29.65**

65. ‖ The uniform 30 mT magnetic field in **FIGURE P29.65** points in the positive z-direction. An electron enters the region of magnetic field with a speed of 5.0×10^6 m/s and at an angle of $30°$ above the xy-plane. Find the radius r and the pitch p of the electron's spiral trajectory.

66. ‖ Particle accelerators, such as the Large Hadron Collider, use magnetic fields to steer charged particles around a ring. Consider a proton ring with 36 identical bending magnets connected by straight segments. The protons move along a 1.0-m-long circular arc as they pass through each magnet. What magnetic field strength is needed in each magnet to steer protons around the ring with a speed of 2.5×10^7 m/s? Assume that the field is uniform inside the magnet, zero outside.

67. ‖ A particle of charge q and mass m moves in the uniform fields $\vec{E} = E_0\hat{k}$ and $\vec{B} = B_0\hat{k}$. At $t = 0$, the particle has velocity $\vec{v}_0 = v_0\hat{i}$. What is the particle's speed at a later time t?

68. ‖ A Hall-effect probe to measure magnetic field strengths needs to be calibrated in a known magnetic field. Although it is not easy to do, magnetic fields can be precisely measured by measuring the cyclotron frequency of protons. A testing laboratory adjusts a magnetic field until the proton's cyclotron frequency is 10.0 MHz. At this field strength, the Hall voltage on the probe is 0.543 mV when the current through the probe is 0.150 mA. Later, when an unknown magnetic field is measured, the Hall voltage at the same current is 1.735 mV. What is the strength of this magnetic field?

69. ‖‖ It is shown in more advanced courses that charged particles
CALC in circular orbits radiate electromagnetic waves, called *cyclotron radiation.* As a result, a particle undergoing cyclotron motion with speed v is actually losing kinetic energy at the rate

$$\frac{dK}{dt} = -\left(\frac{\mu_0 q^4}{6\pi c m^2}\right) B^2 v^2$$

How long does it take (a) an electron and (b) a proton to radiate away half its energy while spiraling in a 2.0 T magnetic field?

70. ‖‖ A proton in a cyclotron gains $\Delta K = 2e\Delta V$ of kinetic energy per
CALC revolution, where ΔV is the potential between the dees. Although the energy gain comes in small pulses, the proton makes so many revolutions that it is reasonable to model the energy as increasing at the constant rate $P = dK/dt = \Delta K/T$, where T is the period of the cyclotron motion. This is *power* input because it is a rate of increase of energy.

   a. Find an expression for $r(t)$, the radius of a proton's orbit in a cyclotron, in terms of m, e, B, P, and t. Assume that $r = 0$ at $t = 0$.
   Hint: Start by finding an expression for the proton's kinetic energy in terms of r.
   b. A relatively small cyclotron is 2.0 m in diameter, uses a 0.55 T magnetic field, and has a 400 V potential difference between the dees. What is the power input to a proton, in W?
   c. How long does it take a proton to spiral from the center out to the edge?

71. ‖ The 10-turn loop of wire shown in **FIGURE P29.71** lies in a horizontal plane, parallel to a uniform horizontal magnetic field, and carries a 2.0 A current. The loop is free to rotate about a nonmagnetic axle through the center. A 50 g mass hangs from one edge of the loop. What magnetic field strength will prevent the loop from rotating about the axle?

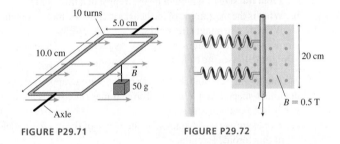

FIGURE P29.71 **FIGURE P29.72**

72. ‖ The two springs in **FIGURE P29.72** each have a spring constant of 10 N/m. They are compressed by 1.0 cm when a current passes through the wire. How big is the current?

73. ‖ **FIGURE P29.73** is an edge view of a 2.0 kg square loop, 2.5 m on each side, with its lower edge resting on a frictionless, horizontal surface. A 25 A current is flowing around the loop in the direction shown. What is the strength of a uniform, horizontal magnetic field for which the loop is in static equilibrium at the angle shown?

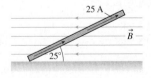

FIGURE P29.73

74. ‖ Magnetic fields are sometimes measured by balancing magnetic forces against known mechanical forces. Your task is to measure the strength of a horizontal magnetic field using a 12-cm-long rigid metal rod that hangs from two nonmagnetic springs, one at each end, with spring constants 1.3 N/m. You first position the rod to be level and perpendicular to the field, whose direction you determined with a compass. You then connect the ends of the rod to wires that run parallel to the field and thus experience no forces. Finally, you measure the downward deflection of the rod, stretching the springs, as you pass current through it. Your data are as follows:

Current (A)	Deflection (mm)
1.0	4
2.0	9
3.0	12
4.0	15
5.0	21

Use an appropriate graph of the data to determine the magnetic field strength.

75. ‖ A conducting bar of length l and mass m rests at the left end of the two frictionless rails of length d in FIGURE P29.75. A uniform magnetic field of strength B points upward.
   a. In which direction, into or out of the page, will a current through the conducting bar cause the bar to experience a force to the right?
   b. Find an expression for the bar's speed as it leaves the rails at the right end.

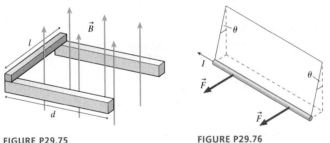

FIGURE P29.75   FIGURE P29.76

76. ‖ a. In FIGURE P29.76, a long, straight, current-carrying wire of linear mass density μ is suspended by threads. A magnetic field perpendicular to the wire exerts a horizontal force that deflects the wire to an equilibrium angle θ. Find an expression for the strength and direction of the magnetic field $\vec{B}$.
   b. What $\vec{B}$ deflects a 55 g/m wire to a 12° angle when the current is 10 A?

77. ‖‖ A wire along the x-axis carries current I in the negative x-
CALC direction through the magnetic field

$$\vec{B} = \begin{cases} B_0 \dfrac{x}{l} \hat{k} & 0 \le x \le l \\ 0 & \text{elsewhere} \end{cases}$$

   a. Draw a graph of B versus x over the interval $-\frac{3}{2}l < x < \frac{3}{2}l$.
   b. Find an expression for the net force $\vec{F}_{net}$ on the wire.
   c. Find an expression for the net torque on the wire about the point $x = 0$.

78. ‖ A *nonuniform* magnetic field exerts a net force on a current loop of radius R. FIGURE P29.78 shows a magnetic field that is diverging from the end of a bar magnet. The magnetic field at the position of the current loop makes an angle θ with respect to the vertical.
   a. Find an expression for the net magnetic force on the current.
   b. Calculate the force if $R = 2.0$ cm, $I = 0.50$ A, $B = 200$ mT, and $\theta = 20°$.

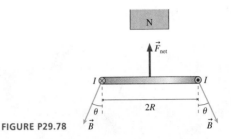

FIGURE P29.78

Challenge Problems

79. ‖‖‖ You have a 1.0-m-long copper wire. You want to make an N-turn current loop that generates a 1.0 mT magnetic field at the center when the current is 1.0 A. You must use the entire wire. What will be the diameter of your coil?

80. ‖‖‖ a. Derive an expression for the magnetic field strength at
CALC distance d from the center of a straight wire of finite length l that carries current I.
   b. Determine the field strength at the center of a current-carrying *square* loop having sides of length $2R$.
   c. Compare your answer to part b to the field at the center of a *circular* loop of diameter $2R$. Do so by computing the ratio B_{square}/B_{circle}.

81. ‖‖‖ A flat, circular disk of radius R is uniformly charged with
CALC total charge Q. The disk spins at angular velocity ω about an axis through its center. What is the magnetic field strength at the center of the disk?

82. ‖‖‖ A long, straight conducting wire of radius R has a nonuniform
CALC current density $J = J_0 r/R$, where J_0 is a constant. The wire carries total current I.
   a. Find an expression for J_0 in terms of I and R.
   b. Find an expression for the magnetic field strength inside the wire at radius r.
   c. At the boundary, $r = R$, does your solution match the known field outside a long, straight current-carrying wire?

83. ‖‖‖ An infinitely wide flat sheet of charge flows out of the page in FIGURE CP29.83. The current per unit width along the sheet (amps per meter) is given by the linear current density J_s.
   a. What is the *shape* of the magnetic field? To answer this question, you may find it helpful to approximate the current sheet as many parallel, closely spaced current-carrying wires. Give your answer as a picture showing magnetic field vectors.
   b. Find the magnetic field strength at distance d above or below the current sheet.

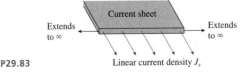

FIGURE CP29.83

30 Electromagnetic Induction

Electromagnetic induction is the physics that underlies many modern technologies, from the generation of electricity to data storage.

IN THIS CHAPTER, you will learn what electromagnetic induction is and how it is used.

What is an induced current?

A magnetic field can create a current in a loop of wire, *but only if the amount of field through the loop is changing*.

- This is called an induced current.
- The process is called **electromagnetic induction**.

« LOOKING BACK Chapter 29 Magnetic fields

What is magnetic flux?

A key idea will be the amount of magnetic field passing through a loop or coil. This is called magnetic flux. Magnetic flux depends on the strength of the magnetic field, the area of the loop, and the angle between them.

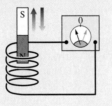

« LOOKING BACK Section 24.3 Electric flux

What is Lenz's law?

Lenz's law says that a current is induced in a closed loop if and only if the magnetic flux through the loop is changing. Simply having a flux does nothing; the flux has to change. You'll learn how to use Lenz's law to determine the direction of an induced current around a loop.

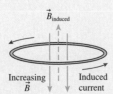

What is Faraday's law?

Faraday's law is the most important law connecting electric and magnetic fields, laying the groundwork for electromagnetic waves. Just as a battery has an emf that drives current, a loop of wire has an induced emf determined by the rate of change of magnetic flux through the loop.

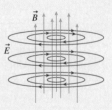

« LOOKING BACK Section 26.4 Sources of potential

What is an induced field?

At its most fundamental level, Faraday's law tells us that a changing magnetic field creates an induced electric field. This is an entirely new way to create an electric field, independent of charges. It is the induced electric field that drives the induced current around a conducting loop.

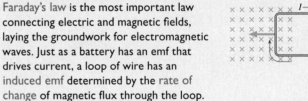

How is electromagnetic induction used?

Electromagnetic induction is one of the most important applications of electricity and magnetism. Generators use electromagnetic induction to turn the mechanical energy of a spinning turbine into electric energy. Inductors are important circuit elements that rely on electromagnetic induction. All forms of telecommunication are based on electromagnetic induction. And, not least, electromagnetic induction is the basis for light and other electromagnetic waves.

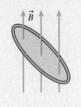

30.1 Induced Currents

Oersted's 1820 discovery that a current creates a magnetic field generated enormous excitement. One question scientists hoped to answer was whether the converse of Oersted's discovery was true: that is, can a magnet be used to create a current?

The breakthrough came in 1831 when the American science teacher Joseph Henry and the English scientist Michael Faraday each discovered the process we now call *electromagnetic induction*. Faraday—whom you met in Chapter 22 as the inventor of the concept of a *field*—was the first to publish his findings, so today we study Faraday's law rather than Henry's law.

Faraday's 1831 discovery, like Oersted's, was a happy combination of an unplanned event and a mind that was ready to recognize its significance. Faraday was experimenting with two coils of wire wrapped around an iron ring, as shown in **FIGURE 30.1**. He had hoped that the magnetic field generated in the coil on the left would induce a magnetic field in the iron, and that the magnetic field in the iron might then somehow create a current in the circuit on the right.

Like all his previous attempts, this technique failed to generate a current. But Faraday happened to notice that the needle of the current meter jumped ever so slightly at the instant he closed the switch in the circuit on the left. After the switch was closed, the needle immediately returned to zero. The needle again jumped when he later opened the switch, but this time in the opposite direction. Faraday recognized that the motion of the needle indicated a current in the circuit on the right, but a momentary current only during the brief interval when the current on the left was starting or stopping.

Faraday's observations, coupled with his mental picture of field lines, led him to suggest that a current is generated only if the magnetic field through the coil is *changing*. This explains why all the previous attempts to generate a current with static magnetic fields had been unsuccessful. Faraday set out to test this hypothesis.

FIGURE 30.1 Faraday's discovery.

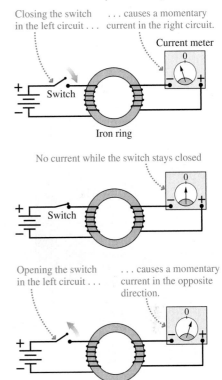

Closing the switch in the left circuit . . . . . . causes a momentary current in the right circuit.

No current while the switch stays closed

Opening the switch in the left circuit . . . . . . causes a momentary current in the opposite direction.

Faraday investigates electromagnetic induction

Faraday placed one coil directly above the other, without the iron ring. There was no current in the lower circuit while the switch was in the closed position, but a momentary current appeared whenever the switch was opened or closed.

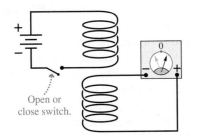

Opening or closing the switch creates a momentary current.

He pushed a bar magnet into a coil of wire. This action caused a momentary deflection of the current-meter needle, although *holding* the magnet inside the coil had no effect. A quick withdrawal of the magnet deflected the needle in the other direction.

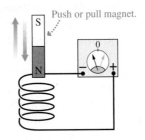

Pushing the magnet into the coil or pulling it out creates a momentary current.

Must the magnet move? Faraday created a momentary current by rapidly pulling a coil of wire out of a magnetic field. Pushing the coil *into* the magnet caused the needle to deflect in the opposite direction.

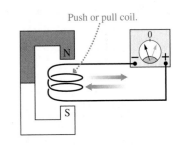

Pushing the coil into the magnet or pulling it out creates a momentary current.

Faraday found that there is a current in a coil of wire if and only if the magnetic field passing through the coil is *changing*. This is an informal statement of what we'll soon call *Faraday's law*. The current in a circuit due to a changing magnetic field is called an **induced current.** An induced current is not caused by a battery; it is a completely new way to generate a current.

30.2 Motional emf

We'll start our investigation of electromagnetic induction by looking at situations in which the magnetic field is fixed while the circuit moves or changes. Consider a conductor of length l that moves with velocity $\vec{v}$ through a perpendicular uniform magnetic field $\vec{B}$, as shown in **FIGURE 30.2**. The charge carriers inside the wire—assumed to be positive—also move with velocity $\vec{v}$, so they each experience a magnetic force $\vec{F}_B = q\vec{v} \times \vec{B}$ of strength $F_B = qvB$. This force causes the charge carriers to move, separating the positive and negative charges. The separated charges then create an electric field inside the conductor.

FIGURE 30.2 The magnetic force on the charge carriers in a moving conductor creates an electric field inside the conductor.

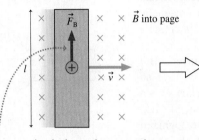
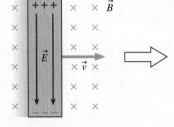
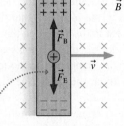

Charge carriers in the conductor experience a force of magnitude $F_B = qvB$. Positive charges are free to move and drift upward.

The resulting charge separation creates an electric field in the conductor. $\vec{E}$ increases as more charge flows.

The charge flow continues until the electric and magnetic forces balance. For a positive charge carrier, the upward magnetic force $\vec{F}_B$ is equal to the downward electric force $\vec{F}_E$.

The charge carriers continue to separate until the electric force $F_E = qE$ exactly balances the magnetic force $F_B = qvB$, creating an equilibrium situation. This balance happens when the electric field strength is

$$E = vB \tag{30.1}$$

In other words, **the magnetic force on the charge carriers in a moving conductor creates an electric field $E = vB$ inside the conductor.**

The electric field, in turn, creates an electric potential difference between the two ends of the moving conductor. **FIGURE 30.3a** defines a coordinate system in which $\vec{E} = -vB\,\hat{\jmath}$. Using the connection between the electric field and the electric potential,

$$\Delta V = V_{\text{top}} - V_{\text{bottom}} = -\int_0^l E_y\,dy = -\int_0^l (-vB)\,dy = vlB \tag{30.2}$$

Thus **the motion of the wire through a magnetic field *induces* a potential difference vlB between the ends of the conductor.** The potential difference depends on the strength of the magnetic field and on the wire's speed through the field.

There's an important analogy between this potential difference and the potential difference of a battery. **FIGURE 30.3b** reminds you that a battery uses a nonelectric force—the charge escalator—to separate positive and negative charges. The emf $\mathcal{E}$ of the battery was defined as the work performed per charge (W/q) to separate the charges. An isolated battery, with no current, has a potential difference $\Delta V_{\text{bat}} = \mathcal{E}$. We could refer to a battery, where the charges are separated by chemical reactions, as a source of *chemical emf.*

The moving conductor develops a potential difference because of the work done by magnetic forces to separate the charges. You can think of the moving conductor as a "battery" that stays charged only as long as it keeps moving but "runs down" if it stops. The emf of the conductor is due to its motion, rather than to chemical reactions inside, so we can define the **motional emf** of a conductor moving with velocity $\vec{v}$ perpendicular to a magnetic field $\vec{B}$ to be

$$\mathcal{E} = vlB \tag{30.3}$$

FIGURE 30.3 Generating an emf.

(a) Magnetic forces separate the charges and cause a potential difference between the ends. This is a motional emf.

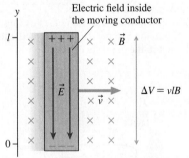

Electric field inside the moving conductor

$\Delta V = vlB$

(b) Chemical reactions separate the charges and cause a potential difference between the ends. This is a chemical emf.

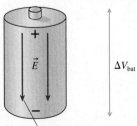

ΔV_{bat}

Electric field inside the battery

STOP TO THINK 30.1 A square conductor moves through a uniform magnetic field. Which of the figures shows the correct charge distribution on the conductor?

(a) $\vec{v}$

(b) $\vec{v}$

(c) $\vec{v}$

(d) $\vec{v}$

EXAMPLE 30.1 | Measuring the earth's magnetic field

It is known that the earth's magnetic field over northern Canada points straight down. The crew of a Boeing 747 aircraft flying at 260 m/s over northern Canada finds a 0.95 V potential difference between the wing tips. The wing span of a Boeing 747 is 65 m. What is the magnetic field strength there?

MODEL The wing is a conductor moving through a magnetic field, so there is a motional emf.

SOLVE The magnetic field is perpendicular to the velocity, so we can use Equation 30.3 to find

$$B = \frac{\mathcal{E}}{vL} = \frac{0.95 \text{ V}}{(260 \text{ m/s})(65 \text{ m})} = 5.6 \times 10^{-5} \text{ T}$$

ASSESS Chapter 29 noted that the earth's magnetic field is roughly 5×10^{-5} T. The field is somewhat stronger than this near the magnetic poles, somewhat weaker near the equator.

EXAMPLE 30.2 | Potential difference along a rotating bar

A metal bar of length l rotates with angular velocity ω about a pivot at one end of the bar. A uniform magnetic field $\vec{B}$ is perpendicular to the plane of rotation. What is the potential difference between the ends of the bar?

VISUALIZE FIGURE 30.4 is a pictorial representation of the bar. The magnetic forces on the charge carriers will cause the outer end to be positive with respect to the pivot.

FIGURE 30.4 Pictorial representation of a metal bar rotating in a magnetic field.

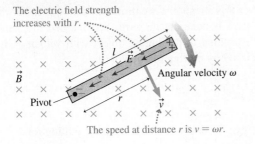

The electric field strength increases with r.

$\vec{B}$
Pivot
Angular velocity ω
The speed at distance r is $v = \omega r$.

SOLVE Even though the bar is rotating, rather than moving in a straight line, the velocity of each charge carrier is perpendicular to $\vec{B}$. Consequently, the electric field created inside the bar is exactly that given in Equation 30.1, $E = vB$. But v, the speed of the charge carrier, now depends on its distance from the pivot. Recall that in rotational motion the tangential speed at radius r from the center of rotation is $v = \omega r$. Thus the electric field at distance r from the pivot is $E = \omega rB$. The electric field increases in strength as you move outward along the bar.

The electric field $\vec{E}$ points toward the pivot, so its radial component is $E_r = -\omega rB$. If we integrate outward from the center, the potential difference between the ends of the bar is

$$\Delta V = V_{\text{tip}} - V_{\text{pivot}} = -\int_0^l E_r \, dr$$

$$= -\int_0^l (-\omega rB) \, dr = \omega B \int_0^l r \, dr = \tfrac{1}{2}\omega l^2 B$$

ASSESS $\tfrac{1}{2}\omega l$ is the speed at the midpoint of the bar. Thus ΔV is $v_{\text{mid}} lB$, which seems reasonable.

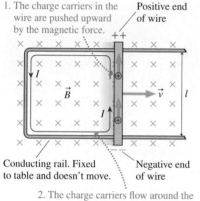

FIGURE 30.5 A current is induced in the circuit as the wire moves through a magnetic field.

1. The charge carriers in the wire are pushed upward by the magnetic force.

Positive end of wire

Conducting rail. Fixed to table and doesn't move.

Negative end of wire

2. The charge carriers flow around the conducting loop as an induced current.

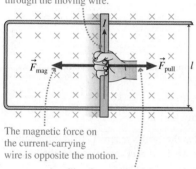

FIGURE 30.6 A pulling force is needed to move the wire to the right.

The induced current flows through the moving wire.

The magnetic force on the current-carrying wire is opposite the motion.

A pulling force to the right must balance the magnetic force to keep the wire moving at constant speed.

Induced Current in a Circuit

The moving conductor of Figure 30.2 had an emf, but it couldn't sustain a current because the charges had nowhere to go. It's like a battery that is disconnected from a circuit. We can change this by including the moving conductor in a circuit.

FIGURE 30.5 shows a conducting wire sliding with speed v along a U-shaped conducting rail. We'll assume that the rail is attached to a table and cannot move. The wire and the rail together form a closed conducting loop—a circuit.

Suppose a magnetic field $\vec{B}$ is perpendicular to the plane of the circuit. Charges in the moving wire will be pushed to the ends of the wire by the magnetic force, just as they were in Figure 30.2, but now the charges can continue to flow around the circuit. That is, the moving wire acts like a battery in a circuit.

The current in the circuit is an *induced current*. In this example, the induced current is counterclockwise (ccw). If the total resistance of the circuit is R, the induced current is given by Ohm's law as

$$I = \frac{\mathcal{E}}{R} = \frac{vlB}{R} \tag{30.4}$$

In this situation, the induced current is due to magnetic forces on moving charges.

We've assumed that the wire is moving along the rail at constant speed. It turns out that we must apply a continuous pulling force $\vec{F}_{pull}$ to make this happen. **FIGURE 30.6** shows why. The moving wire, which now carries induced current I, is in a magnetic field. You learned in Chapter 29 that a magnetic field exerts a force on a current-carrying wire. According to the right-hand rule, the magnetic force $\vec{F}_{mag}$ on the moving wire points to the left. This "magnetic drag" will cause the wire to slow down and stop *unless* we exert an equal but opposite pulling force $\vec{F}_{pull}$ to keep the wire moving.

The magnitude of the magnetic force on a current-carrying wire was found in Chapter 29 to be $F_{mag} = IlB$. Using that result, along with Equation 30.4 for the induced current, we find that the force required to pull the wire with a constant speed v is

$$F_{pull} = F_{mag} = IlB = \left(\frac{vlB}{R}\right)lB = \frac{vl^2B^2}{R} \tag{30.5}$$

STOP TO THINK 30.2 Is there an induced current in this circuit? If so, what is its direction?

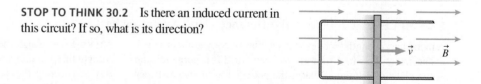

Energy Considerations

The environment must do work on the wire to pull it. What happens to the energy transferred to the wire by this work? Is energy conserved as the wire moves along the rail? It will be easier to answer this question if we think about power rather than work. Power is the *rate* at which work is done on the wire. You learned in Chapter 9 that the power exerted by a force pushing or pulling an object with velocity v is $P = Fv$. The power provided to the circuit by pulling on the wire is

$$P_{input} = F_{pull}v = \frac{v^2l^2B^2}{R} \tag{30.6}$$

This is the rate at which energy is added to the circuit by the pulling force.

FIGURE 29.42b shows a wire *perpendicular* to the magnetic field. By the right-hand rule, each charge in the current has a force of magnitude qvB directed to the left. Consequently, the entire length of wire within the magnetic field experiences a force to the left, perpendicular to both the current direction and the field direction.

A current is moving charge, and the magnetic force on a current-carrying wire is simply the net magnetic force on all the charge carriers in the wire. FIGURE 29.43 shows a wire carrying current I and a segment of length l in which the charge carriers—moving with drift velocity $\vec{v}_d$—have total charge Q. Because the magnetic force is proportional to q, the *net* force on all the charge carriers in the wire is the force on the net charge: $\vec{F} = Q\vec{v}_d \times \vec{B}$. But we need to express this in terms of the current.

By definition, the current I is the amount of charge Q divided by the time t it takes the charge to flow through this segment: $I = Q/t$. The charge carriers have drift speed v_d, so they move distance l in $t = l/v_d$. Combining these equations, we have

$$I = \frac{Qv_d}{l}$$

and thus $Qv_d = Il$. If we define vector $\vec{l}$ to point in the direction of $\vec{v}_d$, the current direction, then $Q\vec{v}_d = I\vec{l}$. Substituting $I\vec{l}$ for $Q\vec{v}_d$ in the force equation, we find that the magnetic force on length l of a current-carrying wire is

$$\vec{F}_{wire} = I\vec{l} \times \vec{B} = (IlB\sin\alpha, \text{ direction of right-hand rule}) \qquad (29.26)$$

where α is the angle between $\vec{l}$ (the direction of the current) and $\vec{B}$. As an aside, you can see from Equation 29.26 that the magnetic field B must have units of N/A m. This is why we defined 1 T = 1 N/A m in Section 29.3.

NOTE The familiar right-hand rule applies to a current-carrying wire. Point your right thumb in the direction of the current (parallel to $\vec{l}$) and your index finger in the direction of $\vec{B}$. Your middle finger is then pointing in the direction of the force $\vec{F}$ on the wire.

FIGURE 29.42 Magnetic force on a current-carrying wire.

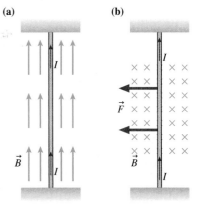

There's no force on a current parallel to a magnetic field.

There is a magnetic force in the direction of the right-hand rule.

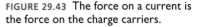

FIGURE 29.43 The force on a current is the force on the charge carriers.

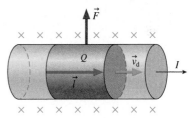

EXAMPLE 29.13 | **Magnetic levitation**

The 0.10 T uniform magnetic field of FIGURE 29.44 is horizontal, parallel to the floor. A straight segment of 1.0-mm-diameter copper wire, also parallel to the floor, is perpendicular to the magnetic field. What current through the wire, and in which direction, will allow the wire to "float" in the magnetic field?

FIGURE 29.44 Magnetic levitation.

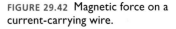

MODEL The wire will float in the magnetic field if the magnetic force on the wire points upward and has magnitude mg, allowing it to balance the downward gravitational force.

SOLVE We can use the right-hand rule to determine which current direction experiences an upward force. With $\vec{B}$ pointing away from us, the direction of the current needs to be from left to right. The forces will balance when

$$F = IlB = mg = \rho(\pi r^2 l)g$$

where $\rho = 8920$ kg/m³ is the density of copper. The length of the wire cancels, leading to

$$I = \frac{\rho\pi r^2 g}{B} = \frac{(8920 \text{ kg/m}^3)\pi(0.00050 \text{ m})^2(9.80 \text{ m/s}^2)}{0.10 \text{ T}}$$

$$= 0.69 \text{ A}$$

A 0.69 A current from left to right will levitate the wire in the magnetic field.

ASSESS A 0.69 A current is quite reasonable, but this idea is useful only if we can get the current into and out of this segment of wire. In practice, we could do so with wires that come in from below the page. These input and output wires would be parallel to $\vec{B}$ and not experience a magnetic force. Although this example is very simple, it is the basis for applications such as magnetic levitation trains.

Force Between Two Parallel Wires

Now consider Ampère's experimental arrangement of two parallel wires of length l, distance d apart. FIGURE 29.45a on the next page shows the currents I_1 and I_2 in the same direction; FIGURE 29.45b shows the currents in opposite directions. We will assume that the wires are sufficiently long to allow us to use the earlier result for the magnetic field of a long, straight wire: $B = \mu_0 I/2\pi r$.

FIGURE 30.9 Eddy currents.

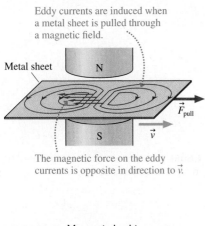

FIGURE 30.9 Eddy currents.

Eddy currents are induced when a metal sheet is pulled through a magnetic field.

Metal sheet

N

$\vec{F}_{\text{pull}}$

$\vec{v}$

S

The magnetic force on the eddy currents is opposite in direction to $\vec{v}$.

FIGURE 30.10 Magnetic braking system.

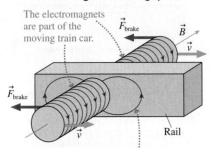

The electromagnets are part of the moving train car.

$\vec{F}_{\text{brake}}$

$\vec{B}$

$\vec{v}$

$\vec{F}_{\text{brake}}$

$\vec{v}$

Rail

Eddy currents are induced in the rail. Magnetic forces between the eddy currents and the electromagnets slow the train.

FIGURE 30.11 The amount of air flowing through a loop depends on the effective area of the loop.

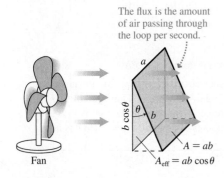

The flux is the amount of air passing through the loop per second.

a

$b \cos\theta$

θ

b

$A = ab$

Fan

$A_{\text{eff}} = ab \cos\theta$

Eddy Currents

These ideas have interesting implications. Consider pulling a *sheet* of metal through a magnetic field, as shown in **FIGURE 30.9a**. The metal, we will assume, is not a magnetic material, so it experiences no magnetic force if it is at rest. The charge carriers in the metal experience a magnetic force as the sheet is dragged between the pole tips of the magnet. A current is induced, just as in the loop of wire, but here the currents do not have wires to define their path. As a consequence, two "whirlpools" of current begin to circulate in the metal. These spread-out current whirlpools in a solid metal are called **eddy currents.**

As the eddy current passes between the pole tips, it experiences a magnetic force to the left—a retarding force. Thus **an external force is required to pull a metal through a magnetic field.** If the pulling force ceases, the retarding magnetic force quickly causes the metal to decelerate until it stops. Similarly, a force is required to push a sheet of metal *into* a magnetic field.

Eddy currents are often undesirable. The power dissipation of eddy currents can cause unwanted heating, and the magnetic forces on eddy currents mean that extra energy must be expended to move metals in magnetic fields. But eddy currents also have important useful applications. A good example is magnetic braking.

The moving train car has an electromagnet that straddles the rail, as shown in **FIGURE 30.10**. During normal travel, there is no current through the electromagnet and no magnetic field. To stop the car, a current is switched into the electromagnet. The current creates a strong magnetic field that passes *through* the rail, and the motion of the rail relative to the magnet induces eddy currents in the rail. The magnetic force between the electromagnet and the eddy currents acts as a braking force on the magnet and, thus, on the car. Magnetic braking systems are very efficient, and they have the added advantage that they heat the rail rather than the brakes.

STOP TO THINK 30.3 A square loop of copper wire is pulled through a region of magnetic field. Rank in order, from strongest to weakest, the pulling forces $\vec{F}_a$, $\vec{F}_b$, $\vec{F}_c$, and $\vec{F}_d$ that must be applied to keep the loop moving at constant speed.

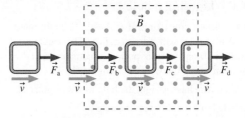

30.3 Magnetic Flux

Faraday found that a current is induced when the amount of magnetic field passing through a coil or a loop of wire changes. And that's exactly what happens as the slide wire moves down the rail in Figure 30.5! As the circuit expands, more magnetic field passes through. It's time to define more clearly what we mean by "the amount of field passing through a loop."

Imagine holding a rectangular loop in front of the fan shown in **FIGURE 30.11**. The amount of air flowing *through* the loop—the *flux*—depends on the angle of the loop. The flow is maximum if the loop is perpendicular to the flow, zero if the loop is rotated to be parallel to the flow. In general, the amount of air flowing through is proportional to the *effective area* of the loop (i.e., the area facing the fan):

$$A_{\text{eff}} = ab\cos\theta = A\cos\theta \tag{30.8}$$

where $A = ab$ is the area of the loop and θ is the tilt angle of the loop. A loop perpendicular to the flow, with $\theta = 0°$, has $A_{\text{eff}} = A$, the full area of the loop.

We can apply this idea to a magnetic field passing through a loop. FIGURE 30.12 shows a loop of area $A = ab$ in a uniform magnetic field. Think of the field vectors, seen here from behind, as if they were arrows shot into the page. The density of arrows (arrows per m^2) is proportional to the strength B of the magnetic field; a stronger field would be represented by arrows packed closer together. The number of arrows passing through a loop of wire depends on two factors:

1. The density of arrows, which is proportional to B, and
2. The effective area $A_{eff} = A \cos \theta$ of the loop.

The angle θ is the angle between the magnetic field and the axis of the loop. The maximum number of arrows passes through the loop when it is perpendicular to the magnetic field ($\theta = 0°$). No arrows pass through the loop if it is tilted 90°.

FIGURE 30.12 Magnetic field through a loop that is tilted at various angles.

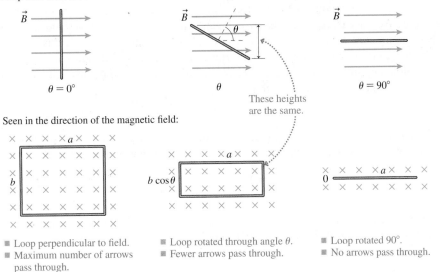

Loop seen from the side:

$\theta = 0°$ θ $\theta = 90°$

These heights are the same.

Seen in the direction of the magnetic field:

- Loop perpendicular to field.
- Maximum number of arrows pass through.

- Loop rotated through angle θ.
- Fewer arrows pass through.

- Loop rotated 90°.
- No arrows pass through.

With this in mind, let's define the **magnetic flux** Φ_m as

$$\Phi_m = A_{eff}B = AB \cos \theta \qquad (30.9)$$

The magnetic flux measures the amount of magnetic field passing through a loop of area A if the loop is tilted at angle θ from the field. The SI unit of magnetic flux is the **weber.** From Equation 30.9 you can see that

$$1 \text{ weber} = 1 \text{ Wb} = 1 \text{ T m}^2$$

Equation 30.9 is reminiscent of the vector dot product: $\vec{A} \cdot \vec{B} = AB \cos \theta$. With that in mind, let's define an **area vector** $\vec{A}$ to be a vector *perpendicular* to the loop, with magnitude equal to the area A of the loop. Vector $\vec{A}$ has units of m^2. FIGURE 30.13a shows the area vector $\vec{A}$ for a circular loop of area A.

FIGURE 30.13b shows a magnetic field passing through a loop. The angle between vectors $\vec{A}$ and $\vec{B}$ is the same angle used in Equations 30.8 and 30.9 to define the effective area and the magnetic flux. So Equation 30.9 really is a dot product, and we can define the magnetic flux more concisely as

$$\Phi_m = \vec{A} \cdot \vec{B} \qquad (30.10)$$

Writing the flux as a dot product helps make clear how angle θ is defined: θ is the angle between the magnetic field and the axis of the loop.

FIGURE 30.13 Magnetic flux can be defined in terms of an area vector $\vec{A}$.

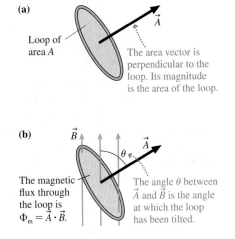

(a)

Loop of area A

The area vector is perpendicular to the loop. Its magnitude is the area of the loop.

(b)

The magnetic flux through the loop is $\Phi_m = \vec{A} \cdot \vec{B}$.

The angle θ between $\vec{A}$ and $\vec{B}$ is the angle at which the loop has been tilted.

EXAMPLE 30.4 | A circular loop in a magnetic field

FIGURE 30.14 is an edge view of a 10-cm-diameter circular loop in a uniform 0.050 T magnetic field. What is the magnetic flux through the loop?

SOLVE Angle θ is the angle between the loop's area vector $\vec{A}$, which is perpendicular to the plane of the loop, and the magnetic field $\vec{B}$. In this case, $\theta = 60°$, not the 30° angle shown in the figure. Vector $\vec{A}$ has magnitude $A = \pi r^2 = 7.85 \times 10^{-3} \text{ m}^2$. Thus the magnetic flux is

$$\Phi_m = \vec{A} \cdot \vec{B} = AB\cos\theta = 2.0 \times 10^{-4} \text{ Wb}$$

FIGURE 30.14 A circular loop in a magnetic field.

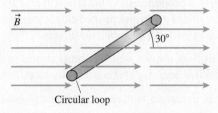

Magnetic Flux in a Nonuniform Field

Equation 30.10 for the magnetic flux assumes that the field is uniform over the area of the loop. We can calculate the flux in a nonuniform field, one where the field strength changes from one edge of the loop to the other, but we'll need to use calculus.

FIGURE 30.15 shows a loop in a nonuniform magnetic field. Imagine dividing the loop into many small pieces of area dA. The infinitesimal flux $d\Phi_m$ through one such area, where the magnetic field is $\vec{B}$, is

$$d\Phi_m = \vec{B} \cdot d\vec{A} \tag{30.11}$$

The total magnetic flux through the loop is the sum of the fluxes through each of the small areas. We find that sum by integrating. Thus the total magnetic flux through the loop is

$$\Phi_m = \int_{\text{area of loop}} \vec{B} \cdot d\vec{A} \tag{30.12}$$

Equation 30.12 is a more general definition of magnetic flux. It may look rather formidable, so we'll illustrate its use with an example.

FIGURE 30.15 A loop in a nonuniform magnetic field.

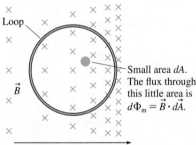

Loop

$\vec{B}$

Small area dA. The flux through this little area is $d\Phi_m = \vec{B} \cdot d\vec{A}$.

Increasing field strength

EXAMPLE 30.5 | Magnetic flux from the current in a long straight wire

The 1.0 cm $\times$ 4.0 cm rectangular loop of **FIGURE 30.16** is 1.0 cm away from a long, straight wire. The wire carries a current of 1.0 A. What is the magnetic flux through the loop?

FIGURE 30.16 A loop next to a current-carrying wire.

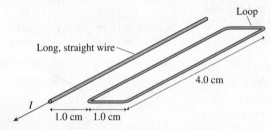

Loop

Long, straight wire

4.0 cm

I

1.0 cm   1.0 cm

MODEL Model the wire as if it were infinitely long. The magnetic field strength of a wire decreases with distance from the wire, so the field is *not* uniform over the area of the loop.

VISUALIZE Using the right-hand rule, we see that the field, as it circles the wire, is perpendicular to the plane of the loop. **FIGURE 30.17** redraws the loop with the field coming out of the page and establishes a coordinate system.

SOLVE Let the loop have dimensions a and b, as shown, with the near edge at distance c from the wire. The magnetic field varies

FIGURE 30.17 Calculating the magnetic flux through the loop.

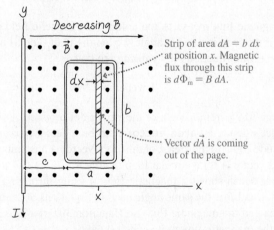

Decreasing B

Strip of area $dA = b\,dx$ at position x. Magnetic flux through this strip is $d\Phi_m = B\,dA$.

dx

b

Vector $d\vec{A}$ is coming out of the page.

c a

with distance x from the wire, but the field is constant along a line parallel to the wire. This suggests dividing the loop into many narrow rectangular strips of length b and width dx, each forming a small area $dA = b\,dx$. The magnetic field has the same strength at all points within this small area. One such strip is shown in the figure at position x.

The area vector $d\vec{A}$ is perpendicular to the strip (coming out of the page), which makes it parallel to $\vec{B}$ ($\theta = 0°$). Thus the infinitesimal flux through this little area is

$$d\Phi_m = \vec{B} \cdot d\vec{A} = B \, dA = B(b \, dx) = \frac{\mu_0 Ib}{2\pi x} dx$$

where, from Chapter 29, we've used $B = \mu_0 I/2\pi x$ as the magnetic field at distance x from a long, straight wire. Integrating "over the area of the loop" means to integrate from the near edge of the loop at $x = c$ to the far edge at $x = c + a$. Thus

$$\Phi_m = \frac{\mu_0 Ib}{2\pi} \int_c^{c+a} \frac{dx}{x} = \frac{\mu_0 Ib}{2\pi} \ln x \Big|_c^{c+a} = \frac{\mu_0 Ib}{2\pi} \ln\left(\frac{c+a}{c}\right)$$

Evaluating for $a = c = 0.010$ m, $b = 0.040$ m, and $I = 1.0$ A gives

$$\Phi_m = 5.5 \times 10^{-9} \text{ Wb}$$

ASSESS The flux measures how much of the wire's magnetic field passes through the loop, but we had to integrate, rather than simply using Equation 30.10, because the field is stronger at the near edge of the loop than at the far edge.

30.4 Lenz's Law

We started out by looking at a situation in which a moving wire caused a loop to expand in a magnetic field. This is one way to change the magnetic flux through the loop. But Faraday found that a current can be induced by any change in the magnetic flux, no matter how it's accomplished.

For example, a momentary current is induced in the loop of FIGURE 30.18 as the bar magnet is pushed toward the loop, increasing the flux through the loop. Pulling the magnet back out of the loop causes the current meter to deflect in the opposite direction. The conducting wires aren't moving, so this is not a motional emf. Nonetheless, the induced current is very real.

The German physicist Heinrich Lenz began to study electromagnetic induction after learning of Faraday's discovery. Three years later, in 1834, Lenz announced a rule for determining the direction of the induced current. We now call his rule **Lenz's law,** and it can be stated as follows:

> **Lenz's law** There is an induced current in a closed, conducting loop if and only if the magnetic flux through the loop is changing. The direction of the induced current is such that the induced magnetic field opposes the *change* in the flux.

Lenz's law is rather subtle, and it takes some practice to see how to apply it.

NOTE One difficulty with Lenz's law is the term *flux*. In everyday language, the word *flux* already implies that something is changing. Think of the phrase, "The situation is in flux." Not so in physics, where *flux*, the root of the word *flow*, means "passes through." A steady magnetic field through a loop creates a steady, *un*changing magnetic flux.

Lenz's law tells us to look for situations where the flux is *changing*. This can happen in three ways.

1. The magnetic field through the loop changes (increases or decreases),
2. The loop changes in area or angle, or
3. The loop moves into or out of a magnetic field.

Lenz's law depends on the idea that an induced current generates its own magnetic field $\vec{B}_{induced}$. This is the *induced magnetic field* of Lenz's law. You learned in Chapter 29 how to use the right-hand rule to determine the direction of this induced magnetic field.

In Figure 30.18, pushing the bar magnet toward the loop causes the magnetic flux to *increase* in the downward direction. To oppose the *change* in flux, which is what Lenz's law requires, the loop itself needs to generate the *upward*-pointing magnetic field of FIGURE 30.19. The induced magnetic field at the center of the loop will point upward if the current is ccw. Thus pushing the north end of a bar magnet toward the loop induces a ccw current around the loop. The induced current ceases as soon as the magnet stops moving.

FIGURE 30.18 Pushing a bar magnet toward the loop induces a current.

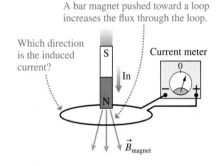

A bar magnet pushed toward a loop increases the flux through the loop.

Which direction is the induced current?

Current meter

$\vec{B}_{magnet}$

FIGURE 30.19 The induced current is ccw.

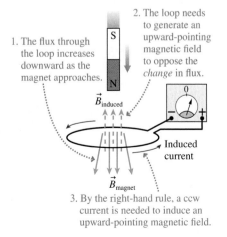

2. The loop needs to generate an upward-pointing magnetic field to oppose the *change* in flux.

1. The flux through the loop increases downward as the magnet approaches.

$\vec{B}_{induced}$

Induced current

$\vec{B}_{magnet}$

3. By the right-hand rule, a ccw current is needed to induce an upward-pointing magnetic field.

Now suppose the bar magnet is pulled back away from the loop, as shown in FIGURE 30.20a. There is a downward magnetic flux through the loop, but the flux *decreases* as the magnet moves away. According to Lenz's law, the induced magnetic field of the loop *opposes this decrease*. To do so, the induced field needs to point in the *downward* direction, as shown in FIGURE 30.20b. Thus as the magnet is withdrawn, the induced current is clockwise (cw), opposite to the induced current of Figure 30.19.

FIGURE 30.20 Pulling the magnet away induces a cw current.

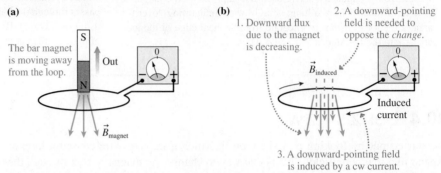

NOTE Notice that the magnetic field of the bar magnet is pointing downward in both Figures 30.19 and 30.20. It is not the *flux* due to the magnet that the induced current opposes, but the *change* in the flux. This is a subtle but critical distinction. If the induced current opposed the flux itself, the current in both Figures 30.19 and 30.20 would be ccw to generate an upward magnetic field. But that's not what happens. When the field of the magnet points down and is increasing, the induced current opposes the increase by generating an upward field. When the field of the magnet points down but is decreasing, the induced current opposes the decrease by generating a downward field.

Using Lenz's Law

FIGURE 30.21 shows six basic situations. The magnetic field can point either up or down through the loop. For each, the flux can either increase, hold steady, or decrease in strength. These observations form the basis for a set of rules about using Lenz's law.

FIGURE 30.21 The induced current for six different situations.

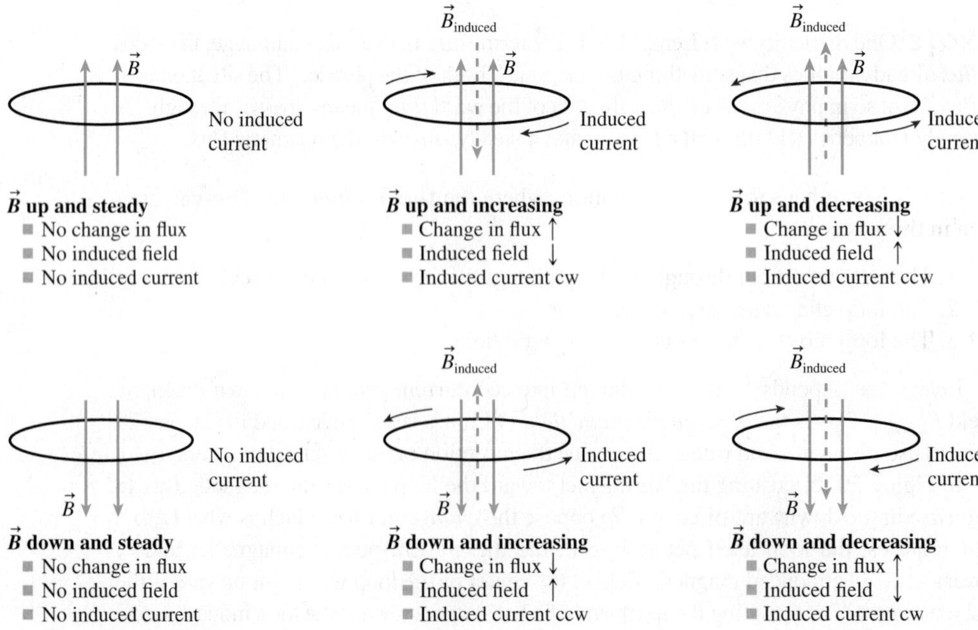

$\vec{B}$ up and steady
- No change in flux
- No induced field
- No induced current

$\vec{B}$ up and increasing
- Change in flux ↑
- Induced field ↓
- Induced current cw

$\vec{B}$ up and decreasing
- Change in flux ↓
- Induced field ↑
- Induced current ccw

$\vec{B}$ down and steady
- No change in flux
- No induced field
- No induced current

$\vec{B}$ down and increasing
- Change in flux ↓
- Induced field ↑
- Induced current ccw

$\vec{B}$ down and decreasing
- Change in flux ↑
- Induced field ↓
- Induced current cw

TACTICS BOX 30.1

Using Lenz's law

❶ **Determine the direction of the applied magnetic field.** The field must pass through the loop.

❷ **Determine how the flux is changing.** Is it increasing, decreasing, or staying the same?

❸ **Determine the direction of an induced magnetic field that will oppose the *change* in the flux.**

- ■ Increasing flux: the induced magnetic field points opposite the applied magnetic field.
- ■ Decreasing flux: the induced magnetic field points in the same direction as the applied magnetic field.
- ■ Steady flux: there is no induced magnetic field.

❹ **Determine the direction of the induced current.** Use the right-hand rule to determine the current direction in the loop that generates the induced magnetic field you found in step 3.

Exercises 10–14

Let's look at two examples.

EXAMPLE 30.6 | Lenz's law 1

FIGURE 30.22 shows two loops, one above the other. The upper loop has a battery and a switch that has been closed for a long time. How does the lower loop respond when the switch is opened in the upper loop?

MODEL We'll use the right-hand rule to find the magnetic fields of current loops.

SOLVE FIGURE 30.23 shows the four steps of using Lenz's law. Opening the switch induces a ccw current in the lower loop. This is a momentary current, lasting only until the magnetic field of the upper loop drops to zero.

ASSESS The conclusion is consistent with Figure 30.21.

FIGURE 30.22 The two loops of Example 30.6.

FIGURE 30.23 Applying Lenz's law.

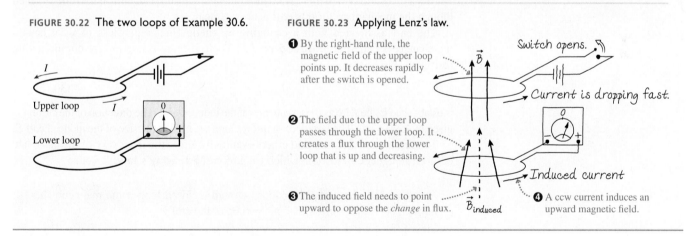

❶ By the right-hand rule, the magnetic field of the upper loop points up. It decreases rapidly after the switch is opened.

❷ The field due to the upper loop passes through the lower loop. It creates a flux through the lower loop that is up and decreasing.

❸ The induced field needs to point upward to oppose the *change* in flux.

❹ A ccw current induces an upward magnetic field.

Switch opens.
Current is dropping fast.
Induced current

EXAMPLE 30.7 | Lenz's law 2

FIGURE 30.24 shows two coils wrapped side by side on a cylinder. When the switch for coil 1 is closed, does the induced current in coil 2 pass from right to left or from left to right through the current meter?

MODEL We'll use the right-hand rule to find the magnetic field of a coil.

VISUALIZE It is very important to look at the *direction* in which a coil is wound around the cylinder. Notice that the two coils in Figure 30.24 are wound in opposite directions.

FIGURE 30.24 The two solenoids of Example 30.7.

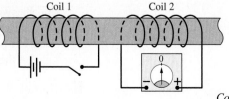

Coil 1   Coil 2

Continued

SOLVE FIGURE 30.25 shows the four steps of using Lenz's law. Closing the switch induces a current that passes from right to left through the current meter. The induced current is only momentary. It lasts only until the field from coil 1 reaches full strength and is no longer changing.

ASSESS The conclusion is consistent with Figure 30.21.

FIGURE 30.25 Applying Lenz's law.

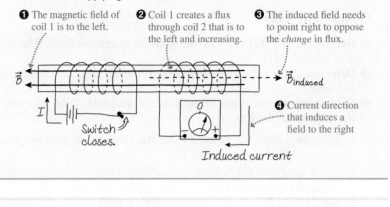

❶ The magnetic field of coil 1 is to the left.

❷ Coil 1 creates a flux through coil 2 that is to the left and increasing.

❸ The induced field needs to point right to oppose the *change* in flux.

❹ Current direction that induces a field to the right

Induced current

STOP TO THINK 30.4 A current-carrying wire is pulled away from a conducting loop in the direction shown. As the wire is moving, is there a cw current around the loop, a ccw current, or no current?

30.5 Faraday's Law

Charges don't start moving spontaneously. A current requires an emf to provide the energy. We started our analysis of induced currents with circuits in which a *motional emf* can be understood in terms of magnetic forces on moving charges. But we've also seen that a current can be induced by changing the magnetic field through a stationary circuit, a circuit in which there is no motion. There *must* be an emf in this circuit, even though the mechanism for this emf is not yet clear.

The emf associated with a changing magnetic flux, regardless of what causes the change, is called an **induced emf** $\mathcal{E}$. Then, if there is a complete circuit having resistance R, a current

$$I_{\text{induced}} = \frac{\mathcal{E}}{R} \tag{30.13}$$

is established in the wire as a *consequence* of the induced emf. The direction of the current is given by Lenz's law. The last piece of information we need is the size of the induced emf $\mathcal{E}$.

The research of Faraday and others eventually led to the discovery of the basic law of electromagnetic induction, which we now call **Faraday's law.** It states:

> **Faraday's law** An emf $\mathcal{E}$ is induced around a closed loop if the magnetic flux through the loop changes. The magnitude of the emf is
>
> $$\mathcal{E} = \left| \frac{d\Phi_{\text{m}}}{dt} \right| \tag{30.14}$$
>
> and the direction of the emf is such as to drive an induced current in the direction given by Lenz's law.

In other words, the induced emf is the *rate of change* of the magnetic flux through the loop.

As a corollary to Faraday's law, an N-turn coil of wire in a changing magnetic field acts like N batteries in series. The induced emf of each of the coils adds, so the induced emf of the entire coil is

$$\mathcal{E}_{\text{coil}} = N \left| \frac{d\Phi_{\text{per coil}}}{dt} \right| \quad \text{(Faraday's law for an } N\text{-turn coil)} \tag{30.15}$$

As a first example of using Faraday's law, return to the situation of Figure 30.5, where a wire moves through a magnetic field by sliding on a U-shaped conducting rail. FIGURE 30.26 shows the circuit again. The magnetic field $\vec{B}$ is perpendicular to the plane of the conducting loop, so $\theta = 0°$ and the magnetic flux is $\Phi = AB$, where A is the area of the loop. If the slide wire is distance x from the end, the area is $A = xl$ and the flux at that instant of time is

$$\Phi_m = AB = xlB \tag{30.16}$$

The flux through the loop increases as the wire moves. According to Faraday's law, the induced emf is

$$\mathcal{E} = \left| \frac{d\Phi_m}{dt} \right| = \frac{d}{dt}(xlB) = \frac{dx}{dt}lB = vlB \tag{30.17}$$

where the wire's velocity is $v = dx/dt$. We can now use Equation 30.13 to find that the induced current is

$$I = \frac{\mathcal{E}}{R} = \frac{vlB}{R} \tag{30.18}$$

The flux is increasing into the loop, so the induced magnetic field opposes this increase by pointing out of the loop. This requires a ccw induced current in the loop. Faraday's law leads us to the conclusion that the loop will have a ccw induced current $I = vlB/R$. This is exactly the conclusion we reached in Section 30.2, where we analyzed the situation from the perspective of magnetic forces on moving charge carriers. Faraday's law confirms what we already knew but, at least in this case, doesn't seem to offer anything new.

FIGURE 30.26 The magnetic flux through the loop increases as the slide wire moves.

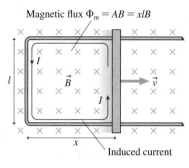

Magnetic flux $\Phi_m = AB = xlB$

Using Faraday's Law

Most electromagnetic induction problems can be solved with a four-step strategy.

PROBLEM-SOLVING STRATEGY 30.1   (MP)

Electromagnetic induction

MODEL Make simplifying assumptions about wires and magnetic fields.

VISUALIZE Draw a picture or a circuit diagram. Use Lenz's law to determine the direction of the induced current.

SOLVE The mathematical representation is based on Faraday's law

$$\mathcal{E} = \left| \frac{d\Phi_m}{dt} \right|$$

For an N-turn coil, multiply by N. The size of the induced current is $I = \mathcal{E}/R$.

ASSESS Check that your result has correct units and significant figures, is reasonable, and answers the question.

Exercise 18

EXAMPLE 30.8 | Electromagnetic induction in a solenoid

A 2.0-cm-diameter loop of wire with a resistance of 0.010 Ω is placed in the center of the solenoid seen in FIGURE 30.27a on the next page. The solenoid is 4.0 cm in diameter, 20 cm long, and wrapped with 1000 turns of wire. FIGURE 30.27b shows the current through the solenoid as a function of time as the solenoid is "powered up." A positive current is defined to be cw when seen from the left. Find the current in the loop as a function of time and show the result as a graph.

MODEL The solenoid's length is much greater than its diameter, so the field near the center should be nearly uniform.

VISUALIZE The magnetic field of the solenoid creates a magnetic flux through the loop of wire. The solenoid current is always positive, meaning that it is cw as seen from the left. Consequently, from the right-hand rule, the magnetic field inside the solenoid always points to the right. During the first second, while the solenoid current is

Continued

FIGURE 30.27 A loop inside a solenoid.

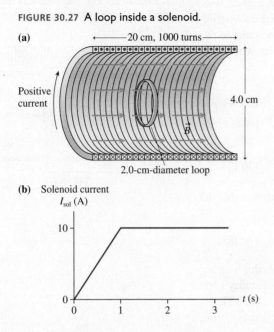

(a)

Positive current

20 cm, 1000 turns

4.0 cm

$\vec{B}$

2.0-cm-diameter loop

(b) Solenoid current

I_{sol} (A)

increasing, the flux through the loop is to the right and increasing. To oppose the *change* in the flux, the loop's induced magnetic field must point to the left. Thus, again using the right-hand rule, the induced current must flow ccw as seen from the left. This is a *negative* current. There's no *change* in the flux for $t > 1$ s, so the induced current is zero.

SOLVE Now we're ready to use Faraday's law to find the magnitude of the current. Because the field is uniform inside the solenoid and perpendicular to the loop ($\theta = 0°$), the flux is $\Phi_m = AB$, where $A = \pi r^2 = 3.14 \times 10^{-4}$ m^2 is the area of the loop (*not* the area of the solenoid). The field of a long solenoid of length l was found in Chapter 29 to be

$$B = \frac{\mu_0 N I_{sol}}{l}$$

The flux when the solenoid current is I_{sol} is thus

$$\Phi_m = \frac{\mu_0 A N I_{sol}}{l}$$

The changing flux creates an induced emf $\mathcal{E}$ that is given by Faraday's law:

$$\mathcal{E} = \left|\frac{d\Phi_m}{dt}\right| = \frac{\mu_0 A N}{l}\left|\frac{dI_{sol}}{dt}\right| = 2.0 \times 10^{-6}\left|\frac{dI_{sol}}{dt}\right|$$

From the slope of the graph, we find

$$\left|\frac{dI_{sol}}{dt}\right| = \begin{cases} 10 \text{ A/s} & 0.0 \text{ s} < t < 1.0 \text{ s} \\ 0 & 1.0 \text{ s} < t < 3.0 \text{ s} \end{cases}$$

Thus the induced emf is

$$\mathcal{E} = \begin{cases} 2.0 \times 10^{-5} \text{ V} & 0.0 \text{ s} < t < 1.0 \text{ s} \\ 0 \text{ V} & 1.0 \text{ s} < t < 3.0 \text{ s} \end{cases}$$

Finally, the current induced in the loop is

$$I_{loop} = \frac{\mathcal{E}}{R} = \begin{cases} -2.0 \text{ mA} & 0.0 \text{ s} < t < 1.0 \text{ s} \\ 0 \text{ mA} & 1.0 \text{ s} < t < 3.0 \text{ s} \end{cases}$$

where the negative sign comes from Lenz's law. This result is shown in **FIGURE 30.28**.

FIGURE 30.28 The induced current in the loop.

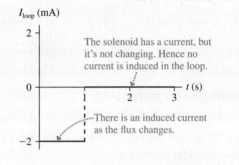

I_{loop} (mA)

The solenoid has a current, but it's not changing. Hence no current is induced in the loop.

There is an induced current as the flux changes.

EXAMPLE 30.9 | **Current induced by an MRI machine**

The body is a conductor, so rapid magnetic field changes in an MRI machine can induce currents in the body. To estimate the size of these currents, and any biological hazard they might impose, consider the "loop" of muscle tissue shown in **FIGURE 30.29**. This might be muscle circling the bone of your arm or thigh. Although muscle is not a great conductor—its resistivity is 1.5 Ω m—we can consider it to be a conducting loop with a rather high resistance. Suppose the magnetic field along the axis of the loop drops from

FIGURE 30.29 Edge view of a loop of muscle tissue in a magnetic field.

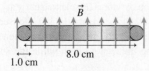

$\vec{B}$

8.0 cm

1.0 cm

1.6 T to 0 T in 0.30 s, which is about the largest possible rate of change for an MRI solenoid. What current will be induced?

MODEL Model the muscle as a conducting loop. Assume that B decreases linearly with time.

SOLVE The magnetic field is parallel to the axis of the loop, with $\theta = 0°$, so the magnetic flux through the loop is $\Phi_m = AB = \pi r^2 B$. The flux changes with time because B changes. According to Faraday's law, the magnitude of the induced emf is

$$\mathcal{E} = \left|\frac{d\Phi_m}{dt}\right| = \pi r^2\left|\frac{dB}{dt}\right|$$

The rate at which the magnetic field changes is

$$\frac{dB}{dt} = \frac{\Delta B}{\Delta t} = \frac{-1.60 \text{ T}}{0.30 \text{ s}} = -5.3 \text{ T/s}$$

But the circuit also dissipates energy by transforming electric energy into the thermal energy of the wires and components, heating them up. The power dissipated by current I as it passes through resistance R is $P = I^2R$. Equation 30.4 for the induced current I gives us the power dissipated by the circuit of Figure 30.5:

$$P_{\text{dissipated}} = I^2R = \frac{v^2l^2B^2}{R} \tag{30.7}$$

You can see that Equations 30.6 and 30.7 are identical. **The rate at which work is done on the circuit exactly balances the rate at which energy is dissipated.** Thus *energy is conserved.*

If you have to *pull* on the wire to get it to move to the right, you might think that it would spring back to the left on its own. FIGURE 30.7 shows the same circuit with the wire moving to the left. In this case, you must *push* the wire to the left to keep it moving. The magnetic force is always opposite to the wire's direction of motion.

In both Figure 30.6, where the wire is pulled, and Figure 30.7, where it is pushed, a mechanical force is used to create a current. In other words, we have a conversion of *mechanical* energy to *electric* energy. A device that converts mechanical energy to electric energy is called a **generator.** The slide-wire circuits of Figures 30.6 and 30.7 are simple examples of a generator. We will look at more practical examples of generators later in the chapter.

We can summarize our analysis as follows:

1. Pulling or pushing the wire through the magnetic field at speed v creates a motional emf $\mathcal{E}$ in the wire and induces a current $I = \mathcal{E}/R$ in the circuit.
2. To keep the wire moving at constant speed, a pulling or pushing force must balance the magnetic force on the wire. This force does work on the circuit.
3. The work done by the pulling or pushing force exactly balances the energy dissipated by the current as it passes through the resistance of the circuit.

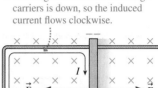

FIGURE 30.7 A pushing force is needed to move the wire to the left.

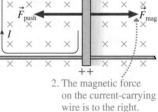

1. The magnetic force on the charge carriers is down, so the induced current flows clockwise.

2. The magnetic force on the current-carrying wire is to the right.

EXAMPLE 30.3 | **Lighting a bulb**

FIGURE 30.8 shows a circuit consisting of a flashlight bulb, rated 3.0 V/1.5 W, and ideal wires with no resistance. The right wire of the circuit, which is 10 cm long, is pulled at constant speed v through a perpendicular magnetic field of strength 0.10 T.

a. What speed must the wire have to light the bulb to full brightness?

b. What force is needed to keep the wire moving?

FIGURE 30.8 Circuit of Example 30.3.

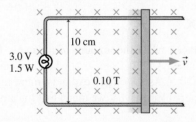

MODEL Treat the moving wire as a source of motional emf.

VISUALIZE The magnetic force on the charge carriers, $\vec{F}_B = q\vec{v} \times \vec{B}$, causes a counterclockwise (ccw) induced current.

SOLVE a. The bulb's rating of 3.0 V/1.5 W means that at full brightness it will dissipate 1.5 W at a potential difference of 3.0 V. Because the power is related to the voltage and current by $P = I\Delta V$,

the current causing full brightness is

$$I = \frac{P}{\Delta V} = \frac{1.5\ \text{W}}{3.0\ \text{V}} = 0.50\ \text{A}$$

The bulb's resistance—the total resistance of the circuit—is

$$R = \frac{\Delta V}{I} = \frac{3.0\ \text{V}}{0.50\ \text{A}} = 6.0\ \Omega$$

Equation 30.4 gives the speed needed to induce this current:

$$v = \frac{IR}{lB} = \frac{(0.50\ \text{A})(6.0\ \Omega)}{(0.10\ \text{m})(0.10\ \text{T})} = 300\ \text{m/s}$$

You can confirm from Equation 30.6 that the input power at this speed is 1.5 W.

b. From Equation 30.5, the pulling force must be

$$F_{\text{pull}} = \frac{vl^2B^2}{R} = 5.0 \times 10^{-3}\ \text{N}$$

You can also obtain this result from $F_{\text{pull}} = P/v$.

ASSESS Example 30.1 showed that high speeds are needed to produce significant potential difference. Thus 300 m/s is not surprising. The pulling force is not very large, but even a small force can deliver large amounts of power $P = Fv$ when v is large.

dB/dt is negative because the field is decreasing, but all we need for Faraday's law is the absolute value. Thus

$$\mathcal{E} = \pi r^2 \left| \frac{dB}{dt} \right| = \pi (0.040 \text{ m})^2 (5.3 \text{ T/s}) = 0.027 \text{ V}$$

To find the current, we need to know the resistance of the loop. Recall, from Chapter 27, that a conductor with resistivity ρ, length L, and cross-section area A has resistance $R = \rho L/A$. The length is the circumference of the loop, calculated to be $L = 0.25$ m, and we can use the 1.0 cm diameter of the "wire" to find $A = 7.9 \times 10^{-5}$ m^2.

With these values, we can compute $R = 4700 \ \Omega$. As a result, the induced current is

$$I = \frac{\mathcal{E}}{R} = \frac{0.027 \text{ V}}{4700 \ \Omega} = 5.7 \times 10^{-6} \text{ A} = 5.7 \ \mu\text{A}$$

ASSESS This is a very small current. Power—the rate of energy dissipation in the muscle—is

$$P = I^2 R = (5.7 \times 10^{-6} \text{ A})^2 (4700 \ \Omega) = 1.5 \times 10^{-7} \text{ W}$$

The current is far too small to notice, and the tiny energy dissipation will certainly not heat the tissue.

What Does Faraday's Law Tell Us?

The induced current in the slide-wire circuit of Figure 30.26 can be understood as a motional emf due to magnetic forces on moving charges. We had not anticipated this kind of current in Chapter 29, but it takes no new laws of physics to understand it. The induced currents in Examples 30.8 and 30.9 are different. We cannot explain these induced currents on the basis of previous laws or principles. This is new physics.

Faraday recognized that all induced currents are associated with a changing magnetic flux. There are two fundamentally different ways to change the magnetic flux through a conducting loop:

1. The loop can expand, contract, or rotate, creating a motional emf.
2. The magnetic field can change.

We can see both of these if we write Faraday's law as

$$\mathcal{E} = \left| \frac{d\Phi_m}{dt} \right| = \left| \vec{B} \cdot \frac{d\vec{A}}{dt} + \vec{A} \cdot \frac{d\vec{B}}{dt} \right| \tag{30.19}$$

The first term on the right side represents a motional emf. The magnetic flux changes because the loop itself is changing. This term includes not only situations like the slide-wire circuit, where the area A changes, but also loops that rotate in a magnetic field. The physical area of a rotating loop does not change, but the area *vector* $\vec{A}$ does. The loop's motion causes magnetic forces on the charge carriers in the loop.

The second term on the right side is the new physics in Faraday's law. It says that an emf can also be created simply by changing a magnetic field, even if nothing is moving. This was the case in Examples 30.8 and 30.9. Faraday's law tells us that the induced emf is simply the rate of change of the magnetic flux through the loop, *regardless* of what causes the flux to change.

STOP TO THINK 30.5 A conducting loop is halfway into a magnetic field. Suppose the magnetic field begins to increase rapidly in strength. What happens to the loop?

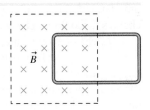

a. The loop is pushed upward, toward the top of the page.
b. The loop is pushed downward, toward the bottom of the page.
c. The loop is pulled to the left, into the magnetic field.
d. The loop is pushed to the right, out of the magnetic field.
e. The tension in the wires increases but the loop does not move.

FIGURE 30.30 An induced electric field creates a current in the loop.

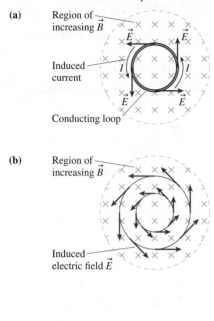

FIGURE 30.31 Two ways to create an electric field.

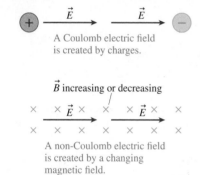

30.6 Induced Fields

Faraday's law is a tool for calculating the strength of an induced current, but one important piece of the puzzle is still missing. What *causes* the current? That is, what *force* pushes the charges around the loop against the resistive forces of the metal? The agents that exert forces on charges are electric fields and magnetic fields. Magnetic forces are responsible for motional emfs, but magnetic forces cannot explain the current induced in a *stationary* loop by a changing magnetic field.

FIGURE 30.30a shows a conducting loop in an increasing magnetic field. According to Lenz's law, there is an induced current in the ccw direction. Something has to act on the charge carriers to make them move, so we infer that there must be an *electric* field tangent to the loop at all points. This electric field is *caused* by the changing magnetic field and is called an **induced electric field.** The induced electric field is the mechanism that creates a current inside a stationary loop when there's a changing magnetic field.

The conducting loop isn't necessary. The space in which the magnetic field is changing is filled with the pinwheel pattern of induced electric fields shown in FIGURE 30.30b. Charges will move if a conducting path is present, but the induced electric field is there as a direct consequence of the changing magnetic field.

But this is a rather peculiar electric field. All the electric fields we have examined until now have been created by charges. Electric field vectors pointed away from positive charges and toward negative charges. An electric field created by charges is called a **Coulomb electric field.** The induced electric field of Figure 30.30b is caused not by charges but by a changing magnetic field. It is called a **non-Coulomb electric field.**

So it appears that there are two different ways to create an electric field:

1. A Coulomb electric field is created by positive and negative charges.
2. A non-Coulomb electric field is created by a changing magnetic field.

Both exert a force $\vec{F} = q\vec{E}$ on a charge, and both create a current in a conductor. However, the origins of the fields are very different. FIGURE 30.31 is a quick summary of the two ways to create an electric field.

We first introduced the idea of a field as a way of thinking about how two charges exert long-range forces on each other through the emptiness of space. The field may have seemed like a useful pictorial representation of charge interactions, but we had little evidence that fields are *real,* that they actually exist. Now we do. The electric field has shown up in a completely different context, independent of charges, as the explanation of the very real existence of induced currents.

The electric field is not just a pictorial representation; it is real.

Calculating the Induced Field

The induced electric field is peculiar in another way: It is nonconservative. Recall that a force is conservative if it does no net work on a particle moving around a closed path. "Uphills" are balanced by "downhills." We can associate a potential energy with a conservative force, hence we have gravitational potential energy for the conservative gravitational force and electric potential energy for the conservative electric force of charges (a Coulomb electric field).

But a charge moving around a closed path in the induced electric field of Figure 30.30 is always being pushed *in the same direction* by the electric force $\vec{F} = q\vec{E}$. There's never any negative work to balance the positive work, so the net work done in going around a closed path is not zero. Because it's nonconservative, we cannot associate an electric potential with an induced electric field. Only the Coulomb field of charges has an electric potential.

However, we can associate the induced field with the emf of Faraday's law. The emf was defined as the work required per unit charge to separate the charge. That is,

$$\mathcal{E} = \frac{W}{q} \tag{30.20}$$

In batteries, a familiar source of emf, this work is done by chemical forces. But the emf that appears in Faraday's law arises when work is done by the force of an induced electric field.

If a charge q moves through a small displacement $d\vec{s}$, the small amount of work done by the electric field is $dW = \vec{F} \cdot d\vec{s} = q\vec{E} \cdot d\vec{s}$. The emf of Faraday's law is an emf around a *closed curve* through which the magnetic flux Φ_m is changing. The work done by the induced electric field as charge q moves around a closed curve is

$$W_{\text{closed curve}} = q \oint \vec{E} \cdot d\vec{s} \tag{30.21}$$

where the integration symbol with the circle is the same as the one we used in Ampère's law to indicate an integral around a closed curve. If we use this work in Equation 30.20, we find that the emf around a closed loop is

$$\mathcal{E} = \frac{W_{\text{closed curve}}}{q} = \oint \vec{E} \cdot d\vec{s} \tag{30.22}$$

If we restrict ourselves to situations such as Figure 30.30 where the loop is perpendicular to the magnetic field and only the field is changing, we can write Faraday's law as $\mathcal{E} = |d\Phi_m/dt| = A|dB/dt|$. Consequently

$$\oint \vec{E} \cdot d\vec{s} = A \left| \frac{dB}{dt} \right| \tag{30.23}$$

Equation 30.23 is an alternative statement of Faraday's law that relates the induced electric field to the changing magnetic field.

The solenoid in **FIGURE 30.32a** provides a good example of the connection between $\vec{E}$ and $\vec{B}$. If there were a conducting loop inside the solenoid, we could use Lenz's law to determine that the direction of the induced current would be clockwise. But Faraday's law, in the form of Equation 30.23, tells us that **an induced electric field is present whether there's a conducting loop or not.** The electric field is induced simply due to the fact that $\vec{B}$ is changing.

FIGURE 30.32 The induced electric field circulates around the changing magnetic field inside a solenoid.

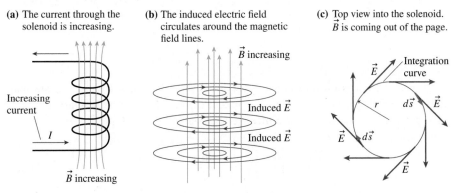

(a) The current through the solenoid is increasing.

(b) The induced electric field circulates around the magnetic field lines.

(c) Top view into the solenoid. $\vec{B}$ is coming out of the page.

The shape and direction of the induced electric field have to be such that it *could* drive a current around a conducting loop, if one were present, and it has to be consistent with the cylindrical symmetry of the solenoid. The only possible choice, shown in **FIGURE 30.32b**, is an electric field that circulates clockwise around the magnetic field lines.

NOTE Circular electric field lines violate the Chapter 23 rule that electric field lines have to start and stop on charges. However, that rule applied only to Coulomb fields created by source charges. An induced electric field is a non-Coulomb field created not by source charges but by a changing magnetic field. Without source charges, induced electric field lines *must* form closed loops.

To use Faraday's law, choose a *clockwise* circle of radius r as the closed curve for evaluating the integral. **FIGURE 30.32c** shows that the electric field vectors are everywhere tangent to the curve, so the line integral of $\vec{E}$ is

$$\oint \vec{E} \cdot d\vec{s} = El = 2\pi rE \tag{30.24}$$

where $l = 2\pi r$ is the length of the closed curve. This is exactly like the integrals we did for Ampère's law in Chapter 29.

If we stay inside the solenoid $(r < R)$, the flux passes through area $A = \pi r^2$ and Equation 30.24 becomes

$$\oint \vec{E} \cdot d\vec{s} = 2\pi rE = A\left|\frac{dB}{dt}\right| = \pi r^2 \left|\frac{dB}{dt}\right| \tag{30.25}$$

Thus the strength of the induced electric field inside the solenoid is

$$E_{\text{inside}} = \frac{r}{2}\left|\frac{dB}{dt}\right| \tag{30.26}$$

This result shows very directly that the induced electric field is created by a *changing* magnetic field. A constant $\vec{B}$, with $dB/dt = 0$, would give $E = 0$.

EXAMPLE 30.10 | An induced electric field

A 4.0-cm-diameter solenoid is wound with 2000 turns per meter. The current through the solenoid oscillates at 60 Hz with an amplitude of 2.0 A. What is the maximum strength of the induced electric field inside the solenoid?

MODEL Assume that the magnetic field inside the solenoid is uniform.

VISUALIZE The electric field lines are concentric circles around the magnetic field lines, as was shown in Figure 30.32b. They reverse direction twice every period as the current oscillates.

SOLVE You learned in Chapter 29 that the magnetic field strength inside a solenoid with n turns per meter is $B = \mu_0 nI$. In this case, the current through the solenoid is $I = I_0 \sin\omega t$, where $I_0 = 2.0$ A is the peak current and $\omega = 2\pi(60\text{ Hz}) = 377$ rad/s. Thus the

induced electric field strength at radius r is

$$E = \frac{r}{2}\left|\frac{dB}{dt}\right| = \frac{r}{2}\frac{d}{dt}(\mu_0 nI_0 \sin\omega t) = \tfrac{1}{2}\mu_0 nr\omega I_0 \cos\omega t$$

The field strength is maximum at maximum radius $(r = R)$ *and* at the instant when $\cos\omega t = 1$. That is,

$$E_{\text{max}} = \tfrac{1}{2}\mu_0 nR\omega I_0 = 0.019 \text{ V/m}$$

ASSESS This field strength, although not large, is similar to the field strength that the emf of a battery creates in a wire. Hence this induced electric field can drive a substantial induced current through a conducting loop *if* a loop is present. But the induced electric field exists inside the solenoid whether or not there is a conducting loop.

Occasionally it is useful to have a version of Faraday's law without the absolute value signs. The essence of Lenz's law is that the emf $\mathcal{E}$ opposes the *change* in Φ_m. Mathematically, this means that $\mathcal{E}$ must be opposite in sign to dB/dt. Consequently, we can write Faraday's law as

$$\mathcal{E} = \oint \vec{E} \cdot d\vec{s} = -\frac{d\Phi_m}{dt} \tag{30.27}$$

For practical applications, it's always easier to calculate just the magnitude of the emf with Faraday's law and to use Lenz's law to find the direction of the emf or the induced current. However, the mathematically rigorous version of Faraday's law in Equation 30.27 will prove to be useful when we combine it with other equations, in Chapter 31, to predict the existence of electromagnetic waves.

Maxwell's Theory of Electromagnetic Waves

In 1855, less than two years after receiving his undergraduate degree, the Scottish physicist James Clerk Maxwell presented a paper titled "On Faraday's Lines of Force." In this paper, he began to sketch out how Faraday's pictorial ideas about fields could be given a rigorous mathematical basis. Maxwell was troubled by a certain lack of symmetry. Faraday had found that a changing magnetic field creates an induced electric field, a non-Coulomb electric field not tied to charges. But what, Maxwell began to wonder, about a changing *electric* field?

To complete the symmetry, Maxwell proposed that a changing electric field creates an **induced magnetic field,** a new kind of magnetic field not tied to the existence of currents. FIGURE 30.33 shows a region of space where the *electric* field is increasing. This region of space, according to Maxwell, is filled with a pinwheel pattern of induced magnetic fields. The induced magnetic field looks like the induced electric field, with $\vec{E}$ and $\vec{B}$ interchanged, except that—for technical reasons explored in the next chapter—the induced $\vec{B}$ points the opposite way from the induced $\vec{E}$. Although there was no experimental evidence that induced magnetic fields existed, Maxwell went ahead and included them in his electromagnetic field theory. This was an inspired hunch, soon to be vindicated.

Maxwell soon realized that it might be possible to establish self-sustaining electric and magnetic fields that would be entirely independent of any charges or currents. That is, a changing electric field $\vec{E}$ creates a magnetic field $\vec{B}$, which then changes in just the right way to recreate the electric field, which then changes in just the right way to again recreate the magnetic field, and so on. The fields are continually recreated through electromagnetic induction without any reliance on charges or currents.

Maxwell was able to predict that electric and magnetic fields would be able to sustain themselves, free from charges and currents, if they took the form of an **electromagnetic wave.** The wave would have to have a very specific geometry, shown in FIGURE 30.34, in which $\vec{E}$ and $\vec{B}$ are perpendicular to each other as well as perpendicular to the direction of travel. That is, an electromagnetic wave would be a *transverse* wave.

Furthermore, Maxwell's theory predicted that the wave would travel with speed

$$v_{\text{em wave}} = \frac{1}{\sqrt{\epsilon_0 \mu_0}}$$

where ϵ_0 is the permittivity constant from Coulomb's law and μ_0 is the permeability constant from the law of Biot and Savart. Maxwell computed that an electromagnetic wave, if it existed, would travel with speed $v_{\text{em wave}} = 3.00 \times 10^8$ m/s.

We don't know Maxwell's immediate reaction, but it must have been both shock and excitement. His predicted speed for electromagnetic waves, a prediction that came directly from his theory, was none other than the speed of light! This agreement could be just a coincidence, but Maxwell didn't think so. Making a bold leap of imagination, Maxwell concluded that **light is an electromagnetic wave.**

It took 25 more years for Maxwell's predictions to be tested. In 1886, the German physicist Heinrich Hertz discovered how to generate and transmit radio waves. Two years later, in 1888, he was able to show that radio waves travel at the speed of light. Maxwell, unfortunately, did not live to see his triumph. He had died in 1879, at the age of 48.

30.7 Induced Currents: Three Applications

There are many applications of Faraday's law and induced currents in modern technology. In this section we will look at three: generators, transformers, and metal detectors.

Generators

A generator is a device that transforms mechanical energy into electric energy. FIGURE 30.35 on the next page shows a generator in which a coil of wire, perhaps spun by a windmill, rotates in a magnetic field. Both the field and the area of the loop are constant, but the magnetic flux through the loop changes continuously as the loop rotates. The induced current is removed from the rotating loop by *brushes* that press up against rotating *slip rings*.

The flux through the coil is

$$\Phi_{\text{m}} = \vec{A} \cdot \vec{B} = AB \cos \theta = AB \cos \omega t \tag{30.28}$$

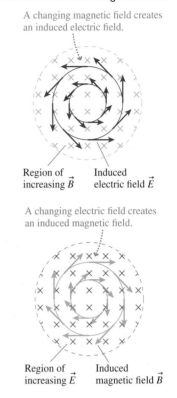

FIGURE 30.33 Maxwell hypothesized the existence of induced magnetic fields.

A changing magnetic field creates an induced electric field.

Region of increasing $\vec{B}$ Induced electric field $\vec{E}$

A changing electric field creates an induced magnetic field.

Region of increasing $\vec{E}$ Induced magnetic field $\vec{B}$

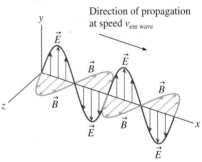

FIGURE 30.34 A self-sustaining electromagnetic wave.

Direction of propagation at speed $v_{\text{em wave}}$

A generator inside a hydroelectric dam uses electromagnetic induction to convert the mechanical energy of a spinning turbine into electric energy.

where ω is the angular frequency $(\omega = 2\pi f)$ with which the coil rotates. The induced emf is given by Faraday's law,

$$\mathcal{E}_{coil} = -N\frac{d\Phi_m}{dt} = -ABN\frac{d}{dt}(\cos\omega t) = \omega ABN\sin\omega t \qquad (30.29)$$

where N is the number of turns on the coil. Here it's best to use the signed version of Faraday's law to see how $\mathcal{E}_{coil}$ alternates between positive and negative.

Because the emf alternates in sign, the current through resistor R alternates back and forth in direction. Hence the generator of Figure 30.35 is an alternating-current generator, producing what we call an *AC voltage*.

FIGURE 30.35 An alternating-current generator.

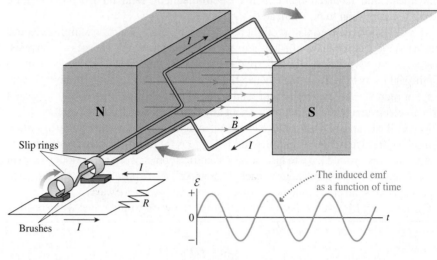

EXAMPLE 30.11 | **An AC generator**

A coil with area 2.0 m² rotates in a 0.010 T magnetic field at a frequency of 60 Hz. How many turns are needed to generate a peak voltage of 160 V?

SOLVE The coil's maximum voltage is found from Equation 30.29:

$$\mathcal{E}_{max} = \omega ABN = 2\pi f ABN$$

The number of turns needed to generate $\mathcal{E}_{max} = 160$ V is

$$N = \frac{\mathcal{E}_{max}}{2\pi f AB} = \frac{160 \text{ V}}{2\pi(60 \text{ Hz})(2.0 \text{ m}^2)(0.010 \text{ T})} = 21 \text{ turns}$$

ASSESS A 0.010 T field is modest, so you can see that generating large voltages is not difficult with large (2 m²) coils. Commercial generators use water flowing through a dam, rotating windmill blades, or turbines spun by expanding steam to rotate the generator coils. Work is required to rotate the coil, just as work was required to pull the slide wire in Section 30.2, because the magnetic field exerts retarding forces on the currents in the coil. Thus a generator is a device that turns motion (mechanical energy) into a current (electric energy). A generator is the opposite of a motor, which turns a current into motion.

Transformers

FIGURE 30.36 A transformer.

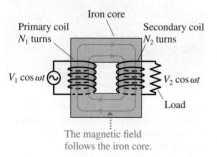

The magnetic field follows the iron core.

FIGURE 30.36 shows two coils wrapped on an iron core. The left coil is called the **primary coil.** It has N_1 turns and is driven by an oscillating voltage $V_1 \cos\omega t$. The magnetic field of the primary follows the iron core and passes through the right coil, which has N_2 turns and is called the **secondary coil.** The alternating current through the primary coil causes an oscillating magnetic flux through the secondary coil and, hence, an induced emf. The induced emf of the secondary coil is delivered to the load as the oscillating voltage $V_2 \cos\omega t$.

The changing magnetic field inside the iron core is inversely proportional to the number of turns on the primary coil: $B \propto 1/N_1$. (This relation is a consequence of the coil's inductance, an idea discussed in the next section.) According to Faraday's law, the emf induced in the secondary coil is directly proportional to its number of turns:

$\mathcal{E}_{\text{sec}} \propto N_2$. Combining these two proportionalities, the secondary voltage of an ideal transformer is related to the primary voltage by

$$V_2 = \frac{N_2}{N_1}V_1 \qquad (30.30)$$

Depending on the ratio N_2/N_1, the voltage V_2 across the load can be *transformed* to a higher or a lower voltage than V_1. Consequently, this device is called a **transformer.** Transformers are widely used in the commercial generation and transmission of electricity. A *step-up transformer*, with $N_2 \gg N_1$, boosts the voltage of a generator up to several hundred thousand volts. Delivering power with smaller currents at higher voltages reduces losses due to the resistance of the wires. High-voltage transmission lines carry electric power to urban areas, where *step-down transformers* $(N_2 \ll N_1)$ lower the voltage to 120 V.

Transformers are essential for transporting electric energy from the power plant to cities and homes.

Metal Detectors

Metal detectors, such as those used in airports for security, seem fairly mysterious. How can they detect the presence of *any* metal—not just magnetic materials such as iron—but not detect plastic or other materials? Metal detectors work because of induced currents.

A metal detector, shown in **FIGURE 30.37**, consists of two coils: a *transmitter coil* and a *receiver coil*. A high-frequency alternating current in the transmitter coil generates an alternating magnetic field along the axis. This magnetic field creates a changing flux through the receiver coil and causes an alternating induced current. The transmitter and receiver are similar to a transformer.

Suppose a piece of metal is placed between the transmitter and the receiver. The alternating magnetic field through the metal induces eddy currents in a plane parallel to the transmitter and receiver coils. The receiver coil then responds to the *superposition* of the transmitter's magnetic field and the magnetic field of the eddy currents. Because the eddy currents attempt to prevent the flux from changing, in accordance with Lenz's law, the net field at the receiver *decreases* when a piece of metal is inserted between the coils. Electronic circuits detect the current decrease in the receiver coil and set off an alarm. Eddy currents can't flow in an insulator, so this device detects only metals.

FIGURE 30.37 A metal detector.

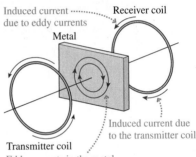

Induced current due to eddy currents
Receiver coil
Metal
Induced current due to the transmitter coil
Transmitter coil
Eddy currents in the metal reduce the induced current in the receiver coil.

30.8 Inductors

Capacitors are useful circuit elements because they store potential energy U_C in the electric field. Similarly, a coil of wire can be a useful circuit element because it stores energy in the magnetic field. In circuits, a coil is called an **inductor** because, as you'll see, the potential difference across an inductor is an *induced* emf. An *ideal inductor* is one for which the wire forming the coil has no electric resistance. The circuit symbol for an inductor is —.

We define the **inductance** L of a coil to be its flux-to-current ratio:

$$L = \frac{\Phi_m}{I} \qquad (30.31)$$

Strictly speaking, this is called *self-inductance* because the flux we're considering is the magnetic flux the solenoid creates in itself when there is a current. The SI unit of inductance is the **henry,** named in honor of Joseph Henry, defined as

$$1 \text{ henry} = 1 \text{ H} \equiv 1 \text{ Wb/A} = 1 \text{ T}\,\text{m}^2/\text{A}$$

Practical inductances are typically millihenries (mH) or microhenries (μH).

It's not hard to find the inductance of a solenoid. In Chapter 29 we found that the magnetic field inside an ideal solenoid having N turns and length l is

$$B = \frac{\mu_0 NI}{l}$$

The magnetic flux through one turn of the coil is $\Phi_{\text{per turn}} = AB$, where A is the cross-section area of the solenoid. The total magnetic flux through all N turns is

$$\Phi_{\text{m}} = N\Phi_{\text{per turn}} = \frac{\mu_0 N^2 A}{l} I \tag{30.32}$$

Thus the inductance of the solenoid, using the definition of Equation 30.31, is

$$L_{\text{solenoid}} = \frac{\Phi_{\text{m}}}{I} = \frac{\mu_0 N^2 A}{l} \tag{30.33}$$

The inductance of a solenoid depends only on its geometry, not at all on the current. You may recall that the capacitance of two parallel plates depends only on their geometry, not at all on their potential difference.

EXAMPLE 30.12 | **The length of an inductor**

An inductor is made by tightly wrapping 0.30-mm-diameter wire around a 4.0-mm-diameter cylinder. What length cylinder has an inductance of 10 μH?

SOLVE The cross-section area of the solenoid is $A = \pi r^2$. If the wire diameter is d, the number of turns of wire on a cylinder of length l is $N = l/d$. Thus the inductance is

$$L = \frac{\mu_0 N^2 A}{l} = \frac{\mu_0 (l/d)^2 \pi r^2}{l} = \frac{\mu_0 \pi r^2 l}{d^2}$$

The length needed to give inductance $L = 1.0 \times 10^{-5}$ H is

$$l = \frac{d^2 L}{\mu_0 \pi r^2} = \frac{(0.00030 \text{ m})^2 (1.0 \times 10^{-5} \text{ H})}{(4\pi \times 10^{-7} \text{ T m/A}) \pi (0.0020 \text{ m})^2}$$

$$= 0.057 \text{ m} = 5.7 \text{ cm}$$

The Potential Difference Across an Inductor

An inductor is not very interesting when the current through it is steady. If the inductor is ideal, with $R = 0 \ \Omega$, the potential difference due to a steady current is zero. **Inductors become important circuit elements when currents are changing.** FIGURE 30.38a shows a steady current into the left side of an inductor. The solenoid's magnetic field passes through the coils of the solenoid, establishing a flux.

In FIGURE 30.38b, the current into the solenoid is increasing. This creates an increasing flux to the left. According to Lenz's law, an induced current in the coils will oppose this increase by creating an induced magnetic field pointing to the right. This requires the induced current to be *opposite* the current into the solenoid. This induced current will carry positive charge carriers to the left until a potential difference is established across the solenoid.

You saw a similar situation in Section 30.2. The induced current in a conductor moving through a magnetic field carried positive charge carriers to the top of the wire and established a potential difference across the conductor. The induced current in the moving wire was due to magnetic forces on the moving charges. Now, in Figure 30.38b, the induced current is due to the non-Coulomb electric field induced by the changing magnetic field. Nonetheless, the outcome is the same: a potential difference across the conductor.

We can use Faraday's law to find the potential difference. The emf induced in a coil is

$$\mathcal{E}_{\text{coil}} = N \left| \frac{d\Phi_{\text{per turn}}}{dt} \right| = \left| \frac{d\Phi_{\text{m}}}{dt} \right| \tag{30.34}$$

where $\Phi_{\text{m}} = N\Phi_{\text{per turn}}$ is the total flux through all the coils. The inductance was defined such that $\Phi_{\text{m}} = LI$, so Equation 30.34 becomes

$$\mathcal{E}_{\text{coil}} = L \left| \frac{dI}{dt} \right| \tag{30.35}$$

FIGURE 30.38 Increasing the current through an inductor.

(a)

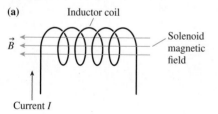

Inductor coil

$\vec{B}$

Solenoid magnetic field

Current I

(b)

The induced current is opposite the solenoid current.

The induced magnetic field opposes the change in flux.

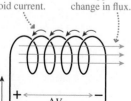

$+$ ΔV_{L} $-$

Increasing current

The induced current carries positive charge carriers to the left and establishes a potential difference across the inductor.

The induced emf is directly proportional to the *rate of change* of current through the coil. We'll consider the appropriate sign in a moment, but Equation 30.35 gives us the size of the potential difference that is developed across a coil as the current through the coil changes. Note that $\mathcal{E}_{coil} = 0$ for a steady, unchanging current.

FIGURE 30.39 shows the same inductor, but now the current (still *in* to the left side) is decreasing. To oppose the decrease in flux, the induced current is in the *same* direction as the input current. The induced current carries charge to the right and establishes a potential difference opposite that in Figure 30.38b.

NOTE Notice that the induced current does not oppose the current through the inductor, which is from left to right in both Figures 30.38 and 30.39. Instead, in accordance with Lenz's law, the induced current opposes the *change* in the current in the solenoid. The practical result is that it is hard to change the current through an inductor. Any effort to increase or decrease the current is met with opposition in the form of an opposing induced current. You can think of the current in an inductor as having inertia, trying to continue what it was doing without change.

Before we can use inductors in a circuit we need to establish a rule about signs that is consistent with our earlier circuit analysis. FIGURE 30.40 first shows current I passing through a resistor. You learned in Chapter 28 that the potential difference across a resistor is $\Delta V_{res} = -\Delta V_R = -IR$, where the minus sign indicates that the potential *decreases* in the direction of the current.

We'll use the same convention for an inductor. The potential difference across an inductor, *measured along the direction of the current*, is

$$\Delta V_L = -L\frac{dI}{dt} \tag{30.36}$$

If the current is increasing ($dI/dt > 0$), the input side of the inductor is more positive than the output side and the potential decreases in the direction of the current ($\Delta V_L < 0$). This was the situation in Figure 30.38b. If the current is decreasing ($dI/dt < 0$), the input side is more negative and the potential increases in the direction of the current ($\Delta V_L > 0$). This was the situation in Figure 30.39.

The potential difference across an inductor can be very large if the current changes very abruptly (large dI/dt). FIGURE 30.41 shows an inductor connected across a battery. There is a large current through the inductor, limited only by the internal resistance of the battery. Suppose the switch is suddenly opened. A very large induced voltage is created across the inductor as the current rapidly drops to zero. This potential difference (plus ΔV_{bat}) appears across the gap of the switch as it is opened. A large potential difference across a small gap often creates a spark.

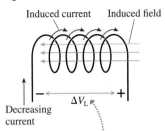

FIGURE 30.39 Decreasing the current through an inductor.

Induced current   Induced field

Decreasing current

ΔV_L

The induced current carries positive charge carriers to the right. The potential difference is opposite that of Figure 30.38b.

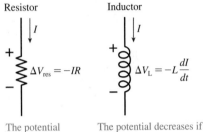

FIGURE 30.40 The potential difference across a resistor and an inductor.

Resistor

$\Delta V_{res} = -IR$

Inductor

$\Delta V_L = -L\dfrac{dI}{dt}$

The potential always decreases.

The potential decreases if the current is increasing.

The potential increases if the current is decreasing.

FIGURE 30.41 Creating sparks.

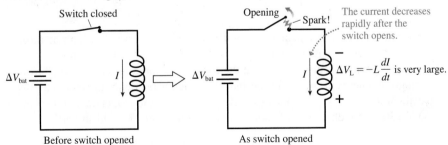

Switch closed

ΔV_{bat} I

Before switch opened

Opening   Spark!

The current decreases rapidly after the switch opens.

ΔV_{bat} I

$\Delta V_L = -L\dfrac{dI}{dt}$ is very large.

As switch opened

Indeed, this is exactly how the spark plugs in your car work. The car's generator sends a current through the *coil,* which is a big inductor. When a switch is suddenly opened, breaking the current, the induced voltage, typically a few thousand volts, appears across the terminals of the spark plug, creating the spark that ignites the gasoline. Older cars use a *distributor* to open and close an actual switch; more recent cars have *electronic ignition* in which the mechanical switch has been replaced by a transistor.

EXAMPLE 30.13 | **Large voltage across an inductor**

A 1.0 A current passes through a 10 mH inductor coil. What potential difference is induced across the coil if the current drops to zero in 5.0 μs?

MODEL Assume this is an ideal inductor, with $R = 0 \; \Omega$, and that the current decrease is linear with time.

SOLVE The rate of current decrease is

$$\frac{dI}{dt} \approx \frac{\Delta I}{\Delta t} = \frac{-1.0 \; \text{A}}{5.0 \times 10^{-6} \; \text{s}} = -2.0 \times 10^5 \; \text{A/s}$$

The induced voltage is

$$\Delta V_{\text{L}} = -L\frac{dI}{dt} \approx -(0.010 \; \text{H})(-2.0 \times 10^5 \; \text{A/s}) = 2000 \; \text{V}$$

ASSESS Inductors may be physically small, but they can pack a punch if you try to change the current through them too quickly.

STOP TO THINK 30.6 The potential at a is higher than the potential at b. Which of the following statements about the inductor current I could be true?

$V_{\text{a}} > V_{\text{b}}$

a ⟶⦚⦚⦚⦚⟶ b

a. I is from a to b and steady.
c. I is from a to b and decreasing.
e. I is from b to a and increasing.

b. I is from a to b and increasing.
d. I is from b to a and steady.
f. I is from b to a and decreasing.

Energy in Inductors and Magnetic Fields

Recall that electric power is $P_{\text{elec}} = I\Delta V$. As current passes through an inductor, for which $\Delta V_{\text{L}} = -L(dI/dt)$, the electric power is

$$P_{\text{elec}} = I\Delta V_{\text{L}} = -LI\frac{dI}{dt} \tag{30.37}$$

P_{elec} is negative because a circuit with an increasing current is *losing* electric energy. That energy is being transferred to the inductor, which is *storing* energy U_{L} at the rate

$$\frac{dU_{\text{L}}}{dt} = +LI\frac{dI}{dt} \tag{30.38}$$

where we've noted that power is the rate of change of energy.

We can find the total energy stored in an inductor by integrating Equation 30.38 from $I = 0$, where $U_{\text{L}} = 0$, to a final current I. Doing so gives

$$U_{\text{L}} = L\int_0^I I \, dI = \tfrac{1}{2}LI^2 \tag{30.39}$$

The potential energy stored in an inductor depends on the square of the current through it. Notice the analogy with the energy $U_{\text{C}} = \tfrac{1}{2}C(\Delta V)^2$ stored in a capacitor.

In working with circuits we say that the energy is "stored in the inductor." Strictly speaking, the energy is stored in the inductor's magnetic field, analogous to how a capacitor stores energy in the electric field. We can use the inductance of a solenoid, Equation 30.33, to relate the inductor's energy to the magnetic field strength:

$$U_{\text{L}} = \tfrac{1}{2}LI^2 = \frac{\mu_0 N^2 A}{2l}I^2 = \frac{1}{2\mu_0}Al\left(\frac{\mu_0 NI}{l}\right)^2 \tag{30.40}$$

We made the last rearrangement in Equation 30.40 because $\mu_0 NI/l$ is the magnetic field inside the solenoid. Thus

$$U_{\text{L}} = \frac{1}{2\mu_0}AlB^2 \tag{30.41}$$

Energy in electric and magnetic fields

Electric fields	Magnetic fields
A capacitor stores energy $U_{\text{C}} = \tfrac{1}{2}C(\Delta V)^2$	An inductor stores energy $U_{\text{L}} = \tfrac{1}{2}LI^2$
Energy density in the field is $u_{\text{E}} = \dfrac{\epsilon_0}{2}E^2$	Energy density in the field is $u_{\text{B}} = \dfrac{1}{2\mu_0}B^2$

But Al is the volume inside the solenoid. Dividing by Al, the magnetic field *energy density* inside the solenoid (energy per m³) is

$$u_B = \frac{1}{2\mu_0}B^2 \qquad (30.42)$$

We've derived this expression for energy density based on the properties of a solenoid, but it turns out to be the correct expression for the energy density anywhere there's a magnetic field. Compare this to the energy density of an electric field $u_E = \frac{1}{2}\epsilon_0 E^2$ that we found in Chapter 26.

EXAMPLE 30.14 | **Energy stored in an inductor**

The 10 μH inductor of Example 30.12 was 5.7 cm long and 4.0 mm in diameter. Suppose it carries a 100 mA current. What are the energy stored in the inductor, the magnetic energy density, and the magnetic field strength?

SOLVE The stored energy is

$$U_L = \frac{1}{2}LI^2 = \frac{1}{2}(1.0 \times 10^{-5}\,\text{H})(0.10\,\text{A})^2 = 5.0 \times 10^{-8}\,\text{J}$$

The solenoid volume is $(\pi r^2)l = 7.16 \times 10^{-7}\,\text{m}^3$. Using this gives the energy density of the magnetic field:

$$u_B = \frac{5.0 \times 10^{-8}\,\text{J}}{7.16 \times 10^{-7}\,\text{m}^3} = 0.070\,\text{J/m}^3$$

From Equation 30.42, the magnetic field with this energy density is

$$B = \sqrt{2\mu_0 u_B} = 4.2 \times 10^{-4}\,\text{T}$$

30.9 *LC* Circuits

Telecommunication—radios, televisions, cell phones—is based on electromagnetic signals that *oscillate* at a well-defined frequency. These oscillations are generated and detected by a simple circuit consisting of an inductor and a capacitor. This is called an **LC circuit.** In this section we will learn why an *LC* circuit oscillates and determine the oscillation frequency.

FIGURE 30.42 shows a capacitor with initial charge Q_0, an inductor, and a switch. The switch has been open for a long time, so there is no current in the circuit. Then, at $t = 0$, the switch is closed. How does the circuit respond? Let's think it through qualitatively before getting into the mathematics.

As **FIGURE 30.43** shows, the inductor provides a conducting path for discharging the capacitor. However, the discharge current has to pass through the inductor, and, as we've seen, an inductor resists changes in current. Consequently, the current doesn't stop when the capacitor charge reaches zero.

FIGURE 30.42 An *LC* circuit.

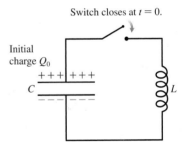

FIGURE 30.43 The capacitor charge oscillates much like a block attached to a spring.

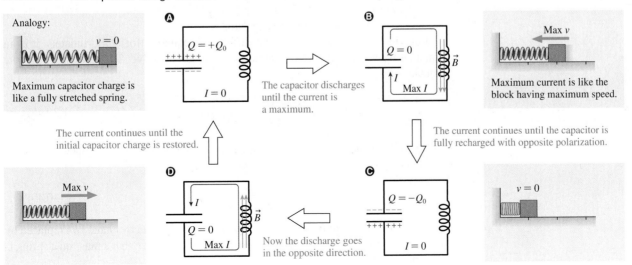

A cell phone is actually a very sophisticated two-way radio that communicates with the nearest base station via high-frequency radio waves—roughly 1000 MHz. As in any radio or communications device, the transmission frequency is established by the oscillating current in an *LC* circuit.

A block attached to a stretched spring is a useful mechanical analogy. Closing the switch to discharge the capacitor is like releasing the block. The block doesn't stop when it reaches the origin; its inertia keeps it going until the spring is fully compressed. Likewise, the current continues until it has recharged the capacitor with the opposite polarization. This process repeats over and over, charging the capacitor first one way, then the other. That is, the charge and current *oscillate*.

The goal of our circuit analysis will be to find expressions showing how the capacitor charge Q and the inductor current I change with time. As always, our starting point for circuit analysis is Kirchhoff's voltage law, which says that all the potential differences around a closed loop must sum to zero. Choosing a cw direction for I, Kirchhoff's law is

$$\Delta V_C + \Delta V_L = 0 \tag{30.43}$$

The potential difference across a capacitor is $\Delta V_C = Q/C$, and we found the potential difference across an inductor in Equation 30.36. Using these, Kirchhoff's law becomes

$$\frac{Q}{C} - L\frac{dI}{dt} = 0 \tag{30.44}$$

Equation 30.44 has two unknowns, Q and I. We can eliminate one of the unknowns by finding another relation between Q and I. Current is the rate at which charge moves, $I = dq/dt$, but the charge flowing through the inductor is charge that was *removed* from the capacitor. That is, an infinitesimal charge dq flows through the inductor when the capacitor charge changes by $dQ = -dq$. Thus the current through the inductor is related to the charge on the capacitor by

$$I = -\frac{dQ}{dt} \tag{30.45}$$

Now I is positive when Q is decreasing, as we would expect. This is a subtle but important step in the reasoning.

Equations 30.44 and 30.45 are two equations in two unknowns. To solve them, we'll first take the time derivative of Equation 30.45:

$$\frac{dI}{dt} = \frac{d}{dt}\left(-\frac{dQ}{dt}\right) = -\frac{d^2Q}{dt^2} \tag{30.46}$$

We can substitute this result into Equation 30.44:

$$\frac{Q}{C} + L\frac{d^2Q}{dt^2} = 0 \tag{30.47}$$

Now we have an equation for the capacitor charge Q.

Equation 30.47 is a second-order differential equation for Q. Fortunately, it is an equation we've seen before and already know how to solve. To see this, we rewrite Equation 30.47 as

$$\frac{d^2Q}{dt^2} = -\frac{1}{LC}Q \tag{30.48}$$

Recall, from Chapter 15, that the equation of motion for an undamped mass on a spring is

$$\frac{d^2x}{dt^2} = -\frac{k}{m}x \tag{30.49}$$

Equation 30.48 is *exactly the same equation*, with x replaced by Q and k/m replaced by $1/LC$. This should be no surprise because we've already seen that a mass on a spring is a mechanical analog of the *LC* circuit.

We know the solution to Equation 30.49. It is simple harmonic motion $x(t) = x_0 \cos \omega t$ with angular frequency $\omega = \sqrt{k/m}$. Thus the solution to Equation 30.48 must be

$$Q(t) = Q_0 \cos \omega t \qquad (30.50)$$

where Q_0 is the initial charge, at $t = 0$, and the angular frequency is

$$\omega = \sqrt{\frac{1}{LC}} \qquad (30.51)$$

The charge on the upper plate of the capacitor oscillates back and forth between $+Q_0$ and $-Q_0$ (the opposite polarization) with period $T = 2\pi/\omega$.

As the capacitor charge oscillates, so does the current through the inductor. Using Equation 30.45 gives the current through the inductor:

$$I = -\frac{dQ}{dt} = \omega Q_0 \sin \omega t = I_{max} \sin \omega t \qquad (30.52)$$

where $I_{max} = \omega Q_0$ is the maximum current.

An *LC* circuit is an *electric oscillator*, oscillating at frequency $f = \omega/2\pi$. FIGURE 30.44 shows graphs of the capacitor charge Q and the inductor current I as functions of time. Notice that Q and I are 90° out of phase. The current is zero when the capacitor is fully charged, as expected, and the charge is zero when the current is maximum.

FIGURE 30.44 The oscillations of an *LC* circuit.

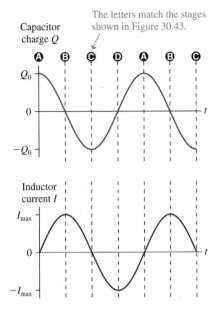

EXAMPLE 30.15 | **An AM radio oscillator**

You have a 1.0 mH inductor. What capacitor should you choose to make an oscillator with a frequency of 920 kHz? (This frequency is near the center of the AM radio band.)

SOLVE The angular frequency is $\omega = 2\pi f = 5.78 \times 10^6$ rad/s. Using Equation 30.51 for ω gives the required capacitor:

$$C = \frac{1}{\omega^2 L} = 3.0 \times 10^{-11} \text{ F} = 30 \text{ pF}$$

An *LC* circuit, like a mass on a spring, wants to respond only at its natural oscillation frequency $\omega = 1/\sqrt{LC}$. In Chapter 15 we defined a strong response at the natural frequency as a *resonance,* and resonance is the basis for all telecommunications. The input circuit in radios, televisions, and cell phones is an *LC* circuit driven by the signal picked up by the antenna. This signal is the superposition of hundreds of sinusoidal waves at different frequencies, one from each transmitter in the area, but the circuit responds only to the *one* signal that matches the circuit's natural frequency. That particular signal generates a large-amplitude current that can be further amplified and decoded to become the output that you hear.

30.10 *LR* Circuits

A circuit consisting of an inductor, a resistor, and (perhaps) a battery is called an **LR circuit**. FIGURE 30.45a is an example of an *LR* circuit. We'll assume that the switch has been in position a for such a long time that the current is steady and unchanging. There's no potential difference across the inductor, because $dI/dt = 0$, so it simply acts like a piece of wire. The current flowing around the circuit is determined entirely by the battery and the resistor: $I_0 = \Delta V_{bat}/R$.

What happens if, at $t = 0$, the switch is suddenly moved to position b? With the battery no longer in the circuit, you might expect the current to stop immediately. But the inductor won't let that happen. The current will continue for some period of time as the inductor's magnetic field drops to zero. In essence, the energy stored in the inductor allows it to act like a battery for a short period of time. Our goal is to determine how the current decays after the switch is moved.

FIGURE 30.45 An *LR* circuit.

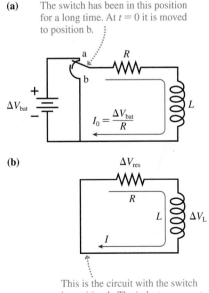

(a) The switch has been in this position for a long time. At $t = 0$ it is moved to position b.

(b)

This is the circuit with the switch in position b. The inductor prevents the current from stopping instantly.

NOTE It's important not to open switches in inductor circuits because they'll spark, as Figure 30.41 showed. The unusual switch in Figure 30.45 is designed to make the new contact just before breaking the old one.

FIGURE 30.45b shows the circuit after the switch is changed. Our starting point, once again, is Kirchhoff's voltage law. The potential differences around a closed loop must sum to zero. For this circuit, Kirchhoff's law is

$$\Delta V_{res} + \Delta V_L = 0 \tag{30.53}$$

The potential differences in the direction of the current are $\Delta V_{res} = -IR$ for the resistor and $\Delta V_L = -L(dI/dt)$ for the inductor. Substituting these into Equation 30.53 gives

$$-RI - L\frac{dI}{dt} = 0 \tag{30.54}$$

We're going to need to integrate to find the current I as a function of time. Before doing so, we rearrange Equation 30.54 to get all the current terms on one side of the equation and all the time terms on the other:

$$\frac{dI}{I} = -\frac{R}{L}dt = -\frac{dt}{(L/R)} \tag{30.55}$$

We know that the current at $t = 0$, when the switch was moved, was I_0. We want to integrate from these starting conditions to current I at the unspecified time t. That is,

$$\int_{I_0}^{I}\frac{dI}{I} = -\frac{1}{(L/R)}\int_0^t dt \tag{30.56}$$

Both are common integrals, giving

$$\ln I\Big|_{I_0}^{I} = \ln I - \ln I_0 = \ln\left(\frac{I}{I_0}\right) = -\frac{t}{(L/R)} \tag{30.57}$$

We can solve for the current I by taking the exponential of both sides, then multiplying by I_0. Doing so gives I, the current as a function of time:

$$I = I_0 e^{-t/(L/R)} \tag{30.58}$$

Notice that $I = I_0$ at $t = 0$, as expected.

The argument of the exponential function must be dimensionless, so L/R must have dimensions of time. If we define the **time constant** τ of the LR circuit to be

$$\tau = \frac{L}{R} \tag{30.59}$$

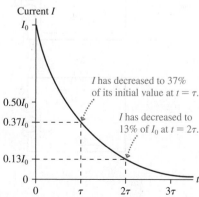

FIGURE 30.46 The current decay in an LR circuit.

Current I

I has decreased to 37% of its initial value at $t = \tau$.

I has decreased to 13% of I_0 at $t = 2\tau$.

then we can write Equation 30.58 as

$$I = I_0 e^{-t/\tau} \tag{30.60}$$

The time constant is the time at which the current has decreased to e^{-1} (about 37%) of its initial value. We can see this by computing the current at the time $t = \tau$:

$$I(\text{at } t = \tau) = I_0 e^{-\tau/\tau} = e^{-1}I_0 = 0.37I_0 \tag{30.61}$$

Thus the time constant for an LR circuit functions in exactly the same way as the time constant for the RC circuit we analyzed in Chapter 29. At time $t = 2\tau$, the current has decreased to $e^{-2}I_0$, or about 13% of its initial value.

The current is graphed in **FIGURE 30.46**. You can see that the current decays exponentially. The *shape* of the graph is always the same, regardless of the specific value of the time constant τ.

EXAMPLE 30.16 | Exponential decay in an *LR* circuit

The switch in **FIGURE 30.47** has been in position a for a long time. It is changed to position b at $t = 0$ s.

a. What is the current in the circuit at $t = 5.0\ \mu s$?

b. At what time has the current decayed to 1% of its initial value?

FIGURE 30.47 The *LR* circuit of Example 30.16.

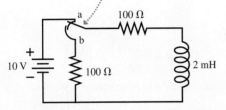

The switch moves from a to b at $t = 0$.

MODEL This is an *LR* circuit. We'll assume ideal wires and an ideal inductor.

VISUALIZE The two resistors will be in series after the switch is thrown.

SOLVE Before the switch is thrown, while $\Delta V_L = 0$, the current is $I_0 = (10\ \text{V})/(100\ \Omega) = 0.10\ \text{A} = 100\ \text{mA}$. This will be the initial current after the switch is thrown because the current through an inductor can't change instantaneously. The circuit resistance after the switch is thrown is $R = 200\ \Omega$, so the time constant is

$$\tau = \frac{L}{R} = \frac{2.0 \times 10^{-3}\ \text{H}}{200\ \Omega} = 1.0 \times 10^{-5}\ \text{s} = 10\ \mu s$$

a. The current at $t = 5.0\ \mu s$ is

$$I = I_0 e^{-t/\tau} = (100\ \text{mA})e^{-(5.0\mu s)/(10\mu s)} = 61\ \text{mA}$$

b. To find the time at which a particular current is reached we need to go back to Equation 30.57 and solve for t:

$$t = -\frac{L}{R} \ln\left(\frac{I}{I_0}\right) = -\tau \ln\left(\frac{I}{I_0}\right)$$

The time at which the current has decayed to 1 mA (1% of I_0) is

$$t = -(10\ \mu s) \ln\left(\frac{1\ \text{mA}}{100\ \text{mA}}\right) = 46\ \mu s$$

ASSESS For all practical purposes, the current has decayed away in $\approx 50\ \mu s$. The inductance in this circuit is not large, so a short decay time is not surprising.

STOP TO THINK 30.7 Rank in order, from largest to smallest, the time constants τ_a, τ_b, and τ_c of these three circuits.

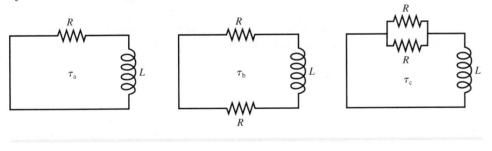

CHALLENGE EXAMPLE 30.17 | Induction heating

Induction heating uses induced currents to heat metal objects to high temperatures for applications such as surface hardening, brazing, or even melting. To illustrate the idea, consider a copper wire formed into a 4.0 cm × 4.0 cm square loop and placed in a magnetic field—perpendicular to the plane of the loop—that oscillates with 0.010 T amplitude at a frequency of 1000 Hz. What is the wire's initial temperature rise, in °C/min?

MODEL The changing magnetic flux through the loop will induce a current that, because of the wire's resistance, will heat the wire. Eventually, when the wire gets hot, heat loss through radiation and/or convection will limit the temperature rise, but initially we can consider the temperature change due only to the heating by the current. Assume that the wire's diameter is much less than the 4.0 cm width of the loop.

FIGURE 30.48 A copper wire being heated by induction.

Cross section A

$\rho_{mass} = 892.0\ \text{kg/m}^3$
$\rho_{elec} = 1.7 \times 10^{-8}\ \Omega\ m$
$c = 385\ \text{J/kg K}$

$L = 4.0\ \text{cm}$

$L = 4.0\ \text{cm}$

VISUALIZE FIGURE 30.48 shows the copper loop in the magnetic field. The wire's cross-section area A is unknown, but our assumption of a thin wire means that the loop has a well-defined area L^2.

Continued

Values of copper's resistivity, density, and specific heat were taken from tables inside the back cover of the book. We've used subscripts to distinguish between mass density ρ_{mass} and resistivity ρ_{elec}, a potentially confusing duplication of symbols.

SOLVE Power dissipation by a current, $P = I^2R$, heats the wire. As long as heat losses are negligible, we can use the heating rate and the wire's specific heat c to calculate the rate of temperature change. Our first task is to find the induced current. According to Faraday's law,

$$I = \frac{\mathcal{E}}{R} = -\frac{1}{R}\frac{d\Phi_m}{dt} = -\frac{L^2}{R}\frac{dB}{dt}$$

where R is the loop's resistance and $\Phi_m = L^2 B$ is the magnetic flux through a loop of area L^2. The oscillating magnetic field can be written $B = B_0 \cos \omega t$, with $B_0 = 0.010$ T and $\omega = 2\pi \times 1000$ Hz $= 6280$ rad/s. Thus

$$\frac{dB}{dt} = -\omega B_0 \sin \omega t$$

from which we find that the induced current oscillates as

$$I = \frac{\omega B_0 L^2}{R} \sin \omega t$$

As the current oscillates, the power dissipation in the wire is

$$P = I^2 R = \frac{\omega^2 B_0^2 L^4}{R} \sin^2 \omega t$$

The power dissipation also oscillates, but very rapidly in comparison to a temperature rise that we expect to occur over seconds or minutes. Consequently, we are justified in replacing the oscillating P with its *average* value P_{avg}. Recall that the time average of the function $\sin^2 \omega t$ is $\frac{1}{2}$, a result that can be proven by integration or justified by noticing that a graph of $\sin^2 \omega t$ oscillates symmetrically between 0 and 1. Thus the average power dissipation in the wire is

$$P_{\text{avg}} = \frac{\omega^2 B_0^2 L^4}{2R}$$

Recall that power is the *rate* of energy transfer. In this case, the power dissipated in the wire is the wire's heating rate: $dQ/dt = P_{\text{avg}}$, where here Q is heat, not charge. Using $Q = mc\,\Delta T$, from thermodynamics, we can write

$$\frac{dQ}{dt} = mc\frac{dT}{dt} = P_{\text{avg}} = \frac{\omega^2 B_0^2 L^4}{2R}$$

To complete the calculation, we need the mass and resistance of the wire. The wire's total length is $4L$, and its cross-section area is A. Thus

$$m = \rho_{\text{mass}} V = 4\rho_{\text{mass}} LA$$

$$R = \frac{\rho_{\text{elec}}(4L)}{A} = \frac{4\rho_{\text{elec}} L}{A}$$

Substituting these into the heating equation, we have

$$4\rho_{\text{mass}} LAc\frac{dT}{dt} = \frac{\omega^2 B_0^2 L^3 A}{8\rho_{\text{elec}}}$$

Interestingly, the wire's cross-section area cancels. The wire's temperature initially increases at the rate

$$\frac{dT}{dt} = \frac{\omega^2 B_0^2 L^2}{32\rho_{\text{elec}}\rho_{\text{mass}} c}$$

All the terms on the right-hand side are known. Evaluating, we find

$$\frac{dT}{dt} = 3.3 \text{ K/s} = 200°\text{C/min}$$

ASSESS This is a rapid but realistic temperature rise for a small object, although the rate of increase will slow as the object begins losing heat to the environment through radiation and/or convection. Induction heating can increase an object's temperature by several hundred degrees in a few minutes.

SUMMARY

The goal of Chapter 30 has been to learn what electromagnetic induction is and how it is used.

GENERAL PRINCIPLES

Lenz's Law

There is an induced current in a closed conducting loop if and only if the magnetic flux through the loop is changing. The direction of the induced current is such that the induced magnetic field opposes the *change* in the flux.

Faraday's Law

An emf is induced around a closed loop if the magnetic flux through the loop changes.

Magnitude: $\mathcal{E} = \left| \dfrac{d\Phi_m}{dt} \right|$

Direction: As given by Lenz's law

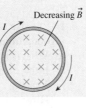
Decreasing $\vec{B}$

Using Electromagnetic Induction

MODEL Make simplifying assumptions.

VISUALIZE Use Lenz's law to determine the direction of the **induced current.**

SOLVE The **induced emf** is

$$\mathcal{E} = \left| \frac{d\Phi_m}{dt} \right|$$

Multiply by N for an N-turn coil.
The size of the induced current is $I = \mathcal{E}/R$.

ASSESS Is the result reasonable?

IMPORTANT CONCEPTS

Magnetic flux

Magnetic flux measures the amount of magnetic field passing through a surface.

$$\Phi_m = \vec{A} \cdot \vec{B} = AB\cos\theta$$

Loop of area A

Three ways to change the flux

1. A loop moves into or out of a magnetic field.

2. The loop changes area or rotates.
3. The magnetic field through the loop increases or decreases.

Two ways to create an induced current

1. A **motional emf** is due to magnetic forces on moving charge carriers.

$$\mathcal{E} = vlB$$

2. An **induced electric field** is due to a changing magnetic field.

$$\oint \vec{E} \cdot d\vec{s} = -\frac{d\Phi_m}{dt}$$

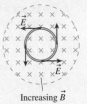
Increasing $\vec{B}$

APPLICATIONS

Inductors

Solenoid inductance $L_{solenoid} = \dfrac{\mu_0 N^2 A}{l}$

Potential difference $\Delta V_L = -L\dfrac{dI}{dt}$

Energy stored $U_L = \frac{1}{2}LI^2$

Magnetic energy density $u_B = \dfrac{1}{2\mu_0}B^2$

LC circuit

Oscillates at $\omega = \sqrt{\dfrac{1}{LC}}$

LR circuit

Exponential change with $\tau = \dfrac{L}{R}$

TERMS AND NOTATION

electromagnetic induction	area vector, $\vec{A}$	induced magnetic field	henry, H
induced current	Lenz's law	electromagnetic wave	*LC* circuit
motional emf	induced emf, $\mathcal{E}$	primary coil	*LR* circuit
generator	Faraday's law	secondary coil	time constant, τ
eddy current	induced electric field	transformer	
magnetic flux, Φ_m	Coulomb electric field	inductor	
weber, Wb	non-Coulomb electric field	inductance, *L*	

CONCEPTUAL QUESTIONS

1. What is the direction of the induced current in **FIGURE Q30.1**?

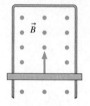

FIGURE Q30.1

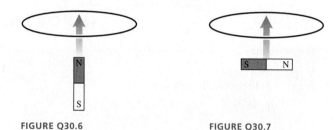

FIGURE Q30.2

2. You want to insert a loop of copper wire between the two permanent magnets in **FIGURE Q30.2**. Is there an attractive magnetic force that tends to *pull* the loop in, like a magnet pulls on a paper clip? Or do you need to *push* the loop in against a repulsive force? Explain.

3. A vertical, rectangular loop of copper wire is half in and half out of the horizontal magnetic field in **FIGURE Q30.3**. (The field is zero beneath the dashed line.) The loop is released and starts to fall. Is there a net magnetic force on the loop? If so, in which direction? Explain.

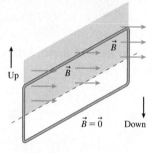

FIGURE Q30.3

4. Does the loop of wire in **FIGURE Q30.4** have a clockwise current, a counterclockwise current, or no current under the following circumstances? Explain.
   a. The magnetic field points out of the page and is increasing.
   b. The magnetic field points out of the page and is constant.
   c. The magnetic field points out of the page and is decreasing.

FIGURE Q30.4

FIGURE Q30.5

5. The two loops of wire in **FIGURE Q30.5** are stacked one above the other. Does the upper loop have a clockwise current, a counterclockwise current, or no current at the following times? Explain.
   a. Before the switch is closed.
   b. Immediately after the switch is closed.
   c. Long after the switch is closed.
   d. Immediately after the switch is reopened.

6. **FIGURE Q30.6** shows a bar magnet being pushed toward a conducting loop from below, along the axis of the loop.
   a. What is the current direction in the loop? Explain.
   b. Is there a magnetic force on the loop? If so, in which direction? Explain.
      Hint: A current loop is a magnetic dipole.
   c. Is there a force on the magnet? If so, in which direction?

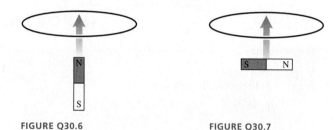

FIGURE Q30.6   **FIGURE Q30.7**

7. A bar magnet is pushed toward a loop of wire as shown in **FIGURE Q30.7**. Is there a current in the loop? If so, in which direction? If not, why not?

8. **FIGURE Q30.8** shows a bar magnet, a coil of wire, and a current meter. Is the current through the meter right to left, left to right, or zero for the following circumstances? Explain.
   a. The magnet is inserted into the coil.
   b. The magnet is held at rest inside the coil.
   c. The magnet is withdrawn from the left side of the coil.

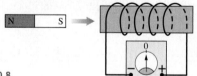

FIGURE Q30.8

9. Is the magnetic field strength in **FIGURE Q30.9** increasing, decreasing, or steady? Explain.

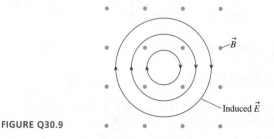

FIGURE Q30.9

10. An inductor with a 2.0 A current stores energy. At what current will the stored energy be twice as large?

11. a. Can you tell which of the inductors in FIGURE Q30.11 has the larger current through it? If so, which one? Explain.
    b. Can you tell through which inductor the current is changing more rapidly? If so, which one? Explain.
    c. If the current enters the inductor from the bottom, can you tell if the current is increasing, decreasing, or staying the same? If so, which? Explain.

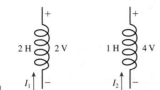

FIGURE Q30.11

12. An *LC* circuit oscillates at a frequency of 2000 Hz. What will the frequency be if the inductance is quadrupled?

13. Rank in order, from largest to smallest, the three time constants τ_a to τ_c for the three circuits in FIGURE Q30.13. Explain.

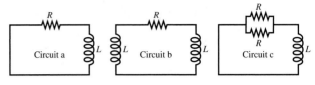

FIGURE Q30.13

14. For the circuit of FIGURE Q30.14:
    a. What is the battery current immediately after the switch closes? Explain.
    b. What is the battery current after the switch has been closed a long time? Explain.

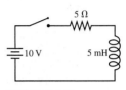

FIGURE Q30.14

EXERCISES AND PROBLEMS

Problems labeled ▨ integrate material from earlier chapters.

Exercises

Section 30.2 Motional emf

1. | The earth's magnetic field strength is 5.0×10^{-5} T. How fast would you have to drive your car to create a 1.0 V motional emf along your 1.0-m-tall radio antenna? Assume that the motion of the antenna is perpendicular to $\vec{B}$.

2. | A potential difference of 0.050 V is developed across the 10-cm-long wire of FIGURE EX30.2 as it moves through a magnetic field perpendicular to the page. What are the strength and direction (in or out) of the magnetic field?

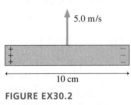

FIGURE EX30.2

3. | A 10-cm-long wire is pulled along a U-shaped conducting rail in a perpendicular magnetic field. The total resistance of the wire and rail is 0.20 Ω. Pulling the wire at a steady speed of 4.0 m/s causes 4.0 W of power to be dissipated in the circuit.
   a. How big is the pulling force?
   b. What is the strength of the magnetic field?

Section 30.3 Magnetic Flux

4. ‖ What is the magnetic flux through the loop shown in FIGURE EX30.4?

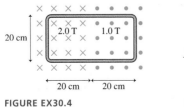

FIGURE EX30.4

5. ‖ FIGURE EX30.5 shows a 10 cm ×10 cm square bent at a 90° angle. A uniform 0.050 T magnetic field points downward at a 45° angle. What is the magnetic flux through the loop?

6. ‖ An equilateral triangle 8.0 cm on a side is in a 5.0 mT uniform magnetic field. The magnetic flux through the triangle is 6.0μWb. What is the angle between the magnetic field and an axis perpendicular to the plane of the triangle?

7. | What is the magnetic flux through the loop shown in FIGURE EX30.7?

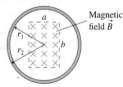

FIGURE EX30.7

8. ‖ FIGURE EX30.8 shows a 2.0-cm-diameter solenoid passing through the center of a 6.0-cm-diameter loop. The magnetic field inside the solenoid is 0.20 T. What is the magnetic flux through the loop when it is perpendicular to the solenoid and when it is tilted at a 60° angle?

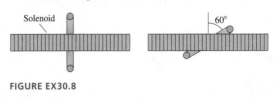

FIGURE EX30.8

Section 30.4 Lenz's Law

9. | There is a cw induced current in the conducting loop shown in FIGURE EX30.9. Is the magnetic field inside the loop increasing in strength, decreasing in strength, or steady?

FIGURE EX30.9

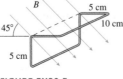

FIGURE EX30.5

10. | A solenoid is wound as shown in **FIGURE EX30.10**.
   a. Is there an induced current as magnet 1 is moved away from the solenoid? If so, what is the current direction through resistor R?
   b. Is there an induced current as magnet 2 is moved away from the solenoid? If so, what is the current direction through resistor R?

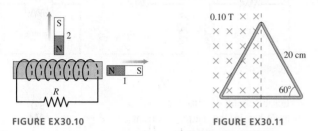

FIGURE EX30.10

FIGURE EX30.11

11. | The metal equilateral triangle in **FIGURE EX30.11**, 20 cm on each side, is halfway into a 0.10 T magnetic field.
   a. What is the magnetic flux through the triangle?
   b. If the magnetic field strength decreases, what is the direction of the induced current in the triangle?
12. || The current in the solenoid of **FIGURE EX30.12** is increasing. The solenoid is surrounded by a conducting loop. Is there a current in the loop? If so, is the loop current cw or ccw?

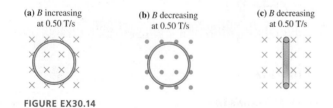

FIGURE EX30.12

Section 30.5 Faraday's Law

13. | The loop in **FIGURE EX30.13** is being pushed into the 0.20 T magnetic field at 50 m/s. The resistance of the loop is 0.10 Ω. What are the direction and the magnitude of the current in the loop?

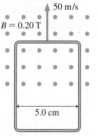

FIGURE EX30.13

14. | **FIGURE EX30.14** shows a 10-cm-diameter loop in three different magnetic fields. The loop's resistance is 0.20 Ω. For each, what are the size and direction of the induced current?

(a) B increasing at 0.50 T/s

(b) B decreasing at 0.50 T/s

(c) B decreasing at 0.50 T/s

FIGURE EX30.14

15. | The resistance of the loop in **FIGURE EX30.15** is 0.20 Ω. Is the magnetic field strength increasing or decreasing? At what rate (T/s)?

150 mA

8.0 cm

8.0 cm

FIGURE EX30.15

16. || A 1000-turn coil of wire 1.0 cm in diameter is in a magnetic field that increases from 0.10 T to 0.30 T in 10 ms. The axis of the coil is parallel to the field. What is the emf of the coil?
17. || A 5.0-cm-diameter coil has 20 turns and a resistance of
CALC 0.50 Ω. A magnetic field perpendicular to the coil is $B = 0.020t + 0.010t^2$, where B is in tesla and t is in seconds.
   a. Find an expression for the induced current $I(t)$ as a function of time.
   b. Evaluate I at $t = 5$ s and $t = 10$ s.

Section 30.6 Induced Fields

18. || **FIGURE EX30.18** shows the current as a function of time through a 20-cm-long, 4.0-cm-diameter solenoid with 400 turns. Draw a graph of the induced electric field strength as a function of time at a point 1.0 cm from the axis of the solenoid.

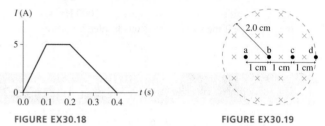

FIGURE EX30.18

FIGURE EX30.19

19. || The magnetic field in **FIGURE EX30.19** is decreasing at the rate 0.10 T/s. What is the acceleration (magnitude and direction) of a proton initially at rest at points a to d?
20. || The magnetic field inside a 5.0-cm-diameter solenoid is 2.0 T and decreasing at 4.0 T/s. What is the electric field strength inside the solenoid at a point (a) on the axis and (b) 2.0 cm from the axis?
21. || Scientists studying an anomalous magnetic field find that it is inducing a circular electric field in a plane perpendicular to the magnetic field. The electric field strength 1.5 m from the center of the circle is 4.0 mV/m. At what rate is the magnetic field changing?

Section 30.7 Induced Currents: Three Applications

22. | Electricity is distributed from electrical substations to neighborhoods at 15,000 V. This is a 60 Hz oscillating (AC) voltage. Neighborhood transformers, seen on utility poles, step this voltage down to the 120 V that is delivered to your house.
   a. How many turns does the primary coil on the transformer have if the secondary coil has 100 turns?
   b. No energy is lost in an ideal transformer, so the output power P_{out} from the secondary coil equals the input power P_{in} to the primary coil. Suppose a neighborhood transformer delivers 250 A at 120 V. What is the current in the 15,000 V line from the substation?
23. | The charger for your electronic devices is a transformer. Suppose a 60 Hz outlet voltage of 120 V needs to be reduced to a device voltage of 3.0 V. The side of the transformer attached to the electronic device has 60 turns of wire. How many turns are on the side that plugs into the outlet?

Section 30.8 Inductors

24. | The maximum allowable potential difference across a 200 mH inductor is 400 V. You need to raise the current through the inductor from 1.0 A to 3.0 A. What is the minimum time you should allow for changing the current?

46. ‖ **FIGURE P30.46** shows a 4.0-cm-diameter loop with resistance 0.10 Ω around a 2.0-cm-diameter solenoid. The solenoid is 10 cm long, has 100 turns, and carries the current shown in the graph. A positive current is cw when seen from the left. Find the current in the loop at (a) $t = 0.5$ s, (b) $t = 1.5$ s, and (c) $t = 2.5$ s.

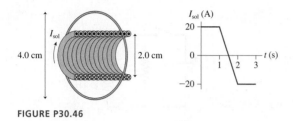

FIGURE P30.46

47. ‖‖ **FIGURE P30.47** shows a 1.0-cm-diameter loop with $R = 0.50$ Ω inside a 2.0-cm-diameter solenoid. The solenoid is 8.0 cm long, has 120 turns, and carries the current shown in the graph. A positive current is cw when seen from the left. Determine the current in the loop at $t = 0.010$ s.

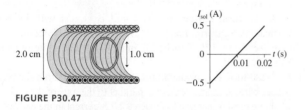

FIGURE P30.47

48. ‖ **FIGURE P30.48** shows two 20-turn coils tightly wrapped on the same 2.0-cm-diameter cylinder with 1.0-mm-diameter wire. The current through coil 1 is shown in the graph. Determine the current in coil 2 at (a) $t = 0.05$ s and (b) $t = 0.25$ s. A positive current is into the page at the top of a loop. Assume that the magnetic field of coil 1 passes entirely through coil 2.

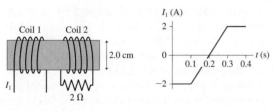

FIGURE P30.48

49. ‖ An electric generator has an 18-cm-diameter, 120-turn coil that **CALC** rotates at 60 Hz in a uniform magnetic field that is perpendicular to the rotation axis. What magnetic field strength is needed to generate a peak voltage of 170 V?

50. ‖ A 40-turn, 4.0-cm-diameter coil with $R = 0.40$ Ω surrounds **CALC** a 3.0-cm-diameter solenoid. The solenoid is 20 cm long and has 200 turns. The 60 Hz current through the solenoid is $I = I_0 \sin(2\pi ft)$. What is I_0 if the maximum induced current in the coil is 0.20 A?

51. ‖‖ A small, 2.0-mm-diameter circular loop with $R = 0.020$ Ω is at the center of a large 100-mm-diameter circular loop. Both loops lie in the same plane. The current in the outer loop changes from +1.0 A to −1.0 A in 0.10 s. What is the induced current in the inner loop?

52. ‖ A rectangular metal loop with 0.050 Ω resistance is placed next to one wire of the *RC* circuit shown in **FIGURE P30.52**. The capacitor is charged to 20 V with the polarity shown, then the switch is closed at $t = 0$ s.
   a. What is the direction of current in the loop for $t > 0$ s?
   b. What is the current in the loop at $t = 5.0$ μs? Assume that only the circuit wire next to the loop is close enough to produce a significant magnetic field.

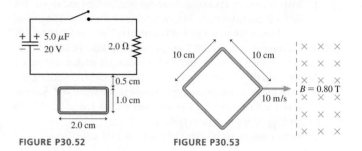

FIGURE P30.52   **FIGURE P30.53**

53. ‖ The square loop shown in **FIGURE P30.53** moves into a 0.80 T **CALC** magnetic field at a constant speed of 10 m/s. The loop has a resistance of 0.10 Ω, and it enters the field at $t = 0$ s.
   a. Find the induced current in the loop as a function of time. Give your answer as a graph of I versus t from $t = 0$ s to $t = 0.020$ s.
   b. What is the maximum current? What is the position of the loop when the current is maximum?

54. ‖ The L-shaped conductor in **FIGURE P30.54** moves at 10 m/s **CALC** across and touches a stationary L-shaped conductor in a 0.10 T magnetic field. The two vertices overlap, so that the enclosed area is zero, at $t = 0$ s. The conductor has a resistance of 0.010 ohms *per meter*.
   a. What is the direction of the induced current?
   b. Find expressions for the induced emf and the induced current as functions of time.
   c. Evaluate $\mathcal{E}$ and I at $t = 0.10$ s.

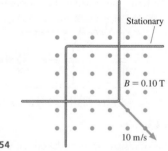

FIGURE P30.54

55. ‖ A 20-cm-long, zero-resistance slide wire moves outward, on zero-resistance rails, at a steady speed of 10 m/s in a 0.10 T magnetic field. (See Figure 30.26.) On the opposite side, a 1.0 Ω carbon resistor completes the circuit by connecting the two rails. The mass of the resistor is 50 mg.
   a. What is the induced current in the circuit?
   b. How much force is needed to pull the wire at this speed?
   c. If the wire is pulled for 10 s, what is the temperature increase of the carbon? The specific heat of carbon is 710 J/kg K.

56. ‖ Your camping buddy has an idea for a light to go inside your
CALC tent. He happens to have a powerful (and heavy!) horseshoe
magnet that he bought at a surplus store. This magnet creates
a 0.20 T field between two pole tips 10 cm apart. His idea is
to build the hand-cranked generator shown in FIGURE P30.56.
He thinks you can make enough current to fully light a 1.0 Ω
lightbulb rated at 4.0 W. That's not super bright, but it should be
plenty of light for routine activities in the tent.
 a. Find an expression for the induced current as a function of
 time if you turn the crank at frequency f. Assume that the
 semicircle is at its highest point at $t = 0$ s.
 b. With what frequency will you have to turn the crank for the
 maximum current to fully light the bulb? Is this feasible?

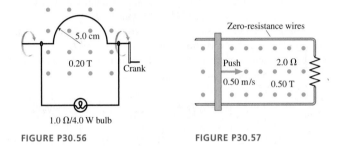

FIGURE P30.56   **FIGURE P30.57**

57. ‖ The 10-cm-wide, zero-resistance slide wire shown in FIGURE
P30.57 is pushed toward the 2.0 Ω resistor at a steady speed of
0.50 m/s. The magnetic field strength is 0.50 T.
 a. How big is the pushing force?
 b. How much power does the pushing force supply to the wire?
 c. What are the direction and magnitude of the induced current?
 d. How much power is dissipated in the resistor?

58. ‖ You've decided to make the magnetic projectile launcher
shown in FIGURE P30.58 for your science project. An aluminum
bar of length l slides along metal rails through a magnetic field B.
The switch closes at $t = 0$ s, while the bar is at rest, and a battery
of emf $\mathcal{E}_{bat}$ starts a current flowing around the loop. The battery
has internal resistance r. The resistances of the rails and the bar
are effectively zero.
 a. Show that the bar reaches a terminal speed v_{term}, and find an
 expression for v_{term}.
 b. Evaluate v_{term} for $\mathcal{E}_{bat} = 1.0$ V, $r = 0.10$ Ω, $l = 6.0$ cm, and
 $B = 0.50$ T.

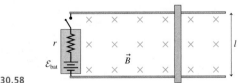

FIGURE P30.58

59. ‖ FIGURE P30.59 shows a U-shaped conducting rail that is
oriented vertically in a horizontal magnetic field. The rail has
no electric resistance and does not move. A slide wire with mass
m and resistance R can slide up and down without friction while
maintaining electrical contact with the rail. The slide wire is
released from rest.
 a. Show that the slide wire reaches a terminal speed v_{term}, and
 find an expression for v_{term}.
 b. Determine the value of v_{term} if $l = 20$ cm, $m = 10$ g, $R =$
 0.10 Ω, and $B = 0.50$ T.

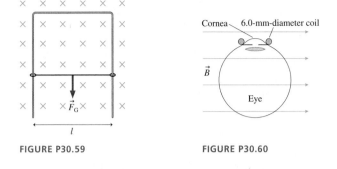

FIGURE P30.59   **FIGURE P30.60**

60. ‖‖‖ Experiments to study vision often need to track the movements
BIO of a subject's eye. One way of doing so is to have the subject sit in
a magnetic field while wearing special contact lenses with a coil
of very fine wire circling the edge. A current is induced in the
coil each time the subject rotates his eye. Consider the experiment
of FIGURE P30.60 in which a 20-turn, 6.0-mm-diameter coil of
wire circles the subject's cornea while a 1.0 T magnetic field is
directed as shown. The subject begins by looking straight ahead.
What emf is induced in the coil if the subject shifts his gaze by
5° in 0.20 s?

61. ‖ A 10-turn coil of wire having a diameter of 1.0 cm and
CALC a resistance of 0.20 Ω is in a 1.0 mT magnetic field, with the
coil oriented for maximum flux. The coil is connected to an
uncharged 1.0 μF capacitor rather than to a current meter. The
coil is quickly pulled out of the magnetic field. Afterward, what
is the voltage across the capacitor?
 Hint: Use $I = dq/dt$ to relate the *net* change of flux to the amount
 of charge that flows to the capacitor.

62. ‖‖‖ The magnetic field at one place on the earth's surface is
55 μT in strength and tilted 60° down from horizontal. A 200-
turn coil having a diameter of 4.0 cm and a resistance of 2.0 Ω
is connected to a 1.0 μF capacitor rather than to a current
meter. The coil is held in a horizontal plane and the capacitor is
discharged. Then the coil is quickly rotated 180° so that the side
that had been facing up is now facing down. Afterward, what is
the voltage across the capacitor? See the Hint in Problem 61.

63. ‖ Equation 30.26 is an expression for the induced electric field
CALC inside a solenoid ($r < R$). Find an expression for the induced
electric field outside a solenoid ($r > R$) in which the magnetic
field is changing at the rate dB/dt.

64. ‖ A solenoid inductor has an emf of 0.20 V when the current
through it changes at the rate 10.0 A/s. A steady current of 0.10
A produces a flux of 5.0 μWb per turn. How many turns does the
inductor have?

65. ‖ One possible concern with MRI (see Exercise 28) is turning the
BIO magnetic field on or off too quickly. Bodily fluids are conductors,
and a changing magnetic field could cause electric currents to flow
through the patient. Suppose a typical patient has a maximum cross-
section area of 0.060 m². What is the smallest time interval in which
a 5.0 T magnetic field can be turned on or off if the induced emf
around the patient's body must be kept to less than 0.10 V?

66. ‖ FIGURE P30.66 shows the current through a 10 mH inductor.
Draw a graph showing the potential difference ΔV_L across the
inductor for these 6 ms.

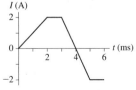

FIGURE P30.66

67. ‖ FIGURE P30.67 shows the potential difference across a 50 mH inductor. The current through the inductor at $t = 0$ s is 0.20 A. Draw a graph showing the current through the inductor from $t = 0$ s to $t = 40$ ms.

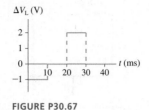

FIGURE P30.67

68. ‖ A 3.6 mH inductor with negligible resistance has a 1.0 A current through it. The current starts to increase at $t = 0$ s, creating a constant 5.0 mV voltage across the inductor. How much charge passes through the inductor between $t = 0$ s and $t = 5.0$ s?

69. ‖ The current through inductance L is given by $I = I_0 \sin \omega t$.
CALC a. Find an expression for the potential difference ΔV_L across the inductor.
   b. The maximum voltage across the inductor is 0.20 V when $L = 50$ μH and $f = 500$ kHz. What is I_0?

70. ‖ The current through inductance L is given by $I = I_0 e^{-t/\tau}$.
CALC a. Find an expression for the potential difference ΔV_L across the inductor.
   b. Evaluate ΔV_L at $t = 0$, 1.0, and 3.0 ms if $L = 20$ mH, $I_0 = 50$ mA, and $\tau = 1.0$ ms.

71. ‖ An LC circuit is built with a 20 mH inductor and an 8.0 pF capacitor. The capacitor voltage has its maximum value of 25 V at $t = 0$ s.
   a. How long is it until the capacitor is first fully discharged?
   b. What is the inductor current at that time?

72. ‖ An electric oscillator is made with a 0.10 μF capacitor and a 1.0 mH inductor. The capacitor is initially charged to 5.0 V. What is the maximum current through the inductor as the circuit oscillates?

73. ‖ For your final exam in electronics, you're asked to build an LC circuit that oscillates at 10 kHz. In addition, the maximum current must be 0.10 A and the maximum energy stored in the capacitor must be 1.0×10^{-5} J. What values of inductance and capacitance must you use?

74. ‖ The inductor in FIGURE P30.74 is a 9.0-cm-long, 2.0-cm-diameter solenoid wrapped with 300 turns. What is the current in the circuit 10 μs after the switch is moved from a to b?

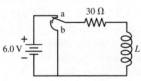

FIGURE P30.74

75. ‖ The 300 μF capacitor in FIGURE P30.75 is initially charged to 100 V, the 1200 μF capacitor is uncharged, and the switches are both open.
   a. What is the maximum voltage to which you can charge the 1200 μF capacitor by the proper closing and opening of the two switches?
   b. How would you do it? Describe the sequence in which you would close and open switches and the times at which you would do so. The first switch is closed at $t = 0$ s.

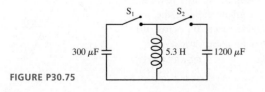

FIGURE P30.75

76. ‖ The switch in FIGURE P30.76 has been open for a long time. It is closed at $t = 0$ s. What is the current through the 20 Ω resistor
   a. Immediately after the switch is closed?
   b. After the switch has been closed a long time?
   c. Immediately after the switch is reopened?

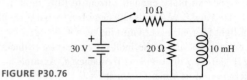

FIGURE P30.76

77. ‖ The switch in FIGURE P30.77 has been open for a long time. It is closed at $t = 0$ s.
   a. After the switch has been closed for a long time, what is the current in the circuit? Call this current I_0.
   b. Find an expression for the current I as a function of time. Write your expression in terms of I_0, R, and L.
   c. Sketch a current-versus-time graph from $t = 0$ s until the current is no longer changing.

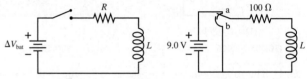

FIGURE P30.77   FIGURE P30.78

78. ‖ To determine the inductance of an unmarked inductor, you set up the circuit shown in FIGURE P30.78. After moving the switch from a to b at $t = 0$ s, you monitor the resistor voltage with an oscilloscope. Your data are shown in the table:

Time (μs)	Voltage (V)
0	9.0
10	6.7
20	4.6
30	3.2
40	2.5

Use an appropriate graph of the data to determine the inductance.

79. ‖‖ 5.0 μs after the switch of FIGURE P30.79 is moved from a to b, the magnetic energy stored in the inductor has decreased by half. What is the value of the inductance L?

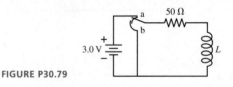

FIGURE P30.79

Challenge Problems

80. ‖‖‖ The rectangular loop in FIGURE CP30.80 has 0.020 Ω resistance. What is the induced current in the loop at this instant?
CALC

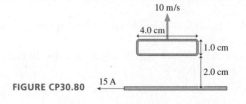

FIGURE CP30.80

81. ‖ In recent years it has been possible to buy a 1.0 F capacitor. This is an enormously large amount of capacitance. Suppose you want to build a 1.0 Hz oscillator with a 1.0 F capacitor. You have a spool of 0.25-mm-diameter wire and a 4.0-cm-diameter plastic cylinder. How long must your inductor be if you wrap it with 2 layers of closely spaced turns?

82. ‖ CALC The metal wire in FIGURE CP30.82 moves with speed v parallel to a straight wire that is carrying current I. The distance between the two wires is d. Find an expression for the potential difference between the two ends of the moving wire.

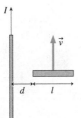

FIGURE CP30.82

83. ‖ CALC Let's look at the details of eddy-current braking. A square loop, length l on each side, is shot with velocity v_0 into a uniform magnetic field B. The field is perpendicular to the plane of the loop. The loop has mass m and resistance R, and it enters the field at $t = 0$ s. Assume that the loop is moving to the right along the x-axis and that the field begins at $x = 0$ m.

   a. Find an expression for the loop's velocity as a function of time as it enters the magnetic field. You can ignore gravity, and you can assume that the back edge of the loop has not entered the field.
   b. Calculate and draw a graph of v over the interval $0\ \text{s} \le t \le 0.04$ s for the case that $v_0 = 10$ m/s, $l = 10$ cm, $m = 1.0$ g, $R = 0.0010\ \Omega$, and $B = 0.10$ T. The back edge of the loop does not reach the field during this time interval.

84. ‖ CALC An 8.0 cm × 8.0 cm square loop is halfway into a magnetic field perpendicular to the plane of the loop. The loop's mass is 10 g and its resistance is 0.010 Ω. A switch is closed at $t = 0$ s, causing the magnetic field to increase from 0 to 1.0 T in 0.010 s.
   a. What is the induced current in the square loop?
   b. With what speed is the loop "kicked" away from the magnetic field?
   Hint: What is the impulse on the loop?

85. ‖ CALC A 2.0-cm-diameter solenoid is wrapped with 1000 turns per meter. 0.50 cm from the axis, the strength of an induced electric field is 5.0×10^{-4} V/m. What is the rate dI/dt with which the current through the solenoid is changing?

86. ‖ CALC High-frequency signals are often transmitted along a *coaxial cable,* such as the one shown in FIGURE CP30.86. For example, the cable TV hookup coming into your home is a coaxial cable. The signal is carried on a wire of radius r_1 while the outer conductor of radius r_2 is grounded. A soft, flexible insulating material fills the space between them, and an insulating plastic coating goes around the outside.
   a. Find an expression for the inductance per meter of a coaxial cable. To do so, consider the flux through a rectangle of length l that spans the gap between the inner and outer conductors.
   b. Evaluate the inductance per meter of a cable having $r_1 = 0.50$ mm and $r_2 = 3.0$ mm.

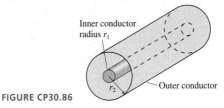

FIGURE CP30.86

31 Electromagnetic Fields and Waves

Sugar crystals seen in polarized light. The crystals rotate the plane of polarization, and the different colors represent portions of the crystals of different thicknesses.

IN THIS CHAPTER, you will study the properties of electromagnetic fields and waves.

How do fields transform?

Whether the field at a point is electric or magnetic depends, surprisingly, on your motion relative to the charges and currents. You'll learn how to transform the fields measured in one reference frame to a second reference frame moving relative to the first.

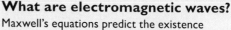

« LOOKING BACK Section 4.3 Relative motion

What is Maxwell's theory of electromagnetism?

Electricity and magnetism can be summarized in four equations for the fields, called Maxwell's equations, and one equation that tells us how charges respond to fields.

- Gauss's law: Charges create electric fields.
- Gauss's law for magnetism: There are no isolated magnetic poles.
- Faraday's law: Electric fields can also be created by changing magnetic fields.
- Ampère-Maxwell law: Magnetic fields can be created either by currents or by changing magnetic fields.

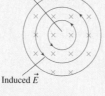

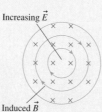

« LOOKING BACK Section 24.4 Gauss's law

« LOOKING BACK Section 29.6 Ampère's law

« LOOKING BACK Section 30.5 Faraday's law

What are electromagnetic waves?

Maxwell's equations predict the existence of self-sustaining oscillations of the electric and magnetic fields—electromagnetic waves—that travel through space without the presence of charges or currents.

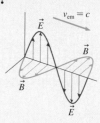

- In a vacuum, all electromagnetic waves—from radio waves to x rays—travel with the *same speed* $v_{em} = 1/\sqrt{\epsilon_0\mu_0} = c$, the speed of light.
- The fields $\vec{E}$ and $\vec{B}$ are perpendicular to each other and to the direction of travel.
- Electromagnetic waves are launched by an oscillating dipole, called an antenna.
- Electromagnetic waves transfer energy.
- Electromagnetic waves also transfer momentum and exert radiation pressure.

« LOOKING BACK Section 16.4 The wave equation

What is polarization?

An electromagnetic wave is polarized if the electric field always oscillates in the same plane—the plane of polarization. Polarizers both create and analyze polarized light. You will learn to calculate the intensity of light transmitted through a polarizer and will see that light is completely blocked by crossed polarizers. Polarization is used in many types of modern optical instrumentation.

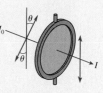

31.1 *E or B?* It Depends on Your Perspective

Our story thus far has been that charges create electric fields and that moving charges, or currents, create magnetic fields. But consider **FIGURE 31.1a**, where Brittney, carrying charge q, runs past Alec with velocity $\vec{v}$. Alec sees a moving charge, and he knows that this charge creates a magnetic field. But from Brittney's perspective, the charge is at rest. Stationary charges don't create magnetic fields, so Brittney claims that the magnetic field is zero. Is there, or is there not, a magnetic field?

Or what about the situation in **FIGURE 31.1b**? Now Brittney is carrying the charge through a magnetic field that Alec has created. Alec sees a charge moving in a magnetic field, so he knows there's a force $\vec{F} = q\vec{v} \times \vec{B}$ on the charge. But for Brittney the charge is still at rest. Stationary charges don't experience magnetic forces, so Brittney claims that $\vec{F} = \vec{0}$.

Now, we may be a bit uncertain about magnetic fields, but surely there can be no disagreement over forces. After all, forces cause observable and measurable effects, so Alec and Brittney should be able to agree on whether or not the charge experiences a force. Further, if Brittney runs with constant velocity, then both Alec and Brittney are in *inertial reference frames*. You learned in Chapter 4 that these are the reference frames in which Newton's laws are valid, so we can't say that there's anything abnormal or unusual about Alec's and Brittney's observations.

This paradox has arisen because magnetic fields and forces depend on velocity, but we haven't looked at the issue of velocity *with respect to what* or velocity *as measured by whom*. The resolution of this paradox will lead us to the conclusion that $\vec{E}$ and $\vec{B}$ are not, as we've been assuming, separate and independent entities. They are closely intertwined.

Reference Frames

We introduced reference frames and relative motion in Chapter 4. To remind you, **FIGURE 31.2** shows two reference frames labeled A and B. You can think of these as the reference frames in which Alec and Brittney, respectively, are at rest. Frame B moves with velocity $\vec{v}_{BA}$ with respect to frame A. That is, an observer (Alec) at rest in A sees the origin of B (Brittney) go past with velocity $\vec{v}_{BA}$. Of course, Brittney would say that Alec has velocity $\vec{v}_{AB} = -\vec{v}_{BA}$ relative to her reference frame. We will stipulate that both reference frames are inertial reference frames, so $\vec{v}_{BA}$ is constant.

Figure 31.2 also shows a particle C. Experimenters in frame A measure the motion of the particle and find that its velocity *relative to frame A* is $\vec{v}_{CA}$. At the same instant, experimenters in B find that the particle's velocity *relative to frame B* is $\vec{v}_{CB}$. In Chapter 4, we found that $\vec{v}_{CA}$ and $\vec{v}_{CB}$ are related by

$$\vec{v}_{CA} = \vec{v}_{CB} + \vec{v}_{BA} \tag{31.1}$$

Equation 31.1, the *Galilean transformation of velocity*, tells us that the velocity of the particle relative to reference frame A is its velocity relative to frame B plus (vector addition!) the velocity of frame B relative to frame A.

Suppose the particle in Figure 31.2 is accelerating. How does its acceleration $\vec{a}_{CA}$, as measured in frame A, compare to the acceleration $\vec{a}_{CB}$ measured in frame B? We can answer this question by taking the time derivative of Equation 31.1:

$$\frac{d\vec{v}_{CA}}{dt} = \frac{d\vec{v}_{CB}}{dt} + \frac{d\vec{v}_{BA}}{dt}$$

The derivatives of $\vec{v}_{CA}$ and $\vec{v}_{CB}$ are the particle's accelerations $\vec{a}_{CA}$ and $\vec{a}_{CB}$ in frames A and B, respectively. But $\vec{v}_{BA}$ is a *constant* velocity, so $d\vec{v}_{BA}/dt = \vec{0}$. Thus the Galilean transformation of acceleration is simply

$$\vec{a}_{CA} = \vec{a}_{CB} \tag{31.2}$$

FIGURE 31.1 Brittney carries a charge past Alec.

(a)

Charge q moves with velocity $\vec{v}$ relative to Alec.

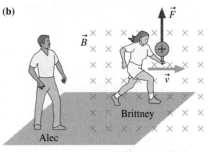

(b)

Charge q moves through a magnetic field established by Alec.

FIGURE 31.2 Reference frames A and B.

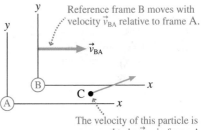

Reference frame B moves with velocity $\vec{v}_{BA}$ relative to frame A.

The velocity of this particle is measured to be $\vec{v}_{CA}$ in frame A and $\vec{v}_{CB}$ in frame B.

Brittney and Alec may measure different positions and velocities for a particle, but they *agree* on its acceleration. And if they agree on its acceleration, they must, by using Newton's second law, agree on the force acting on the particle. That is, **experimenters in all inertial reference frames agree about the force acting on a particle.**

The Transformation of Electric and Magnetic Fields

Imagine that Alec has measured the electric field $\vec{E}_A$ and the magnetic field $\vec{B}_A$ in reference frame A. Our investigations thus far give us no reason to think that Brittney's measurements of the fields will differ from Alec's. After all, it seems like the fields are just "there," waiting to be measured.

To find out if this is true, Alec establishes a region of space with a uniform magnetic field $\vec{B}_A$ but no electric field ($\vec{E}_A = \vec{0}$). Then, as shown in FIGURE 31.3, he shoots a positive charge q through the magnetic field. At an instant when q is moving horizontally with velocity $\vec{v}_{CA}$, Alec observes that the particle experiences force $\vec{F}_A = q\vec{v}_{CA} \times \vec{B}_A$. The direction of the force is straight up.

Suppose that Brittney, in frame B, runs alongside the charge with the same velocity: $\vec{v}_{BA} = \vec{v}_{CA}$. To her, in frame B, the charge is at rest. Nonetheless, because both experimenters must agree about forces, Brittney *must* observe the same upward force on the charge that Alec observed. But there is *no* magnetic force on a stationary charge, so how can this be?

Because Brittney sees a stationary charge being acted on by an upward force, her only possible conclusion is that there is an upward-pointing *electric field.* After all, the electric field was initially defined in terms of the force experienced by a stationary charge. If the electric field in frame B is $\vec{E}_B$, then the force on the charge is $\vec{F}_B = q\vec{E}_B$. But we know that $\vec{F}_B = \vec{F}_A$, and Alec has already measured $\vec{F}_A = q\vec{v}_{CA} \times \vec{B}_A = q\vec{v}_{BA} \times \vec{B}_A$. Thus we're led to the conclusion that

$$\vec{E}_B = \vec{v}_{BA} \times \vec{B}_A \tag{31.3}$$

As Brittney runs past Alec, she finds that at least part of Alec's magnetic field has become an electric field! **Whether a field is seen as "electric" or "magnetic" depends on the motion of the reference frame relative to the sources of the field.**

FIGURE 31.4 shows the situation from Brittney's perspective. There is a force on charge q, the same force that Alec measured in Figure 31.3, but Brittney attributes this force to an electric field rather than a magnetic field. (Brittney needs a moving charge to measure magnetic forces, so we'll need a different experiment to see whether or not there's a magnetic field in frame B.)

More generally, suppose that an experimenter in reference frame A creates both an electric field $\vec{E}_A$ and a magnetic field $\vec{B}_A$. A charge moving in A with velocity $\vec{v}_{CA}$ experiences the force $\vec{F}_A = q(\vec{E}_A + \vec{v}_{CA} \times \vec{B}_A)$ shown in FIGURE 31.5a. The charge is at rest in a reference frame B that moves with velocity $\vec{v}_{BA} = \vec{v}_{CA}$ so the force in B can be due only to an electric field: $\vec{F}_B = q\vec{E}_B$. Equating the forces, because experimenters in all inertial reference frames agree about forces, we find that

$$\vec{E}_B = \vec{E}_A + \vec{v}_{BA} \times \vec{B}_A \tag{31.4}$$

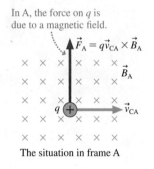

FIGURE 31.3 A charged particle moves through a magnetic field in reference frame A and experiences a magnetic force.

In A, the force on q is due to a magnetic field.

$\vec{F}_A = q\vec{v}_{CA} \times \vec{B}_A$

$\vec{B}_A$

q $\vec{v}_{CA}$

The situation in frame A

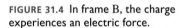
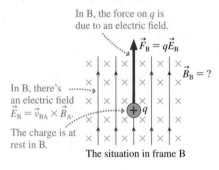

FIGURE 31.4 In frame B, the charge experiences an electric force.

In B, the force on q is due to an electric field.

$\vec{F}_B = q\vec{E}_B$

$\vec{B}_B = ?$

In B, there's an electric field $\vec{E}_B = \vec{v}_{BA} \times \vec{B}_A$.

q

The charge is at rest in B.

The situation in frame B

FIGURE 31.5 A charge in reference frame A experiences electric and magnetic forces. The charge experiences the same force in frame B, but it is due only to an electric field.

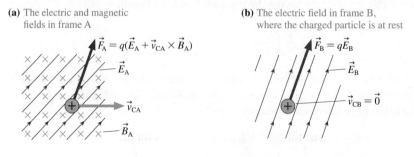

(a) The electric and magnetic fields in frame A

$\vec{F}_A = q(\vec{E}_A + \vec{v}_{CA} \times \vec{B}_A)$

$\vec{E}_A$

$\vec{v}_{CA}$

$\vec{B}_A$

(b) The electric field in frame B, where the charged particle is at rest

$\vec{F}_B = q\vec{E}_B$

$\vec{E}_B$

$\vec{v}_{CB} = \vec{0}$

Equation 31.4 transforms the electric and magnetic fields measured in reference frame A into the electric field measured in a frame B that moves relative to A with velocity $\vec{v}_{BA}$. **FIGURE 31.5b** shows the outcome. Although we used a charge as a probe to find Equation 31.4, the equation is strictly about fields in different reference frames; it makes no mention of charges.

EXAMPLE 31.1 | Transforming the electric field

A laboratory experimenter has created the parallel electric and magnetic fields $\vec{E} = 10,000\,\hat{\imath}$ V/m and $\vec{B} = 0.10\,\hat{\imath}$ T. A proton is shot into these fields with velocity $\vec{v} = 1.0 \times 10^5\,\hat{\jmath}$ m/s. What is the electric field in the proton's reference frame?

MODEL Let the laboratory be reference frame A and a frame moving with the proton be reference frame B. The relative velocity is $\vec{v}_{BA} = 1.0 \times 10^5\,\hat{\jmath}$ m/s.

VISUALIZE **FIGURE 31.6** shows the geometry. The laboratory fields, now labeled A, are parallel to the x-axis while $\vec{v}_{BA}$ is in the y-direction. Thus $\vec{v}_{BA} \times \vec{B}_A$ points in the negative z-direction.

SOLVE $\vec{v}_{BA}$ and $\vec{B}_A$ are perpendicular, so the magnitude of $\vec{v}_{BA} \times \vec{B}_A$ is $(1.0 \times 10^5$ m/s$)(0.10$ T$)(\sin 90°) = 10,000$ V/m. Thus the electric field in frame B, the proton's frame, is

$$\vec{E}_B = \vec{E}_A + \vec{v}_{BA} \times \vec{B}_A = (10,000\,\hat{\imath} - 10,000\,\hat{k})\ \text{V/m}$$

$$= (14,000\ \text{V/m, } 45° \text{ below the x-axis})$$

ASSESS The force on the proton is the same in both reference frames. But in the proton's reference frame that force is due entirely to an electric field tilted 45° below the x-axis.

FIGURE 31.6 Finding electric field $\vec{E}_B$.

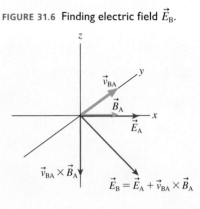

To find a transformation equation for the magnetic field, **FIGURE 31.7a** shows charge q at rest in reference frame A. Alec measures the fields of a stationary point charge:

$$\vec{E}_A = \frac{1}{4\pi\epsilon_0}\frac{q}{r^2}\hat{r} \qquad \vec{B}_A = \vec{0}$$

What are the fields at this point in space as measured by Brittney in frame B? We can use Equation 31.4 to find $\vec{E}_B$. Because $\vec{B}_A = \vec{0}$, the electric field in frame B is

$$\vec{E}_B = \vec{E}_A = \frac{1}{4\pi\epsilon_0}\frac{q}{r^2}\hat{r} \tag{31.5}$$

In other words, **Coulomb's law is still valid in a frame in which the point charge is moving.**

But Brittney also measures a magnetic field $\vec{B}_B$, because, as seen in **FIGURE 31.7b**, charge q is moving in reference frame B. The magnetic field of a moving point charge is given by the Biot-Savart law:

$$\vec{B}_B = \frac{\mu_0}{4\pi}\frac{q}{r^2}\vec{v}_{CB} \times \hat{r} = -\frac{\mu_0}{4\pi}\frac{q}{r^2}\vec{v}_{BA} \times \hat{r} \tag{31.6}$$

where we used the fact that the charge's velocity in frame B is $\vec{v}_{CB} = -\vec{v}_{BA}$.

It will be useful to rewrite Equation 31.6 as

$$\vec{B}_B = -\frac{\mu_0}{4\pi}\frac{q}{r^2}\vec{v}_{BA} \times \hat{r} = -\epsilon_0\mu_0\vec{v}_{BA} \times \left(\frac{1}{4\pi\epsilon_0}\frac{q}{r^2}\hat{r}\right)$$

The expression in parentheses is simply $\vec{E}_A$, the electric field in frame A, so we have

$$\vec{B}_B = -\epsilon_0\mu_0\vec{v}_{BA} \times \vec{E}_A \tag{31.7}$$

Thus we find the remarkable idea that **the Biot-Savart law for the magnetic field of a moving point charge is nothing other than the Coulomb electric field of a stationary point charge transformed into a moving reference frame.**

We will assert without proof that if the experimenters in frame A create a magnetic field $\vec{B}_A$ in addition to the electric field $\vec{E}_A$, then the magnetic field $\vec{B}_B$ is

$$\vec{B}_B = \vec{B}_A - \epsilon_0\mu_0\vec{v}_{BA} \times \vec{E}_A \tag{31.8}$$

This is a general transformation matching Equation 31.4 for the electric field $\vec{E}_B$.

FIGURE 31.7 A charge at rest in frame A is moving in frame B.

(a) In frame A, the static charge creates an electric field but no magnetic field.

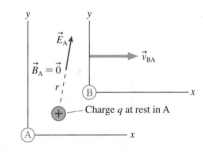

(b) In frame B, the moving charge creates both an electric and a magnetic field.

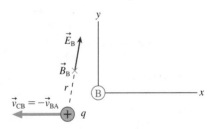

Notice something interesting. The constant μ_0 has units of T m/A; those of ϵ_0 are $C^2/N m^2$. By definition, 1 T = 1 N/A m and 1 A = 1 C/s. Consequently, the units of $\epsilon_0\mu_0$ turn out to be s^2/m^2. In other words, the quantity $1/\sqrt{\epsilon_0\mu_0}$, with units of m/s, is a speed. But what speed? The constants are well known from measurements of static electric and magnetic fields, so we can compute

$$\frac{1}{\sqrt{\epsilon_0\mu_0}} = \frac{1}{\sqrt{(8.85 \times 10^{-12} \text{ C}^2/\text{N m}^2)(1.26 \times 10^{-6} \text{ T m/A})}} = 3.00 \times 10^8 \text{ m/s}$$

Of all the possible values you might get from evaluating $1/\sqrt{\epsilon_0\mu_0}$, what are the chances it would turn out to be c, the speed of light? It is not a random coincidence. In Section 31.5 we'll show that electric and magnetic fields can exist as a *traveling wave,* and that the wave speed is predicted by the theory to be none other than

$$v_{\text{em}} = c = \frac{1}{\sqrt{\epsilon_0\mu_0}} \tag{31.9}$$

For now, we'll go ahead and write $\epsilon_0\mu_0 = 1/c^2$. With this, our **Galilean field transformation equations** are

$$\vec{E}_B = \vec{E}_A + \vec{v}_{BA} \times \vec{B}_A$$

$$\vec{B}_B = \vec{B}_A - \frac{1}{c^2}\vec{v}_{BA} \times \vec{E}_A \tag{31.10}$$

where $\vec{v}_{BA}$ is the velocity of reference frame B relative to frame A and where, to reiterate, the fields are measured *at the same point in space* by experimenters *at rest* in each reference frame.

NOTE We'll see shortly that these equations are valid only if $v_{BA} \ll c$.

We can no longer believe that electric and magnetic fields have a separate, independent existence. Changing from one reference frame to another mixes and rearranges the fields. Different experimenters watching an event will agree on the outcome, such as the deflection of a charged particle, but they will ascribe it to different combinations of fields. Our conclusion is that **there is a single electromagnetic field that presents different faces, in terms of $\vec{E}$ and $\vec{B}$, to different viewers.**

EXAMPLE 31.2 | **Two views of a magnet**

The 1.0 T field of a large laboratory magnet points straight up. A rocket flies past the laboratory, parallel to the ground, at 1000 m/s. What are the fields between the magnet's pole tips as measured—very quickly!—by scientists on the rocket?

MODEL Let the laboratory be reference frame A and a frame moving with the rocket be reference frame B.

VISUALIZE FIGURE 31.8 shows the magnet and establishes the coordinate systems. The relative velocity is $\vec{v}_{BA} = 1000\,\hat{\imath}$ m/s.

FIGURE 31.8 The rocket and the magnet.

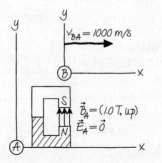

SOLVE The fields in the laboratory reference frame are $\vec{E}_A = \vec{0}$ and $\vec{B}_A = 1.0\,\hat{\jmath}$ T. Transforming the fields to the rocket's reference frame gives first, for the electric field,

$$\vec{E}_B = \vec{E}_A + \vec{v}_{BA} \times \vec{B}_A = \vec{v}_{BA} \times \vec{B}_A$$

From the right-hand rule, $\vec{v}_{BA} \times \vec{B}_A$ is out of the page, in the z-direction. $\vec{v}_{BA}$ and $\vec{B}_A$ are perpendicular, so

$$\vec{E}_B = v_{BA}B_A\,\hat{k} = 1000\,\hat{k} \text{ V/m}$$

Similarly, for the magnetic field,

$$\vec{B}_B = \vec{B}_A - \frac{1}{c^2}\vec{v}_{BA} \times \vec{E}_A = \vec{B}_A = 1.0\,\hat{\jmath} \text{ T}$$

Thus the rocket scientists measure

$$\vec{E}_B = 1000\,\hat{k} \text{ V/m} \quad \text{and} \quad \vec{B}_B = 1.0\,\hat{\jmath} \text{ T}$$

25. | What is the potential difference across a 10 mH inductor if the current through the inductor drops from 150 mA to 50 mA in 10 μs? What is the direction of this potential difference? That is, does the potential increase or decrease along the direction of the current?

26. ‖ A 100 mH inductor whose windings have a resistance of 4.0 Ω is connected across a 12 V battery having an internal resistance of 2.0 Ω. How much energy is stored in the inductor?

27. ‖ How much energy is stored in a 3.0-cm-diameter, 12-cm-long solenoid that has 200 turns of wire and carries a current of 0.80 A?

28. | MRI (magnetic resonance imaging) is a medical technique that
BIO produces detailed "pictures" of the interior of the body. The patient is placed into a solenoid that is 40 cm in diameter and 1.0 m long. A 100 A current creates a 5.0 T magnetic field inside the solenoid. To carry such a large current, the solenoid wires are cooled with liquid helium until they become superconducting (no electric resistance).
    a. How much magnetic energy is stored in the solenoid? Assume that the magnetic field is uniform within the solenoid and quickly drops to zero outside the solenoid.
    b. How many turns of wire does the solenoid have?

Section 30.9 *LC* Circuits

29. ‖ An FM radio station broadcasts at a frequency of 100 MHz. What inductance should be paired with a 10 pF capacitor to build a receiver circuit for this station?

30. ‖ A 2.0 mH inductor is connected in parallel with a variable capacitor. The capacitor can be varied from 100 pF to 200 pF. What is the range of oscillation frequencies for this circuit?

31. ‖ An MRI machine needs to detect signals that oscillate at very
BIO high frequencies. It does so with an *LC* circuit containing a 15 mH coil. To what value should the capacitance be set to detect a 450 MHz signal?

32. ‖ An *LC* circuit has a 10 mH inductor. The current has its maximum value of 0.60 A at $t = 0$ s. A short time later the capacitor reaches its maximum potential difference of 60 V. What is the value of the capacitance?

33. ‖ The switch in FIGURE EX30.33 has been in position 1 for a long time. It is changed to position 2 at $t = 0$ s.
    a. What is the maximum current through the inductor?
    b. What is the first time at which the current is maximum?

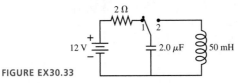

FIGURE EX30.33

Section 30.10 *LR* Circuits

34. | What value of resistor R gives the circuit in FIGURE EX30.34 a time constant of 25 μs?

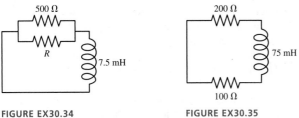

FIGURE EX30.34   FIGURE EX30.35

35. ‖ At $t = 0$ s, the current in the circuit in FIGURE EX30.35 is I_0. At what time is the current $\frac{1}{2}I_0$?

36. | The switch in FIGURE EX30.36 has been open for a long time. It is closed at $t = 0$ s.
    a. What is the current through the battery immediately after the switch is closed?
    b. What is the current through the battery after the switch has been closed a long time?

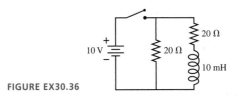

FIGURE EX30.36

Problems

37. ‖ A 20 cm × 20 cm square loop has a resistance of 0.10 Ω. A
CALC magnetic field perpendicular to the loop is $B = 4t - 2t^2$, where B is in tesla and t is in seconds. What is the current in the loop at $t = 0.0$ s, $t = 1.0$ s, and $t = 2.0$ s?

38. ‖ A 100-turn, 2.0-cm-diameter coil is at rest with its axis vertical. A uniform magnetic field 60° away from vertical increases from 0.50 T to 1.50 T in 0.60 s. What is the induced emf in the coil?

39. ‖ A 100-turn, 8.0-cm-diameter coil is made of 0.50-mm-diameter copper wire. A magnetic field is parallel to the axis of the coil. At what rate must B increase to induce a 2.0 A current in the coil?

40. ‖ A circular loop made from a flexible, conducting wire is
CALC shrinking. Its radius as a function of time is $r = r_0 e^{-\beta t}$. The loop is perpendicular to a steady, uniform magnetic field B. Find an expression for the induced emf in the loop at time t.

41. ‖ A 10 cm × 10 cm square loop lies in the *xy*-plane. The
CALC magnetic field in this region of space is $\vec{B} = (0.30t\,\hat{\imath} + 0.50t^2\,\hat{k})$ T, where t is in s. What is the emf induced in the loop at (a) $t = 0.5$ s and (b) $t = 1.0$ s?

42. ‖ A spherical balloon with a volume of 2.5 L is in a 45 mT
CALC uniform, vertical magnetic field. A horizontal elastic but conducting wire with 2.5 Ω resistance circles the balloon at its equator. Suddenly the balloon starts expanding at 0.75 L/s. What is the current in the wire 2.0 s later?

43. ‖‖ A 3.0-cm-diameter, 10-turn coil of wire, located at $z = 0$
CALC in the *xy*-plane, carries a current of 2.5 A. A 2.0-mm-diameter conducting loop with 2.0×10^{-4} Ω resistance is also in the *xy*-plane at the center of the coil. At $t = 0$ s, the loop begins to move along the *z*-axis with a constant speed of 75 m/s. What is the induced current in the conducting loop at $t = 200\,\mu$s? The diameter of the conducting loop is much smaller than that of the coil, so you can assume that the magnetic field through the loop is everywhere the on-axis field of the coil.

44. ‖‖ A 20 cm × 20 cm square loop of wire lies in the *xy*-plane
CALC with its bottom edge on the *x*-axis. The resistance of the loop is 0.50 Ω. A magnetic field parallel to the *z*-axis is given by $B = 0.80y^2 t$, where B is in tesla, y in meters, and t in seconds. What is the size of the induced current in the loop at $t = 0.50$ s?

45. ‖‖ A 2.0 cm × 2.0 cm square loop of wire with resistance 0.010 Ω has one edge parallel to a long straight wire. The near edge of the loop is 1.0 cm from the wire. The current in the wire is increasing at the rate of 100 A/s. What is the current in the loop?

Almost Relativity

FIGURE 31.9a shows two positive charges moving side by side through frame A with velocity $\vec{v}_{CA}$. Charge q_1 creates an electric field and a magnetic field at the position of charge q_2. These are

$$\vec{E}_A = \frac{1}{4\pi\epsilon_0}\frac{q_1}{r^2}\hat{j} \quad \text{and} \quad \vec{B}_A = \frac{\mu_0}{4\pi}\frac{q_1 v_{CA}}{r^2}\hat{k}$$

where r is the distance between the charges, and we've used $\hat{r} = \hat{j}$ and $\vec{v} \times \hat{r} = v\hat{k}$.

How are the fields seen in frame B, which moves with $\vec{v}_{BA} = \vec{v}_{CA}$ and in which the charges are at rest? From the field transformation equations,

$$\vec{B}_B = \vec{B}_A - \frac{1}{c^2}\vec{v}_{BA} \times \vec{E}_A = \frac{\mu_0}{4\pi}\frac{q_1 v_{CA}}{r^2}\hat{k} - \frac{1}{c^2}\left(v_{CA}\hat{i} \times \frac{1}{4\pi\epsilon_0}\frac{q_1}{r^2}\hat{j}\right)$$

$$= \frac{\mu_0}{4\pi}\frac{q_1 v_{CA}}{r^2}\left(1 - \frac{1}{\epsilon_0\mu_0 c^2}\right)\hat{k} \tag{31.11}$$

where we used $\hat{i} \times \hat{j} = \hat{k}$. But $\epsilon_0\mu_0 = 1/c^2$, so the term in parentheses is zero and thus $\vec{B}_B = \vec{0}$. This result was expected because q_1 is at rest in frame B and shouldn't create a magnetic field.

The transformation of the electric field is similar:

$$\vec{E}_B = \vec{E}_A + \vec{v}_{BA} \times \vec{B}_A = \frac{1}{4\pi\epsilon_0}\frac{q_1}{r^2}\hat{j} + v_{BA}\hat{i} \times \frac{\mu_0}{4\pi}\frac{q_1 v_{CA}}{r^2}\hat{k}$$

$$= \frac{1}{4\pi\epsilon_0}\frac{q_1}{r^2}(1 - \epsilon_0\mu_0 v_{BA}^2)\hat{j} = \frac{1}{4\pi\epsilon_0}\frac{q_1}{r^2}\left(1 - \frac{v_{BA}^2}{c^2}\right)\hat{j} \tag{31.12}$$

where we used $\hat{i} \times \hat{k} = -\hat{j}$, $\vec{v}_{CA} = \vec{v}_{BA}$, and $\epsilon_0\mu_0 = 1/c^2$. **FIGURE 31.9b** shows the charges and fields in frame B.

But now we have a problem. In frame B where the two charges are at rest and separated by distance r, the electric field due to charge q_1 should be simply

$$\vec{E}_B = \frac{1}{4\pi\epsilon_0}\frac{q_1}{r^2}\hat{j}$$

The field transformation equations have given a "wrong" result for the electric field $\vec{E}_B$.

It turns out that the field transformations of Equations 31.10, which are based on Galilean relativity, aren't quite right. We would need Einstein's relativity—a topic that we'll take up in Chapter 36—to give the correct transformations. However, the Galilean field transformations in Equations 31.10 are equivalent to the relativistically correct transformations when $v \ll c$, in which case $v^2/c^2 \ll 1$. You can see that the two expressions for $\vec{E}_B$ do, in fact, agree if v_{BA}^2/c^2 can be neglected.

Thus our use of the field transformation equations has an additional rule: Set v^2/c^2 to zero. This is an acceptable rule for speeds $v < 10^7$ m/s. Even with this limitation, our investigation has provided us with a deeper understanding of electric and magnetic fields.

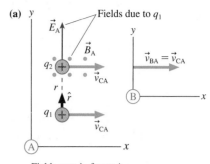

FIGURE 31.9 Two charges moving parallel to each other.

(a)

Fields seen in frame A

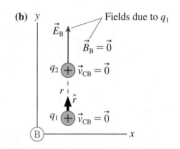

(b)

Fields seen in frame B

STOP TO THINK 31.1 The first diagram shows electric and magnetic fields in reference frame A. Which diagram shows the fields in frame B?

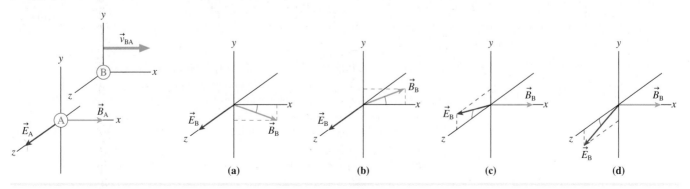

(a)   (b)   (c)   (d)

Faraday's Law Revisited

The transformation of electric and magnetic fields gives us new insight into Faraday's law. **FIGURE 31.10a** shows a reference frame A in which a conducting loop is moving with velocity $\vec{v}$ into a magnetic field. You learned in Chapter 30 that the magnetic field exerts a magnetic force $\vec{F}_B = q\vec{v} \times \vec{B} = (qvB, \text{upward})$ on the charges in the leading edge of the wire, creating an emf $\mathcal{E} = vLB$ and an induced current in the loop. We called this a *motional emf.*

How do things appear to an experimenter who is in frame B that moves with the loop at velocity $\vec{v}_{BA} = \vec{v}$ and for whom the loop is at rest? We have learned the important lesson that experimenters in different inertial reference frames agree about the outcome of any experiment; hence an experimenter in frame B agrees that there is an induced current in the loop. But the charges are at rest in frame B so there cannot be any magnetic force on them. How is the emf established in frame B?

We can use the field transformations to determine that the fields in frame B are

$$\vec{E}_B = \vec{E}_A + \vec{v} \times \vec{B}_A = \vec{v} \times \vec{B}$$
$$\vec{B}_B = \vec{B}_A - \frac{1}{c^2}\vec{v} \times \vec{E}_A = \vec{B} \tag{31.13}$$

where we used the fact that $\vec{E}_A = \vec{0}$ in frame A.

An experimenter in the loop's frame sees not only a magnetic field but also the electric field $\vec{E}_B$ shown in **FIGURE 31.10b**. The magnetic field exerts no force on the charges, because they're at rest in this frame, but the electric field does. The force on charge q is $\vec{F}_E = q\vec{E}_B = q\vec{v} \times \vec{B} = (qvB, \text{upward})$. This is the same force as was measured in the laboratory frame, so it will cause the same emf and the same current. The outcome is identical, as we knew it had to be, but the experimenter in B attributes the emf to an electric field whereas the experimenter in A attributes it to a magnetic field.

Field $\vec{E}_B$ is, in fact, the *induced electric field* of Faraday's law. Faraday's law, fundamentally, is a statement that **a changing magnetic field creates an electric field.** But only in frame B, the frame of the loop, is the magnetic field changing. Thus the induced electric field is seen in the loop's frame but not in the laboratory frame.

31.2 The Field Laws Thus Far

Let's remind ourselves where we are in terms of discovering laws about the electromagnetic field. Gauss's law, which you studied in Chapter 24, states a very general property of the electric field. It says that charges create electric fields in such a way that the electric flux of the field is the same through *any* closed surface surrounding the charges. **FIGURE 31.11** illustrates this idea by showing the field lines passing through a Gaussian surface enclosing a charge.

The mathematical statement of Gauss's law for the electric field says that for any *closed* surface enclosing total charge Q_{in}, the net electric flux through the surface is

$$(\Phi_e)_{\text{closed surface}} = \oint \vec{E} \cdot d\vec{A} = \frac{Q_{in}}{\epsilon_0} \tag{31.14}$$

The circle on the integral sign indicates that the integration is over a closed surface. Gauss's law is the first of what will turn out to be four *field equations.*

There's an analogous equation for magnetic fields, an equation we implied in Chapter 29—where we noted that isolated north or south poles do not exist—but didn't explicitly write it down. **FIGURE 31.12** shows a Gaussian surface around a magnetic dipole. Magnetic field lines form continuous curves, without starting or stopping, so every field line leaving the surface at some point must reenter it at another. Consequently, the net magnetic flux over a *closed* surface is zero.

FIGURE 31.10 A motional emf as seen in two different reference frames.

(a) Laboratory frame A

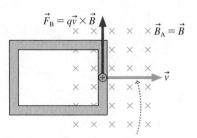

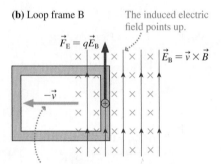

The loop is moving to the right.

(b) Loop frame B

The induced electric field points up.

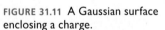
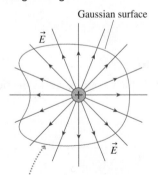

The magnetic field is moving to the left.

FIGURE 31.11 A Gaussian surface enclosing a charge.

Gaussian surface

There is a net electric flux through this surface that encloses a charge.

We've shown only one surface and one magnetic field, but this conclusion turns out to be a general property of magnetic fields. Because every north pole is accompanied by a south pole, we can't enclose a "net pole" within a surface. Thus Gauss's law for magnetic fields is

$$(\Phi_m)_{\text{closed surface}} = \oint \vec{B} \cdot d\vec{A} = 0 \tag{31.15}$$

Equation 31.14 is the mathematical statement that Coulomb electric field lines start and stop on charges. Equation 31.15 is the mathematical statement that magnetic field lines form closed loops; they don't start or stop (i.e., there are no isolated magnetic poles). These two versions of Gauss's law are important statements about what types of fields can and cannot exist. They will become two of Maxwell's equations.

The third field law we've established is Faraday's law:

$$\mathcal{E} = \oint \vec{E} \cdot d\vec{s} = -\frac{d\Phi_m}{dt} \tag{31.16}$$

where the line integral of $\vec{E}$ is around the closed curve that bounds the surface through which the magnetic flux Φ_m is calculated. Equation 31.16 is the mathematical statement that an electric field can also be created by a changing magnetic field. The correct use of Faraday's law requires a convention for determining when fluxes are positive and negative. The sign convention will be given in the next section, where we discuss the fourth and last field equation—an analogous equation for magnetic fields.

31.3 The Displacement Current

We introduced Ampère's law in Chapter 29 as an alternative to the Biot-Savart law for calculating the magnetic field of a current. Whenever total current I_{through} passes through an area bounded by a closed curve, the line integral of the magnetic field around the curve is

$$\oint \vec{B} \cdot d\vec{s} = \mu_0 I_{\text{through}} \tag{31.17}$$

FIGURE 31.13 illustrates the geometry of Ampère's law. The sign of each current can be determined by using Tactics Box 31.1. In this case, $I_{\text{through}} = I_1 - I_2$.

TACTICS BOX 31.1   (MP)

Determining the signs of flux and current

❶ For a surface S bounded by a closed curve C, choose either the clockwise (cw) or counterclockwise (ccw) direction around C.

❷ Curl the fingers of your *right* hand around the curve in the chosen direction, with your thumb perpendicular to the surface. Your thumb defines the positive direction.

■ A flux Φ through the surface is positive if the field is in the same direction as your thumb, negative if the field is in the opposite direction.

■ A current through the surface in the direction of your thumb is positive, in the direction opposite your thumb is negative.

Exercises 4–6   ✐

Ampère's law is the formal statement that **currents create magnetic fields.** Although Ampère's law can be used to calculate magnetic fields in situations with a high degree of symmetry, it is more important as a statement about what types of magnetic field can and cannot exist.

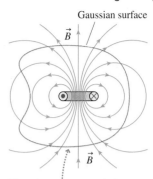

FIGURE 31.12 There is no net flux through a Gaussian surface around a magnetic dipole.

Gaussian surface

There is no net magnetic flux through this closed surface.

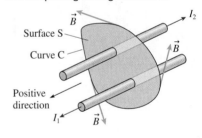

FIGURE 31.13 Ampère's law relates the line integral of $\vec{B}$ around curve C to the current passing through surface S.

Surface S
Curve C
Positive direction

Something Is Missing

Nothing restricts the bounded surface of Ampère's law to being flat. It's not hard to see that any current passing through surface S_1 in **FIGURE 31.14** must also pass through the curved surface S_2. To interpret Ampère's law properly, we have to say that the current $I_{through}$ is the net current passing through *any* surface S that is bounded by curve C.

FIGURE 31.14 The *net* current passing through the flat surface S_1 also passes through the curved surface S_2.

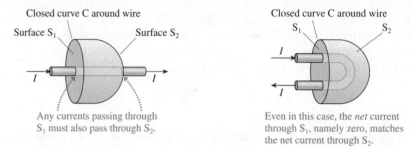

Any currents passing through S_1 must also pass through S_2.

Even in this case, the *net* current through S_1, namely zero, matches the net current through S_2.

FIGURE 31.15 There is no current through surface S_2 as the capacitor charges, but there is a changing electric flux.

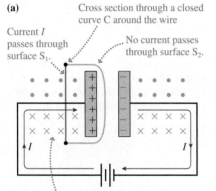

(a)

Cross section through a closed curve C around the wire

Current I passes through surface S_1.

No current passes through surface S_2.

This is the magnetic field of the current I that is charging the capacitor.

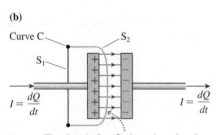

(b)

Curve C

The electric flux Φ_e through surface S_2 increases as the capacitor charges.

But this leads to an interesting puzzle. **FIGURE 31.15a** shows a capacitor being charged. Current I, from the left, brings positive charge to the left capacitor plate. The same current carries charges away from the right capacitor plate, leaving the right plate negatively charged. This is a perfectly ordinary current in a conducting wire, and you can use the right-hand rule to verify that its magnetic field is as shown.

Curve C is a closed curve encircling the wire on the left. The current passes through surface S_1, a flat surface across C, and we could use Ampère's law to find that the magnetic field is that of a straight wire. But what happens if we try to use surface S_2 to determine $I_{through}$? Ampère's law says that we can consider *any* surface bounded by curve C, and surface S_2 certainly qualifies. But *no* current passes through S_2. Charges are brought to the left plate of the capacitor and charges are removed from the right plate, but *no* charge moves across the gap between the plates. Surface S_1 has $I_{through} = I$, but surface S_2 has $I_{through} = 0$. Another dilemma!

It would appear that Ampère's law is either wrong or incomplete. Maxwell was the first to recognize the seriousness of this problem. He noted that there may be no current passing through S_2, but, as **FIGURE 31.15b** shows, there is an electric flux Φ_e through S_2 due to the electric field inside the capacitor. Furthermore, this flux is *changing* with time as the capacitor charges and the electric field strength grows. Faraday had discovered the significance of a changing magnetic flux, but no one had considered a changing electric flux.

The current I passes through S_1, so Ampère's law applied to S_1 gives

$$\oint \vec{B} \cdot d\vec{s} = \mu_0 I_{through} = \mu_0 I$$

We believe this result because it gives the correct magnetic field for a current-carrying wire. Now the line integral depends only on the magnetic field at points on curve C, so its value won't change if we choose a different surface S to evaluate the current. The problem is with the right side of Ampère's law, which would incorrectly give zero if applied to surface S_2. We need to modify the right side of Ampère's law to recognize that an electric flux rather than a current passes through S_2.

The electric flux between two capacitor plates of surface area A is

$$\Phi_e = EA$$

The capacitor's electric field is $E = Q/\epsilon_0 A$; hence the flux is actually independent of the plate size:

$$\Phi_e = \frac{Q}{\epsilon_0 A} A = \frac{Q}{\epsilon_0} \tag{31.18}$$

The *rate* at which the electric flux is changing is

$$\frac{d\Phi_e}{dt} = \frac{1}{\epsilon_0}\frac{dQ}{dt} = \frac{I}{\epsilon_0} \qquad (31.19)$$

where we used $I = dQ/dt$. The flux is changing with time at a rate directly proportional to the charging current I.

Equation 31.19 suggests that the quantity $\epsilon_0(d\Phi_e/dt)$ is in some sense "equivalent" to current I. Maxwell called the quantity

$$I_{\text{disp}} = \epsilon_0\frac{d\Phi_e}{dt} \qquad (31.20)$$

the **displacement current.** He had started with a fluid-like model of electric and magnetic fields, and the displacement current was analogous to the displacement of a fluid. The fluid model has since been abandoned, but the name lives on despite the fact that nothing is actually being displaced.

Maxwell hypothesized that the displacement current was the "missing" piece of Ampère's law, so he modified Ampère's law to read

$$\oint \vec{B} \cdot d\vec{s} = \mu_0(I_{\text{through}} + I_{\text{disp}}) = \mu_0\left(I_{\text{through}} + \epsilon_0\frac{d\Phi_e}{dt}\right) \qquad (31.21)$$

Equation 31.21 is now known as the Ampère-Maxwell law. When applied to Figure 31.15b, the Ampère-Maxwell law gives

$$S_1: \quad \oint \vec{B} \cdot d\vec{s} = \mu_0\left(I_{\text{through}} + \epsilon_0\frac{d\Phi_e}{dt}\right) = \mu_0(I + 0) = \mu_0 I$$

$$S_2: \quad \oint \vec{B} \cdot d\vec{s} = \mu_0\left(I_{\text{through}} + \epsilon_0\frac{d\Phi_e}{dt}\right) = \mu_0(0 + I) = \mu_0 I$$

where, for surface S_2, we used Equation 31.19 for $d\Phi_e/dt$. Surfaces S_1 and S_2 now both give the same result for the line integral of $\vec{B} \cdot d\vec{s}$ around the closed curve C.

> **NOTE** The displacement current I_{disp} between the capacitor plates is numerically equal to the current I in the wires to and from the capacitor, so in some sense it allows "current" to be continuous all the way through the capacitor. Nonetheless, the displacement current is *not* a flow of charge. The displacement current is equivalent to a real current in that it creates the same magnetic field, but it does so with a changing electric flux rather than a flow of charge.

The Induced Magnetic Field

Ordinary Coulomb electric fields are created by charges, but a second way to create an electric field is by having a changing magnetic field. That's Faraday's law. Ordinary magnetic fields are created by currents, but now we see that a second way to create a magnetic field is by having a changing electric field. Just as the electric field created by a changing $\vec{B}$ is called an induced electric field, the magnetic field created by a changing $\vec{E}$ is called an *induced magnetic field.*

FIGURE 31.16 shows the close analogy between induced electric fields, governed by Faraday's law, and induced magnetic fields, governed by the second term in the Ampère-Maxwell law. An increasing solenoid current causes an increasing magnetic field. The changing magnetic field, in turn, induces a circular electric field. The negative sign in Faraday's law dictates that the induced electric field direction is ccw when seen looking along the magnetic field direction.

An increasing capacitor charge causes an increasing electric field. The changing electric field, in turn, induces a circular magnetic field. But the sign of the Ampère-Maxwell law is positive, the opposite of the sign of Faraday's law, so the induced magnetic field direction is cw when you're looking along the electric field direction.

FIGURE 31.16 The close analogy between an induced electric field and an induced magnetic field.

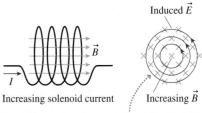

Faraday's law describes an induced electric field.

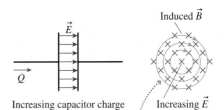

The Ampère-Maxwell law describes an induced magnetic field.

EXAMPLE 31.3 | **The fields inside a charging capacitor**

A 2.0-cm-diameter parallel-plate capacitor with a 1.0 mm spacing is being charged at the rate 0.50 C/s. What is the magnetic field strength inside the capacitor at a point 0.50 cm from the axis?

MODEL The electric field inside a parallel-plate capacitor is uniform. As the capacitor is charged, the changing electric field induces a magnetic field.

VISUALIZE FIGURE 31.17 shows the fields. The induced magnetic field lines are circles concentric with the capacitor.

FIGURE 31.17 The magnetic field strength is found by integrating around a closed curve of radius r.

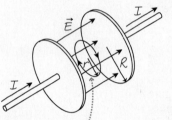

The magnetic field line is a circle concentric with the capacitor. The electric flux through this circle is $\pi r^2 E$.

SOLVE The electric field of a parallel-plate capacitor is $E = Q/\epsilon_0 A = Q/\epsilon_0 \pi R^2$. The electric flux through the circle of radius r (not the full flux of the capacitor) is

$$\Phi_e = \pi r^2 E = \pi r^2 \frac{Q}{\epsilon_0 \pi R^2} = \frac{r^2}{R^2} \frac{Q}{\epsilon_0}$$

Thus the Ampère-Maxwell law is

$$\oint \vec{B} \cdot d\vec{s} = \epsilon_0 \mu_0 \frac{d\Phi_e}{dt} = \epsilon_0 \mu_0 \frac{d}{dt}\left(\frac{r^2}{R^2}\frac{Q}{\epsilon_0}\right) = \mu_0 \frac{r^2}{R^2}\frac{dQ}{dt}$$

The magnetic field is everywhere tangent to the circle of radius r, so the integral of $\vec{B} \cdot d\vec{s}$ around the circle is simply $BL = 2\pi rB$. With this value for the line integral, the Ampère-Maxwell law becomes

$$2\pi rB = \mu_0 \frac{r^2}{R^2}\frac{dQ}{dt}$$

and thus

$$B = \frac{\mu_0}{2\pi}\frac{r}{R^2}\frac{dQ}{dt} = (2.0 \times 10^{-7}\,\text{T m/A})\frac{0.0050\,\text{m}}{(0.010\,\text{m})^2}(0.50\,\text{C/s})$$

$$= 5.0 \times 10^{-6}\,\text{T}$$

ASSESS Charging a capacitor at 0.5 C/s requires a 0.5 A charging current. We've seen many previous examples in which a current-carrying wire with $I \approx 1$ A generates a nearby magnetic field of a few microtesla, so the result seems reasonable.

If a changing magnetic field can induce an electric field and a changing electric field can induce a magnetic field, what happens when both fields change simultaneously? That is the question that Maxwell was finally able to answer after he modified Ampère's law to include the displacement current, and it is the subject to which we turn next.

STOP TO THINK 31.2 The electric field in four identical capacitors is shown as a function of time. Rank in order, from largest to smallest, the magnetic field strength at the outer edge of the capacitor at time T.

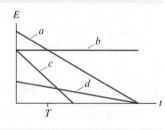

31.4 Maxwell's Equations

James Clerk Maxwell was a young, mathematically brilliant Scottish physicist. In 1855, barely 24 years old, he presented a paper to the Cambridge Philosophical Society entitled "On Faraday's Lines of Force." It had been 30 years and more since the major discoveries of Oersted, Ampère, Faraday, and others, but electromagnetism remained a loose collection of facts and "rules of thumb" without a consistent theory to link these ideas together.

Maxwell's goal was to synthesize this body of knowledge and to form a *theory* of electromagnetic fields. The critical step along the way was his recognition of the need to include a displacement-current term in Ampère's law.

Maxwell's theory of electromagnetism is embodied in four equations that we today call **Maxwell's equations.** These are

$$\oint \vec{E} \cdot d\vec{A} = \frac{Q_{in}}{\epsilon_0}$$ Gauss's law

$$\oint \vec{B} \cdot d\vec{A} = 0$$ Gauss's law for magnetism

$$\oint \vec{E} \cdot d\vec{s} = -\frac{d\Phi_m}{dt}$$ Faraday's law

$$\oint \vec{B} \cdot d\vec{s} = \mu_0 I_{through} + \epsilon_0 \mu_0 \frac{d\Phi_e}{dt}$$ Ampère-Maxwell law

Maxwell's claim is that these four equations are a *complete* description of electric and magnetic fields. They tell us how fields are created by charges and currents, and also how fields can be induced by the changing of other fields. We need one more equation for completeness, an equation that tells us how matter responds to electromagnetic fields. The general force equation

$$\vec{F} = q(\vec{E} + \vec{v} \times \vec{B})$$ (Lorentz force law)

is known as the *Lorentz force law.* **Maxwell's equations for the fields, together with the Lorentz force law to tell us how matter responds to the fields, form the complete theory of electromagnetism.**

Maxwell's equations bring us to the pinnacle of classical physics. When combined with Newton's three laws of motion, his law of gravity, and the first and second laws of thermodynamics, we have all of classical physics—a total of just 11 equations.

While some physicists might quibble over whether all 11 are truly fundamental, the important point is not the exact number but how few equations we need to describe the overwhelming majority of our experience of the physical world. It seems as if we could have written them all on page 1 of this book and been finished, but it doesn't work that way. Each of these equations is the synthesis of a tremendous number of physical phenomena and conceptual developments. To know physics isn't just to know the equations, but to know what the equations *mean* and how they're used. That's why it's taken us so many chapters and so much effort to get to this point. Each equation is a shorthand way to summarize a book's worth of information!

Let's summarize the physical meaning of the five electromagnetic equations:

Classical physics

Newton's first law
Newton's second law
Newton's third law
Newton's law of gravity
Gauss's law
Gauss's law for magnetism
Faraday's law
Ampère-Maxwell law
Lorentz force law
First law of thermodynamics
Second law of thermodynamics

- **Gauss's law:** Charged particles create an electric field.
- **Faraday's law:** An electric field can also be created by a changing magnetic field.
- **Gauss's law for magnetism:** There are no isolated magnetic poles.
- **Ampère-Maxwell law, first half:** Currents create a magnetic field.
- **Ampère-Maxwell law, second half:** A magnetic field can also be created by a changing electric field.
- **Lorentz force law, first half:** An electric force is exerted on a charged particle in an electric field.
- **Lorentz force law, second half:** A magnetic force is exerted on a charge moving in a magnetic field.

These are the *fundamental ideas* of electromagnetism. Other important ideas, such as Ohm's law, Kirchhoff's laws, and Lenz's law, despite their practical importance, are not fundamental ideas. They can be derived from Maxwell's equations, sometimes with the addition of empirically based concepts such as resistance.

It's true that Maxwell's equations are mathematically more complex than Newton's laws and that their solution, for many problems of practical interest, requires advanced mathematics. Fortunately, we have the mathematical tools to get just far enough into Maxwell's equations to discover their most startling and revolutionary implication—the prediction of electromagnetic waves.

31.5 ADVANCED TOPIC Electromagnetic Waves

> **NOTE** This optional section goes through the mathematics of showing that Maxwell's equations predict electromagnetic waves. The key results of this section are summarized at the beginning of Section 31.6, so this section may be omitted with no loss of continuity.

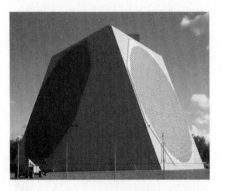

Large radar installations like this one are used to track rockets and missiles.

Maxwell developed his four equations as a mathematical summary of what was known about electricity and magnetism in the mid-19th century: the properties of *static* electric and magnetic fields plus Faraday's discovery of electromagnetic induction. Maxwell introduced the idea of *displacement current*—that a changing electric flux creates a magnetic field—on purely theoretical grounds; there was no experimental evidence at the time. But this new concept was the key to Maxwell's success because it soon allowed him to make the remarkable and totally unexpected prediction of **electromagnetic waves**—self-sustaining oscillations of the electric and magnetic fields that propagate through space without the need for charges or currents.

Our goals in this section are to show that Maxwell's equations lead to a *wave equation* for the electric and magnetic fields and to discover that all electromagnetic waves, regardless of frequency, travel through vacuum at the same speed, a speed we now call the *speed of light*. A completely general derivation of the wave equation is too mathematically advanced for this textbook, so we will make a small number of assumptions—but assumptions that will seem quite reasonable after our study of induced fields.

To begin, we will assume that electric and magnetic fields can exist independently of charges and currents in a *source-free* region of space. This is a very important assumption because it makes the statement that **fields are real entities;** they're not just cute pictures that tell us about charges and currents. The source-free Maxwell's equations, with no charges or currents, are

$$\oint \vec{E} \cdot d\vec{A} = 0 \qquad \oint \vec{E} \cdot d\vec{s} = -\frac{d\Phi_m}{dt}$$

$$\oint \vec{B} \cdot d\vec{A} = 0 \qquad \oint \vec{B} \cdot d\vec{s} = \epsilon_0 \mu_0 \frac{d\Phi_e}{dt} \tag{31.22}$$

Any electromagnetic wave traveling in empty space must be consistent with these equations.

The Structure of Electromagnetic Waves

Faraday discovered that a changing magnetic field creates an induced electric field, and Maxwell's postulated displacement current says that a changing electric field creates an induced magnetic field. The idea behind electromagnetic waves, illustrated in FIGURE 31.18, is that the fields can exist in a self-sustaining mode *if* a changing magnetic field creates an electric field that, in turn, happens to change in just the right way to recreate the original magnetic field. Notice that it has to be an *electromagnetic* wave, with changing electric *and* magnetic fields. A purely electric or purely magnetic wave cannot exist.

You saw in Section 30.6 that an induced electric field, which can drive an induced current around a conducting loop, is *perpendicular* to the changing magnetic field.

FIGURE 31.18 Induced fields can be self-sustaining.

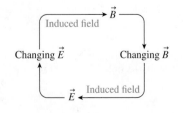

And earlier in this chapter, when we introduced the displacement current, the induced magnetic field in a charging capacitor was *perpendicular* to the changing electric field. Thus we'll make the assumption that $\vec{E}$ **and** $\vec{B}$ **are perpendicular to each other** in an electromagnetic wave. Furthermore—we'll justify this shortly—$\vec{E}$ **and** $\vec{B}$ **are each perpendicular to the direction of travel.** Thus an electromagnetic wave is a *transverse wave,* analogous to a wave on a string, rather than a sound-like longitudinal wave.

We will also assume, to keep the mathematics as simple as possible, that an electromagnetic wave can travel as a *plane wave,* which you will recall from Chapter 16 is a wave for which the fields are same *everywhere* in a plane perpendicular to the direction of travel. FIGURE 31.19a shows an electromagnetic plane wave propagating at speed v_{em} along the x-axis. $\vec{E}$ and $\vec{B}$ are perpendicular to each other, as we've assumed, *and* to the direction of travel. We've defined the y- and z-axes to be, respectively, parallel to $\vec{E}$ and $\vec{B}$. Notice how the fields are the same at every point in a yz-plane slicing the x-axis.

Because a wave is a traveling disturbance, FIGURE 31.19b shows that the fields— at one instant of time—*do* change along the x-axis. These changing fields are the disturbance that is moving down the x-axis at speed v_{em}, so $\vec{E}$ and $\vec{B}$ of a plane wave are functions of the two variables x and t. We're not assuming that the wave has any particular shape—the shape of the wave is what we want to predict from Maxwell's equations—simply that it's a transverse wave moving along the x-axis.

Now that we know something about the structure of the wave, we can start to check its consistency with Maxwell's equations. FIGURE 31.20 shows an imaginary box, a Gaussian surface, centered on the x-axis. Both electric and magnetic field vectors exist at each point in space, but the figure shows them separately for clarity. $\vec{E}$ oscillates along the y-axis, so all electric field lines enter and leave the box through the top and bottom surfaces; no electric field lines pass through the sides of the box.

FIGURE 31.19 An electromagnetic plane wave.

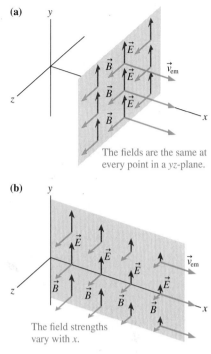

(a)

The fields are the same at every point in a yz-plane.

(b)

The field strengths vary with x.

FIGURE 31.20 A closed surface can be used to check Gauss's law for the electric and magnetic fields.

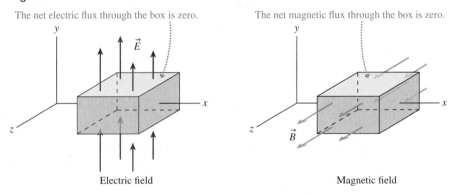

The net electric flux through the box is zero.

The net magnetic flux through the box is zero.

Electric field

Magnetic field

Because this is a plane wave, the magnitude of each electric field vector entering the bottom of the box is exactly matched by an electric field vector leaving the top. The electric flux through the top of the box is equal in magnitude but opposite in sign to the flux through the bottom, and the flux through the sides is zero. Thus the *net* electric flux is $\Phi_e = 0$. There is no charge inside the box, because there are no sources in this region of space, so we also have $Q_{in} = 0$. Hence the electric field of a plane wave is consistent with the first of the source-free Maxwell's equations, Gauss's law.

The exact same argument applies to the magnetic field. The net magnetic flux is $\Phi_m = 0$; thus the magnetic field is consistent with the second of Maxwell's equations.

Suppose that $\vec{E}$ or $\vec{B}$ had a component along the x-axis, the direction of travel. The fields *change* along the x-axis—that's what a traveling wave is—so it would not be possible for the flux through the right face to exactly cancel the flux through the left face at every instant of time. An x-component of either field would violate Gauss's law by creating a net flux when there are no enclosed sources. Thus our claim that an electromagnetic wave must be a transverse wave, with the fields perpendicular to the direction of travel, is a requirement of Gauss's law.

Faraday's Law

Gauss's law tells us that an electromagnetic wave has to be a transverse wave. What does Faraday's law have to say? Faraday's law is concerned with the changing magnetic flux through a closed curve, so let's apply Faraday's law to the narrow rectangle in the xy-plane shown in **FIGURE 31.21**. We'll assume that Δx is so small that $\vec{B}$ is essentially constant over the width of the rectangle.

FIGURE 31.21 Applying Faraday's law.

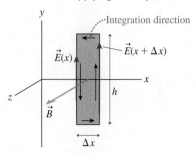

The magnetic field $\vec{B}$ is perpendicular to the rectangle, so the magnetic flux is $\Phi_{\mathrm{m}} = B_z A_{\text{rectangle}} = B_z h \Delta x$. As the wave moves, the flux *changes* at the rate

$$\frac{d\Phi_{\mathrm{m}}}{dt} = \frac{d}{dt}(B_z h \Delta x) = \frac{\partial B_z}{\partial t} h \Delta x \tag{31.23}$$

The ordinary derivative dB_z/dt, which is the full rate of change of B from all possible causes, becomes a partial derivative $\partial B_z/\partial t$ in this situation because the change in magnetic flux is due entirely to the change of B_z with time and not at all to the spatial variation of B_z.

According to our sign convention, we need to go around the rectangle in a counterclockwise direction to make the flux positive. Thus we must also use a counterclockwise direction to evaluate the line integral:

$$\oint \vec{E} \cdot d\vec{s} = \int_{\text{right}} \vec{E} \cdot d\vec{s} + \int_{\text{top}} \vec{E} \cdot d\vec{s} + \int_{\text{left}} \vec{E} \cdot d\vec{s} + \int_{\text{bottom}} \vec{E} \cdot d\vec{s} \tag{31.24}$$

The electric field $\vec{E}$ points in the y-direction; hence $\vec{E} \cdot d\vec{s} = 0$ at all points on the top and bottom edges and these two integrals are zero.

Along the left edge of the loop, at position x, $\vec{E}$ has the same value at every point. Figure 31.21 shows that the direction of $\vec{E}$ is *opposite* to $d\vec{s}$; thus $\vec{E} \cdot d\vec{s} = -E_y(x)ds$. On the right edge of the loop, at position $x + \Delta x$, $\vec{E}$ is *parallel* to $d\vec{s}$, and $\vec{E} \cdot d\vec{s} = E_y(x + \Delta x)ds$. Thus the line integral of $\vec{E} \cdot d\vec{s}$ around the rectangle is

$$\oint \vec{E} \cdot d\vec{s} = -E_y(x)h + E_y(x + \Delta x)h = [E_y(x + \Delta x) - E_y(x)]h \tag{31.25}$$

NOTE $E_y(x)$ indicates that E_y is a function of the position x. It is *not* E_y multiplied by x.

You learned in calculus that the derivative of the function $f(x)$ is

$$\frac{df}{dx} = \lim_{\Delta x \to 0}\left[\frac{f(x + \Delta x) - f(x)}{\Delta x}\right]$$

We've assumed that Δx is very small. If we now let the width of the rectangle go to zero, $\Delta x \to 0$, Equation 31.25 becomes

$$\oint \vec{E} \cdot d\vec{s} = \frac{\partial E_y}{\partial x} h \Delta x \tag{31.26}$$

We've used a partial derivative because E_y is a function of both position x and time t.

Now, using Equations 31.23 and 31.26, we can write Faraday's law as

$$\oint \vec{E} \cdot d\vec{s} = \frac{\partial E_y}{\partial x} h \Delta x = -\frac{d\Phi_{\mathrm{m}}}{dt} = -\frac{\partial B_z}{\partial t} h \Delta x$$

The area $h \Delta x$ of the rectangle cancels, and we're left with

$$\frac{\partial E_y}{\partial x} = -\frac{\partial B_z}{\partial t} \tag{31.27}$$

Equation 31.27, which compares the rate at which E_y varies with position to the rate at which B_z varies with time, is a *required condition* that an electromagnetic wave must satisfy to be consistent with Maxwell's equations.

The Ampère-Maxwell Law

We have only one equation to go, but this one will now be easier. The Ampère-Maxwell law is concerned with the changing electric flux through a closed curve. **FIGURE 31.22** shows a very narrow rectangle in the xz-plane. The electric field is perpendicular to this rectangle; hence the electric flux through it is $\Phi_e = E_y A_{\text{rectangle}} = E_y l \Delta x$. This flux is changing at the rate

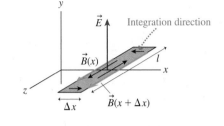

FIGURE 31.22 Applying the Ampère-Maxwell law.

$$\frac{d\Phi_e}{dt} = \frac{d}{dt}(E_y l \Delta x) = \frac{\partial E_y}{\partial t} l \Delta x \tag{31.28}$$

The line integral of $\vec{B} \cdot d\vec{s}$ around this closed rectangle is calculated just like the line integral of $\vec{E} \cdot d\vec{s}$ in Figure 31.21. $\vec{B}$ is perpendicular to $d\vec{s}$ on the narrow ends, so $\vec{B} \cdot d\vec{s} = 0$. The field at *all* points on the left edge is $\vec{B}(x)$, and this field is parallel to $d\vec{s}$ to make $\vec{B} \cdot d\vec{s} = B_z(x)\,ds$. Similarly, $\vec{B} \cdot d\vec{s} = -B_z(x + \Delta x)\,ds$ at all points on the right edge, where $\vec{B}$ is opposite to $d\vec{s}$. Thus, if we let $\Delta x \to 0$,

$$\oint \vec{B} \cdot d\vec{s} = B_z(x)l - B_z(x + \Delta x)l = -[B_z(x + \Delta x) - B_z(x)]l = -\frac{\partial B_z}{\partial x} l \Delta x \tag{31.29}$$

Equations 31.28 and 31.29 can now be used in the Ampère-Maxwell law:

$$\oint \vec{B} \cdot d\vec{s} = -\frac{\partial B_z}{\partial x} l \Delta x = \epsilon_0 \mu_0 \frac{d\Phi_e}{dt} = \epsilon_0 \mu_0 \frac{\partial E_y}{\partial t} l \Delta x$$

The area of the rectangle cancels, and we're left with

$$\frac{\partial B_z}{\partial x} = -\epsilon_0 \mu_0 \frac{\partial E_y}{\partial t} \tag{31.30}$$

Equation 31.30 is a second required condition that the fields must satisfy.

The Wave Equation

In « Section 16.4, during our study of traveling waves, we derived the *wave equation:*

$$\frac{\partial^2 D}{\partial t^2} = v^2 \frac{\partial^2 D}{\partial x^2} \tag{31.31}$$

There we learned that any physical system that obeys this equation for some type of displacement D can have traveling waves that propagate along the x-axis with speed v.

If we start with Equation 31.27, the Faraday's law requirement for any electromagnetic wave, we can take the second derivative with respect to x to find

$$\frac{\partial^2 E_y}{\partial x^2} = -\frac{\partial^2 B_z}{\partial x \partial t} \tag{31.32}$$

You've learned in calculus that the order of differentiation doesn't matter, so $\partial^2 B_z / \partial x \partial t = \partial^2 B_z / \partial t \partial x$. And from Equation 31.30,

$$\frac{\partial^2 B_z}{\partial t \partial x} = -\epsilon_0 \mu_0 \frac{\partial^2 E_y}{\partial t^2} \tag{31.33}$$

Substituting Equation 31.33 into Equation 31.32 and taking the constants to the other side, we have

$$\frac{\partial^2 E_y}{\partial t^2} = \frac{1}{\epsilon_0 \mu_0} \frac{\partial^2 E_y}{\partial x^2} \qquad \text{(the wave equation for electromagnetic waves)} \tag{31.34}$$

Equation 31.34 is the wave equation! And it's easy to show, by taking second derivatives of B_z rather than E_y, that the magnetic field B_z obeys exactly the same wave equation.

As we anticipated, Maxwell's equations have led to a prediction of electromagnetic waves. Referring to the general wave equation, Equation 31.31, we see that an electromagnetic wave must travel (in vacuum) with speed

$$v_{em} = \frac{1}{\sqrt{\epsilon_0 \mu_0}} \qquad (31.35)$$

The constants ϵ_0 and μ_0 are known from electrostatics and magnetostatics, where they determined the size of $\vec{E}$ and $\vec{B}$ due to point charges. Thus we can calculate

$$v_{em} = \frac{1}{\sqrt{\epsilon_0 \mu_0}} = 3.00 \times 10^8 \text{ m/s} = c \qquad (31.36)$$

This is a remarkable conclusion. Coulomb's law and the Biot-Savart law, in which ϵ_0 and μ_0 first appeared, have nothing to do with waves. Yet Maxwell's theory of electromagnetism ends up predicting that electric and magnetic fields can form a self-sustaining electromagnetic wave *if* that wave travels with speed $v_{em} = 1/\sqrt{\epsilon_0 \mu_0}$. No other speed will satisfy Maxwell's equations.

Laboratory measurements had already determined that light travels at 3.0×10^8 m/s, so Maxwell was entirely justified in concluding that light is an electromagnetic wave. Furthermore, we've made no assumption about the frequency of the wave, so apparently electromagnetic waves of any frequency, from radio waves to x rays, travel (in vacuum) with speed c, the speed of light.

Connecting E and B

The electric and magnetic fields of an electromagnetic wave both oscillate, but not independently of each other. The two field strengths are related. E_y and B_z both satisfy the same wave equation, so the traveling waves—just like a wave on a string—are

$$E_y = E_0 \sin(kx - \omega t) = E_0 \sin\left[2\pi\left(\frac{x}{\lambda} - ft\right)\right]$$

$$B_z = B_0 \sin(kx - \omega t) = B_0 \sin\left[2\pi\left(\frac{x}{\lambda} - ft\right)\right] \qquad (31.37)$$

where E_0 and B_0 are the amplitudes of the electric and magnetic portions of the wave and, as for any sinusoidal wave, $k = 2\pi/\lambda$, $\omega = 2\pi f$, and $\lambda f = v = c$. These waves have to satisfy Equation 31.27; thus

$$\frac{\partial E_y}{\partial x} = \frac{2\pi E_0}{\lambda} \cos\left[2\pi\left(\frac{x}{\lambda} - ft\right)\right] = -\frac{\partial B_z}{\partial t} = 2\pi f B_0 \cos\left[2\pi\left(\frac{x}{\lambda} - ft\right)\right] \qquad (31.38)$$

Equation 31.38 is true only if $E_0 = \lambda f B_0 = cB_0$. And since the electric and magnetic fields oscillate together, this relationship between their amplitudes has to hold true at any point on the wave. Thus $E = cB$ **at any point on the wave.**

STOP TO THINK 31.3 An electromagnetic wave is propagating in the positive x-direction. At this instant of time, what is the direction of $\vec{E}$ at the center of the rectangle?

a. In the positive x-direction
b. In the negative x-direction
c. In the positive y-direction
d. In the negative y-direction
e. In the positive z-direction
f. In the negative z-direction

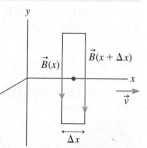

31.6 Properties of Electromagnetic Waves

It had been known since the early 19th century, from experiments with interference and diffraction, that light is a wave, but no one understood what was "waving." Faraday speculated that light was somehow connected to electricity and magnetism, but Maxwell was the first to understand not only that light is an electromagnetic wave but also that electromagnetic waves can exist at any frequency, not just the frequencies of visible light.

In the previous section, we used Maxwell's equations to discover that:

1. Maxwell's equations predict the existence of sinusoidal electromagnetic waves that travel through empty space independent of any charges or currents.
2. The waves are transverse waves, with $\vec{E}$ and $\vec{B}$ perpendicular to the direction of propagation $\vec{v}_{em}$.
3. $\vec{E}$ and $\vec{B}$ are perpendicular to each other in a manner such that $\vec{E} \times \vec{B}$ is in the direction of $\vec{v}_{em}$.
4. All electromagnetic waves, regardless of frequency or wavelength, travel in vacuum at speed $v_{em} = 1/\sqrt{\epsilon_0\mu_0} = c$, the speed of light.
5. The field strengths are related by $E = cB$ at every point on the wave.

FIGURE 31.23 illustrates many of these characteristics of electromagnetic waves. It's a useful picture, and one that you'll see in any textbook, but a picture that can be very misleading if you don't think about it carefully. First and foremost, $\vec{E}$ and $\vec{B}$ are *not* spatial vectors. That is, they don't stretch spatially in the y- or z-direction for a certain distance. Instead, these vectors are showing the field strengths and directions along a single line, the x-axis. An $\vec{E}$ vector pointing in the y-direction is saying, "At this point on the x-axis, where the tail is, this is the direction and strength of the electric field." Nothing is "reaching" to a point in space above the x-axis.

Second, we're assuming this is a *plane wave,* which, you'll recall, is a wave for which the fields are the same *everywhere* in any plane perpendicular to $\vec{v}_{em}$. Figure 31.23 shows the fields only along one line. But whatever the fields are doing at a point on the x-axis, they are doing the same thing everywhere in the yz-plane that slices the x-axis at that point. With this in mind, let's explore some other properties of electromagnetic waves.

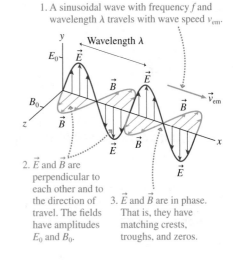

FIGURE 31.23 A sinusoidal electro-magnetic wave.

1. A sinusoidal wave with frequency f and wavelength λ travels with wave speed v_{em}.

2. $\vec{E}$ and $\vec{B}$ are perpendicular to each other and to the direction of travel. The fields have amplitudes E_0 and B_0.

3. $\vec{E}$ and $\vec{B}$ are in phase. That is, they have matching crests, troughs, and zeros.

Energy and Intensity

Waves transfer energy. Ocean waves erode beaches, sound waves set your eardrums vibrating, and light from the sun warms the earth. The energy flow of an electromagnetic wave is described by the **Poynting vector** $\vec{S}$, defined as

$$\vec{S} \equiv \frac{1}{\mu_0}\vec{E} \times \vec{B} \qquad (31.39)$$

The Poynting vector, shown in **FIGURE 31.24**, has two important properties:

1. The Poynting vector points in the direction in which an electromagnetic wave is traveling. You can see this by looking back at Figure 31.23.
2. It is straightforward to show that the units of S are W/m^2, or power (joules per second) per unit area. Thus the magnitude S of the Poynting vector measures the rate of energy transfer per unit area of the wave.

Because $\vec{E}$ and $\vec{B}$ of an electromagnetic wave are perpendicular to each other, and $E = cB$, the magnitude of the Poynting vector is

$$S = \frac{EB}{\mu_0} = \frac{E^2}{c\mu_0} = c\epsilon_0 E^2$$

The Poynting vector is a function of time, oscillating from zero to $S_{max} = E_0^2/c\mu_0$ and back to zero twice during each period of the wave's oscillation. That is, the energy

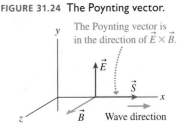

FIGURE 31.24 The Poynting vector.

The Poynting vector is in the direction of $\vec{E} \times \vec{B}$.

flow in an electromagnetic wave is not smooth. It "pulses" as the electric and magnetic fields oscillate in intensity. We're unaware of this pulsing because the electromagnetic waves that we can sense—light waves—have such high frequencies.

Of more interest is the *average* energy transfer, averaged over one cycle of oscillation, which is the wave's **intensity** *I*. In our earlier study of waves, we defined the intensity of a wave to be $I = P/A$, where *P* is the power (energy transferred per second) of a wave that impinges on area *A*. Because $E = E_0 \sin[2\pi(x/\lambda - ft)]$, and the average over one period of $\sin^2[2\pi(x/\lambda - ft)]$ is $\frac{1}{2}$, the intensity of an electromagnetic wave is

$$I = \frac{P}{A} = S_{\text{avg}} = \frac{1}{2c\mu_0}E_0^2 = \frac{c\epsilon_0}{2}E_0^2 \qquad (31.40)$$

Equation 31.40 relates the intensity of an electromagnetic wave, a quantity that is easily measured, to the amplitude E_0 of the wave's electric field.

The intensity of a plane wave, with constant electric field amplitude E_0, would not change with distance. But a plane wave is an idealization; there are no true plane waves in nature. You learned in Chapter 16 that, to conserve energy, the intensity of a wave far from its source decreases with the inverse square of the distance. If a source with power P_{source} emits electromagnetic waves *uniformly* in all directions, the electromagnetic wave intensity at distance *r* from the source is

$$I = \frac{P_{\text{source}}}{4\pi r^2} \qquad (31.41)$$

Equation 31.41 simply expresses the recognition that the energy of the wave is spread over a sphere of surface area $4\pi r^2$.

EXAMPLE 31.4 | Fields of a cell phone

A digital cell phone broadcasts a 0.60 W signal at a frequency of 1.9 GHz. What are the amplitudes of the electric and magnetic fields at a distance of 10 cm, about the distance to the center of the user's brain?

MODEL Treat the cell phone as a point source of electromagnetic waves.

SOLVE The intensity of a 0.60 W point source at a distance of 10 cm is

$$I = \frac{P_{\text{source}}}{4\pi r^2} = \frac{0.60 \text{ W}}{4\pi(0.10 \text{ m})^2} = 4.78 \text{ W/m}^2$$

We can find the electric field amplitude from the intensity:

$$E_0 = \sqrt{\frac{2I}{c\epsilon_0}} = \sqrt{\frac{2(4.78 \text{ W/m}^2)}{(3.00 \times 10^8 \text{ m/s})(8.85 \times 10^{-12} \text{ C}^2/\text{N m}^2)}}$$
$$= 60 \text{ V/m}$$

The amplitudes of the electric and magnetic fields are related by the speed of light. This allows us to compute

$$B_0 = \frac{E_0}{c} = 2.0 \times 10^{-7} \text{ T}$$

ASSESS The electric field amplitude is modest; the magnetic field amplitude is very small. This implies that the interaction of electromagnetic waves with matter is mostly due to the electric field.

STOP TO THINK 31.4 An electromagnetic wave is traveling in the positive *y*-direction. The electric field at one instant of time is shown at one position. The magnetic field at this position points

a. In the positive *x*-direction.
b. In the negative *x*-direction.
c. In the positive *y*-direction.
d. In the negative *y*-direction.
e. Toward the origin.
f. Away from the origin.

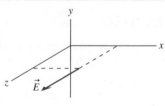

Radiation Pressure

Electromagnetic waves transfer not only energy but also momentum. An object gains momentum when it absorbs electromagnetic waves, much as a ball at rest gains momentum when struck by a ball in motion.

Suppose we shine a beam of light on an object that completely absorbs the light energy. If the object absorbs energy during a time interval Δt, its momentum changes by

$$\Delta p = \frac{\text{energy absorbed}}{c}$$

This is a consequence of Maxwell's theory, which we'll state without proof.

The momentum change implies that the light is exerting a force on the object. Newton's second law, in terms of momentum, is $F = \Delta p/\Delta t$. The radiation force due to the beam of light is

$$F = \frac{\Delta p}{\Delta t} = \frac{(\text{energy absorbed})/\Delta t}{c} = \frac{P}{c}$$

where P is the power (joules per second) of the light.

It's more interesting to consider the force exerted on an object per unit area, which is called the **radiation pressure** p_{rad}. The radiation pressure on an object that absorbs all the light is

$$p_{rad} = \frac{F}{A} = \frac{P/A}{c} = \frac{I}{c} \qquad (31.42)$$

where I is the intensity of the light wave. The subscript on p_{rad} is important in this context to distinguish the radiation pressure from the momentum p.

Artist's conception of a future spacecraft powered by radiation pressure from the sun.

EXAMPLE 31.5 | Solar sailing

A low-cost way of sending spacecraft to other planets would be to use the radiation pressure on a solar sail. The intensity of the sun's electromagnetic radiation at distances near the earth's orbit is about 1300 W/m². What size sail would be needed to accelerate a 10,000 kg spacecraft toward Mars at 0.010 m/s²?

MODEL Assume that the solar sail is perfectly absorbing.

SOLVE The force that will create a 0.010 m/s² acceleration is $F = ma = 100$ N. We can use Equation 31.42 to find the sail

area that, by absorbing light, will receive a 100 N force from the sun:

$$A = \frac{cF}{I} = \frac{(3.00 \times 10^8 \text{ m/s})(100 \text{ N})}{1300 \text{ W/m}^2} = 2.3 \times 10^7 \text{ m}^2$$

ASSESS If the sail is a square, it would need to be 4.8 km × 4.8 km, or roughly 3 mi × 3 mi. This is large, but not entirely out of the question with thin films that can be unrolled in space. But how will the crew return from Mars?

Antennas

We've seen that an electromagnetic wave is self-sustaining, independent of charges or currents. However, charges and currents are needed at the *source* of an electromagnetic wave. We'll take a brief look at how an electromagnetic wave is generated by an antenna.

FIGURE 31.25 is the electric field of an electric dipole. If the dipole is vertical, the electric field $\vec{E}$ at points along a horizontal line is also vertical. Reversing the dipole, by switching the charges, reverses $\vec{E}$. If the charges were to oscillate back and forth,

FIGURE 31.25 An electric dipole creates an electric field that reverses direction if the dipole charges are switched.

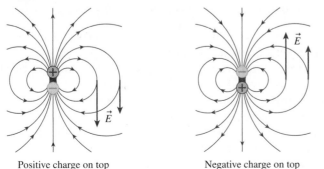

Positive charge on top   Negative charge on top

FIGURE 31.26 An antenna generates a self-sustaining electromagnetic wave.

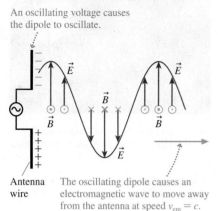

An oscillating voltage causes the dipole to oscillate.

Antenna wire

The oscillating dipole causes an electromagnetic wave to move away from the antenna at speed $v_{em} = c$.

switching position at frequency f, then $\vec{E}$ would oscillate in a vertical plane. The changing $\vec{E}$ would then create an induced magnetic field $\vec{B}$, which could then create an $\vec{E}$, which could then create a $\vec{B}$, . . . , and an electromagnetic wave at frequency f would radiate out into space.

This is exactly what an **antenna** does. **FIGURE 31.26** shows two metal wires attached to the terminals of an oscillating voltage source. The figure shows an instant when the top wire is negative and the bottom is positive, but these will reverse in half a cycle. The wire is basically an oscillating dipole, and it creates an oscillating electric field. The oscillating $\vec{E}$ induces an oscillating $\vec{B}$, and they take off as an electromagnetic wave at speed $v_{em} = c$. The wave does need oscillating charges as a *wave source,* but once created it is self-sustaining and independent of the source. The antenna might be destroyed, but the wave could travel billions of light years across the universe, bearing the legacy of James Clerk Maxwell.

STOP TO THINK 31.5 The amplitude of the oscillating electric field at your cell phone is 4.0 μV/m when you are 10 km east of the broadcast antenna. What is the electric field amplitude when you are 20 km east of the antenna?

a. 1.0 μV/m

b. 2.0 μV/m

c. 4.0 μV/m

d. There's not enough information to tell.

31.7 Polarization

The plane of the electric field vector $\vec{E}$ and the Poynting vector $\vec{S}$ (the direction of propagation) is called the **plane of polarization** of an electromagnetic wave. **FIGURE 31.27** shows two electromagnetic waves moving along the x-axis. The electric field in Figure 31.27a oscillates vertically, so we would say that this wave is *vertically polarized.* Similarly the wave in Figure 31.27b is *horizontally polarized.* Other polarizations are possible, such as a wave polarized 30° away from horizontal.

FIGURE 31.27 The plane of polarization is the plane in which the electric field vector oscillates.

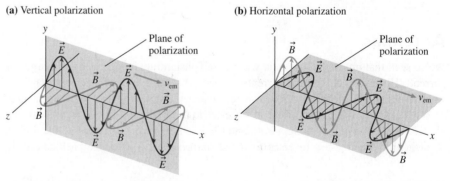

(a) Vertical polarization

(b) Horizontal polarization

NOTE This use of the term "polarization" is completely independent of the idea of *charge polarization* that you learned about in Chapter 22.

Some wave sources, such as lasers and radio antennas, emit *polarized* electromagnetic waves with a well-defined plane of polarization. By contrast, most natural sources of electromagnetic radiation are unpolarized, emitting waves whose electric fields oscillate randomly with all possible orientations.

A few natural sources are *partially polarized,* meaning that one direction of polarization is more prominent than others. The light of the sky at right angles to the sun is partially polarized because of how the sun's light scatters from air molecules to create skylight. Bees and other insects make use of this partial polarization to navigate. Light reflected from a flat, horizontal surface, such as a road or the surface of a lake, has a predominantly horizontal polarization. This is the rationale for using polarizing sunglasses.

The most common way of artificially generating polarized visible light is to send unpolarized light through a *polarizing filter.* The first widely used polarizing filter was invented by Edwin Land in 1928, while he was still an undergraduate student. He developed an improved version, called Polaroid, in 1938. Polaroid, as shown in **FIGURE 31.28,** is a plastic sheet containing very long organic molecules known as polymers. The sheets are formed in such a way that the polymers are all aligned to form a grid, rather like the metal bars in a barbecue grill. The sheet is then chemically treated to make the polymer molecules somewhat conducting.

As a light wave travels through Polaroid, the component of the electric field oscillating parallel to the polymer grid drives the conduction electrons up and down the molecules. The electrons absorb energy from the light wave, so the parallel component of $\vec{E}$ is absorbed in the filter. But the conduction electrons can't oscillate perpendicular to the molecules, so the component of $\vec{E}$ perpendicular to the polymer grid passes through without absorption. Thus the light wave emerging from a polarizing filter is polarized perpendicular to the polymer grid. The direction of the transmitted polarization is called the *polarizer axis.*

FIGURE 31.28 A polarizing filter.

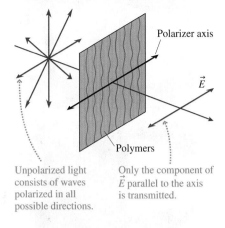

Unpolarized light consists of waves polarized in all possible directions.

Only the component of $\vec{E}$ parallel to the axis is transmitted.

Malus's Law

Suppose a *polarized* light wave of intensity I_0 approaches a polarizing filter. What is the intensity of the light that passes through the filter? **FIGURE 31.29** shows that an oscillating electric field can be decomposed into components parallel and perpendicular to the polarizer axis. If we call the polarizer axis the *y*-axis, then the incident electric field is

$$\vec{E}_{\text{incident}} = E_\perp \hat{\imath} + E_\| \hat{\jmath} = E_0 \sin\theta\,\hat{\imath} + E_0 \cos\theta\,\hat{\jmath} \qquad (31.43)$$

where θ is the angle between the incident plane of polarization and the polarizer axis.

If the polarizer is ideal, meaning that light polarized parallel to the axis is 100% transmitted and light perpendicular to the axis is 100% blocked, then the electric field of the light transmitted by the filter is

$$\vec{E}_{\text{transmitted}} = E_\| \hat{\jmath} = E_0 \cos\theta\,\hat{\jmath} \qquad (31.44)$$

Because the intensity depends on the square of the electric field amplitude, you can see that the transmitted intensity is related to the incident intensity by

$$I_{\text{transmitted}} = I_0 \cos^2\theta \qquad \text{(incident light polarized)} \qquad (31.45)$$

This result, which was discovered experimentally in 1809, is called **Malus's law.**

FIGURE 31.30a shows that Malus's law can be demonstrated with two polarizing filters. The first, called the *polarizer,* is used to produce polarized light of intensity I_0. The second, called the *analyzer,* is rotated by angle θ relative to the polarizer. As the photographs of **FIGURE 31.30b** show, the transmission of the analyzer is (ideally) 100% when $\theta = 0°$ and steadily decreases to zero when $\theta = 90°$. Two polarizing filters with perpendicular axes, called *crossed polarizers,* block all the light.

FIGURE 31.29 An incident electric field can be decomposed into components parallel and perpendicular to a polarizer axis.

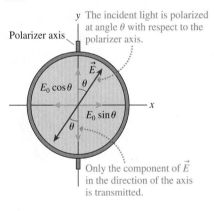

The incident light is polarized at angle θ with respect to the polarizer axis.

Only the component of $\vec{E}$ in the direction of the axis is transmitted.

FIGURE 31.30 The intensity of the transmitted light depends on the angle between the polarizing filters.

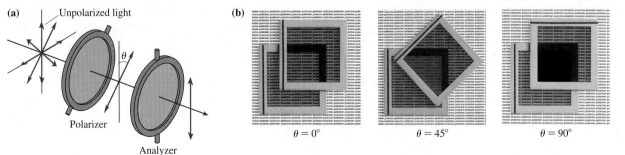

(a) Unpolarized light

Polarizer

Analyzer

(b)

$\theta = 0°$ $\theta = 45°$ $\theta = 90°$

The vertical polarizer blocks the horizontally polarized glare from the surface of the water.

Suppose the light incident on a polarizing filter is *unpolarized,* as is the light incident from the left on the polarizer in Figure 31.30a. The electric field of unpolarized light varies randomly through all possible values of θ. Because the *average* value of $\cos^2\theta$ is $\frac{1}{2}$, the intensity transmitted by a polarizing filter is

$$I_{\text{transmitted}} = \tfrac{1}{2}I_0 \qquad \text{(incident light unpolarized)} \qquad (31.46)$$

In other words, a polarizing filter passes 50% of unpolarized light and blocks 50%.

In polarizing sunglasses, the polymer grid is aligned horizontally (when the glasses are in the normal orientation) so that the glasses transmit vertically polarized light. Most natural light is unpolarized, so the glasses reduce the light intensity by 50%. But *glare*—the reflection of the sun and the skylight from roads and other horizontal surfaces—has a strong horizontal polarization. This light is almost completely blocked by the Polaroid, so the sunglasses "cut glare" without affecting the main scene you wish to see.

You can test whether your sunglasses are polarized by holding them in front of you and rotating them as you look at the glare reflecting from a horizontal surface. Polarizing sunglasses substantially reduce the glare when the glasses are "normal" but not when the glasses are 90° from normal. (You can also test them against a pair of sunglasses known to be polarizing by seeing if all light is blocked when the lenses of the two pairs are crossed.)

STOP TO THINK 31.6 Unpolarized light of equal intensity is incident on four pairs of polarizing filters. Rank in order, from largest to smallest, the intensities I_a to I_d transmitted through the second polarizer of each pair.

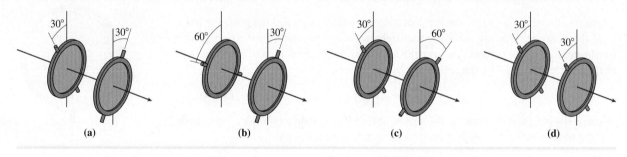

(a)　　　　　(b)　　　　　(c)　　　　　(d)

CHALLENGE EXAMPLE 31.6 | Light propulsion

Future space rockets might propel themselves by firing laser beams, rather than exhaust gases, out the back. The acceleration would be small, but it could continue for months or years in the vacuum of space. Consider a 1200 kg unmanned space probe powered by a 15 MW laser. After one year, how far will it have traveled and how fast will it be going?

MODEL Assume the laser efficiency is so high that it can be powered for a year with a negligible mass of fuel.

SOLVE Light waves transfer not only energy but also momentum, which is how they exert a radiation-pressure force. We found that the radiation force of a light beam of power P is

$$F = \frac{P}{c}$$

From Newton's third law, the emitted light waves must exert an equal-but-opposite reaction force on the source of the light. In this case, the emitted light exerts a force of this magnitude on the space

probe to which the laser is attached. This reaction force causes the probe to accelerate at

$$a = \frac{F}{m} = \frac{P}{mc} = \frac{15 \times 10^6 \text{ W}}{(1200 \text{ kg})(3.0 \times 10^8 \text{ m/s})}$$
$$= 4.2 \times 10^{-5} \text{ m/s}^2$$

As expected, the acceleration is extremely small. But one year is a large amount of time: $\Delta t = 3.15 \times 10^7$ s. After one year of acceleration,

$$v = a\,\Delta t = 1300 \text{ m/s}$$
$$d = \tfrac{1}{2}a(\Delta t)^2 = 2.1 \times 10^{10} \text{ m}$$

The space probe will have traveled 2.1×10^{10} m and will be going 1300 m/s.

ASSESS Even after a year, the speed is not exceptionally fast— only about 2900 mph. But the probe will have traveled a substantial distance, about 25% of the distance to Mars.

SUMMARY

The goal of Chapter 31 has been to study the properties of electromagnetic fields and waves.

GENERAL PRINCIPLES

Maxwell's Equations

These equations govern electromagnetic fields:

$$\oint \vec{E} \cdot d\vec{A} = \frac{Q_{in}}{\epsilon_0} \qquad \text{Gauss's law}$$

$$\oint \vec{B} \cdot d\vec{A} = 0 \qquad \text{Gauss's law for magnetism}$$

$$\oint \vec{E} \cdot d\vec{s} = -\frac{d\Phi_m}{dt} \qquad \text{Faraday's law}$$

$$\oint \vec{B} \cdot d\vec{s} = \mu_0 I_{through} + \epsilon_0 \mu_0 \frac{d\Phi_e}{dt} \qquad \textit{Ampère-Maxwell law}$$

Maxwell's equations tell us that:

An electric field can be created by

- Charged particles
- A changing magnetic field

A magnetic field can be created by

- A current
- A changing electric field

Lorentz Force

This force law governs the interaction of charged particles with electromagnetic fields:

$$\vec{F} = q(\vec{E} + \vec{v} \times \vec{B})$$

- An electric field exerts a force on any charged particle.
- A magnetic field exerts a force on a moving charged particle.

Field Transformations

Fields measured in reference frame A to be $\vec{E}_A$ and $\vec{B}_A$ are found in frame B to be

$$\vec{E}_B = \vec{E}_A + \vec{v}_{BA} \times \vec{B}_A$$

$$\vec{B}_B = \vec{B}_A - \frac{1}{c^2} \vec{v}_{BA} \times \vec{E}_A$$

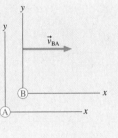

IMPORTANT CONCEPTS

Induced fields

An induced electric field is created by a changing magnetic field.

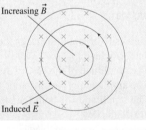

An induced magnetic field is created by a changing electric field.

These fields can exist independently of charges and currents.

An **electromagnetic wave** is a self-sustaining electromagnetic field.

- An em wave is a transverse wave with $\vec{E}$, $\vec{B}$, and $\vec{v}_{em}$ mutually perpendicular.
- An em wave propagates with speed $v_{em} = c = 1/\sqrt{\epsilon_0 \mu_0}$.
- The electric and magnetic field strengths are related by $E = cB$.
- The **Poynting vector** $\vec{S} = (\vec{E} \times \vec{B})/\mu_0$ is the energy transfer in the direction of travel.
- The wave **intensity** is $I = P/A = (1/2c\mu_0)E_0^2 = (c\epsilon_0/2)E_0^2$.

APPLICATIONS

Polarization

The electric field and the Poynting vector define the **plane of polarization.** The intensity of polarized light transmitted through a polarizing filter is given by Malus's law:

$$I = I_0 \cos^2 \theta$$

where θ is the angle between the electric field and the polarizer axis.

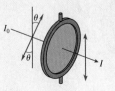

TERMS AND NOTATION

Galilean field transformation equations	Maxwell's equations	intensity, I	plane of polarization
	electromagnetic wave	radiation pressure, p_{rad}	Malus's law
displacement current	Poynting vector, $\vec{S}$	antenna	

CONCEPTUAL QUESTIONS

1. Andre is flying his spaceship to the left through the laboratory magnetic field of FIGURE Q31.1.
   a. Does Andre see a magnetic field? If so, in which direction does it point?
   b. Does Andre see an electric field? If so, in which direction does it point?

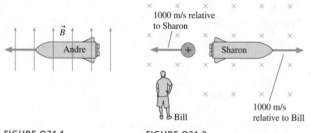

FIGURE Q31.1   FIGURE Q31.2

2. Sharon drives her rocket through the magnetic field of FIGURE Q31.2 traveling to the right at a speed of 1000 m/s as measured by Bill. As she passes Bill, she shoots a positive charge backward at a speed of 1000 m/s relative to her.
   a. According to Bill, what kind of force or forces act on the charge? In which directions? Explain.
   b. According to Sharon, what kind of force or forces act on the charge? In which directions? Explain.

3. If you curl the fingers of your right hand as shown, are the electric fluxes in FIGURE Q31.3 positive or negative?

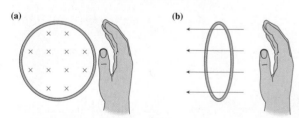

(a)   (b)

FIGURE Q31.3

4. What is the current through surface S in FIGURE Q31.4 if you curl your right fingers in the direction of the arrow?

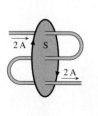

FIGURE Q31.4

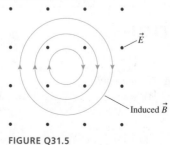

FIGURE Q31.5

5. Is the electric field strength in FIGURE Q31.5 increasing, decreasing, or not changing? Explain.
6. Do the situations in FIGURE Q31.6 represent possible electromagnetic waves? If not, why not?

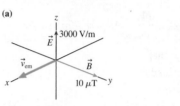

(a)   (b)

FIGURE Q31.6

7. In what directions are the electromagnetic waves traveling in FIGURE Q31.7 ?

(a)   (b)

FIGURE Q31.7

8. The intensity of an electromagnetic wave is 10 W/m^2. What will the intensity be if:
   a. The amplitude of the electric field is doubled?
   b. The amplitude of the magnetic field is doubled?
   c. The amplitudes of both the electric and the magnetic fields are doubled?
   d. The frequency is doubled?
9. Older televisions used a *loop antenna* like the one in FIGURE Q31.9. How does this antenna work?

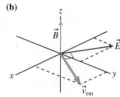

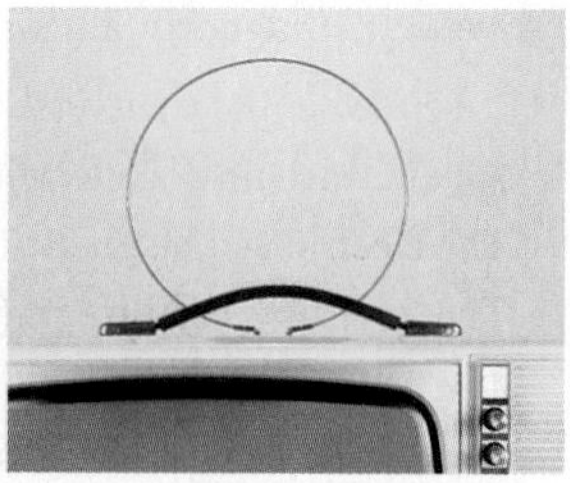

FIGURE Q31.9

10. A vertically polarized electromagnetic wave passes through the five polarizers in FIGURE Q31.10. Rank in order, from largest to smallest, the transmitted intensities I_a to I_e.

a   b   c   d   e

FIGURE Q31.10

EXERCISES AND PROBLEMS

Problems labeled [gray] integrate material from earlier chapters.

Exercises

Section 31.1 *E* or *B*? It Depends on Your Perspective

1. | **FIGURE EX31.1** shows the electric and magnetic field in frame A. A rocket in frame B travels parallel to one of the axes of the A coordinate system. Along which axis must the rocket travel, and in which direction, in order for the rocket scientists to measure (a) $B_B > B_A$, (b) $B_B = B_A$, and (c) $B_B < B_A$?

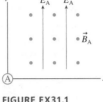

FIGURE EX31.1

2. || A rocket cruises past a laboratory at 1.00×10^6 m/s in the positive *x*-direction just as a proton is launched with velocity (in the laboratory frame) $\vec{v} = (1.41 \times 10^6 \hat{\imath} + 1.41 \times 10^6 \hat{\jmath})$ m/s. What are the proton's speed and its angle from the *y*-axis in (a) the laboratory frame and (b) the rocket frame?

3. || Laboratory scientists have created the electric and magnetic fields shown in **FIGURE EX31.3**. These fields are also seen by scientists who zoom past in a rocket traveling in the *x*-direction at 1.0×10^6 m/s. According to the rocket scientists, what angle does the electric field make with the axis of the rocket?

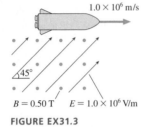

FIGURE EX31.3

4. || Scientists in the laboratory create a uniform electric field $\vec{E} = 1.0 \times 10^6 \hat{k}$ V/m in a region of space where $\vec{B} = \vec{0}$. What are the fields in the reference frame of a rocket traveling in the positive *x*-direction at 1.0×10^6 m/s?

5. || A rocket zooms past the earth at $v = 2.0 \times 10^6$ m/s. Scientists on the rocket have created the electric and magnetic fields shown in **FIGURE EX31.5**. What are the fields measured by an earthbound scientist?

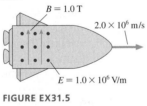

FIGURE EX31.5

Section 31.2 The Field Laws Thus Far

Section 31.3 The Displacement Current

6. || The magnetic field is uniform over each face of the box shown in **FIGURE EX31.6**. What are the magnetic field strength and direction on the front surface?

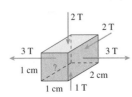

FIGURE EX31.6

7. | Show that the quantity $\epsilon_0(d\Phi_e/dt)$ has units of current.

8. || Show that the displacement current inside a parallel-plate capacitor can be written $C(dV_C/dt)$.

9. | What capacitance, in μF, has its potential difference increasing at 1.0×10^6 V/s when the displacement current in the capacitor is 1.0 A?

10. || A 5.0-cm-diameter parallel-plate capacitor has a 0.50 mm gap. What is the displacement current in the capacitor if the potential difference across the capacitor is increasing at 500,000 V/s?

11. ||| A 10-cm-diameter parallel-plate capacitor has a 1.0 mm spacing. The electric field between the plates is increasing at the rate 1.0×10^6 V/m s. What is the magnetic field strength (a) on the axis, (b) 3.0 cm from the axis, and (c) 7.0 cm from the axis?

Section 31.5 Electromagnetic Waves

12. | What is the magnetic field amplitude of an electromagnetic wave whose electric field amplitude is 10 V/m?

13. | What is the electric field amplitude of an electromagnetic wave whose magnetic field amplitude is 2.0 mT?

14. | The magnetic field of an electromagnetic wave in a vacuum is $B_z = (3.00 \, \mu\text{T}) \sin[(1.00 \times 10^7)x - \omega t]$, where *x* is in m and *t* is in s. What are the wave's (a) wavelength, (b) frequency, and (c) electric field amplitude?

15. | The electric field of an electromagnetic wave in a vacuum is $E_y = (20.0 \text{ V/m}) \cos[(6.28 \times 10^8)x - \omega t]$, where *x* is in m and *t* is in s. What are the wave's (a) wavelength, (b) frequency, and (c) magnetic field amplitude?

Section 31.6 Properties of Electromagnetic Waves

16. | a. What is the magnetic field amplitude of an electromagnetic wave whose electric field amplitude is 100 V/m?
    b. What is the intensity of the wave?

17. | A radio wave is traveling in the negative *y*-direction. What is the direction of $\vec{E}$ at a point where $\vec{B}$ is in the positive *x*-direction?

18. || A radio receiver can detect signals with electric field amplitudes as small as 300 μV/m. What is the intensity of the smallest detectable signal?

19. || A helium-neon laser emits a 1.0-mm-diameter laser beam with a power of 1.0 mW. What are the amplitudes of the electric and magnetic fields of the light wave?

20. || A radio antenna broadcasts a 1.0 MHz radio wave with 25 kW of power. Assume that the radiation is emitted uniformly in all directions.
    a. What is the wave's intensity 30 km from the antenna?
    b. What is the electric field amplitude at this distance?

21. || A 200 MW laser pulse is focused with a lens to a diameter of 2.0 μm.
    a. What is the laser beam's electric field amplitude at the focal point?
    b. What is the ratio of the laser beam's electric field to the electric field that keeps the electron bound to the proton of a hydrogen atom? The radius of the electron orbit is 0.053 nm.

22. | A 1000 W carbon-dioxide laser emits light with a wavelength of 10 μm into a 3.0-mm-diameter laser beam. What force does the laser beam exert on a completely absorbing target?

23. ‖ At what distance from a 10 W point source of electromagnetic waves is the magnetic field amplitude 1.0 μT?

Section 31.7 Polarization

24. ‖ Only 25% of the intensity of a polarized light wave passes through a polarizing filter. What is the angle between the electric field and the axis of the filter?

25. ‖ FIGURE EX31.25 shows a vertically polarized radio wave of frequency 1.0×10^6 Hz traveling into the page. The maximum electric field strength is 1000 V/m. What are

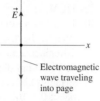

FIGURE EX31.25

   a. The maximum magnetic field strength?
   b. The magnetic field strength and direction at a point where $\vec{E} = (500$ V/m, down$)$?

26. ‖ Unpolarized light with intensity 350 W/m² passes first through a polarizing filter with its axis vertical, then through a second polarizing filter. It emerges from the second filter with intensity 131 W/m². What is the angle from vertical of the axis of the second polarizing filter?

27. ‖ A 200 mW vertically polarized laser beam passes through a polarizing filter whose axis is 35° from horizontal. What is the power of the laser beam as it emerges from the filter?

Problems

28. ‖ What are the electric field strength and direction at the position of the electron in FIGURE P31.28 ?

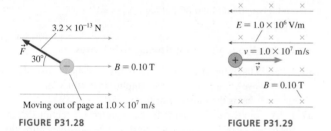

FIGURE P31.28   FIGURE P31.29

29. ‖ What is the force (magnitude and direction) on the proton in FIGURE P31.29? Give the direction as an angle cw or ccw from vertical.

30. ‖ What electric field strength and direction will allow the proton in FIGURE P31.30 to pass through this region of space without being deflected?

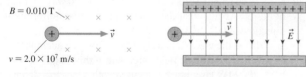

FIGURE P31.30   FIGURE P31.31

31. ‖ A proton is fired with a speed of 1.0×10^6 m/s through the parallel-plate capacitor shown in FIGURE P31.31. The capacitor's electric field is $\vec{E} = (1.0 \times 10^5$ V/m, down$)$.
   a. What magnetic field $\vec{B}$, both strength and direction, must be applied to allow the proton to pass through the capacitor with no change in speed or direction?
   b. Find the electric and magnetic fields in the proton's reference frame.

c. How does an experimenter in the proton's frame explain that the proton experiences no force as the charged plates fly by?

32. ‖ An electron travels with $\vec{v} = 5.0 \times 10^6\,\hat{\imath}$ m/s through a point in space where $\vec{E} = (2.0 \times 10^5\,\hat{\imath} - 2.0 \times 10^5\,\hat{\jmath})$ V/m and $\vec{B} = -0.10\,\hat{k}$ T. What is the force on the electron?

33. ‖ In FIGURE P31.33, a circular loop of radius r travels with speed v along a charged wire having linear charge density λ. The wire is at rest in the laboratory frame, and it passes through the center of the loop.
   a. What are $\vec{E}$ and $\vec{B}$ at a point on the loop as measured by a scientist in the laboratory? Include both strength and direction.
   b. What are the fields $\vec{E}$ and $\vec{B}$ at a point on the loop as measured by a scientist in the frame of the loop?
   c. Show that an experimenter in the loop's frame sees a current $I = \lambda v$ passing through the center of the loop.
   d. What electric and magnetic fields would an experimenter in the loop's frame calculate at distance r from the current of part c?
   e. Show that your fields of parts b and d are the same.

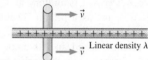

FIGURE P31.33

34. ‖ The magnetic field inside a 4.0-cm-diameter superconducting solenoid varies sinusoidally between 8.0 T and 12.0 T at a frequency of 10 Hz.
   a. What is the maximum electric field strength at a point 1.5 cm from the solenoid axis?
   b. What is the value of B at the instant E reaches its maximum value?

35. ‖ A simple series circuit consists of a 150 Ω resistor, a 25 V battery, a switch, and a 2.5 pF parallel-plate capacitor (initially uncharged) with plates 5.0 mm apart. The switch is closed at $t = 0$ s.
   a. After the switch is closed, find the maximum electric flux and the maximum displacement current through the capacitor.
   b. Find the electric flux and the displacement current at $t = 0.50$ ns.

36. ‖ A wire with conductivity σ carries current I. The current is increasing at the rate dI/dt.
   a. Show that there is a displacement current in the wire equal to $(\epsilon_0/\sigma)(dI/dt)$.
   b. Evaluate the displacement current for a copper wire in which the current is increasing at 1.0×10^6 A/s.

37. ‖ A 10 A current is charging a 1.0-cm-diameter parallel-plate capacitor.
   a. What is the magnetic field strength at a point 2.0 mm radially from the center of the wire leading to the capacitor?
   b. What is the magnetic field strength at a point 2.0 mm radially from the center of the capacitor?

38. ‖ FIGURE P31.38 shows the voltage across a 0.10 μF capacitor. Draw a graph showing the displacement current through the capacitor as a function of time.

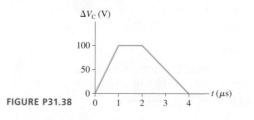

FIGURE P31.38

39. ‖ **FIGURE P31.39** shows the electric field inside a cylinder of radius $R = 3.0$ mm. The field strength is increasing with time as $E = 1.0 \times 10^8 t^2$ V/m, where t is in s. The electric field outside the cylinder is always zero, and the field inside the cylinder was zero for $t < 0$.

FIGURE P31.39

   a. Find an expression for the electric flux Φ_e through the entire cylinder as a function of time.
   b. Draw a picture showing the magnetic field lines inside and outside the cylinder. Be sure to include arrowheads showing the field's direction.
   c. Find an expression for the magnetic field strength as a function of time at a distance $r < R$ from the center. Evaluate the magnetic field strength at $r = 2.0$ mm, $t = 2.0$ s.
   d. Find an expression for the magnetic field strength as a function of time at a distance $r > R$ from the center. Evaluate the magnetic field strength at $r = 4.0$ mm, $t = 2.0$ s.

40. ‖ A long, thin superconducting wire carrying a 15 A current passes through the center of a thin, 2.0-cm-diameter ring. A uniform electric field of increasing strength also passes through the ring, parallel to the wire. The magnetic field through the ring is zero.
   a. At what rate is the electric field strength increasing?
   b. Is the electric field in the direction of the current or opposite to the current?

41. ‖ A $1.0\,\mu$F capacitor is discharged, starting at $t = 0$ s. The displacement current through the plates is $I_{disp} = (10\ \text{A})\exp(-t/2.0\,\mu\text{s})$. What was the capacitor's initial voltage $(\Delta V_C)_0$?

42. ‖‖ At one instant, the electric and magnetic fields at one point of an electromagnetic wave are $\vec{E} = (200\,\hat{\imath} + 300\,\hat{\jmath} - 50\,\hat{k})$ V/m and $\vec{B} = B_0(7.3\,\hat{\imath} - 7.3\,\hat{\jmath} + a\,\hat{k})\,\mu$T.
   a. What are the values of a and B_0?
   b. What is the Poynting vector at this time and position?

43. ‖ a. Show that u_E and u_B, the energy densities of the electric and magnetic fields, are equal to each other in an electromagnetic wave. In other words, show that the wave's energy is divided equally between the electric field and the magnetic field.
   b. What is the total energy density in an electromagnetic wave of intensity 1000 W/m²?

44. ‖ The intensity of sunlight reaching the earth is 1360 W/m².
   a. What is the power output of the sun?
   b. What is the intensity of sunlight reaching Mars?

45. ‖ Assume that a 7.0-cm-diameter, 100 W lightbulb radiates all its energy as a single wavelength of visible light. Estimate the electric and magnetic field strengths at the surface of the bulb.

46. ‖ The electric field of a 450 MHz radio wave has a maximum rate of change of 4.5×10^{11} (V/m)/s. What is the wave's magnetic field amplitude?

47. ‖ When the Voyager 2 spacecraft passed Neptune in 1989, it was 4.5×10^9 km from the earth. Its radio transmitter, with which it sent back data and images, broadcast with a mere 21 W of power. Assuming that the transmitter broadcast equally in all directions,
   a. What signal intensity was received on the earth?
   b. What electric field amplitude was detected?
   The received signal was somewhat stronger than your result because the spacecraft used a directional antenna, but not by much.

48. ‖ In reading the instruction manual that came with your garage-door opener, you see that the transmitter unit in your car produces a 250 mW, 350 MHz signal and that the receiver unit is supposed to respond to a radio wave at this frequency if the electric field amplitude exceeds 0.10 V/m. You wonder if this is really true. To find out, you put fresh batteries in the transmitter and start walking away from your garage while opening and closing the door. Your garage door finally fails to respond when you're 42 m away. What is the electric field amplitude at the receiver when it first fails?

49. ‖ The maximum electric field strength in air is 3.0 MV/m. Stronger electric fields ionize the air and create a spark. What is the maximum power that can be delivered by a 1.0-cm-diameter laser beam propagating through air?

50. ‖‖ A LASIK vision-correction system uses a laser that emits
    BIO 10-ns-long pulses of light, each with 2.5 mJ of energy. The laser beam is focused to a 0.85-mm-diameter circle on the cornea. What is the electric field amplitude of the light wave at the cornea?

51. ‖ The intensity of sunlight reaching the earth is 1360 W/m². Assuming all the sunlight is absorbed, what is the radiation-pressure force on the earth? Give your answer (a) in newtons and (b) as a fraction of the sun's gravitational force on the earth.

52. ‖ For radio and microwaves, the depth of penetration into
    BIO the human body is proportional to $\lambda^{1/2}$. If 27 MHz radio waves penetrate to a depth of 14 cm, how far do 2.4 GHz microwaves penetrate?

53. ‖ A laser beam shines straight up onto a flat, black foil of mass m.
   a. Find an expression for the laser power P needed to levitate the foil.
   b. Evaluate P for a foil with a mass of 25 μg.

54. ‖ For a science project, you would like to horizontally suspend an 8.5 by 11 inch sheet of black paper in a vertical beam of light whose dimensions exactly match the paper. If the mass of the sheet is 1.0 g, what light intensity will you need?

55. ‖ You've recently read about a chemical laser that generates a 20-cm-diameter, 25 MW laser beam. One day, after physics class, you start to wonder if you could use the radiation pressure from this laser beam to launch small payloads into orbit. To see if this might be feasible, you do a quick calculation of the acceleration of a 20-cm-diameter, 100 kg, perfectly absorbing block. What speed would such a block have if pushed *horizontally* 100 m along a frictionless track by such a laser?

56. ‖ Unpolarized light of intensity I_0 is incident on three polarizing filters. The axis of the first is vertical, that of the second is 45° from vertical, and that of the third is horizontal. What light intensity emerges from the third filter?

57. ‖ Unpolarized light of intensity I_0 is incident on two polarizing filters. The transmitted light intensity is $I_0/10$. What is the angle between the axes of the two filters?

58. ‖ Unpolarized light of intensity I_0 is incident on a stack of 7 polarizing filters, each with its axis rotated 15° cw with respect to the previous filter. What light intensity emerges from the last filter?

Challenge Problems

59. ‖‖ A cube of water 10 cm on a side is placed in a microwave beam having $E_0 = 11$ kV/m. The microwaves illuminate one face of the cube, and the water absorbs 80% of the incident energy. How long will it take to raise the water temperature by 50°C? Assume that the water has no heat loss during this time.

60. ||| An 80 kg astronaut has gone outside his space capsule to do some repair work. Unfortunately, he forgot to lock his safety tether in place, and he has drifted 5.0 m away from the capsule. Fortunately, he has a 1000 W portable laser with fresh batteries that will operate it for 1.0 h. His only chance is to accelerate himself toward the space capsule by firing the laser in the opposite direction. He has a 10-h supply of oxygen. How long will it take him to reach safety?

61. ||| An electron travels with $\vec{v} = 5.0 \times 10^6 \,\hat{\imath}$ m/s through a point in space where $\vec{B} = 0.10 \,\hat{\jmath}$ T. The force on the electron at this point is $\vec{F} = (9.6 \times 10^{-14}\,\hat{\imath} - 9.6 \times 10^{-14}\,\hat{k})$ N. What is the electric field?

62. ||| The radar system at an airport broadcasts 11 GHz microwaves with 150 kW of power. An approaching airplane with a 31 m² cross section is 30 km away. Assume that the radar broadcasts uniformly in all directions and that the airplane scatters microwaves uniformly in all directions. What is the electric field strength of the microwave signal received back at the airport 200 μs later?

63. ||| Large quantities of dust should have been left behind after the creation of the solar system. Larger dust particles, comparable in size to soot and sand grains, are common. They create shooting stars when they collide with the earth's atmosphere. But very small dust particles are conspicuously absent. Astronomers believe that the very small dust particles have been blown out of the solar system by the sun. By comparing the forces on dust particles, determine the diameter of the smallest dust particles that can remain in the solar system over long periods of time. Assume that the dust particles are spherical, black, and have a density of 2000 kg/m³. The sun emits electromagnetic radiation with power 3.9×10^{26} W.

64. ||| Consider current I passing through a resistor of radius r,
CALC length L, and resistance R.
   a. Determine the electric and magnetic fields at the surface of the resistor. Assume that the electric field is uniform throughout, including at the surface.
   b. Determine the strength and direction of the Poynting vector at the surface of the resistor.
   c. Show that the flux of the Poynting vector (i.e., the integral of $\vec{S} \cdot d\vec{A}$) over the surface of the resistor is I^2R. Then give an interpretation of this result.

32 AC Circuits

Transmission lines carry alternating current at voltages as high as 500,000 V.

IN THIS CHAPTER, you will learn about and analyze AC circuits.

What is an AC circuit?

The circuits of Chapter 28, with a steady current in one direction, are called DC circuits—*direct current*. A circuit with an oscillating emf is called an AC circuit, for *alternating current*. The wires that transport electricity across the country—the grid—use alternating current.

« LOOKING BACK Chapter 28 Circuits

How do circuit elements act in an AC circuit?

Resistors in an AC circuit act as they do in a DC circuit. But you'll learn that capacitors and inductors are more useful in AC circuits than in DC circuits.

- The voltage across and the current through a capacitor or inductor are 90° out of phase. One is peaking when the other is zero, and vice versa.
- The peak voltage V and peak current I have an Ohm's-law-like relationship $V = IX$, where X, which depends on frequency, is called the reactance.
- Unlike resistors, capacitors and inductors do not dissipate energy.

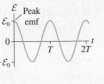

The current and voltage are 90° out of phase with each other.

« LOOKING BACK Section 26.5 Capacitors
« LOOKING BACK Section 30.8 Inductors

What is a phasor?

AC voltages oscillate sinusoidally, so the mathematics of AC circuits is that of SHM. You'll learn a new way to represent oscillating quantities—as a rotating vector called a phasor. The instantaneous value of a phasor quantity is its horizontal projection.

« LOOKING BACK Chapter 15 Simple harmonic motion and resonance

What is an *RLC* circuit?

A circuit with a resistor, inductor, and capacitor in series is called an *RLC* circuit. An *RLC* circuit has a resonance—a large current over a narrow range of frequencies—that allows it to be tuned to a specific frequency. As a result, *RLC* circuits are very important in communications.

Why are AC circuits important?

AC circuits are the backbone of our technological society. Generators automatically produce an oscillating emf, AC power is easily transported over large distances, and transformers allow engineers to shift the AC voltage up or down. The circuits of radio, television, and cell phones are also AC circuits because they work with oscillating voltages and currents—at much higher frequencies than the grid, but the physical principles are the same.

FIGURE 32.1 An oscillating emf can be represented as a graph or as a phasor diagram.

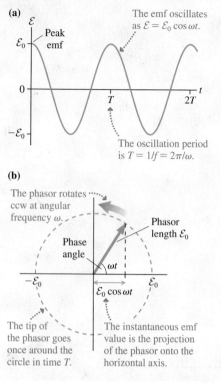

The emf oscillates as $\mathcal{E} = \mathcal{E}_0 \cos \omega t$.

Peak emf

The oscillation period is $T = 1/f = 2\pi/\omega$.

The phasor rotates ccw at angular frequency ω.

Phasor length $\mathcal{E}_0$

Phase angle

The tip of the phasor goes once around the circle in time T.

The instantaneous emf value is the projection of the phasor onto the horizontal axis.

FIGURE 32.2 The correspondence between a phasor and points on a graph.

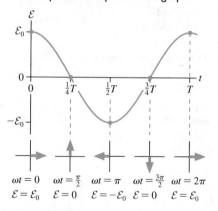

$$
\begin{array}{ccccc}
\omega t = 0 & \omega t = \frac{\pi}{2} & \omega t = \pi & \omega t = \frac{3\pi}{2} & \omega t = 2\pi \\
\mathcal{E} = \mathcal{E}_0 & \mathcal{E} = 0 & \mathcal{E} = -\mathcal{E}_0 & \mathcal{E} = 0 & \mathcal{E} = \mathcal{E}_0
\end{array}
$$

32.1 AC Sources and Phasors

One of the examples of Faraday's law cited in Chapter 30 was an electric generator. A turbine, which might be powered by expanding steam or falling water, causes a coil of wire to rotate in a magnetic field. As the coil spins, the emf and the induced current oscillate sinusoidally. The emf is alternately positive and negative, causing the charges to flow in one direction and then, a half cycle later, in the other. The oscillation frequency of the *grid* in North and South America is $f = 60$ Hz, whereas most of the rest of the world uses a 50 Hz oscillation.

The generator's peak emf—the peak voltage—is a fixed, unvarying quantity, so it might seem logical to call a generator an *alternating-voltage source*. Nonetheless, circuits powered by a sinusoidal emf are called **AC circuits,** where AC stands for *alternating current*. By contrast, the steady-current circuits you studied in Chapter 28 are called **DC circuits,** for *direct current*.

AC circuits are not limited to the use of 50 Hz or 60 Hz power-line voltages. Audio, radio, television, and telecommunication equipment all make extensive use of AC circuits, with frequencies ranging from approximately 10^2 Hz in audio circuits to approximately 10^9 Hz in cell phones. These devices use *electrical oscillators* rather than generators to produce a sinusoidal emf, but the basic principles of circuit analysis are the same.

You can think of an AC generator or oscillator as a battery whose output voltage undergoes sinusoidal oscillations. The instantaneous emf of an AC generator or oscillator, shown graphically in FIGURE 32.1a, can be written

$$\mathcal{E} = \mathcal{E}_0 \cos \omega t \tag{32.1}$$

where $\mathcal{E}_0$ is the peak or maximum emf and $\omega = 2\pi f$ is the angular frequency in radians per second. Recall that the units of emf are volts. As you can imagine, the mathematics of AC circuit analysis are going to be very similar to the mathematics of simple harmonic motion.

An alternative way to represent the emf and other oscillatory quantities is with the *phasor diagram* of FIGURE 32.1b. A **phasor** is a vector that rotates *counterclockwise* (ccw) around the origin at angular frequency ω. The length or magnitude of the phasor is the maximum value of the quantity. For example, the length of an emf phasor is $\mathcal{E}_0$. The angle ωt is the *phase angle*, an idea you learned about in Chapter 15, where we made a connection between circular motion and simple harmonic motion.

The quantity's instantaneous value, the value you would measure at time t, is the projection of the phasor onto the horizontal axis. This is also analogous to the connection between circular motion and simple harmonic motion. FIGURE 32.2 helps you visualize the phasor rotation by showing how the phasor corresponds to the more familiar graph at several specific points in the cycle.

STOP TO THINK 32.1   The magnitude of the instantaneous value of the emf represented by this phasor is

a. Increasing.
b. Decreasing.
c. Constant.
d. It's not possible to tell without knowing t.

Resistor Circuits

In Chapter 28 you learned to analyze a circuit in terms of the current I, voltage V, and potential difference ΔV. Now, because the current and voltage are oscillating, we will use lowercase i to represent the *instantaneous* current through a circuit element and v for the circuit element's *instantaneous* voltage.

FIGURE 32.3 shows the instantaneous current i_R through a resistor R. The potential difference across the resistor, which we call the *resistor voltage* v_R, is given by Ohm's law:

$$v_R = i_R R \qquad (32.2)$$

FIGURE 32.4 shows a resistor R connected across an AC emf $\mathcal{E}$. Notice that the circuit symbol for an AC generator is 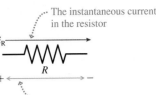. We can analyze this circuit in exactly the same way we analyzed a DC resistor circuit. Kirchhoff's loop law says that the sum of all the potential differences around a closed path is zero:

$$\sum \Delta V = \Delta V_{source} + \Delta V_{res} = \mathcal{E} - v_R = 0 \qquad (32.3)$$

The minus sign appears, just as it did in the equation for a DC circuit, because the potential *decreases* when we travel through a resistor in the direction of the current. We find from the loop law that $v_R = \mathcal{E} = \mathcal{E}_0 \cos \omega t$. This isn't surprising because the resistor is connected directly across the terminals of the emf.

The resistor voltage in an AC circuit can be written

$$v_R = V_R \cos \omega t \qquad (32.4)$$

where V_R is the peak or maximum voltage. You can see that $V_R = \mathcal{E}_0$ in the single-resistor circuit of Figure 32.4. Thus the current through the resistor is

$$i_R = \frac{v_R}{R} = \frac{V_R \cos \omega t}{R} = I_R \cos \omega t \qquad (32.5)$$

where $I_R = V_R/R$ is the peak current.

NOTE Ohm's law applies to both the instantaneous *and* peak currents and voltages of a resistor.

The resistor's instantaneous current and voltage are in phase, both oscillating as $\cos \omega t$. FIGURE 32.5 shows the voltage and the current simultaneously on a graph and as a phasor diagram. The fact that the current phasor is shorter than the voltage phasor has no significance. Current and voltage are measured in different units, so you can't compare the length of one to the length of the other. Showing the two different quantities on a single graph—a tactic that can be misleading if you're not careful—illustrates that they oscillate in phase and that their phasors rotate together at the same angle and frequency.

FIGURE 32.3 Instantaneous current i_R through a resistor.

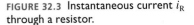

The instantaneous current in the resistor

The instantaneous resistor voltage is $v_R = i_R R$. The potential decreases in the direction of the current.

FIGURE 32.4 An AC resistor circuit.

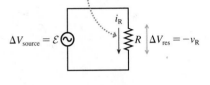

This is the current direction when $\mathcal{E} > 0$. A half cycle later it will be in the opposite direction.

FIGURE 32.5 Graph and phasor diagrams of the resistor current and voltage.

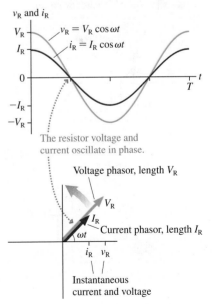

$$v_R = V_R \cos \omega t$$
$$i_R = I_R \cos \omega t$$

The resistor voltage and current oscillate in phase.

Voltage phasor, length V_R

Current phasor, length I_R

Instantaneous current and voltage

EXAMPLE 32.1 | **Finding resistor voltages**

In the circuit of FIGURE 32.6, what are (a) the peak voltage across each resistor and (b) the instantaneous resistor voltages at $t = 20$ ms?

VISUALIZE Figure 32.6 shows the circuit diagram. The two resistors are in series.

SOLVE a. The equivalent resistance of the two series resistors is $R_{eq} = 5\ \Omega + 15\ \Omega = 20\ \Omega$. The instantaneous current through the equivalent resistance is

$$i_R = \frac{v_R}{R_{eq}} = \frac{\mathcal{E}_0 \cos \omega t}{R_{eq}} = \frac{(100\ \text{V}) \cos\left(2\pi(60\ \text{Hz})t\right)}{20\ \Omega}$$

$$= (5.0\ \text{A}) \cos\left(2\pi(60\ \text{Hz})t\right)$$

FIGURE 32.6 An AC resistor circuit.

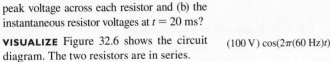

$(100\ \text{V}) \cos(2\pi(60\ \text{Hz})t)$

$5\ \Omega$

$15\ \Omega$

Continued

The peak current is $I_R = 5.0$ A, and this is also the peak current through the two resistors that form the 20 Ω equivalent resistance. Hence the peak voltage across each resistor is

$$V_R = I_R R = \begin{cases} 25 \text{ V} & 5 \text{ Ω resistor} \\ 75 \text{ V} & 15 \text{ Ω resistor} \end{cases}$$

b. The instantaneous current at $t = 0.020$ s is

$$i_R = (5.0 \text{ A}) \cos(2\pi(60 \text{ Hz})(0.020 \text{ s})) = 1.55 \text{ A}$$

The resistor voltages at this time are

$$v_R = i_R R = \begin{cases} 7.7 \text{ V} & 5 \text{ Ω resistor} \\ 23.2 \text{ V} & 15 \text{ Ω resistor} \end{cases}$$

ASSESS The sum of the instantaneous voltages, 30.9 V, is what you would find by calculating $\mathcal{E}$ at $t = 20$ ms. This self-consistency gives us confidence in the answer.

STOP TO THINK 32.2 The resistor whose voltage and current phasors are shown here has resistance R

a. > 1 Ω
b. < 1 Ω
c. It's not possible to tell.

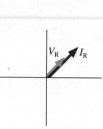

32.2 Capacitor Circuits

FIGURE 32.7 An AC capacitor circuit.

(a)

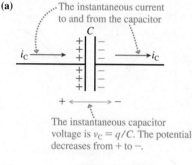

The instantaneous current to and from the capacitor

The instantaneous capacitor voltage is $v_C = q/C$. The potential decreases from + to −.

(b)

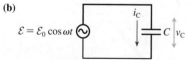

FIGURE 32.7a shows current i_C charging a capacitor with capacitance C. The instantaneous capacitor voltage is $v_C = q/C$, where $\pm q$ is the charge on the two capacitor plates at this instant. It is useful to compare Figure 32.7a to Figure 32.3 for a resistor.

FIGURE 32.7b, where capacitance C is connected across an AC source of emf $\mathcal{E}$, is the most basic capacitor circuit. The capacitor is in parallel with the source, so the capacitor voltage equals the emf: $v_C = \mathcal{E} = \mathcal{E}_0 \cos \omega t$. It will be useful to write

$$v_C = V_C \cos \omega t \tag{32.6}$$

where V_C is the peak or maximum voltage across the capacitor. You can see that $V_C = \mathcal{E}_0$ in this single-capacitor circuit.

To find the current to and from the capacitor, we first write the charge

$$q = Cv_C = CV_C \cos \omega t \tag{32.7}$$

The current is the *rate* at which charge flows through the wires, $i_C = dq/dt$, thus

$$i_C = \frac{dq}{dt} = \frac{d}{dt}(CV_C \cos \omega t) = -\omega CV_C \sin \omega t \tag{32.8}$$

We can most easily see the relationship between the capacitor voltage and current if we use the trigonometric identity $-\sin(x) = \cos(x + \pi/2)$ to write

$$i_C = \omega CV_C \cos\left(\omega t + \frac{\pi}{2}\right) \tag{32.9}$$

In contrast to a resistor, a capacitor's current and voltage are *not* in phase. In **FIGURE 32.8a**, a graph of the instantaneous voltage v_C and current i_C, you can see that the current peaks one-quarter of a period *before* the voltage peaks. The phase angle of the current phasor on the phasor diagram of FIGURE 32.8b is $\pi/2$ rad—a quarter of a circle—larger than the phase angle of the voltage phasor.

We can summarize this finding:

The AC current of a capacitor *leads* the capacitor voltage by $\pi/2$ rad, or 90°.

The current reaches its peak value I_C at the instant the capacitor is fully discharged and $v_C = 0$. The current is zero at the instant the capacitor is fully charged.

A simple harmonic oscillator provides a mechanical analogy of the 90° phase difference between current and voltage. You learned in Chapter 15 that the position and velocity of a simple harmonic oscillator are

$$x = A \cos \omega t$$

$$v = \frac{dx}{dt} = -\omega A \sin \omega t = -v_{max} \sin \omega t = v_{max} \cos\left(\omega t + \frac{\pi}{2}\right)$$

You can see that the velocity of an oscillator leads the position by 90° in the same way that the capacitor current leads the voltage.

Capacitive Reactance

We can use Equation 32.9 to see that the peak current to and from a capacitor is $I_C = \omega C V_C$. This relationship between the peak voltage and peak current looks much like Ohm's law for a resistor if we define the **capacitive reactance** X_C to be

$$X_C \equiv \frac{1}{\omega C} \tag{32.10}$$

With this definition,

$$I_C = \frac{V_C}{X_C} \quad \text{or} \quad V_C = I_C X_C \tag{32.11}$$

The units of reactance, like those of resistance, are ohms.

> **NOTE** Reactance relates the *peak* voltage V_C and current I_C. But reactance differs from resistance in that it does *not* relate the instantaneous capacitor voltage and current because they are out of phase. That is, $v_C \neq i_C X_C$.

A resistor's resistance R is independent of the emf frequency. In contrast, as FIGURE 32.9 shows, a capacitor's reactance X_C depends inversely on the frequency. The reactance becomes very large at low frequencies (i.e., the capacitor is a large impediment to current). This makes sense because $\omega = 0$ would be a nonoscillating DC circuit, and we know that a steady DC current cannot pass through a capacitor. The reactance decreases as the frequency increases until, at very high frequencies, $X_C \approx 0$ and the capacitor begins to act like an ideal wire. This result has important consequences for how capacitors are used in many circuits.

FIGURE 32.8 Graph and phasor diagrams of the capacitor current and voltage.

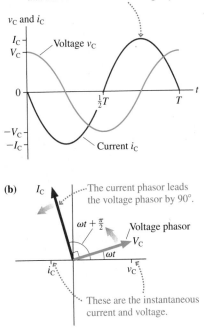

(a) i_C peaks $\frac{1}{4}T$ before v_C peaks. We say that the current *leads* the voltage by 90°.

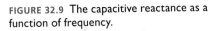

(b)

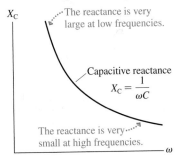

FIGURE 32.9 The capacitive reactance as a function of frequency.

EXAMPLE 32.2 | **Capacitive reactance**

What is the capacitive reactance of a 0.10 μF capacitor at 100 Hz (an audio frequency) and at 100 MHz (an FM-radio frequency)?

SOLVE At 100 Hz,

$$X_C(\text{at } 100 \text{ Hz}) = \frac{1}{\omega C} = \frac{1}{2\pi(100 \text{ Hz})(1.0 \times 10^{-7} \text{ F})} = 16,000 \ \Omega$$

Increasing the frequency by a factor of 10^6 decreases X_C by a factor of 10^6, giving

$$X_C(\text{at } 100 \text{ MHz}) = 0.016 \ \Omega$$

ASSESS A capacitor with a substantial reactance at audio frequencies has virtually no reactance at FM-radio frequencies.

EXAMPLE 32.3 | **Capacitor current**

A 10 μF capacitor is connected to a 1000 Hz oscillator with a peak emf of 5.0 V. What is the peak current to the capacitor?

VISUALIZE Figure 32.7b showed the circuit diagram. It is a simple one-capacitor circuit.

SOLVE The capacitive reactance at $\omega = 2\pi f = 6280$ rad/s is

$$X_C = \frac{1}{\omega C} = \frac{1}{(6280 \text{ rad/s})(10 \times 10^{-6} \text{ F})} = 16 \ \Omega$$

The peak voltage across the capacitor is $V_C = \mathcal{E}_0 = 5.0$ V; hence the peak current is

$$I_C = \frac{V_C}{X_C} = \frac{5.0 \text{ V}}{16 \ \Omega} = 0.31 \text{ A}$$

ASSESS Using reactance is just like using Ohm's law, but don't forget it applies to only the *peak* current and voltage, not the instantaneous values.

STOP TO THINK 32.3 What is the capacitive reactance of "no capacitor," just a continuous wire?

a. 0   b. ∞ c. Undefined

32.3 *RC* Filter Circuits

You learned in Chapter 28 that a resistance R causes a capacitor to be charged or discharged with time constant $\tau = RC$. We called this an *RC* circuit. Now that we've looked at resistors and capacitors individually, let's explore what happens if an *RC* circuit is driven continuously by an alternating current source.

FIGURE 32.10 shows a circuit in which a resistor R and capacitor C are in series with an emf $\mathcal{E}$ oscillating at angular frequency ω. Before launching into a formal analysis, let's try to understand qualitatively how this circuit will respond as the frequency is varied. If the frequency is very low, the capacitive reactance will be very large, and thus the peak current I_C will be very small. The peak current through the resistor is the same as the peak current to and from the capacitor (just as in DC circuits, conservation of charge requires $I_R = I_C$); hence we expect the resistor's peak voltage $V_R = I_R R$ to be very small at very low frequencies.

On the other hand, suppose the frequency is very high. Then the capacitive reactance approaches zero and the peak current, determined by the resistance alone, will be $I_R = \mathcal{E}_0/R$. The resistor's peak voltage $V_R = IR$ will approach the peak source voltage $\mathcal{E}_0$ at very high frequencies.

This reasoning leads us to expect that V_R will *increase* steadily from 0 to $\mathcal{E}_0$ as ω is increased from 0 to very high frequencies. Kirchhoff's loop law has to be obeyed, so the capacitor voltage V_C will *decrease* from $\mathcal{E}_0$ to 0 during the same change of frequency. A quantitative analysis will show us how this behavior can be used as a *filter*.

The goal of a quantitative analysis is to determine the peak current I and the two peak voltages V_R and V_C as functions of the emf amplitude $\mathcal{E}_0$ and frequency ω. Our analytic procedure is based on the fact that the instantaneous current i is the same for two circuit elements in series.

FIGURE 32.10 An *RC* circuit driven by an AC source.

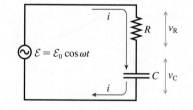

Using phasors to analyze an *RC* circuit

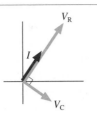

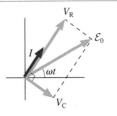

Begin by drawing a current phasor of length I. This is the starting point because the series circuit elements have the same current i. The angle at which the phasor is drawn is not relevant.

The current and voltage of a resistor are in phase, so draw a resistor voltage phasor of length V_R parallel to the current phasor I. The capacitor current leads the capacitor voltage by 90°, so draw a capacitor voltage phasor of length V_C that is 90° behind [i.e., clockwise (cw) from] the current phasor.

The series resistor and capacitor are in parallel with the emf, so their *instantaneous* voltages satisfy $v_R + v_C = \mathcal{E}$. This is a *vector* addition of phasors, so draw the emf phasor as the vector sum of the two voltage phasors. The emf is $\mathcal{E} = \mathcal{E}_0 \cos \omega t$, hence the emf phasor is at angle ωt.

The length of the emf phasor, $\mathcal{E}_0$, is the hypotenuse of a right triangle formed by the resistor and capacitor phasors. Thus $\mathcal{E}_0^2 = V_R^2 + V_C^2$.

The relationship $\mathcal{E}_0^2 = V_R^2 + V_C^2$ is based on the peak values, not the instantaneous values, because the peak values are the lengths of the sides of the right triangle. The peak voltages are related to the peak current I via $V_R = IR$ and $V_C = IX_C$, thus

$$\mathcal{E}_0^2 = V_R^2 + V_C^2 = (IR)^2 + (IX_C)^2 = (R^2 + X_C^2)I^2$$
$$= (R^2 + 1/\omega^2 C^2)I^2 \tag{32.12}$$

Consequently, the peak current in the *RC* circuit is

$$I = \frac{\mathcal{E}_0}{\sqrt{R^2 + X_C^2}} = \frac{\mathcal{E}_0}{\sqrt{R^2 + 1/\omega^2 C^2}} \tag{32.13}$$

Knowing I gives us the two peak voltages:

$$V_R = IR = \frac{\mathcal{E}_0 R}{\sqrt{R^2 + X_C^2}} = \frac{\mathcal{E}_0 R}{\sqrt{R^2 + 1/\omega^2 C^2}}$$

$$V_C = IX_C = \frac{\mathcal{E}_0 X_C}{\sqrt{R^2 + X_C^2}} = \frac{\mathcal{E}_0/\omega C}{\sqrt{R^2 + 1/\omega^2 C^2}} \tag{32.14}$$

Frequency Dependence

Our goal was to see how the peak current and voltages vary as functions of the frequency ω. Equations 32.13 and 32.14 are rather complex and best interpreted by looking at graphs. **FIGURE 32.11** is a graph of V_R and V_C versus ω.

You can see that our qualitative predictions have been borne out. That is, V_R increases from 0 to $\mathcal{E}_0$ as ω is increased, while V_C decreases from $\mathcal{E}_0$ to 0. The explanation for this behavior is that the capacitive reactance X_C decreases as ω increases. For low frequencies, where $X_C \gg R$, the circuit is primarily capacitive. For high frequencies, where $X_C \ll R$, the circuit is primarily resistive.

The frequency at which $V_R = V_C$ is called the **crossover frequency** ω_c. The *crossover* frequency is easily found by setting the two expressions in Equations 32.14 equal to each other. The denominators are the same and cancel, as does $\mathcal{E}_0$, leading to

$$\omega_c = \frac{1}{RC} \tag{32.15}$$

In practice, $f_c = \omega_c/2\pi$ is also called the crossover frequency.

FIGURE 32.11 Graph of the resistor and capacitor peak voltages as functions of the emf angular frequency ω.

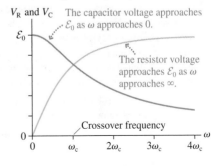

We'll leave it as a homework problem to show that $V_R = V_C = \mathcal{E}_0/\sqrt{2}$ when $\omega = \omega_c$. This may seem surprising. After all, shouldn't V_R and V_C add up to $\mathcal{E}_0$?

No! V_R and V_C are the *peak values* of oscillating voltages, not the instantaneous values. The instantaneous values do, indeed, satisfy $v_R + v_C = \mathcal{E}$ at all instants of time. But the resistor and capacitor voltages are out of phase with each other, as the phasor diagram shows, so the two circuit elements don't reach their peak values at the same time. The peak values are related by $\mathcal{E}_0^2 = V_R^2 + V_C^2$, and you can see that $V_R = V_C = \mathcal{E}_0/\sqrt{2}$ satisfies this equation.

NOTE It's very important in AC circuit analysis to make a clear distinction between instantaneous values and peak values of voltages and currents. Relationships that are true for one set of values may not be true for the other.

Filters

FIGURE 32.12 Low-pass and high-pass filter circuits.

(a) Low-pass filter

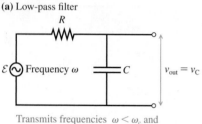

Transmits frequencies $\omega < \omega_c$ and blocks frequencies $\omega > \omega_c$

(b) High-pass filter

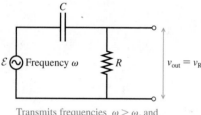

Transmits frequencies $\omega > \omega_c$ and blocks frequencies $\omega < \omega_c$

FIGURE 32.12a is the circuit we've just analyzed; the only difference is that the capacitor voltage v_C is now identified as the *output voltage* v_{out}. This is a voltage you might measure or, perhaps, send to an amplifier for use elsewhere in an electronic instrument. You can see from the capacitor voltage graph in Figure 32.11 that the peak output voltage is $V_{out} \approx \mathcal{E}_0$ if $\omega \ll \omega_c$, but $V_{out} \approx 0$ if $\omega \gg \omega_c$. In other words,

- If the frequency of an input signal is well below the crossover frequency, the input signal is transmitted with little loss to the output.
- If the frequency of an input signal is well above the crossover frequency, the input signal is strongly attenuated and the output is very nearly zero.

This circuit is called a **low-pass filter.**

The circuit of FIGURE 32.12b, which instead uses the resistor voltage v_R for the output v_{out}, is a **high-pass filter.** The output is $V_{out} \approx 0$ if $\omega \ll \omega_c$, but $V_{out} \approx \mathcal{E}_0$ if $\omega \gg \omega_c$. That is, an input signal whose frequency is well above the crossover frequency is transmitted without loss to the output.

Filter circuits are widely used in electronics. For example, a high-pass filter designed to have $f_c = 100$ Hz would pass the audio frequencies associated with speech ($f > 200$ Hz) while blocking 60 Hz "noise" that can be picked up from power lines. Similarly, the high-frequency hiss from old vinyl records can be attenuated with a low-pass filter, allowing the lower-frequency audio signal to pass.

A simple RC filter suffers from the fact that the crossover region where $V_R \approx V_C$ is fairly broad. More sophisticated filters have a sharper transition from off ($V_{out} \approx 0$) to on ($V_{out} \approx \mathcal{E}_0$), but they're based on the same principles as the RC filter analyzed here.

EXAMPLE 32.4 | **Designing a filter**

For a science project, you've built a radio to listen to AM radio broadcasts at frequencies near 1 MHz. The basic circuit is an antenna, which produces a very small oscillating voltage when it absorbs the energy of an electromagnetic wave, and an amplifier. Unfortunately, your neighbor's short-wave radio broadcast at 10 MHz interferes with your reception. Having just finished physics, you decide to solve this problem by placing a filter between the antenna and the amplifier. You happen to have a 500 pF capacitor. What frequency should you select as the filter's crossover frequency? What value of resistance will you need to build this filter?

MODEL You need a low-pass filter to block signals at 10 MHz while passing the lower-frequency AM signal at 1 MHz.

VISUALIZE The circuit will look like the low-pass filter in Figure 32.12a. The oscillating voltage generated by the antenna will be the emf, and v_{out} will be sent to the amplifier.

SOLVE You might think that a crossover frequency near 5 MHz, about halfway between 1 MHz and 10 MHz, would work best. But 5 MHz is a factor of 5 higher than 1 MHz while only a factor of 2 less than 10 MHz. A crossover frequency the *same factor* above 1 MHz as it is below 10 MHz will give the best results. In practice, choosing $f_c = 3$ MHz would be sufficient. You can then use Equation 32.15 to select the proper resistor value:

$$R = \frac{1}{\omega_c C} = \frac{1}{2\pi(3 \times 10^6 \text{ Hz})(500 \times 10^{-12} \text{ F})}$$

$$= 106 \ \Omega \approx 100 \ \Omega$$

ASSESS Rounding to 100 Ω is appropriate because the crossover frequency was determined to only one significant figure. Such "sloppy design" is adequate when the two frequencies you need to distinguish are well separated.

STOP TO THINK 32.4 Rank in order, from largest to smallest, the crossover frequencies $(\omega_c)_a$ to $(\omega_c)_d$ of these four circuits.

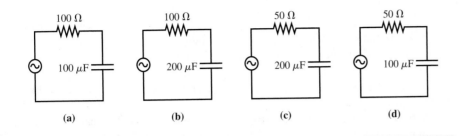

(a)   (b)   (c)   (d)

32.4 Inductor Circuits

FIGURE 32.13a shows the instantaneous current i_L through an inductor. If the current is changing, the instantaneous inductor voltage is

$$v_L = L\frac{di_L}{dt} \tag{32.16}$$

You learned in Chapter 30 that the potential decreases in the direction of the current if the current is increasing $(di_L/dt > 0)$ and increases if the current is decreasing $(di_L/dt < 0)$.

FIGURE 32.13b is the simplest inductor circuit. The inductor L is connected across the AC source, so the inductor voltage equals the emf: $v_L = \mathcal{E} = \mathcal{E}_0 \cos \omega t$. We can write

$$v_L = V_L \cos \omega t \tag{32.17}$$

where V_L is the peak or maximum voltage across the inductor. You can see that $V_L = \mathcal{E}_0$ in this single-inductor circuit.

We can find the inductor current i_L by integrating Equation 32.17. First, we use Equation 32.17 to write Equation 32.16 as

$$di_L = \frac{v_L}{L}dt = \frac{V_L}{L}\cos \omega t\, dt \tag{32.18}$$

Integrating gives

$$i_L = \frac{V_L}{L}\int \cos \omega t\, dt = \frac{V_L}{\omega L}\sin \omega t = \frac{V_L}{\omega L}\cos\left(\omega t - \frac{\pi}{2}\right)$$
$$= I_L \cos\left(\omega t - \frac{\pi}{2}\right) \tag{32.19}$$

where $I_L = V_L/\omega L$ is the peak or maximum inductor current.

NOTE Mathematically, Equation 32.19 could have an integration constant i_0. An integration constant would represent a constant DC current through the inductor, but there is no DC source of potential in an AC circuit. Hence, on physical grounds, we set $i_0 = 0$ for an AC circuit.

We define the **inductive reactance,** analogous to the capacitive reactance, to be

$$X_L \equiv \omega L \tag{32.20}$$

FIGURE 32.13 Using an inductor in an AC circuit.

(a) The instantaneous current through the inductor

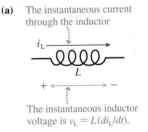

The instantaneous inductor voltage is $v_L = L(di_L/dt)$.

(b)

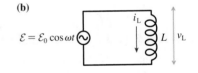

Then the peak current $I_L = V_L/\omega L$ and the peak voltage are related by

$$I_L = \frac{V_L}{X_L} \quad \text{or} \quad V_L = I_L X_L \qquad (32.21)$$

FIGURE 32.14 The inductive reactance as a function of frequency.

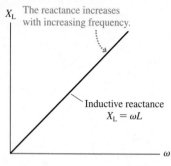

FIGURE 32.14 shows that the inductive reactance increases as the frequency increases. This makes sense. Faraday's law tells us that the induced voltage across a coil increases as the time rate of change of $\vec{B}$ increases, and $\vec{B}$ is directly proportional to the inductor current. For a given peak current $I_L, \vec{B}$ changes more rapidly at higher frequencies than at lower frequencies, and thus V_L is larger at higher frequencies than at lower frequencies.

FIGURE 32.15a is a graph of the inductor voltage and current. You can see that the current peaks one-quarter of a period *after* the voltage peaks. The angle of the current phasor on the phasor diagram of FIGURE 32.15b is $\pi/2$ rad less than the angle of the voltage phasor. We can summarize this finding:

The AC current through an inductor *lags* the inductor voltage by $\pi/2$ rad, or 90°.

FIGURE 32.15 Graph and phasor diagrams of the inductor current and voltage.

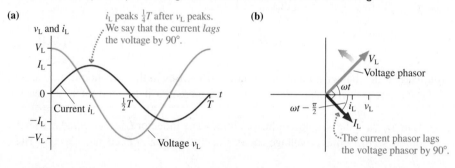

EXAMPLE 32.5 | **Current and voltage of an inductor**

A 25 μH inductor is used in a circuit that oscillates at 100 kHz. The current through the inductor reaches a peak value of 20 mA at $t = 5.0\ \mu$s. What is the peak inductor voltage, and when, closest to $t = 5.0\ \mu$s, does it occur?

MODEL The inductor current lags the voltage by 90°, or, equivalently, the voltage reaches its peak value one-quarter period *before* the current.

VISUALIZE The circuit looks like Figure 32.15b.

SOLVE The inductive reactance at $f = 100$ kHz is

$$X_L = \omega L = 2\pi(1.0 \times 10^5\ \text{Hz})(25 \times 10^{-6}\ \text{H}) = 16\ \Omega$$

Thus the peak voltage is $V_L = I_L X_L = (20\ \text{mA})(16\ \Omega) = 320$ mV. The voltage peak occurs one-quarter period before the current peaks, and we know that the current peaks at $t = 5.0\ \mu$s. The period of a 100 kHz oscillation is 10.0 μs, so the voltage peaks at

$$t = 5.0\ \mu\text{s} - \frac{10.0\ \mu\text{s}}{4} = 2.5\ \mu\text{s}$$

32.5 The Series *RLC* Circuit

FIGURE 32.16 A series *RLC* circuit.

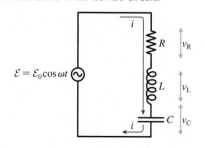

The circuit of FIGURE 32.16, where a resistor, inductor, and capacitor are in series, is called a **series *RLC* circuit.** The series *RLC* circuit has many important applications because, as you will see, it exhibits resonance behavior.

The analysis, which is very similar to our analysis of the *RC* circuit in Section 32.3, will be based on a phasor diagram. Notice that the three circuit elements are in series with each other and, together, are in parallel with the emf. We can draw two conclusions that form the basis of our analysis:

1. The instantaneous current of all three elements is the same: $i = i_R = i_L = i_C$.
2. The sum of the instantaneous voltages matches the emf: $\mathcal{E} = v_R + v_L + v_C$.

Using phasors to analyze an *RLC* circuit

Begin by drawing a current phasor of length I. This is the starting point because the series circuit elements have the same current i.

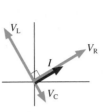

The current and voltage of a resistor are in phase, so draw a resistor voltage phasor parallel to the current phasor I. The capacitor current leads the capacitor voltage by 90°, so draw a capacitor voltage phasor that is 90° behind the current phasor. The inductor current *lags* the voltage by 90°, so draw an inductor voltage phasor 90° ahead of the current phasor.

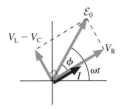

The instantaneous voltages satisfy $\mathcal{E} = v_R + v_L + v_C$. In terms of phasors, this is a *vector* addition. We can do the addition in two steps. Because the capacitor and inductor phasors are in opposite directions, their vector sum has length $V_L - V_C$. Adding the resistor phasor, at right angles, then gives the emf phasor $\mathcal{E}$ at angle ωt.

The length $\mathcal{E}_0$ of the emf phasor is the hypotenuse of a right triangle. Thus
$$\mathcal{E}_0^2 = V_R^2 + (V_L - V_C)^2$$

If $V_L > V_C$, which we've assumed, then the instantaneous current i lags the emf by a phase angle ϕ. We can write the current, in terms of ϕ, as

$$i = I\cos(\omega t - \phi) \tag{32.22}$$

Of course, there's no guarantee that V_L will be larger than V_C. If the opposite is true, $V_L < V_C$, the emf phasor is on the other side of the current phasor. Our analysis is still valid if we consider ϕ to be negative when i is ccw from $\mathcal{E}$. Thus ϕ can be anywhere between $-90°$ and $+90°$.

Now we can continue much as we did with the *RC* circuit. Based on the right triangle, $\mathcal{E}_0^2$ is

$$\mathcal{E}_0^2 = V_R^2 + (V_L - V_C)^2 = \left[R^2 + (X_L - X_C)^2\right]I^2 \tag{32.23}$$

where we wrote each of the peak voltages in terms of the peak current I and a resistance or a reactance. Consequently, the peak current in the *RLC* circuit is

$$I = \frac{\mathcal{E}_0}{\sqrt{R^2 + (X_L - X_C)^2}} = \frac{\mathcal{E}_0}{\sqrt{R^2 + (\omega L - 1/\omega C)^2}} \tag{32.24}$$

The three peak voltages are then found from $V_R = IR$, $V_L = IX_L$, and $V_C = IX_C$.

Impedance

The denominator of Equation 32.24 is called the **impedance** Z of the circuit:

$$Z = \sqrt{R^2 + (X_L - X_C)^2} \tag{32.25}$$

Impedance, like resistance and reactance, is measured in ohms. The circuit's peak current can be written in terms of the source emf and the circuit impedance as

$$I = \frac{\mathcal{E}_0}{Z} \tag{32.26}$$

Equation 32.26 is a compact way to write I, but it doesn't add anything new to Equation 32.24.

FIGURE 32.17 The current is not in phase with the emf.

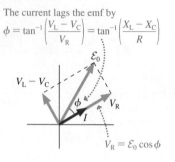

The current lags the emf by
$$\phi = \tan^{-1}\left(\frac{V_L - V_C}{V_R}\right) = \tan^{-1}\left(\frac{X_L - X_C}{R}\right)$$

$\mathcal{E}_0$

$V_L - V_C$

ϕ

V_R

I

$V_R = \mathcal{E}_0 \cos \phi$

Phase Angle

It is often useful to know the phase angle ϕ between the emf and the current. You can see from **FIGURE 32.17** that

$$\tan \phi = \frac{V_L - V_C}{V_R} = \frac{(X_L - X_C)I}{RI}$$

The current I cancels, and we're left with

$$\phi = \tan^{-1}\left(\frac{X_L - X_C}{R}\right) \tag{32.27}$$

We can check that Equation 32.27 agrees with our analyses of single-element circuits. A resistor-only circuit has $X_L = X_C = 0$ and thus $\phi = \tan^{-1}(0) = 0$ rad. In other words, as we discovered previously, the emf and current are in phase. An AC inductor circuit has $R = X_C = 0$ and thus $\phi = \tan^{-1}(\infty) = \pi/2$ rad, agreeing with our earlier finding that the inductor current lags the voltage by 90°.

Other relationships can be found from the phasor diagram and written in terms of the phase angle. For example, we can write the peak resistor voltage as

$$V_R = \mathcal{E}_0 \cos \phi \tag{32.28}$$

Notice that the resistor voltage oscillates in phase with the emf only if $\phi = 0$ rad.

Resonance

Suppose we vary the emf frequency ω while keeping everything else constant. There is very little current at very low frequencies because the capacitive reactance $X_C = 1/\omega C$ (and thus Z) is very large. Similarly, there is very little current at very high frequencies because the inductive reactance $X_L = \omega L$ becomes very large.

If I approaches zero at very low and very high frequencies, there should be some intermediate frequency where I is a maximum. Indeed, you can see from Equation 32.24 that the denominator will be a minimum, making I a maximum, when $X_L = X_C$, or

$$\omega L = \frac{1}{\omega C} \tag{32.29}$$

The frequency ω_0 that satisfies Equation 32.29 is called the **resonance frequency:**

$$\omega_0 = \frac{1}{\sqrt{LC}} \tag{32.30}$$

This is the frequency for *maximum current* in the series *RLC* circuit. The maximum current

$$I_{\max} = \frac{\mathcal{E}_0}{R} \tag{32.31}$$

is that of a purely resistive circuit because the impedance is $Z = R$ at resonance.

You'll recognize ω_0 as the oscillation frequency of the *LC* circuit we analyzed in Chapter 30. The current in an ideal *LC* circuit oscillates forever as energy is transferred back and forth between the capacitor and the inductor. This is analogous to an ideal, frictionless simple harmonic oscillator in which the energy is transformed back and forth between kinetic and potential.

Adding a resistor to the circuit is like adding damping to a mechanical oscillator. The emf is then a sinusoidal driving force, and the series *RLC* circuit is directly analogous to the driven, damped oscillator that you studied in Chapter 15. A mechanical oscillator exhibits *resonance* by having a large-amplitude response when the driving frequency matches the system's natural frequency. Equation 32.30 is the natural frequency of the series *RLC* circuit, the frequency at which the current would like to oscillate. Consequently, the circuit has a large current response when the oscillating emf matches this frequency.

FIGURE 32.18 shows the peak current I of a series *RLC* circuit as the emf frequency ω is varied. Notice how the current increases until reaching a maximum at frequency ω_0, then decreases. This is the hallmark of a resonance.

As R decreases, causing the damping to decrease, the maximum current becomes larger and the curve in Figure 32.18 becomes narrower. You saw exactly the same behavior for a driven mechanical oscillator. The emf frequency must be very close to ω_0 in order for a lightly damped system to respond, but the response at resonance is very large.

For a different perspective, **FIGURE 32.19** graphs the instantaneous emf $\mathcal{E} = \mathcal{E}_0 \cos \omega t$ and current $i = I \cos(\omega t - \phi)$ for frequencies below, at, and above ω_0. The current and the emf are in phase at resonance ($\phi = 0$ rad) because the capacitor and inductor essentially cancel each other to give a purely resistive circuit. Away from resonance, the current decreases *and* begins to get out of phase with the emf. You can see, from Equation 32.27, that the phase angle ϕ is negative when $X_L < X_C$ (i.e., the frequency is below resonance) and positive when $X_L > X_C$ (the frequency is above resonance).

Resonance circuits are widely used in radio, television, and communication equipment because of their ability to respond to one particular frequency (or very narrow range of frequencies) while suppressing others. The selectivity of a resonance circuit improves as the resistance decreases, but the inherent resistance of the wires and the inductor coil keeps R from being 0 Ω.

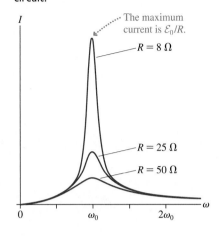

FIGURE 32.18 A graph of the current I versus emf frequency for a series *RLC* circuit.

FIGURE 32.19 Graphs of the emf $\mathcal{E}$ and the current i at frequencies below, at, and above the resonance frequency ω_0.

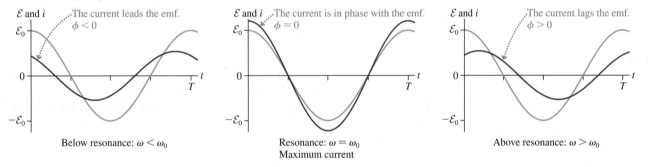

EXAMPLE 32.6 | Designing a radio receiver

An AM radio antenna picks up a 1000 kHz signal with a peak voltage of 5.0 mV. The tuning circuit consists of a 60 μH inductor in series with a variable capacitor. The inductor coil has a resistance of 0.25 Ω, and the resistance of the rest of the circuit is negligible.

a. To what value should the capacitor be tuned to listen to this radio station?
b. What is the peak current through the circuit at resonance?
c. A stronger station at 1050 kHz produces a 10 mV antenna signal. What is the current at this frequency when the radio is tuned to 1000 kHz?

MODEL The inductor's 0.25 Ω resistance can be modeled as a resistance in series with the inductance, hence we have a series *RLC* circuit. The antenna signal at $\omega = 2\pi \times 1000$ kHz is the emf.

VISUALIZE The circuit looks like Figure 32.16.

SOLVE a. The capacitor needs to be tuned to where it and the inductor are resonant at $\omega_0 = 2\pi \times 1000$ kHz. The appropriate value is

$$C = \frac{1}{L\omega_0^2} = \frac{1}{(60 \times 10^{-6} \text{ H})(6.28 \times 10^6 \text{ rad/s})^2}$$

$$= 4.2 \times 10^{-10} \text{ F} = 420 \text{ pF}$$

b. $X_L = X_C$ at resonance, so the peak current is

$$I = \frac{\mathcal{E}_0}{R} = \frac{5.0 \times 10^{-3} \text{ V}}{0.25 \ \Omega} = 0.020 \text{ A} = 20 \text{ mA}$$

c. The 1050 kHz signal is "off resonance," so we need to compute $X_L = \omega L = 396 \ \Omega$ and $X_C = 1/\omega C = 361 \ \Omega$ at $\omega = 2\pi \times 1050$ kHz. The peak voltage of this signal is $\mathcal{E}_0 = 10$ mV. With these values, Equation 32.24 for the peak current is

$$I = \frac{\mathcal{E}_0}{\sqrt{R^2 + (X_L - X_C)^2}} = 0.28 \text{ mA}$$

ASSESS These are realistic values for the input stage of an AM radio. You can see that the signal from the 1050 kHz station is strongly suppressed when the radio is tuned to 1000 kHz.

32.6 Power in AC Circuits

A primary role of the emf is to supply energy. Some circuit devices, such as motors and lightbulbs, use the energy to perform useful tasks. Other circuit devices dissipate the energy as an increased thermal energy in the components and the surrounding air. Chapter 28 examined the topic of power in DC circuits. Now we can perform a similar analysis for AC circuits.

The emf supplies energy to a circuit at the rate

$$p_{source} = i\mathcal{E} \tag{32.32}$$

where i and $\mathcal{E}$ are the instantaneous current from and potential difference across the emf. We've used a lowercase p to indicate that this is the instantaneous power. We need to look at the power losses in individual circuit elements.

Resistors

A resistor dissipates energy at the rate

$$p_R = i_R v_R = i_R^2 R \tag{32.33}$$

We can use $i_R = I_R \cos \omega t$ to write the resistor's instantaneous power loss as

$$p_R = i_R^2 R = I_R^2 R \cos^2 \omega t \tag{32.34}$$

FIGURE 32.20 shows the instantaneous power graphically. You can see that, because the cosine is squared, the power oscillates twice during every cycle of the emf. The energy dissipation peaks both when $i_R = I_R$ and when $i_R = -I_R$.

In practice, we're more interested in the *average power* than in the instantaneous power. The **average power** P is the total energy dissipated per second. We can find P_R for a resistor by using the identity $\cos^2(x) = \frac{1}{2}(1 + \cos 2x)$ to write

$$P_R = I_R^2 R \cos^2 \omega t = I_R^2 R \left[\tfrac{1}{2}(1 + \cos 2\omega t) \right] = \tfrac{1}{2} I_R^2 R + \tfrac{1}{2} I_R^2 R \cos 2\omega t$$

The $\cos 2\omega t$ term oscillates positive and negative twice during each cycle of the emf. Its average, over one cycle, is zero. Thus the average power loss in a resistor is

$$P_R = \tfrac{1}{2} I_R^2 R \qquad \text{(average power loss in a resistor)} \tag{32.35}$$

It is useful to write Equation 32.25 as

$$P_R = \left(\frac{I_R}{\sqrt{2}} \right)^2 R = (I_{rms})^2 R \tag{32.36}$$

where the quantity

$$I_{rms} = \frac{I_R}{\sqrt{2}} \tag{32.37}$$

is called the **root-mean-square current**, or rms current, I_{rms}. Technically, an rms quantity is the square root of the average, or mean, of the quantity squared. For a sinusoidal oscillation, the rms value turns out to be the peak value divided by $\sqrt{2}$.

The rms current allows us to compare Equation 32.36 directly to the energy dissipated by a resistor in a DC circuit: $P = I^2 R$. You can see that the average power loss of a resistor in an AC circuit with $I_{rms} = 1$ A is the same as in a DC circuit with $I = 1$ A. **As far as power is concerned, an rms current is equivalent to an equal DC current.**

FIGURE 32.20 The instantaneous power loss in a resistor.

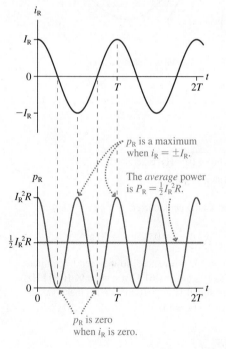

Similarly, we can define the root-mean-square voltage and emf:

$$V_{rms} = \frac{V_R}{\sqrt{2}} \qquad \mathcal{E}_{rms} = \frac{\mathcal{E}_0}{\sqrt{2}} \qquad (32.38)$$

The resistor's average power loss in terms of the rms quantities is

$$P_R = (I_{rms})^2 R = \frac{(V_{rms})^2}{R} = I_{rms} V_{rms} \qquad (32.39)$$

and the average power supplied by the emf is

$$P_{source} = I_{rms} \mathcal{E}_{rms} \qquad (32.40)$$

The single-resistor circuit that we analyzed in Section 32.1 had $V_R = \mathcal{E}$ or, equivalently, $V_{rms} = \mathcal{E}_{rms}$. You can see from Equations 32.39 and 32.40 that the power loss in the resistor exactly matches the power supplied by the emf. This must be the case in order to conserve energy.

NOTE Voltmeters, ammeters, and other AC measuring instruments are calibrated to give the rms value. An AC voltmeter would show that the "line voltage" of an electrical outlet in the United States is 120 V. This is $\mathcal{E}_{rms}$. The peak voltage $\mathcal{E}_0$ is larger by a factor of $\sqrt{2}$, or $\mathcal{E}_0 = 170$ V. The power-line voltage is sometimes specified as "120 V/60 Hz," showing the rms voltage and the frequency.

The power rating on a lightbulb is its average power at $V_{rms} = 120$ V.

EXAMPLE 32.7 | **Lighting a bulb**

A 100 W incandescent lightbulb is plugged into a 120 V/60 Hz outlet. What is the resistance of the bulb's filament? What is the peak current through the bulb?

MODEL The filament in a lightbulb acts as a resistor.

VISUALIZE FIGURE 32.21 is a simple one-resistor circuit.

FIGURE 32.21 An AC circuit with a lightbulb as a resistor.

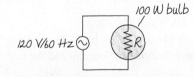

SOLVE A bulb labeled 100 W is designed to dissipate an average 100 W at $V_{rms} = 120$ V. We can use Equation 32.39 to find

$$R = \frac{(V_{rms})^2}{P_R} = \frac{(120\ \text{V})^2}{100\ \text{W}} = 144\ \Omega$$

The rms current is then found from

$$I_{rms} = \frac{P_R}{V_{rms}} = \frac{100\ \text{W}}{120\ \text{V}} = 0.833\ \text{A}$$

The peak current is $I_R = \sqrt{2} I_{rms} = 1.18$ A.

ASSESS Calculations with rms values are just like the calculations for DC circuits.

Capacitors and Inductors

In Section 32.2, we found that the instantaneous current to a capacitor is $i_C = -\omega C V_C \sin \omega t$. Thus the instantaneous energy dissipation in a capacitor is

$$p_C = v_C i_C = (V_C \cos \omega t)(-\omega C V_C \sin \omega t) = -\tfrac{1}{2} \omega C V_C^2 \sin 2\omega t \qquad (32.41)$$

where we used $\sin(2x) = 2\sin(x)\cos(x)$.

FIGURE 32.22 on the next page shows Equation 32.41 graphically. Energy is transferred into the capacitor (positive power) as it is charged, but, instead of being dissipated, as it would be by a resistor, the energy is stored as potential energy in the capacitor's electric field. Then, as the capacitor discharges, this energy is given back to the circuit. Power is the rate at which energy is *removed* from the circuit, hence p is negative as the capacitor transfers energy back into the circuit.

Using a mechanical analogy, a capacitor is like an ideal, frictionless simple harmonic oscillator. Kinetic and potential energy are constantly being exchanged,

FIGURE 32.22 Energy flows into and out of a capacitor as it is charged and discharged.

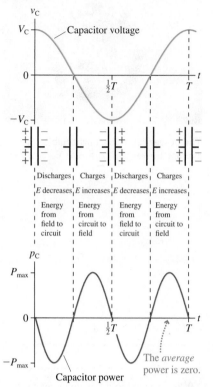

but there is no dissipation because none of the energy is transformed into thermal energy. The important conclusion is that **a capacitor's average power loss is zero: $P_C = 0$.**

The same is true of an inductor. An inductor alternately stores energy in the magnetic field, as the current is increasing, then transfers energy back to the circuit as the current decreases. The instantaneous power oscillates between positive and negative, but **an inductor's average power loss is zero: $P_L = 0$.**

> **NOTE** We're assuming ideal capacitors and inductors. Real capacitors and inductors inevitably have a small amount of resistance and dissipate a small amount of energy. However, their energy dissipation is negligible compared to that of the resistors in most practical circuits.

The Power Factor

In an *RLC* circuit, energy is supplied by the emf and dissipated by the resistor. But an *RLC* circuit is unlike a purely resistive circuit in that the current is not in phase with the potential difference of the emf.

We found in Equation 32.22 that the instantaneous current in an *RLC* circuit is $i = I\cos(\omega t - \phi)$, where ϕ is the angle by which the current lags the emf. Thus the instantaneous power supplied by the emf is

$$p_{source} = i\mathcal{E} = (I\cos(\omega t - \phi))(\mathcal{E}_0 \cos \omega t) = I\mathcal{E}_0 \cos \omega t \cos(\omega t - \phi) \quad (32.42)$$

We can use the expression $\cos(x - y) = \cos(x)\cos(y) + \sin(x)\sin(y)$ to write the power as

$$p_{source} = (I\mathcal{E}_0 \cos \phi) \cos^2 \omega t + (I\mathcal{E}_0 \sin \phi) \sin \omega t \cos \omega t \quad (32.43)$$

In our analysis of the power loss in a resistor and a capacitor, we found that the average of $\cos^2 \omega t$ is $\frac{1}{2}$ and the average of $\sin \omega t \cos \omega t$ is zero. Thus we can immediately write that the *average* power supplied by the emf is

$$P_{source} = \frac{1}{2} I\mathcal{E}_0 \cos \phi = I_{rms} \mathcal{E}_{rms} \cos \phi \quad (32.44)$$

The rms values, you will recall, are $I/\sqrt{2}$ and $\mathcal{E}_0/\sqrt{2}$.

The term $\cos \phi$, called the **power factor,** arises because the current and the emf in a series *RLC* circuit are not in phase. Because the current and the emf aren't pushing and pulling together, the source delivers less energy to the circuit.

We'll leave it as a homework problem for you to show that the peak current in an *RLC* circuit can be written $I = I_{max} \cos \phi$, where $I_{max} = \mathcal{E}_0/R$ was given in Equation 32.31. In other words, the current term in Equation 32.44 is a function of the power factor. Consequently, the average power is

$$P_{source} = P_{max} \cos^2 \phi \quad (32.45)$$

where $P_{max} = \frac{1}{2} I_{max} \mathcal{E}_0$ is the *maximum* power the source can deliver to the circuit.

The source delivers maximum power only when $\cos \phi = 1$. This is the case when $X_L - X_C = 0$, requiring either a purely resistive circuit or an *RLC* circuit operating at the resonance frequency ω_0. The average power loss is zero for a purely capacitive or purely inductive load with, respectively, $\phi = -90°$ or $\phi = +90°$, as found above.

Motors of various types, especially large industrial motors, use a significant fraction of the electric energy generated in industrialized nations. Motors operate most efficiently, doing the maximum work per second, when the power factor is as close to 1 as possible. But motors are inductive devices, due to their electromagnet coils, and if too many motors are attached to the electric grid, the power factor is pulled away from 1. To compensate, the electric company places large capacitors throughout the

Industrial motors use a significant fraction of the electric energy generated in the United States.

transmission system. The capacitors dissipate no energy, but they allow the electric system to deliver energy more efficiently by keeping the power factor close to 1.

Finally, we found in Equation 32.28 that the resistor's peak voltage in an RLC circuit is related to the emf peak voltage by $V_R = \mathcal{E}_0 \cos\phi$ or, dividing both sides by $\sqrt{2}$, $V_{rms} = \mathcal{E}_{rms} \cos\phi$. We can use this result to write the energy loss in the resistor as

$$P_R = I_{rms}V_{rms} = I_{rms}\mathcal{E}_{rms}\cos\phi \qquad (32.46)$$

But this expression is P_{source}, as we found in Equation 32.44. Thus we see that the energy supplied to an RLC circuit by the emf is ultimately dissipated by the resistor.

STOP TO THINK 32.6 The emf and the current in a series RLC circuit oscillate as shown. Which of the following (perhaps more than one) would increase the rate at which energy is supplied to the circuit?

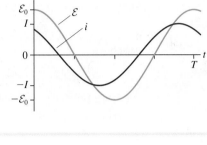

a. Increase $\mathcal{E}_0$
b. Increase L
c. Increase C
d. Decrease $\mathcal{E}_0$
e. Decrease L
f. Decrease C

CHALLENGE EXAMPLE 32.8 | Power in an RLC circuit

An audio amplifier drives a series RLC circuit consisting of an 8.0 Ω loudspeaker, a 160 μF capacitor, and a 1.5 mH inductor. The amplifier output is 15.0 V rms at 500 Hz.

a. What power is delivered to the speaker?
b. What maximum power could the amplifier deliver, and how would the capacitor have to be changed for this to happen?

MODEL The emf and voltage of an RLC circuit are not in phase, and that affects the power delivered to the circuit. All the power is dissipated by the circuit's resistance, which in this case is the loudspeaker.

VISUALIZE The circuit looks like Figure 32.16.

SOLVE a. The emf delivers power $P_{source} = I_{rms}\mathcal{E}_{rms}\cos\phi$, where ϕ is the phase angle between the emf and the current. The rms current is $I_{rms} = \mathcal{E}_{rms}/Z$, where Z is the impedance. To calculate Z, we need the reactances of the capacitor and inductor, and these, in turn, depend on the frequency. At 500 Hz, the angular frequency is $\omega = 2\pi(500 \text{ Hz}) = 3140 \text{ rad/s}$. With this, we can find

$$X_C = \frac{1}{\omega C} = \frac{1}{(3140 \text{ rad/s})(160 \times 10^{-6} \text{ F})} = 1.99 \ \Omega$$

$$X_L = \omega L = (3140 \text{ rad/s})(0.0015 \text{ H}) = 4.71 \ \Omega$$

Now we can calculate the impedance:

$$Z = \sqrt{R^2 + (X_L - X_C)^2} = 8.45 \ \Omega$$

and thus

$$I_{rms} = \frac{\mathcal{E}_{rms}}{Z} = \frac{15.0 \text{ V}}{8.45 \ \Omega} = 1.78 \text{ A}$$

Lastly, we need the phase angle between the emf and the current:

$$\phi = \tan^{-1}\left(\frac{X_L - X_C}{R}\right) = 18.8°$$

The power factor is $\cos(18.8°) = 0.947$, and thus the power delivered by the emf is

$$P_{source} = I_{rms}\mathcal{E}_{rms}\cos\phi = (1.78 \text{ A})(15.0 \text{ V})(0.947) = 25 \text{ W}$$

b. Maximum power is delivered when the current is in phase with the emf, making the power factor 1.00. This occurs when $X_C = X_L$, making the impedance $Z = R = 8.0 \ \Omega$ and the current $I_{rms} = \mathcal{E}_{rms}/R = 1.88$ A. Then

$$P_{source} = I_{rms}\mathcal{E}_{rms}\cos\phi = (1.88 \text{ A})(15.0 \text{ V})(1.00) = 28 \text{ W}$$

To deliver maximum power, we need to change the capacitance to make $X_C = X_L = 4.71 \ \Omega$. The required capacitance is

$$C = \frac{1}{(3140 \text{ rad/s})(4.71 \ \Omega)} = 68 \ \mu\text{F}$$

So delivering maximum power requires lowering the capacitance from 160 μF to 68 μF.

ASSESS Changing the capacitor not only increases the power factor, it also increases the current. Both contribute to the higher power.

SUMMARY

The goal of Chapter 32 has been to learn about and analyze AC circuits.

IMPORTANT CONCEPTS

AC circuits are driven by an emf

$$\mathcal{E} = \mathcal{E}_0 \cos \omega t$$

that oscillates with angular frequency $\omega = 2\pi f$.

Phasors can be used to represent the oscillating emf, current, and voltage.

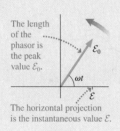

The length of the phasor is the peak value $\mathcal{E}_0$.

The horizontal projection is the instantaneous value $\mathcal{E}$.

Basic circuit elements

Element	i and v	Resistance/reactance	I and V	Power
Resistor	In phase	R is fixed	$V = IR$	$I_{rms}V_{rms}$
Capacitor	i leads v by 90°	$X_C = 1/\omega C$	$V = IX_C$	0
Inductor	i lags v by 90°	$X_L = \omega L$	$V = IX_L$	0

For many purposes, especially calculating power, the **root-mean-square** (rms) quantities

$$V_{rms} = V/\sqrt{2} \qquad I_{rms} = I/\sqrt{2} \qquad \mathcal{E}_{rms} = \mathcal{E}_0/\sqrt{2}$$

are equivalent to the corresponding DC quantities.

KEY SKILLS

Using phasor diagrams

- Start with a phasor (v or i) common to two or more circuit elements.

- The sum of instantaneous quantities is vector addition.

- Use the Pythagorean theorem to relate peak quantities.

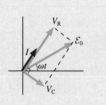

For an RC circuit, shown here,

$$v_R + v_C = \mathcal{E}$$
$$V_R^2 + V_C^2 = \mathcal{E}_0^2$$

Instantaneous and peak quantities

The instantaneous quantities v and i vary sinusoidally. The peak quantities V and I are the maximum values of v and i. For capacitors and inductors, the peak quantities are related by $V = IX$, where X is the reactance, but this relationship does *not* apply to v and i.

Kirchhoff's loop law says that the sum of the potential differences around a loop is zero.

Charge conservation says that circuit elements in series all have the same instantaneous current i and the same peak current I.

APPLICATIONS

RC filter circuits

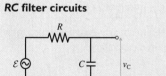

$$v_C = \frac{\mathcal{E}_0 X_C}{\sqrt{R^2 + X_C^2}}$$

$$v_C \to \mathcal{E}_0 \text{ as } \omega \to 0$$

A **low-pass filter** transmits low frequencies and blocks high frequencies.

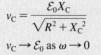

$$v_R = \frac{\mathcal{E}_0 R}{\sqrt{R^2 + X_C^2}}$$

$$v_R \to \mathcal{E}_0 \text{ as } \omega \to \infty$$

A **high-pass filter** transmits high frequencies and blocks low frequencies.

Series *RLC* circuits

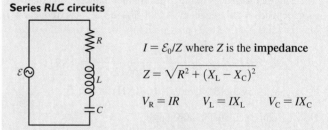

$$I = \mathcal{E}_0/Z \text{ where } Z \text{ is the } \mathbf{impedance}$$

$$Z = \sqrt{R^2 + (X_L - X_C)^2}$$

$$V_R = IR \qquad V_L = IX_L \qquad V_C = IX_C$$

When $\omega = \omega_0 = 1/\sqrt{LC}$ (the **resonance frequency**), the current in the circuit is a maximum $I_{max} = \mathcal{E}_0/R$.

In general, the current i lags behind $\mathcal{E}$ by the **phase angle** $\phi = \tan^{-1}\big((X_L - X_C)/R\big)$.

The power supplied by the emf is $P_{source} = I_{rms}\mathcal{E}_{rms}\cos\phi$, where $\cos\phi$ is called the **power factor**.

The power lost in the resistor is $P_R = I_{rms}V_{rms} = (I_{rms})^2 R$.

TERMS AND NOTATION

AC circuit	crossover frequency, ω_c	series RLC circuit	root-mean-square current, I_{rms}
DC circuit	low-pass filter	impedance, Z	power factor, $\cos\phi$
phasor	high-pass filter	resonance frequency, ω_0	
capacitive reactance, X_C	inductive reactance, X_L	average power, P	

CONCEPTUAL QUESTIONS

1. **FIGURE Q32.1** shows emf phasors a, b, and c.
   a. For each, what is the instantaneous value of the emf?
   b. At this instant, is the magnitude of each emf increasing, decreasing, or holding constant?

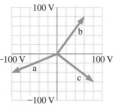

FIGURE Q32.1

2. A resistor is connected across an oscillating emf. The peak current through the resistor is 2.0 A. What is the peak current if:
   a. The resistance R is doubled?
   b. The peak emf $\mathcal{E}_0$ is doubled?
   c. The frequency ω is doubled?

3. A capacitor is connected across an oscillating emf. The peak current through the capacitor is 2.0 A. What is the peak current if:
   a. The capacitance C is doubled?
   b. The peak emf $\mathcal{E}_0$ is doubled?
   c. The frequency ω is doubled?

4. A low-pass RC filter has a crossover frequency $f_c = 200$ Hz. What is f_c if:
   a. The resistance R is doubled?
   b. The capacitance C is doubled?
   c. The peak emf $\mathcal{E}_0$ is doubled?

5. An inductor is connected across an oscillating emf. The peak current through the inductor is 2.0 A. What is the peak current if:
   a. The inductance L is doubled?
   b. The peak emf $\mathcal{E}_0$ is doubled?
   c. The frequency ω is doubled?

6. The resonance frequency of a series RLC circuit is 1000 Hz. What is the resonance frequency if:
   a. The resistance R is doubled?
   b. The inductance L is doubled?
   c. The capacitance C is doubled?
   d. The peak emf $\mathcal{E}_0$ is doubled?

7. In the series RLC circuit represented by the phasors of **FIGURE Q32.7**, is the emf frequency less than, equal to, or greater than the resonance frequency ω_0? Explain.

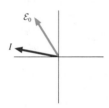

FIGURE Q32.7

8. The resonance frequency of a series RLC circuit is less than the emf frequency. Does the current lead or lag the emf? Explain.

9. The current in a series RLC circuit lags the emf by 20°. You cannot change the emf. What two different things could you do to the circuit that would increase the power delivered to the circuit by the emf?

10. The average power dissipated by a resistor is 4.0 W. What is P_R if:
   a. The resistance R is doubled while $\mathcal{E}_0$ is held fixed?
   b. The peak emf $\mathcal{E}_0$ is doubled while R is held fixed?
   c. Both are doubled simultaneously?

EXERCISES AND PROBLEMS

Problems labeled ▓ integrate material from earlier chapters.

Exercises

Section 32.1 AC Sources and Phasors

1. ‖ The emf phasor in **FIGURE EX32.1** is shown at $t = 2.0$ ms.
   a. What is the angular frequency ω? Assume this is the first rotation.
   b. What is the peak value of the emf?

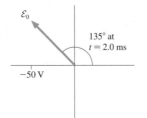

FIGURE EX32.1

2. ‖ The emf phasor in **FIGURE EX32.2** is shown at $t = 15$ ms.
   a. What is the angular frequency ω? Assume this is the first rotation.
   b. What is the instantaneous value of the emf?

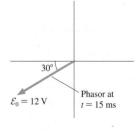

FIGURE EX32.2

3. ‖ A 110 Hz source of emf has a peak voltage of 50 V. Draw the emf phasor at $t = 3.0$ ms.

4. ‖ Draw the phasor for the emf $\mathcal{E} = (170 \text{ V})\cos\big((2\pi \times 60 \text{ Hz})t\big)$ at $t = 60$ ms.

5. ‖ **FIGURE EX32.5** shows voltage and current graphs for a resistor.
   a. What is the emf frequency f?
   b. What is the value of the resistance R?
   c. Draw the resistor's voltage and current phasors at $t = 15$ ms.

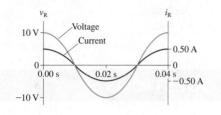

FIGURE EX32.5

6. ‖ A 200 Ω resistor is connected to an AC source with $\mathcal{E}_0 = 10$ V. What is the peak current through the resistor if the emf frequency is (a) 100 Hz? (b) 100 kHz?

Section 32.2 Capacitor Circuits

7. ‖ The peak current to and from a capacitor is 10 mA. What is the peak current if
   a. The emf frequency is doubled?
   b. The emf peak voltage is doubled (at the original frequency)?

8. ‖ A 0.30 μF capacitor is connected across an AC generator that produces a peak voltage of 10 V. What is the peak current to and from the capacitor if the emf frequency is (a) 100 Hz? (b) 100 kHz?

9. ‖ **FIGURE EX32.9** shows voltage and current graphs for a capacitor.
   a. What is the emf frequency f?
   b. What is the value of the capacitance C?

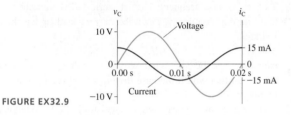

FIGURE EX32.9

10. ‖ A 20 nF capacitor is connected across an AC generator that produces a peak voltage of 5.0 V.
    a. At what frequency f is the peak current 50 mA?
    b. What is the instantaneous value of the emf at the instant when $i_C = I_C$?

11. ‖ A capacitor is connected to a 15 kHz oscillator. The peak current is 65 mA when the rms voltage is 6.0 V. What is the value of the capacitance C?

12. ‖ A capacitor has a peak current of 330 μA when the peak voltage at 250 kHz is 2.2 V.
    a. What is the capacitance?
    b. If the peak voltage is held constant, what is the peak current at 500 kHz?

13. ‖ a. Evaluate V_C in **FIGURE EX32.13** at emf frequencies 1, 3, 10, 30, and 100 kHz.
    b. Graph V_C versus frequency. Draw a smooth curve through your five points.

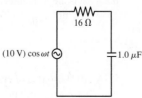

FIGURE EX32.13

14. ‖ a. Evaluate V_R in **FIGURE EX32.14** at emf frequencies 100, 300, 1000, 3000, and 10,000 Hz.
    b. Graph V_R versus frequency. Draw a smooth curve through your five points.

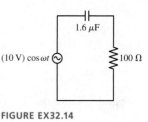

FIGURE EX32.14

Section 32.3 RC Filter Circuits

15. ‖ A high-pass RC filter is connected to an AC source with a peak voltage of 10.0 V. The peak capacitor voltage is 6.0 V. What is the resistor voltage?

16. ‖ A high-pass RC filter with a crossover frequency of 1000 Hz uses a 100 Ω resistor. What is the value of the capacitor?

17. ‖ A low-pass RC filter with a crossover frequency of 1000 Hz uses a 100 Ω resistor. What is the value of the capacitor?

18. ‖ A low-pass filter consists of a 100 μF capacitor in series with a 159 Ω resistor. The circuit is driven by an AC source with a peak voltage of 5.00 V.
    a. What is the crossover frequency f_c?
    b. What is V_C when $f = \frac{1}{2}f_c, f_c$, and $2f_c$?

19. ‖ What are V_R and V_C if the emf frequency in **FIGURE EX32.19** is 10 kHz?

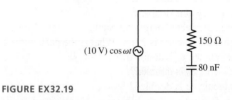

FIGURE EX32.19

20. ‖ A high-pass filter consists of a 1.59 μF capacitor in series with a 100 Ω resistor. The circuit is driven by an AC source with a peak voltage of 5.00 V.
    a. What is the crossover frequency f_c?
    b. What is V_R when $f = \frac{1}{2}f_c, f_c$, and $2f_c$?

21. ‖ An electric circuit, whether it's a simple lightbulb or a complex amplifier, has two input terminals that are connected to the two output terminals of the voltage source. The impedance between the two input terminals (often a function of frequency) is the circuit's *input impedance*. Most circuits are designed to have a large input impedance. To see why, suppose you need to amplify the output of a high-pass filter that is constructed with a 1.2 kΩ resistor and a 15 μF capacitor. The amplifier you've chosen has a purely resistive input impedance. For a 60 Hz signal, what is the ratio $V_{R\,load}/V_{R\,no\,load}$ of the filter's peak voltage output with (load) and without (no load) the amplifier connected if the amplifier's input impedance is (a) 1.5 kΩ and (b) 150 kΩ?

Section 32.4 Inductor Circuits

22. ‖ A 20 mH inductor is connected across an AC generator that produces a peak voltage of 10 V. What is the peak current through the inductor if the emf frequency is (a) 100 Hz? (b) 100 kHz?

23. ‖ The peak current through an inductor is 10 mA. What is the peak current if
    a. The emf frequency is doubled?
    b. The emf peak voltage is doubled (at the original frequency)?

24. ‖ **FIGURE EX32.24** shows voltage and current graphs for an inductor.
   a. What is the emf frequency f?
   b. What is the value of the inductance L?

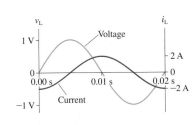

FIGURE EX32.24

25. ‖ An inductor is connected to a 15 kHz oscillator. The peak current is 65 mA when the rms voltage is 6.0 V. What is the value of the inductance L?

26. ‖ An inductor has a peak current of 330 μA when the peak voltage at 45 MHz is 2.2 V.
   a. What is the inductance?
   b. If the peak voltage is held constant, what is the peak current at 90 MHz?

Section 32.5 The Series *RLC* Circuit

27. | A series *RLC* circuit has a 200 kHz resonance frequency. What is the resonance frequency if the capacitor value is doubled and, at the same time, the inductor value is halved?

28. | A series *RLC* circuit has a 200 kHz resonance frequency. What is the resonance frequency if
   a. The resistor value is doubled?
   b. The capacitor value is doubled?

29. ‖ What capacitor in series with a 100 Ω resistor and a 20 mH inductor will give a resonance frequency of 1000 Hz?

30. ‖ A series *RLC* circuit consists of a 50 Ω resistor, a 3.3 mH inductor, and a 480 nF capacitor. It is connected to an oscillator with a peak voltage of 5.0 V. Determine the impedance, the peak current, and the phase angle at frequencies (a) 3000 Hz, (b) 4000 Hz, and (c) 5000 Hz.

31. ‖ At what frequency f do a 1.0 μF capacitor and a 1.0 μH inductor have the same reactance? What is the value of the reactance at this frequency?

32. | For the circuit of **FIGURE EX32.32**,
   a. What is the resonance frequency, in both rad/s and Hz?
   b. Find V_R and V_L at resonance.
   c. How can V_L be larger than $\mathcal{E}_0$? Explain.

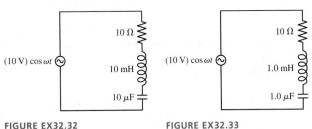

FIGURE EX32.32   **FIGURE EX32.33**

33. | For the circuit of **FIGURE EX32.33**,
   a. What is the resonance frequency, in both rad/s and Hz?
   b. Find V_R and V_C at resonance.
   c. How can V_C be larger than $\mathcal{E}_0$? Explain.

Section 32.6 Power in AC Circuits

34. | The heating element of a hair drier dissipates 1500 W when connected to a 120 V/60 Hz power line. What is its resistance?

35. ‖ A resistor dissipates 2.0 W when the rms voltage of the emf is 10.0 V. At what rms voltage will the resistor dissipate 10.0 W?

36. | For what absolute value of the phase angle does a source deliver 75% of the maximum possible power to an *RLC* circuit?

37. | The motor of an electric drill draws a 3.5 A rms current at the power-line voltage of 120 V rms. What is the motor's power if the current lags the voltage by 20°?

38. ‖ A series *RLC* circuit attached to a 120 V/60 Hz power line draws a 2.4 A rms current with a power factor of 0.87. What is the value of the resistor?

39. ‖ A series *RLC* circuit with a 100 Ω resistor dissipates 80 W when attached to a 120 V/60 Hz power line. What is the power factor?

Problems

40. ‖ a. For an *RC* circuit, find an expression for the angular frequency at which $V_R = \frac{1}{2}\mathcal{E}_0$.
   b. What is V_C at this frequency?

41. ‖ a. For an *RC* circuit, find an expression for the angular frequency at which $V_C = \frac{1}{2}\mathcal{E}_0$.
   b. What is V_R at this frequency?

42. ‖ For an *RC* filter circuit, show that $V_R = V_C = \mathcal{E}_0/\sqrt{2}$ at $\omega = \omega_c$.

43. ‖ A series *RC* circuit is built with a 12 kΩ resistor and a parallel-plate capacitor with 15-cm-diameter electrodes. A 12 V, 36 kHz source drives a peak current of 0.65 mA through the circuit. What is the spacing between the capacitor plates?

44. ‖ Show that Equation 32.27 for the phase angle ϕ of a series *RLC* circuit gives the correct result for a capacitor-only circuit.

45. ‖ a. What is the peak current supplied by the emf in **FIGURE P32.45**?
   b. What is the peak voltage across the 3.0 μF capacitor?

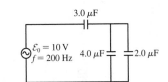

FIGURE P32.45

46. ‖ You have a resistor and a capacitor of unknown values. First, you charge the capacitor and discharge it through the resistor. By monitoring the capacitor voltage on an oscilloscope, you see that the voltage decays to half its initial value in 2.5 ms. You then use the resistor and capacitor to make a low-pass filter. What is the crossover frequency f_c?

47. ‖ **FIGURE P32.47** shows a parallel *RC* circuit.
   a. Use a phasor-diagram analysis to find expressions for the peak currents I_R and I_C.
   Hint: What do the resistor and capacitor have in common? Use that as the initial phasor.
   b. Complete the phasor analysis by finding an expression for the peak emf current I.

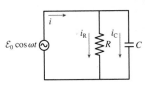

FIGURE P32.47

48. ‖ The small transformers that power many consumer products produce a 12.0 V rms, 60 Hz emf. Design a circuit using resistors and capacitors that uses the transformer voltage as an input and produces a 6.0 V rms output that leads the input voltage by 45°.

49. ‖ Use a phasor diagram to analyze the *RL* circuit of **FIGURE P32.49**. In particular,

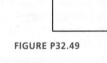

a. Find expressions for I, V_R, and V_L.
b. What is V_R in the limits $\omega \to 0$ and $\omega \to \infty$?
c. If the output is taken from the resistor, is this a low-pass or a high-pass filter? Explain.

FIGURE P32.49

d. Find an expression for the crossover frequency ω_c.

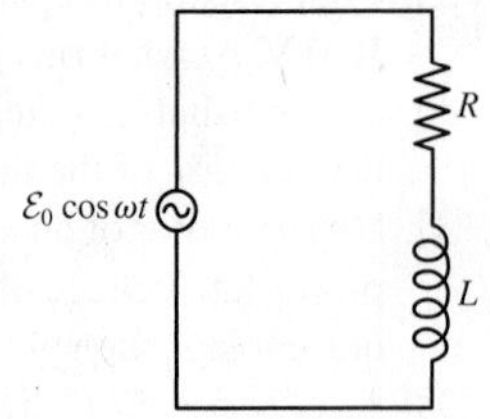

50. ‖ A series *RL* circuit is built with a 110 Ω resistor and a 5.0-cm-long, 1.0-cm-diameter solenoid with 800 turns of wire. What is the peak magnetic flux through the solenoid if the circuit is driven by a 12 V, 5.0 kHz source?

51. ‖‖ A series *RLC* circuit consists of a 75 Ω resistor, a 0.12 H inductor, and a 30 μF capacitor. It is attached to a 120 V/60 Hz power line. What are (a) the peak current I, (b) the phase angle ϕ, and (c) the average power loss?

52. ‖ A series *RLC* circuit consists of a 25 Ω resistor, a 0.10 H inductor, and a 100 μF capacitor. It draws a 2.5 A rms current when attached to a 60 Hz source. What are (a) the emf $\mathcal{E}_{rms}$, (b) the phase angle ϕ, and (c) the average power loss?

53. ‖ A series *RLC* circuit consists of a 550 Ω resistor, a 2.1 mH inductor, and a 550 nF capacitor. It is connected to a 50 V rms oscillating voltage source with an adjustable frequency. An oscillating magnetic field is observed at a point 2.5 mm from the center of one of the circuit wires, which is long and straight. What is the maximum magnetic field amplitude that can be generated?

54. ‖ In **FIGURE P32.54**, what is the current supplied by the emf when (a) the frequency is very small and (b) the frequency is very large?

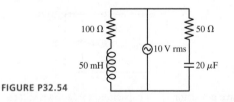

FIGURE P32.54

55. ‖ The current lags the emf by 30° in a series *RLC* circuit with $\mathcal{E}_0 = 10$ V and $R = 50$ Ω. What is the peak current through the circuit?

56. ‖ A series *RLC* circuit consists of a 50 Ω resistor, a 3.3 mH inductor, and a 480 nF capacitor. It is connected to a 5.0 kHz oscillator with a peak voltage of 5.0 V. What is the instantaneous current i when
a. $\mathcal{E} = \mathcal{E}_0$?
b. $\mathcal{E} = 0$ V and is decreasing?

57. ‖ A series *RLC* circuit consists of a 50 Ω resistor, a 3.3 mH inductor, and a 480 nF capacitor. It is connected to a 3.0 kHz oscillator with a peak voltage of 5.0 V. What is the instantaneous emf $\mathcal{E}$ when
a. $i = I$?
b. $i = 0$ A and is decreasing?
c. $i = -I$?

58. ‖ Show that the power factor of a series *RLC* circuit is $\cos\phi = R/Z$.

59. ‖ For a series *RLC* circuit, show that
a. The peak current can be written $I = I_{max}\cos\phi$.
b. The average power can be written $P_{source} = P_{max}\cos^2\phi$.

60. ‖ The tuning circuit in an FM radio receiver is a series *RLC* circuit with a 0.200 μH inductor.
a. The receiver is tuned to a station at 104.3 MHz. What is the value of the capacitor in the tuning circuit?
b. FM radio stations are assigned frequencies every 0.2 MHz, but two nearby stations cannot use adjacent frequencies. What is the maximum resistance the tuning circuit can have if the peak current at a frequency of 103.9 MHz, the closest frequency that can be used by a nearby station, is to be no more than 0.10% of the peak current at 104.3 MHz? The radio is still tuned to 104.3 MHz, and you can assume the two stations have equal strength.

61. ‖ A television channel is assigned the frequency range from 54 MHz to 60 MHz. A series *RLC* tuning circuit in a TV receiver resonates in the middle of this frequency range. The circuit uses a 16 pF capacitor.
a. What is the value of the inductor?
b. In order to function properly, the current throughout the frequency range must be at least 50% of the current at the resonance frequency. What is the minimum possible value of the circuit's resistance?

62. ‖ Lightbulbs labeled 40 W, 60 W, and 100 W are connected to a 120 V/60 Hz power line as shown in **FIGURE P32.62**. What is the rate at which energy is dissipated in each bulb?

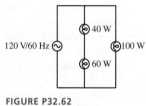

FIGURE P32.62

63. ‖ A generator consists of a 12-cm by 16-cm rectangular loop with 500 turns of wire spinning at 60 Hz in a 25 mT uniform magnetic field. The generator output is connected to a series *RC* circuit consisting of a 120 Ω resistor and a 35 μF capacitor. What is the average power delivered to the circuit?

64. ‖ Commercial electricity is generated and transmitted as *three-phase electricity*. Instead of a single emf, three separate wires carry currents for the emfs $\mathcal{E}_1 = \mathcal{E}_0\cos\omega t$, $\mathcal{E}_2 = \mathcal{E}_0\cos(\omega t + 120°)$, and $\mathcal{E}_3 = \mathcal{E}_0\cos(\omega t - 120°)$ over three parallel wires, each of which supplies one-third of the power. This is why the long-distance transmission lines you see in the countryside have three wires. Suppose the transmission lines into a city supply a total of 450 MW of electric power, a realistic value.
a. What would be the current in each wire if the transmission voltage were $\mathcal{E}_0 = 120$ V rms?
b. In fact, transformers are used to step the transmission-line voltage up to 500 kV rms. What is the current in each wire?
c. Big transformers are expensive. Why does the electric company use step-up transformers?

65. ‖ You're the operator of a 15,000 V rms, 60 Hz electrical substation. When you get to work one day, you see that the station is delivering 6.0 MW of power with a power factor of 0.90.
a. What is the rms current leaving the station?
b. How much series capacitance should you add to bring the power factor up to 1.0?
c. How much power will the station then be delivering?

66. ‖ Commercial electricity is generated and transmitted as *three-phase electricity*. Instead of a single emf $\mathcal{E} = \mathcal{E}_0 \cos \omega t$, three separate wires carry currents for the emfs $\mathcal{E}_1 = \mathcal{E}_0 \cos \omega t, \mathcal{E}_2 = \mathcal{E}_0 \cos(\omega t + 120°)$, and $\mathcal{E}_3 = \mathcal{E}_0 \cos(\omega t - 120°)$. This is why the long-distance transmission lines you see in the countryside have three parallel wires, as do many distribution lines within a city.
    a. Draw a phasor diagram showing phasors for all three phases of a three-phase emf.
    b. Show that the sum of the three phases is zero, producing what is referred to as *neutral*. In *single-phase* electricity, provided by the familiar 120 V/60 Hz electric outlets in your home, one side of the outlet is neutral, as established at a nearby electrical substation. The other, called the *hot side*, is one of the three phases. (The round opening is connected to ground.)
    c. Show that the potential difference between any two of the phases has the rms value $\sqrt{3}\,\mathcal{E}_{\text{rms}}$, where $\mathcal{E}_{\text{rms}}$ is the familiar single-phase rms voltage. Evaluate this potential difference for $\mathcal{E}_{\text{rms}} = 120$ V. Some high-power home appliances, especially electric clothes dryers and hot-water heaters, are designed to operate between two of the phases rather than between one phase and neutral. Heavy-duty industrial motors are designed to operate from all three phases, but full three-phase power is rare in residential or office use.

67. ‖ A motor attached to a 120 V/60 Hz power line draws an 8.0 A current. Its average energy dissipation is 800 W.
    a. What is the power factor?
    b. What is the rms resistor voltage?
    c. What is the motor's resistance?
    d. How much series capacitance needs to be added to increase the power factor to 1.0?

Challenge Problems

68. ‖‖ **FIGURE CP32.68** shows voltage and current graphs for a series *RLC* circuit.
    a. What is the resistance R?
    b. If $L = 200$ μH, what is the resonance frequency in Hz?

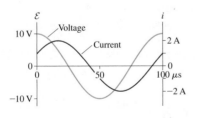

FIGURE CP32.68

69. ‖‖ a. Show that the average power loss in a series *RLC* circuit is
$$P_{\text{avg}} = \frac{\omega^2 (\mathcal{E}_{\text{rms}})^2 R}{\omega^2 R^2 + L^2 (\omega^2 - \omega_0^2)^2}$$
    b. Prove that the energy dissipation is a maximum at $\omega = \omega_0$.

70. ‖‖ a. Show that the peak inductor voltage in a series *RLC* circuit is maximum at frequency
$$\omega_{\text{L}} = \left(\frac{1}{\omega_0^2} - \frac{1}{2} R^2 C^2 \right)^{-1/2}$$
    b. A series *RLC* circuit with $\mathcal{E}_0 = 10.0$ V consists of a 1.0 Ω resistor, a 1.0 μH inductor, and a 1.0 μF capacitor. What is V_{L} at $\omega = \omega_0$ and at $\omega = \omega_{\text{L}}$?

71. ‖‖ The telecommunication circuit shown in **FIGURE CP32.71** has a parallel inductor and capacitor in series with a resistor.
    a. Use a phasor diagram to show that the peak current through the resistor is
$$I = \frac{\mathcal{E}_0}{\sqrt{R^2 + \left(\dfrac{1}{X_{\text{L}}} - \dfrac{1}{X_{\text{C}}} \right)^{-2}}}$$
    Hint: Start with the inductor phasor v_{L}.
    b. What is I in the limits $\omega \to 0$ and $\omega \to \infty$?
    c. What is the resonance frequency ω_0? What is I at this frequency?

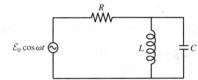
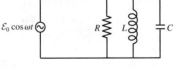

FIGURE CP32.71

72. ‖‖ Consider the parallel *RLC* circuit shown in **FIGURE CP32.72**.
    a. Show that the current drawn from the emf is
$$I = \mathcal{E}_0 \sqrt{\frac{1}{R^2} + \left(\frac{1}{\omega L} - \omega C \right)^2}$$
    Hint: Start with a phasor that is common to all three circuit elements.
    b. What is I in the limits $\omega \to 0$ and $\omega \to \infty$?
    c. Find the frequency for which I is a minimum.
    d. Sketch a graph of I versus ω.

FIGURE CP32.72

Electricity and Magnetism

KEY FINDINGS   What are the overarching findings of Part VI?

- **Charge,** like mass, is a fundamental property of matter. There are two kinds of charge: positive and negative.

- Charges interact via the **electric field:**
  - Source charges create an electric field.
  - A charge in the field experiences an electric force.

- Moving charges (currents) interact with other moving charges via **magnetic fields** and forces.

- Fields are also created when other fields change:
  - A changing magnetic field induces an electric field.
  - A changing electric field induces a magnetic field.

LAWS   What laws of physics govern electricity and magnetism?

Maxwell's equations are the formal statements of the laws of electricity and magnetism, but most problem solving is based on simpler versions of these equations:

Coulomb's law $\vec{E}_{\text{point charge}} = \dfrac{1}{4\pi\epsilon_0}\dfrac{q}{r^2}\hat{r}$
    Gauss's law $\oint \vec{E}\cdot d\vec{A} = \dfrac{Q_{\text{in}}}{\epsilon_0}$

Biot-Savart law $\vec{B}_{\text{point charge}} = \dfrac{\mu_0}{4\pi}\dfrac{q\vec{v}\times\hat{r}}{r^2}$
    Ampère's law $\oint \vec{B}\cdot d\vec{s} = \mu_0 I_{\text{through}}$

Lorentz force law $\vec{F}_{\text{on }q} = q\vec{E} + q\vec{v}\times\vec{B}$

Faraday's law $\mathcal{E} = \left| d\Phi_{\text{m}}/dt \right|$ $I_{\text{induced}} = \mathcal{E}/R$ in the direction of Lenz's law

MODELS   What are the most important models of electricity and magnetism?

Charge model

- Two types of charge: positive and negative.
  - Like charges repel, opposite charges attract.
  - Charge is conserved but can be transferred.
- Two types of materials: conductors and insulators.
- Neutral objects (no net charge) can be polarized.

Electric field model

- Charges interact via the electric field.
  - Source charges create an electric field.
  - Charge q experiences $\vec{E}_{\text{on }q} = q\vec{E}$.
- Important electric field models
  - Point charge
  - Long charged wire
  - Charged sphere
  - Charged plane

Magnetic field models

- Moving charges and currents interact via the magnetic field.
  - Current creates a magnetic field.
  - Charge q experiences $\vec{F}_{\text{on }q} = q\vec{v}\times\vec{B}$.
  - Current I experiences $\vec{F}_{\text{on }I} = I\vec{l}\times\vec{B}$.
- Important magnetic field models
  - Straight, current-carrying wire
  - Current loop   - Solenoid

Electromagnetic waves

- An **electromagnetic wave** is a self-sustaining electromagnetic field.
  - $\vec{E}$ and $\vec{B}$ are perpendicular to $\vec{v}$.
  - Wave speed $v_{\text{em}} = 1/\sqrt{\epsilon_0\mu_0} = c$.

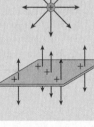
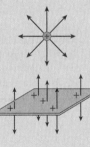
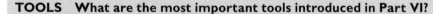
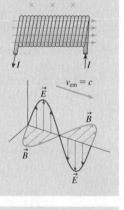

TOOLS   What are the most important tools introduced in Part VI?

Electric potential and potential energy

- Electric interactions can also be described by an **electric potential:**

  $V_{\text{point charge}} = q/4\pi\epsilon_0 r$

  - A charge q in a potential has potential energy $U = qV$.
  - Mechanical energy is conserved.

- Field and potential energy are closely related.
  - The field is perpendicular to equipotential surfaces.
  - The field points "downhill" toward lower potential.

  Decreasing $V \longrightarrow$

Circuits

- Batteries are sources of emf.
- **Current** is the rate of flow of charge: $I = dQ/dt$.
- Circuit elements:
  - Capacitors: $Q = C\Delta V$
  - Resistors: $I = \Delta V/R$ (Ohm's law)
  - Capacitors and resistors can be combined in series or in parallel.
- **Kirchhoff's laws:**
  - Loop law: The sum of potential differences around a loop is zero.
  - Junction law: The sum of currents at a junction is zero.

Uniform fields

- A parallel-plate capacitor creates a uniform electric field.
- A solenoid creates a uniform magnetic field.

Induced currents

- **Lenz's law:** An induced electric current flows in the direction such that the induced magnetic field opposes the *change* in the magnetic flux.

Decreasing B

Induced current

Optics

OVERVIEW

The Story of Light

Optics is the area of physics concerned with the properties and applications of light, including how light interacts with matter. But what is light? This is an ancient question with no simple answer. Rather than an all-encompassing theory of light, we will develop three different *models* of light, each of which describes and explains the behavior of light within a certain range of physical situations.

The *wave model* is based on the well-known fact that light is a wave—specifically, an electromagnetic wave. This model builds on what you learned about waves in Part IV of this text. Light waves, like any wave, are diffuse, spread out through space, and exhibit superposition and interference. The wave model of light has many important applications, including

■ Precision measurements.
■ Optical coatings on lenses, sensors, and windows.
■ Optical computing.

> **NOTE** Although light is an electromagnetic wave, your understanding of these chapters depends on nothing more than the "waviness" of light. You can begin these chapters either before or after your study of electricity and magnetism in Part VI. The electromagnetic aspects of light waves are discussed in Chapter 31.

Another well-known fact—that light travels in straight lines—is the basis for the *ray model*. The ray model will allow us to understand

■ Image formation with lenses and mirrors.
■ Optical fibers.
■ Optical instruments ranging from cameras to telescopes.

One of the most amazing optical instruments is your eye. We'll investigate optics of the eye and learn how glasses or contact lenses can correct some defects of vision.

The *photon model*, part of quantum physics, is mentioned here for completeness but won't be developed until Part VIII of this text. In the quantum world, light consists of tiny packets of energy—photons—that have both wave-like and particle-like properties. Photons will help us understand how atoms emit and absorb light.

For the most part, these three models are mutually exclusive; hence we'll pay close attention to establishing guidelines for when each model is valid.

These optical fibers—thin, flexible threads of glass that channel laser light much like water flowing through a pipe —are what make high-speed internet a reality.

33 Wave Optics

The vivid colors of this peacock feather—which change as you see it from different angles—are not due to pigments. Instead, the colors are due to the interference of light waves.

IN THIS CHAPTER, you will learn about and apply the wave model of light.

What is light?

You will learn that light has aspects of both waves and particles. We will introduce three models of light.

- The wave model of light—the subject of this chapter—describes how light waves spread out and how the superposition of multiple light waves causes interference.
- The ray model of light, in which light travels in straight lines, will explain how mirrors and lenses work. It is the subject of Chapter 34.
- The photon model of light, which will be discussed in Part VIII, is an important part of quantum physics.

One of our tasks will be to learn when each model is appropriate.

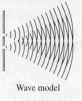

Wave model

Ray model

Photon model

« LOOKING BACK Sections 16.5, 16.7, and 16.8 Light waves, wave fronts, phase, and intensity

What is diffraction?

Diffraction is the ability of a wave to spread out after passing through a small hole or going around a corner. The diffraction of light indicates that light is a wave. One interesting finding will be that a *smaller* hole causes *more* spreading.

Does light exhibit interference?

Yes. You previously studied the thin-film interference of light reflecting from two surfaces. In this chapter we will examine the interference fringes that are seen after light passes through two narrow, closely spaced slits in an opaque screen.

Double-slit interference

« LOOKING BACK Section 17.7 Interference

What is a diffraction grating?

A diffraction grating is a periodic array of closely spaced slits or grooves. Different wavelengths are sent in different directions when light passes through a diffraction grating. Two similar wavelengths can be distinguished because the fringes of each are very narrow and precisely located.

Diffraction-grating fringes

How is interference used?

Diffraction gratings are the basis of spectroscopy, a tool for analyzing the composition of materials by the wavelengths they emit. Interferometers make precise measurements, ranging from the vibrations of wings to the movements of continents, with the controlled use of interference. And interference plays a key role in optical computers.

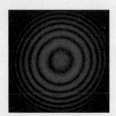

Interferometer fringes

33.1 Models of Light

The study of light is called **optics.** But what is light? The first Greek scientists did not make a distinction between light and vision. Light, to them, was inseparable from seeing. But gradually there arose a view that light actually "exists," that light is some sort of physical entity that is present regardless of whether or not someone is looking. But if light is a physical entity, what is it?

Newton, in addition to his pioneering work in mathematics and mechanics in the 1660s, investigated the nature of light. Newton knew that a water wave, after passing through an opening, *spreads out* to fill the space behind the opening. You can see this in FIGURE 33.1a, where plane waves, approaching from the left, spread out in circular arcs after passing through a hole in a barrier. This inexorable spreading of waves is the phenomenon called **diffraction.** Diffraction is a sure sign that whatever is passing through the hole is a wave.

In contrast, FIGURE 33.1b shows that sunlight makes sharp-edged shadows. We don't see the light spreading out in circular arcs after passing through an opening. This behavior is what you would expect if light consists of noninteracting particles traveling in straight lines. Some particles would pass through the openings to make bright areas on the floor, others would be blocked and cause the well-defined shadow. This reasoning led Newton to the conclusion that light consists of very small, light, fast particles that he called *corpuscles.*

The situation changed dramatically in 1801, when the English scientist Thomas Young announced that he had produced *interference* of light. Young's experiment, which we will analyze in the next section, quickly settled the debate in favor of a wave theory of light because interference is a distinctly wave-like phenomenon. But if light is a wave, what is waving? It was ultimately established that light is an *electromagnetic wave,* an oscillation of the electromagnetic field requiring no material medium in which to travel.

But this satisfying conclusion was soon undermined by new discoveries at the start of the 20th century. Albert Einstein's introduction of the concept of the *photon*—a wave having certain particle-like characteristics—marked the end of *classical physics* and the beginning of a new era called *quantum physics.* Equally important, Einstein's theory marked yet another shift in our age-old effort to understand light.

> **NOTE** Optics, as we will study it, was developed before it was known that light is an electromagnetic wave. This chapter requires an understanding of waves, from Chapters 16 and 17, but does not require a knowledge of electromagnetic fields. Thus you can study Part VII either before or after your study of electricity and magnetism in Part VI. Light polarization, the one aspect of optics that does require some familiarity with electromagnetic waves, is covered in Chapter 31.

Three Views

Light is a real physical entity, but the nature of light is elusive. Light is the chameleon of the physical world. Under some circumstances, light acts like particles traveling in straight lines. But change the circumstances, and light shows the same kinds of wave-like behavior as sound waves or water waves. Change the circumstances yet again, and light exhibits behavior that is neither wave-like nor particle-like but has characteristics of both.

Rather than an all-encompassing "theory of light," it will be better to develop three **models of light.** Each model successfully explains the behavior of light within a certain domain—that is, within a certain range of physical situations. Our task will be twofold:

1. To develop clear and distinct models of light.
2. To learn the conditions and circumstances for which each model is valid.

We'll begin with a brief summary of all three models.

FIGURE 33.1 Water waves spread out behind a small hole in a barrier, but light passing through an archway makes a sharp-edged shadow.

(a) Plane waves approach from the left.

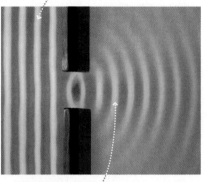

Circular waves spread out on the right.

(b)

A beam of sunlight has a sharp edge.

Three models of light

| The Wave Model | The Ray Model | The Photon Model |

The Wave Model

The wave model of light is responsible for the well-known "fact" that light is a *wave*. Indeed, under many circumstances light exhibits the same behavior as sound or water waves. Lasers and electro-optical devices are best described by the wave model of light. Some aspects of the wave model were introduced in Chapters 16 and 17, and it is the primary focus of this chapter.

The Ray Model

An equally well-known "fact" is that light travels in straight lines. These straight-line paths are called *light rays*. The properties of prisms, mirrors, and lenses are best understood in terms of light rays. Unfortunately, it's difficult to reconcile "light travels in straight lines" with "light is a wave." For the most part, waves and rays are mutually exclusive models of light. One of our important tasks will be to learn when each model is appropriate. Ray optics is the subject of Chapters 34 and 35.

The Photon Model

Modern technology is increasingly reliant on quantum physics. In the quantum world, light behaves like neither a wave nor a particle. Instead, light consists of *photons* that have both wave-like and particle-like properties. Much of the quantum theory of light is beyond the scope of this textbook, but we will take a peek at some of the important ideas in Part VIII.

FIGURE 33.2 Light, just like a water wave, does spread out behind an opening if the opening is sufficiently narrow.

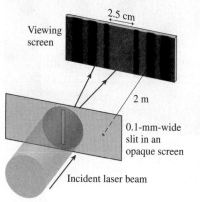

33.2 The Interference of Light

Newton might have reached a different conclusion about the nature of light had he seen the experiment depicted in FIGURE 33.2. Here light of a single wavelength (or color) passes through a narrow slit that is only 0.1 mm wide, about twice the width of a human hair. The image shows how the light appears on a viewing screen 2 m behind the slit. If light travels in straight lines, as Newton thought, we should see a narrow strip of light, about 0.1 mm wide, with dark shadows on either side. Instead, we see a band of light extending over about 2.5 cm, a distance much wider than the aperture, with dimmer patches of light extending even farther on either side.

If you compare Figure 33.2 to the water wave of Figure 33.1a, you see that *the light is spreading out* behind the opening. The light is exhibiting diffraction, the sure sign of waviness. We will look at diffraction in more detail later in the chapter. For now, we merely need the *observation* that light does, indeed, spread out behind an opening that is sufficiently narrow. Light is acting as a wave, so we should—as Thomas Young did—be able to observe the interference of light waves.

A Brief Review of Interference

Waves obey the *principle of superposition:* If two waves overlap, their displacements add—whether it's the displacement of air molecules in a sound wave or the electric field in a light wave. You learned in **« Section 17.5–17.7** that *constructive interference* occurs when the crests of two waves overlap, adding to produce a wave with a larger amplitude and thus a greater intensity. Conversely, the overlap of the crest of one wave with the trough of another wave produces a wave with reduced amplitude (perhaps even zero) and thus a lesser intensity. This is *destructive interference*.

FIGURE 33.3 shows two in-phase sources of sinusoidal waves. The circular rings, you will recall, are the *wave crests*, and they are spaced one wavelength λ apart. Although Figure 33.3 is a snapshot, frozen in time, you should envision both sets of rings propagating outward at the wave speed as the sources oscillate.

We can measure distances simply by counting rings. For example, you can see that the dot on the right side is exactly three wavelengths from the top source and exactly four wavelengths from the bottom source, making $r_1 = 3\lambda$ and $r_2 = 4\lambda$. At this point, the *path-length difference* is $\Delta r = r_2 - r_1 = \lambda$. **Points where Δr is an integer number of wavelengths are points of constructive interference.** The waves may have traveled difference distances, but crests align with crests and troughs align with troughs to produce a wave with a larger amplitude.

FIGURE 33.3 Constructive and destructive interference occur along antinodal and nodal lines.

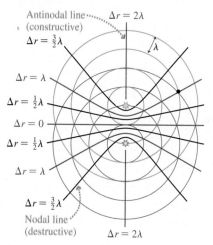

If the path-length difference is a half-integer number of wavelengths, then the crest of one wave will arrive with the trough of the other. Thus **points where Δr is a half-integer number of wavelengths are points of destructive interference.** Mathematically, the conditions for interference are:

$$
\begin{array}{ll}
\text{Constructive interference:} & \Delta r = m\lambda \\
\text{Destructive interference:} & \Delta r = \left(m + \tfrac{1}{2}\right)\lambda
\end{array}
\qquad m = 0, 1, 2, \ldots \qquad (33.1)
$$

Destructive interference is *perfect destructive interference*, with zero intensity, only if both waves have exactly the same amplitude.

The set of all points for which $\Delta r = m\lambda$ is called an *antinodal line*. Maximum constructive interference is happening at every point along this line. If these are sound waves, you will hear maximum loudness by standing on an antinodal line. If these are light waves, you will see maximum brightness everywhere an antinodal line touches a viewing screen.

Similarly, maximum destructive interference occurs along *nodal lines*—analogous to the *nodes* of a one-dimensional standing wave. If these are light waves, a viewing screen will be dark everywhere it intersects a nodal line.

Young's Double-Slit Experiment

Let's now see how these ideas about interference apply to light waves. FIGURE 33.4a shows an experiment in which a laser beam is aimed at an opaque screen containing two long, narrow slits that are very close together. This pair of slits is called a **double slit,** and in a typical experiment they are ≈ 0.1 mm wide and spaced ≈ 0.5 mm apart. We will assume that the laser beam illuminates both slits equally and that any light passing through the slits impinges on a viewing screen. This is the essence of Young's experiment of 1801, although he used sunlight rather than a laser.

What should we expect to see on the screen? FIGURE 33.4b is a view from above the experiment, looking down on the top ends of the slits and the top edge of the viewing screen. Because the slits are very narrow, **light spreads out behind each slit** exactly as it did in Figure 33.2. These two spreading waves overlap and interfere with each other, just as if they were sound waves emitted by two

FIGURE 33.4 A double-slit interference experiment.

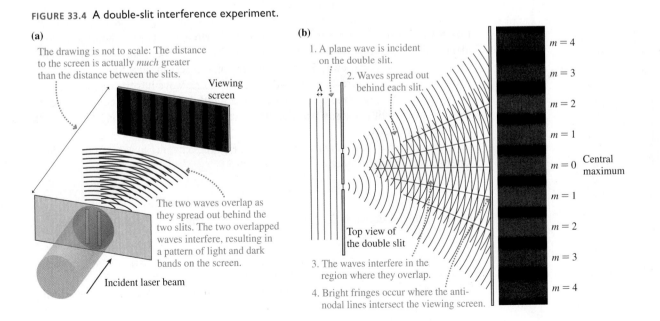

(a)

The drawing is not to scale: The distance to the screen is actually *much* greater than the distance between the slits.

Viewing screen

The two waves overlap as they spread out behind the two slits. The two overlapped waves interfere, resulting in a pattern of light and dark bands on the screen.

Incident laser beam

(b)

1. A plane wave is incident on the double slit.

2. Waves spread out behind each slit.

λ

Top view of the double slit

3. The waves interfere in the region where they overlap.

4. Bright fringes occur where the anti-nodal lines intersect the viewing screen.

$m = 4$
$m = 3$
$m = 2$
$m = 1$
$m = 0$ Central maximum
$m = 1$
$m = 2$
$m = 3$
$m = 4$

loudspeakers. Notice how the antinodal lines of constructive interference are just like those in Figure 33.3.

The image in Figure 33.4b shows how the screen looks. As expected, the light is intense at points where an antinodal line intersects the screen. There is no light at all at points where a nodal line intersects the screen. These alternating bright and dark bands of light, due to constructive and destructive interference, are called **interference fringes.** The fringes are numbered $m = 0, 1, 2, 3, \ldots$, going outward from the center. The brightest fringe, at the midpoint of the viewing screen, with $m = 0$, is called the **central maximum.**

STOP TO THINK 33.1 Suppose the viewing screen in Figure 33.4 is moved closer to the double slit. What happens to the interference fringes?

a. They get brighter but otherwise do not change.
b. They get brighter and closer together.
c. They get brighter and farther apart.
d. They get out of focus.
e. They fade out and disappear.

Analyzing Double-Slit Interference

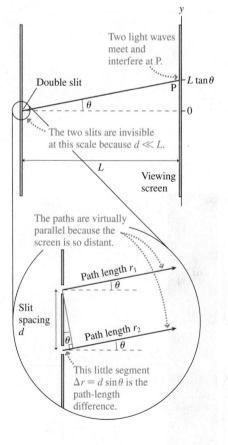

FIGURE 33.5 Geometry of the double-slit experiment.

Figure 33.4 showed qualitatively how interference is produced behind a double slit by the overlap of the light waves spreading out behind each slit. Now let's analyze the experiment more carefully. **FIGURE 33.5** shows a double-slit experiment in which the spacing between the two slits is d and the distance to the viewing screen is L. We will assume that L is *very* much larger than d. Consequently, we don't see the individual slits in the upper part of Figure 33.5.

Let P be a point on the screen at angle θ. Our goal is to determine whether the interference at P is constructive, destructive, or in between. The insert to Figure 33.5 shows the individual slits and the paths from these slits to point P. Because P is so far away on this scale, the two paths are virtually parallel, both at angle θ. Both slits are illuminated by the *same* wave front from the laser; hence the slits act as sources of identical, in-phase waves ($\Delta\phi_0 = 0$). Thus the interference at point P is constructive, producing a bright fringe, if $\Delta r = m\lambda$ at that point.

The midpoint on the viewing screen at $y = 0$ is equally distant from both slits ($\Delta r = 0$) and thus is a point of constructive interference. This is the bright fringe identified as the central maximum in Figure 33.4b. The path-length difference increases as you move away from the center of the screen, and the $m = 1$ fringes occur at the points where $\Delta r = 1\lambda$—that is, where one wave has traveled exactly one wavelength farther than the other. In general, **the mth bright fringe occurs where the wave from one slit travels m wavelengths farther than the wave from the other slit and thus $\Delta r = m\lambda$.**

You can see from the magnified portion of Figure 33.5 that the wave from the lower slit travels an extra distance

$$\Delta r = d\sin\theta \qquad (33.2)$$

If we use this in Equation 33.1, we find that bright fringes (constructive interference) occur at angles θ_m such that

$$\Delta r = d\sin\theta_m = m\lambda \qquad m = 0, 1, 2, 3, \ldots \qquad (33.3)$$

We added the subscript m to denote that θ_m is the angle of the mth bright fringe, starting with $m = 0$ at the center.

In practice, the angle θ in a double-slit experiment is very small ($<1°$). We can use the small-angle approximation $\sin\theta \approx \theta$, where θ must be in radians, to write Equation 33.3 as

$$\theta_m = m\frac{\lambda}{d} \qquad m = 0, 1, 2, 3, \ldots \qquad \text{(angles of bright fringes)} \qquad (33.4)$$

This gives the angular positions *in radians* of the bright fringes in the interference pattern.

Positions of the Fringes

It's usually easier to measure distances rather than angles, so we can also specify point P by its position on a y-axis with the origin directly across from the midpoint between the slits. You can see from Figure 33.5 that

$$y = L\tan\theta \qquad (33.5)$$

Using the small-angle approximation once again, this time in the form $\tan\theta \approx \theta$, we can substitute θ_m from Equation 33.4 for $\tan\theta_m$ in Equation 33.5 to find that the mth bright fringe occurs at position

$$y_m = \frac{m\lambda L}{d} \qquad m = 0, 1, 2, 3, \ldots \qquad \text{(positions of bright fringes)} \qquad (33.6)$$

The interference pattern is symmetrical, so there is an mth bright fringe at the same distance on both sides of the center. You can see this in Figure 33.4b. As we've noted, **the $m = 1$ fringes occur at points on the screen where the light from one slit travels exactly one wavelength farther than the light from the other slit.**

> **NOTE** Equations 33.4 and 33.6 do *not* apply to the interference of sound waves from two loudspeakers. The approximations we've used (small angles, $L \gg d$) are usually not valid for the much longer wavelengths of sound waves.

Equation 33.6 predicts that **the interference pattern is a series of equally spaced bright lines** on the screen, exactly as shown in Figure 33.4b. How do we know the fringes are equally spaced? The **fringe spacing** between the m fringe and the $m + 1$ fringe is

$$\Delta y = y_{m+1} - y_m = \frac{(m + 1)\lambda L}{d} - \frac{m\lambda L}{d} = \frac{\lambda L}{d} \qquad (33.7)$$

Because Δy is independent of m, *any* two adjacent bright fringes have the same spacing.

The dark fringes in the image are bands of destructive interference. They occur at positions where the path-length difference of the waves is a half-integer number of wavelengths:

$$\Delta r = \left(m + \tfrac{1}{2}\right)\lambda \qquad m = 0, 1, 2, \ldots \qquad \begin{array}{l}\text{(destructive}\\\text{interference)}\end{array} \qquad (33.8)$$

We can use Equation 33.2 for Δr and the small-angle approximation to find that the dark fringes are located at positions

$$y'_m = \left(m + \tfrac{1}{2}\right)\frac{\lambda L}{d} \qquad m = 0, 1, 2, \ldots \qquad \begin{array}{l}\text{(positions of}\\\text{dark fringes)}\end{array} \qquad (33.9)$$

We have used y'_m, with a prime, to distinguish the location of the mth minimum from the mth maximum at y_m. You can see from Equation 33.9 that **the dark fringes are located exactly halfway between the bright fringes.**

As a quick example, suppose that light from a helium-neon laser ($\lambda = 633$ nm) illuminates two slits spaced 0.40 mm apart and that a viewing screen is 2.0 m behind the slits. The $m = 2$ bright fringe is located at position

$$y_2 = \frac{2\lambda L}{d} = \frac{2(633 \times 10^{-9}\text{ m})(2.0\text{ m})}{4.0 \times 10^{-4}\text{ m}} = 6.3\text{ mm}$$

Similarly, the $m = 2$ dark fringe is found at $y_2' = \left(2 + \frac{1}{2}\right)\lambda L/d = 7.9$ mm. Because the fringes are counted outward from the center, the $m = 2$ bright fringe occurs *before* the $m = 2$ dark fringe.

EXAMPLE 33.1 | **Measuring the wavelength of light**

A double-slit interference pattern is observed on a screen 1.0 m behind two slits spaced 0.30 mm apart. Ten bright fringes span a distance of 1.7 cm. What is the wavelength of the light?

MODEL It is not always obvious which fringe is the central maximum. Slight imperfections in the slits can make the interference fringe pattern less than ideal. However, you do not need to identify the $m = 0$ fringe because you can make use of the fact that the fringe spacing Δy is uniform. Ten bright fringes have *nine* spaces between them (not ten—be careful!).

VISUALIZE The interference pattern looks like the image of Figure 33.4b.

SOLVE The fringe spacing is

$$\Delta y = \frac{1.7\text{ cm}}{9} = 1.89 \times 10^{-3}\text{ m}$$

Using this fringe spacing in Equation 33.7, we find that the wavelength is

$$\lambda = \frac{d}{L}\Delta y = 5.7 \times 10^{-7}\text{ m} = 570\text{ nm}$$

It is customary to express the wavelengths of visible light in nanometers. Be sure to do this as you solve problems.

ASSESS Young's double-slit experiment not only demonstrated that light is a wave, it provided a means for measuring the wavelength. You learned in Chapter 16 that the wavelengths of visible light span the range 400–700 nm. These lengths are smaller than we can easily comprehend. A wavelength of 570 nm, which is in the middle of the visible spectrum, is only about 1% of the diameter of a human hair.

STOP TO THINK 33.2 Light of wavelength λ_1 illuminates a double slit, and interference fringes are observed on a screen behind the slits. When the wavelength is changed to λ_2, the fringes get closer together. Is λ_2 larger or smaller than λ_1?

Intensity of the Double-Slit Interference Pattern

Equations 33.6 and 33.9 locate the positions of maximum and zero intensity. To complete our analysis we need to calculate the light *intensity* at every point on the screen. In « Chapter 17, where interference was introduced, you learned that the net amplitude E of two superimposed waves is

$$E = \left|2e\cos\left(\frac{\Delta\phi}{2}\right)\right| \tag{33.10}$$

where, for light waves, e is the electric field amplitude of each individual wave. Because the sources (i.e., the two slits) are in phase, the phase difference $\Delta\phi$ at the point where the two waves are combined is due only to the path-length difference: $\Delta\phi = 2\pi(\Delta r/\lambda)$. Using Equation 33.2 for Δr, along with the small-angle approximation and Equation 33.5 for y, we find the phase difference at position y on the screen to be

$$\Delta\phi = 2\pi\frac{\Delta r}{\lambda} = 2\pi\frac{d\sin\theta}{\lambda} \approx 2\pi\frac{d\tan\theta}{\lambda} = \frac{2\pi d}{\lambda L}y \tag{33.11}$$

Substituting Equation 33.11 into Equation 33.10, we find the wave amplitude at position y to be

$$E = \left| 2e \cos\left(\frac{\pi d}{\lambda L} y\right) \right| \qquad (33.12)$$

The light *intensity*, which is what we see, is proportional to the square of the amplitude. The intensity of a single slit, with amplitude e, is $I_1 = Ce^2$, where C is a proportionality constant. For the double slit, the intensity at position y on the screen is

$$I = CE^2 = 4Ce^2 \cos^2\left(\frac{\pi d}{\lambda L} y\right) \qquad (33.13)$$

Replacing Ce^2 with I_1, we see that the intensity of an *ideal* double-slit interference pattern at position y is

$$I_{\text{double}} = 4I_1 \cos^2\left(\frac{\pi d}{\lambda L} y\right) \qquad (33.14)$$

We've said "ideal" because we've assumed that e, the electric field amplitude of each wave, is constant across the screen.

FIGURE 33.6a is a graph of the ideal double-slit intensity versus position y. Notice the unusual orientation of the graph, with the intensity increasing toward the *left* so that the y-axis can match the experimental layout. You can see that the intensity oscillates between dark fringes ($I_{\text{double}} = 0$) and bright fringes ($I_{\text{double}} = 4I_1$). The maxima occur at points where $y_m = m\lambda L/d$. This is what we found earlier for the positions of the bright fringes, so Equation 33.14 is consistent with our initial analysis.

One curious feature is that the light intensity at the maxima is $I = 4I_1$, four times the intensity of the light from each slit alone. You might think that two slits would make the light twice as intense as one slit, but interference leads to a different result. Mathematically, two slits make the *amplitude* twice as big at points of constructive interference ($E = 2e$), so the intensity increases by a factor of $2^2 = 4$. Physically, this is conservation of energy. The line labeled $2I_1$ in Figure 33.6a is the uniform intensity that two slits would produce *if* the waves did not interfere. Interference does not change the amount of light energy coming through the two slits, but it does redistribute the light energy on the viewing screen. You can see that the *average* intensity of the oscillating curve is $2I_1$, but the intensity of the bright fringes gets pushed up from $2I_1$ to $4I_1$ in order for the intensity of the dark fringes to drop from $2I_1$ to 0.

Equation 33.14 predicts that, ideally, all interference fringes are equally bright, but you saw in Figure 33.4b that the fringes decrease in brightness as you move away from the center. The erroneous prediction stems from our assumption that the amplitude e of the wave from each slit is constant across the screen. A more detailed calculation, which we will do in Section 33.5, must consider the varying intensity of the light that has diffracted through each of the slits. We'll find that Equation 33.14 is still correct if I_1 slowly decreases as y increases.

FIGURE 33.6b summarizes this analysis by graphing the light intensity (Equation 33.14) with I_1 slowly decreasing as y increases. Comparing this graph to the image, you can see that the wave model of light has provided an excellent description of Young's double-slit interference experiment.

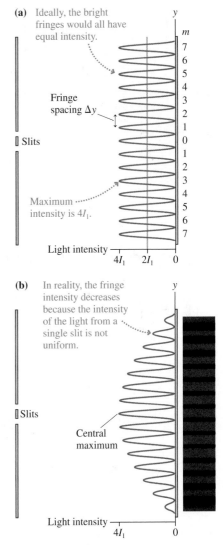

FIGURE 33.6 Intensity of the interference fringes in a double-slit experiment.

(a) Ideally, the bright fringes would all have equal intensity.

Fringe spacing Δy

Slits

Maximum intensity is $4I_1$.

Light intensity

$4I_1$ $2I_1$ 0

(b) In reality, the fringe intensity decreases because the intensity of the light from a single slit is not uniform.

Slits

Central maximum

Light intensity

$4I_1$ 0

33.3 The Diffraction Grating

Suppose we were to replace the double slit with an opaque screen that has N closely spaced slits. When illuminated from one side, each of these slits becomes the source of a light wave that diffracts, or spreads out, behind the slit. Such a multi-slit device is called a **diffraction grating.** The light intensity pattern on a screen behind a diffraction grating is due to the interference of N overlapped waves.

FIGURE 33.7 Top view of a diffraction grating with $N = 10$ slits.

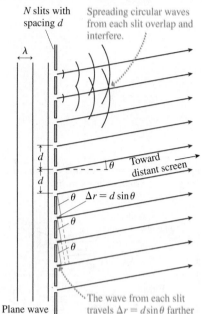

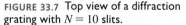

FIGURE 33.8 Angles of constructive interference.

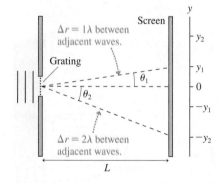

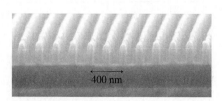

A microscopic side-on look at a diffraction grating.

NOTE The terms "interference" and "diffraction" have historical roots that predate our modern understanding of wave optics, and thus their use can be confusing. Physically, both arise from the superposition of overlapped waves. *Interference* usually describes a superposition of waves coming from distinct sources—as in double-slit interference. *Diffraction* usually describes how the superposition of different portions of a single wave front causes light to bend or spread after encountering an obstacle, such as a narrow slit. Diffraction is important in a diffraction grating because light does spread out behind each slit, but the intensity pattern seen behind a diffraction grating results from the interference of light waves coming from each of the slits. It might more logically be called an *interference grating,* but it's not. Don't let the name confuse you.

FIGURE 33.7 shows a diffraction grating in which N slits are equally spaced a distance d apart. This is a top view of the grating, as we look down on the experiment, and the slits extend above and below the page. Only 10 slits are shown here, but a practical grating will have hundreds or even thousands of slits. Suppose a plane wave of wavelength λ approaches from the left. The crest of a plane wave arrives *simultaneously* at each of the slits, causing the wave emerging from each slit to be in phase with the wave emerging from every other slit. Each of these emerging waves spreads out, just like the light wave in Figure 33.2, and after a short distance they all overlap with each other and interfere.

We want to know how the interference pattern will appear on a screen behind the grating. The light wave at the screen is the superposition of N waves, from N slits, as they spread and overlap. As we did with the double slit, we'll assume that the distance L to the screen is very large in comparison with the slit spacing d; hence the path followed by the light from one slit to a point on the screen is *very nearly* parallel to the path followed by the light from neighboring slits. The paths cannot be perfectly parallel, of course, or they would never meet to interfere, but the slight deviation from perfect parallelism is too small to notice. You can see in Figure 33.7 that the wave from one slit travels distance $\Delta r = d \sin \theta$ more than the wave from the slit above it and $\Delta r = d \sin \theta$ less than the wave below it. This is the same reasoning we used in Figure 33.5 to analyze the double-slit experiment.

Figure 33.7 is a magnified view of the slits. FIGURE 33.8 steps back to where we can see the viewing screen. If the angle θ is such that $\Delta r = d \sin \theta = m\lambda$, where m is an integer, then the light wave arriving at the screen from one slit will be *exactly in phase* with the light waves arriving from the two slits next to it. But each of those waves is in phase with waves from the slits next to them, and so on until we reach the end of the grating. In other words, **N light waves, from N different slits, will *all* be in phase with each other when they arrive at a point on the screen at angle θ_m such that**

$$d \sin \theta_m = m\lambda \qquad m = 0, 1, 2, 3, \ldots \qquad (33.15)$$

The screen will have bright constructive-interference fringes at the values of θ_m given by Equation 33.15. We say that the light is "diffracted at angle θ_m."

Because it's usually easier to measure distances rather than angles, the position y_m of the mth maximum is

$$y_m = L \tan \theta_m \qquad \text{(positions of bright fringes)} \qquad (33.16)$$

The integer m is called the **order** of the diffraction. For example, light diffracted at θ_2 would be the second-order diffraction. Practical gratings, with very small values for d, display only a few orders. Because d is usually very small, it is customary to characterize a grating by the number of *lines per millimeter.* Here "line" is synonymous with "slit," so the number of lines per millimeter is simply the inverse of the slit spacing d in millimeters.

NOTE The condition for constructive interference in a grating of N slits is identical to Equation 33.4 for just two slits. Equation 33.15 is simply the requirement that the path-length difference between adjacent slits, be they two or N, is $m\lambda$. But unlike the angles in double-slit interference, the angles of constructive interference from a diffraction grating are generally *not* small angles. The reason is that the slit spacing d in a diffraction grating is so small that λ/d is not a small number. Thus you *cannot* use the small-angle approximation to simplify Equations 33.15 and 33.16.

The wave amplitude at the points of constructive interference is Ne because N waves of amplitude e combine in phase. Because the intensity depends on the square of the amplitude, the intensities of the bright fringes of a diffraction grating are

$$I_{max} = N^2 I_1 \tag{33.17}$$

where, as before, I_1 is the intensity of the wave from a single slit. You can see that the fringe intensities increase rapidly as the number of slits increases.

Not only do the fringes get brighter as N increases, they also get narrower. This is again a matter of conservation of energy. If the light waves did not interfere, the intensity from N slits would be NI_1. Interference increases the intensity of the bright fringes by an extra factor of N, so to conserve energy the width of the bright fringes must be proportional to $1/N$. For a realistic diffraction grating, with $N > 100$, the interference pattern consists of a small number of *very* bright and *very* narrow fringes while most of the screen remains dark. **FIGURE 33.9a** shows the interference pattern behind a diffraction grating both graphically and with a simulation of the viewing screen. A comparison with Figure 33.6b shows that the bright fringes of a diffraction grating are much sharper and more distinct than the fringes of a double slit.

Because the bright fringes are so distinct, diffraction gratings are used for measuring the wavelengths of light. Suppose the incident light consists of two slightly different wavelengths. Each wavelength will be diffracted at a slightly different angle and, if N is sufficiently large, we'll see two distinct fringes on the screen. **FIGURE 33.9b** illustrates this idea. By contrast, the bright fringes in a double-slit experiment are too broad to distinguish the fringes of one wavelength from those of the other.

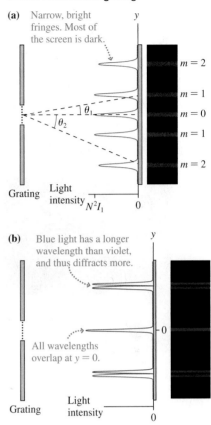

FIGURE 33.9 The interference pattern behind a diffraction grating.

(a) Narrow, bright fringes. Most of the screen is dark.

$m = 2$
$m = 1$
$m = 0$
$m = 1$
$m = 2$

Grating   Light intensity $N^2 I_1$ 0

(b) Blue light has a longer wavelength than violet, and thus diffracts more.

All wavelengths overlap at $y = 0$.

Grating   Light intensity   0

EXAMPLE 33.2 | **Measuring wavelengths emitted by sodium atoms**

Light from a sodium lamp passes through a diffraction grating having 1000 slits per millimeter. The interference pattern is viewed on a screen 1.000 m behind the grating. Two bright yellow fringes are visible 72.88 cm and 73.00 cm from the central maximum. What are the wavelengths of these two fringes?

VISUALIZE This is the situation shown in Figure 33.9b. The two fringes are very close together, so we expect the wavelengths to be only slightly different. No other yellow fringes are mentioned, so we will assume these two fringes are the first-order diffraction ($m = 1$).

SOLVE The distance y_m of a bright fringe from the central maximum is related to the diffraction angle by $y_m = L\tan\theta_m$. Thus the diffraction angles of these two fringes are

$$\theta_1 = \tan^{-1}\left(\frac{y_1}{L}\right) = \begin{cases} 36.08° & \text{fringe at 72.88 cm} \\ 36.13° & \text{fringe at 73.00 cm} \end{cases}$$

These angles must satisfy the interference condition $d\sin\theta_1 = \lambda$, so the wavelengths are $\lambda = d\sin\theta_1$. What is d? If a 1 mm length of the grating has 1000 slits, then the spacing from one slit to the next must be 1/1000 mm, or $d = 1.000 \times 10^{-6}$ m. Thus the wavelengths creating the two bright fringes are

$$\lambda = d\sin\theta_1 = \begin{cases} 589.0 \text{ nm} & \text{fringe at 72.88 cm} \\ 589.6 \text{ nm} & \text{fringe at 73.00 cm} \end{cases}$$

ASSESS We had data accurate to four significant figures, and all four were necessary to distinguish the two wavelengths.

The science of measuring the wavelengths of atomic and molecular emissions is called **spectroscopy**. The two sodium wavelengths in this example are called the *sodium doublet,* a name given to two closely spaced wavelengths emitted by the atoms of one element. This doublet is an identifying characteristic of sodium. Because no other element emits these two wavelengths, the doublet can be used to identify the presence of sodium in a sample of unknown composition, even if sodium is only a very minor constituent. This procedure is called *spectral analysis*.

FIGURE 33.10 Reflection gratings.

FIGURE 33.10 Reflection gratings.

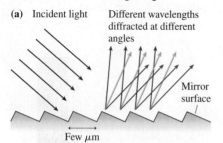

(a) Incident light   Different wavelengths
diffracted at different
angles

Mirror
surface

Few μm

A reflection grating can be made by cutting
parallel grooves in a mirror surface.

(b)

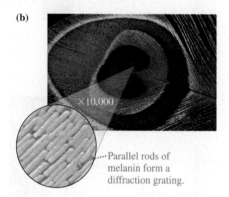

×10,000

⋯Parallel rods of
melanin form a
diffraction grating.

Reflection Gratings

We have analyzed what is called a *transmission grating,* with many parallel slits. In practice, most diffraction gratings are manufactured as *reflection gratings.* The simplest reflection grating, shown in FIGURE 33.10a, is a mirror with hundreds or thousands of narrow, parallel grooves cut into the surface. The grooves divide the surface into many parallel reflective stripes, each of which, when illuminated, becomes the source of a spreading wave. Thus an incident light wave is divided into N overlapped waves. The interference pattern is exactly the same as the interference pattern of light transmitted through N parallel slits.

Naturally occurring reflection gratings are responsible for some forms of color in nature. As the micrograph of FIGURE 33.10b shows, a peacock feather consists of nearly parallel rods of melanin. These act as a reflection grating and create the ever-changing, multicolored hues of iridescence as the angle between the grating and your eye changes. The iridescence of some insects is due to diffraction from parallel microscopic ridges on the shell.

The rainbow of colors reflected from the surface of a DVD is a similar display of interference. The surface of a DVD is smooth plastic with a mirror-like reflective coating in which millions of microscopic holes, each about 1 μm in diameter, encode digital information. From an optical perspective, the array of holes in a shiny surface is a two-dimensional version of the reflection grating shown in Figure 33.10a. Reflection gratings can be manufactured at very low cost simply by stamping holes or grooves into a reflective surface, and these are widely sold as toys and novelty items. Rainbows of color are seen as each wavelength of white light is diffracted at a unique angle.

STOP TO THINK 33.3 White light passes through a diffraction grating and forms rainbow patterns on a screen behind the grating. For each rainbow,

   a. The red side is on the right, the violet side on the left.
   b. The red side is on the left, the violet side on the right.
   c. The red side is closest to the center of the screen, the violet side is farthest from the center.
   d. The red side is farthest from the center of the screen, the violet side is closest to the center.

33.4 Single-Slit Diffraction

We opened this chapter with a photograph (Figure 33.1a) of a water wave passing through a hole in a barrier, then spreading out on the other side. You then saw an image (Figure 33.2) showing that light, after passing through a very narrow slit, also spreads out on the other side. This is called *diffraction.*

FIGURE 33.11 shows the experimental arrangement for observing the diffraction of light through a narrow slit of width a. Diffraction through a tall, narrow slit is known as **single-slit diffraction.** A viewing screen is placed distance L behind the slit, and we will assume that $L \gg a$. The light pattern on the viewing screen consists of a *central maximum* flanked by a series of weaker **secondary maxima** and dark fringes. Notice that the central maximum is significantly broader than the secondary maxima. It is also significantly brighter than the secondary maxima, although that is hard to tell here because this image has been overexposed to make the secondary maxima show up better.

FIGURE 33.11 A single-slit diffraction
experiment.

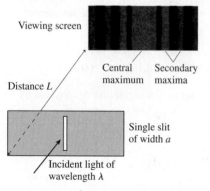

Viewing screen

Central   Secondary
maximum   maxima

Distance L

Single slit
of width a

Incident light of
wavelength λ

Huygens' Principle

Our analysis of the superposition of waves from distinct sources, such as two loudspeakers or the two slits in a double-slit experiment, has tacitly assumed that the sources are *point sources,* with no measurable extent. To understand diffraction, we need to think about the propagation of an *extended* wave front. This is a problem first considered by the Dutch scientist Christiaan Huygens, a contemporary of Newton.

Huygens lived before a mathematical theory of waves had been developed, so he developed a geometrical model of wave propagation. His idea, which we now call **Huygens' principle,** has two steps:

1. Each point on a wave front is the source of a spherical *wavelet* that spreads out at the wave speed.
2. At a later time, the shape of the wave front is the line tangent to all the wavelets.

FIGURE 33.12 illustrates Huygens' principle for a plane wave and a spherical wave. As you can see, the line tangent to the wavelets of a plane wave is a plane that has propagated to the right. The line tangent to the wavelets of a spherical wave is a larger sphere.

Huygens' principle is a visual device, not a theory of waves. Nonetheless, the full mathematical theory of waves, as it developed in the 19th century, justifies Huygens' basic idea, although it is beyond the scope of this textbook to prove it.

Analyzing Single-Slit Diffraction

FIGURE 33.13a shows a wave front passing through a narrow slit of width a. According to Huygens' principle, each point on the wave front can be thought of as the source of a spherical wavelet. These wavelets overlap and interfere, producing the diffraction pattern seen on the viewing screen. The full mathematical analysis, using *every* point on the wave front, is a fairly difficult problem in calculus. We'll be satisfied with a geometrical analysis based on just a few wavelets.

FIGURE 33.13b shows the paths of several wavelets that travel straight ahead to the central point on the screen. (The screen is *very* far to the right in this magnified view of the slit.) The paths are very nearly parallel to each other, thus all the wavelets travel the same distance and arrive at the screen *in phase* with each other. The *constructive interference* between these wavelets produces the central maximum of the diffraction pattern at $\theta = 0$.

The situation is different at points away from the center. Wavelets 1 and 2 in **FIGURE 33.13c** start from points that are distance $a/2$ apart. If the angle is such that Δr_{12}, the extra distance traveled by wavelet 2, happens to be $\lambda/2$, then wavelets 1 and 2 arrive out of phase and interfere destructively. But if Δr_{12} is $\lambda/2$, then the difference Δr_{34} between paths 3 and 4 and the difference Δr_{56} between paths 5 and 6 are also $\lambda/2$.

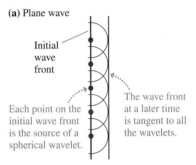

FIGURE 33.12 Huygens' principle applied to the propagation of plane waves and spherical waves.

(a) Plane wave

Initial wave front

Each point on the initial wave front is the source of a spherical wavelet.

The wave front at a later time is tangent to all the wavelets.

(b) Spherical wave

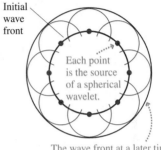

Initial wave front

Each point is the source of a spherical wavelet.

The wave front at a later time is tangent to all the wavelets.

FIGURE 33.13 Each point on the wave front is a source of spherical wavelets. The superposition of these wavelets produces the diffraction pattern on the screen.

(a) Greatly magnified view of slit

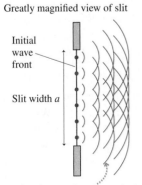

Initial wave front

Slit width a

The wavelets from each point on the initial wave front overlap and interfere, creating a diffraction pattern on the screen.

(b)

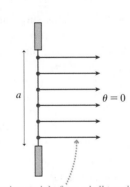

a

$\theta = 0$

The wavelets going straight forward all travel the same distance to the screen. Thus they arrive in phase and interfere constructively to produce the central maximum.

(c)

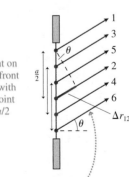

Each point on the wave front is paired with another point distance $a/2$ away.

$\frac{a}{2}$

θ

1
3
5
2
4
6
Δr_{12}

θ

These wavelets all meet on the screen at angle θ. Wavelet 2 travels distance $\Delta r_{12} = (a/2)\sin\theta$ farther than wavelet 1.

Those pairs of wavelets also interfere destructively. The superposition of all the wavelets produces perfect destructive interference.

Figure 33.13c shows six wavelets, but our conclusion is valid for any number of wavelets. The key idea is that **every point on the wave front can be paired with another point distance a/2 away.** If the path-length difference is $\lambda/2$, the wavelets from these two points arrive at the screen out of phase and interfere destructively. When we sum the displacements of all N wavelets, they will—pair by pair—add to zero. The viewing screen at this position will be dark. This is the main idea of the analysis, one worth thinking about carefully.

You can see from Figure 33.13c that $\Delta r_{12} = (a/2)\sin\theta$. This path-length difference will be $\lambda/2$, the condition for destructive interference, if

$$\Delta r_{12} = \frac{a}{2}\sin\theta_1 = \frac{\lambda}{2} \tag{33.18}$$

or, equivalently, if $a\sin\theta_1 = \lambda$.

> **NOTE** Equation 33.18 cannot be satisfied if the slit width a is less than the wavelength λ. If a wave passes through an opening smaller than the wavelength, the central maximum of the diffraction pattern expands to where it *completely* fills the space behind the opening. There are no minima or dark spots at any angle. This situation is uncommon for light waves, because λ is so small, but quite common in the diffraction of sound and water waves.

We can extend this idea to find other angles of perfect destructive interference. Suppose each wavelet is paired with another wavelet from a point $a/4$ away. If Δr between these wavelets is $\lambda/2$, then all N wavelets will again cancel in pairs to give complete destructive interference. The angle θ_2 at which this occurs is found by replacing $a/2$ in Equation 33.18 with $a/4$, leading to the condition $a\sin\theta_2 = 2\lambda$. This process can be continued, and we find that the general condition for complete destructive interference is

$$a\sin\theta_p = p\lambda \qquad p = 1, 2, 3, \ldots \tag{33.19}$$

When $\theta_p \ll 1$ rad, which is almost always true for light waves, we can use the small-angle approximation to write

$$\theta_p = p\frac{\lambda}{a} \qquad p = 1, 2, 3, \ldots \qquad \text{(angles of dark fringes)} \tag{33.20}$$

Equation 33.20 gives the angles *in radians* to the dark minima in the diffraction pattern of Figure 33.11. Notice that $p = 0$ is explicitly *excluded*. $p = 0$ corresponds to the straight-ahead position at $\theta = 0$, but you saw in Figures 33.11 and 33.13b that $\theta = 0$ is the central *maximum,* not a minimum.

> **NOTE** It is perhaps surprising that Equations 33.19 and 33.20 are *mathematically* the same as the condition for the mth *maximum* of the double-slit interference pattern. But the physical meaning here is quite different. Equation 33.20 locates the *minima* (dark fringes) of the single-slit diffraction pattern.

You might think that we could use this method of pairing wavelets from different points on the wave front to find the maxima in the diffraction pattern. Why not take two points on the wave front that are distance $a/2$ apart, find the angle at which their wavelets are in phase and interfere constructively, then sum over all points on the wave front? There is a subtle but important distinction. **FIGURE 33.14** shows six vector arrows. The arrows in Figure 33.14a are arranged in pairs such that the two members of each pair cancel. The sum of all six vectors is clearly the zero vector $\vec{0}$, representing destructive interference. This is the procedure we used in Figure 33.13c to arrive at Equation 33.18.

FIGURE 33.14 Destructive interference by pairs leads to net destructive interference, but constructive interference by pairs does *not* necessarily lead to net constructive interference.

(a)

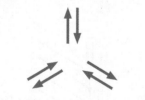

Each pair of vectors interferes destructively. The vector sum of all six vectors is zero.

(b)

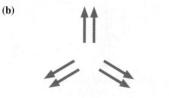

Each pair of vectors interferes constructively. Even so, the vector sum of all six vectors is zero.

The arrows in Figure 33.14b are arranged in pairs such that the two members of each pair point in the same direction—constructive interference! Nonetheless, the sum of all six vectors is still $\vec{0}$. To have N waves interfere constructively requires more than simply having constructive interference between pairs. Each pair must also be in phase with every other pair, a condition not satisfied in Figure 33.14b. Constructive interference by pairs does *not* necessarily lead to net constructive interference. It turns out that there is no simple formula to locate the maxima of a single-slit diffraction pattern.

Optional Section 33.5 will calculate the full intensity pattern of a single slit. The results are shown graphically in FIGURE 33.15. You can see the bright central maximum at $\theta = 0$, the weaker secondary maxima, and the dark points of destructive interference at the angles given by Equation 33.20. Compare this graph to the image of Figure 33.11 and make sure you see the agreement between the two.

FIGURE 33.15 A graph of the intensity of a single-slit diffraction pattern.

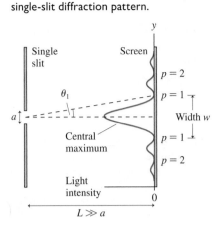

EXAMPLE 33.3 | Diffraction of a laser through a slit

Light from a helium-neon laser ($\lambda = 633$ nm) passes through a narrow slit and is seen on a screen 2.0 m behind the slit. The first minimum in the diffraction pattern is 1.2 cm from the central maximum. How wide is the slit?

MODEL A narrow slit produces a single-slit diffraction pattern. A displacement of only 1.2 cm in a distance of 200 cm means that angle θ_1 is certainly a small angle.

VISUALIZE The intensity pattern will look like Figure 33.15.

SOLVE We can use the small-angle approximation to find that the angle to the first minimum is

$$\theta_1 = \frac{1.2 \text{ cm}}{200 \text{ cm}} = 0.00600 \text{ rad} = 0.344°$$

The first minimum is at angle $\theta_1 = \lambda/a$, from which we find that the slit width is

$$a = \frac{\lambda}{\theta_1} = \frac{633 \times 10^{-9} \text{ m}}{6.00 \times 10^{-3} \text{ rad}} = 1.1 \times 10^{-4} \text{ m} = 0.11 \text{ mm}$$

ASSESS This is typical of the slit widths used to observe single-slit diffraction. You can see that the small-angle approximation is well satisfied.

The Width of a Single-Slit Diffraction Pattern

We'll find it useful, as we did for the double slit, to measure *positions* on the screen rather than angles. The position of the pth dark fringe, at angle θ_p, is $y_p = L \tan \theta_p$, where L is the distance from the slit to the viewing screen. Using Equation 33.20 for θ_p and the small-angle approximation $\tan \theta_p \approx \theta_p$, we find that the dark fringes in the single-slit diffraction pattern are located at

$$y_p = \frac{p \lambda L}{a} \qquad p = 1, 2, 3, \ldots \qquad \text{(positions of dark fringes)} \qquad (33.21)$$

A diffraction pattern is dominated by the central maximum, which is much brighter than the secondary maxima. The width w of the central maximum, shown in Figure 33.15, is defined as the distance between the two $p = 1$ minima on either side of the central maximum. Because the pattern is symmetrical, the width is simply $w = 2y_1$. This is

$$w = \frac{2 \lambda L}{a} \qquad \text{(single slit)} \qquad (33.22)$$

The width of the central maximum is *twice* the spacing $\lambda L/a$ between the dark fringes on either side. The farther away the screen (larger L), the wider the pattern of light on it becomes. In other words, the light waves are *spreading out* behind the slit, and they fill a wider and wider region as they travel farther.

An important implication of Equation 33.22, one contrary to common sense, is that a narrower slit (smaller a) causes a *wider* diffraction pattern. **The smaller the opening you squeeze a wave through, the *more* it spreads out on the other side.**

The central maximum of this single-slit diffraction pattern appears white because it is overexposed. The width of the central maximum is clear.

EXAMPLE 33.4 | Determining the wavelength

Light passes through a 0.12-mm-wide slit and forms a diffraction pattern on a screen 1.00 m behind the slit. The width of the central maximum is 0.85 cm. What is the wavelength of the light?

SOLVE From Equation 33.22, the wavelength is

$$\lambda = \frac{aw}{2L} = \frac{(1.2 \times 10^{-4}\text{ m})(0.0085\text{ m})}{2(1.00\text{ m})}$$

$$= 5.1 \times 10^{-7}\text{ m} = 510\text{ nm}$$

STOP TO THINK 33.4 The figure shows two single-slit diffraction patterns. The distance between the slit and the viewing screen is the same in both cases. Which of the following (perhaps more than one) could be true?

a. The slits are the same for both; $\lambda_1 > \lambda_2$.
b. The slits are the same for both; $\lambda_2 > \lambda_1$.
c. The wavelengths are the same for both; $a_1 > a_2$.
d. The wavelengths are the same for both; $a_2 > a_1$.
e. The slits and the wavelengths are the same for both; $p_1 > p_2$.
f. The slits and the wavelengths are the same for both; $p_2 > p_1$.

33.5 ADVANCED TOPIC A Closer Look at Diffraction

Interference and diffraction are manifestations of superposition. Mathematically, the superposition of waves at a fixed point in space (r_1 and r_2 constant) involves sums such as $e\cos(\omega t) + e\cos(\omega t + \Delta\phi)$, where e is the electric field amplitude of each wave and $\Delta\phi$ is the phase difference between them due to the fact that the waves have traveled different distances. Interestingly, we can use geometry to compute the sums that are relevant to interference and diffraction.

FIGURE 33.16a shows two vectors, each with amplitude e, rotating in the xy-plane with angular frequency ω. At any instant, the angle from the x-axis is the vector's phase, ωt or $\omega t + \Delta\phi$. A rotating *vector* that encodes amplitude and *phase* information is called a **phasor,** and Figure 33.16a is a *phasor diagram*. Notice two key features. First, the vectors rotate together, keeping a fixed angle $\Delta\phi$ between them. Second, the projections of the phasors onto the x-axis are $e\cos(\omega t)$ and $e\cos(\omega t + \Delta\phi)$, exactly what we add in a superposition calculation.

To see how this works, let's return to Young's double-slit experiment. The bright and dark interference fringes arise from the superposition of two waves, one from each slit, with a phase difference $\Delta\phi = 2\pi\,\Delta r/\lambda$ due to the path-length difference Δr. **FIGURE 33.16b** represents each wave as a phasor with amplitude e. We're interested only in their superposition, not the rapid oscillation at frequency ω, so we can draw the first phasor horizontally. If we add the phasors as vectors, using the tip-to-tail method, **the magnitude E of their vector sum is the electric field amplitude of the superposition** of the two waves.

We can use geometry and trigonometry to determine E. We have an isosceles triangle whose large angle, complementary to $\Delta\phi$, is $180° - \Delta\phi$. Consequently, the two smaller, equal angles are each $\Delta\phi/2$, and thus the base of the isosceles triangle has length $E = 2e\cos(\Delta\phi/2)$. The figure shows a triangle for which $\cos(\Delta\phi/2)$ is positive, but $\cos(\Delta\phi/2)$ can be negative at some points in the double-slit pattern. Amplitude, however, must be a positive number, so in general

$$E = \left| 2e\cos\left(\frac{\Delta\phi}{2}\right) \right| \tag{33.23}$$

FIGURE 33.16 Phasor diagrams for double-slit interference.

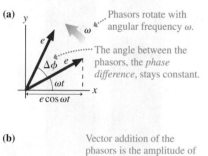

(a) Phasors rotate with angular frequency ω.
The angle between the phasors, the *phase difference*, stays constant.

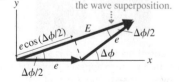

(b) Vector addition of the phasors is the amplitude of the wave superposition.

Equation 33.23 is identical to Equation 33.10, which we found previously to be the amplitude of the double-slit interference pattern. The intensity of the interference pattern is proportional to E^2.

The Single Slit Revisited

Let's use phasors to find the intensity of the diffraction pattern of a single slit. FIGURE 33.17a shows a slit of width a with N point sources of Huygens' wavelets, each separated by distance a/N. (We'll soon consider the entire wave front by letting $N \rightarrow \infty$.) We need to calculate the superposition of these N wavelets at a point on a distant screen, so far away in comparison with the slit width a that the directions are all essentially parallel at angle θ.

In the double-slit experiment, two waves from slits separated by distance d had a phase difference $\Delta\phi = 2\pi d \sin\theta/\lambda$. By exactly the same reasoning, the phase difference between two adjacent wavelets separated by a/N is $\Delta\phi_{adj} = 2\pi(a/N)\sin\theta/\lambda$. This is the phase difference for *every* pair of adjacent wavelets.

FIGURE 33.17b analyzes the diffraction at $\theta = 0$, the center of the diffraction pattern. Here all the wavelets travel straight ahead, and the phase difference between adjacent wavelets is $\Delta\phi_{adj} = 0$. Consequently, the phasor diagram shows N phasors in a straight line with amplitude $E_0 = Ne$.

FIGURE 33.17c is the phasor diagram for superposition at an arbitrary point on the screen with $\theta \neq 0$. All N phasors have the same length e—so the length of the chain of phasors is still $E_0 = Ne$—but each is rotated by angle $\Delta\phi_{adj}$ with respect to the preceding phasor. The angle of the last phasor, after N rotations, is

$$\beta = N\Delta\phi_{adj} = \frac{2\pi a \sin\theta}{\lambda} \tag{33.24}$$

Notice that β is independent of N. It is the phase difference between a wavelet originating at the top edge of the slit and one originating at the bottom edge, distance a away.

E is the amplitude of the superposition of the N wavelets. To determine E, let $N \rightarrow \infty$. This makes our calculation exact, because now we're considering every point on the wave front. It also makes our calculation easier, because now the chain of phasors, with length E_0, is simply the arc of a circle.

FIGURE 33.18 shows the geometry. The triangle at the upper right is a right triangle, so $\alpha + \beta = 90°$. But α and the angle subtending the arc also add up to $90°$, so the angle subtending the arc must be β. We've divided it into two angles, each $\beta/2$, in order to create two right triangles along E. You can see that $E = 2R\sin(\beta/2)$. In addition, the arc length, if β is in radians, is $E_0 = \beta R$. Eliminating R, we find that the amplitude of the superposition is

$$E = E_0 \frac{\sin(\beta/2)}{\beta/2} = E_0 \frac{\sin(\pi a \sin\theta/\lambda)}{\pi a \sin\theta/\lambda} \tag{33.25}$$

The diffraction-pattern intensity is proportional to E^2, thus

$$I_{slit} = I_0 \left[\frac{\sin(\pi a \sin\theta/\lambda)}{\pi a \sin\theta/\lambda} \right]^2 \tag{33.26}$$

where I_0 (proportional to E_0^2) is the intensity at the center of the central maximum, $\theta = 0$. (Recall, from l'Hôpital's rule, that $\sin x/x \rightarrow 1$ as $x \rightarrow 0$.) FIGURE 33.19 is a graph of Equation 33.26. You can see that it is, indeed, exactly what we observe for single-slit diffraction—a bright central maximum flanked by weaker secondary maxima. The minima occur where the numerator of Equation 33.26 is zero. This requires $\sin\theta_p = p\lambda/a$ for $p = 1, 2, 3, \ldots$, which is exactly the result for the dark fringes that we found previously.

We usually measure positions on a screen rather than angles. For a screen at distance L, a point on the screen at distance y from the center of the pattern is at

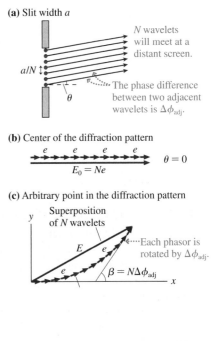

FIGURE 33.17 Phasor diagrams for single-slit diffraction.

(a) Slit width a

a/N

N wavelets will meet at a distant screen.

The phase difference between two adjacent wavelets is $\Delta\phi_{adj}$.

(b) Center of the diffraction pattern

$\theta = 0$

$E_0 = Ne$

(c) Arbitrary point in the diffraction pattern

Superposition of N wavelets

E

Each phasor is rotated by $\Delta\phi_{adj}$.

$\beta = N\Delta\phi_{adj}$

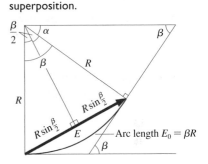

FIGURE 33.18 Calculating the superposition.

$\frac{\beta}{2}$ α β

β R

R $R\sin\frac{\beta}{2}$

$R\sin\frac{\beta}{2}$ E

Arc length $E_0 = \beta R$

β

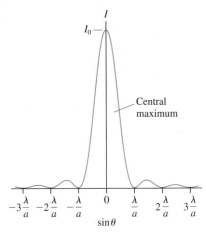

FIGURE 33.19 The single-slit diffraction pattern.

I

I_0

Central maximum

$-3\frac{\lambda}{a}$ $-2\frac{\lambda}{a}$ $-\frac{\lambda}{a}$ 0 $\frac{\lambda}{a}$ $2\frac{\lambda}{a}$ $3\frac{\lambda}{a}$

$\sin\theta$

angle $\theta = \tan^{-1}(y/L)$. For very small angles, which is also typical, the small-angle approximation $\tan\theta \approx \sin\theta \approx \theta$ allows us to write the intensity at position y as

$$I_{\text{slit}} = I_0 \left[\frac{\sin(\pi ay/\lambda L)}{\pi ay/\lambda L} \right]^2 \qquad \text{(small angles)} \qquad (33.27)$$

In this case the minima are at $y_p = p\lambda L/a$, also as we found previously.

You might think that the maximum would occur where the numerator in Equation 33.27 is 1. However, y also appears in the denominator, and that affects the maxima. Setting the derivative of Equation 33.27 to zero, to locate the maxima, leads to a *transcendental equation,* one that cannot be solved algebraically. It can be solved numerically, leading to the result that, for small angles, the first two maxima occur at $y_{\text{max 1}} = 1.43\lambda L/a$ and $y_{\text{max 2}} = 2.46\lambda L/a$.

EXAMPLE 33.5 | **Single-slit intensity**

Light with a wavelength of 500 nm passes through a 150-μm-wide slit and is viewed on a screen 2.5 m behind the slit. At what distance from the center of the diffraction pattern is the intensity 50% of maximum?

MODEL The slit produces a single-slit diffraction pattern. Assume that the diffraction angles are small enough to justify the small-angle approximation.

VISUALIZE Figure 33.19 showed a graph of the intensity distribution. The intensity falls to 50% of maximum before the first minimum and never returns to 50% after the first minimum.

SOLVE From Equation 33.27, the intensity at position y is

$$I = I_0 \left[\frac{\sin(\pi ay/\lambda L)}{\pi ay/\lambda L} \right]^2$$

The intensity will have fallen to 50% of maximum I_0 when

$$\frac{\sin(\pi ay/\lambda L)}{\pi ay/\lambda L} = \sqrt{\frac{1}{2}} = 0.707$$

This is a transcendental equation; there is no exact solution. However, it can easily be solved on a calculator with only a small amount of trial and error. If we let $x = \pi ay/\lambda L$, then the equation we want to solve is $\sin x/x = 0.707$, where x is in radians. Put your calculator in radian mode, guess a value of x, compute $\sin x/x$, and then use the result to make an improved guess. The first minimum, $y_1 = \lambda L/a$, has $x = \pi$ rad, and we know that the solution is less than this.

First try: $x = 1.0$ rad gives $\sin x/x = 0.841$.
Second try: $x = 1.5$ rad gives $\sin x/x = 0.665$.

With just two guesses we've narrowed the range to 1.0 rad $< x <$ 1.5 rad. It only takes about three more tries to arrive at $x = 1.39$ rad as the answer to three significant figures. Thus the intensity has dropped to 50% of maximum at

$$y = \frac{1.39}{\pi} \frac{\lambda L}{a} = 3.7 \text{ mm}$$

ASSESS Diffraction patterns seen in the laboratory are typically a centimeter or two wide. This point is within the central maximum, so ≈ 4 mm from the center is reasonable. And ≈ 4 mm from center at a distance of 2.5 m certainly justifies our use of the small-angle approximation.

The Complete Double-Slit Intensity

Figure 33.4 showed double-slit interference occurring between two overlapping waves as they "spread out behind the two slits." The waves are spreading because light has passed through two narrow slits, and each slit is causing single-slit diffraction. What we see in double-slit interference is actually interference between two overlapping single-slit diffraction patterns. Interference produces the fringes, but the diffraction pattern—the amount of light reaching the screen—determines *how bright* the fringes are.

We earlier calculated the ideal double-slit intensity, $I_{\text{double}} = 4I_1 \cos^2(\pi dy/\lambda L)$, for two slits separated by distance d. But each slit has width a, so the double-slit pattern is *modulated* by the single-slit diffraction intensity for a slit of width a. Thus a realistic double-slit intensity, for small angles, is

$$I_{\text{double}} = I_0 \left[\frac{\sin(\pi ay/\lambda L)}{\pi ay/\lambda L} \right]^2 \cos^2(\pi dy/\lambda L) \qquad (33.28)$$

The cosine term produces the fringe oscillations, but now the overall intensity is determined by the diffraction of the individual slits. If the slits are extremely narrow ($a \ll d$), which we tacitly assumed before, then the central maximum of the single-slit pattern is very broad and we see many fringes with only a slow decline in the fringe intensity. This was the case in Figure 33.6b.

But many double slits have a width a that is only slightly smaller than the slit spacing d, and this leads to a complex interplay between diffraction and interference. **FIGURE 33.20** shows a double-slit interference pattern for two 0.055-mm-wide slits separated by 0.35 mm and, for comparison, the diffraction pattern of a single 0.055-mm-wide slit. Diffraction of the individual slits determines the overall brightness on the screen—it is the spreading of the light behind the slit—and within this we see the interference between light waves from the two slits. It seems there could be no better proof that light is a wave!

Notice that the $m = 7$ interference fringe is missing (and $m = 8$ is so weak as to be almost invisible). If an interference maximum falls exactly on a minimum (a zero) in the single-slit diffraction pattern, then we have what is called a **missing order.** Interference maxima occur at $y_m = m\lambda L/d$ and diffraction minima are at $y_p = p\lambda L/a$, where m and p are integers. Equating these, to find where they overlap, we see that order m is missing if

$$m_{\text{missing}} = p\frac{d}{a} \qquad p = 1, 2, 3, \ldots \qquad (33.29)$$

m has to be an integer, so the order is truly missing only for certain slit-spacing-to-width ratios d/a. In the case of Figure 33.20, $d/a = 7$ and so the $m = 7$ order is missing (it falls on the $p = 1$ diffraction minimum), as is $m = 14$. In practice, one or more interference maxima may be too weak to be seen even if they don't satisfy Equation 33.29 exactly.

33.6 Circular-Aperture Diffraction

Diffraction occurs if a wave passes through an opening of any shape. Diffraction by a single slit establishes the basic ideas of diffraction, but a common situation of practical importance is diffraction of a wave by a **circular aperture.** Circular diffraction is mathematically more complex than diffraction from a slit, and we will present results without derivation.

Consider some examples. A loudspeaker cone generates sound by the rapid oscillation of a diaphragm, but the sound wave must pass through the circular aperture defined by the outer edge of the speaker cone before it travels into the room beyond. This is diffraction by a circular aperture. Telescopes and microscopes are the reverse. Light waves from outside need to enter the instrument. To do so, they must pass through a circular lens. In fact, the performance limit of optical instruments is determined by the diffraction of the circular openings through which the waves must pass. This is an issue we'll look at in Chapter 35.

FIGURE 33.21 shows a circular aperture of diameter D. Light waves passing through this aperture spread out to generate a *circular* diffraction pattern. You should compare this to Figure 33.11 for a single slit to note the similarities and differences. The diffraction pattern still has a *central maximum,* now circular, and it is surrounded by a series of secondary bright fringes.

FIGURE 33.20 The overall intensity of a double-slit interference pattern is governed by the single-slit diffraction through each slit.

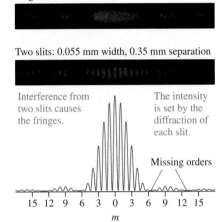

FIGURE 33.21 The diffraction of light by a circular opening.

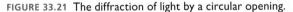
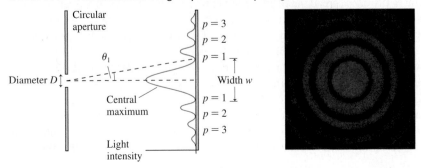

Angle θ_1 locates the first minimum in the intensity, where there is perfect destructive interference. A mathematical analysis of circular diffraction finds

$$\theta_1 = \frac{1.22\lambda}{D} \qquad (33.30)$$

where D is the *diameter* of the circular opening. Equation 33.30 has assumed the small-angle approximation, which is almost always valid for the diffraction of light but usually is *not* valid for the diffraction of longer-wavelength sound waves.

Within the small-angle approximation, the width of the central maximum is

$$w = 2y_1 = 2L\tan\theta_1 \approx \frac{2.44\lambda L}{D} \qquad \text{(circular aperture)} \qquad (33.31)$$

The diameter of the diffraction pattern increases with distance L, showing that light spreads out behind a circular aperture, but it decreases if the size D of the aperture is increased.

EXAMPLE 33.6 | **Shining a laser through a circular hole**

Light from a helium-neon laser ($\lambda = 633$ nm) passes through a 0.50-mm-diameter hole. How far away should a viewing screen be placed to observe a diffraction pattern whose central maximum is 3.0 mm in diameter?

SOLVE Equation 33.31 gives us the appropriate screen distance:

$$L = \frac{wD}{2.44\lambda} = \frac{(3.0 \times 10^{-3}\text{ m})(5.0 \times 10^{-4}\text{ m})}{2.44(633 \times 10^{-9}\text{ m})} = 0.97\text{ m}$$

33.7 The Wave Model of Light

We opened this chapter by noting that there are three models of light, each useful within a certain range of circumstances. We are now at a point where we can establish an important condition that separates the wave model of light from the ray model of light.

When light passes through an opening of size a, the angle of the first diffraction minimum is

$$\theta_1 = \sin^{-1}\left(\frac{\lambda}{a}\right) \qquad (33.32)$$

Equation 33.32 is for a slit, but the result is very nearly the same if a is the diameter of a circular aperture. Regardless of the shape of the opening, **the factor that determines how much a wave spreads out behind an opening is the ratio λ/a, the size of the wavelength compared to the size of the opening.**

FIGURE 33.22 illustrates the difference between a wave whose wavelength is much smaller than the size of the opening and a second wave whose wavelength is comparable to the opening. A wave with $\lambda/a \approx 1$ quickly spreads to fill the region behind the opening. Light waves, because of their very short wavelength, almost always have $\lambda/a \ll 1$ and diffract to produce a slowly spreading "beam" of light.

Now we can better appreciate Newton's dilemma. With everyday-sized openings, sound and water waves have $\lambda/a \approx 1$ and diffract to fill the space behind the opening. Consequently, this is what we come to expect for the behavior of waves. We see now that light really does spread out behind an opening, but the very small λ/a ratio usually makes the diffraction pattern too small to see. Diffraction begins to be discernible only when the size of the opening is a fraction of a millimeter or less. If we wanted the diffracted light wave to *fill* the space behind the opening ($\theta_1 \approx 90°$), as a sound wave does, we would need to reduce the size of the opening to $a \approx 0.001$ mm!

FIGURE 33.23 shows light passing through a hole of diameter D. According to the ray model, light rays passing through the hole travel straight ahead to create a bright circular spot of diameter D on a viewing screen. This is the *geometric image* of the slit. In reality, diffraction causes the light to spread out behind the slit, but—and this is the important point—**we will not notice the spreading if it is less than the diameter**

FIGURE 33.22 The diffraction of a long-wavelength wave and a short-wavelength wave through the same opening.

Long wavelength, $\lambda \approx a$. This wave quickly fills the region behind the opening.

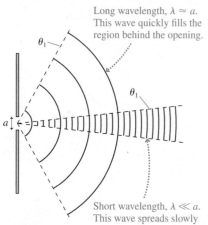

Short wavelength, $\lambda \ll a$. This wave spreads slowly and remains a well-defined beam.

D of the geometric image. That is, we will not be aware of diffraction unless the bright spot on the screen increases in diameter.

This idea provides a reasonable criterion for when to use ray optics and when to use wave optics:

- If the spreading due to diffraction is less than the size of the opening, use the ray model and think of light as traveling in straight lines.
- If the spreading due to diffraction is greater than the size of the opening, use the wave model of light.

The crossover point between these two regimes occurs when the spreading due to diffraction is equal to the size of the opening. The central-maximum width of a circular-aperture diffraction pattern is $w = 2.44\lambda L/D$. If we equate this diffraction width to the diameter of the aperture itself, we have

$$\frac{2.44\lambda L}{D_c} = D_c \tag{33.33}$$

where the subscript c on D_c indicates that this is the crossover between the ray model and the wave model. Because we're making an estimate—the change from the ray model to the wave model is gradual, not sudden—to one significant figure, we find

$$D_c \approx \sqrt{2\lambda L} \tag{33.34}$$

This is the diameter of a circular aperture whose diffraction pattern, at distance L, has width $w \approx D$. We know that visible light has $\lambda \approx 500$ nm, and a typical distance in laboratory work is $L \approx 1$ m. For these values,

$$D_c \approx 1 \text{ mm}$$

Thus diffraction is significant, and you should use the wave model, when light passes through openings smaller than about 1 mm. The ray model, which we'll study in the next chapter, is the more appropriate model when light passes through openings larger than about 1 mm. Lenses and mirrors, in particular, are almost always larger than 1 mm, so they will be analyzed with the ray model.

We can now pull all these ideas together in a more complete presentation of the wave model of light.

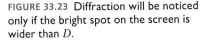

FIGURE 33.23 Diffraction will be noticed only if the bright spot on the screen is wider than *D*.

If light travels in straight lines, the image on the screen is the same size as the hole. Diffraction will not be noticed unless the light spreads over a diameter larger than *D*.

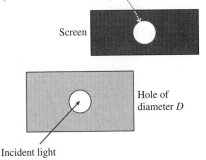

Screen

Hole of diameter *D*

Incident light

MODEL 33.1

Wave model of light

For use when diffraction is significant.

- Light is an electromagnetic wave.
  - Light travels through vacuum at speed *c*.
  - Wavelength λ and frequency *f* are related by $\lambda f = c$.
  - Most of optics depends only on the waviness of light, not on its electromagnetic properties.
- Light exhibits diffraction and interference.
  - Light spreads out after passing through an opening. The amount of spread is inversely proportional to the size of the opening.
  - Two equal-wavelength light waves interfere. Constructive and destructive interference depend on the path-length difference.
- Limitations:
  - The *ray model* is a better description in situations with no diffraction.
    Use the wave model with openings < 1 mm in size.
    Use the ray model with openings > 1 mm in size.
  - The *photon model* is a better description of extremely weak light or the light emitted in atomic transitions.

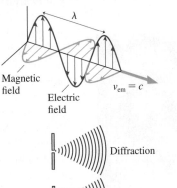

λ

Magnetic field

Electric field

$v_{em} = c$

Diffraction

Interference

33.8 Interferometers

Scientists and engineers have devised many ingenious methods for using interference to control the flow of light and to make very precise measurements with light waves. A device that makes practical use of interference is called an **interferometer.**

Interference requires two waves of *exactly* the same wavelength. One way of guaranteeing that two waves have exactly equal wavelengths is to divide one wave into two parts of smaller amplitude. Later, at a different point in space, the two parts are recombined. **Interferometers are based on the division and recombination of a single wave.**

The Michelson Interferometer

Albert Michelson, the first American scientist to receive a Nobel Prize, invented an optical interferometer, shown in **FIGURE 33.24**, in which an incoming light wave is divided by a **beam splitter,** a partially silvered mirror that reflects half the light but transmits the other half. The two waves then travel toward mirrors M_1 and M_2. Half of the wave reflected from M_1 is transmitted through the beam splitter, where it recombines with the reflected half of the wave returning from M_2. The superimposed waves travel on to a light detector, originally a human observer but now more likely an electronic photodetector.

After separating, the two waves travel distances $r_1 = 2L_1$ and $r_2 = 2L_2$ before recombining, with the factors of 2 appearing because the waves travel to the mirrors and back again. Thus the path-length difference between the two waves is

$$\Delta r = 2L_2 - 2L_1 \tag{33.35}$$

The conditions for constructive and destructive interference between the two recombined beams are the same as for double-slit interference: $\Delta r = m\lambda$ and $\Delta r = \left(m + \frac{1}{2}\lambda\right)$, respectively. Thus constructive and destructive interference occur when

$$\text{Constructive:} \quad L_2 - L_1 = m\frac{\lambda}{2}$$
$$\text{Destructive:} \quad L_2 - L_1 = \left(m + \tfrac{1}{2}\right)\frac{\lambda}{2} \qquad m = 0, 1, 2, \ldots \tag{33.36}$$

You might expect the interferometer output to be either "bright" or "dark." Instead, a viewing screen shows the pattern of circular interference fringes seen in **FIGURE 33.25**. Our analysis was for light waves that impinge on the mirrors exactly perpendicular to the surface. In an actual experiment, some of the light waves enter the interferometer at slightly different angles and, as a result, the recombined waves have slightly altered path-length differences Δr. These waves cause the alternating bright and dark fringes as you move outward from the center of the pattern. Their analysis will be left to more advanced courses in optics. Equations 33.36 are valid at the *center* of the circular pattern; thus there is a bright or dark central spot when one of the conditions in Equations 33.36 is true.

Mirror M_2 can be moved forward or backward by turning a precision screw, causing the central spot to alternate in a bright-dark-bright-dark-bright cycle that is easily seen or monitored by a photodetector. Suppose the interferometer is adjusted to produce a bright central spot. The next bright spot will appear when M_2 has moved *half* a wavelength, increasing the path-length difference by one full wavelength. The number Δm of maxima appearing as M_2 moves through distance ΔL_2 is

$$\Delta m = \frac{\Delta L_2}{\lambda/2} \tag{33.37}$$

Very precise wavelength measurements can be made by moving the mirror while counting the number of new bright spots appearing at the center of the pattern. The number Δm is counted and known exactly. The only limitation on how precisely λ

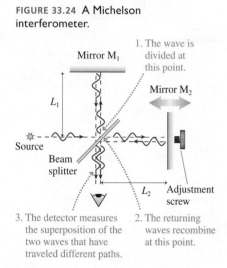

FIGURE 33.24 A Michelson interferometer.

Mirror M_1

1. The wave is divided at this point.

Mirror M_2

L_1

Source

Beam splitter

L_2 Adjustment screw

3. The detector measures the superposition of the two waves that have traveled different paths.

2. The returning waves recombine at this point.

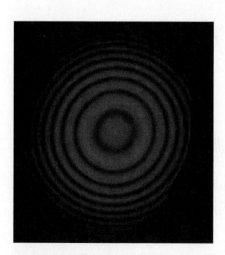

FIGURE 33.25 Photograph of the interference fringes produced by a Michelson interferometer.

can be measured this way is the precision with which distance ΔL_2 can be measured. Unlike λ, which is microscopic, ΔL_2 is typically a few millimeters, a macroscopic distance that can be measured very accurately using precision screws, micrometers, and other techniques. Michelson's invention provided a way to transfer the precision of macroscopic distance measurements to an equal precision for the wavelength of light.

EXAMPLE 33.7 | **Measuring the wavelength of light**

An experimenter uses a Michelson interferometer to measure one of the wavelengths of light emitted by neon atoms. She slowly moves mirror M_2 until 10,000 new bright central spots have appeared. (In a modern experiment, a photodetector and computer would eliminate the possibility of experimenter error while counting.) She then measures that the mirror has moved a distance of 3.164 mm. What is the wavelength of the light?

MODEL An interferometer produces a new maximum each time L_2 increases by $\lambda/2$.

SOLVE The mirror moves $\Delta L_2 = 3.164 \text{ mm} = 3.164 \times 10^{-3} \text{ m}$. We can use Equation 33.37 to find

$$\lambda = \frac{2 \Delta L_2}{\Delta m} = 6.328 \times 10^{-7} \text{ m} = 632.8 \text{ nm}$$

ASSESS A measurement of ΔL_2 accurate to four significant figures allowed us to determine λ to four significant figures. This happens to be the neon wavelength that is emitted as the laser beam in a helium-neon laser.

STOP TO THINK 33.5 A Michelson interferometer using light of wavelength λ has been adjusted to produce a bright spot at the center of the interference pattern. Mirror M_1 is then moved distance λ toward the beam splitter while M_2 is moved distance λ away from the beam splitter. How many bright-dark-bright fringe shifts are seen?

a. 0
b. 1
c. 2
d. 4
e. 8
f. It's not possible to say without knowing λ.

Holography

No discussion of wave optics would be complete without mentioning holography, which has both scientific and artistic applications. The basic idea is a simple extension of interferometry.

FIGURE 33.26a shows how a **hologram** is made. A beam splitter divides a laser beam into two waves. One wave illuminates the object of interest. The light scattered by this object is a very complex wave, but it is the wave you would see if you looked at the object from the position of the film. The other wave, called the *reference beam,* is reflected directly toward the film. The scattered light and the reference beam meet at the film and interfere. The film records their interference pattern.

The interference patterns we've looked at in this chapter have been simple patterns of stripes and circles because the light waves have been well-behaved plane waves and

FIGURE 33.26 Holography is an important application of wave optics.

(a) Recording a hologram

The interference between the scattered light and the reference beam is recorded on the film.

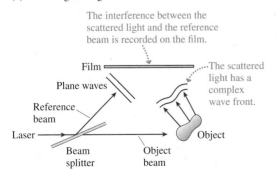

The scattered light has a complex wave front.

(b) A hologram

An enlarged photo of the developed film. This is the hologram.

(c) Playing a hologram

The diffraction of the laser beam through the light and dark patches of the film reconstructs the original scattered wave.

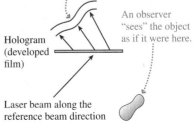

An observer "sees" the object as if it were here.

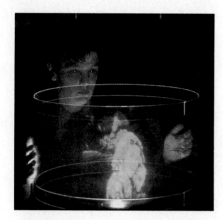

A hologram.

spherical waves. The light wave scattered by the object in Figure 33.26a is exceedingly complex. As a result, the interference pattern recorded on the film—the hologram—is a seemingly random pattern of whorls and blotches. **FIGURE 33.26b** is an enlarged photograph of a portion of a hologram. It's certainly not obvious that information is stored in this pattern, but it is.

The hologram is "played" by sending just the reference beam through it, as seen in **FIGURE 33.26c**. The reference beam diffracts through the transparent parts of the hologram, just as it would through the slits of a diffraction grating. Amazingly, the diffracted wave is *exactly the same* as the light wave that had been scattered by the object! In other words, the diffracted reference beam *reconstructs* the original scattered wave. As you look at this diffracted wave, from the far side of the hologram, you "see" the object exactly as if it were there. The view is three dimensional because, by moving your head with respect to the hologram, you can see different portions of the wave front.

CHALLENGE EXAMPLE 33.8 | **Measuring the index of refraction of a gas**

A Michelson interferometer uses a helium-neon laser with wavelength $\lambda_{vac} = 633$ nm. In one arm, the light passes through a 4.00-cm-thick glass cell. Initially the cell is evacuated, and the interferometer is adjusted so that the central spot is a bright fringe. The cell is then slowly filled to atmospheric pressure with a gas. As the cell fills, 43 bright-dark-bright fringe shifts are seen and counted. What is the index of refraction of the gas at this wavelength?

MODEL Adding one additional wavelength to the round trip causes one bright-dark-bright fringe shift. Changing the length of the arm is one way to add wavelengths, but not the only way. Increasing the index of refraction also adds wavelengths because light has a shorter wavelength when traveling through a material with a larger index of refraction.

VISUALIZE **FIGURE 33.27** shows a Michelson interferometer with a cell of thickness d in one arm.

FIGURE 33.27 Measuring the index of refraction.

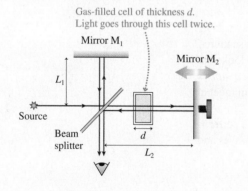

Gas-filled cell of thickness d.
Light goes through this cell twice.

Mirror M_1

L_1

Mirror M_2

Source

Beam splitter

d

L_2

SOLVE To begin, all the air is pumped out of the cell. As light travels from the beam splitter to the mirror and back, the number of wavelengths inside the cell is

$$m_1 = \frac{2d}{\lambda_{vac}}$$

where the 2 appears because the light passes through the cell twice.

The cell is then filled with gas at 1 atm pressure. Light travels slower in the gas, $v = c/n$, and you learned in Chapter 16 that the reduction in speed decreases the wavelength to λ_{vac}/n. With the cell filled, the number of wavelengths spanning distance d is

$$m_2 = \frac{2d}{\lambda} = \frac{2d}{\lambda_{vac}/n}$$

The physical distance has not changed, but the number of wavelengths along the path has. Filling the cell has increased the path by

$$\Delta m = m_2 - m_1 = (n - 1)\frac{2d}{\lambda_{vac}}$$

wavelengths. Each increase of one wavelength causes one bright-dark-bright fringe shift at the output. Solving for n, we find

$$n = 1 + \frac{\lambda_{vac}\,\Delta m}{2d} = 1 + \frac{(6.33 \times 10^{-7}\ \text{m})(43)}{2(0.0400\ \text{m})} = 1.00034$$

ASSESS This may seem like a six-significant-figure result, but there are really only two. What we're measuring is not n but $n - 1$. We know the fringe count to two significant figures, and that has allowed us to compute $n - 1 = \lambda_{vac}\,\Delta m/2d = 3.4 \times 10^{-4}$.

SUMMARY

The goal of Chapter 33 has been to learn about and apply the wave model of light.

GENERAL PRINCIPLES

Huygens' principle says that each point on a wave front is the source of a spherical wavelet. The wave front at a later time is tangent to all the wavelets.

Diffraction is the spreading of a wave after it passes through an opening.

Constructive and destructive **interference** are due to the overlap of two or more waves as they spread behind openings.

IMPORTANT CONCEPTS

The **wave model** of light considers light to be a wave propagating through space. Diffraction and interference are important.

The **ray model** of light considers light to travel in straight lines like little particles. Diffraction and interference are not important.

Diffraction is important when the width of the diffraction pattern of an aperture equals or exceeds the size of the aperture. For a circular aperture, the crossover between the ray and wave models occurs for an opening of diameter $D_c \approx \sqrt{2\lambda L}$.

In practice, $D_c \approx 1$ mm for visible light. Thus

- Use the wave model when light passes through openings < 1 mm in size. Diffraction effects are usually important.
- Use the ray model when light passes through openings > 1 mm in size. Diffraction is usually not important.

APPLICATIONS

Single slit of width a.
A bright **central maximum** of width

$$w = \frac{2\lambda L}{a}$$

is flanked by weaker **secondary maxima.**
Dark fringes are located at angles such that

$$a \sin\theta_p = p\lambda \qquad p = 1, 2, 3, \ldots$$

If $\lambda/a \ll 1$, then from the small-angle approximation

$$\theta_p = \frac{p\lambda}{a} \qquad y_p = \frac{p\lambda L}{a}$$

Circular aperture of diameter D.
A bright central maximum of diameter

$$w = \frac{2.44\lambda L}{D}$$

is surrounded by circular secondary maxima.
The first dark fringe is located at

$$\theta_1 = \frac{1.22\lambda}{D} \qquad y_1 = \frac{1.22\lambda L}{D}$$

For an aperture of any shape, a smaller opening causes a more rapid spreading of the wave behind the opening.

Interference due to wave-front division

Waves overlap as they spread out behind slits. Constructive interference occurs along antinodal lines. Bright fringes are seen where the antinodal lines intersect the viewing screen.

Double slit with separation d.
Equally spaced bright fringes are located at

$$\theta_m = \frac{m\lambda}{d} \qquad y_m = \frac{m\lambda L}{d} \qquad m = 0, 1, 2, \ldots$$

The **fringe spacing** is $\Delta y = \dfrac{\lambda L}{d}$

Diffraction grating with slit spacing d.
Very bright and narrow fringes are located at angles and positions

$$d \sin\theta_m = m\lambda \qquad y_m = L \tan\theta_m$$

Interference due to amplitude division

An interferometer divides a wave, lets the two waves travel different paths, then recombines them. Interference is constructive if one wave travels an integer number of wavelengths more or less than the other wave. The difference can be due to an actual path-length difference or to a different index of refraction.

Michelson interferometer

The number of bright-dark-bright fringe shifts as mirror M_2 moves distance ΔL_2 is

$$\Delta m = \frac{\Delta L_2}{\lambda/2}$$

TERMS AND NOTATION

optics	double slit	spectroscopy	circular aperture
diffraction	interference fringes	single-slit diffraction	interferometer
models of light	central maximum	secondary maxima	beam splitter
wave model	fringe spacing, Δy	Huygens' principle	hologram
ray model	diffraction grating	phasor	
photon model	order, m	missing order	

CONCEPTUAL QUESTIONS

1. **FIGURE Q33.1** shows light waves passing through two closely spaced, narrow slits. The graph shows the intensity of light on a screen behind the slits. Reproduce these graph axes, including the zero and the tick marks locating the double-slit fringes, then draw a graph to show how the light-intensity pattern will appear if the right slit is blocked, allowing light to go through only the left slit. Explain your reasoning.

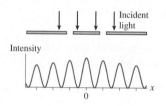

FIGURE Q33.1

2. In a double-slit interference experiment, which of the following actions (perhaps more than one) would cause the fringe spacing to increase? (a) Increasing the wavelength of the light. (b) Increasing the slit spacing. (c) Increasing the distance to the viewing screen. (d) Submerging the entire experiment in water.

3. **FIGURE Q33.3** shows the viewing screen in a double-slit experiment. Fringe C is the central maximum. What will happen to the fringe spacing if
   a. The wavelength of the light is decreased?
   b. The spacing between the slits is decreased?
   c. The distance to the screen is decreased?
   d. Suppose the wavelength of the light is 500 nm. How much farther is it from the dot on the screen in the center of fringe E to the left slit than it is from the dot to the right slit?

FIGURE Q33.3

4. **FIGURE Q33.3** is the interference pattern seen on a viewing screen behind 2 slits. Suppose the 2 slits were replaced by 20 slits having the same spacing d between adjacent slits.
   a. Would the number of fringes on the screen increase, decrease, or stay the same?
   b. Would the fringe spacing increase, decrease, or stay the same?
   c. Would the width of each fringe increase, decrease, or stay the same?
   d. Would the brightness of each fringe increase, decrease, or stay the same?

5. **FIGURE Q33.5** shows the light intensity on a viewing screen behind a single slit of width a. The light's wavelength is λ. Is $\lambda < a$, $\lambda = a$, $\lambda > a$, or is it not possible to tell? Explain.

FIGURE Q33.5   FIGURE Q33.6

6. **FIGURE Q33.6** shows the light intensity on a viewing screen behind a circular aperture. What happens to the width of the central maximum if
   a. The wavelength of the light is increased?
   b. The diameter of the aperture is increased?
   c. How will the screen appear if the aperture diameter is less than the light wavelength?

7. Narrow, bright fringes are observed on a screen behind a diffraction grating. The entire experiment is then immersed in water. Do the fringes on the screen get closer together, get farther apart, remain the same, or disappear? Explain.

8. a. Green light shines through a 100-mm-diameter hole and is observed on a screen. If the hole diameter is increased by 20%, does the circular spot of light on the screen decrease in diameter, increase in diameter, or stay the same? Explain.
   b. Green light shines through a 100-μm-diameter hole and is observed on a screen. If the hole diameter is increased by 20%, does the circular spot of light on the screen decrease in diameter, increase in diameter, or stay the same? Explain.

9. A Michelson interferometer using 800 nm light is adjusted to have a bright central spot. One mirror is then moved 200 nm forward, the other 200 nm back. Afterward, is the central spot bright, dark, or in between? Explain.

10. A Michelson interferometer is set up to display constructive interference (a bright central spot in the fringe pattern of Figure 33.25) using light of wavelength λ. If the wavelength is changed to $\lambda/2$, does the central spot remain bright, does the central spot become dark, or do the fringes disappear? Explain. Assume the fringes are viewed by a detector sensitive to both wavelengths.

EXERCISES AND PROBLEMS

Problems labeled 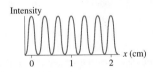 integrate material from earlier chapters.

Exercises

Section 33.2 The Interference of Light

1. | A double slit is illuminated simultaneously with orange light of wavelength 620 nm and light of an unknown wavelength. The $m = 4$ bright fringe of the unknown wavelength overlaps the $m = 3$ bright orange fringe. What is the unknown wavelength?

2. || Two narrow slits 80 μm apart are illuminated with light of wavelength 620 nm. What is the angle of the $m = 3$ bright fringe in radians? In degrees?

3. | A double-slit experiment is performed with light of wavelength 630 nm. The bright interference fringes are spaced 1.8 mm apart on the viewing screen. What will the fringe spacing be if the light is changed to a wavelength of 420 nm?

4. | Light of wavelength 550 nm illuminates a double slit, and the interference pattern is observed on a screen. At the position of the $m = 2$ bright fringe, how much farther is it to the more distant slit than to the nearer slit?

5. || Light of 630 nm wavelength illuminates two slits that are 0.25 mm apart. FIGURE EX33.5 shows the intensity pattern seen on a screen behind the slits. What is the distance to the screen?

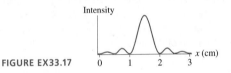

FIGURE EX33.5

6. || In a double-slit experiment, the slit separation is 200 times the wavelength of the light. What is the angular separation (in degrees) between two adjacent bright fringes?

7. || Light from a sodium lamp ($\lambda = 589$ nm) illuminates two narrow slits. The fringe spacing on a screen 150 cm behind the slits is 4.0 mm. What is the spacing (in mm) between the two slits?

8. || A double-slit interference pattern is created by two narrow slits spaced 0.25 mm apart. The distance between the first and the fifth minimum on a screen 60 cm behind the slits is 5.5 mm. What is the wavelength (in nm) of the light used in this experiment?

Section 33.3 The Diffraction Grating

9. | A 4.0-cm-wide diffraction grating has 2000 slits. It is illuminated by light of wavelength 550 nm. What are the angles (in degrees) of the first two diffraction orders?

10. || Light of wavelength 620 nm illuminates a diffraction grating. The second-order maximum is at angle 39.5°. How many lines per millimeter does this grating have?

11. || A diffraction grating produces a first-order maximum at an angle of 20.0°. What is the angle of the second-order maximum?

12. | A diffraction grating is illuminated simultaneously with red light of wavelength 660 nm and light of an unknown wavelength. The fifth-order maximum of the unknown wavelength exactly overlaps the third-order maximum of the red light. What is the unknown wavelength?

13. || The two most prominent wavelengths in the light emitted by a hydrogen discharge lamp are 656 nm (red) and 486 nm (blue). Light from a hydrogen lamp illuminates a diffraction grating with 500 lines/mm, and the light is observed on a screen 1.50 m behind the grating. What is the distance between the first-order red and blue fringes?

14. || A helium-neon laser ($\lambda = 633$ nm) illuminates a diffraction grating. The distance between the two $m = 1$ bright fringes is 32 cm on a screen 2.0 m behind the grating. What is the spacing between slits of the grating?

Section 33.4 Single-Slit Diffraction

15. || In a single-slit experiment, the slit width is 200 times the wavelength of the light. What is the width (in mm) of the central maximum on a screen 2.0 m behind the slit?

16. || A helium-neon laser ($\lambda = 633$ nm) illuminates a single slit and is observed on a screen 1.5 m behind the slit. The distance between the first and second minima in the diffraction pattern is 4.75 mm. What is the width (in mm) of the slit?

17. || Light of 630 nm wavelength illuminates a single slit of width 0.15 mm. FIGURE EX33.17 shows the intensity pattern seen on a screen behind the slit. What is the distance to the screen?

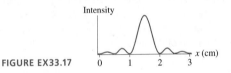

FIGURE EX33.17

18. | A 0.50-mm-wide slit is illuminated by light of wavelength 500 nm. What is the width (in mm) of the central maximum on a screen 2.0 m behind the slit?

19. || You need to use your cell phone, which broadcasts an 800 MHz signal, but you're behind two massive, radio-wave-absorbing buildings that have only a 15 m space between them. What is the angular width, in degrees, of the electromagnetic wave after it emerges from between the buildings?

20. | For what slit-width-to-wavelength ratio does the first minimum of a single-slit diffraction pattern appear at (a) 30°, (b) 60°, and (c) 90°?

21. | Light from a helium-neon laser ($\lambda = 633$ nm) is incident on a single slit. What is the largest slit width for which there are no minima in the diffraction pattern?

Section 33.5 A Closer Look at Diffraction

22. | A laser beam illuminates a single, narrow slit, and the diffraction pattern is observed on a screen behind the slit. The first secondary maximum is 26 mm from the center of the diffraction pattern. How far is the first minimum from the center of the diffraction pattern?

23. || Two 50-μm-wide slits spaced 0.25 mm apart are illuminated by blue laser light with a wavelength of 450 nm. The interference pattern is observed on a screen 2.0 m behind the slits. How many bright fringes are seen in the *central maximum* that spans the distance between the first missing order on one side and the first missing order on the other side?

24. ‖ A laser beam with a wavelength of 480 nm illuminates two 0.12-mm-wide slits separated by 0.30 mm. The interference pattern is observed on a screen 2.3 m behind the slits. What is the light intensity, as a fraction of the maximum intensity I_0, at a point halfway between the center and the first minimum?

Section 33.6 Circular-Aperture Diffraction

25. ‖ A 0.50-mm-diameter hole is illuminated by light of wavelength 550 nm. What is the width (in mm) of the central maximum on a screen 2.0 m behind the slit?
26. ‖ Infrared light of wavelength 2.5 μm illuminates a 0.20-mm-diameter hole. What is the angle of the first dark fringe in radians? In degrees?
27. ‖ You want to photograph a circular diffraction pattern whose central maximum has a diameter of 1.0 cm. You have a helium-neon laser ($\lambda = 633$ nm) and a 0.12-mm-diameter pinhole. How far behind the pinhole should you place the screen that's to be photographed?
28. ‖ Your artist friend is designing an exhibit inspired by circular-aperture diffraction. A pinhole in a red zone is going to be illuminated with a red laser beam of wavelength 670 nm, while a pinhole in a violet zone is going to be illuminated with a violet laser beam of wavelength 410 nm. She wants all the diffraction patterns seen on a distant screen to have the same size. For this to work, what must be the ratio of the red pinhole's diameter to that of the violet pinhole?
29. ‖ Light from a helium-neon laser ($\lambda = 633$ nm) passes through a circular aperture and is observed on a screen 4.0 m behind the aperture. The width of the central maximum is 2.5 cm. What is the diameter (in mm) of the hole?

Section 33.8 Interferometers

30. ‖ A Michelson interferometer uses red light with a wavelength of 656.45 nm from a hydrogen discharge lamp. How many bright-dark-bright fringe shifts are observed if mirror M_2 is moved exactly 1 cm?
31. ‖ Moving mirror M_2 of a Michelson interferometer a distance of 100 μm causes 500 bright-dark-bright fringe shifts. What is the wavelength of the light?
32. ‖ A Michelson interferometer uses light from a sodium lamp. Sodium atoms emit light having wavelengths 589.0 nm and 589.6 nm. The interferometer is initially set up with both arms of equal length ($L_1 = L_2$), producing a bright spot at the center of the interference pattern. How far must mirror M_2 be moved so that one wavelength has produced one more new maximum than the other wavelength?

Problems

33. ‖ FIGURE P33.33 shows the light intensity on a screen 2.5 m behind an aperture. The aperture is illuminated with light of wavelength 620 nm.
   a. Is the aperture a single slit or a double slit? Explain.
   b. If the aperture is a single slit, what is its width? If it is a double slit, what is the spacing between the slits?

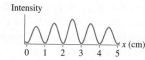

FIGURE P33.33

34. ‖ FIGURE P33.34 shows the light intensity on a screen 2.5 m behind an aperture. The aperture is illuminated with light of wavelength 620 nm.
   a. Is the aperture a single slit or a double slit? Explain.
   b. If the aperture is a single slit, what is its width? If it is a double slit, what is the spacing between the slits?

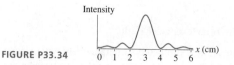

FIGURE P33.34

35. ‖ Light from a helium-neon laser ($\lambda = 633$ nm) is used to illuminate two narrow slits. The interference pattern is observed on a screen 3.0 m behind the slits. Twelve bright fringes are seen, spanning a distance of 52 mm. What is the spacing (in mm) between the slits?
36. ‖ FIGURE P33.36 shows the light intensity on a screen behind a double slit. The slit spacing is 0.20 mm and the wavelength of the light is 620 nm. What is the distance from the slits to the screen?

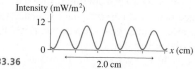

FIGURE P33.36

37. ‖ FIGURE P33.36 shows the light intensity on a screen behind a double slit. The slit spacing is 0.20 mm and the screen is 2.0 m behind the slits. What is the wavelength (in nm) of the light?
38. ‖ FIGURE P33.36 shows the light intensity on a screen behind a double slit. Suppose one slit is covered. What will be the light intensity at the center of the screen due to the remaining slit?
39. ‖ A diffraction grating having 500 lines/mm diffracts visible light at 30°. What is the light's wavelength?
40. ‖ Helium atoms emit light at several wavelengths. Light from a helium lamp illuminates a diffraction grating and is observed on a screen 50.00 cm behind the grating. The emission at wavelength 501.5 nm creates a first-order bright fringe 21.90 cm from the central maximum. What is the wavelength of the bright fringe that is 31.60 cm from the central maximum?
41. ‖ A triple-slit experiment consists of three narrow slits, equally spaced by distance d and illuminated by light of wavelength λ. Each slit alone produces intensity I_1 on the viewing screen at distance L.
   a. Consider a point on the distant viewing screen such that the path-length difference between any two adjacent slits is λ. What is the intensity at this point?
   b. What is the intensity at a point where the path-length difference between any two adjacent slits is $\lambda/2$?
42. ‖ Because sound is a wave, it's possible to make a diffraction grating for sound from a large board of sound-absorbing material with several parallel slits cut for sound to go through. When 10 kHz sound waves pass through such a grating, listeners 10 m from the grating report "loud spots" 1.4 m on both sides of center. What is the spacing between the slits? Use 340 m/s for the speed of sound.
43. ‖ A diffraction grating with 600 lines/mm is illuminated with light of wavelength 510 nm. A very wide viewing screen is 2.0 m behind the grating.
   a. What is the distance between the two $m = 1$ bright fringes?
   b. How many bright fringes can be seen on the screen?

44. ‖ A 500 line/mm diffraction grating is illuminated by light of wavelength 510 nm. How many bright fringes are seen on a 2.0-m-wide screen located 2.0 m behind the grating?

45. ‖ White light (400–700 nm) incident on a 600 line/mm diffraction grating produces rainbows of diffracted light. What is the width of the first-order rainbow on a screen 2.0 m behind the grating?

46. ‖ A chemist identifies compounds by identifying bright lines in their spectra. She does so by heating the compounds until they glow, sending the light through a diffraction grating, and measuring the positions of first-order spectral lines on a detector 15.0 cm behind the grating. Unfortunately, she has lost the card that gives the specifications of the grating. Fortunately, she has a known compound that she can use to calibrate the grating. She heats the known compound, which emits light at a wavelength of 461 nm, and observes a spectral line 9.95 cm from the center of the diffraction pattern. What are the wavelengths emitted by compounds A and B that have spectral lines detected at positions 8.55 cm and 12.15 cm, respectively?

47. ‖‖‖ a. Find an expression for the positions y_1 of the first-order fringes of a diffraction grating if the line spacing is large enough for the small-angle approximation $\tan\theta \approx \sin\theta \approx \theta$ to be valid. Your expression should be in terms of d, L, and λ.
   b. Use your expression from part a to find an expression for the separation Δy on the screen of two fringes that differ in wavelength by $\Delta\lambda$.
   c. Rather than a viewing screen, modern spectrometers use detectors—similar to the one in your digital camera—that are divided into *pixels*. Consider a spectrometer with a 333 line/mm grating and a detector with 100 pixels/mm located 12 cm behind the grating. The *resolution* of a spectrometer is the smallest wavelength separation $\Delta\lambda_{min}$ that can be measured reliably. What is the resolution of this spectrometer for wavelengths near 550 nm, in the center of the visible spectrum? You can assume that the fringe due to one specific wavelength is narrow enough to illuminate only one column of pixels.

48. ‖ For your science fair project you need to design a diffraction grating that will disperse the visible spectrum (400–700 nm) over 30.0° in first order.
   a. How many lines per millimeter does your grating need?
   b. What is the first-order diffraction angle of light from a sodium lamp ($\lambda = 589$ nm)?

49. ‖ FIGURE P33.49 shows the interference pattern on a screen 1.0 m behind an 800 line/mm diffraction grating. What is the wavelength (in nm) of the light?

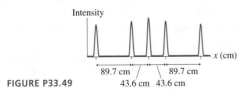

FIGURE P33.49

50. ‖ FIGURE P33.49 shows the interference pattern on a screen 1.0 m behind a diffraction grating. The wavelength of the light is 620 nm. How many lines per millimeter does the grating have?

51. ‖ Light from a sodium lamp ($\lambda = 589$ nm) illuminates a narrow slit and is observed on a screen 75 cm behind the slit. The distance between the first and third dark fringes is 7.5 mm. What is the width (in mm) of the slit?

52. ‖ The wings of some beetles have closely spaced parallel lines of melanin, causing the wing to act as a reflection grating. Suppose sunlight shines straight onto a beetle wing. If the melanin lines on the wing are spaced 2.0 μm apart, what is the first-order diffraction angle for green light ($\lambda = 550$ nm)?

53. ‖ If sunlight shines straight onto a peacock feather, the feather appears bright blue when viewed from 15° on either side of the incident beam of light. The blue color is due to diffraction from parallel rods of melanin in the feather barbules, as was shown in the photograph on page 940. Other wavelengths in the incident light are diffracted at different angles, leaving only the blue light to be seen. The average wavelength of blue light is 470 nm. Assuming this to be the first-order diffraction, what is the spacing of the melanin rods in the feather?

54. ‖ You've found an unlabeled diffraction grating. Before you can use it, you need to know how many lines per mm it has. To find out, you illuminate the grating with light of several different wavelengths and then measure the distance between the two first-order bright fringes on a viewing screen 150 cm behind the grating. Your data are as follows:

Wavelength (nm)	Distance (cm)
430	109.6
480	125.4
530	139.8
580	157.2
630	174.4
680	194.8

Use the best-fit line of an appropriate graph to determine the number of lines per mm.

55. ‖‖‖ A diffraction grating has slit spacing d. Fringes are viewed on a screen at distance L. Find an expression for the wavelength of light that produces a first-order fringe on the viewing screen at distance L from the center of the screen.

56. ‖ FIGURE P33.56 shows the light intensity on a screen behind a single slit. The slit width is 0.20 mm and the screen is 1.5 m behind the slit. What is the wavelength (in nm) of the light?

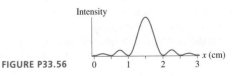

FIGURE P33.56

57. ‖ FIGURE P33.56 shows the light intensity on a screen behind a single slit. The wavelength of the light is 600 nm and the slit width is 0.15 mm. What is the distance from the slit to the screen?

58. ‖ FIGURE P33.56 shows the light intensity on a screen behind a circular aperture. The wavelength of the light is 500 nm and the screen is 1.0 m behind the slit. What is the diameter (in mm) of the aperture?

59. ‖ A student performing a double-slit experiment is using a green laser with a wavelength of 530 nm. She is confused when the $m = 5$ maximum does not appear. She had predicted that this bright fringe would be 1.6 cm from the central maximum on a screen 1.5 m behind the slits.
    a. Explain what prevented the fifth maximum from being observed.
    b. What is the width of her slits?

60. ‖ Scientists shine a laser beam on a 35-μm-wide slit and produce a diffraction pattern on a screen 70 cm behind the slit. Careful measurements show that the intensity first falls to 25% of maximum at a distance of 7.2 mm from the center of the diffraction pattern. What is the wavelength of the laser light?
    Hint: Use the trial-and-error technique demonstrated in Example 33.5 to solve the transcendental equation.

61. ‖ Light from a helium-neon laser ($\lambda = 633$ nm) illuminates a circular aperture. It is noted that the diameter of the central maximum on a screen 50 cm behind the aperture matches the diameter of the geometric image. What is the aperture's diameter (in mm)?

62. ‖ A helium-neon laser ($\lambda = 633$ nm) is built with a glass tube of inside diameter 1.0 mm, as shown in FIGURE P33.62. One mirror is partially transmitting to allow the laser beam out. An electrical discharge in the tube causes it to glow like a neon light. From an optical perspective, the laser beam is a light wave that diffracts out through a 1.0-mm-diameter circular opening.
    a. Can a laser beam be *perfectly* parallel, with no spreading? Why or why not?
    b. The angle θ_1 to the first minimum is called the *divergence angle* of a laser beam. What is the divergence angle of this laser beam?
    c. What is the diameter (in mm) of the laser beam after it travels 3.0 m?
    d. What is the diameter of the laser beam after it travels 1.0 km?

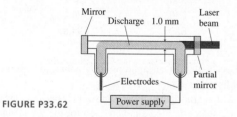

FIGURE P33.62

63. ‖ One day, after pulling down your window shade, you notice that sunlight is passing through a pinhole in the shade and making a small patch of light on the far wall. Having recently studied optics in your physics class, you're not too surprised to see that the patch of light seems to be a circular diffraction pattern. It appears that the central maximum is about 1 cm across, and you estimate that the distance from the window shade to the wall is about 3 m. Estimate (a) the average wavelength of the sunlight (in nm) and (b) the diameter of the pinhole (in mm).

64. ‖ A radar for tracking aircraft broadcasts a 12 GHz microwave beam from a 2.0-m-diameter circular radar antenna. From a wave perspective, the antenna is a circular aperture through which the microwaves diffract.
    a. What is the diameter of the radar beam at a distance of 30 km?
    b. If the antenna emits 100 kW of power, what is the average microwave intensity at 30 km?

65. ‖ Scientists use *laser range-finding* to measure the distance to the moon with great accuracy. A brief laser pulse is fired at the moon, then the time interval is measured until the "echo" is seen by a telescope. A laser beam spreads out as it travels because it diffracts through a circular exit as it leaves the laser. In order for the reflected light to be bright enough to detect, the laser spot on the moon must be no more than 1.0 km in diameter. Staying within this diameter is accomplished by using a special large-diameter laser. If $\lambda = 532$ nm, what is the minimum diameter of the circular opening from which the laser beam emerges? The earth-moon distance is 384,000 km.

66. ‖ Light of wavelength 600 nm passes though two slits separated by 0.20 mm and is observed on a screen 1.0 m behind the slits. The location of the central maximum is marked on the screen and labeled $y = 0$.
    a. At what distance, on either side of $y = 0$, are the $m = 1$ bright fringes?
    b. A very thin piece of glass is then placed in one slit. Because light travels slower in glass than in air, the wave passing through the glass is delayed by 5.0×10^{-16} s in comparison to the wave going through the other slit. What fraction of the period of the light wave is this delay?
    c. With the glass in place, what is the phase difference $\Delta\phi_0$ between the two waves as they leave the slits?
    d. The glass causes the interference fringe pattern on the screen to shift sideways. Which way does the central maximum move (toward or away from the slit with the glass) and by how far?

67. ‖ A 600 line/mm diffraction grating is in an empty aquarium tank. The index of refraction of the glass walls is $n_{glass} = 1.50$. A helium-neon laser ($\lambda = 633$ nm) is outside the aquarium. The laser beam passes through the glass wall and illuminates the diffraction grating.
    a. What is the first-order diffraction angle of the laser beam?
    b. What is the first-order diffraction angle of the laser beam after the aquarium is filled with water ($n_{water} = 1.33$)?

68. ‖ A Michelson interferometer operating at a 600 nm wavelength has a 2.00-cm-long glass cell in one arm. To begin, the air is pumped out of the cell and mirror M_2 is adjusted to produce a bright spot at the center of the interference pattern. Then a valve is opened and air is slowly admitted into the cell. The index of refraction of air at 1.00 atm pressure is 1.00028. How many bright-dark-bright fringe shifts are observed as the cell fills with air?

69. ‖‖‖ Optical computers require microscopic optical switches to turn signals on and off. One device for doing so, which can be implemented in an integrated circuit, is the *Mach-Zender interferometer* seen in FIGURE P33.69. Light from an on-chip infrared laser ($\lambda = 1.000$ μm) is split into two waves that travel equal distances around the arms of the interferometer. One arm passes through an *electro-optic crystal,* a transparent material that can change its index of refraction in response to an applied voltage. Suppose both arms are exactly the same length and the crystal's index of refraction with no applied voltage is 1.522.
    a. With no voltage applied, is the output bright (switch closed, optical signal passing through) or dark (switch open, no signal)? Explain.
    b. What is the first index of refraction of the electro-optic crystal larger than 1.522 that changes the optical switch to the state opposite the state you found in part a?

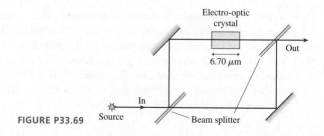

FIGURE P33.69

70. ‖ To illustrate one of the ideas of holography in a simple way, consider a diffraction grating with slit spacing d. The small-angle approximation is usually not valid for diffraction gratings, because d is only slightly larger than λ, but assume that the λ/d ratio of this grating is small enough to make the small-angle approximation valid.

a. Use the small-angle approximation to find an expression for the fringe spacing on a screen at distance L behind the grating.

b. Rather than a screen, suppose you place a piece of film at distance L behind the grating. The bright fringes will expose the film, but the dark spaces in between will leave the film unexposed. After being developed, the film will be a series of alternating light and dark stripes. What if you were to now "play" the film by using it as a diffraction grating? In other words, what happens if you shine the same laser through the film and look at the film's diffraction pattern on a screen at the same distance L? Demonstrate that the film's diffraction pattern is a reproduction of the original diffraction grating.

Challenge Problems

71. ‖‖ A double-slit experiment is set up using a helium-neon laser ($\lambda = 633$ nm). Then a very thin piece of glass ($n = 1.50$) is placed over one of the slits. Afterward, the central point on the screen is occupied by what had been the $m = 10$ dark fringe. How thick is the glass?

72. ‖‖ The intensity at the central maximum of a double-slit interference pattern is $4I_1$. The intensity at the first minimum is zero. At what fraction of the distance from the central maximum to the first minimum is the intensity I_1? Assume an ideal double slit.

73. ‖‖ FIGURE CP33.73 shows two nearly overlapped intensity peaks of the sort you might produce with a diffraction grating (see Figure 33.9b). As a practical matter, two peaks can just barely be resolved if their spacing Δy equals the width w of each peak, where w is measured at half of the peak's height. Two peaks closer together than w will merge into a single peak. We can use this idea to understand the resolution of a diffraction grating.

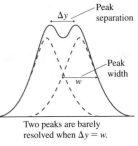

Two peaks are barely resolved when $\Delta y = w$.

FIGURE CP33.73

a. In the small-angle approximation, the position of the $m = 1$ peak of a diffraction grating falls at the same location as the $m = 1$ fringe of a double slit: $y_1 = \lambda L/d$. Suppose two wavelengths differing by $\Delta \lambda$ pass through a grating at the same time. Find an expression for Δy, the separation of their first-order peaks.

b. We noted that the widths of the bright fringes are proportional to $1/N$, where N is the number of slits in the grating. Let's hypothesize that the fringe width is $w = y_1/N$. Show that this is true for the double-slit pattern. We'll then assume it to be true as N increases.

c. Use your results from parts a and b together with the idea that $\Delta y_{min} = w$ to find an expression for $\Delta \lambda_{min}$, the minimum wavelength separation (in first order) for which the diffraction fringes can barely be resolved.

d. Ordinary hydrogen atoms emit red light with a wavelength of 656.45 nm. In deuterium, which is a "heavy" isotope of hydrogen, the wavelength is 656.27 nm. What is the minimum number of slits in a diffraction grating that can barely resolve these two wavelengths in the first-order diffraction pattern?

74. ‖‖‖ FIGURE CP33.74 shows light of wavelength λ incident at angle ϕ on a *reflection* grating of spacing d. We want to find the angles θ_m at which constructive interference occurs.

a. The figure shows paths 1 and 2 along which two waves travel and interfere. Find an expression for the path-length difference $\Delta r = r_2 - r_1$.

b. Using your result from part a, find an equation (analogous to Equation 33.15) for the angles θ_m at which diffraction occurs when the light is incident at angle ϕ. Notice that m can be a negative integer in your expression, indicating that path 2 is shorter than path 1.

c. Show that the zeroth-order diffraction is simply a "reflection." That is, $\theta_0 = \phi$.

d. Light of wavelength 500 nm is incident at $\phi = 40°$ on a reflection grating having 700 reflection lines/mm. Find all angles θ_m at which light is diffracted. Negative values of θ_m are interpreted as an angle left of the vertical.

e. Draw a picture showing a *single* 500 nm light ray incident at $\phi = 40°$ and showing all the diffracted waves at the correct angles.

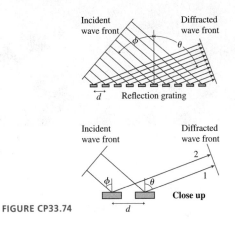

FIGURE CP33.74

75. ‖‖‖ The pinhole camera of FIGURE CP33.75 images distant objects by allowing only a narrow bundle of light rays to pass through the hole and strike the film. If light con-

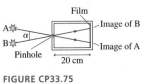

FIGURE CP33.75

sisted of particles, you could make the image sharper and sharper (at the expense of getting dimmer and dimmer) by making the aperture smaller and smaller. In practice, diffraction of light by the circular aperture limits the maximum sharpness that can be obtained. Consider two distant points of light, such as two distant streetlights. Each will produce a circular diffraction pattern on the film. The two images can just barely be resolved if the central maximum of one image falls on the first dark fringe of the other image. (This is called Rayleigh's criterion, and we will explore its implication for optical instruments in Chapter 35.)

a. Optimum sharpness of one image occurs when the diameter of the central maximum equals the diameter of the pinhole. What is the optimum hole size for a pinhole camera in which the film is 20 cm behind the hole? Assume $\lambda = 550$ nm, an average value for visible light.

b. For this hole size, what is the angle α (in degrees) between two distant sources that can barely be resolved?

c. What is the distance between two street lights 1 km away that can barely be resolved?

34 Ray Optics

The observation that light travels in straight lines—*light rays*—will help us understand the physics of lenses and prisms.

IN THIS CHAPTER, you will learn about and apply the ray model of light.

What are light rays?
A light ray is a concept, not a physical thing. It is the line along which light energy flows.

- Rays travel in straight lines. Two rays can cross without disturbing one another.
- Objects are sources of light rays.
- Reflection and refraction by mirrors and lenses create images of objects. Points to which light rays converge are called real images. Points from which light rays diverge are called virtual images.
- The eye sees an object or an image when diverging bundles of rays enter the pupil and are focused to a real image on the retina.

You'll use both graphical and mathematical techniques to analyze how light rays travel and how images are formed.

What is the law of reflection?
Light rays bounce, or reflect, off a surface.

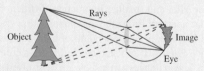

- Specular reflection is mirror like.
- Diffuse reflection is like light reflecting from the page of this book.

The law of reflection says that the angle of reflection equals the angle of incidence. You'll learn how reflection allows images to be seen in both flat and curved mirrors.

What is refraction?
Light rays change direction at the boundary when they move from one medium to another. This is called refraction, and it is the basis for image formation by lenses. Snell's law will allow you to find the angles on both sides of the boundary.

≪ LOOKING BACK Section 16.5 Index of refraction

How do lenses form images?
Lenses form images by refraction.

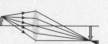

- We'll start with ray tracing, a graphical method of seeing how and where images are formed.
- We'll then develop the thin-lens equation for more quantitative results.

The same methods apply to image formation by curved mirrors.

Why is optics important?
Optics is everywhere, from your smart phone camera and your car headlights to laser pointers and the optical scanners that read bar codes. Our knowledge of the microscopic world and of the cosmos comes through optical instruments. And, of course, your eye is one of the most marvelous optical devices of all. Modern optical engineering is called photonics. Photonics does draw on all three models of light, as needed, but ray optics is usually the foundation on which optical instruments are designed.

34.1 The Ray Model of Light

A flashlight makes a beam of light through the night's darkness. Sunbeams stream into a darkened room through a small hole in the shade. Laser beams are even more well defined. Our everyday experience that light travels in straight lines is the basis of the *ray model* of light.

The ray model is an oversimplification of reality but nonetheless is very useful within its range of validity. In particular, the ray model of light is valid as long as any apertures through which the light passes (lenses, mirrors, and holes) are very large compared to the wavelength of light. In that case, diffraction and other wave aspects of light are negligible and can be ignored. The analysis of Section 33.7 found that the crossover between wave optics and ray optics occurs for apertures ≈ 1 mm in diameter. Lenses and mirrors are almost always larger than 1 mm, so the ray model of light is an excellent basis for the practical optics of image formation.

To begin, let us define a **light ray** as a line in the direction along which light energy is flowing. A light ray is an abstract idea, not a physical entity or a "thing." Any narrow beam of light, such as the laser beam in **FIGURE 34.1**, is actually a bundle of many parallel light rays. You can think of a single light ray as the limiting case of a laser beam whose diameter approaches zero. Laser beams are good approximations of light rays, but any real laser beam is a bundle of many parallel rays.

Chapter 33 briefly introduced the three models of light. Here we expand on the ray model, the subject of this chapter.

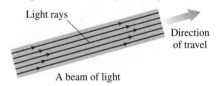

FIGURE 34.1 A laser beam or beam of sunlight is a bundle of parallel light rays.

Light rays

Direction of travel

A beam of light

MODEL 34.1

Ray model of light

For use when diffraction is not significant.

- Light rays travel in straight lines.
  - The speed of light is $v = c/n$, where n is the material's index of refraction.
  - Light rays cross without interacting.
- Light rays travel forever unless they interact with matter.
  - At an interface between two materials, rays can be either reflected or refracted.
  - Within a material, light rays can be either scattered or absorbed.
- An **object** is a source of light rays.
  - Rays originate at *every* point on an object.
  - Rays are sent in *all* directions.
- The eye sees by focusing a diverging bundle of light rays.
  - Diverging rays enter the pupil and are focused on the retina.
  - Your brain perceives the object as being at the point from which the rays are diverging.
- Limitations: Use the wave model if diffraction is significant. The ray model is usually valid if openings are larger than about 1 mm, while the wave model is more appropriate if openings are smaller than about 1 mm.

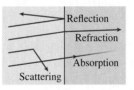

Reflection
Refraction
Absorption
Scattering

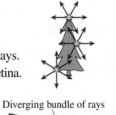

Diverging bundle of rays

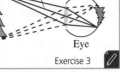

Eye

Exercise 3

Objects

FIGURE 34.2 on the next page illustrates the idea that objects can be either *self-luminous,* such as the sun, flames, and lightbulbs, or *reflective.* Most objects are reflective. A tree, unless it is on fire, is seen or photographed by virtue of reflected sunlight or

FIGURE 34.2 Self-luminous and reflective objects.

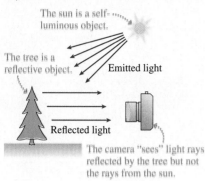

FIGURE 34.4 A ray diagram simplifies the situation by showing only a few rays.

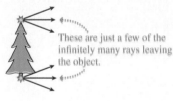

FIGURE 34.5 A camera obscura.

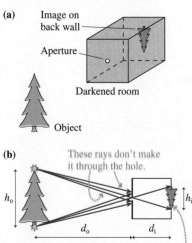

reflected skylight. People, houses, and this page in the book reflect light from self-luminous sources. In this chapter we are concerned not with how the light originates but with how it behaves after leaving the object.

Light rays from an object are emitted in all directions, but you are not *aware* of light rays unless they enter the pupil of your eye. Consequently, most light rays go completely unnoticed. For example, light rays travel from the sun to the tree in Figure 34.2, but you're not aware of these unless the tree reflects some of them into your eye. Or consider a laser beam. You've probably noticed that it's almost impossible to see a laser beam from the side unless there's dust in the air. The dust scatters a few of the light rays toward your eye, but in the absence of dust you would be completely unaware of a very powerful light beam traveling past you. **Light rays exist independently of whether you are seeing them.**

FIGURE 34.3 shows two idealized sets of light rays. The diverging rays from a **point source** are emitted in all directions. It is useful to think of each point on an object as a point source of light rays. A **parallel bundle** of rays could be a laser beam. Alternatively it could represent a *distant object,* an object such as a star so far away that the rays arriving at the observer are essentially parallel to each other.

FIGURE 34.3 Point sources and parallel bundles represent idealized objects.

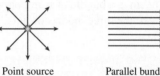

Point source   Parallel bundle

Ray Diagrams

Rays originate from *every* point on an object and travel outward in *all* directions, but a diagram trying to show all these rays would be hopelessly messy and confusing. To simplify the picture, we usually use a **ray diagram** showing only a few rays. For example, **FIGURE 34.4** is a ray diagram showing only a few rays leaving the top and bottom points of the object and traveling to the right. These rays will be sufficient to show us how the object is imaged by lenses or mirrors.

> **NOTE** Ray diagrams are the basis for a *pictorial representation* that we'll use throughout this chapter. Be careful not to think that a ray diagram shows all of the rays. The rays shown on the diagram are just a subset of the infinitely many rays leaving the object.

Apertures

A popular form of entertainment during ancient Roman times was a visit to a **camera obscura,** Latin for "dark room." As **FIGURE 34.5a** shows, a camera obscura was a darkened room with a single, small hole to the outside world. After their eyes became dark adapted, visitors could see a dim but full-color image of the outside world displayed on the back wall of the room. However, the image was upside down! The *pinhole camera* is a miniature version of the camera obscura.

A hole through which light passes is called an **aperture**. **FIGURE 34.5b** uses the ray model of light passing through a small aperture to explain how the camera obscura works. Each point on an object emits light rays in all directions, but only a very few of these rays pass through the aperture and reach the back wall. As the figure illustrates, the geometry of the rays causes the image to be upside down.

Actually, as you may have realized, each *point* on the object illuminates a small but extended *patch* on the wall. This is because the non-zero size of the aperture—needed for the image to be bright enough to see—allows several rays from each point on the object to pass through at slightly different angles. As a result, the image is slightly blurred and out of focus. (Diffraction also becomes an issue if the hole gets too small.) We'll later discover how a modern camera, with a lens, improves on the camera obscura.

You can see from the similar triangles in Figure 34.5b that the object and image heights are related by

$$\frac{h_i}{h_o} = \frac{d_i}{d_o} \qquad (34.1)$$

where d_o is the distance to the object and d_i is the depth of the camera obscura. Any realistic camera obscura has $d_i < d_o$; thus the image is smaller than the object.

We can apply the ray model to more complex apertures, such as the L-shaped aperture in **FIGURE 34.6**. The pattern of light on the screen is found by tracing all the straight-line paths—the ray trajectories—that start from the point source and pass through the aperture. We will see an enlarged L on the screen, with a sharp boundary between the image and the dark shadow.

FIGURE 34.6 Light through an aperture.

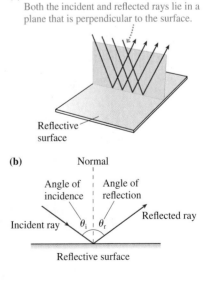

STOP TO THINK 34.1 A long, thin lightbulb illuminates a vertical aperture. Which pattern of light do you see on a viewing screen behind the aperture?

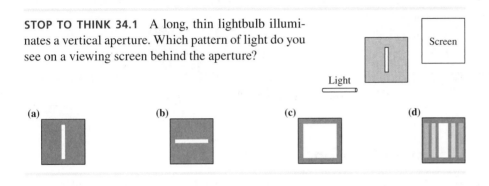

34.2 Reflection

Reflection of light is a familiar, everyday experience. You see your reflection in the bathroom mirror first thing every morning, reflections in your car's rearview mirror as you drive to school, and the sky reflected in puddles of standing water. Reflection from a flat, smooth surface, such as a mirror or a piece of polished metal, is called **specular reflection,** from *speculum,* the Latin word for "mirror."

FIGURE 34.7a shows a bundle of parallel light rays reflecting from a mirror-like surface. You can see that the incident and reflected rays are both in a plane that is normal, or perpendicular, to the reflective surface. A three-dimensional perspective accurately shows the relationship between the light rays and the surface, but figures such as this are hard to draw by hand. Instead, it is customary to represent reflection with the simpler pictorial representation of **FIGURE 34.7b**. In this figure,

FIGURE 34.7 Specular reflection of light.

(a)

Both the incident and reflected rays lie in a plane that is perpendicular to the surface.

Reflective surface

(b)

- The plane of the page is the *plane of incidence,* the plane containing both incident and reflected rays. The reflective surface extends into the page.
- A *single* light ray represents the entire bundle of parallel rays. This is oversimplified, but it keeps the figure and the analysis clear.

The angle θ_i between the ray and a line perpendicular to the surface—the *normal* to the surface—is called the **angle of incidence.** Similarly, the **angle of reflection** θ_r is the angle between the reflected ray and the normal to the surface. The **law of reflection,** easily demonstrated with simple experiments, states that

1. The incident ray and the reflected ray are in the same plane normal to the surface, and
2. The angle of reflection equals the angle of incidence: $\theta_r = \theta_i$.

Normal

Angle of incidence   Angle of reflection

Incident ray θ_i θ_r Reflected ray

Reflective surface

NOTE Optics calculations *always* use the angle measured from the normal, not the angle between the ray and the surface.

EXAMPLE 34.1 | **Light reflecting from a mirror**

A dressing mirror on a closet door is 1.50 m tall. The bottom is 0.50 m above the floor. A bare lightbulb hangs 1.00 m from the closet door, 2.50 m above the floor. How long is the streak of reflected light across the floor?

MODEL Treat the lightbulb as a point source and use the ray model of light.

FIGURE 34.8 Pictorial representation of the light rays reflecting from a mirror.

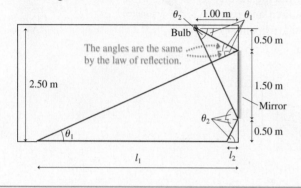

VISUALIZE FIGURE 34.8 is a pictorial representation of the light rays. We need to consider only the two rays that strike the edges of the mirror. All other reflected rays will fall between these two.

SOLVE Figure 34.8 has used the law of reflection to set the angles of reflection equal to the angles of incidence. Other angles have been identified with simple geometry. The two angles of incidence are

$$\theta_1 = \tan^{-1}\left(\frac{0.50 \text{ m}}{1.00 \text{ m}}\right) = 26.6°$$

$$\theta_2 = \tan^{-1}\left(\frac{2.00 \text{ m}}{1.00 \text{ m}}\right) = 63.4°$$

The distances to the points where the rays strike the floor are then

$$l_1 = \frac{2.00 \text{ m}}{\tan\theta_1} = 4.00 \text{ m}$$

$$l_2 = \frac{0.50 \text{ m}}{\tan\theta_2} = 0.25 \text{ m}$$

Thus the length of the light streak is $l_1 - l_2 = 3.75$ m.

Diffuse Reflection

FIGURE 34.9 Diffuse reflection from an irregular surface.

Each ray obeys the law of reflection at that point, but the irregular surface causes the reflected rays to leave in many random directions.

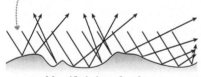

Magnified view of surface

Most objects are seen by virtue of their reflected light. For a "rough" surface, the law of reflection $\theta_r = \theta_i$ is obeyed at each point but the irregularities of the surface cause the reflected rays to leave in many random directions. This situation, shown in FIGURE 34.9, is called **diffuse reflection.** It is how you see this page, the wall, your hand, your friend, and so on.

By a "rough" surface, we mean a surface that is rough or irregular in comparison to the wavelength of light. Because visible-light wavelengths are ≈ 0.5 μm, any surface with texture, scratches, or other irregularities larger than 1 μm will cause diffuse reflection rather than specular reflection. A piece of paper may feel quite smooth to your hand, but a microscope would show that the surface consists of distinct fibers much larger than 1 μm. By contrast, the irregularities on a mirror or a piece of polished metal are much smaller than 1 μm.

The Plane Mirror

One of the most commonplace observations is that you can see yourself in a mirror. How? FIGURE 34.10a shows rays from point source P reflecting from a mirror. Consider the particular ray shown in FIGURE 34.10b. The reflected ray travels along a line that passes through point P' on the "back side" of the mirror. Because $\theta_r = \theta_i$, simple geometry dictates that P' is the same distance behind the mirror as P is in front of the mirror. That is, $s' = s$.

FIGURE 34.10 The light rays reflecting from a plane mirror.

(a)

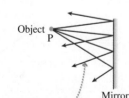

Rays from P reflect from the mirror. Each ray obeys the law of reflection.

(b)

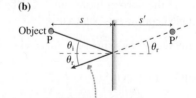

This reflected ray appears to have been traveling along a line that passed through point P'.

(c)

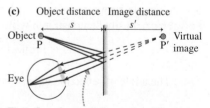

The reflected rays *all* diverge from P', which appears to be the source of the reflected rays. Your eye collects the bundle of diverging rays and "sees" the light coming from P'.

The location of point P′ in Figure 34.10b is independent of the value of θ_i. Consequently, as **FIGURE 34.10c** shows, **the reflected rays all *appear* to be coming from point P′.** For a plane mirror, the distance s′ to point P′ is equal to the object distance s:

$$s' = s \qquad \text{(plane mirror)} \qquad (34.2)$$

If rays diverge from an object point P and interact with a mirror so that the reflected rays diverge from point P′ and *appear* to come from P′, then we call P′ a **virtual image** of point P. The image is "virtual" in the sense that no rays actually leave P′, which is in darkness behind the mirror. But as far as your eye is concerned, the light rays act exactly *as if* the light really originated at P′. So while you may say "I see P in the mirror," what you are actually seeing is the virtual image of P. Distance s′ is the *image distance*.

For an extended object, such as the one in **FIGURE 34.11**, each point on the object from which rays strike the mirror has a corresponding image point an equal distance on the opposite side of the mirror. The eye captures and focuses diverging bundles of rays from each point of the image in order to see the full image in the mirror. Two facts are worth noting:

1. Rays from each point on the object spread out in all directions and strike *every point* on the mirror. Only a very few of these rays enter your eye, but the other rays are very real and might be seen by other observers.
2. Rays from points P and Q enter your eye after reflecting from *different* areas of the mirror. This is why you can't always see the full image of an object in a very small mirror.

FIGURE 34.11 Each point on the extended object has a corresponding image point an equal distance on the opposite side of the mirror.

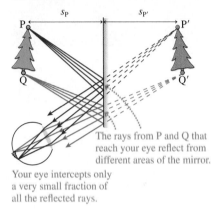

EXAMPLE 34.2 | How high is the mirror?

If your height is h, what is the shortest mirror on the wall in which you can see your full image? Where must the top of the mirror be hung?

MODEL Use the ray model of light.

VISUALIZE FIGURE 34.12 is a pictorial representation of the light rays. We need to consider only the two rays that leave your head and feet and reflect into your eye.

SOLVE Let the distance from your eyes to the top of your head be l_1 and the distance to your feet be l_2. Your height is $h = l_1 + l_2$. A light ray from the top of your head that reflects from the mirror at $\theta_r = \theta_i$ and enters your eye must, by congruent triangles, strike the mirror a distance $\frac{1}{2}l_1$ above your eyes. Similarly, a ray from your foot to your eye strikes the mirror a distance $\frac{1}{2}l_2$ below your eyes. The distance between these two points on the mirror is $\frac{1}{2}l_1 + \frac{1}{2}l_2 = \frac{1}{2}h$. A ray from anywhere else on your body can reach your eye if it strikes the mirror between these two points. Pieces of the mirror outside these two points are irrelevant, not because rays don't strike them but because the reflected rays don't reach

FIGURE 34.12 Pictorial representation of light rays from your head and feet reflecting into your eye.

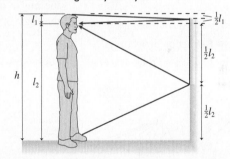

your eye. Thus the shortest mirror in which you can see your full reflection is $\frac{1}{2}h$. But this will work only if the top of the mirror is hung midway between your eyes and the top of your head.

ASSESS It is interesting that the answer does not depend on how far you are from the mirror.

STOP TO THINK 34.2 Two plane mirrors form a right angle. How many images of the ball can you see in the mirrors?

a. 1
b. 2
c. 3
d. 4

Observer

FIGURE 34.13 A light beam refracts twice in passing through a glass prism.

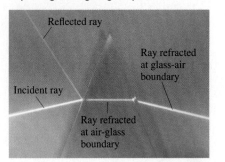

34.3 Refraction

Two things happen when a light ray is incident on a smooth boundary between two transparent materials, such as the boundary between air and glass:

1. Part of the light *reflects* from the boundary, obeying the law of reflection. This is how you see reflections from pools of water or storefront windows.
2. Part of the light continues into the second medium. It is *transmitted* rather than reflected, but the transmitted ray changes direction as it crosses the boundary. The transmission of light from one medium to another, but with a change in direction, is called **refraction.**

The photograph of FIGURE 34.13 shows the refraction of a light beam as it passes through a glass prism. Notice that the ray direction changes as the light enters and leaves the glass. Our goal in this section is to understand refraction, so we will usually ignore the weak reflection and focus on the transmitted light.

> **NOTE** A transparent material through which light travels is called the *medium.* This term has to be used with caution. The material does affect the light speed, but a transparent material differs from the medium of a sound or water wave in that particles of the medium do *not* oscillate as a light wave passes through. For a light wave it is the electromagnetic field that oscillates.

FIGURE 34.14a shows the refraction of light rays in a parallel beam of light, such as a laser beam, and rays from a point source. Our analysis will be easier, however, if we focus on a single light ray. FIGURE 34.14b is a ray diagram showing the refraction of a single ray at a boundary between medium 1 and medium 2. Let the angle between the ray and the normal be θ_1 in medium 1 and θ_2 in medium 2. For the medium in which the ray is approaching the boundary, this is the *angle of incidence* as we've previously defined it. The angle on the transmitted side, *measured from the normal,* is called the **angle of refraction.** Notice that θ_1 is the angle of incidence in Figure 34.14b and the angle of refraction in FIGURE 34.14c, where the ray is traveling in the opposite direction.

FIGURE 34.14 Refraction of light rays.

(a) Refraction of parallel and point-source rays

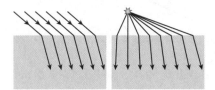

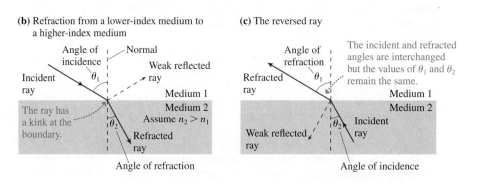

(b) Refraction from a lower-index medium to a higher-index medium

(c) The reversed ray

Refraction was first studied experimentally by the Arab scientist Ibn al-Haitham, in about the year 1000, and later by the Dutch scientist Willebrord Snell. **Snell's law** says that when a ray refracts between medium 1 and medium 2, having indices of refraction n_1 and n_2, the ray angles θ_1 and θ_2 in the two media are related by

$$n_1 \sin\theta_1 = n_2 \sin\theta_2 \qquad \text{(Snell's law of refraction)} \qquad (34.3)$$

Notice that Snell's law does not mention which is the incident angle and which is the refracted angle.

The Index of Refraction

To Snell and his contemporaries, n was simply an "index of the refractive power" of a transparent substance. The relationship between the index of refraction and the speed of light was not recognized until the development of a wave theory of light in the 19th century. Theory predicts, and experiment confirms, that light travels through a transparent medium, such as glass or water, at a speed *less* than its speed c in vacuum. In Section 16.5, we defined the *index of refraction n* of a transparent medium as

$$n = \frac{c}{v_{medium}} \qquad (34.4)$$

where v_{medium} is the light speed in the medium. This implies, of course, that $v_{medium} = c/n$. The index of refraction of a medium is always $n > 1$ except for vacuum.

TABLE 34.1 shows measured values of n for several materials. There are many types of glass, each with a slightly different index of refraction, so we will keep things simple by accepting $n = 1.50$ as a typical value. Notice that cubic zirconia, used to make costume jewelry, has an index of refraction much higher than glass.

We can accept Snell's law as simply an empirical discovery about light. Alternatively, and perhaps surprisingly, we can use the wave model of light to justify Snell's law. The key ideas we need are:

- Wave fronts represent the crests of waves. They are spaced one wavelength apart.
- The wavelength in a medium with index of refraction n is $\lambda = \lambda_{vac}/n$, where λ_{vac} is the vacuum wavelength.
- Wave fronts are perpendicular to the wave's direction of travel.
- The wave fronts stay lined up as a wave crosses from one medium into another.

FIGURE 34.15 shows a wave crossing the boundary between two media, where we're assuming $n_2 > n_1$. **Because the wavelengths differ on opposite sides of the boundary, the wave fronts can stay lined up only if the waves in the two media are traveling in different directions.** In other words, the wave must refract at the boundary to keep the crests of the wave aligned.

To analyze Figure 34.15, consider the segment of boundary of length l between the two wave fronts. This segment is the common hypotenuse of two right triangles. From the upper triangle, which has one side of length λ_1, we see

$$l = \frac{\lambda_1}{\sin\theta_1} \qquad (34.5)$$

where θ_1 is the angle of incidence. Similarly, the lower triangle, where θ_2 is the angle of refraction, gives

$$l = \frac{\lambda_2}{\sin\theta_2} \qquad (34.6)$$

Equating these two expressions for l, and using $\lambda_1 = \lambda_{vac}/n_1$ and $\lambda_2 = \lambda_{vac}/n_2$, we find

$$\frac{\lambda_{vac}}{n_1 \sin\theta_1} = \frac{\lambda_{vac}}{n_2 \sin\theta_2} \qquad (34.7)$$

Equation 34.7 can be true only if

$$n_1 \sin\theta_1 = n_2 \sin\theta_2 \qquad (34.8)$$

which is Snell's law.

Examples of Refraction

Look back at Figure 34.14. As the ray in Figure 34.14b moves from medium 1 to medium 2, where $n_2 > n_1$, it bends closer to the normal. In Figure 34.14c, where the ray moves from medium 2 to medium 1, it bends away from the normal. This is a general conclusion that follows from Snell's law:

- When a ray is transmitted into a material with a higher index of refraction, it bends toward the normal.
- When a ray is transmitted into a material with a lower index of refraction, it bends away from the normal.

TABLE 34.1 Indices of refraction

Medium	n
Vacuum	1.00 exactly
Air (actual)	1.0003
Air (accepted)	1.00
Water	1.33
Ethyl alcohol	1.36
Oil	1.46
Glass (typical)	1.50
Polystyrene plastic	1.59
Cubic zirconia	2.18
Diamond	2.41
Silicon (infrared)	3.50

FIGURE 34.15 Snell's law is a consequence of the wave model of light.

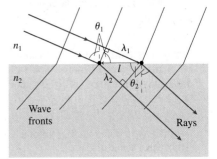

This rule becomes a central idea in a procedure for analyzing refraction problems.

TACTICS BOX 34.1 (MP)

Analyzing refraction

❶ **Draw a ray diagram.** Represent the light beam with one ray.
❷ **Draw a line normal to the boundary.** Do this at each point where the ray intersects a boundary.
❸ **Show the ray bending in the correct direction.** The angle is larger on the side with the smaller index of refraction. This is the qualitative application of Snell's law.
❹ **Label angles of incidence and refraction.** Measure all angles from the normal.
❺ **Use Snell's law.** Calculate the unknown angle or unknown index of refraction.

Exercises 11–15 ✏

EXAMPLE 34.3 | Deflecting a laser beam

A laser beam is aimed at a 1.0-cm-thick sheet of glass at an angle 30° above the glass.

a. What is the laser beam's direction of travel in the glass?
b. What is its direction in the air on the other side?
c. By what distance is the laser beam displaced?

MODEL Represent the laser beam with a single ray and use the ray model of light.

VISUALIZE FIGURE 34.16 is a pictorial representation in which the first four steps of Tactics Box 34.1 are identified. Notice that the angle of incidence is $\theta_1 = 60°$, not the 30° value given in the problem.

FIGURE 34.16 The ray diagram of a laser beam passing through a sheet of glass.

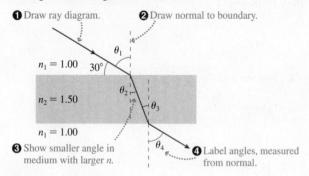

❶ Draw ray diagram.   ❷ Draw normal to boundary.

θ_1
$n_1 = 1.00$ 30°
$n_2 = 1.50$ θ_2
θ_3
$n_1 = 1.00$
❸ Show smaller angle in medium with larger n. θ_4 ❹ Label angles, measured from normal.

SOLVE a. Snell's law, the final step in the Tactics Box, is $n_1 \sin\theta_1 = n_2 \sin\theta_2$. Using $\theta_1 = 60°$, we find that the direction of travel in the glass is

$$\theta_2 = \sin^{-1}\left(\frac{n_1 \sin\theta_1}{n_2}\right) = \sin^{-1}\left(\frac{\sin 60°}{1.5}\right)$$
$$= \sin^{-1}(0.577) = 35.3°$$

b. Snell's law at the second boundary is $n_2 \sin\theta_3 = n_1 \sin\theta_4$. You can see from Figure 34.16 that the interior angles are equal:

$\theta_3 = \theta_2 = 35.3°$. Thus the ray emerges back into the air traveling at angle

$$\theta_4 = \sin^{-1}\left(\frac{n_2 \sin\theta_3}{n_1}\right) = \sin^{-1}(1.5 \sin 35.3°)$$
$$= \sin^{-1}(0.867) = 60°$$

This is the same as θ_1, the original angle of incidence. The glass doesn't change the direction of the laser beam.

c. Although the exiting laser beam is parallel to the initial laser beam, it has been displaced sideways by distance d. FIGURE 34.17 shows the geometry for finding d. From trigonometry, $d = l \sin\phi$. Further, $\phi = \theta_1 - \theta_2$ and $l = t/\cos\theta_2$, where t is the thickness of the glass. Combining these gives

$$d = l \sin\phi = \frac{t}{\cos\theta_2} \sin(\theta_1 - \theta_2)$$
$$= \frac{(1.0 \text{ cm}) \sin 24.7°}{\cos 35.3°} = 0.51 \text{ cm}$$

The glass causes the laser beam to be displaced sideways by 0.51 cm.

FIGURE 34.17 The laser beam is deflected sideways by distance d.

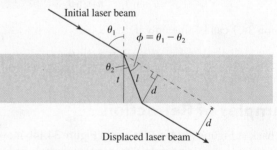

Initial laser beam
θ_1 $\phi = \theta_1 - \theta_2$
θ_2
t l d
d
Displaced laser beam

ASSESS The laser beam exits the glass still traveling in the same direction as it entered. This is a general result for light traveling through a medium with parallel sides. Notice that the displacement d becomes zero in the limit $t \to 0$. This will be an important observation when we get to lenses.

EXAMPLE 34.4 | Measuring the index of refraction

FIGURE 34.18 shows a laser beam deflected by a 30°-60°-90° prism. What is the prism's index of refraction?

FIGURE 34.18 A prism deflects a laser beam.

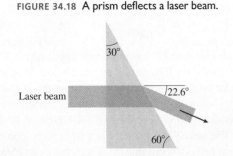

FIGURE 34.19 Pictorial representation of a laser beam passing through the prism.

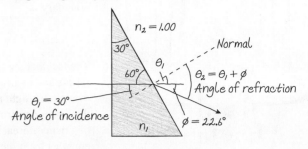

θ_1 and θ_2 are measured from the normal.

MODEL Represent the laser beam with a single ray and use the ray model of light.

VISUALIZE FIGURE 34.19 uses the steps of Tactics Box 34.1 to draw a ray diagram. The ray is incident perpendicular to the front face of the prism ($\theta_{\text{incident}} = 0°$), thus it is transmitted through the first boundary without deflection. At the second boundary it is especially important to *draw the normal to the surface* at the point of incidence and to *measure angles from the normal*.

SOLVE From the geometry of the triangle you can find that the laser's angle of incidence on the hypotenuse of the prism is $\theta_1 = 30°$,

the same as the apex angle of the prism. The ray exits the prism at angle θ_2 such that the deflection is $\phi = \theta_2 - \theta_1 = 22.6°$. Thus $\theta_2 = 52.6°$. Knowing both angles and $n_2 = 1.00$ for air, we can use Snell's law to find n_1:

$$n_1 = \frac{n_2 \sin \theta_2}{\sin \theta_1} = \frac{1.00 \sin 52.6°}{\sin 30°} = 1.59$$

ASSESS Referring to the indices of refraction in Table 34.1, we see that the prism is made of plastic.

Total Internal Reflection

What would have happened in Example 34.4 if the prism angle had been 45° rather than 30°? The light rays would approach the rear surface of the prism at an angle of incidence $\theta_1 = 45°$. When we try to calculate the angle of refraction at which the ray emerges into the air, we find

$$\sin \theta_2 = \frac{n_1}{n_2} \sin \theta_1 = \frac{1.59}{1.00} \sin 45° = 1.12$$

$$\theta_2 = \sin^{-1}(1.12) = ???$$

Angle θ_2 doesn't compute because the sine of an angle can't be larger than 1. The ray is unable to refract through the boundary. Instead, 100% of the light *reflects* from the boundary back into the prism. This process is called **total internal reflection,** often abbreviated TIR. That it really happens is illustrated in FIGURE 34.20. Here three laser beams enter a prism from the left. The bottom two refract out through the right side of the prism. The blue beam, which is incident on the prism's top face, undergoes total internal reflection and then emerges through the right surface.

FIGURE 34.21 shows several rays leaving a point source in a medium with index of refraction n_1. The medium on the other side of the boundary has $n_2 < n_1$. As we've seen, crossing a boundary into a material with a lower index of refraction causes the ray to bend away from the normal. Two things happen as angle θ_1 increases. First, the refraction angle θ_2 approaches 90°. Second, the fraction of the light energy transmitted decreases while the reflected fraction increases.

A **critical angle** is reached when $\theta_2 = 90°$. Because $\sin 90° = 1$, Snell's law $n_1 \sin \theta_c = n_2 \sin 90°$ gives the critical angle of incidence as

$$\theta_c = \sin^{-1}\left(\frac{n_2}{n_1}\right) \qquad (34.9)$$

FIGURE 34.20 The blue laser beam undergoes total internal reflection inside the prism.

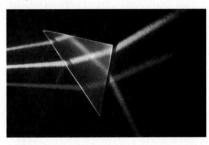

FIGURE 34.21 Refraction and reflection of rays as the angle of incidence increases.

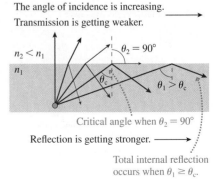

The angle of incidence is increasing. Transmission is getting weaker.

Critical angle when $\theta_2 = 90°$

Reflection is getting stronger.

Total internal reflection occurs when $\theta_1 \geq \theta_c$.

FIGURE 34.22 Binoculars make use of total internal reflection.

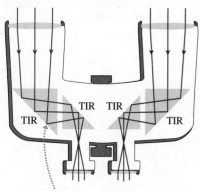

Angles of incidence exceed the critical angle.

The refracted light vanishes at the critical angle and the reflection becomes 100% for any angle $\theta_1 \geq \theta_c$. The critical angle is well defined because of our assumption that $n_2 < n_1$. **There is no critical angle and no total internal reflection if $n_2 > n_1$.**

As a quick example, the critical angle in a typical piece of glass at the glass-air boundary is

$$\theta_{c\ \text{glass}} = \sin^{-1}\left(\frac{1.00}{1.50}\right) = 42°$$

The fact that the critical angle is less than 45° has important applications. For example, **FIGURE 34.22** shows a pair of binoculars. The lenses are much farther apart than your eyes, so the light rays need to be brought together before exiting the eyepieces. Rather than using mirrors, which get dirty and require alignment, binoculars use a pair of prisms on each side. Thus the light undergoes two total internal reflections and emerges from the eyepiece. (The actual arrangement is a little more complex than in Figure 34.22, to avoid left-right reversals, but this illustrates the basic idea.)

EXAMPLE 34.5 | Total internal reflection

A small lightbulb is set in the bottom of a 3.0-m-deep swimming pool. What is the diameter of the circle of light seen on the water's surface from above?

MODEL Use the ray model of light.

VISUALIZE FIGURE 34.23 is a pictorial representation. The lightbulb emits rays at all angles, but only some of the rays refract into the air and are seen from above. Rays striking the surface at greater than the critical angle undergo TIR and remain within the water. The diameter of the circle of light is the distance between the two points at which rays strike the surface at the critical angle.

SOLVE From trigonometry, the circle diameter is $D = 2h\tan\theta_c$, where h is the depth of the water. The critical angle for a water-air boundary is $\theta_c = \sin^{-1}(1.00/1.33) = 48.7°$. Thus

$$D = 2(3.0\ \text{m})\tan 48.7° = 6.8\ \text{m}$$

FIGURE 34.23 Pictorial representation of the rays leaving a lightbulb at the bottom of a swimming pool.

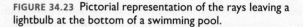

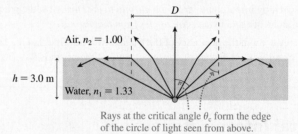

Rays at the critical angle θ_c form the edge of the circle of light seen from above.

Fiber Optics

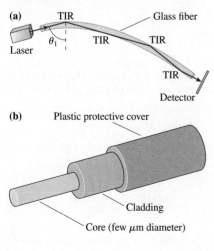

The most important modern application of total internal reflection is the transmission of light through optical fibers. **FIGURE 34.24a** shows a laser beam shining into the end of a long, narrow-diameter glass tube. The light rays pass easily from the air into the glass, but they then impinge on the inside wall of the glass tube at an angle of incidence θ_1 approaching 90°. This is well above the critical angle, so the laser beam undergoes TIR and remains inside the glass. The laser beam continues to "bounce" its way down the tube as if the light were inside a pipe. Indeed, optical fibers are sometimes called "light pipes." The rays are *below* the critical angle ($\theta_1 \approx 0$) when they finally reach the end of the fiber, thus they refract out without difficulty and can be detected.

While a simple glass tube can transmit light, a glass-air boundary is not sufficiently reliable for commercial use. Any small scratch on the side of the tube alters the rays' angle of incidence and allows leakage of light. **FIGURE 34.24b** shows the construction of a practical optical fiber. A small-diameter glass *core* is surrounded by a layer of glass *cladding*. The glasses used for the core and the cladding have $n_{\text{core}} > n_{\text{cladding}}$; thus light undergoes TIR at the core-cladding boundary and remains confined within the core. This boundary is not exposed to the environment and hence retains its integrity even under adverse conditions.

Even glass of the highest purity is not perfectly transparent. Absorption in the glass, even if very small, causes a gradual decrease in light intensity. The glass used for the core of optical fibers has a minimum absorption at a wavelength of 1.3 μm, in the infrared, so this is the laser wavelength used for long-distance signal transmission. Light at this wavelength can travel hundreds of kilometers through a fiber without significant loss.

STOP TO THINK 34.3 A light ray travels from medium 1 to medium 3 as shown. For these media,

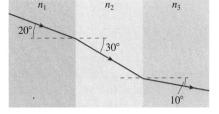

a. $n_3 > n_1$ b. $n_3 = n_1$ c. $n_3 < n_1$
d. We can't compare n_1 to n_3 without knowing n_2.

34.4 Image Formation by Refraction at a Plane Surface

If you see a fish that appears to be swimming close to the front window of the aquarium, but then look through the side of the aquarium, you'll find that the fish is actually farther from the window than you thought. Why is this?

To begin, recall that vision works by focusing a diverging bundle of rays onto the retina. The point from which the rays diverge is where you perceive the object to be. FIGURE 34.25a shows how you would see a fish out of water at distance d.

Now place the fish back into the aquarium at the same distance d. For simplicity, we'll ignore the glass wall of the aquarium and consider the water-air boundary. (The thin glass of a typical window has only a very small effect on the refraction of the rays and doesn't change the conclusions.) Light rays again leave the fish, but this time they refract at the water-air boundary. Because they're going from a higher to a lower index of refraction, the rays refract *away from* the normal. FIGURE 34.25b shows the consequences.

A bundle of diverging rays still enters your eye, but now these rays are diverging from a closer point, at distance d'. As far as your eye and brain are concerned, it's exactly *as if* the rays really originate at distance d', and this is the location at which you see the fish. **The object appears closer than it really is because of the refraction of light at the boundary.**

We found that the rays reflected from a mirror diverge from a point that is not the object point. We called that point a *virtual image*. Similarly, if rays from an object point P refract at a boundary between two media such that the rays then diverge from a point P′ and *appear* to come from P′, we call P′ a virtual image of point P. **The virtual image of the fish is what you see.**

Let's examine this image formation a bit more carefully. FIGURE 34.26 shows a boundary between two transparent media having indices of refraction n_1 and n_2. Point P, a source of light rays, is the object. Point P′, from which the rays *appear* to diverge, is the virtual image of P. Distance s is called the **object distance**. Our goal is to determine distance s', the **image distance**. Both are measured from the boundary.

A line perpendicular to the boundary is called the **optical axis.** Consider a ray leaving the object at angle θ_1 with respect to the optical axis. θ_1 is also the angle of incidence at the boundary, where the ray refracts into the second medium at angle θ_2. By tracing the refracted ray backward, you can see that θ_2 is also the angle between the refracted ray and the optical axis at point P′.

The distance l is common to both the incident and the refracted rays, and you can see that $l = s \tan\theta_1 = s' \tan\theta_2$. However, **it is customary in optics for virtual image distances to be negative.** (The reason will be clear when we get to image formation by lenses.) Hence we will insert a minus sign, finding that

$$s' = -\frac{\tan\theta_1}{\tan\theta_2}s \qquad (34.10)$$

FIGURE 34.25 Refraction of the light rays causes a fish in the aquarium to be seen at distance d'.

(a) A fish out of water

The rays that reach the eye are diverging from this point, the object.

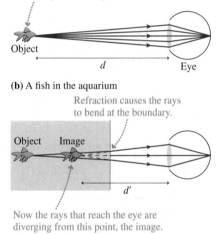

(b) A fish in the aquarium

Refraction causes the rays to bend at the boundary.

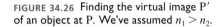

Now the rays that reach the eye are diverging from this point, the image.

FIGURE 34.26 Finding the virtual image P′ of an object at P. We've assumed $n_1 > n_2$.

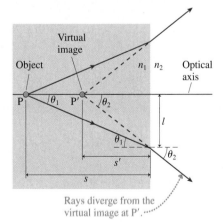

Rays diverge from the virtual image at P′.

Snell's law relates the sines of angles θ_1 and θ_2; that is,

$$\frac{\sin\theta_1}{\sin\theta_2} = \frac{n_2}{n_1} \tag{34.11}$$

In practice, the angle between any of these rays and the optical axis is very small because the size of the pupil of your eye is very much less than the distance between the object and your eye. (The angles in the figure have been greatly exaggerated.) Rays that are nearly *parallel* to the *axis* are called **paraxial rays.** The small-angle approximation $\sin\theta \approx \tan\theta \approx \theta$, where θ is in radians, can be applied to paraxial rays. Consequently,

$$\frac{\tan\theta_1}{\tan\theta_2} \approx \frac{\sin\theta_1}{\sin\theta_2} = \frac{n_2}{n_1} \tag{34.12}$$

Using this result in Equation 34.10, we find that the image distance is

$$s' = -\frac{n_2}{n_1}s \tag{34.13}$$

The minus sign tells us that we have a virtual image.

> **NOTE** The fact that the result for s' is independent of θ_1 implies that *all* paraxial rays appear to diverge from the same point P'. This property of the diverging rays is essential in order to have a well-defined image.

EXAMPLE 34.6 | **An air bubble in a window**

A fish and a sailor look at each other through a 5.0-cm-thick glass porthole in a submarine. There happens to be an air bubble right in the center of the glass. How far behind the surface of the glass does the air bubble appear to the fish? To the sailor?

MODEL Represent the air bubble as a point source and use the ray model of light.

VISUALIZE Paraxial light rays from the bubble refract into the air on one side and into the water on the other. The ray diagram looks like Figure 34.26.

SOLVE The index of refraction of the glass is $n_1 = 1.50$. The bubble is in the center of the window, so the object distance from either side of the window is $s = 2.5$ cm. On the water side, the image distance is

$$s' = -\frac{n_2}{n_1}s = -\frac{1.33}{1.50}(2.5\text{ cm}) = -2.2\text{ cm}$$

The minus sign indicates a virtual image. Physically, the fish sees the bubble 2.2 cm behind the surface. The image distance on the water side is

$$s' = -\frac{n_2}{n_1}s = -\frac{1.00}{1.50}(2.5\text{ cm}) = -1.7\text{ cm}$$

So the sailor sees the bubble 1.7 cm behind the surface.

ASSESS The image distance is *less* for the sailor because of the *larger* difference between the two indices of refraction.

34.5 Thin Lenses: Ray Tracing

A camera obscura or a pinhole camera forms images on a screen, but the images are faint and not perfectly focused. The ability to create a bright, well-focused image is vastly improved by using a lens. A **lens** is a transparent object that uses refraction at *curved* surfaces to form an image from diverging light rays. We will defer a mathematical analysis of lenses until the next section. First, we want to establish a pictorial method of understanding image formation. This method is called **ray tracing.**

FIGURE 34.27 shows parallel light rays entering two different lenses. The left lens, called a **converging lens,** causes the rays to refract *toward* the optical axis. The common point through which initially parallel rays pass is called the **focal point** of the lens. The distance of the focal point from the lens is called the **focal length** f of the lens. The right lens, called a **diverging lens,** refracts parallel rays *away from* the optical axis. This lens also has a focal point, but it is not as obvious.

FIGURE 34.27 Parallel light rays pass through a converging lens and a diverging lens.

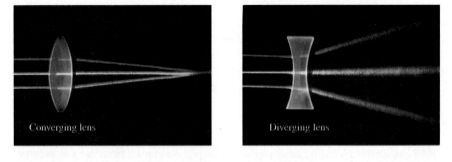

NOTE ▶ A converging lens is thicker in the center than at the edges. A diverging lens is thicker at the edges than at the center. ◀

FIGURE 34.28 clarifies the situation. In the case of a diverging lens, a backward projection of the diverging rays shows that they *appear* to have started from the same point. This is the focal point of a diverging lens, and its distance from the lens is the focal length of the lens. In the next section we'll relate the focal length to the curvature and index of refraction of the lens, but now we'll use the practical definition that **the focal length is the distance from the lens at which rays parallel to the optical axis converge or from which they diverge.**

NOTE ▶ The focal length f is a property *of the lens,* independent of how the lens is used. The focal length characterizes a lens in much the same way that a mass m characterizes an object or a spring constant k characterizes a spring. ◀

Converging Lenses

These basic observations about lenses are enough to understand image formation by a thin lens. A **thin lens** is a lens whose thickness is very small in comparison to its focal length and in comparison to the object and image distances. We'll make the approximation that the thickness of a thin lens is zero and that the lens lies in a plane called the **lens plane.** Within this approximation, **all refraction occurs as the rays cross the lens plane, and all distances are measured from the lens plane.** Fortunately, the thin-lens approximation is quite good for most practical applications of lenses.

NOTE ▶ We'll *draw* lenses as if they have a thickness, because that is how we expect lenses to look, but our analysis will not depend on the shape or thickness of a lens. ◀

FIGURE 34.29 shows three important situations of light rays passing through a thin converging lens. Part a is familiar from Figure 34.28. If the direction of each of the rays in Figure 34.29a is reversed, Snell's law tells us that each ray will exactly retrace its path and emerge from the lens parallel to the optical axis. This leads to Figure 34.29b, which is the "mirror image" of part a. Notice that the lens actually has *two* focal points, located at distances f on either side of the lens.

FIGURE 34.28 The focal lengths of converging and diverging lenses.

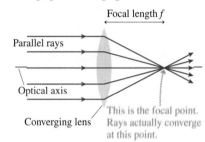

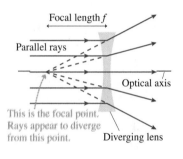

FIGURE 34.29 Three important sets of rays passing through a thin converging lens.

(a)

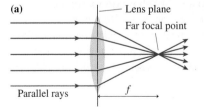

Any ray initially parallel to the optical axis will refract through the focal point on the far side of the lens.

(b)

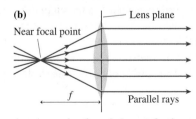

Any ray passing through the near focal point emerges from the lens parallel to the optical axis.

(c)

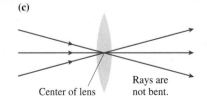

Any ray directed at the center of the lens passes through in a straight line.

Figure 34.29c shows three rays passing through the *center* of the lens. At the center, the two sides of a lens are very nearly parallel to each other. Earlier, in Example 34.3, we found that a ray passing through a piece of glass with parallel sides is *displaced* but *not bent* and that the displacement becomes zero as the thickness approaches zero. Consequently, a ray through the center of a thin lens, with zero thickness, is neither bent nor displaced but travels in a straight line.

These three situations form the basis for ray tracing.

Real Images

FIGURE 34.30 shows a lens and an object whose *object distance s* from the lens is larger than the focal length. Rays from point P on the object are refracted by the lens so as to converge at point P′ on the opposite side of the lens at *image distance s′*. If rays diverge from an object point P and interact with a lens such that the refracted rays *converge* at point P′, actually meeting at P′, then we call P′ a **real image** of point P. Contrast this with our prior definition of a *virtual image* as a point from which rays—which never meet—appear to *diverge*. For a real image, the image distance *s′* is *positive*.

FIGURE 34.30 Rays from an object point P are refracted by the lens and converge to a real image at point P′.

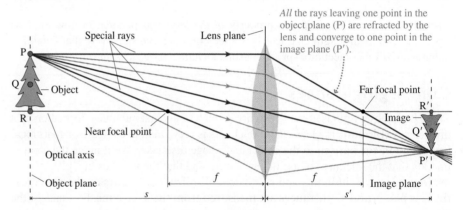

All points on the object that are in the same plane, the **object plane,** converge to image points in the **image plane.** Points Q and R in the object plane of Figure 34.30 have image points Q′ and R′ in the same plane as point P′. Once we locate *one* point in the image plane, such as P′, we know that the full image lies in the same plane.

There are two important observations to make about Figure 34.30. First, the image is upside down with respect to the object. This is called an **inverted image,** and it is a standard characteristic of real-image formation with a converging lens. Second, rays from point P *fill* the entire lens surface, and all portions of the lens contribute to the image. A larger lens will "collect" more rays and thus make a brighter image.

FIGURE 34.31 is a close-up view of the rays very near the image plane. The rays don't stop at P′ unless we place a screen in the image plane. When we do so, we see a sharp, well-focused image on the screen. To focus an image, you must either move the screen to coincide with the image plane or move the lens or object to make the image plane coincide with the screen. For example, the focus knob on a projector moves the lens forward or backward until the image plane matches the screen position.

NOTE The ability to see a real image on a screen sets real images apart from *virtual* images. But keep in mind that we need not *see* a real image in order to *have* an image. A real image exists at a point in space where the rays converge even if there's no viewing screen in the image plane.

FIGURE 34.31 A close-up look at the rays near the image plane.

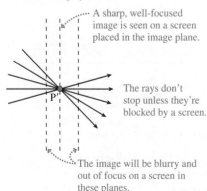

Figure 34.30 highlights three "special rays" based on the three situations of Figure 34.29. These three rays alone are sufficient to locate the image point P′. That is, we don't need to draw all the rays shown in Figure 34.30. The procedure known as *ray tracing* consists of locating the image by the use of just these three rays.

TACTICS BOX 34.2 (MP)

Ray tracing for a converging lens

❶ **Draw an optical axis.** Use graph paper or a ruler! Establish an appropriate scale.

❷ **Center the lens on the axis.** Mark and label the focal points at distance f on either side.

❸ **Represent the object with an upright arrow at distance s.** It's usually best to place the base of the arrow on the axis and to draw the arrow about half the radius of the lens.

❹ **Draw the three "special rays" from the tip of the arrow.** Use a straightedge.
  a. A ray parallel to the axis refracts through the far focal point.
  b. A ray that enters the lens along a line through the near focal point emerges parallel to the axis.
  c. A ray through the center of the lens does not bend.

❺ **Extend the rays until they converge.** This is the image point. Draw the rest of the image in the image plane. If the base of the object is on the axis, then the base of the image will also be on the axis.

❻ **Measure the image distance s'.** Also, if needed, measure the image height relative to the object height.

Exercises 18–23

EXAMPLE 34.7 | Finding the image of a flower

A 4.0-cm-diameter flower is 200 cm from the 50-cm-focal-length lens of a camera. How far should the light detector be placed behind the lens to record a well-focused image? What is the diameter of the image on the detector?

MODEL The flower is in the object plane. Use ray tracing to locate the image.

VISUALIZE FIGURE 34.32 shows the ray-tracing diagram and the steps of Tactics Box 34.2. The image has been drawn in the plane where the three special rays converge. You can see *from the drawing* that the image distance is $s' \approx 67$ cm. This is where the detector needs to be placed to record a focused image.

The heights of the object and image are labeled h and h'. The ray through the center of the lens is a straight line, thus the object and image both subtend the same angle θ. Using similar triangles,

$$\frac{h'}{s'} = \frac{h}{s}$$

Solving for h' gives

$$h' = h\frac{s'}{s} = (4.0 \text{ cm})\frac{67 \text{ cm}}{200 \text{ cm}} = 1.3 \text{ cm}$$

The flower's image has a diameter of 1.3 cm.

ASSESS We've been able to learn a great deal about the image from a simple geometric procedure.

FIGURE 34.32 Ray-tracing diagram for Example 34.7.

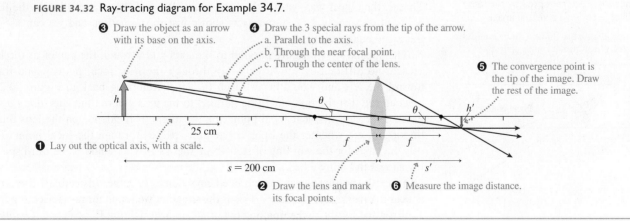

❸ Draw the object as an arrow with its base on the axis.
❹ Draw the 3 special rays from the tip of the arrow.
  a. Parallel to the axis.
  b. Through the near focal point.
  c. Through the center of the lens.
❺ The convergence point is the tip of the image. Draw the rest of the image.
❶ Lay out the optical axis, with a scale.
❷ Draw the lens and mark its focal points.
❻ Measure the image distance.

25 cm
$s = 200$ cm
s'

Lateral Magnification

The image can be either larger or smaller than the object, depending on the location and focal length of the lens. But there's more to a description of the image than just its size. We also want to know its *orientation* relative to the object. That is, is the image upright or inverted? It is customary to combine size and orientation information into a single number. The **lateral magnification** m is defined as

$$m = -\frac{s'}{s} \tag{34.14}$$

You just saw in Example 34.7 that the image-to-object height ratio is $h'/h = s'/s$. Consequently, we interpret the lateral magnification m as follows:

1. A positive value of m indicates that the image is upright relative to the object. A negative value of m indicates that the image is inverted relative to the object.
2. The absolute value of m gives the size ratio of the image and object: $h'/h = |m|$.

The lateral magnification in Example 34.7 would be $m = -0.33$, indicating that the image is inverted and 33% the size of the object.

NOTE The image-to-object height ratio is called *lateral* magnification to distinguish it from angular magnification, which we'll introduce in the next chapter. In practice, m is simply called "magnification" when there's no chance of confusion. Although we usually think that "to magnify" means "to make larger," in optics the magnification can be either > 1 (the image is larger than the object) or < 1 (the image is smaller than the object).

STOP TO THINK 34.4 A lens produces a sharply focused, inverted image on a screen. What will you see on the screen if the lens is removed?

a. The image will be inverted and blurry.
b. The image will be upright and sharp.
c. The image will be upright and blurry.
d. The image will be much dimmer but otherwise unchanged.
e. There will be no image at all.

Virtual Images

The previous section considered a converging lens with the object at distance $s > f$. That is, the object was outside the focal point. What if the object is inside the focal point, at distance $s < f$? FIGURE 34.33 shows just this situation, and we can use ray tracing to analyze it.

The special rays initially parallel to the axis and through the center of the lens present no difficulties. However, a ray through the near focal point would travel toward the left and would never reach the lens! Referring back to Figure 34.29b, you can see that the rays emerging parallel to the axis entered the lens *along a line* passing through the near focal point. It's the angle of incidence on the lens that is important, not whether the light ray actually passes through the focal point. This was the basis for the wording of step 4b in Tactics Box 34.2 and is the third special ray shown in Figure 34.33.

You can see that the three refracted rays don't converge. Instead, all three rays appear to *diverge* from point P′. This is the situation we found for rays reflecting from a mirror and for the rays refracting out of an aquarium. Point P′ is a *virtual image* of

FIGURE 34.33 Rays from an object at distance $s < f$ are refracted by the lens and diverge to form a virtual image.

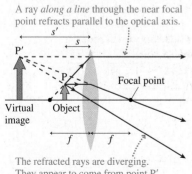

A ray *along a line* through the near focal point refracts parallel to the optical axis.

The refracted rays are diverging. They appear to come from point P′.

the object point P. Furthermore, it is an **upright image,** having the same orientation as the object.

The refracted rays, which are all to the right of the lens, *appear* to come from P′, but none of the rays were ever at that point. No image would appear on a screen placed in the image plane at P′. So what good is a virtual image?

Your eye collects and focuses bundles of diverging rays; thus, as FIGURE 34.34a shows, you can see a virtual image by looking *through* the lens. This is exactly what you do with a magnifying glass, producing a scene like the one in FIGURE 34.34b. In fact, you view a virtual image anytime you look *through* the eyepiece of an optical instrument such as a microscope or binoculars.

As before, **the image distance s′ for a virtual image is defined to be a** *negative* **number (s′ < 0),** indicating that the image is on the opposite side of the lens from a real image. With this choice of sign, the definition of magnification, $m = -s'/s$, is still valid. A virtual image with negative s′ has $m > 0$, thus the image is upright. This agrees with the rays in Figure 34.33 and the photograph of Figure 34.34b.

> **NOTE** A lens thicker in the middle than at the edges is classified as a converging lens. The light rays from an object *can* converge to form a real image after passing through such a lens, but only if the object distance is larger than the focal length of the lens: $s > f$. If $s < f$, the rays diverge to produce a virtual image.

FIGURE 34.34 A converging lens is a magnifying glass when the object distance is less than f.

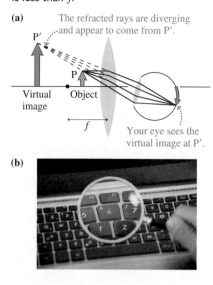

(a)

(b)

EXAMPLE 34.8 | **Magnifying a flower**

To see a flower better, a naturalist holds a 6.0-cm-focal-length magnifying glass 4.0 cm from the flower. What is the magnification?

MODEL The flower is in the object plane. Use ray tracing to locate the image.

VISUALIZE FIGURE 34.35 shows the ray-tracing diagram. The three special rays diverge from the lens, but we can use a straight-edge to extend the rays backward to the point from which they diverge. This point, the image point, is seen to be 12 cm to the left of the lens. Because this is a virtual image, the image distance is a negative s′ = −12 cm. Thus the magnification is

$$m = -\frac{s'}{s} = -\frac{-12 \text{ cm}}{4.0 \text{ cm}} = 3.0$$

The image is three times as large as the object and, because m is positive, upright.

FIGURE 34.35 Ray-tracing diagram for Example 34.8.

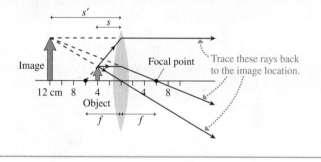

Diverging Lenses

A lens thicker at the edges than in the middle is called a *diverging lens.* FIGURE 34.36 shows three important sets of rays passing through a diverging lens. These are based on Figures 34.27 and 34.28, where you saw that rays initially parallel to the axis diverge after passing through a diverging lens.

FIGURE 34.36 Three important sets of rays passing through a thin diverging lens.

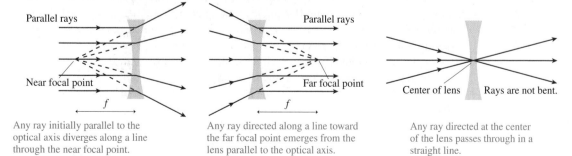

Any ray initially parallel to the optical axis diverges along a line through the near focal point.

Any ray directed along a line toward the far focal point emerges from the lens parallel to the optical axis.

Any ray directed at the center of the lens passes through in a straight line.

Ray tracing follows the steps of Tactics Box 34.2 for a converging lens *except* that two of the three special rays in step 4 are different.

TACTICS BOX 34.3   (MP)

Ray tracing for a diverging lens

❶–❸ **Follow steps 1 through 3 of Tactics Box 34.2.**
❹ **Draw the three "special rays" from the tip of the arrow.** Use a straightedge.
   a. A ray parallel to the axis diverges along a line through the near focal point.
   b. A ray along a line toward the far focal point emerges parallel to the axis.
   c. A ray through the center of the lens does not bend.
❺ **Trace the diverging rays backward.** The point from which they are diverging is the image point, which is always a virtual image.
❻ **Measure the image distance s'.** This will be a negative number.

Exercise 24 🖎

EXAMPLE 34.9 | Demagnifying a flower

A diverging lens with a focal length of 50 cm is placed 100 cm from a flower. Where is the image? What is its magnification?

MODEL The flower is in the object plane. Use ray tracing to locate the image.

VISUALIZE **FIGURE 34.37** shows the ray-tracing diagram. The three special rays (labeled a, b, and c to match the Tactics Box) do not converge. However, they can be traced backward to an intersection ≈ 33 cm to the left of the lens. A virtual image is formed at $s' = -33$ cm with magnification

$$m = -\frac{s'}{s} = -\frac{-33 \text{ cm}}{100 \text{ cm}} = 0.33$$

The image, which can be seen by looking *through* the lens, is one-third the size of the object and upright.

FIGURE 34.37 Ray-tracing diagram for Example 34.9.

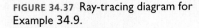
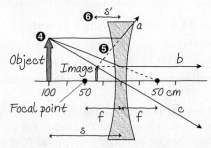

ASSESS Ray tracing with a diverging lens is somewhat trickier than with a converging lens, so this example is worth careful study.

Diverging lenses *always* make virtual images and, for this reason, are rarely used alone. However, they have important applications when used in combination with other lenses. Cameras, eyepieces, and eyeglasses often incorporate diverging lenses.

34.6 Thin Lenses: Refraction Theory

Ray tracing is a powerful visual approach for understanding image formation, but it doesn't provide precise information about the image. We need to develop a quantitative relationship between the object distance s and the image distance s'.

To begin, **FIGURE 34.38** shows a *spherical* boundary between two transparent media with indices of refraction n_1 and n_2. The sphere has radius of curvature R. Consider a ray that leaves object point P at angle α and later, after refracting, reaches point P'. Figure 34.38 has exaggerated the angles to make the picture clear, but we will restrict our analysis to *paraxial rays* traveling nearly parallel to the axis. For paraxial rays, all the angles are small and we can use the small-angle approximation.

The ray from P is incident on the boundary at angle θ_1 and refracts into medium n_2 at angle θ_2, both measured from the normal to the surface at the point of incidence. Snell's law is $n_1 \sin \theta_1 = n_2 \sin \theta_2$, which in the small-angle approximation is

$$n_1 \theta_1 = n_2 \theta_2 \tag{34.15}$$

You can see from the geometry of Figure 34.38 that angles α, β, and ϕ are related by

$$\theta_1 = \alpha + \phi \quad \text{and} \quad \theta_2 = \phi - \beta \tag{34.16}$$

FIGURE 34.38 Image formation due to refraction at a spherical surface. The angles are exaggerated.

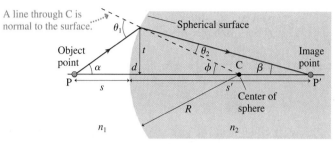

Using these expressions in Equation 34.15, we can write Snell's law as

$$n_1(\alpha + \phi) = n_2(\phi - \beta) \tag{34.17}$$

This is one important relationship between the angles.

The line of height t, from the axis to the point of incidence, is the vertical leg of three different right triangles having vertices at points P, C, and P'. Consequently,

$$\tan \alpha \approx \alpha = \frac{t}{s+d} \qquad \tan \beta \approx \beta = \frac{t}{s'-d} \qquad \tan \phi \approx \phi = \frac{t}{R-d} \tag{34.18}$$

But $d \rightarrow 0$ for paraxial rays, thus

$$\alpha = \frac{t}{s} \qquad \beta = \frac{t}{s'} \qquad \phi = \frac{t}{R} \tag{34.19}$$

This is the second important relationship that comes from Figure 34.38.

If we use Equation 34.19 in Equation 34.17, the t cancels and we find

$$\frac{n_1}{s} + \frac{n_2}{s'} = \frac{n_2 - n_1}{R} \tag{34.20}$$

Equation 34.20 is independent of angle α. Consequently, **all paraxial rays leaving point P later converge at point P'**. If an object is located at distance s from a spherical refracting surface, an image will be formed at distance s' given by Equation 34.20.

Equation 34.20 was derived for a surface that is convex toward the object point, and the image is real. However, the result is also valid for virtual images or for surfaces that are concave toward the object point as long as we adopt the *sign convention* shown in TABLE 34.2.

Section 34.4 considered image formation due to refraction by a plane surface. There we found (in Equation 34.13) an image distance $s' = -(n_2/n_1)s$. A plane can be thought of as a sphere in the limit $R \rightarrow \infty$, so we should be able to reach the same conclusion from Equation 34.20. Indeed, as $R \rightarrow \infty$, the term $(n_2 - n_1)/R \rightarrow 0$ and Equation 34.20 becomes $s' = -(n_2/n_1)s$.

TABLE 34.2 Sign convention for refracting surfaces

	Positive	Negative
R	Convex toward the object	Concave toward the object
s'	Real image, opposite side from object	Virtual image, same side as object

---

EXAMPLE 34.10 | **Image formation inside a glass rod**

One end of a 4.0-cm-diameter glass rod is shaped like a hemisphere. A small lightbulb is 6.0 cm from the end of the rod. Where is the bulb's image located?

MODEL Model the lightbulb as a point source of light and consider the paraxial rays that refract into the glass rod.

FIGURE 34.39 The curved surface refracts the light to form a real image.

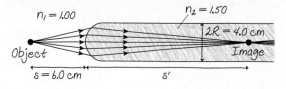

VISUALIZE FIGURE 34.39 shows the situation. $n_1 = 1.00$ for air and $n_2 = 1.50$ for glass.

SOLVE The radius of the surface is half the rod diameter, so $R = 2.0$ cm. Equation 34.20 is

$$\frac{1.00}{6.0 \text{ cm}} + \frac{1.50}{s'} = \frac{1.50 - 1.00}{2.0 \text{ cm}} = \frac{0.50}{2.0 \text{ cm}}$$

Solving for the image distance s' gives

$$\frac{1.50}{s'} = \frac{0.50}{2.0 \text{ cm}} - \frac{1.00}{6.0 \text{ cm}} = 0.0833 \text{ cm}^{-1}$$

$$s' = \frac{1.50}{0.0833} = 18 \text{ cm}$$

ASSESS This is a real image located 18 cm inside the glass rod.

EXAMPLE 34.11 | A goldfish in a bowl

A goldfish lives in a spherical fish bowl 50 cm in diameter. If the fish is 10 cm from the near edge of the bowl, where does the fish appear when viewed from the outside?

MODEL Model the fish as a point source and consider the paraxial rays that refract from the water into the air. The thin glass wall has little effect and will be ignored.

FIGURE 34.40 The curved surface of a fish bowl produces a virtual image of the fish.

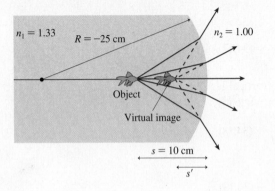

VISUALIZE FIGURE 34.40 shows the rays refracting *away* from the normal as they move from the water into the air. We expect to find a virtual image at a distance less than 10 cm.

SOLVE The object is in the water, so $n_1 = 1.33$ and $n_2 = 1.00$. The inner surface is concave (you can remember "concave" because it's like looking into a cave), so $R = -25$ cm. The object distance is $s = 10$ cm. Thus Equation 34.20 is

$$\frac{1.33}{10 \text{ cm}} + \frac{1.00}{s'} = \frac{1.00 - 1.33}{-25 \text{ cm}} = \frac{0.33}{25 \text{ cm}}$$

Solving for the image distance s' gives

$$\frac{1.00}{s'} = \frac{0.33}{25 \text{ cm}} - \frac{1.33}{10 \text{ cm}} = -0.12 \text{ cm}^{-1}$$

$$s' = \frac{1.00}{-0.12 \text{ cm}^{-1}} = -8.3 \text{ cm}$$

ASSESS The image is virtual, located to the left of the boundary. A person looking into the bowl will see a fish that appears to be 8.3 cm from the edge of the bowl.

STOP TO THINK 34.5 Which of these actions will move the real image point P′ farther from the boundary? More than one may work.

a. Increase the radius of curvature R.
b. Increase the index of refraction n.
c. Increase the object distance s.
d. Decrease the radius of curvature R.
e. Decrease the index of refraction n.
f. Decrease the object distance s.

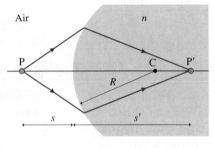

Lenses

The thin-lens approximation assumes rays refract one time, at the lens plane. In fact, as FIGURE 34.41 shows, rays refract *twice*, at spherical surfaces having radii of curvature R_1 and R_2. Let the lens have thickness t and be made of a material with index of refraction n. For simplicity, we'll assume that the lens is surrounded by air.

FIGURE 34.41 Image formation by a lens.

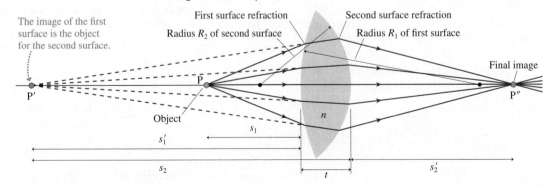
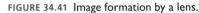

The object at point P is distance s_1 to the left of the lens. The first surface of the lens, of radius R_1, refracts the rays from P to create an image at point P'. We can use Equation 34.20 for a spherical surface to find the image distance s_1':

$$\frac{1}{s_1} + \frac{n}{s_1'} = \frac{n-1}{R_1} \tag{34.21}$$

where we used $n_1 = 1$ for the air and $n_2 = n$ for the lens. We'll assume that the image P' is a virtual image, but this assumption isn't essential to the outcome.

With two refracting surfaces, the image P' of the first surface becomes the object for the second surface. That is, the rays refracting at the second surface appear to have come from P'. Object distance s_2 from P' to the second surface looks like it should be $s_2 = s_1' + t$, but P' is a virtual image, so s_1' is a *negative* number. Thus the distance to the second surface is $s_2 = |s_1'| + t = t - s_1'$. We can find the image of P' by a second application of Equation 34.20, but now the rays are incident on the surface from within the lens, so $n_1 = n$ and $n_2 = 1$. Consequently,

$$\frac{n}{t - s_1'} + \frac{1}{s_2'} = \frac{1-n}{R_2} \tag{34.22}$$

For a *thin lens,* which has $t \to 0$, Equation 34.22 becomes

$$-\frac{n}{s_1'} + \frac{1}{s_2'} = \frac{1-n}{R_2} = -\frac{n-1}{R_2} \tag{34.23}$$

Our goal is to find the distance s_2' to point P", the image produced by the lens as a whole. This goal is easily reached if we simply add Equations 34.21 and 34.23, eliminating s_1' and giving

$$\frac{1}{s_1} + \frac{1}{s_2'} = \frac{n-1}{R_1} - \frac{n-1}{R_2} = (n-1)\left(\frac{1}{R_1} - \frac{1}{R_2}\right) \tag{34.24}$$

The numerical subscripts on s_1 and s_2' no longer serve a purpose. If we replace s_1 by s, the object distance from the lens, and s_2' by s', the image distance, Equation 34.24 becomes the **thin-lens equation:**

$$\frac{1}{s} + \frac{1}{s'} = \frac{1}{f} \qquad \text{(thin-lens equation)} \tag{34.25}$$

where the *focal length* of the lens is

$$\frac{1}{f} = (n-1)\left(\frac{1}{R_1} - \frac{1}{R_2}\right) \qquad \text{(lens maker's equation)} \tag{34.26}$$

Equation 34.26 is known as the **lens maker's equation.** It allows you to determine the focal length from the shape of a thin lens and the material used to make it.

We can verify that this expression for f really is the focal length of the lens by recalling that rays initially parallel to the optical axis pass through the focal point on the far side. In fact, this was our *definition* of the focal length of a lens. Parallel rays must come from an object extremely far away, with object distance $s \to \infty$ and thus $1/s = 0$. In that case, Equation 34.25 tells us that the parallel rays will converge at distance $s' = f$ on the far side of the lens, exactly as expected.

We derived the thin-lens equation and the lens maker's equation from the specific lens geometry shown in Figure 34.41, but the results are valid for any lens as long as all quantities are given appropriate signs. The sign convention used with Equations 34.25 and 34.26 is given in TABLE 34.3.

TABLE 34.3 Sign convention for thin lenses

	Positive	Negative
R_1, R_2	Convex toward the object	Concave toward the object
f	Converging lens, thicker in center	Diverging lens, thinner in center
s'	Real image, opposite side from object	Virtual image, same side as object

NOTE For a *thick lens,* where the thickness t is not negligible, we can solve Equations 34.21 and 34.22 in sequence to find the position of the image point P".

EXAMPLE 34.12 | **Focal length of a meniscus lens**

What is the focal length of the glass *meniscus lens* shown in FIGURE 34.42? Is this a converging or diverging lens?

FIGURE 34.42 A meniscus lens.

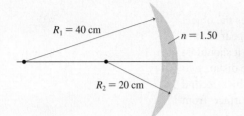

SOLVE If the object is on the left, then the first surface has $R_1 = -40$ cm (concave toward the object) and the second surface has $R_2 = -20$ cm (also concave toward the object). The index of refraction of glass is $n = 1.50$, so the lens maker's equation is

$$\frac{1}{f} = (n-1)\left(\frac{1}{R_1} - \frac{1}{R_2}\right) = (1.50 - 1)\left(\frac{1}{-40\text{ cm}} - \frac{1}{-20\text{ cm}}\right)$$

$$= 0.0125\text{ cm}^{-1}$$

Inverting this expression gives $f = 80$ cm. This is a converging lens, as seen both from the positive value of f and from the fact that the lens is thicker in the center.

Thin-Lens Image Formation

Although the thin-lens equation allows precise calculations, the lessons of ray tracing should not be forgotten. The most powerful tool of optical analysis is a combination of ray tracing, to gain an intuitive understanding of the ray trajectories, and the thin-lens equation.

EXAMPLE 34.13 | **Designing a lens**

The objective lens of a microscope uses a planoconvex glass lens with the flat side facing the specimen. A real image is formed 160 mm behind the lens when the lens is 8.0 mm from the specimen. What is the radius of the lens's curved surface?

MODEL Treat the lens as a thin lens with the specimen as the object. The lens's focal length is given by the lens maker's equation.

VISUALIZE FIGURE 34.43 clarifies the shape of the lens and defines R_2. The index of refraction was taken from Table 34.1.

FIGURE 34.43 A planoconvex microscope lens.

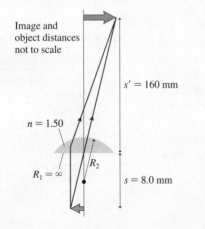

SOLVE We can use the lens maker's equation to solve for R_2 if we know the lens's focal length. Because we know both the object and image distances, we can use the thin-lens equation to find

$$\frac{1}{f} = \frac{1}{s} + \frac{1}{s'} = \frac{1}{8.0\text{ mm}} + \frac{1}{160\text{ mm}} = 0.131\text{ mm}^{-1}$$

The focal length is $f = 1/(0.131\text{ mm}^{-1}) = 7.6$ mm, but $1/f$ is all we need for the lens maker's equation. The front surface of the lens is planar, which we can consider a portion of a sphere with $R_1 \to \infty$. Consequently $1/R_1 = 0$. With this, we can solve the lens maker's equation for R_2:

$$\frac{1}{R_2} = \frac{1}{R_1} - \frac{1}{n-1}\frac{1}{f} = 0 - \left(\frac{1}{1.50 - 1}\right)(0.131\text{ mm}^{-1})$$

$$= -0.262\text{ mm}^{-1}$$

$$R_2 = -3.8\text{ mm}$$

The minus sign appears because the curved surface is concave toward the object. Physically, the radius of the curved surface is 3.8 mm.

ASSESS The actual thickness of the lens has to be less than R_2, probably no more than about 1.0 mm. This thickness is significantly less than the object and image distances, so the thin-lens approximation is justified.

EXAMPLE 34.14 | A magnifying lens

A stamp collector uses a magnifying lens that sits 2.0 cm above the stamp. The magnification is 4.0. What is the focal length of the lens?

FIGURE 34.44 Pictorial representation of a magnifying lens.

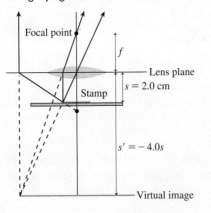

MODEL A magnifying lens is a converging lens with the object distance less than the focal length $(s < f)$. Assume it is a thin lens.

VISUALIZE FIGURE 34.44 shows the lens and a ray-tracing diagram. We do not need to know the actual shape of the lens, so the figure shows a generic converging lens.

SOLVE A virtual image is upright, so $m = +4.0$. The magnification is $m = -s'/s$, thus

$$s' = -4.0s = -(4.0)(2.0 \text{ cm}) = -8.0 \text{ cm}$$

We can use s and s' in the thin-lens equation to find the focal length:

$$\frac{1}{f} = \frac{1}{s} + \frac{1}{s'} = \frac{1}{2.0 \text{ cm}} + \frac{1}{-8.0 \text{ cm}} = 0.375 \text{ cm}^{-1}$$

$$f = 2.7 \text{ cm}$$

ASSESS $f > 2$ cm, as expected.

STOP TO THINK 34.6 A lens forms a real image of a lightbulb, but the image of the bulb on a viewing screen is blurry because the screen is slightly in front of the image plane. To focus the image, should you move the lens toward the bulb or away from the bulb?

34.7 Image Formation with Spherical Mirrors

Curved mirrors—such as those used in telescopes, security and rearview mirrors, and searchlights—can be used to form images, and their images can be analyzed with ray diagrams similar to those used with lenses. We'll consider only the important case of **spherical mirrors**, whose surface is a section of a sphere.

Concave Mirrors

FIGURE 34.45 shows a **concave mirror,** a mirror in which the edges curve *toward* the light source. Rays parallel to the optical axis reflect from the surface of the mirror so as to pass through a single point on the optical axis. This is the focal point of the mirror. The focal length is the distance from the mirror surface to the focal point. A concave mirror is analogous to a converging lens, but it has only one focal point.

Let's begin by considering the case where the object's distance s from the mirror is greater than the focal length $(s > f)$, as shown in FIGURE 34.46 on the next page. We see that the image is *real* (and inverted) because rays from the object point P converge at the image point P'. Although an infinite number of rays from P all meet at P', each ray obeying the law of reflection, you can see that three "special rays" are enough to determine the position and size of the image:

- A ray parallel to the axis reflects through the focal point.
- A ray through the focal point reflects parallel to the axis.
- A ray striking the center of the mirror reflects at an equal angle on the opposite side of the axis.

These three rays also locate the image if $s < f$, but in that case the image is *virtual* and behind the mirror. Once again, virtual images have a *negative* image distance s'.

FIGURE 34.45 The focal point and focal length of a concave mirror.

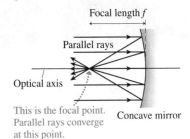

This is the focal point. Parallel rays converge at this point.

FIGURE 34.46 A real image formed by a concave mirror.

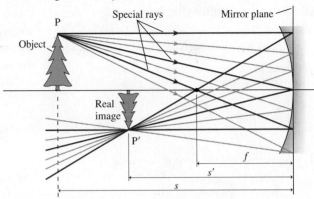

NOTE A thin lens has negligible thickness and all refraction occurs at the lens plane. Similarly, we will assume that mirrors are thin (even though drawings may show a thickness) and thus *all reflection occurs at the mirror plane.*

FIGURE 34.47 The focal point and focal length of a convex mirror.

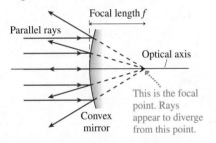

FIGURE 34.48 A city skyline is reflected in this polished sphere.

Convex Mirrors

FIGURE 34.47 shows parallel light rays approaching a mirror in which the edges curve *away from* the light source. This is called a **convex mirror.** In this case, the reflected rays appear to come from a point behind the mirror. This is the focal point for a convex mirror.

A common example of a convex mirror is a silvered ball, such as a tree ornament. You may have noticed that if you look at your reflection in such a ball, your image appears right-side-up but is quite small. As another example, **FIGURE 34.48** shows a city skyline reflected in a polished metal sphere. Let's use ray tracing to understand why the skyscrapers all appear to be so small.

FIGURE 34.49 shows an object in front of a convex mirror. In this case, the reflected rays—each obeying the law of reflection—create an upright image of reduced height behind the mirror. We see that the image is virtual because no rays actually converge at the image point P′. Instead, diverging rays *appear* to come from this point. Once again, three special rays are enough to find the image.

Convex mirrors are used for a variety of safety and monitoring applications, such as passenger-side rearview mirrors and the round mirrors used in stores to keep an eye

FIGURE 34.49 A virtual image formed by a convex mirror.

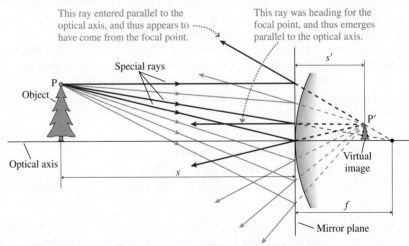

on the customers. When an object is reflected in a convex mirror, the image appears smaller than the object itself. Because the image is, in a sense, a miniature version of the object, you can *see much more of it* within the edges of the mirror than you could with an equal-sized flat mirror.

TACTICS BOX 34.4   (MP)

Ray tracing for a spherical mirror

❶ **Draw an optical axis.** Use graph paper or a ruler! Establish a scale.

❷ **Center the mirror on the axis.** Mark and label the focal point at distance f from the mirror's surface.

❸ **Represent the object with an upright arrow at distance s.** It's usually best to place the base of the arrow on the axis and to draw the arrow about half the radius of the mirror.

❹ **Draw the three "special rays" from the tip of the arrow.** All reflections occur at the mirror plane.
  a. A ray parallel to the axis reflects through (concave) or away from (convex) the focal point.
  b. An incoming ray passing through (concave) or heading toward (convex) the focal point reflects parallel to the axis.
  c. A ray that strikes the center of the mirror reflects at an equal angle on the opposite side of the optical axis.

❺ **Extend the rays forward or backward until they converge.** This is the image point. Draw the rest of the image in the image plane. If the base of the object is on the axis, then the base of the image will also be on the axis.

❻ **Measure the image distance s'.** Also, if needed, measure the image height relative to the object height.

Exercises 28–29   ✎

EXAMPLE 34.15 | **Analyzing a concave mirror**

A 3.0-cm-high object is located 60 cm from a concave mirror. The mirror's focal length is 40 cm. Use ray tracing to find the position and height of the image.

MODEL Use the ray-tracing steps of Tactics Box 34.4.

VISUALIZE FIGURE 34.50 shows the steps of Tactics Box 34.4.

SOLVE We can use a ruler to find that the image position is $s' \approx 120$ cm in front of the mirror and its height is $h' \approx 6$ cm.

ASSESS The image is a *real* image because light rays converge at the image point.

FIGURE 34.50 Ray-tracing diagram for a concave mirror.

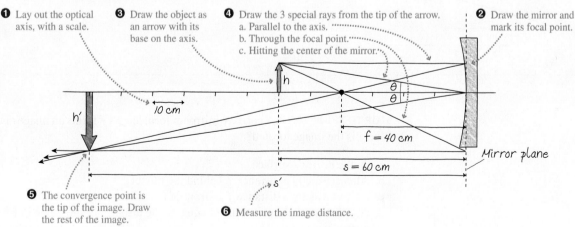

❶ Lay out the optical axis, with a scale.

❸ Draw the object as an arrow with its base on the axis.

❹ Draw the 3 special rays from the tip of the arrow.
  a. Parallel to the axis.
  b. Through the focal point.
  c. Hitting the center of the mirror.

❷ Draw the mirror and mark its focal point.

h

θ
θ

$f = 40$ cm

Mirror plane

h'

10 cm

$s = 60$ cm

s'

❺ The convergence point is the tip of the image. Draw the rest of the image.

❻ Measure the image distance.

The Mirror Equation

The thin-lens equation assumes lenses have negligible thickness (so a single refraction occurs in the lens plane) and the rays are nearly parallel to the optical axis (paraxial rays). If we make the same assumptions about spherical mirrors—the mirror has negligible thickness and so paraxial rays reflect at the mirror plane—then the object and image distances are related exactly as they were for thin lenses:

$$\frac{1}{s} + \frac{1}{s'} = \frac{1}{f} \qquad \text{(mirror equation)} \qquad (34.27)$$

The focal length of the mirror, as you can show as a homework problem, is related to the mirror's radius of curvature by

$$f = \frac{R}{2} \qquad (34.28)$$

TABLE 34.4 shows the sign convention used with spherical mirrors. It differs from the convention for lenses, so you'll want to carefully compare this table to Table 34.3. A concave mirror (analogous to a converging lens) has a positive focal length while a convex mirror (analogous to a diverging lens) has a negative focal length. The lateral magnification of a spherical mirror is computed exactly as for a lens:

$$m = -\frac{s'}{s} \qquad (34.29)$$

TABLE 34.4 Sign convention for spherical mirrors

	Positive	Negative
R, f	Concave toward the object	Convex toward the object
s'	Real image, same side as object	Virtual image, opposite side from object

---

EXAMPLE 34.16 | Analyzing a concave mirror

A 3.0-cm-high object is located 20 cm from a concave mirror. The mirror's radius of curvature is 80 cm. Determine the position, orientation, and height of the image.

MODEL Treat the mirror as a thin mirror.

VISUALIZE The mirror's focal length is $f = R/2 = +40$ cm, where we used the sign convention from Table 34.4. With the focal length known, the three special rays in FIGURE 34.51 show that the image is a magnified, virtual image behind the mirror.

FIGURE 34.51 Pictorial representation of Example 34.16.

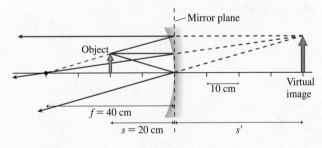

SOLVE The thin-mirror equation is

$$\frac{1}{20 \text{ cm}} + \frac{1}{s'} = \frac{1}{40 \text{ cm}}$$

This is easily solved to give $s' = -40$ cm, in agreement with the ray tracing. The negative sign tells us this is a virtual image behind the mirror. The magnification is

$$m = -\frac{-40 \text{ cm}}{20 \text{ cm}} = +2.0$$

Consequently, the image is 6.0 cm tall and upright.

ASSESS This is a virtual image because light rays diverge from the image point. You could see this enlarged image by standing behind the object and looking into the mirror. In fact, this is how magnifying cosmetic mirrors work.

STOP TO THINK 34.7 A concave mirror of focal length f forms an image of the moon. Where is the image located?

a. At the mirror's surface
b. Almost exactly a distance f behind the mirror
c. Almost exactly a distance f in front of the mirror
d. At a distance behind the mirror equal to the distance of the moon in front of the mirror

CHALLENGE **EXAMPLE 34.17** | Optical fiber imaging

An *endoscope* is a thin bundle of optical fibers that can be inserted through a bodily opening or small incision to view the interior of the body. As **FIGURE 34.52** shows, an *objective* lens forms a real image on the entrance face of the fiber bundle. Individual fibers, using total internal reflection, transport the light to the exit face, where it emerges. The doctor (or a TV camera) observes the object by viewing the exit face through an *eyepiece* lens.

FIGURE 34.52 An endoscope.

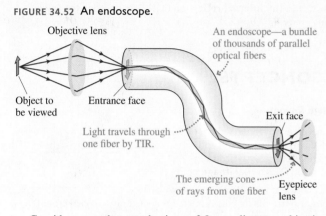

Consider an endoscope having a 3.0-mm-diameter objective lens with a focal length of 1.1 mm. These are typical values. The indices of refraction of the core and the cladding of the optical fibers are 1.62 and 1.50, respectively. To give maximum brightness, the objective lens is positioned so that, for an on-axis object, rays passing through the outer edge of the lens have the maximum angle of incidence for undergoing TIR in the fiber. How far should the objective lens be placed from the object the doctor wishes to view?

MODEL Represent the object as an on-axis point source and use the ray model of light.

VISUALIZE FIGURE 34.53 shows the real image being focused on the entrance face of the endoscope. Inside the fiber, rays that strike the cladding at an angle of incidence greater than the critical angle θ_c undergo TIR and stay in the fiber; rays are lost if their angle of incidence is less than θ_c. For maximum brightness, the lens is positioned so that a ray passing through the outer edge refracts into the fiber at the maximum angle of incidence θ_{max} for which TIR is possible. A smaller-diameter lens would sacrifice light-gathering power, whereas the outer rays from a larger-diameter lens would impinge on the core-cladding boundary at less than θ_c and would not undergo TIR.

FIGURE 34.53 Magnified view of the entrance of an optical fiber.

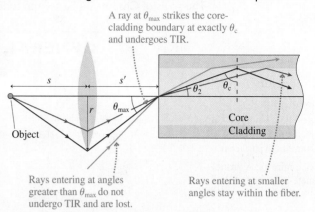

SOLVE We know the focal length of the lens. We can use the geometry of the ray at the critical angle to find the image distance s', then use the thin-lens equation to find the object distance s. The critical angle for TIR inside the fiber is

$$\theta_c = \sin^{-1}\left(\frac{n_{cladding}}{n_{core}}\right) = \sin^{-1}\left(\frac{1.50}{1.62}\right) = 67.8°$$

A ray incident on the core-cladding boundary at exactly the critical angle must have entered the fiber, at the entrance face, at angle $\theta_2 = 90° - \theta_c = 22.2°$. For optimum lens placement, this ray passed through the outer edge of the lens and was incident on the entrance face at angle θ_{max}. Snell's law at the entrance face is

$$n_{air} \sin\theta_{max} = 1.00\sin\theta_{max} = n_{core}\sin\theta_2$$

and thus

$$\theta_{max} = \sin^{-1}(1.62\sin 22.2°) = 37.7°$$

We know the lens radius, $r = 1.5\,\text{mm}$, so the distance of the lens from the fiber—the image distance s'—is

$$s' = \frac{r}{\tan\theta_{max}} = \frac{1.5\ \text{mm}}{\tan(37.7°)} = 1.9\ \text{mm}$$

Now we can use the thin-lens equation to locate the object:

$$\frac{1}{s} = \frac{1}{f} - \frac{1}{s'} = \frac{1}{1.1\ \text{mm}} - \frac{1}{1.9\ \text{mm}}$$

$$s = 2.6\ \text{mm}$$

The doctor, viewing the exit face of the fiber bundle, will see a focused image when the objective lens is 2.6 mm from the object she wishes to view.

ASSESS The object and image distances are both greater than the focal length, which is correct for forming a real image.

SUMMARY

The goals of Chapter 34 have been to learn about and apply the ray model of light.

GENERAL PRINCIPLES

Reflection

Law of reflection: $\theta_r = \theta_i$

Reflection can be **specular** (mirror-like) or **diffuse** (from rough surfaces).

Plane mirrors: A virtual image is formed at P′ with $s' = s$.

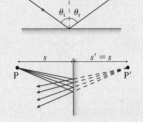

Refraction

Snell's law of refraction:

$$n_1 \sin\theta_1 = n_2 \sin\theta_2$$

Index of refraction is $n = c/v$. The ray is closer to the normal on the side with the larger index of refraction.

If $n_2 < n_1$, **total internal reflection** (TIR) occurs when the angle of incidence $\theta_1 \geq \theta_c = \sin^{-1}(n_2/n_1)$.

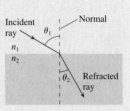

IMPORTANT CONCEPTS

The ray model of light

Light travels along straight lines, called **light rays,** at speed $v = c/n$.

A light ray continues forever unless an interaction with matter causes it to reflect, refract, scatter, or be absorbed.

Light rays come from **objects.** Each point on the object sends rays in all directions.

The eye sees an object (or an image) when diverging rays are collected by the pupil and focused on the retina.

▶ Ray optics is valid when lenses, mirrors, and apertures are larger than ≈ 1 mm.

Image formation

If rays diverge from P and interact with a lens or mirror so that the refracted/reflected rays *converge* at P′, then P′ is a **real image** of P.

If rays diverge from P and interact with a lens or mirror so that the refracted/reflected rays *diverge* from P′ and appear to come from P′, then P′ is a **virtual image** of P.

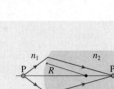

Spherical surface: Object and image distances are related by

$$\frac{n_1}{s} + \frac{n_2}{s'} = \frac{n_2 - n_1}{R}$$

Plane surface: $R \to \infty$, so $s' = -(n_2/n_1)s$.

APPLICATIONS

Ray tracing

3 special rays in 3 basic situations:

Converging lens
Real image

Converging lens
Virtual image

Diverging lens
Virtual image

Magnification $m = -\dfrac{s'}{s}$

m is + for an upright image, − for inverted.
The height ratio is $h'/h = |m|$.

Thin lenses

The image and object distances are related by

$$\frac{1}{s} + \frac{1}{s'} = \frac{1}{f}$$

where the focal length is given by the lens maker's equation:

$$\frac{1}{f} = (n-1)\left(\frac{1}{R_1} - \frac{1}{R_2}\right)$$

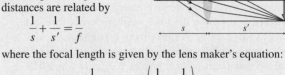

R	+ for surface convex toward object	− for concave
f	+ for a converging lens	− for diverging
s'	+ for a real image	− for virtual

Spherical mirrors

The image and object distances are related by

$$\frac{1}{s} + \frac{1}{s'} = \frac{1}{f}$$

R, f	+ for concave mirror	− for convex
s'	+ for a real image	− for virtual

Focal length $f = R/2$

TERMS AND NOTATION

light ray	diffuse reflection	lens	inverted image
object	virtual image	ray tracing	lateral magnification, m
point source	refraction	converging lens	upright image
parallel bundle	angle of refraction	focal point	thin-lens equation
ray diagram	Snell's law	focal length, f	lens maker's equation
camera obscura	total internal reflection (TIR)	diverging lens	spherical mirror
aperture	critical angle, θ_c	thin lens	concave mirror
specular reflection	object distance, s	lens plane	convex mirror
angle of incidence	image distance, s'	real image	
angle of reflection	optical axis	object plane	
law of reflection	paraxial rays	image plane	

CONCEPTUAL QUESTIONS

1. Suppose you have two pinhole cameras. The first has a small round hole in the front. The second is identical except it has a square hole of the same area as the round hole in the first camera. Would the pictures taken by these two cameras, under the same conditions, be different in any obvious way? Explain.

2. You are looking at the image of a pencil in a mirror, as shown in **FIGURE Q34.2**.
   a. What happens to the image if the top half of the mirror, down to the midpoint, is covered with a piece of cardboard? Explain.
   b. What happens to the image if the bottom half of the mirror is covered with a piece of cardboard? Explain.

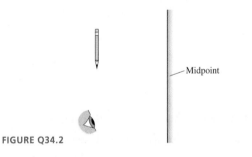

FIGURE Q34.2

3. One problem with using optical fibers for communication is that a light ray passing directly down the center of the fiber takes less time to travel from one end to the other than a ray taking a longer, zig-zag path. Thus light rays starting at the same time but traveling in slightly different directions reach the end of the fiber at different times. This problem can be solved by making the refractive index of the glass change gradually from a higher value in the center to a lower value near the edges of the fiber. Explain how this reduces the difference in travel times.

4. A light beam passing from medium 2 to medium 1 is refracted as shown in **FIGURE Q34.4**. Is n_1 larger than n_2, is n_1 smaller than n_2, or is there not enough information to tell? Explain.

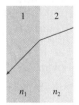

FIGURE Q34.4

5. A fish in an aquarium with flat sides looks out at a hungry cat. To the fish, does the distance to the cat appear to be less than the actual distance, the same as the actual distance, or more than the actual distance? Explain.

6. Consider *one* point on an object near a lens.
   a. What is the minimum number of rays needed to locate its image point? Explain.
   b. How many rays from this point actually strike the lens and refract to the image point?

7. The object and lens in **FIGURE Q34.7** are positioned to form a well-focused, inverted image on a viewing screen. Then a piece of cardboard is lowered just in front of the lens to cover the top half of the lens. Describe what you see on the screen when the cardboard is in place.

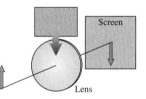

FIGURE Q34.7

8. A converging lens creates the image shown in **FIGURE Q34.8**. Is the object distance less than the focal length f, between f and $2f$, or greater than $2f$? Explain.

FIGURE Q34.8

9. A concave mirror brings the sun's rays to a focus in front of the mirror. Suppose the mirror is submerged in a swimming pool but still pointed up at the sun. Will the sun's rays be focused nearer to, farther from, or at the same distance from the mirror? Explain.

10. You see an upright, magnified image of your face when you look into a magnifying cosmetic mirror. Where is the image? Is it in front of the mirror's surface, on the mirror's surface, or behind the mirror's surface? Explain.

11. When you look at your reflection in the bowl of a spoon, it is upside down. Why?

EXERCISES AND PROBLEMS

Problems labeled ![] integrate material from earlier chapters.

Exercises

Section 34.1 The Ray Model of Light

1. ‖ A point source of light illuminates an aperture 2.0 m away. A 12.0-cm-wide bright patch of light appears on a screen 1.0 m behind the aperture. How wide is the aperture?

2. ‖ a. How long (in ns) does it take light to travel 1.0 m in vacuum?
   b. What distance does light travel in water, glass, and cubic zirconia during the time that it travels 1.0 m in vacuum?

3. ‖ A student has built a 15-cm-long pinhole camera for a science fair project. She wants to photograph her 180-cm-tall friend and have the image on the film be 5.0 cm high. How far should the front of the camera be from her friend?

4. ‖ A 5.0-cm-thick layer of oil is sandwiched between a 1.0-cm-thick sheet of glass and a 2.0-cm-thick sheet of polystyrene plastic. How long (in ns) does it take light incident perpendicular to the glass to pass through this 8.0-cm-thick sandwich?

Section 34.2 Reflection

5. ‖ A light ray leaves point A in **FIGURE EX34.5**, reflects from the mirror, and reaches point B. How far below the top edge does the ray strike the mirror?

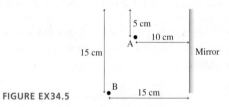

FIGURE EX34.5

6. ‖ The mirror in **FIGURE EX34.6** deflects a horizontal laser beam by 60°. What is the angle ϕ?

FIGURE EX34.6

7. ‖ At what angle ϕ should the laser beam in **FIGURE EX34.7** be aimed at the mirrored ceiling in order to hit the midpoint of the far wall?

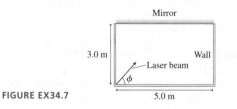

FIGURE EX34.7

8. ‖ A laser beam is incident on the left mirror in **FIGURE EX34.8**. Its initial direction is parallel to a line that bisects the mirrors. What is the angle ϕ of the reflected laser beam?

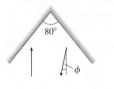

FIGURE EX34.8

9. ‖ It is 165 cm from your eyes to your toes. You're standing 200 cm in front of a tall mirror. How far is it from your eyes to the image of your toes?

Section 34.3 Refraction

10. ‖ A laser beam in air is incident on a liquid at an angle of 53° with respect to the normal. The laser beam's angle in the liquid is 35°. What is the liquid's index of refraction?

11. ‖ A 1.0-cm-thick layer of water stands on a horizontal slab of glass. A light ray in the air is incident on the water 60° from the normal. What is the ray's direction of travel in the glass?

12. ‖ A costume jewelry pendant made of cubic zirconia is submerged in oil. A light ray strikes one face of the zirconia crystal at an angle of incidence of 25°. Once inside, what is the ray's angle with respect to the face of the crystal?

13. ‖ An underwater diver sees the sun 50° above horizontal. How high is the sun above the horizon to a fisherman in a boat above the diver?

14. ‖ The glass core of an optical fiber has an index of refraction 1.60. The index of refraction of the cladding is 1.48. What is the maximum angle a light ray can make with the wall of the core if it is to remain inside the fiber?

15. ‖ A thin glass rod is submerged in oil. What is the critical angle for light traveling inside the rod?

16. ‖ **FIGURE EX34.16** shows a transparent hemisphere with radius R and index of refraction n. What is the maximum distance d for which a light ray parallel to the axis refracts out through the curved surface?

FIGURE EX34.16

Section 34.4 Image Formation by Refraction at a Plane Surface

17. ‖ A fish in a flat-sided aquarium sees a can of fish food on the counter. To the fish's eye, the can looks to be 30 cm outside the aquarium. What is the actual distance between the can and the aquarium? (You can ignore the thin glass wall of the aquarium.)

18. ‖ A biologist keeps a specimen of his favorite beetle embedded in a cube of polystyrene plastic. The hapless bug appears to be 2.0 cm within the plastic. What is the beetle's actual distance beneath the surface?

19. ‖ A 150-cm-tall diver is standing completely submerged on the bottom of a swimming pool full of water. You are sitting on the end of the diving board, almost directly over her. How tall does the diver appear to be?

20. ‖ To a fish in an aquarium, the 4.00-mm-thick walls appear to be only 3.50 mm thick. What is the index of refraction of the walls?

Section 34.5 Thin Lenses: Ray Tracing

21. ‖ An object is 20 cm in front of a converging lens with a focal length of 10 cm. Use ray tracing to determine the location of the image. Is the image upright or inverted?

22. ‖ An object is 30 cm in front of a converging lens with a focal length of 5 cm. Use ray tracing to determine the location of the image. Is the image upright or inverted?

23. ‖ An object is 6 cm in front of a converging lens with a focal length of 10 cm. Use ray tracing to determine the location of the image. Is the image upright or inverted?

24. ‖ An object is 15 cm in front of a diverging lens with a focal length of −15 cm. Use ray tracing to determine the location of the image. Is the image upright or inverted?

Section 34.6 Thin Lenses: Refraction Theory

25. ‖ Find the focal length of the glass lens in **FIGURE EX34.25**.

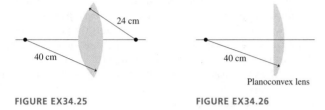

FIGURE EX34.25   **FIGURE EX34.26**

26. ‖ Find the focal length of the planoconvex polystyrene plastic lens in **FIGURE EX34.26**.

27. ‖ Find the focal length of the glass lens in **FIGURE EX34.27**.

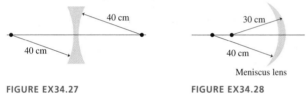

FIGURE EX34.27   **FIGURE EX34.28**

28. ‖ Find the focal length of the meniscus polystyrene plastic lens in **FIGURE EX34.28**.

29. ‖ An air bubble inside an 8.0-cm-diameter plastic ball is 2.0 cm from the surface. As you look at the ball with the bubble turned toward you, how far beneath the surface does the bubble appear to be?

30. ‖ A goldfish lives in a 50-cm-diameter spherical fish bowl. The fish sees a cat watching it. If the cat's face is 20 cm from the edge of the bowl, how far from the edge does the fish see it as being? (You can ignore the thin glass wall of the bowl.)

31. | A 1.0-cm-tall candle flame is 60 cm from a lens with a focal length of 20 cm. What are the image distance and the height of the flame's image?

32. ‖ A 1.0-cm-tall object is 10 cm in front of a converging lens that has a 30 cm focal length.
   a. Use ray tracing to find the position and height of the image. To do this accurately, use a ruler or paper with a grid. Determine the image distance and image height by making measurements on your diagram.
   b. Calculate the image position and height. Compare with your ray-tracing answers in part a.

33. ‖ A 2.0-cm-tall object is 40 cm in front of a converging lens that has a 20 cm focal length.
   a. Use ray tracing to find the position and height of the image. To do this accurately, use a ruler or paper with a grid. Determine the image distance and image height by making measurements on your diagram.
   b. Calculate the image position and height. Compare with your ray-tracing answers in part a.

34. ‖ A 1.0-cm-tall object is 75 cm in front of a converging lens that has a 30 cm focal length.

   a. Use ray tracing to find the position and height of the image. To do this accurately, use a ruler or paper with a grid. Determine the image distance and image height by making measurements on your diagram.
   b. Calculate the image position and height. Compare with your ray-tracing answers in part a.

35. ‖ A 2.0-cm-tall object is 15 cm in front of a converging lens that has a 20 cm focal length.
   a. Use ray tracing to find the position and height of the image. To do this accurately, use a ruler or paper with a grid. Determine the image distance and image height by making measurements on your diagram.
   b. Calculate the image position and height. Compare with your ray-tracing answers in part a.

36. ‖ A 1.0-cm-tall object is 60 cm in front of a diverging lens that has a −30 cm focal length.
   a. Use ray tracing to find the position and height of the image. To do this accurately, use a ruler or paper with a grid. Determine the image distance and image height by making measurements on your diagram.
   b. Calculate the image position and height. Compare with your ray-tracing answers in part a.

37. ‖ A 2.0-cm-tall object is 15 cm in front of a diverging lens that has a −20 cm focal length.
   a. Use ray tracing to find the position and height of the image. To do this accurately, use a ruler or paper with a grid. Determine the image distance and image height by making measurements on your diagram.
   b. Calculate the image position and height. Compare with your ray-tracing answers in part a.

Section 34.7 Image Formation with Spherical Mirrors

38. ‖ An object is 40 cm in front of a concave mirror with a focal length of 20 cm. Use ray tracing to locate the image. Is the image upright or inverted?

39. ‖ An object is 12 cm in front of a concave mirror with a focal length of 20 cm. Use ray tracing to locate the image. Is the image upright or inverted?

40. ‖ An object is 30 cm in front of a convex mirror with a focal length of −20 cm. Use ray tracing to locate the image. Is the image upright or inverted?

41. | A 1.0-cm-tall object is 20 cm in front of a concave mirror that has a 60 cm focal length. Calculate the position and height of the image. State whether the image is in front of or behind the mirror, and whether the image is upright or inverted.

42. | A 1.0-cm-tall object is 20 cm in front of a convex mirror that has a −60 cm focal length. Calculate the position and height of the image. State whether the image is in front of or behind the mirror, and whether the image is upright or inverted.

Problems

43. ‖ An advanced computer sends information to its various parts via infrared light pulses traveling through silicon fibers. To acquire data from memory, the central processing unit sends a light-pulse request to the memory unit. The memory unit processes the request, then sends a data pulse back to the central processing unit. The memory unit takes 0.5 ns to process a request. If the information has to be obtained from memory in 2.0 ns, what is the maximum distance the memory unit can be from the central processing unit?

44. ‖ A red ball is placed at point A in **FIGURE P34.44**.
   a. How many images are seen by an observer at point O?
   b. What are the (x, y) coordinates of each image?

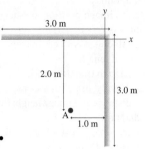

FIGURE P34.44 O

45. ‖ The place you get your hair cut has two nearly parallel mirrors 5.0 m apart. As you sit in the chair, your head is 2.0 m from the nearer mirror. Looking toward this mirror, you first see your face and then, farther away, the back of your head. (The mirrors need to be slightly nonparallel for you to be able to see the back of your head, but you can treat them as parallel in this problem.) How far away does the back of your head appear to be? Neglect the thickness of your head.

46. ‖ A microscope is focused on a black dot. When a 1.00-cm-thick piece of plastic is placed over the dot, the microscope objective has to be raised 0.40 cm to bring the dot back into focus. What is the index of refraction of the plastic?

47. ‖ A light ray in air is incident on a transparent material whose index of refraction is n.
   a. Find an expression for the (non-zero) angle of incidence whose angle of refraction is half the angle of incidence.
   b. Evaluate your expression for light incident on glass.

48. ‖ The meter stick in **FIGURE P34.48** lies on the bottom of a 100-cm-long tank with its zero mark against the left edge. You look into the tank at a 30° angle, with your line of sight just grazing the upper left edge of the tank. What mark do you see on the meter stick if the tank is (a) empty, (b) half full of water, and (c) completely full of water?

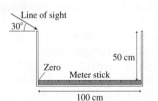

FIGURE P34.48

49. ‖ The 80-cm-tall, 65-cm-wide tank shown in **FIGURE P34.49** is completely filled with water. The tank has marks every 10 cm along one wall, and the 0 cm mark is barely submerged. As you stand beside the opposite wall, your eye is level with the top of the water.
   a. Can you see the marks from the top of the tank (the 0 cm mark) going down, or from the bottom of the tank (the 80 cm mark) coming up? Explain.
   b. Which is the lowest or highest mark, depending on your answer to part a, that you can see?

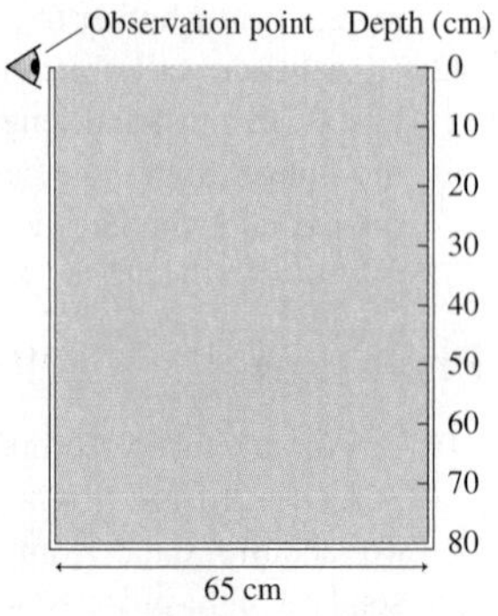

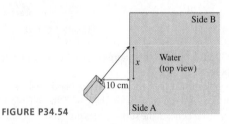

FIGURE P34.49

50. ‖ A horizontal meter stick is centered at the bottom of a 3.0-m-deep, 3.0-m-wide pool of water. How long does the meter stick appear to be as you look at it from the edge of the pool?

51. ‖ A 4.0-m-wide swimming pool is filled to the top. The bottom of the pool becomes completely shaded in the afternoon when the sun is 20° above the horizon. How deep is the pool?

52. ‖ It's nighttime, and you've dropped your goggles into a 3.0-m-deep swimming pool. If you hold a laser pointer 1.0 m above the edge of the pool, you can illuminate the goggles if the laser beam enters the water 2.0 m from the edge. How far are the goggles from the edge of the pool?

53. ‖ An astronaut is exploring an unknown planet when she accidentally drops an oxygen canister into a 1.50-m-deep pool filled with an unknown liquid. Although she dropped the canister 21 cm from the edge, it appears to be 31 cm away when she peers in from the edge. What is the liquid's index of refraction? Assume that the planet's atmosphere is similar to earth's.

54. ‖ Shown from above in **FIGURE P34.54** is one corner of a rectangular box filled with water. A laser beam starts 10 cm from side A of the container and enters the water at position x. You can ignore the thin walls of the container.
   a. If $x = 15$ cm, does the laser beam refract back into the air through side B or reflect from side B back into the water? Determine the angle of refraction or reflection.
   b. Repeat part a for $x = 25$ cm.
   c. Find the minimum value of x for which the laser beam passes through side B and emerges into the air.

Side B

Water
(top view)

x

10 cm

Side A

FIGURE P34.54

55. ‖ A light beam can use reflections to form a closed, N-sided polygon inside a solid, transparent cylinder if N is sufficiently large. What is the minimum possible value of N for light inside a cylinder of (a) water, (b) polystyrene plastic, and (c) cubic zirconia? Assume the cylinder is surrounded by air.

56. ‖ Optical engineers need to know the *cone of acceptance* of an optical fiber. This is the maximum angle that an entering light ray can make with the axis of the fiber if it is to be guided down the fiber. What is the cone of acceptance of an optical fiber for which the index of refraction of the core is 1.55 while that of the cladding is 1.45? You can model the fiber as a cylinder with a flat entrance face.

57. ‖‖ One of the contests at the school carnival is to throw a spear at an underwater target lying flat on the bottom of a pool. The water is 1.0 m deep. You're standing on a small stool that places your eyes 3.0 m above the bottom of the pool. As you look at the target, your gaze is 30° below horizontal. At what angle below horizontal should you throw the spear in order to hit the target? Your raised arm brings the spear point to the level of your eyes as you throw it, and over this short distance you can assume that the spear travels in a straight line rather than a parabolic trajectory.

58. ‖ There's one angle of incidence β onto a prism for which the light inside an isosceles prism travels parallel to the base and emerges at angle β.

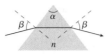

FIGURE P34.58

 a. Find an expression for β in terms of the prism's apex angle α and index of refraction n.

 b. A laboratory measurement finds that $\beta = 52.2°$ for a prism shaped like an equilateral triangle. What is the prism's index of refraction?

59. ‖ You're visiting the shark tank at the aquarium when you see a 2.5-m-long shark that appears to be swimming straight toward you at 2.0 m/s. What is the shark's actual speed through the water? You can ignore the glass wall of the tank.

60. ‖ Paraxial light rays approach a transparent sphere parallel to an optical axis passing through the center of the sphere. The rays come to a focus on the far surface of the sphere. What is the sphere's index of refraction?

61. ‖ To determine the focal length of a lens, you place the lens in front of a small lightbulb and then adjust a viewing screen to get a sharply focused image. Varying the lens position produces the following data:

Bulb to lens (cm)	Lens to screen (cm)
20	61
22	47
24	39
26	37
28	32

Use the best-fit line of an appropriate graph to determine the focal length of the lens.

62. ‖ BIO The illumination lights in an operating room use a concave mirror to focus an image of a bright lamp onto the surgical site. One such light uses a mirror with a 30 cm radius of curvature. If the mirror is 1.2 m from the patient, how far should the lamp be from the mirror?

63. ‖ BIO A dentist uses a curved mirror to view the back side of teeth in the upper jaw. Suppose she wants an upright image with a magnification of 1.5 when the mirror is 1.2 cm from a tooth. Should she use a convex or a concave mirror? What focal length should it have?

64. ‖ BIO A *keratometer* is an optical device used to measure the radius of curvature of the eye's cornea—its entrance surface. This measurement is especially important when fitting contact lenses, which must match the cornea's curvature. Most light incident on the eye is transmitted into the eye, but some light reflects from the cornea, which, due to its curvature, acts like a convex mirror. The keratometer places a small, illuminated ring of known diameter 7.5 cm in front of the eye. The optometrist, using an eyepiece, looks through the center of this ring and sees a small virtual image of the ring that appears to be behind the cornea. The optometrist uses a scale inside the eyepiece to measure the diameter of the image and calculate its magnification. Suppose the optometrist finds that the magnification for one patient is 0.049. What is the absolute value of the radius of curvature of her cornea?

65. ‖ The mirror in **FIGURE P34.65** is covered with a piece of glass. A point source of light is outside the glass. How far from the mirror is the image of this source?

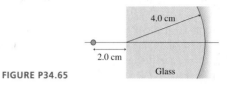

FIGURE P34.65

66. ‖ A 2.0-cm-tall candle flame is 2.0 m from a wall. You happen to have a lens with a focal length of 32 cm. How many places can you put the lens to form a well-focused image of the candle flame on the wall? For each location, what are the height and orientation of the image?

67. ‖ A 25 g rubber ball is dropped from a height of 3.0 m above the center of a horizontal, concave mirror. The ball and its image coincide 0.65 s after the ball is released. What is the mirror's radius of curvature?

68. ‖ In recent years, physicists have learned to create *metamaterials*—engineered materials not found in nature—with negative indices of refraction. It's not yet possible to form a lens from a material with a negative index of refraction, but researchers are optimistic. Suppose you had a planoconvex lens (flat on one side, a 15 cm radius of curvature on the other) that is made from a metamaterial with $n = -1.25$. If you place an object 12 cm from this lens, (a) what type of image will be formed and (b) where will the image be located?

69. ‖ A lightbulb is 3.0 m from a wall. What are the focal length and the position (measured from the bulb) of a lens that will form an image on the wall that is twice the size of the lightbulb?

70. ‖ An old-fashioned slide projector needs to create a 98-cm-high image of a 2.0-cm-tall slide. The screen is 300 cm from the slide.

 a. What focal length does the lens need? Assume that it is a thin lens.

 b. How far should you place the lens from the slide?

71. ‖ Some *electro-optic materials* can change their index of refraction in response to an applied voltage. Suppose a planoconvex lens (flat on one side, a 15.0 cm radius of curvature on the other), made from a material whose normal index of refraction is 1.500, is creating an image of an object that is 50.0 cm from the lens. By how much would the index of refraction need to be increased to move the image 5.0 cm closer to the lens?

72. ‖ A point source of light is distance d from the surface of a 6.0-cm-diameter glass sphere. For what value of d is there an image at the same distance d on the opposite side of the sphere?

73. ‖ A lens placed 10 cm in front of an object creates an upright image twice the height of the object. The lens is then moved along the optical axis until it creates an inverted image twice the height of the object. How far did the lens move?

74. ‖ An object is 60 cm from a screen. What are the radii of a symmetric converging plastic lens (i.e., two equally curved surfaces) that will form an image on the screen twice the height of the object?

75. ‖ CALC A wildlife photographer with a 200-mm-focal-length telephoto lens on his camera is taking a picture of a rhinoceros that is 100 m away. Suddenly, the rhino starts charging straight toward the photographer at a speed of 5.0 m/s. What is the speed, in μm/s, of the image of the rhinoceros? Is the image moving toward or away from the lens?

76. ‖ A concave mirror has a 40 cm radius of curvature. How far from the mirror must an object be placed to create an upright image three times the height of the object?

77. ‖‖ A 2.0-cm-tall object is placed in front of a mirror. A 1.0-cm-tall upright image is formed behind the mirror, 150 cm from the object. What is the focal length of the mirror?

78. ‖ A spherical mirror of radius R has its center at C, as shown in **FIGURE P34.78**. A ray parallel to the axis reflects through F, the focal point. Prove that $f = R/2$ if $\phi \ll 1$ rad.

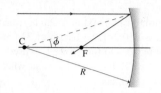

FIGURE P34.78

Challenge Problems

79. ‖‖‖ Consider a lens having index of refraction n_2 and surfaces with radii R_1 and R_2. The lens is immersed in a fluid that has index of refraction n_1.
    a. Derive a generalized lens maker's equation to replace Equation 34.26 when the lens is surrounded by a medium other than air. That is, when $n_1 \neq 1$.
    b. A symmetric converging glass lens (i.e., two equally curved surfaces) has two surfaces with radii of 40 cm. Find the focal length of this lens in air and the focal length of this lens in water.

80. ‖‖‖ **FIGURE CP34.80** shows a light ray that travels from point A
CALC to point B. The ray crosses the boundary at position x, making angles θ_1 and θ_2 in the two media. Suppose that you did *not* know Snell's law.
    a. Write an expression for the *time t* it takes the light ray to travel from A to B. Your expression should be in terms of the distances a, b, and w; the variable x; and the indices of refraction n_1 and n_2.

    b. The time depends on x. There's one value of x for which the light travels from A to B in the shortest possible time. We'll call it x_{min}. Write an expression (but don't try to solve it!) from which x_{min} could be found.
    c. Now, by using the geometry of the figure, derive Snell's law from your answer to part b.
    You've proven that Snell's law is equivalent to the statement that "light traveling between two points follows the path that requires the shortest time." This interesting way of thinking about refraction is called *Fermat's principle*.

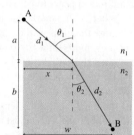

FIGURE CP34.80

81. ‖‖‖ A fortune teller's "crystal ball" (actually just glass) is 10 cm in diameter. Her secret ring is placed 6.0 cm from the edge of the ball.
    a. An image of the ring appears on the opposite side of the crystal ball. How far is the image from the center of the ball?
    b. Draw a ray diagram showing the formation of the image.
    c. The crystal ball is removed and a thin lens is placed where the center of the ball had been. If the image is still in the same position, what is the focal length of the lens?

82. ‖‖‖ Consider an object of thickness ds (parallel to the axis) in
CALC front of a lens or mirror. The image of the object has thickness ds'. Define the *longitudinal magnification* as $M = ds'/ds$. Prove that $M = -m^2$, where m is the lateral magnification.

35 Optical Instruments

The world's greatest collection of telescopes is on the summit of Mauna Kea on the Big Island of Hawaii, towering 4200 m (13,800 ft) over the Pacific Ocean.

IN THIS CHAPTER, you will learn about some common optical instruments and their limitations.

What is an optical instrument?

Optical instruments, such as cameras, microscopes, and telescopes, are used to produce images for viewing or detection. Most use several individual lenses in combination to improve performance. You'll learn how to analyze a system with multiple lenses.

« LOOKING BACK Sections 34.5–34.6 Thin lenses

How does a camera work?

A camera uses a lens—made of several individual lenses—to project a real image onto a light-sensitive detector. The detector in a digital camera uses millions of tiny pixels.

- You'll learn about focusing and zoom.
- You'll also learn how to calculate a lens's f-number, which, along with shutter speed, determines the exposure.

Lens

Aperture

How does vision work?

The human eye is much like a camera; the cornea and lens together focus a real image onto the retina. You'll learn about two defects of vision—myopia (nearsightedness) and hyperopia (farsightedness)—and how they can be corrected with eyeglasses.

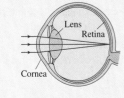

Lens
Retina
Cornea

What optical systems are used to magnify things?

Lenses and mirrors can be used to magnify objects both near and far.

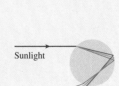
You can magnify text

- A simple magnifying glass has a low magnification of 2× or 3×.
- Microscopes use two sets of lenses to reach magnifications up to 1000×.
- Telescopes magnify distant objects.

Is color important in optics?

Color depends on the wavelength of light.

- The index of refraction is slightly wavelength dependent, so different wavelengths refract at different angles. This is the main reason we have rainbows.
- Many materials absorb or scatter some wavelengths more than others.

Sunlight

Rainbow

What is the resolution of an optical system?

Light passing through a lens undergoes diffraction, just like light passing through a circular hole. Images are not perfect points but are tiny diffraction patterns, and this limits how well two nearby objects can be resolved. You'll learn about Rayleigh's criterion for the resolution of two images.

« LOOKING BACK Section 33.6 Circular diffraction

35.1 Lenses in Combination

Only the simplest magnifiers are built with a single lens of the sort we analyzed in Chapter 34. Optical instruments, such as microscopes and cameras, are invariably built with multiple lenses. The reason, as we'll see, is to improve the image quality.

The analysis of multi-lens systems requires only one new rule: **The image of the first lens acts as the object for the second lens.** To see why this is so, FIGURE 35.1 shows a simple telescope consisting of a large-diameter converging lens, called the *objective,* and a smaller converging lens used as the *eyepiece.* (We'll analyze telescopes more thoroughly later in the chapter.) Highlighted are the three special rays you learned to use in Chapter 34:

- ▨ A ray parallel to the optical axis refracts through the focal point.
- ▨ A ray through the focal point refracts parallel to the optical axis.
- ▨ A ray through the center of the lens is undeviated.

FIGURE 35.1 Ray-tracing diagram of a simple astronomical telescope.

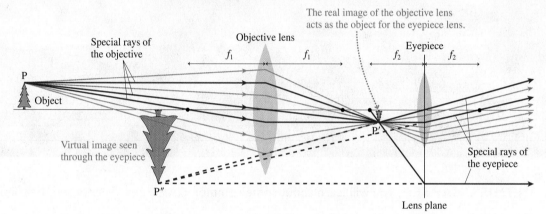

The rays passing through the objective converge to a real image at P′, **but they don't stop there.** Instead, light rays *diverge* from P′ as they approach the second lens. **As far as the eyepiece is concerned, the rays are coming from P′, and thus P′ acts as the object for the second lens.** The three special rays passing through the objective lens are sufficient to locate the image P′, but these rays are generally *not* the special rays for the second lens. However, other rays converging at P′ leave at the correct angles to be the special rays for the eyepiece. That is, a new set of special rays is drawn from P′ to the second lens and used to find the final image point P″.

> **NOTE** One ray seems to "miss" the eyepiece lens, but this isn't a problem. All rays passing through the lens converge to (or diverge from) a single point, and the purpose of the special rays is to locate that point. To do so, we can let the special rays refract as they cross the *lens plane,* regardless of whether the physical lens really extends that far.

EXAMPLE 35.1 | A camera lens

The "lens" on a camera is usually a combination of two or more single lenses. Consider a camera in which light passes first through a diverging lens, with $f_1 = -120$ mm, then a converging lens, with $f_2 = 42$ mm, spaced 60 mm apart. A reasonable definition of the *effective focal length* of this lens combination is the focal length of a *single* lens that could produce an image in the same location if placed at the midpoint of the lens combination. A 10-cm-tall object is 500 mm from the first lens.

a. What are the location, size, and orientation of the image?

b. What is the effective focal length of the double-lens system used in this camera?

MODEL Each lens is a thin lens. The image of the first lens is the object for the second.

VISUALIZE The ray-tracing diagram of FIGURE 35.2 shows the production of a real, inverted image ≈ 55 mm behind the second lens.

SOLVE a. $s_1 = 500$ mm is the object distance of the first lens. Its image, a virtual image, is found from the thin-lens equation:

$$\frac{1}{s_1'} = \frac{1}{f_1} - \frac{1}{s_1} = \frac{1}{-120 \text{ mm}} - \frac{1}{500 \text{ mm}} = -0.0103 \text{ mm}^{-1}$$

$$s_1' = -97 \text{ mm}$$

This is consistent with the ray-tracing diagram. The image of the first lens now acts as the object for the second lens. Because the lenses are 60 mm apart, the object distance is $s_2 = 97$ mm + 60 mm = 157 mm. A second application of the thin-lens equation yields

$$\frac{1}{s_2'} = \frac{1}{f_2} - \frac{1}{s_2} = \frac{1}{42 \text{ mm}} - \frac{1}{157 \text{ mm}} = 0.0174 \text{ mm}^{-1}$$

$$s_2' = 57 \text{ mm}$$

The image of the lens combination is 57 mm behind the second lens. The lateral magnifications of the two lenses are

$$m_1 = -\frac{s_1'}{s_1} = -\frac{-97 \text{ cm}}{500 \text{ cm}} = 0.194$$

$$m_2 = -\frac{s_2'}{s_2} = -\frac{57 \text{ cm}}{157 \text{ cm}} = -0.363$$

The second lens magnifies the image of the first lens, which magnifies the object, so **the total magnification is the product of the individual magnifications:**

$$m = m_1 m_2 = -0.070$$

Thus the image is 57 mm behind the second lens, inverted (m is negative), and 0.70 cm tall.

b. If a single lens midway between these two lenses produced an image in the same plane, its object and image distances would be $s = 500$ mm + 30 mm = 530 mm and $s' = 57$ mm + 30 mm = 87 mm. A final application of the thin-lens equation gives the effective focal length:

$$\frac{1}{f_{\text{eff}}} = \frac{1}{s} + \frac{1}{s'} = \frac{1}{530 \text{ mm}} + \frac{1}{87 \text{ mm}} = 0.0134 \text{ mm}^{-1}$$

$$f_{\text{eff}} = 75 \text{ mm}$$

ASSESS This combination lens would be sold as a "75 mm lens."

FIGURE 35.2 Pictorial representation of a combination lens.

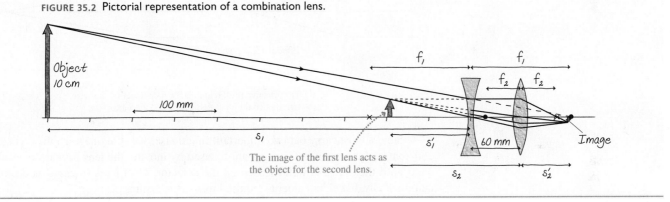

The image of the first lens acts as the object for the second lens.

STOP TO THINK 35.1 The second lens in this optical instrument

a. Causes the light rays to focus closer than they would with the first lens acting alone.
b. Causes the light rays to focus farther away than they would with the first lens acting alone.
c. Inverts the image but does not change where the light rays focus.
d. Prevents the light rays from reaching a focus.

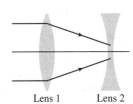

Lens 1   Lens 2

35.2 The Camera

A **camera,** shown in FIGURE 35.3, "takes a picture" by using a lens to form a real, inverted image on a light-sensitive detector in a light-tight box. Film was the detector of choice for well over a hundred years, but today's digital cameras use an electronic detector called a *charge-coupled device,* or CCD.

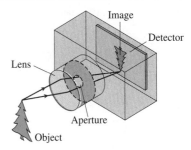

FIGURE 35.3 A camera.

The camera "lens" is always a combination of two or more individual lenses. The simplest such lens, shown in FIGURE 35.4, consists of a converging lens and a somewhat weaker diverging lens. This combination of positive and negative lenses corrects some of the defects inherent in single lenses, as we'll discuss later in the chapter. As Example 35.1 suggested, we can model a combination lens as a single lens with an **effective focal length** (usually called simply "the focal length") f. A *zoom lens* changes the effective focal length by changing the spacing between the converging lens and the diverging lens; this is what happens when the lens barrel on your digital camera moves in and out as you use the zoom. A typical digital camera has a lens whose effective focal length can be varied from 6 mm to 18 mm, giving, as we'll see, a 3× zoom.

FIGURE 35.4 A simple camera lens is a combination lens.

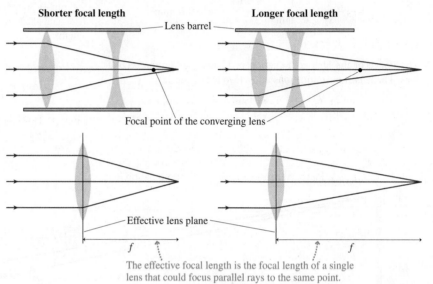

A camera must carry out two important functions: focus the image on the detector and control the exposure. Cameras are focused by moving the lens forward or backward until the image is well focused on the detector. Most modern cameras do this automatically, but older cameras required manual focusing.

EXAMPLE 35.2 | Focusing a camera

Your digital camera lens, with an effective focal length of 10.0 mm, is focused on a flower 20.0 cm away. You then turn to take a picture of a distant landscape. How far, and in which direction, must the lens move to bring the landscape into focus?

MODEL Model the camera's combination lens as a single thin lens with $f = 10.0$ mm. Image and object distances are measured from the effective lens plane. Assume all the lenses in the combination move together as the camera refocuses.

SOLVE The flower is at object distance $s = 20.0$ cm = 200 mm. When the camera is focused, the image distance between the effective lens plane and the detector is found by solving the thin-lens equation $1/s + 1/s' = 1/f$ to give

$$s' = \left(\frac{1}{f} - \frac{1}{s}\right)^{-1} = \left(\frac{1}{10.0\text{ mm}} - \frac{1}{200\text{ mm}}\right)^{-1} = 10.5\text{ mm}$$

The distant landscape is effectively at object distance $s = \infty$, so its image distance is $s' = f = 10.0$ mm. To refocus as you shift scenes, the lens must move 0.5 mm closer to the detector.

ASSESS The required motion of the lens is very small, about the diameter of the lead used in a mechanical pencil.

Zoom Lenses

For objects more than 10 focal lengths from the lens (roughly $s > 20$ cm for a typical digital camera), the approximation $s \gg f$ (and thus $1/s \ll 1/f$) leads to $s' \approx f$. In other words, objects more than about 10 focal lengths away are essentially "at infinity," and we know that the parallel rays from an infinitely distant object are focused one focal length behind the lens. For such an object, the lateral magnification of the image is

$$m = -\frac{s'}{s} \approx -\frac{f}{s} \qquad (35.1)$$

The magnification is much less than 1, because $s \gg f$, so the image on the detector is much smaller than the object itself. This comes as no surprise. More important, **the size of the image is directly proportional to the focal length of the lens.** We saw in Figure 35.4 that the effective focal length of a combination lens is easily changed by varying the distance between the individual lenses, and this is exactly how a zoom lens works. A lens that can be varied from $f_{min} = 6$ mm to $f_{max} = 18$ mm gives magnifications spanning a factor of 3, and that is why you see it specified as a 3× zoom lens.

Controlling the Exposure

The camera also must control the amount of light reaching the detector. Too little light results in photos that are *underexposed;* too much light gives *overexposed* pictures. Both the shutter and the lens diameter help control the exposure.

The *shutter* is "opened" for a selected amount of time as the image is recorded. Older cameras used a spring-loaded mechanical shutter that literally opened and closed; digital cameras electronically control the amount of time the detector is active. Either way, the exposure—the amount of light captured by the detector—is directly proportional to the time the shutter is open. Typical exposure times range from 1/1000 s or less for a sunny scene to 1/30 s or more for dimly lit or indoor scenes. The exposure time is generally referred to as the *shutter speed*.

The amount of light passing through the lens is controlled by an adjustable **aperture,** also called an *iris* because it functions much like the iris of your eye. The aperture sets the effective diameter D of the lens. The full area of the lens is used when the aperture is fully open, but a *stopped-down* aperture allows light to pass through only the central portion of the lens.

The light intensity on the detector is directly proportional to the area of the lens; a lens with twice as much area will collect and focus twice as many light rays from the object to make an image twice as bright. The lens area is proportional to the square of its diameter, so the intensity I is proportional to D^2. The light intensity—power per square meter—is also *inversely* proportional to the area of the image. That is, the light reaching the detector is more intense if the rays collected from the object are focused into a small area than if they are spread out over a large area. The lateral size of the image is proportional to the focal length of the lens, as we saw in Equation 35.1, so the *area* of the image is proportional to f^2 and thus I is proportional to $1/f^2$. Altogether, $I \propto D^2/f^2$.

By long tradition, the light-gathering ability of a lens is specified by its **f-number,** defined as

$$f\text{-number} = \frac{f}{D} \qquad (35.2)$$

An iris can change the effective diameter of a lens and thus the amount of light reaching the detector.

The *f*-number of a lens may be written either as *f*/4.0, to mean that the *f*-number is 4.0, or as F4.0. The instruction manuals with some digital cameras call this the *aperture value* rather than the *f*-number. A digital camera in fully automatic mode does not display shutter speed or *f*-number, but that information is displayed if you set your camera to any of the other modes. For example, the display 1/125 F5.6 means that your camera is going to achieve the correct exposure by adjusting the diameter of the lens aperture to give $f/D = 5.6$ and by opening the shutter for 1/125 s. If your lens's effective focal length is 10 mm, the diameter of the lens aperture will be

$$D = \frac{f}{f\text{-number}} = \frac{10 \text{ mm}}{5.6} = 1.8 \text{ mm}$$

Focal length and f-number information is stamped on a camera lens. This lens is labeled 5.8–23.2 mm 1:2.6–5.5. The first numbers are the range of focal lengths. They span a factor of 4, so this is a 4× zoom lens. The second numbers show that the minimum f-number ranges from $f/2.6$ (for the $f = 5.8$ mm focal length) to $f/5.5$ (for the $f = 23.2$ mm focal length).

NOTE The f in f-number is not the focal length f; it's just a name. And the / in $f/4$ does not mean division; it's just a notation. These both derive from the long history of photography.

Because the aperture diameter is in the denominator of the f-number, a *larger-diameter* aperture, which gathers more light and makes a brighter image, has a *smaller* f-number. The light intensity on the detector is related to the lens's f-number by

$$I \propto \frac{D^2}{f^2} = \frac{1}{(f\text{-number})^2} \tag{35.3}$$

Historically, a lens's f-numbers could be adjusted in the sequence 2.0, 2.8, 4.0, 5.6, 8.0, 11, 16. Each differs from its neighbor by a factor of $\sqrt{2}$, so changing the lens by one "f stop" changed the light intensity by a factor of 2. A modern digital camera is able to adjust the f-number continuously.

The exposure, the total light reaching the detector while the shutter is open, depends on the product $I \Delta t_{\text{shutter}}$. A small f-number (large aperture diameter D) and short $\Delta t_{\text{shutter}}$ can produce the same exposure as a larger f-number (smaller aperture) and a longer $\Delta t_{\text{shutter}}$. It might not make any difference for taking a picture of a distant mountain, but action photography needs very short shutter times to "freeze" the action. Thus action photography requires a large-diameter lens with a small f-number.

EXAMPLE 35.3 | **Capturing the action**

Before a race, a photographer finds that she can make a perfectly exposed photo of the track while using a shutter speed of 1/250 s and a lens setting of $f/8.0$. To freeze the sprinters as they go past, she plans to use a shutter speed of 1/1000 s. To what f-number must she set her lens?

MODEL The exposure depends on $I \Delta t_{\text{shutter}}$, and the light intensity depends inversely on the square of the f-number.

SOLVE Changing the shutter speed from 1/250 s to 1/1000 s will reduce the light reaching the detector by a factor of 4. To compensate, she needs to let 4 times as much light through the lens. Because $I \propto 1/(f\text{-number})^2$, the intensity will increase by a factor of 4 if she *decreases* the f-number by a factor of 2. Thus the correct lens setting is $f/4.0$.

The Detector

For traditional cameras, the light-sensitive detector is film. Today's digital cameras use an electronic light-sensitive surface called a *charge-coupled device* or **CCD.** A CCD consists of a rectangular array of many millions of small detectors called **pixels.** When light hits one of these pixels, it generates an electric charge proportional to the light intensity. Thus an image is recorded on the CCD in terms of little packets of charge. After the CCD has been exposed, the charges are read out, the signal levels are digitized, and the picture is stored in the digital memory of the camera.

FIGURE 35.5a shows a CCD "chip" and, schematically, the magnified appearance of the pixels on its surface. To record color information, different pixels are covered by red, green, or blue filters. A pixel covered by a green filter, for instance, records only the intensity of the green light hitting it. Later, the camera's microprocessor interpolates nearby colors to give each pixel an overall true color. The pixels are so small that the picture looks "smooth" even after some enlargement, but, as you can see in **FIGURE 35.5b**, sufficient magnification reveals the individual pixels.

FIGURE 35.5 The CCD detector used in a digital camera.

(a) 4600 × 3500 pixels

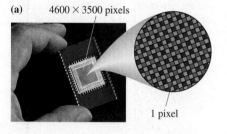

1 pixel

(b)

40×

STOP TO THINK 35.2 A photographer has adjusted his camera for a correct exposure with a short-focal-length lens. He then decides to zoom in by increasing the focal length. To maintain a correct exposure without changing the shutter speed, the diameter of the lens aperture should

a. Be increased.   b. Be decreased.   c. Stay the same.

35.3 Vision

The human eye is a marvelous and intricate organ. If we leave the biological details to biologists and focus on the eye's optical properties, we find that it functions very much like a camera. Like a camera, the eye has refracting surfaces that focus incoming light rays, an adjustable iris to control the light intensity, and a light-sensitive detector.

FIGURE 35.6 shows the basic structure of the eye. It is roughly spherical, about 2.4 cm in diameter. The transparent **cornea,** which is somewhat more sharply curved, and the *lens* are the eye's refractive elements. The eye is filled with a clear, jellylike fluid called the *aqueous humor* (in front of the lens) and the *vitreous humor* (behind the lens). The indices of refraction of the aqueous and vitreous humors are 1.34, only slightly different from water. The lens, although not uniform, has an average index of 1.44. The **pupil,** a variable-diameter aperture in the **iris,** automatically opens and closes to control the light intensity. A fully dark-adapted eye can open to $\approx$ 8 mm, and the pupil closes down to $\approx$ 1.5 mm in bright sun. This corresponds to *f*-numbers from roughly *f*/3 to *f*/16, very similar to a camera.

FIGURE 35.6 The human eye.

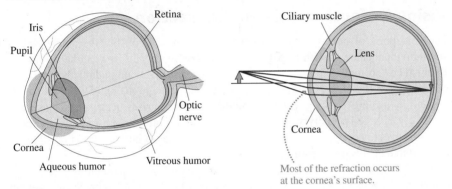

Most of the refraction occurs at the cornea's surface.

The eye's detector, the **retina,** consists of specialized light-sensitive cells called *rods* and *cones*. The rods, sensitive mostly to light and dark, are most important in very dim lighting. Color vision, which requires somewhat more light, is due to the cones, of which there are three types. **FIGURE 35.7** shows the wavelength responses of the cones. They have overlapping ranges, especially the red- and green-sensitive cones, so two or even all three cones respond to light of any particular wavelength. The relative response of the different cones is interpreted by your brain as light of a particular color. Color is a *perception,* a response of our sensory and nervous systems, not something inherent in the light itself. Other animals, with slightly different retinal cells, can see ultraviolet or infrared wavelengths that we cannot see.

FIGURE 35.7 Wavelength sensitivity of the three types of cones in the human retina.

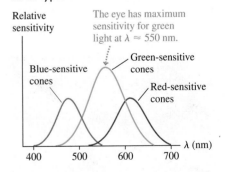

Focusing and Accommodation

The eye, like a camera, focuses light rays to an inverted image on the retina. Perhaps surprisingly, most of the refractive power of the eye is due to the cornea, not the lens. The cornea is a sharply curved, spherical surface, and you learned in Chapter 34 that images are formed by refraction at a spherical surface. The rather large difference between the index of refraction of air and that of the aqueous humor causes a significant refraction of light rays at the cornea. In contrast, there is much less difference between the indices of the lens and its surrounding fluid, so refraction at the lens surfaces is weak. The lens is important for fine-tuning, but the air-cornea boundary is responsible for the majority of the refraction.

FIGURE 35.8 Normal vision of far and near objects.

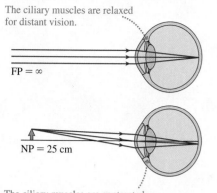

The ciliary muscles are relaxed for distant vision.

FP = ∞

NP = 25 cm

The ciliary muscles are contracted for near vision, causing the lens to curve more.

You can recognize the power of the cornea if you open your eyes underwater. Everything is very blurry! When light enters the cornea through water, rather than through air, there's almost no difference in the indices of refraction at the surface. Light rays pass through the cornea with almost no refraction, so what little focusing ability you have while underwater is due to the lens alone.

A camera focuses by moving the lens. The eye focuses by changing the focal length of the lens, a feat it accomplishes by using the *ciliary muscles* to change the curvature of the lens surface. The ciliary muscles are relaxed when you look at a distant scene. Thus the lens surface is relatively flat and the lens has its longest focal length. As you shift your gaze to a nearby object, the ciliary muscles contract and cause the lens to bulge. This process, called **accommodation,** decreases the lens's radius of curvature and thus decreases its focal length.

The farthest distance at which a relaxed eye can focus is called the eye's **far point** (FP). The far point of a normal eye is infinity; that is, the eye can focus on objects extremely far away. The closest distance at which an eye can focus, using maximum accommodation, is the eye's **near point** (NP). (Objects can be *seen* closer than the near point, but they're not sharply focused on the retina.) Both situations are shown in **FIGURE 35.8**.

Vision Defects and Their Correction

The near point of normal vision is considered to be 25 cm, but the near point of any individual changes with age. The near point of young children can be as little as 10 cm. The "normal" 25 cm near point is characteristic of young adults, but the near point of most individuals begins to move outward by age 40 or 45 and can reach 200 cm by age 60. This loss of accommodation, which arises because the lens loses flexibility, is called **presbyopia.** Even if their vision is otherwise normal, individuals with presbyopia need reading glasses to bring their near point back to 25 or 30 cm, a comfortable distance for reading.

Presbyopia is known as a *refractive error* of the eye. Two other common refractive errors are *hyperopia* and *myopia*. All three can be corrected with lenses—either eyeglasses or contact lenses—that assist the eye's focusing. Corrective lenses are prescribed not by their focal length but by their **power.** The power of a lens is the inverse of its focal length:

$$\text{Power of a lens} = P = \frac{1}{f} \tag{35.4}$$

The optometrist's prescription is −2.25 D for the right eye (top) and −2.50 D for the left (bottom), the minus sign indicating that these are diverging lenses. The optometrist doesn't write the D because the lens maker already knows that prescriptions are in diopters. Most people's eyes are not exactly the same, so each eye usually gets a different lens.

A lens with more power (shorter focal length) causes light rays to refract through a larger angle. The SI unit of lens power is the **diopter,** abbreviated D, defined as 1 D = 1 m⁻¹. Thus a lens with $f = 50$ cm = 0.50 m has power $P = 2.0$ D.

A person who is *farsighted* can see faraway objects (but even then must use some accommodation rather than a relaxed eye), but his near point is larger than 25 cm, often much larger, so he cannot focus on nearby objects. The cause of farsightedness—called **hyperopia**—is an eyeball that is too short for the refractive power of the cornea and lens. As **FIGURES 35.9a** and b show, no amount of accommodation allows the eye to focus on an object 25 cm away, the normal near point.

With hyperopia, the eye needs assistance to focus the rays from a near object onto the closer-than-normal retina. This assistance is obtained by adding refractive power with the positive (i.e., converging) lens shown in **FIGURE 35.9c**. To understand why this works, recall that the image of a first lens acts as the object for a second lens. The goal is to allow the person to focus on an object 25 cm away. If a corrective lens forms an upright, virtual image at the person's actual near point, that virtual image acts as an object for the eye itself and, with maximum accommodation, the eye can focus these rays onto the retina. Presbyopia, the loss of accommodation with age, is corrected in the same way.

FIGURE 35.9 Hyperopia.

FIGURE 35.10 Myopia.

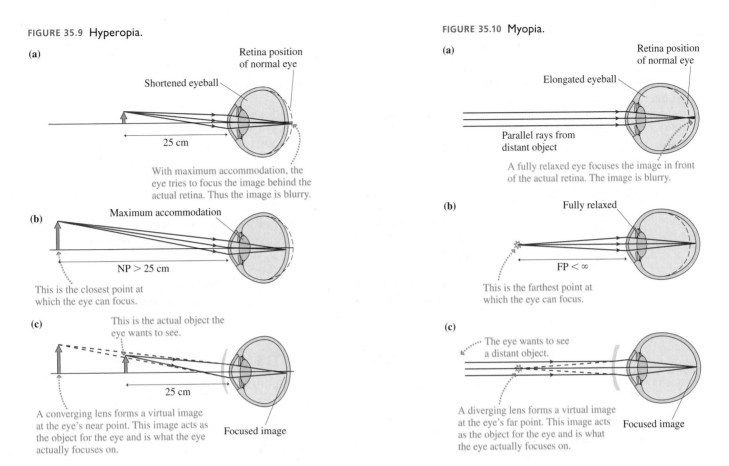

NOTE Figures 35.9 and 35.10 show the corrective lenses as they are actually shaped—called *meniscus lenses*—rather than with our usual lens shape. Nonetheless, the lens in Figure 35.9c is a converging lens because it's thicker in the center than at the edges. The lens in Figure 35.10c is a diverging lens because it's thicker at the edges than in the center.

A person who is *nearsighted* can clearly see nearby objects when the eye is relaxed (and extremely close objects by using accommodation), but no amount of relaxation allows her to see distant objects. Nearsightedness—called **myopia**—is caused by an eyeball that is too long. As **FIGURE 35.10a** shows, rays from a distant object come to a focus in front of the retina and have begun to diverge by the time they reach the retina. The eye's far point, shown in **FIGURE 35.10b**, is less than infinity.

To correct myopia, we needed a diverging lens, as shown in **FIGURE 35.10c**, to slightly defocus the rays and move the image point back to the retina. To focus on a very distant object, the person needs a corrective lens that forms an upright, virtual image at her actual far point. That virtual image acts as an object for the eye itself and, when fully relaxed, the eye can focus these rays onto the retina.

EXAMPLE 35.4 | Correcting hyperopia

Sanjay has hyperopia. The near point of his left eye is 150 cm. What prescription lens will restore normal vision?

MODEL Normal vision will allow Sanjay to focus on an object 25 cm away. In measuring distances, we'll ignore the small space between the lens and his eye.

SOLVE Because Sanjay can see objects at 150 cm, using maximum accommodation, we want a lens that creates a virtual image at

position $s' = -150$ cm (negative because it's a virtual image) of an object held at $s = 25$ cm. From the thin-lens equation,

$$\frac{1}{f} = \frac{1}{s} + \frac{1}{s'} = \frac{1}{0.25 \text{ m}} + \frac{1}{-1.50 \text{ m}} = 3.3 \text{ m}^{-1}$$

$1/f$ is the lens power, and m^{-1} are diopters. Thus the prescription is for a lens with power $P = 3.3$ D.

ASSESS Hyperopia is always corrected with a converging lens.

EXAMPLE 35.5 | Correcting myopia

Martina has myopia. The far point of her left eye is 200 cm. What prescription lens will restore normal vision?

MODEL Normal vision will allow Martina to focus on a very distant object. In measuring distances, we'll ignore the small space between the lens and her eye.

SOLVE Because Martina can see objects at 200 cm with a fully relaxed eye, we want a lens that will create a virtual image at position

$s' = -200$ cm (negative because it's a virtual image) of a distant object at $s = \infty$ cm. From the thin-lens equation,

$$\frac{1}{f} = \frac{1}{s} + \frac{1}{s'} = \frac{1}{\infty \, \text{m}} + \frac{1}{-2.0 \, \text{m}} = -0.5 \, \text{m}^{-1}$$

Thus the prescription is for a lens with power $P = -0.5$ D.

ASSESS Myopia is always corrected with a diverging lens.

STOP TO THINK 35.3 You need to improvise a magnifying glass to read some very tiny print. Should you borrow the eyeglasses from your hyperopic friend or from your myopic friend?

a. The hyperopic friend
b. The myopic friend
c. Either will do.
d. Neither will work.

35.4 Optical Systems That Magnify

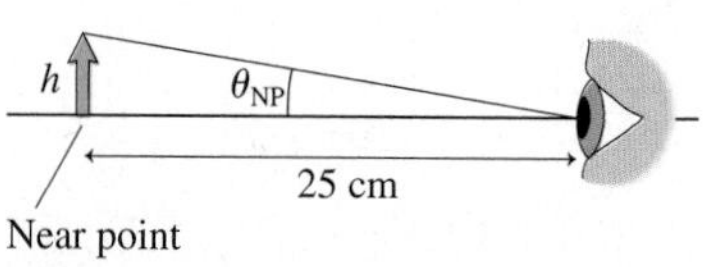

The camera, with its fast shutter speed, allows us to capture images of events that take place too quickly for our unaided eye to resolve. Another use of optical systems is to magnify—to see objects smaller or closer together than our eye can see.

The easiest way to magnify an object requires no extra optics at all; simply get closer! The closer you get, the bigger the object appears. Obviously the actual size of the object is unchanged as you approach it, so what exactly is getting "bigger"? Consider the green arrow in FIGURE 35.11a. We can determine the size of its image on the retina by tracing the rays that are undeviated as they pass through the center of the lens. (Here we're modeling the eye's optical system as one thin lens.) If we get closer to the arrow, now shown as red, we find the arrow makes a larger image on the retina. Our brain interprets the larger image as a larger-appearing object. The object's actual size doesn't change, but its *apparent size* gets larger as it gets closer.

Technically, we say that closer objects look larger because they subtend a larger angle θ, called the **angular size** of the object. The red arrow has a larger angular size than the green arrow, $\theta_2 > \theta_1$, so the red arrow looks larger and we can see more detail. But you can't keep increasing an object's angular size because you can't focus on the object if it's closer than your near point, which we'll take to be a normal 25 cm. FIGURE 35.11b defines the angular size θ_{NP} of an object at your near point. If the object's height is h and if we assume the small-angle approximation $\tan\theta \approx \theta$, the maximum angular size viewable by your unaided eye is

$$\theta_{NP} = \frac{h}{25 \, \text{cm}} \tag{35.5}$$

Suppose we view the same object, of height h, through the single converging lens in FIGURE 35.12. If the object's distance from the lens is less than the lens's focal length, we'll see an enlarged, upright image. Used in this way, the lens is called a **magnifier** or *magnifying glass*. The eye sees the virtual image subtending angle θ, and it can focus on this virtual image as long as the image distance is more than 25 cm. Within the small-angle approximation, the image subtends angle $\theta = h/s$. In practice, we usually want the image to be at distance $s' \approx \infty$ so that we can view it with a relaxed eye as a "distant object." This will be true if the object is very near the focal point: $s \approx f$. In this case, the image subtends angle

$$\theta = \frac{h}{s} \approx \frac{h}{f} \tag{35.6}$$

FIGURE 35.11 Angular size.

(a) Same object at two different distances

As the object gets closer, the angle it subtends becomes larger. Its *angular size* has increased.

θ_1
θ_2

Further, the size of the image on the retina gets larger. The object's *apparent size* has increased.

(b)

h θ_{NP}

25 cm

Near point

FIGURE 35.12 The magnifier.

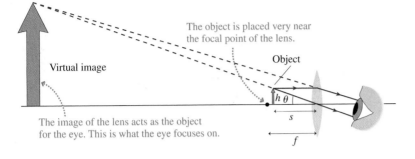

The object is placed very near the focal point of the lens.

Object

Virtual image

The image of the lens acts as the object for the eye. This is what the eye focuses on.

h θ

s

f

Let's define the **angular magnification** M as

$$M = \frac{\theta}{\theta_{NP}} \qquad (35.7)$$

Angular magnification is the increase in the *apparent size* of the object that you achieve by using a magnifying lens rather than simply holding the object at your near point. Substituting from Equations 35.5 and 35.6, we find the angular magnification of a magnifying glass is

$$M = \frac{25 \text{ cm}}{f} \qquad (35.8)$$

The angular magnification depends on the focal length of the lens but not on the size of the object. Although it would appear we could increase angular magnification without limit by using lenses with shorter and shorter focal lengths, the inherent limitations of lenses we discuss later in the chapter limit the magnification of a simple lens to about 4×. Slightly more complex magnifiers with two lenses reach 20×, but beyond that one would use a microscope.

NOTE Don't confuse angular magnification with lateral magnification. Lateral magnification m compares the height of an object to the height of its image. The lateral magnification of a magnifying glass is $\approx \infty$ because the virtual image is at $s' \approx \infty$, but that doesn't make the object seem infinitely big. Its apparent size is determined by the angle subtended on your retina, and that angle remains finite. Thus angular magnification tells us how much bigger things appear.

The Microscope

A microscope, whose major parts are shown in FIGURE 35.13, can attain a magnification of up to 1000× by a *two-step* magnification process. A specimen to be observed is placed on the *stage* of the microscope, directly beneath the **objective,** a converging lens with a relatively short focal length. The objective creates a magnified real image that is further enlarged by the **eyepiece.** Both the objective and the eyepiece are complex combination lenses, but we'll model them as single thin lenses. It's common for a prism to bend the rays so that the eyepiece is at a comfortable viewing angle. However, we'll consider a simplified version of a microscope in which the light travels along a straight tube.

FIGURE 35.14 on the next page is a simple two-lens model of a microscope. The object is placed just outside the focal point of the objective, which creates a highly magnified real image with lateral magnification $m = -s'/s$. The object is so close to the focal point that $s \approx f_{obj}$ is an excellent approximation. In addition, the microscope is designed so that the image distance s' is approximately the tube length L, so $s' \approx L$. With these approximations, the lateral magnification of the objective is

$$m_{obj} = -\frac{s'}{s} \approx -\frac{L}{f_{obj}} \qquad (35.9)$$

FIGURE 35.13 A microscope.

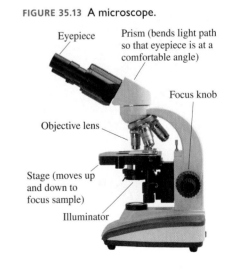

Eyepiece

Prism (bends light path so that eyepiece is at a comfortable angle)

Focus knob

Objective lens

Stage (moves up and down to focus sample)

Illuminator

FIGURE 35.14 The optics of a microscope.

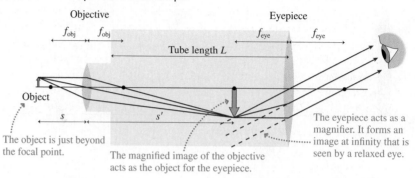

The image of the objective acts as the object for the eyepiece, which functions as a simple magnifier. The angular magnification of the eyepiece is given by Equation 35.8, $M_{eye} = (25\ cm)/f_{eye}$. Together, the objective and eyepiece produce a total angular magnification

$$M = m_{obj}M_{eye} \approx -\frac{L}{f_{obj}}\frac{25\ cm}{f_{eye}} \tag{35.10}$$

The minus sign shows that the image seen in a microscope is inverted.

In practice, the magnifications of the objective (without the minus sign) and the eyepiece are stamped on the barrels. A set of objectives on a rotating turret might include 10×, 20×, 40×, and 100×. When combined with a 10× eyepiece, the microscope's total angular magnification ranges from 100× to 1000×. In addition, most biological microscopes are standardized with a tube length $L = 160$ mm. Thus a 40× objective has focal length $f_{obj} = 160$ mm/40 = 4.0 mm.

EXAMPLE 35.6 | **Viewing blood cells**

A pathologist inspects a sample of 7-μm-diameter human blood cells under a microscope. She selects a 40× objective and a 10× eyepiece. What size object, viewed from 25 cm, has the same apparent size as a blood cell seen through the microscope?

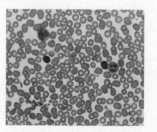

MODEL Angular magnification compares the magnified angular size to the angular size seen at the near-point distance of 25 cm.

SOLVE The microscope's angular magnification is $M = -(40) \times (10) = -400$. The magnified cells will have the same apparent size as an object $400 \times 7\ \mu m \approx 3$ mm in diameter seen from a distance of 25 cm.

ASSESS 3 mm is about the size of a capital O in this textbook, so a blood cell seen through the microscope will have about the same apparent size as an O seen from a comfortable reading distance.

STOP TO THINK 35.4   A biologist rotates the turret of a microscope to replace a 20× objective with a 10× objective. To keep the same overall magnification, the focal length of the eyepiece must be

a. Doubled.   b. Halved.   c. Kept the same.
d. The magnification cannot be kept the same if the objective is changed.

The Telescope

A microscope magnifies small, nearby objects to look large. A telescope magnifies distant objects, which might be quite large, so that we can see details that are blended together when seen by eye.

FIGURE 35.15 shows the optical layout of a simple telescope. A large-diameter objective lens (larger lenses collect more light and thus can see fainter objects) collects the parallel rays from a distant object ($s = \infty$) and forms a real, inverted image at distance $s' = f_{obj}$. Unlike a microscope, which uses a short-focal-length objective, the focal length of a telescope objective is very nearly the length of the telescope tube. Then, just as in the microscope, the eyepiece functions as a simple magnifier. The viewer observes an inverted image, but that's not a serious problem in astronomy. Telescopes for use on earth have a somewhat different design to obtain an upright image.

FIGURE 35.15 A refracting telescope.

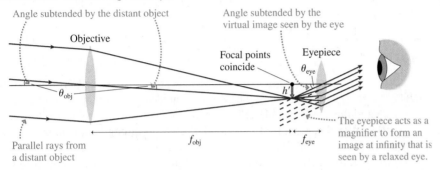

Suppose the distant object, as seen by the objective lens, subtends angle θ_{obj}. If the image seen through the eyepiece subtends a larger angle θ_{eye}, then the angular magnification is $M = \theta_{eye}/\theta_{obj}$. We can see from the undeviated ray passing through the center of the objective lens that (using the small-angle approximation)

$$\theta_{obj} \approx -\frac{h'}{f_{obj}}$$

where the minus sign indicates the inverted image. The image of height h' acts as the object for the eyepiece, and we can see that the final image observed by the viewer subtends angle

$$\theta_{eye} = \frac{h'}{f_{eye}}$$

Consequently, the angular magnification of a telescope is

$$M = \frac{\theta_{eye}}{\theta_{obj}} = -\frac{f_{obj}}{f_{eye}} \qquad (35.11)$$

The angular magnification is simply the ratio of the objective focal length to the eyepiece focal length.

Because the stars and galaxies are so distant, light-gathering power is more important to astronomers than magnification. Large light-gathering power requires a large-diameter objective lens, but large lenses are not practical; they begin to sag under their own weight. Thus **refracting telescopes,** with two lenses, are relatively small. Serious astronomy is done with a **reflecting telescope,** such as the one shown in FIGURE 35.16.

A large-diameter mirror (the *primary mirror*) focuses the rays to form a real image, but, for practical reasons, a *secondary mirror* reflects the rays sideways before they reach a focus. This moves the primary mirror's image out to the edge of the telescope where it can be viewed by an eyepiece on the side. None of these changes affects the overall analysis of the telescope, and its angular magnification is given by Equation 35.11 if f_{obj} is replaced by f_{pri}, the focal length of the primary mirror.

FIGURE 35.16 A reflecting telescope.

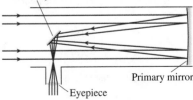

35.5 Color and Dispersion

One of the most obvious visual aspects of light is the phenomenon of color. Yet color, for all its vivid sensation, is not inherent in the light itself. Color is a *perception,* not a physical quantity. Color is associated with the wavelength of light, but the fact that we see light with a wavelength of 650 nm as "red" tells us how our visual system responds to electromagnetic waves of this wavelength. There is no "redness" associated with the light wave itself.

Most of the results of optics do not depend on color; a microscope works the same with red light and blue light. Even so, indices of refraction are slightly wavelength dependent, which will be important in the next section where we consider the resolution of optical instruments. And color in nature is an interesting subject, one worthy of a short digression.

Color

It has been known since antiquity that irregularly shaped glass and crystals cause sunlight to be broken into various colors. A common idea was that the glass or crystal somehow altered the properties of the light by *adding* color to the light. Newton suggested a different explanation. He first passed a sunbeam through a prism, producing the familiar rainbow of light. We say that the prism *disperses* the light. Newton's novel idea, shown in **FIGURE 35.17a**, was to use a second prism, inverted with respect to the first, to "reassemble" the colors. He found that the light emerging from the second prism was a beam of pure, white light.

But the emerging light beam is white only if *all* the rays are allowed to move between the two prisms. Blocking some of the rays with small obstacles, as in **FIGURE 35.17b,** causes the emerging light beam to have color. This suggests that color is associated with the light itself, not with anything that the prism is doing to the light. Newton tested this idea by inserting a small aperture between the prisms to pass only the rays of a particular color, such as green. If the prism alters the properties of light, then the second prism should change the green light to other colors. Instead, the light emerging from the second prism is unchanged from the green light entering the prism.

These and similar experiments show that

1. What we perceive as white light is a mixture of all colors. White light can be dispersed into its various colors and, equally important, mixing all the colors produces white light.
2. The index of refraction of a transparent material differs slightly for different colors of light. Glass has a slightly larger index of refraction for violet light than for green light or red light. Consequently, different colors of light refract at slightly different angles. A prism does not alter the light or add anything to the light; it simply causes the different colors that are inherent in white light to follow slightly different trajectories.

Dispersion

It was Thomas Young, with his two-slit interference experiment, who showed that different colors are associated with light of different wavelengths. The longest wavelengths are perceived as red light and the shortest as violet light. **TABLE 35.1** is a brief summary of the *visible spectrum* of light. Visible-light wavelengths are used so frequently that it is well worth committing this short table to memory.

The slight variation of index of refraction with wavelength is known as **dispersion.** **FIGURE 35.18** shows the *dispersion curves* of two common glasses. Notice that **n is *larger* when the wavelength is *shorter,*** thus violet light refracts more than red light.

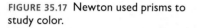

FIGURE 35.17 Newton used prisms to study color.

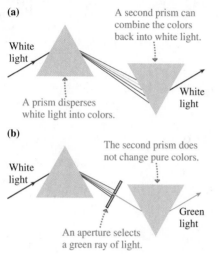

(a)

A second prism can combine the colors back into white light.

White light

A prism disperses white light into colors.

White light

(b)

The second prism does not change pure colors.

White light

An aperture selects a green ray of light.

Green light

TABLE 35.1 A brief summary of the visible spectrum of light

Color	Approximate wavelength
Deepest red	700 nm
Red	650 nm
Green	550 nm
Blue	450 nm
Deepest violet	400 nm

FIGURE 35.18 Dispersion curves show how the index of refraction varies with wavelength.

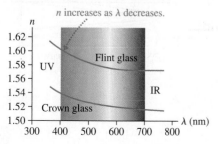

n increases as λ decreases.

n

Flint glass

UV

IR

Crown glass

λ (nm)

EXAMPLE 35.7 | **Dispersing light with a prism**

Example 34.4 found that a ray incident on a 30° prism is deflected by 22.6° if the prism's index of refraction is 1.59. Suppose this is the index of refraction of deep violet light and deep red light has an index of refraction of 1.54.

a. What is the deflection angle for deep red light?

b. If a beam of white light is dispersed by this prism, how wide is the rainbow spectrum on a screen 2.0 m away?

VISUALIZE Figure 34.19 showed the geometry. A ray of any wavelength is incident on the hypotenuse of the prism at $\theta_1 = 30°$.

SOLVE a. If $n_1 = 1.54$ for deep red light, the refraction angle is

$$\theta_2 = \sin^{-1}\left(\frac{n_1 \sin \theta_1}{n_2}\right) = \sin^{-1}\left(\frac{1.54 \sin 30°}{1.00}\right) = 50.4°$$

Example 34.4 showed that the deflection angle is $\phi = \theta_2 - \theta_1$, so deep red light is deflected by $\phi_{red} = 20.4°$. This angle is slightly smaller than the previously observed $\phi_{violet} = 22.6°$.

b. The entire spectrum is spread between $\phi_{red} = 20.4°$ and $\phi_{violet} = 22.6°$. The angular spread is

$$\delta = \phi_{violet} - \phi_{red} = 2.2° = 0.038 \text{ rad}$$

At distance r, the spectrum spans an arc length

$$s = r\delta = (2.0 \text{ m})(0.038 \text{ rad}) = 0.076 \text{ m} = 7.6 \text{ cm}$$

ASSESS The angle is so small that there's no appreciable difference between arc length and a straight line. The spectrum will be 7.6 cm wide at a distance of 2.0 m.

Rainbows

One of the most interesting sources of color in nature is the rainbow. The details get somewhat complicated, but **FIGURE 35.19a** shows that the basic cause of the rainbow is a combination of refraction, reflection, and dispersion.

Figure 35.19a might lead you to think that the top edge of a rainbow is violet. In fact, the top edge is red, and violet is on the bottom. The rays leaving the drop in Figure 35.19a are spreading apart, so they can't all reach your eye. As **FIGURE 35.19b** shows, a ray of red light reaching your eye comes from a drop *higher* in the sky than a ray of violet light. In other words, the colors you see in a rainbow refract toward your eye from different raindrops, not from the same drop. You have to look higher in the sky to see the red light than to see the violet light.

FIGURE 35.19 Light seen in a rainbow has undergone refraction + reflection + refraction in a raindrop.

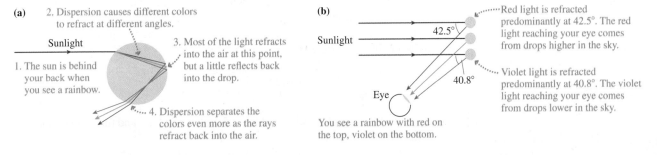

(a)
2. Dispersion causes different colors to refract at different angles.

Sunlight

1. The sun is behind your back when you see a rainbow.

3. Most of the light refracts into the air at this point, but a little reflects back into the drop.

4. Dispersion separates the colors even more as the rays refract back into the air.

(b)
Sunlight

42.5°

40.8°

Eye

You see a rainbow with red on the top, violet on the bottom.

Red light is refracted predominantly at 42.5°. The red light reaching your eye comes from drops higher in the sky.

Violet light is refracted predominantly at 40.8°. The violet light reaching your eye comes from drops lower in the sky.

Colored Filters and Colored Objects

White light passing through a piece of green glass emerges as green light. A possible explanation would be that the green glass *adds* "greenness" to the white light, but Newton found otherwise. Green glass is green because it *absorbs* any light that is "not green." We can think of a piece of colored glass or plastic as a *filter* that removes all wavelengths except a chosen few.

If a green filter and a red filter are overlapped, as in **FIGURE 35.20**, *no* light gets through. The green filter transmits only green light, which is then absorbed by the red filter because it is "not red."

This behavior is true not just for glass filters, which transmit light, but for *pigments* that absorb light of some wavelengths but *reflect* light at other wavelengths. For

FIGURE 35.20 No light at all passes through both a green and a red filter.

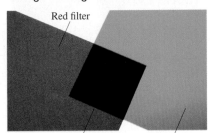

Red filter

Black where filters overlap   Green filter

FIGURE 35.21 The absorption curve of chlorophyll.

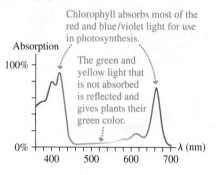

Chlorophyll absorbs most of the red and blue/violet light for use in photosynthesis.

The green and yellow light that is not absorbed is reflected and gives plants their green color.

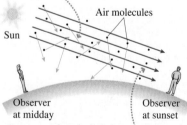

Sunsets are red because all the blue light has scattered as the sunlight passes through the atmosphere.

FIGURE 35.22 Rayleigh scattering by molecules in the air gives the sky and sunsets their color.

At midday the scattered light is mostly blue because molecules preferentially scatter shorter wavelengths.

At sunset, when the light has traveled much farther through the atmosphere, the light is mostly red because the shorter wavelengths have been lost to scattering.

example, red paint contains pigments reflecting light at wavelengths near 650 nm while absorbing all other wavelengths. Pigments in paints, inks, and natural objects are responsible for most of the color we observe in the world, from the red of lipstick to the blue of a blueberry.

As an example, FIGURE 35.21 shows the absorption curve of *chlorophyll*. Chlorophyll is essential for photosynthesis in green plants. The chemical reactions of photosynthesis are able to use red light and blue/violet light, thus chlorophyll absorbs red light and blue/violet light from sunlight and puts it to use. But green and yellow light are not absorbed. Instead, to conserve energy, these wavelengths are mostly *reflected* to give the object a greenish-yellow color. When you look at the green leaves on a tree, you're seeing the light that was reflected because it *wasn't* needed for photosynthesis.

Light Scattering: Blue Skies and Red Sunsets

In the ray model of Section 34.1 we noted that light within a medium can be scattered or absorbed. As we've now seen, the absorption of light can be wavelength dependent and can create color in objects. What are the effects of scattering?

Light can scatter from small particles that are suspended in a medium. If the particles are large compared to the wavelengths of light—even though they may be microscopic and not readily visible to the naked eye—the light essentially reflects off the particles. The law of reflection doesn't depend on wavelength, so all colors are scattered equally. White light scattered from many small particles makes the medium appear cloudy and white. Two well-known examples are clouds, where micrometer-size water droplets scatter the light, and milk, which is a colloidal suspension of microscopic droplets of fats and proteins.

A more interesting aspect of scattering occurs at the atomic level. The atoms and molecules of a transparent medium are much smaller than the wavelengths of light, so they can't scatter light simply by reflection. Instead, the oscillating electric field of the light wave interacts with the electrons in each atom in such a way that the light is scattered. This atomic-level scattering is called **Rayleigh scattering.**

Unlike the scattering by small particles, Rayleigh scattering from atoms and molecules *does* depend on the wavelength. A detailed analysis shows that the intensity of scattered light depends inversely on the fourth power of the wavelength: $I_{\text{scattered}} \propto \lambda^{-4}$. This wavelength dependence explains why the sky is blue and sunsets are red.

As sunlight travels through the atmosphere, the λ^{-4} dependence of Rayleigh scattering causes the shorter wavelengths to be preferentially scattered. If we take 650 nm as a typical wavelength for red light and 450 nm for blue light, the intensity of scattered blue light relative to scattered red light is

$$\frac{I_{\text{blue}}}{I_{\text{red}}} = \left(\frac{650}{450}\right)^4 \approx 4$$

Four times more blue light is scattered toward us than red light and thus, as FIGURE 35.22 shows, the sky appears blue.

Because of the earth's curvature, sunlight has to travel much farther through the atmosphere when we see it at sunrise or sunset than it does during the midday hours. In fact, the path length through the atmosphere at sunset is so long that essentially all the short wavelengths have been lost due to Rayleigh scattering. Only the longer wavelengths remain—orange and red—and they make the colors of the sunset.

35.6 The Resolution of Optical Instruments

A camera *could* focus light with a single lens. A microscope objective *could* be built with a single lens. So why would anyone ever use a lens combination in place of a single lens? There are two primary reasons.

First, as you learned in the previous section, any lens has dispersion. That is, its index of refraction varies slightly with wavelength. Because the index of refraction for violet light is larger than for red light, a lens's focal length is shorter for violet light than for red light. Consequently, different colors of light come to a focus at slightly different distances from the lens. If red light is sharply focused on a viewing screen, then blue and violet wavelengths are not well focused. This imaging error, illustrated in FIGURE 35.23a, is called **chromatic aberration.**

Second, our analysis of thin lenses was based on paraxial rays traveling nearly parallel to the optical axis. A more exact analysis, taking all the rays into account, finds that rays incident on the outer edges of a spherical surface are not focused at exactly the same point as rays incident near the center. This imaging error, shown in FIGURE 35.23b, is called **spherical aberration.** Spherical aberration, which causes the image to be slightly blurred, gets worse as the lens diameter increases.

Fortunately, the chromatic and spherical aberrations of a converging lens and a diverging lens are in opposite directions. When a converging lens and a diverging lens are used in combination, their aberrations tend to cancel. A combination lens, such as the one in FIGURE 35.23c, can produce a much sharper focus than a single lens with the equivalent focal length. Consequently, most optical instruments use combination lenses rather than single lenses.

Diffraction Again

According to the ray model of light, a perfect lens (one with no aberrations) should be able to form a perfect image. But the ray model of light, though a very good model for lenses, is not an absolutely correct description of light. If we look closely, the wave aspects of light haven't entirely disappeared. In fact, the performance of optical equipment is limited by the diffraction of light.

FIGURE 35.24a shows a plane wave, with parallel light rays, being focused by a lens of diameter D. According to the ray model of light, a perfect lens would focus parallel rays to a perfect point. Notice, though, that only a piece of each wave front passes *through* the lens and gets focused. In effect, **the lens itself acts as a circular aperture** in an opaque barrier, allowing through only a portion of each wave front. Consequently, **the lens diffracts the light wave.** The diffraction is usually very small because D is usually much greater than the wavelength of the light; nonetheless, this small amount of diffraction is the limiting factor in how well the lens can focus the light.

FIGURE 35.24b separates the diffraction from the focusing by modeling the lens as an actual aperture of diameter D followed by an "ideal" diffractionless lens. You learned in Chapter 33 that a circular aperture produces a diffraction pattern with a bright central maximum surrounded by dimmer fringes. A converging lens brings this diffraction pattern to a focus in the image plane, as shown in FIGURE 35.24c. As a result, a perfect lens focuses parallel light rays not to a perfect point of light, as we expected, but to a small, circular diffraction pattern.

FIGURE 35.23 Chromatic aberration and spherical aberration prevent simple lenses from forming perfect images.

(a) Chromatic aberration

(b) Spherical aberration

(c) Correcting aberrations

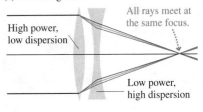

FIGURE 35.24 A lens both focuses and diffracts the light passing through.

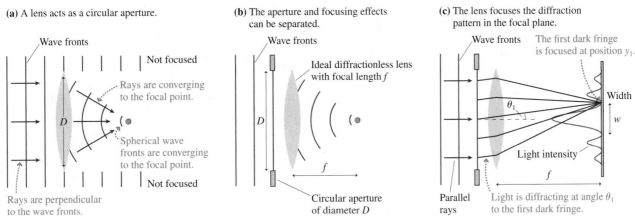

(a) A lens acts as a circular aperture.

(b) The aperture and focusing effects can be separated.

(c) The lens focuses the diffraction pattern in the focal plane.

The angle to the first minimum of a circular diffraction pattern is $\theta_1 = 1.22\lambda/D$. The ray that passes through the center of a lens is not bent, so Figure 35.24c uses this ray to show that the position of the dark fringe is $y_1 = f\tan\theta_1 \approx f\theta_1$. Thus the width of the central maximum in the focal plane is

$$w_{min} \approx 2f\theta_1 = \frac{2.44\lambda f}{D} \qquad \text{(minimum spot size)} \qquad (35.12)$$

This is the **minimum spot size** to which a lens can focus light.

Lenses are often limited by aberrations, so not all lenses can focus parallel light rays to a spot this small. A well-crafted lens, for which Equation 35.12 is the minimum spot size, is called a *diffraction-limited lens*. No optical design can overcome the spreading of light due to diffraction, and it is because of this spreading that the image point has a minimum spot size. The image of an actual object, rather than of parallel rays, becomes a mosaic of overlapping diffraction patterns, so even the most perfect lens inevitably forms an image that is slightly fuzzy.

For various reasons, it is difficult to produce a diffraction-limited lens having a focal length that is much less than its diameter. The very best microscope objectives have $f \approx 0.5D$. This implies that **the smallest diameter to which you can focus a spot of light, no matter how hard you try, is $w_{min} \approx \lambda$.** This is a fundamental limit on the performance of optical equipment. Diffraction has very real consequences!

One example of these consequences is found in the manufacturing of integrated circuits. Integrated circuits are made by creating a "mask" showing all the components and their connections. A lens images this mask onto the surface of a semiconductor wafer that has been coated with a substance called *photoresist*. Bright areas in the mask expose the photoresist, and subsequent processing steps chemically etch away the exposed areas while leaving behind areas that had been in the shadows of the mask. This process is called *photolithography*.

The power of a microprocessor and the amount of memory in a memory chip depend on how small the circuit elements can be made. Diffraction dictates that a circuit element can be no smaller than the smallest spot to which light can be focused, which is roughly the wavelength of the light. If the mask is projected with ultraviolet light having $\lambda \approx 200$ nm, then the smallest elements on a chip are about 200 nm wide. This is, in fact, just about the current limit of technology.

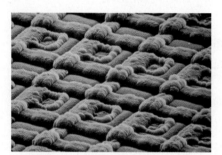

The size of the features in an integrated circuit is limited by the diffraction of light.

EXAMPLE 35.8 | Seeing stars

A 12-cm-diameter telescope lens has a focal length of 1.0 m. What is the diameter of the image of a star in the focal plane if the lens is diffraction limited *and* if the earth's atmosphere is not a limitation?

MODEL Stars are so far away that they appear as points in space. An ideal diffractionless lens would focus their light to arbitrarily small points. Diffraction prevents this. Model the telescope lens as a 12-cm-diameter aperture in front of an ideal lens with a 1.0 m focal length.

SOLVE The minimum spot size in the focal plane of this lens is

$$w = \frac{2.44\lambda f}{D}$$

where D is the lens diameter. What is λ? Because stars emit white light, the *longest* wavelengths spread the most and determine the size of the image that is seen. If we use $\lambda = 700$ nm as the approximate upper limit of visible wavelengths, we find $w = 1.4 \times 10^{-5}$ m $= 14$ μm.

ASSESS This is certainly small, and it would appear as a point to your unaided eye. Nonetheless, the spot size would be easily noticed if it were recorded on film and enlarged. Turbulence and temperature effects in the atmosphere, the causes of the "twinkling" of stars, prevent ground-based telescopes from being this good, but space-based telescopes really are diffraction limited.

Resolution

Suppose you point a telescope at two nearby stars in a galaxy far, far away. If you use the best possible detector, will you be able to distinguish separate images for the two stars, or will they blur into a single blob of light? A similar question could be asked of a microscope. Can two microscopic objects, very close together, be distinguished if sufficient magnification is used? Or is there some size limit at which their images will blur together and never be separated? These are important questions about the *resolution* of optical instruments.

Because of diffraction, the image of a distant star is not a point but a circular diffraction pattern. Our question, then, really is: How close together can two diffraction patterns be before you can no longer distinguish them? One of the major scientists of the 19th century, Lord Rayleigh, studied this problem and suggested a reasonable rule that today is called **Rayleigh's criterion.**

FIGURE 35.25 shows two distant point sources being imaged by a lens of diameter D. The angular separation between the objects, as seen from the lens, is α. Rayleigh's criterion states that

- The two objects are resolvable if $\alpha > \theta_{min}$, where $\theta_{min} = \theta_1 = 1.22\lambda/D$ is the angle of the first dark fringe in the circular diffraction pattern.
- The two objects are not resolvable if $\alpha < \theta_{min}$ because their diffraction patterns are too overlapped.
- The two objects are marginally resolvable if $\alpha = \theta_{min}$. The central maximum of one image falls exactly on top of the first dark fringe of the other image. This is the situation shown in the figure.

FIGURE 35.26 shows enlarged photographs of the images of two point sources. The images are circular diffraction patterns, not points. The two images are close but distinct where the objects are separated by $\alpha > \theta_{min}$. Two objects really were recorded in the photo at the bottom, but their separation is $\alpha < \theta_{min}$ and their images have blended together. In the middle photo, with $\alpha = \theta_{min}$, you can see that the two images are just barely resolved.

The angle

$$\theta_{min} = \frac{1.22\lambda}{D} \qquad \text{(angular resolution of a lens)} \qquad (35.13)$$

is called the **angular resolution** of a lens. The angular resolution of a telescope depends on the diameter of the objective lens (or the primary mirror) and the wavelength of the light; magnification is not a factor. Two images will remain overlapped and unresolved no matter what the magnification if their angular separation is less than θ_{min}. For visible light, where λ is pretty much fixed, the only parameter over which the astronomer has any control is the diameter of the lens or mirror of the telescope. The urge to build ever-larger telescopes is motivated, in part, by a desire to improve the angular resolution. (Another motivation is to increase the light-gathering power so as to see objects farther away.)

The performance of a microscope is also limited by the diffraction of light passing through the objective lens. Just as light cannot be focused to a spot smaller than about a wavelength, the most perfect microscope cannot resolve the features of objects separated by less than one wavelength, or roughly 500 nm. (Some clever tricks with the phase of the light can improve the resolution to about 200 nm, but diffraction is still the limiting factor.) Because atoms are approximately 0.1 nm in diameter, vastly smaller than the wavelength of visible or even ultraviolet light, there is no hope of ever seeing atoms with an optical microscope. This limitation is not simply a matter of needing a better design or more precise components; it is a fundamental limit set by the wave nature of the light with which we see.

FIGURE 35.25 Two images that are marginally resolved.

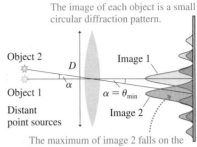

The image of each object is a small circular diffraction pattern.

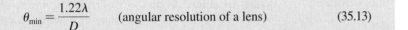

The maximum of image 2 falls on the first dark fringe of image 1. The images are marginally resolved.

FIGURE 35.26 Enlarged photographs of the images of two point sources.

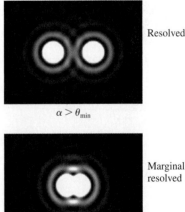

Resolved

$\alpha > \theta_{min}$

Marginally resolved

$\alpha = \theta_{min}$

Not resolved

$\alpha < \theta_{min}$

STOP TO THINK 35.5 Four diffraction-limited lenses focus plane waves of light with the same wavelength λ. Rank in order, from largest to smallest, the spot sizes w_a to w_d.

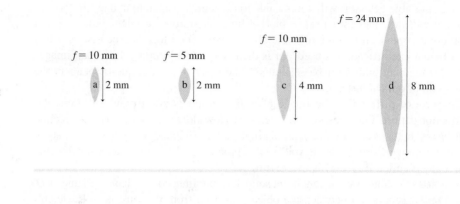

CHALLENGE EXAMPLE 35.9 | Visual acuity

The normal human eye has maximum visual acuity with a pupil diameter of about 3 mm. For larger pupils, acuity decreases due to increasing aberrations; for smaller pupils, acuity decreases due to increasing diffraction. If your pupil diameter is 2.0 mm, as it would be in bright light, what is the smallest-diameter circle that you should be able to see as a circle, rather than just an unresolved blob, on an eye chart at the standard distance of 20 ft? The index of refraction inside the eye is 1.33.

MODEL Assume that a 2.0-mm-diameter pupil is diffraction limited. Then the angular resolution is given by Rayleigh's criterion. Diffraction increases with wavelength, so the eye's acuity will be affected more by longer wavelengths than by shorter wavelengths. Consequently, assume that the light's wavelength in air is 600 nm.

VISUALIZE Let the diameter of the circle be d. FIGURE 35.27 shows the circle at distance $s = 20$ ft $= 6.1$ m. "Seeing the circle," shown edge-on, requires resolving the top and bottom lines as distinct.

FIGURE 35.27 Viewing a circle of diameter d.

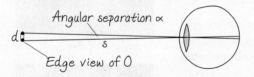

SOLVE The angular separation of the top and bottom lines of the circle is $\alpha = d/s$. Rayleigh's criterion says that a perfect lens with aperture D can just barely resolve these two lines if

$$\alpha = \frac{d}{s} = \theta_{min} = \frac{1.22\lambda_{eye}}{D} = \frac{1.22\lambda_{air}}{n_{eye}D}$$

The diffraction takes place inside the eye, where the wavelength is shortened to $\lambda_{eye} = \lambda_{air}/n_{eye}$. Thus the circle diameter that can barely be resolved with perfect vision is

$$d = \frac{1.22\lambda_{air}s}{n_{eye}D} = \frac{1.22(600 \times 10^{-9} \text{ m})(6.1 \text{ m})}{(1.33)(0.0020 \text{ m})} \approx 2 \text{ mm}$$

That's about the height of a capital O in this book, so in principle you should—in very bright light—just barely be able to recognize it as an O at 20 feet.

ASSESS On an eye chart, the O on the line for 20/20 vision—the standard of excellent vision—is about 7 mm tall, so the calculated 2 mm, although in the right range, is a bit too small. There are two reasons. First, although aberrations of the eye are reduced with a smaller pupil, they haven't vanished. And second, for a 2-mm-tall object at 20 ft, the size of the image on the retina is barely larger than the spacing between the cone cells, so the resolution of the "detector" is also a factor. Your eye is a very good optical instrument, but not perfect.

SUMMARY

The goal of Chapter 35 has been to learn about some common optical instruments and their limitations.

IMPORTANT CONCEPTS

Lens Combinations

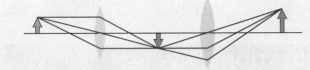

The image of the first lens acts as the object for the second lens.

Lens **power**: $P = \dfrac{1}{f}$ diopters, 1 D = 1 m^{-1}

Resolution

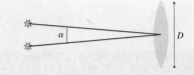

The **angular resolution** of a lens of diameter D is

$$\theta_{min} = 1.22\lambda/D$$

Rayleigh's criterion states that two objects separated by an angle α are marginally resolvable if $\alpha = \theta_{min}$.

APPLICATIONS

Cameras

Form a real, inverted image on a detector. The lens's **f-number** is

$$f\text{-number} = \frac{f}{D}$$

The light intensity on the detector is

$$I \propto \frac{1}{(f\text{-number})^2}$$

Vision

Refraction at the cornea is responsible for most of the focusing. The lens provides fine-tuning by changing its shape (**accommodation**).

In normal vision, the eye can focus from a far point (FP) at ∞ (relaxed eye) to a near point (NP) at ≈ 25 cm (maximum accommodation).

- **Hyperopia** (farsightedness) is corrected with a converging lens.
- **Myopia** (nearsightedness) is corrected with a diverging lens.

Focusing and spatial resolution

The minimum spot size to which a lens of focal length f and diameter D can focus light is limited by diffraction to

$$w_{min} = \frac{2.44\lambda f}{D}$$

With the best lenses that can be manufactured, $w_{min} \approx \lambda$.

Magnifiers

For relaxed-eye viewing, the angular magnification is

$$M = \frac{25 \text{ cm}}{f}$$

For microscopes and telescopes, angular magnification, not lateral magnification, is the important characteristic. The eyepiece acts as a magnifier to view the image formed by the objective lens.

Microscopes

The object is very close to the focal point of the objective.

The total angular magnification is $M \approx -\dfrac{L}{f_{obj}} \dfrac{25 \text{ cm}}{f_{eye}}$.

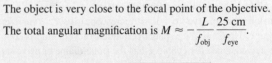

The best possible spatial resolution of a microscope, limited by diffraction, is about one wavelength of light.

Telescopes

The object is very far from the objective.

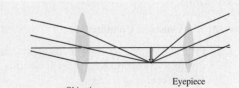

The total angular magnification is $M = -\dfrac{f_{obj}}{f_{eye}}$.

TERMS AND NOTATION

camera	iris	hyperopia	reflecting telescope
effective focal length, f	retina	myopia	dispersion
aperture	accommodation	angular size	Rayleigh scattering
f-number	far point	magnifier	chromatic aberration
CCD	near point	angular magnification, M	spherical aberration
pixel	presbyopia	objective	minimum spot size, w_{min}
cornea	power, P	eyepiece	Rayleigh's criterion
pupil	diopter, D	refracting telescope	angular resolution

CONCEPTUAL QUESTIONS

1. Suppose a camera's exposure is correct when the lens has a focal length of 8.0 mm. Will the picture be overexposed, underexposed, or still correct if the focal length is "zoomed" to 16.0 mm without changing the diameter of the lens aperture? Explain.

2. A camera has a circular aperture immediately behind the lens. Reducing the aperture diameter to half its initial value will
   A. Make the image blurry.
   B. Cut off the outer half of the image and leave the inner half unchanged.
   C. Make the image less bright.
   D. All the above.
   Explain your choice.

3. Suppose you wanted special glasses designed to let you see underwater without a face mask. Should the glasses use a converging or diverging lens? Explain.

4. A red card is illuminated by red light. What color will the card appear? What if it's illuminated by blue light?

5. The center of the galaxy is filled with low-density hydrogen gas that scatters light rays. An astronomer wants to take a picture of the center of the galaxy. Will the view be better using ultraviolet light, visible light, or infrared light? Explain.

6. A friend lends you the eyepiece of his microscope to use on your own microscope. He claims the spatial resolution of your microscope will be halved, since his eyepiece has the same diameter as yours but twice the magnification. Is his claim valid? Explain.

7. A diffraction-limited lens can focus light to a 10-μm-diameter spot on a screen. Do the following actions make the spot diameter larger, make it smaller, or leave it unchanged?
   A. Decreasing the wavelength of the light.
   B. Decreasing the lens diameter.
   C. Decreasing the lens focal length.
   D. Decreasing the lens-to-screen distance.

8. To focus parallel light rays to the smallest possible spot, should you use a lens with a small f-number or a large f-number? Explain.

9. An astronomer is trying to observe two distant stars. The stars are marginally resolved when she looks at them through a filter that passes green light with a wavelength near 550 nm. Which of the following actions would improve the resolution? Assume that the resolution is not limited by the atmosphere.
   A. Changing the filter to a different wavelength. If so, should she use a shorter or a longer wavelength?
   B. Using a telescope with an objective lens of the same diameter but a different focal length. If so, should she select a shorter or a longer focal length?
   C. Using a telescope with an objective lens of the same focal length but a different diameter. If so, should she select a larger or a smaller diameter?
   D. Using an eyepiece with a different magnification. If so, should she select an eyepiece with more or less magnification?

EXERCISES AND PROBLEMS

Problems labeled ▓ integrate material from earlier chapters.

Exercises

Section 35.1 Lenses in Combination

1. ‖ Two converging lenses with focal lengths of 40 cm and 20 cm are 10 cm apart. A 2.0-cm-tall object is 15 cm in front of the 40-cm-focal-length lens.
   a. Use ray tracing to find the position and height of the image. Do this accurately using a ruler or paper with a grid, then make measurements on your diagram.
   b. Calculate the image position and height. Compare with your ray-tracing answers in part a.

2. ‖ A converging lens with a focal length of 40 cm and a diverging lens with a focal length of −40 cm are 160 cm apart. A 2.0-cm-tall object is 60 cm in front of the converging lens.
   a. Use ray tracing to find the position and height of the image. Do this accurately using a ruler or paper with a grid, then make measurements on your diagram.
   b. Calculate the image position and height. Compare with your ray-tracing answers in part a.

3. ‖ A 2.0-cm-tall object is 20 cm to the left of a lens with a focal length of 10 cm. A second lens with a focal length of 15 cm is 30 cm to the right of the first lens.
   a. Use ray tracing to find the position and height of the image. Do this accurately using a ruler or paper with a grid, then make measurements on your diagram.
   b. Calculate the image position and height. Compare with your ray-tracing answers in part a.

4. ‖ A 2.0-cm-tall object is 20 cm to the left of a lens with a focal length of 10 cm. A second lens with a focal length of 5 cm is 30 cm to the right of the first lens.
   a. Use ray tracing to find the position and height of the image. Do this accurately using a ruler or paper with a grid, then make measurements on your diagram.
   b. Calculate the image position and height. Compare with your ray-tracing answers in part a.

5. ‖ A 2.0-cm-tall object is 20 cm to the left of a lens with a focal length of 10 cm. A second lens with a focal length of −5 cm is 30 cm to the right of the first lens.
   a. Use ray tracing to find the position and height of the image. Do this accurately using a ruler or paper with a grid, then make measurements on your diagram.
   b. Calculate the image position and height. Compare with your ray-tracing answers in part a.

Section 35.2 The Camera

6. | A 2.0-m-tall man is 10 m in front of a camera with a 15-mm-focal-length lens. How tall is his image on the detector?

7. | What is the f-number of a lens with a 35 mm focal length and a 7.0-mm-diameter aperture?

8. | What is the aperture diameter of a 12-mm-focal-length lens set to f/4.0?

9. | A camera takes a properly exposed photo at f/5.6 and 1/125 s. What shutter speed should be used if the lens is changed to f/4.0?

10. ‖ A camera takes a properly exposed photo with a 3.0-mm-diameter aperture and a shutter speed of 1/125 s. What is the appropriate aperture diameter for a 1/500 s shutter speed?

Section 35.3 Vision

11. ‖ Ramon has contact lenses with the prescription +2.0 D.
BIO a. What eye condition does Ramon have?
     b. What is his near point without the lenses?

12. | Ellen wears eyeglasses with the prescription −1.0 D.
BIO a. What eye condition does Ellen have?
     b. What is her far point without the glasses?

13. | What is the f-number of a relaxed eye with the pupil fully
BIO dilated to 8.0 mm? Model the eye as a single lens 2.4 cm in front of the retina.

Section 35.4 Optical Systems That Magnify

14. ‖ A magnifier has a magnification of 5×. How far from the lens should an object be held so that its image is seen at the near-point distance of 25 cm? Assume that your eye is immediately behind the lens.

15. ‖ A microscope has a 20 cm tube length. What focal-length objective will give total magnification 500× when used with an eyepiece having a focal length of 5.0 cm?

16. ‖ A standardized biological microscope has an 8.0-mm-focal-length objective. What focal-length eyepiece should be used to achieve a total magnification of 100×?

17. ‖ A 6.0-mm-diameter microscope objective has a focal length of 9.0 mm. What object distance gives a lateral magnification of −40?

18. ‖ A 20× telescope has a 12-cm-diameter objective lens. What minimum diameter must the eyepiece lens have to collect all the light rays from an on-axis distant source?

19. ‖ A reflecting telescope is built with a 20-cm-diameter mirror having a 1.00 m focal length. It is used with a 10× eyepiece. What are (a) the magnification and (b) the f-number of the telescope?

Section 35.5 Color and Dispersion

20. ‖ A sheet of glass has $n_{red} = 1.52$ and $n_{violet} = 1.55$. A narrow beam of white light is incident on the glass at 30°. What is the angular spread of the light inside the glass?

21. | A narrow beam of white light is incident on a sheet of quartz. The beam disperses in the quartz, with red light ($\lambda \approx 700$ nm) traveling at an angle of 26.3° with respect to the normal and violet light ($\lambda \approx 400$ nm) traveling at 25.7°. The index of refraction of quartz for red light is 1.45. What is the index of refraction of quartz for violet light?

22. ‖ A hydrogen discharge lamp emits light with two prominent wavelengths: 656 nm (red) and 486 nm (blue). The light enters a flint-glass prism perpendicular to one face and then refracts through the hypotenuse back into the air. The angle between these two faces is 35°.
   a. Use Figure 35.18 to estimate to ±0.002 the index of refraction of flint glass at these two wavelengths.
   b. What is the angle (in degrees) between the red and blue light as it leaves the prism?

23. ‖ Infrared telescopes, which use special infrared detectors, are able to peer farther into star-forming regions of the galaxy because infrared light is not scattered as strongly as is visible light by the tenuous clouds of hydrogen gas from which new stars are created. For what wavelength of light is the scattering only 1% that of light with a visible wavelength of 500 nm?

Section 35.6 The Resolution of Optical Instruments

24. ‖ A scientist needs to focus a helium-neon laser beam ($\lambda = 633$ nm) to a 10-μm-diameter spot 8.0 cm behind a lens.
   a. What focal-length lens should she use?
   b. What minimum diameter must the lens have?

25. ‖ Two lightbulbs are 1.0 m apart. From what distance can these lightbulbs be marginally resolved by a small telescope with a 4.0-cm-diameter objective lens? Assume that the lens is diffraction limited and $\lambda = 600$ nm.

Problems

26. ‖ In FIGURE P35.26, parallel rays from the left are focused to a point at two locations on the optical axis. Find the position of each location, giving your answer as a distance left or right of the lens.

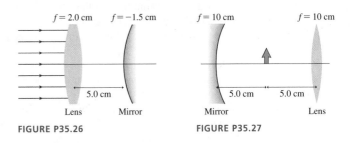

FIGURE P35.26   FIGURE P35.27

27. ‖ The rays leaving the two-component optical system of FIGURE P35.27 produce two distinct images of the 1.0-cm-tall object. What are the position (relative to the lens), orientation, and height of each image?

28. | A common optical instrument in a laser laboratory is a *beam expander*. One type of beam expander is shown in FIGURE P35.28. The parallel rays of a laser beam of width w_1 enter from the left.
   a. For what lens spacing d does a parallel laser beam exit from the right?
   b. What is the width w_2 of the exiting laser beam?

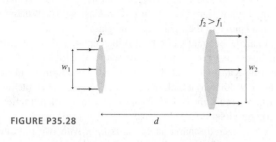

FIGURE P35.28

29. | A common optical instrument in a laser laboratory is a *beam expander*. One type of beam expander is shown in FIGURE P35.29. The parallel rays of a laser beam of width w_1 enter from the left.
   a. For what lens spacing d does a parallel laser beam exit from the right?
   b. What is the width w_2 of the exiting laser beam?

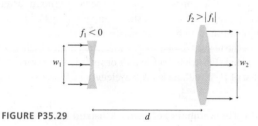

FIGURE P35.29

30. || In FIGURE P35.30, what are the position, height, and orientation of the final image? Give the position as a distance to the right or left of the lens.

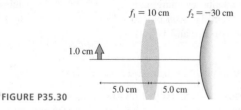

FIGURE P35.30

31. || A 1.0-cm-tall object is 110 cm from a screen. A diverging lens with focal length −20 cm is 20 cm in front of the object. What are the focal length and distance from the screen of a second lens that will produce a well-focused, 2.0-cm-tall image on the screen?

32. ||| A 15-cm-focal-length converging lens is 20 cm to the right of a 7.0-cm-focal-length converging lens. A 1.0-cm-tall object is distance L to the left of the first lens.
   a. For what value of L is the final image of this two-lens system halfway between the two lenses?
   b. What are the height and orientation of the final image?

33. || Yang can focus on objects 150 cm away with a relaxed eye.
BIO With full accommodation, she can focus on objects 20 cm away. After her eyesight is corrected for distance vision, what will her near point be while wearing her glasses?

34. ||| The cornea, a boundary between the air and the aqueous
BIO humor, has a 3.0 cm focal length when acting alone. What is its radius of curvature?

35. | The objective lens of a telescope is a symmetric glass lens with 100 cm radii of curvature. The eyepiece lens is also a symmetric glass lens. What are the radii of curvature of the eyepiece lens if the telescope's magnification is 20×?

36. || Mars (6800 km diameter) is viewed through a telescope on a night when it is 1.1×10^8 km from the earth. Its angular size as seen through the eyepiece is 0.50°, the same size as the full moon seen by the naked eye. If the eyepiece focal length is 25 mm, how long is the telescope?

37. || You've been asked to build a telescope from a 2.0× magnifying lens and a 5.0× magnifying lens.
   a. What is the maximum magnification you can achieve?
   b. Which lens should be used as the objective? Explain.
   c. What will be the length of your telescope?

38. | Marooned on a desert island and with a lot of time on your hands, you decide to disassemble your glasses to make a crude telescope with which you can scan the horizon for rescuers. Luckily you're farsighted, and, like most people, your two eyes have different lens prescriptions. Your left eye uses a lens of power +4.5 D, and your right eye's lens is +3.0 D.
   a. Which lens should you use for the objective and which for the eyepiece? Explain.
   b. What will be the magnification of your telescope?
   c. How far apart should the two lenses be when you focus on distant objects?

39. || A microscope with a tube length of 180 mm achieves a total magnification of 800× with a 40× objective and a 20× eyepiece. The microscope is focused for viewing with a relaxed eye. How far is the sample from the objective lens?

40. || Modern microscopes are more likely to use a camera than
BIO human viewing. This is accomplished by replacing the eyepiece in Figure 35.14 with a *photo-ocular* that focuses the image of the objective to a real image on the sensor of a digital camera. A typical sensor is 22.5 mm wide and consists of 5625 4.0-μm-wide pixels. Suppose a microscopist pairs a 40× objective with a 2.5× photo-ocular.
   a. What is the field of view? That is, what width on the microscope stage, in mm, fills the sensor?
   b. The photo of a cell is 120 pixels in diameter. What is the cell's actual diameter, in μm?

41. || White light is incident onto a 30° prism at the 40° angle shown in FIGURE P35.41. Violet light emerges perpendicular to the rear face of the prism. The index of refraction of violet light in this glass is 2.0% larger than the index of refraction of red light. At what angle ϕ does red light emerge from the rear face?

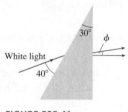

FIGURE P35.41

42. || A beam of white light enters a transparent material.
CALC Wavelengths for which the index of refraction is n are refracted at angle θ_2. Wavelengths for which the index of refraction is $n + \delta n$, where $\delta n \ll n$, are refracted at angle $\theta_2 + \delta\theta$.
   a. Show that the angular separation of the two wavelengths, in radians, is $\delta\theta = -(\delta n/n) \tan\theta_2$.
   b. A beam of white light is incident on a piece of glass at 30.0°. Deep violet light is refracted 0.28° more than deep red light. The index of refraction for deep red light is known to be 1.552. What is the index of refraction for deep violet light?

43. ‖ High-power lasers are used to cut and weld materials by focusing the laser beam to a very small spot. This is like using a magnifying lens to focus the sun's light to a small spot that can burn things. As an engineer, you have designed a laser cutting device in which the material to be cut is placed 5.0 cm behind the lens. You have selected a high-power laser with a wavelength of 1.06 μm. Your calculations indicate that the laser must be focused to a 5.0-μm-diameter spot in order to have sufficient power to make the cut. What is the minimum diameter of the lens you must install?

44. ‖ Once dark adapted, the pupil of your eye is approximately

BIO   7 mm in diameter. The headlights of an oncoming car are 120 cm apart. If the lens of your eye is diffraction limited, at what distance are the two headlights marginally resolved? Assume a wavelength of 600 nm and that the index of refraction inside the eye is 1.33. (Your eye is not really good enough to resolve headlights at this distance, due both to aberrations in the lens and to the size of the receptors in your retina, but it comes reasonably close.)

45. ‖ The resolution of a digital camera is limited by two factors: diffraction by the lens, a limit of any optical system, and the fact that the sensor is divided into discrete pixels. Consider a typical point-and-shoot camera that has a 20-mm-focal-length lens and a sensor with 2.5-μm-wide pixels.
   a. First, assume an ideal, diffractionless lens. At a distance of 100 m, what is the smallest distance, in cm, between two point sources of light that the camera can barely resolve? In answering this question, consider what has to happen on the sensor to show two image points rather than one. You can use $s' = f$ because $s \gg f$.
   b. You can achieve the pixel-limited resolution of part a only if the diffraction width of each image point is no greater than 1 pixel in diameter. For what lens diameter is the minimum spot size equal to the width of a pixel? Use 600 nm for the wavelength of light.
   c. What is the f-number of the lens for the diameter you found in part b? Your answer is a quite realistic value of the f-number at which a camera transitions from being pixel limited to being diffraction limited. For f-numbers smaller than this (larger-diameter apertures), the resolution is limited by the pixel size and does not change as you change the aperture. For f-numbers larger than this (smaller-diameter apertures), the resolution is limited by diffraction, and it gets worse as you "stop down" to smaller apertures.

46. ‖ The Hubble Space Telescope has a mirror diameter of 2.4 m. Suppose the telescope is used to photograph stars near the center of our galaxy, 30,000 light years away, using red light with a wavelength of 650 nm.
   a. What's the distance (in km) between two stars that are marginally resolved? The resolution of a reflecting telescope is calculated exactly the same as for a refracting telescope.
   b. For comparison, what is this distance as a multiple of the distance of Jupiter from the sun?

47. ‖ Alpha Centauri, the nearest star to our solar system, is 4.3 light years away. Assume that Alpha Centauri has a planet with an advanced civilization. Professor Dhg, at the planet's Astronomical Institute, wants to build a telescope with which he can find out whether any planets are orbiting our sun.
   a. What is the minimum diameter for an objective lens that will just barely resolve Jupiter and the sun? The radius of Jupiter's orbit is 780 million km. Assume $\lambda = 600$ nm.
   b. Building a telescope of the necessary size does not appear to be a major problem. What practical difficulties might prevent Professor Dhg's experiment from succeeding?

Challenge Problems

48. ‖‖ Your task in physics laboratory is to make a microscope from two lenses. One lens has a focal length of 2.0 cm, the other 1.0 cm. You plan to use the more powerful lens as the objective, and you want the eyepiece to be 16 cm from the objective.
   a. For viewing with a relaxed eye, how far should the sample be from the objective lens?
   b. What is the magnification of your microscope?

49. ‖‖ The lens shown in FIGURE CP35.49 is called an *achromatic doublet*, meaning that it has no chromatic aberration. The left side is flat, and all other surfaces have radii of curvature R.
   a. For parallel light rays coming from the left, show that the effective focal length of this two-lens system is $f = R/(2n_2 - n_1 - 1)$, where n_1 and n_2 are, respectively, the indices of refraction of the diverging and the converging lenses. Don't forget to make the thin-lens approximation.

   FIGURE CP35.49

   b. Because of dispersion, either lens alone would focus red rays and blue rays at different points. Define Δn_1 and Δn_2 as $n_{blue} - n_{red}$ for the two lenses. What value of the ratio $\Delta n_1/\Delta n_2$ makes $f_{blue} = f_{red}$ for the two-lens system? That is, the two-lens system does *not* exhibit chromatic aberration.
   c. Indices of refraction for two types of glass are given in the table. To make an achromatic doublet, which glass should you use for the converging lens and which for the diverging lens? Explain.

	n_{blue}	n_{red}
Crown glass	1.525	1.517
Flint glass	1.632	1.616

   d. What value of R gives a focal length of 10.0 cm?

50. ‖‖ FIGURE CP35.50 shows a simple zoom lens in which the magnitudes of both focal lengths are f. If the spacing $d < f$, the image of the converging lens falls on the right side of the diverging lens. Our procedure of letting the image of the first lens act as the object of the second lens will continue to work in this case if we use a *negative* object distance for the second lens. This is called a *virtual object*. Consider an object very far to the left ($s \approx \infty$) of the converging lens. Define the effective focal length as the distance from the midpoint between the lenses to the final image.
   a. Show that the effective focal length is

$$f_{eff} = \frac{f^2 - fd + \frac{1}{2}d^2}{d}$$

   b. What is the zoom for a lens that can be adjusted from $d = \frac{1}{2}f$ to $d = \frac{1}{4}f$?

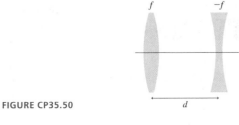

FIGURE CP35.50

KEY FINDINGS What are the overarching findings of Part VII?

Part VII has looked at two models of light: light waves and light rays.

- **Light waves**
  - Are electromagnetic waves.
  - Spread out after passing through openings.
  - Exhibit interference.

A third model of light, the photon model, will be introduced in Part VIII.

- **Light rays**
  - Travel in straight lines.
  - Do not interact.
  - Form images.

LAWS What laws of physics govern optics?

Superposition/interference Constructive interference occurs where crests overlap crests, destructive interference where crests overlap troughs.

Law of reflection $\theta_r = \theta_i$

$\theta_i \quad \theta_r$

Law of refraction (Snell's law) $n_1 \sin \theta_1 = n_2 \sin \theta_2$

θ_1

n_1

n_2

θ_2

Rayleigh's criterion Two objects can be resolved by a lens of diameter D if their angular separation exceeds $1.22\lambda/D$.

MODELS What are the most important models of Part VII?

Wave model

- Light is an electromagnetic wave.
  - Light travels through vacuum at speed c.
  - Wavelength and frequency are related by $\lambda f = c$.

- Light exhibits **diffraction** and **interference.**
  - Light spreads out after passing through an opening.
  - Equal-wavelength light waves interfere. Interference depends on the path-length difference.

- The wave model is usually appropriate for openings smaller than about 1 mm.

Ray model

- Light rays travel in straight lines.
  - The speed is $v = c/n$, where n is the index of refraction.
- Light rays travel forever unless they interact with matter.
  - **Reflection** and **refraction**
  - Scattering and absorption
- An object is a source of rays.
  - Rays originate at every point.
- The eye sees by focusing a diverging bundle of rays.
- The ray model is usually appropriate for openings larger than about 1 mm.

Diverging bundle of rays

Object

Eye

TOOLS What are the most important tools introduced in Part VII?

Diffraction

- Dark fringes in a single-slit diffraction pattern are at $\theta_p = p\lambda/a$, $p = 1,2,\ldots$

Slit width a

- The **central maximum** of the diffraction has width $w = 2\lambda L/a$.
- A circular hole has a central maximum of width $w = 2.44\lambda L/D$.

Double-slit interference

- Bright equally spaced fringes are located at

$$y_m = mL\lambda/d \quad m = 0,1,2,\ldots$$

 Slit spacing d

Diffraction gratings

- Very narrow bright fringes are at

$d \sin \theta_m = m\lambda \quad y_m = L \tan \theta_m$

Ray tracing

- For lenses and mirrors, three special rays locate the image.
  - Parallel to the axis
  - Through the focal point
  - Through the center

$s \qquad s'$

Images

- If rays converge at P′, then P′ is a **real image** and s' is positive.
- If rays diverge from P′, then P′ is a **virtual image** and s' is negative.

Thin lenses and mirrors

- The thin-lens and thin-mirror equation for focal length f is

$$\frac{1}{s} + \frac{1}{s'} = \frac{1}{f}$$

- Lateral magnification is $m = -s'/s$.

Optical instruments

- With multiple lenses, the image of one lens is the object for the next.

Vision

- **Hyperopia** occurs when the eye's near point is too far away. It is corrected with a converging lens.
- **Myopia** occurs when the eye's far point is too close. It is corrected with a diverging lens.

Resolution

- Diffraction limits optical instruments.
  - The smallest spot to which light can be focused is $w_{min} = 2.44\lambda f/D$.
  - Two objects can be resolved if their angular separation exceeds $1.22\lambda/D$.

OVERVIEW

Contemporary Physics

Our journey into physics is nearing its end. We began roughly 350 years ago with Newton's discovery of the laws of motion. Parts VI and VII have brought us to the end of the 19th century, just over 100 years ago. Along the way you've learned about the motion of particles, the conservation of energy, the physics of waves, and the electromagnetic interactions that hold atoms together and generate light waves. We begin the last phase of our journey with confidence.

Newton's mechanics and Maxwell's electromagnetism were the twin pillars of science at the end of the 19th century and remain the basis for much of engineering and applied science in the 21st century. Despite the successes of these theories, a series of discoveries starting around 1900 and continuing into the first few decades of the 20th century profoundly altered our understanding of the universe at the most fundamental level.

- Einstein's theory of relativity forced scientists to completely revise their concepts of space and time. Our exploration of these fascinating ideas will end with perhaps the most famous equation in physics: Einstein's $E = mc^2$.
- Experimenters found that the classical distinction between *particles* and *waves* breaks down at the atomic level. Light sometimes acts like a particle, while electrons and even entire atoms sometimes act like waves. We will need a new theory of light and matter—quantum physics—to explain these phenomena.

These two theories form the basis for physics—and, increasingly, engineering—as it is practiced today.

The complete theory of quantum physics, as it was developed in the 1920s, describes atomic particles in terms of an entirely new concept called a *wave function*. One of our most important tasks in Part VIII will be to learn what a wave function is, what laws govern its behavior, and how to relate wave functions to experimental measurements. We will concentrate on one-dimensional models that, while not perfect, will be adequate for understanding the essential features of scanning tunneling microscopes, various semiconductor devices, radioactive decay, and other applications.

We'll complete our study of quantum physics with an introduction to atomic and nuclear physics. You will learn where the electron-shell model of chemistry comes from, how atoms emit and absorb light, what's inside the nucleus, and why some nuclei undergo radioactive decay.

The quantum world with its wave functions and probabilities can seem strange and mysterious, yet quantum physics gives the most definitive and accurate predictions of any physical theory ever devised. The contemporary perspective of quantum physics will be a fitting end to our journey.

NOTE This edition of *Physics for Scientists and Engineers* contains only Chapter 36, Relativity. The complete Part VIII may be found either in the hardbound *Physics for Scientists and Engineers with Modern Physics* or in the softbound *Volume 3: Relativity and Quantum Physics*.

This plot shows the instantaneous intensity of a focused laser beam.

36 Relativity

The Compact Muon Solenoid detector at the Large Hadron Collider, where protons are accelerated to 99.999999% of the speed of light.

IN THIS CHAPTER, you will learn how relativity changes our concepts of space and time.

What is an inertial reference frame?

Inertial reference frames are reference frames that move relative to each other with constant velocity.

- You'll learn to work with the positions and times of events.
- All the clocks in an inertial reference frame are synchronized.

《 LOOKING BACK Section 4.3 Relative motion

What is relativity about?

Einstein's theory of relativity is based on a simple-sounding principle: The laws of physics are the same in all inertial reference frames. This leads to these conclusions:

- Light travels at the same speed c in all inertial reference frames.
- No object or information can travel faster than the speed of light.

How does relativity affect time?

Time is relative. Two reference frames moving relative to each other measure different time intervals between two events.

- Time dilation is the idea that moving clocks run slower than clocks at rest.
- We'll examine the famous twin paradox.

How does relativity affect space?

Distances are also relative. Two reference frames moving relative to each other find different distances between two events.

- Length contraction is the idea that the length of an object is less when the object is moving than when it is at rest.

How does relativity affect mass and energy?

Einstein's most famous equation, $E = mc^2$, says that mass can be transformed into energy, and energy into mass, as long as the total energy is conserved.

- Nuclear fission converts mass into energy.
- Collisions between high-speed particles create new particles from energy.

Does relativity have applications?

Abstract though it may seem, relativity is important in technologies such as medical PET scans (positron-emission tomography) and nuclear energy. And the global GPS system, a technology we use every day, functions only when the signals from precision clocks in orbiting satellites are corrected for relativistic time dilation.

36.1 Relativity: What's It All About?

What do you think of when you hear the phrase "theory of relativity"? A white-haired Einstein? $E = mc^2$? Black holes? Time travel? Perhaps you've heard that the theory of relativity is so complicated and abstract that only a handful of people in the whole world really understand it.

There is, without doubt, a certain mystique associated with relativity, an aura of the strange and exotic. The good news is that understanding the ideas of relativity is well within your grasp. Einstein's *special theory of relativity,* the portion of relativity we'll study, is not mathematically difficult at all. The challenge is conceptual because relativity questions deeply held assumptions about the nature of space and time. In fact, that's what relativity is all about—space and time.

What's Special About Special Relativity?

Einstein's first paper on relativity, in 1905, dealt exclusively with inertial reference frames, reference frames that move relative to each other with constant velocity. Ten years later, Einstein published a more encompassing theory of relativity that considered accelerated motion and its connection to gravity. The second theory, because it's more general in scope, is called *general relativity.* General relativity is the theory that describes black holes, curved spacetime, and the evolution of the universe. It is a fascinating theory but, alas, very mathematical and outside the scope of this textbook.

Motion at constant velocity is a "special case" of motion—namely, motion for which the acceleration is zero. Hence Einstein's first theory of relativity has come to be known as **special relativity.** It is special in the sense of being a restricted, special case of his more general theory, not special in the everyday sense meaning distinctive or exceptional. Special relativity, with its conclusions about time dilation and length contraction, is what we will study.

Albert Einstein (1879–1955) was one of the most influential thinkers in history.

36.2 Galilean Relativity

Relativity is the process of relating measurements in one reference frame to those in a different reference frame moving *relative to* the first. To appreciate and understand what is new in Einstein's theory, we need a firm grasp of the ideas of relativity that are embodied in Newtonian mechanics. Thus we begin with *Galilean relativity.*

Reference Frames

Suppose you're passing me as we both drive in the same direction along a freeway. My car's speedometer reads 55 mph while your speedometer shows 60 mph. Is 60 mph your "true" speed? That is certainly your speed relative to someone standing beside the road, but your speed relative to me is only 5 mph. Your speed is 120 mph relative to a driver approaching from the other direction at 60 mph.

An object does not have a "true" speed or velocity. The very definition of velocity, $v = \Delta x/\Delta t$, assumes the existence of a coordinate system in which, during some time interval Δt, the displacement Δx is measured. The best we can manage is to specify an object's velocity relative to, or with respect to, the coordinate system in which it is measured.

Let's define a **reference frame** to be a coordinate system in which experimenters equipped with meter sticks, stopwatches, and any other needed equipment make position and time measurements on moving objects. Three ideas are implicit in our definition of a reference frame:

- A reference frame extends infinitely far in all directions.
- The experimenters are at rest in the reference frame.
- The number of experimenters and the quality of their equipment are sufficient to measure positions and velocities to any level of accuracy needed.

The first two ideas are especially important. It is often convenient to say "the laboratory reference frame" or "the reference frame of the rocket." These are shorthand expressions for "a reference frame, infinite in all directions, in which the laboratory (or the rocket) and a set of experimenters happen to be at rest."

NOTE A reference frame is not the same thing as a "point of view." That is, each person or each experimenter does not have his or her own private reference frame. **All experimenters at rest relative to each other share the same reference frame.**

FIGURE 36.1 shows two reference frames called S and S′. The coordinate axes in S are x, y, z and those in S′ are x', y', z'. Reference frame S′ moves with velocity v relative to S or, equivalently, S moves with velocity $-v$ relative to S′. There's no implication that either reference frame is "at rest." Notice that the zero of time, when experimenters start their stopwatches, is the instant that the origins of S and S′ coincide.

We will restrict our attention to *inertial reference frames,* implying that the relative velocity v is constant. You should recall from Chapter 5 that an **inertial reference frame** is a reference frame in which Newton's first law, the law of inertia, is valid. In particular, an inertial reference frame is one in which an isolated particle, one on which there are no forces, either remains at rest or moves in a straight line at constant speed.

Any reference frame moving at constant velocity with respect to an inertial reference frame is itself an inertial reference frame. Conversely, a reference frame accelerating with respect to an inertial reference frame is *not* an inertial reference frame. Our restriction to reference frames moving with respect to each other at constant velocity—with no acceleration—is the "special" part of special relativity.

NOTE An inertial reference frame is an idealization. A true inertial reference frame would need to be floating in deep space, far from any gravitational influence. In practice, an earthbound laboratory is a good approximation of an inertial reference frame because the accelerations associated with the earth's rotation and motion around the sun are too small to influence most experiments.

STOP TO THINK 36.1 Which of these is an inertial reference frame (or a very good approximation)?

a. Your bedroom
b. A car rolling down a steep hill
c. A train coasting along a level track
d. A rocket being launched
e. A roller coaster going over the top of a hill
f. A sky diver falling at terminal speed

FIGURE 36.1 The standard reference frames S and S′.

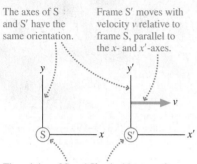

The axes of S and S′ have the same orientation.

Frame S′ moves with velocity v relative to frame S, parallel to the x- and x'-axes.

The origins of S and S′ coincide at $t = 0$.

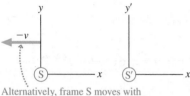

Alternatively, frame S moves with velocity $-v$ relative to frame S′.

FIGURE 36.2 The position of an exploding firecracker is measured in reference frames S and S′.

At time t, the origin of S′ has moved distance vt to the right. Thus $x = x' + vt$.

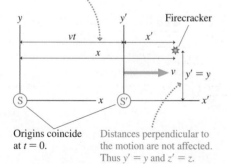

Origins coincide at $t = 0$.

Distances perpendicular to the motion are not affected. Thus $y' = y$ and $z' = z$.

The Galilean Transformations

Suppose a firecracker explodes at time t. The experimenters in reference frame S determine that the explosion happened at position x. Similarly, the experimenters in S′ find that the firecracker exploded at x' in their reference frame. What is the relationship between x and x'?

FIGURE 36.2 shows the explosion and the two reference frames. You can see from the figure that $x = x' + vt$, thus

$$
\begin{aligned}
x &= x' + vt & x' &= x - vt \\
y &= y' & \text{or} \qquad y' &= y \\
z &= z' & z' &= z
\end{aligned}
\qquad (36.1)
$$

These are the *Galilean transformations of position*. If you know a position measured by the experimenters in one inertial reference frame, you can calculate the position that would be measured by experimenters in any other inertial reference frame.

Suppose the experimenters in both reference frames now track the motion of the object in FIGURE 36.3 by measuring its position at many instants of time. The experimenters in S find that the object's velocity is $\vec{u}$. During the *same time interval* Δt, the experimenters in S′ measure the velocity to be $\vec{u}'$.

> **NOTE** In this chapter, we will use v to represent the velocity of one reference frame relative to another. We will use $\vec{u}$ and $\vec{u}'$ to represent the velocities of objects with respect to reference frames S and S′.

We can find the relationship between $\vec{u}$ and $\vec{u}'$ by taking the time derivatives of Equations 36.1 and using the definition $u_x = dx/dt$:

$$u_x = \frac{dx}{dt} = \frac{dx'}{dt} + v = u'_x + v$$

$$u_y = \frac{dy}{dt} = \frac{dy'}{dt} = u'_y$$

The equation for u_z is similar. The net result is

$$
\begin{array}{lll}
u_x = u'_x + v & & u'_x = u_x - v \\
u_y = u'_y & \text{or} & u'_y = u_y \\
u_z = u'_z & & u'_z = u_z
\end{array}
\qquad (36.2)
$$

Equations 36.2 are the *Galilean transformations of velocity*. If you know the velocity of a particle in one inertial reference frame, you can find the velocity that would be measured by experimenters in any other inertial reference frame.

> **NOTE** In Section 4.3 you learned the Galilean transformation of velocity as $\vec{v}_{CB} = \vec{v}_{CA} + \vec{v}_{AB}$, where $\vec{v}_{AB}$ means "the velocity of A relative to B." Equations 36.2 are equivalent for relative motion parallel to the x-axis but are written in a more formal notation that will be useful for relativity.

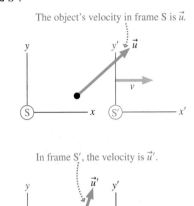

The object's velocity in frame S is $\vec{u}$.

In frame S′, the velocity is $\vec{u}'$.

EXAMPLE 36.1 | The speed of sound

An airplane is flying at speed 200 m/s with respect to the ground. Sound wave 1 is approaching the plane from the front, sound wave 2 is catching up from behind. Both waves travel at 340 m/s relative to the ground. What is the speed of each wave relative to the plane?

MODEL Assume that the earth (frame S) and the airplane (frame S′) are inertial reference frames. Frame S′, in which the airplane is at rest, moves at $v = 200$ m/s relative to frame S.

VISUALIZE FIGURE 36.4 shows the airplane and the sound waves.

FIGURE 36.4 Experimenters in the plane measure different speeds for the waves than do experimenters on the ground.

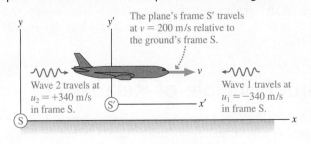

The plane's frame S′ travels at $v = 200$ m/s relative to the ground's frame S.

Wave 2 travels at $u_2 = +340$ m/s in frame S.

Wave 1 travels at $u_1 = -340$ m/s in frame S.

SOLVE The speed of a mechanical wave, such as a sound wave or a wave on a string, is its speed *relative to its medium*. Thus the *speed of sound* is the speed of a sound wave through a reference frame in which the air is at rest. This is reference frame S, where wave 1 travels with velocity $u_1 = -340$ m/s and wave 2 travels with velocity $u_2 = +340$ m/s. Notice that the Galilean transformations use *velocities*, with appropriate signs, not just speeds.

The airplane travels to the right with reference frame S′ at velocity v. We can use the Galilean transformations of velocity to find the velocities of the two sound waves in frame S′:

$$u'_1 = u_1 - v = -340 \text{ m/s} - 200 \text{ m/s} = -540 \text{ m/s}$$
$$u'_2 = u_2 - v = 340 \text{ m/s} - 200 \text{ m/s} = 140 \text{ m/s}$$

ASSESS This isn't surprising. If you're driving at 50 mph, a car coming the other way at 55 mph is approaching you at 105 mph. A car coming up behind you at 55 mph is gaining on you at the rate of only 5 mph. Wave speeds behave the same. Notice that a mechanical wave appears to be stationary to a person moving at the wave speed. To a surfer, the crest of the ocean wave remains at rest under his or her feet.

STOP TO THINK 36.2 Ocean waves are approaching the beach at 10 m/s. A boat heading out to sea travels at 6 m/s. How fast are the waves moving in the boat's reference frame?

a. 16 m/s   b. 10 m/s   c. 6 m/s   d. 4 m/s

The Galilean Principle of Relativity

FIGURE 36.5 Experimenters in both reference frames test Newton's second law by measuring the force on a particle and its acceleration.

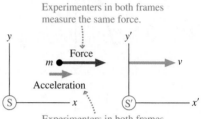

Experimenters in reference frames S and S′ measure different values for position and velocity. What about the force on and the acceleration of the particle in FIGURE 36.5? The strength of a force can be measured with a spring scale. The experimenters in reference frames S and S′ both see the *same reading* on the scale (assume the scale has a bright digital display easily seen by all experimenters), so both conclude that the force is the same. That is, $F' = F$.

We can compare the accelerations measured in the two reference frames by taking the time derivative of the velocity transformation equation $u' = u - v$. (We'll assume, for simplicity, that the velocities and accelerations are all in the x-direction.) The relative velocity v between the two reference frames is *constant*, with $dv/dt = 0$, thus

$$a' = \frac{du'}{dt} = \frac{du}{dt} = a \tag{36.3}$$

Experimenters in reference frames S and S′ measure different values for an object's position and velocity, but they *agree* on its acceleration.

If $F = ma$ in reference frame S, then $F' = ma'$ in reference frame S′. Stated another way, if Newton's second law is valid in one inertial reference frame, then it is valid in all inertial reference frames. Because other laws of mechanics, such as the conservation laws, follow from Newton's laws of motion, we can state this conclusion as the *Galilean principle of relativity*:

> **Galilean principle of relativity** The laws of mechanics are the same in all inertial reference frames.

FIGURE 36.6 Total momentum measured in two reference frames.

(a) Collision seen in frame S

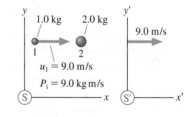

(b) Collision seen in frame S′

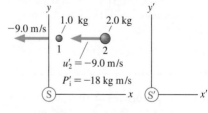

The Galilean principle of relativity is easy to state, but to understand it we must understand what is and is not "the same." To take a specific example, consider the law of conservation of momentum. FIGURE 36.6a shows two particles about to collide. Their total momentum in frame S, where particle 2 is at rest, is $P_i = 9.0$ kg m/s. This is an isolated system, hence the law of conservation of momentum tells us that the momentum after the collision will be $P_f = 9.0$ kg m/s.

FIGURE 36.6b has used the velocity transformation to look at the same particles in frame S′ in which particle 1 is at rest. The initial momentum in S′ is $P_i' = -18$ kg m/s. Thus it is not the *value* of the momentum that is the same in all inertial reference frames. Instead, the Galilean principle of relativity tells us that the *law* of momentum conservation is the same in all inertial reference frames. If $P_f = P_i$ in frame S, then it must be true that $P_f' = P_i'$ in frame S′. Consequently, we can conclude that P_f' will be -18 kg m/s after the collision in S′.

36.3 Einstein's Principle of Relativity

The 19th century was an era of optics and electromagnetism. Thomas Young demonstrated in 1801 that light is a wave, and by midcentury scientists had devised techniques for measuring the speed of light. Faraday discovered electromagnetic induction in 1831, setting in motion a series of events leading to Maxwell's prediction, in 1864, that light waves travel with speed

$$c = \frac{1}{\sqrt{\epsilon_0 \mu_0}} = 3.00 \times 10^8 \text{ m/s}$$

This is a quite specific prediction with no wiggle room. But in what reference frame is this the speed of light? And, if light travels with speed c in some inertial reference frame S, then surely, as FIGURE 36.7 shows, the light speed must be more or less than c in a reference frame S′ that moves with respect to S. There could even be reference frames in which light is frozen and doesn't move at all!

It was in this muddled state of affairs that a young Albert Einstein made his mark on the world. Even as a teenager, Einstein had wondered how a light wave would look to someone "surfing" the wave, traveling alongside the wave at the wave speed. You can do that with a water wave or a sound wave, but light waves seemed to present a logical difficulty. An electromagnetic wave sustains itself by virtue of the fact that a changing magnetic field induces an electric field and a changing electric field induces a magnetic field. But to someone moving with the wave, *the fields would not change.* How could there be an electromagnetic wave under these circumstances?

Several years of thinking about the connection between electromagnetism and reference frames led Einstein to the conclusion that *all* the laws of physics, not just the laws of mechanics, should obey the principle of relativity. In other words, the principle of relativity is a fundamental statement about the nature of the physical universe. Thus we can remove the restriction in the Galilean principle of relativity and state a much more general principle:

> **Principle of relativity** All the laws of physics are the same in all inertial reference frames.

All the results of Einstein's theory of relativity flow from this one simple statement.

The Constancy of the Speed of Light

If Maxwell's equations of electromagnetism are laws of physics, and there's every reason to think they are, then, according to the principle of relativity, Maxwell's equations must be true in *every* inertial reference frame. On the surface this seems to be an innocuous statement, equivalent to saying that the law of conservation of momentum is true in every inertial reference frame. But follow the logic:

1. Maxwell's equations are true in all inertial reference frames.
2. Maxwell's equations predict that electromagnetic waves, including light, travel at speed $c = 3.00 \times 10^8$ m/s.
3. Therefore, **light travels at speed c in all inertial reference frames.**

FIGURE 36.8 shows the implications of this conclusion. *All* experimenters, regardless of how they move with respect to each other, find that *all* light waves, regardless of the source, travel in their reference frame with the *same* speed c. If Cathy's velocity toward Bill and away from Amy is $v = 0.9c$, Cathy finds, by making measurements in her reference frame, that the light from Bill approaches her at speed c, not at $c + v = 1.9c$. And the light from Amy, which left Amy at speed c, catches up from behind at speed c *relative to Cathy,* not the $c - v = 0.1c$ you would have expected.

Although this prediction goes against all shreds of common sense, the experimental evidence for it is strong. Laboratory experiments are difficult because even the highest laboratory speed is insignificant in comparison to c. In the 1930s, however, physicists R. J. Kennedy and E. M. Thorndike realized that they could use the earth itself as a laboratory. The earth's speed as it circles the sun is about 30,000 m/s. The *relative*

FIGURE 36.7 It seems as if the speed of light should differ from c in a reference frame moving through the ether.

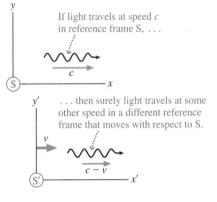

If light travels at speed c in reference frame S, . . .

. . . then surely light travels at some other speed in a different reference frame that moves with respect to S.

FIGURE 36.8 Light travels at speed c in all inertial reference frames, regardless of how the reference frames are moving with respect to the light source.

This light wave leaves Amy at speed c relative to Amy. It approaches Cathy at speed c relative to Cathy.

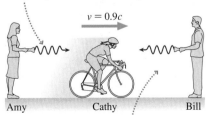

$v = 0.9c$

Amy   Cathy   Bill

This light wave leaves Bill at speed c relative to Bill. It approaches Cathy at speed c relative to Cathy.

FIGURE 36.9 Experiments find that the photons travel through the laboratory with speed c, not the speed $1.99975c$ that you might expect.

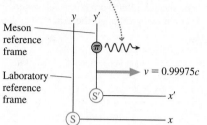

A photon is emitted at speed c relative to the π meson. Measurements find that the photon's speed in the laboratory reference frame is also c.

FIGURE 36.9 Experiments find that the photons travel through the laboratory with speed c, not the speed $1.99975c$ that you might expect.

velocity of the earth in January differs by 60,000 m/s from its velocity in July, when the earth is moving in the opposite direction. Kennedy and Thorndike were able to use a very sensitive and stable interferometer to show that the numerical values of the speed of light in January and July differ by less than 2 m/s.

More recent experiments have used unstable elementary particles, called π mesons, that decay into high-energy photons of light. The π mesons, created in a particle accelerator, move through the laboratory at 99.975% the speed of light, or $v = 0.99975c$, as they emit photons at speed c in the π meson's reference frame. As **FIGURE 36.9** shows, you would expect the photons to travel through the laboratory with speed $c + v = 1.99975c$. Instead, the measured speed of the photons in the laboratory was, within experimental error, 3.00×10^8 m/s.

In summary, *every* experiment designed to compare the speed of light in different reference frames has found that light travels at 3.00×10^8 m/s in every inertial reference frame, regardless of how the reference frames are moving with respect to each other.

How Can This Be?

You're in good company if you find this impossible to believe. Suppose I shot a ball forward at 50 m/s while driving past you at 30 m/s. You would certainly see the ball traveling at 80 m/s relative to you and the ground. What we're saying with regard to light is equivalent to saying that the ball travels at 50 m/s relative to my car and *at the same time* travels at 50 m/s relative to the ground, even though the car is moving across the ground at 30 m/s. It seems logically impossible.

You might think that this is merely a matter of semantics. If we can just get our definitions and use of words straight, then the mystery and confusion will disappear. Or perhaps the difficulty is a confusion between what we "see" versus what "really happens." In other words, a better analysis, one that focuses on what really happens, would find that light "really" travels at different speeds in different reference frames.

Alas, what "really happens" is that light travels at 3.00×10^8 m/s in every inertial reference frame, regardless of how the reference frames are moving with respect to each other. It's not a trick. There remains only one way to escape the logical contradictions.

The definition of velocity is $u = \Delta x / \Delta t$, the ratio of a distance traveled to the time interval in which the travel occurs. Suppose you and I both make measurements on an object as it moves, but you happen to be moving relative to me. Perhaps I'm standing on the corner, you're driving past in your car, and we're both trying to measure the velocity of a bicycle. Further, suppose we have agreed in advance to measure the position of the bicycle first as it passes the tree in **FIGURE 36.10**, then later as it passes the lamppost. Your $\Delta x'$, the bicycle's displacement, differs from my Δx because of your motion relative to me, causing you to calculate a bicycle velocity u' in your reference frame that differs from its velocity u in my reference frame. This is just the Galilean transformations showing up again.

Now let's repeat the measurements, but this time let's measure the velocity of a light wave as it travels from the tree to the lamppost. Once again, your $\Delta x'$ differs from my Δx, and the obvious conclusion is that your light speed u' differs from my light speed u. The difference will be very small if you're driving past in your car, very large if you're flying past in a rocket traveling at nearly the speed of light. Although this conclusion seems obvious, it is wrong. Experiments show that, for a light wave, we'll get the *same* values: $u' = u$.

The only way this can be true is if your Δt is not the same as my Δt. If the time it takes the light to move from the tree to the lamppost in your reference frame, a time we'll now call $\Delta t'$, differs from the time Δt it takes the light to move from the tree to the lamppost in my reference frame, then we might find that $\Delta x'/\Delta t' = \Delta x/\Delta t$. That is, $u' = u$ even though you are moving with respect to me.

FIGURE 36.10 Measuring the velocity of an object by appealing to the basic definition $u = \Delta x / \Delta t$.

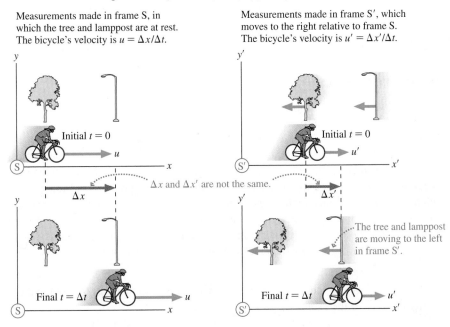

Measurements made in frame S, in which the tree and lamppost are at rest. The bicycle's velocity is $u = \Delta x / \Delta t$.

Measurements made in frame S', which moves to the right relative to frame S. The bicycle's velocity is $u' = \Delta x' / \Delta t$.

Δx and $\Delta x'$ are not the same.

The tree and lamppost are moving to the left in frame S'.

We've assumed, since the beginning of this textbook, that time is simply time. It flows along like a river, and all experimenters in all reference frames simply use it. For example, suppose the tree and the lamppost both have big clocks that we both can see. Shouldn't we be able to agree on the time interval Δt the light needs to move from the tree to the lamppost?

Perhaps not. It's demonstrably true that $\Delta x' \neq \Delta x$. It's experimentally verified that $u' = u$ for light waves. Something must be wrong with *assumptions* that we've made about the nature of time. The principle of relativity has painted us into a corner, and our only way out is to reexamine our understanding of time.

36.4 Events and Measurements

To question some of our most basic assumptions about space and time requires extreme care. We need to be certain that no assumptions slip into our analysis unnoticed. Our goal is to describe the motion of a particle in a clear and precise way, making the barest minimum of assumptions.

Events

The fundamental element of relativity is called an **event.** An event is a physical activity that takes place at a definite point in space and at a definite instant of time. An exploding firecracker is an event. A collision between two particles is an event. A light wave hitting a detector is an event.

Events can be observed and measured by experimenters in different reference frames. An exploding firecracker is as clear to you as you drive by in your car as it is to me standing on the street corner. We can quantify where and when an event occurs with four numbers: the coordinates (x, y, z) and the instant of time t. These four numbers, illustrated in FIGURE 36.11, are called the **spacetime coordinates** of the event.

The spatial coordinates of an event measured in reference frames S and S' may differ. It now appears that the instant of time recorded in S and S' may also differ. Thus the spacetime coordinates of an event measured by experimenters in frame S are (x, y, z, t) and the spacetime coordinates of the *same event* measured by experimenters in frame S' are (x', y', z', t').

FIGURE 36.11 The location and time of an event are described by its spacetime coordinates.

An event has spacetime coordinates (x, y, z, t) in frame S and different spacetime coordinates (x', y', z', t') in frame S'.

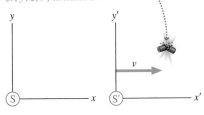

The motion of a particle can be described as a sequence of two or more events. We introduced this idea in the preceding section when we agreed to measure the velocity of a bicycle and then of a light wave by making measurements when the object passed the tree (first event) and when the object passed the lamppost (second event).

Measurements

Events are what "really happen," but how do we learn about an event? That is, how do the experimenters in a reference frame determine the spacetime coordinates of an event? This is a problem of *measurement*.

We defined a reference frame to be a coordinate system in which experimenters can make position and time measurements. That's a good start, but now we need to be more precise as to *how* the measurements are made. Imagine that a reference frame is filled with a cubic lattice of meter sticks, as shown in **FIGURE 36.12**. At every intersection is a clock, and all the clocks in a reference frame are *synchronized*. We'll return in a moment to consider how to synchronize the clocks, but assume for the moment it can be done.

Now, with our meter sticks and clocks in place, we can use a two-part measurement scheme:

- The (x, y, z) coordinates of an event are determined by the intersection of the meter sticks closest to the event.
- The event's time t is the time displayed on the clock nearest the event.

You can imagine, if you wish, that each event is accompanied by a flash of light to illuminate the face of the nearest clock and make its reading known.

Several important issues need to be noted:

1. The clocks and meter sticks in each reference frame are imaginary, so they have no difficulty passing through each other.
2. Measurements of position and time made in one reference frame must use only the clocks and meter sticks in that reference frame.
3. There's nothing special about the sticks being 1 m long and the clocks 1 m apart. The lattice spacing can be altered to achieve whatever level of measurement accuracy is desired.
4. We'll assume that the experimenters in each reference frame have assistants sitting beside every clock to record the position and time of nearby events.
5. Perhaps most important, t is the time at which the event *actually happens,* not the time at which an experimenter sees the event or at which information about the event reaches an experimenter.
6. All experimenters in one reference frame agree on the spacetime coordinates of an event. In other words, **an event has a unique set of spacetime coordinates in each reference frame.**

FIGURE 36.12 The spacetime coordinates of an event are measured by a lattice of meter sticks and clocks.

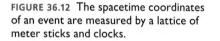
The spacetime coordinates of this event are measured by the nearest meter stick intersection and the nearest clock.

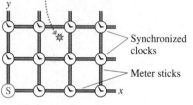
Synchronized clocks

Meter sticks

Reference frame S

Reference frame S' has its own meter sticks and its own clocks.

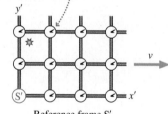

Reference frame S'

STOP TO THINK 36.3   A carpenter is working on a house two blocks away. You notice a slight delay between seeing the carpenter's hammer hit the nail and hearing the blow. At what time does the event "hammer hits nail" occur?

a. At the instant you hear the blow
b. At the instant you see the hammer hit
c. Very slightly before you see the hammer hit
d. Very slightly after you see the hammer hit

Clock Synchronization

It's important that all the clocks in a reference frame be **synchronized,** meaning that all clocks in the reference frame have the same reading at any one instant of time. Thus we need a method of synchronization. One idea that comes to mind is to designate the clock at the origin as the *master clock*. We could then carry this clock around to every clock in the lattice, adjust that clock to match the master clock, and finally return the master clock to the origin.

This would be a perfectly good method of clock synchronization in Newtonian mechanics, where time flows along smoothly, the same for everyone. But we've been driven to reexamine the nature of time by the possibility that time is different in reference frames moving relative to each other. Because the master clock would *move,* we cannot assume that the moving master clock would keep time in the same way as the stationary clocks.

We need a synchronization method that does not require moving the clocks. Fortunately, such a method is easy to devise. Each clock is resting at the intersection of meter sticks, so by looking at the meter sticks, the assistant knows, or can calculate, exactly how far each clock is from the origin. Once the distance is known, the assistant can calculate exactly how long a light wave will take to travel from the origin to each clock. For example, light will take 1.00 μs to travel to a clock 300 m from the origin.

NOTE It's handy for many relativity problems to know that the speed of light is $c = 300$ m/μs.

To synchronize the clocks, the assistants begin by setting each clock to display the light travel time from the origin, but they don't start the clocks. Next, as FIGURE 36.13 shows, a light flashes at the origin and, simultaneously, the clock at the origin starts running from $t = 0$ s. The light wave spreads out in all directions at speed c. A photodetector on each clock recognizes the arrival of the light wave and, without delay, starts the clock. The clock had been preset with the light travel time, so each clock as it starts reads exactly the same as the clock at the origin. Thus all the clocks will be synchronized after the light wave has passed by.

Events and Observations

We noted above that t is the time the event *actually happens*. This is an important point, one that bears further discussion. Light waves take time to travel. Messages, whether they're transmitted by light pulses, telephone, or courier on horseback, take time to be delivered. An experimenter *observes* an event, such as an exploding firecracker, only *at a later time* when light waves reach his or her eyes. But our interest is in the event itself, not the experimenter's observation of the event. The time at which the experimenter sees the event or receives information about the event is not when the event actually occurred.

Suppose at $t = 0$ s a firecracker explodes at $x = 300$ m. The flash of light from the firecracker will reach an experimenter at the origin at $t_1 = 1.0$ μs. The sound of the explosion will reach a sightless experimenter at the origin at $t_2 = 0.88$ s. Neither of these is the time t_{event} of the explosion, although the experimenter can work backward from these times, using known wave speeds, to determine t_{event}. In this example, the spacetime coordinates of the event—the explosion—are (300 m, 0 m, 0 m, 0 s).

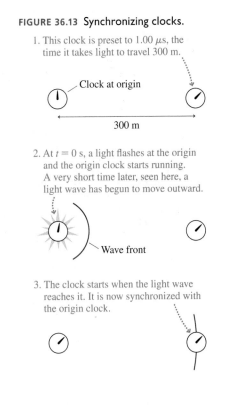

FIGURE 36.13 Synchronizing clocks.

1. This clock is preset to 1.00 μs, the time it takes light to travel 300 m.

Clock at origin

300 m

2. At $t = 0$ s, a light flashes at the origin and the origin clock starts running. A very short time later, seen here, a light wave has begun to move outward.

Wave front

3. The clock starts when the light wave reaches it. It is now synchronized with the origin clock.

EXAMPLE 36.2 | **Finding the time of an event**

Experimenter A in reference frame S stands at the origin looking in the positive x-direction. Experimenter B stands at $x = 900$ m looking in the negative x-direction. A firecracker explodes somewhere between them. Experimenter B sees the light flash at $t = 3.0$ μs.

Experimenter A sees the light flash at $t = 4.0$ μs. What are the spacetime coordinates of the explosion?

MODEL Experimenters A and B are in the same reference frame and have synchronized clocks.

Continued

VISUALIZE FIGURE 36.14 shows the two experimenters and the explosion at unknown position x.

FIGURE 36.14 The light wave reaches the experimenters at different times. Neither of these is the time at which the event actually happened.

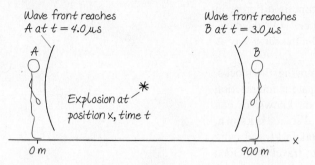

Wave front reaches A at $t = 4.0 \mu s$

Wave front reaches B at $t = 3.0 \mu s$

Explosion at position x, time t

0 m

900 m

x

SOLVE The two experimenters observe light flashes at two different instants, but there's only one event. Light travels at 300 m/μs, so the additional 1.0 μs needed for the light to reach experimenter A implies that distance $(x - 0 \text{ m})$ is 300 m longer than distance $(900 \text{ m} - x)$. That is,

$$(x - 0 \text{ m}) = (900 \text{ m} - x) + 300 \text{ m}$$

This is easily solved to give $x = 600$ m as the position coordinate of the explosion. The light takes 1.0 μs to travel 300 m to experimenter B, 2.0 μs to travel 600 m to experimenter A. The light is received at 3.0 μs and 4.0 μs, respectively; hence it was emitted by the explosion at $t = 2.0 \mu s$. The spacetime coordinates of the explosion are (600 m, 0 m, 0 m, 2.0 μs).

ASSESS Although the experimenters *see* the explosion at different times, they agree that the explosion actually *happened* at $t = 2.0 \mu s$.

Simultaneity

Two events 1 and 2 that take place at different positions x_1 and x_2 but at the *same time* $t_1 = t_2$, as measured in some reference frame, are said to be **simultaneous** in that reference frame. Simultaneity is determined by when the events actually happen, not when they are seen or observed. In general, simultaneous events are *not* seen at the same time because of the difference in light travel times from the events to an experimenter.

EXAMPLE 36.3 | **Are the explosions simultaneous?**

An experimenter in reference frame S stands at the origin looking in the positive x-direction. At $t = 3.0 \mu s$ she sees firecracker 1 explode at $x = 600$ m. A short time later, at $t = 5.0 \mu s$, she sees firecracker 2 explode at $x = 1200$ m. Are the two explosions simultaneous? If not, which firecracker exploded first?

MODEL Light from both explosions travels toward the experimenter at 300 m/μs.

SOLVE The experimenter *sees* two different explosions, but perceptions of the events are not the events themselves. When did the explosions *actually* occur? Using the fact that light travels at 300 m/μs, we can see that firecracker 1 exploded at $t_1 = 1.0 \mu s$ and firecracker 2 also exploded at $t_2 = 1.0 \mu s$. The events *are* simultaneous.

STOP TO THINK 36.4 A tree and a pole are 3000 m apart. Each is suddenly hit by a bolt of lightning. Mark, who is standing at rest midway between the two, sees the two lightning bolts at the same instant of time. Nancy is at rest under the tree. Define event 1 to be "lightning strikes tree" and event 2 to be "lightning strikes pole." For Nancy, does event 1 occur before, after, or at the same time as event 2?

36.5 The Relativity of Simultaneity

We've now established a means for measuring the time of an event in a reference frame, so let's begin to investigate the nature of time. The following "thought experiment" is very similar to one suggested by Einstein.

FIGURE 36.15 shows a long railroad car traveling to the right with a velocity v that may be an appreciable fraction of the speed of light. A firecracker is tied to each end of the car, just above the ground. Each firecracker is powerful enough so that, when it explodes, it will make a burn mark on the ground at the position of the explosion.

Ryan is standing on the ground, watching the railroad car go by. Peggy is standing in the exact center of the car with a special box at her feet. This box has two light detectors, one facing each way, and a signal light on top. The box works as follows:

1. If a flash of light is received at the detector facing right, as seen by Ryan, before a flash is received at the left detector, then the light on top of the box will turn green.
2. If a flash of light is received at the left detector before a flash is received at the right detector, or if two flashes arrive simultaneously, the light on top will turn red.

The firecrackers explode as the railroad car passes Ryan, and he sees the two light flashes from the explosions simultaneously. He then measures the distances to the two burn marks and finds that he was standing exactly halfway between the marks. Because light travels equal distances in equal times, Ryan concludes that the two explosions were simultaneous in his reference frame, the reference frame of the ground. Further, because he was midway between the two ends of the car, he was directly opposite Peggy when the explosions occurred.

FIGURE 36.16a shows the sequence of events in Ryan's reference frame. Light travels at speed c in all inertial reference frames, so, although the firecrackers were moving, the light waves are spheres centered on the burn marks. Ryan determines that the light wave coming from the right reaches Peggy and the box before the light wave coming from the left. Thus, according to Ryan, the signal light on top of the box turns green.

FIGURE 36.15 A railroad car traveling to the right with velocity v.

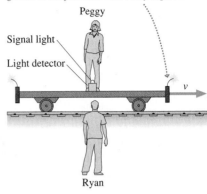

The firecrackers will make burn marks on the ground at the positions where they explode.

FIGURE 36.16 Exploding firecrackers seen in two different reference frames.

(a) The events in Ryan's frame

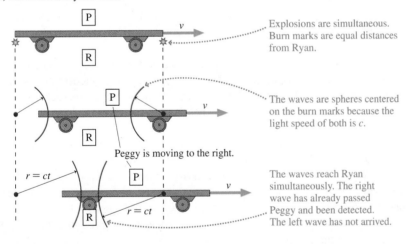

Explosions are simultaneous. Burn marks are equal distances from Ryan.

The waves are spheres centered on the burn marks because the light speed of both is c.

Peggy is moving to the right.

The waves reach Ryan simultaneously. The right wave has already passed Peggy and been detected. The left wave has not arrived.

$r = ct$

(b) Are these the events in Peggy's frame?

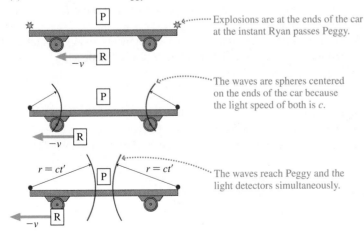

Explosions are at the ends of the car at the instant Ryan passes Peggy.

The waves are spheres centered on the ends of the car because the light speed of both is c.

The waves reach Peggy and the light detectors simultaneously.

$r = ct'$

How do things look in Peggy's reference frame, a reference frame moving to the right at velocity v relative to the ground? As **FIGURE 36.16b** shows, Peggy sees Ryan moving to the left with speed v. Light travels at speed c in all inertial reference frames, so the light waves are spheres centered on the ends of the car. If the explosions are simultaneous, as Ryan has determined, the two light waves reach her and the box simultaneously. Thus, according to Peggy, the signal light on top of the box turns red!

Now the light on top must be either green or red. *It can't be both!* Later, after the railroad car has stopped, Ryan and Peggy can place the box in front of them. Either it has a red light or a green light. Ryan can't see one color while Peggy sees the other. Hence we have a paradox. It's impossible for Peggy and Ryan both to be right. But who is wrong, and why?

What do we know with absolute certainty?

1. Ryan detected the flashes simultaneously.
2. Ryan was halfway between the firecrackers when they exploded.
3. The light from the two explosions traveled toward Ryan at equal speeds.

The conclusion that the explosions were simultaneous in Ryan's reference frame is unassailable. The light is green.

Resolving the Paradox

Peggy, however, made an assumption. It's a perfectly ordinary assumption, one that seems sufficiently obvious that you probably didn't notice, but an assumption nonetheless. Peggy assumed that the explosions were simultaneous.

Didn't Ryan find them to be simultaneous? Indeed, he did. Suppose we call Ryan's reference frame S, the explosion on the right event R, and the explosion on the left event L. Ryan found that $t_R = t_L$. But Peggy has to use a different set of clocks, the clocks in her reference frame S$'$, to measure the times t_R' and t_L' at which the explosions occurred. The fact that $t_R = t_L$ in frame S does *not* allow us to conclude that $t_R' = t_L'$ in frame S$'$.

In fact, in frame S$'$ the right firecracker must explode *before* the left firecracker. Figure 36.16b, with its assumption about simultaneity, was incorrect. **FIGURE 36.17** shows the situation in Peggy's reference frame, with the right firecracker exploding first. Now the wave from the right reaches Peggy and the box first, as Ryan had concluded, and the light on top turns green.

One of the most disconcerting conclusions of relativity is that **two events occurring simultaneously in reference frame S are *not* simultaneous in any reference frame S$'$ moving relative to S.** This is called the **relativity of simultaneity**.

The two firecrackers *really* explode at the same instant of time in Ryan's reference frame. And the right firecracker *really* explodes first in Peggy's reference frame. It's not a matter of when they see the flashes. Our conclusion refers to the times at which the explosions actually occur.

The paradox of Peggy and Ryan contains the essence of relativity, and it's worth careful thought. First, review the logic until you're certain that there *is* a paradox, a logical impossibility. Then convince yourself that the only way to resolve the paradox is to abandon the assumption that the explosions are simultaneous in Peggy's reference frame. If you understand the paradox and its resolution, you've made a big step toward understanding what relativity is all about.

FIGURE 36.17 The real sequence of events in Peggy's reference frame.

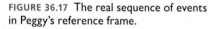
The right firecracker explodes first.

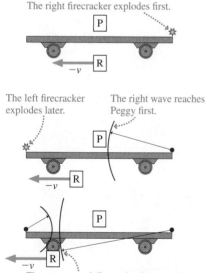
The left firecracker explodes later.

The right wave reaches Peggy first.

The waves reach Ryan simultaneously. The left wave has not reached Peggy.

STOP TO THINK 36.5 A tree and a pole are 3000 m apart. Each is hit by a bolt of lightning. Mark, who is standing at rest midway between the two, sees the two lightning bolts at the same instant of time. Nancy is flying her rocket at $v = 0.5c$ in the direction from the tree toward the pole. The lightning hits the tree just as she passes by it. Define event 1 to be "lightning strikes tree" and event 2 to be "lightning strikes pole." For Nancy, does event 1 occur before, after, or at the same time as event 2?

36.6 Time Dilation

The principle of relativity has driven us to the logical conclusion that time is not the same for two reference frames moving relative to each other. Our analysis thus far has been mostly qualitative. It's time to start developing some quantitative tools that will allow us to compare measurements in one reference frame to measurements in another reference frame.

FIGURE 36.18a shows a special clock called a *light clock*. The light clock is a box with a light source at the bottom and a mirror at the top, separated by distance h. The light source emits a very short pulse of light that travels to the mirror and reflects back to a light detector beside the source. The clock advances one "tick" each time the detector receives a light pulse, and it immediately, with no delay, causes the light source to emit the next light pulse.

Our goal is to compare two measurements of the interval between two ticks of the clock: one taken by an experimenter standing next to the clock and the other by an experimenter moving with respect to the clock. To be specific, **FIGURE 36.18b** shows the clock at rest in reference frame S'. We call this the **rest frame** of the clock. Reference frame S', with the clock, moves to the right with velocity v relative to reference frame S.

Relativity requires us to measure *events,* so let's define event 1 to be the emission of a light pulse and event 2 to be the detection of that light pulse. Experimenters in both reference frames are able to measure where and when these events occur *in their frame*. In frame S, the time interval $\Delta t = t_2 - t_1$ is one tick of the clock. Similarly, one tick in frame S' is $\Delta t' = t'_2 - t'_1$.

To be sure we have a clear understanding of the relativity result, let's first do a classical analysis. In frame S', the clock's rest frame, the light travels straight up and down, a total distance $2h$, at speed c. The time interval is $\Delta t' = 2h/c$.

FIGURE 36.19a shows the operation of the light clock as seen in frame S. The clock is moving to the right at speed v in S, thus the mirror moves distance $\frac{1}{2}v(\Delta t)$ during the time $\frac{1}{2}(\Delta t)$ in which the light pulse moves from the source to the mirror. The distance traveled by the light during this interval is $\frac{1}{2}u_{\text{light}}(\Delta t)$, where u_{light} is the speed of light in frame S. You can see from the vector addition in **FIGURE 36.19b** that the speed of light in frame S is $u_{\text{light}} = (c^2 + v^2)^{1/2}$. (Remember, this is a classical analysis in which the speed of light *does* depend on the motion of the reference frame relative to the light source.)

The Pythagorean theorem applied to the right triangle in Figure 36.19a is

$$h^2 + \left(\tfrac{1}{2}v\,\Delta t\right)^2 = \left(\tfrac{1}{2}u_{\text{light}}\,\Delta t\right)^2 = \left(\tfrac{1}{2}\sqrt{c^2+v^2}\,\Delta t\right)^2$$
$$= \left(\tfrac{1}{2}c\,\Delta t\right)^2 + \left(\tfrac{1}{2}v\,\Delta t\right)^2 \tag{36.4}$$

The term $\left(\tfrac{1}{2}v\,\Delta t\right)^2$ is common to both sides and cancels. Solving for Δt gives $\Delta t = 2h/c$, identical to $\Delta t'$. In other words, a classical analysis finds that the clock ticks at exactly the same rate in both frame S and frame S'. This shouldn't be surprising. There's only one kind of time in classical physics, measured the same by all experimenters independent of their motion.

The principle of relativity changes only one thing, but that change has profound consequences. According to the principle of relativity, light travels at the same speed in *all* inertial reference frames. In frame S', the rest frame of the clock, the light simply goes straight up and back. The time of one tick,

$$\Delta t' = \frac{2h}{c} \tag{36.5}$$

is unchanged from the classical analysis.

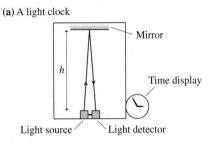

FIGURE 36.18 The ticking of a light clock can be measured by experimenters in two different reference frames.

(a) A light clock

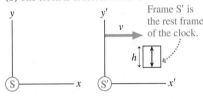

(b) The clock is at rest in frame S'.

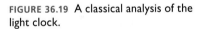

FIGURE 36.19 A classical analysis of the light clock.

(a)

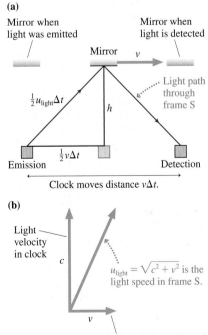

(b)

FIGURE 36.20 A light clock analysis in which the speed of light is the same in all reference frames.

Light speed is the same in both frames.

Mirror

v

$\frac{1}{2}c\,\Delta t$

Light path through frame S

h

Emission

$\frac{1}{2}v\,\Delta t$

Detection

Clock moves distance $v\Delta t$.

FIGURE 36.20 shows the light clock as seen in frame S. The difference from Figure 36.19a is that the light now travels along the hypotenuse at speed c. We can again use the Pythagorean theorem to write

$$h^2 + \left(\tfrac{1}{2}v\,\Delta t\right)^2 = \left(\tfrac{1}{2}c\,\Delta t\right)^2 \tag{36.6}$$

Solving for Δt gives

$$\Delta t = \frac{2h/c}{\sqrt{1 - v^2/c^2}} = \frac{\Delta t'}{\sqrt{1 - v^2/c^2}} \tag{36.7}$$

The time interval between two ticks in frame S is *not* the same as in frame S′.

It's useful to define $\beta = v/c$, the velocity as a fraction of the speed of light. For example, a reference frame moving with $v = 2.4 \times 10^8$ m/s has $\beta = 0.80$. In terms of β, Equation 36.7 is

$$\Delta t = \frac{\Delta t'}{\sqrt{1 - \beta^2}} \tag{36.8}$$

NOTE The expression $(1 - v^2/c^2)^{1/2} = (1 - \beta^2)^{1/2}$ occurs frequently in relativity. The value of the expression is 1 when $v = 0$, and it steadily decreases to 0 as $v \to c$ (or $\beta \to 1$). The square root is an imaginary number if $v > c$, which would make Δt imaginary in Equation 36.8. Time intervals certainly have to be real numbers, suggesting that $v > c$ is not physically possible. One of the predictions of the theory of relativity, as you've undoubtedly heard, is that nothing can travel faster than the speed of light. Now you can begin to see why. We'll examine this topic more closely in Section 36.9. In the meantime, we'll require v to be less than c.

Proper Time

Frame S′ has one important distinction. It is the *one and only* inertial reference frame in which the light clock is at rest. Consequently, it is the one and only inertial reference frame in which the times of both events—the emission of the light and the detection of the light—are measured by the *same* reference-frame clock. You can see that the light pulse in Figure 36.18a starts and ends at the same position. In Figure 36.20, the emission and detection take place at different positions in frame S and must be measured by different reference-frame clocks, one at each position.

The time interval between two events that occur at the *same position* is called the **proper time** $\Delta\tau$. Only one inertial reference frame measures the proper time, and it does so with *a single clock that is present at both events*. An inertial reference frame moving with velocity $v = \beta c$ relative to the proper-time frame must use two clocks to measure the time interval: one at the position of the first event, the other at the position of the second event. The time interval in the frame where two clocks are required is

$$\Delta t = \frac{\Delta\tau}{\sqrt{1 - \beta^2}} \geq \Delta\tau \qquad \text{(time dilation)} \tag{36.9}$$

The "stretching out" of the time interval implied by Equation 36.9 is called **time dilation.** Time dilation is sometimes described by saying that "moving clocks run slow," but this statement has to be interpreted carefully. The whole point of relativity is that all inertial frames are equally valid, so there's no absolute sense in which a clock is "moving" or "at rest."

To illustrate, **FIGURE 36.21** shows two firecracker explosions—two events—that occur at different positions in the ground reference frame. Experimenters on the ground need two clocks to measure the time interval Δt. In the train reference frame,

FIGURE 36.21 The time interval between two events is measured in two different reference frames.

The ground reference frame needs two clocks, A and B, to measure the time interval Δt between events 1 and 2.

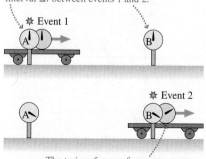

☀ Event 1

A

B

☀ Event 2

A

B

The train reference frame measures the proper time $\Delta\tau$ because one clock is present at both events.

however, a single clock is present at both events; hence the time interval measured in the train reference frame is the proper time $\Delta\tau$. You can see that $\Delta\tau < \Delta t$, so less time has elapsed in the train reference frame.

In this sense, the "moving clock," the one that is present at both events, "runs slower" than clocks that are stationary with respect to the events. More generally, **the time interval between two events is smallest in the reference frame in which the two events occur at the same position.**

> **NOTE** Equation 36.9 was derived using a light clock because the operation of a light clock is clear and easy to analyze. But the conclusion is really about time itself. *Any* clock, regardless of how it operates, behaves the same.

EXAMPLE 36.4 | From the sun to Saturn

Saturn is 1.43×10^{12} m from the sun. A rocket travels along a line from the sun to Saturn at a constant speed of $0.9c$ relative to the solar system. How long does the journey take as measured by an experimenter on earth? As measured by an astronaut on the rocket?

MODEL Let the solar system be in reference frame S and the rocket be in reference frame S′ that travels with velocity $v = 0.9c$ relative to S. Relativity problems must be stated in terms of *events*. Let event 1 be "the rocket and the sun coincide" (the experimenter on earth says that the rocket passes the sun; the astronaut on the rocket says that the sun passes the rocket) and event 2 be "the rocket and Saturn coincide."

FIGURE 36.22 Pictorial representation of the trip as seen in frames S and S′.

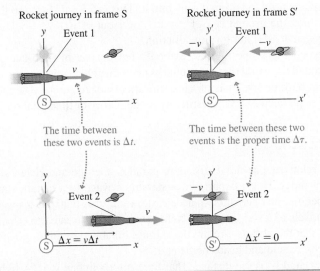

Rocket journey in frame S

The time between these two events is Δt.

Rocket journey in frame S′

The time between these two events is the proper time $\Delta\tau$.

VISUALIZE **FIGURE 36.22** shows the two events as seen from the two reference frames. Notice that the two events occur at the *same position* in S′, the position of the rocket, and consequently can be measured by *one* clock carried on board the rocket.

SOLVE The time interval measured in the solar system reference frame, which includes the earth, is simply

$$\Delta t = \frac{\Delta x}{v} = \frac{1.43 \times 10^{12} \text{ m}}{0.9 \times (3.00 \times 10^8 \text{ m/s})} = 5300 \text{ s}$$

Relativity hasn't abandoned the basic definition $v = \Delta x/\Delta t$, although we do have to be sure that Δx and Δt are measured in just one reference frame and refer to the same two events.

How are things in the rocket's reference frame? The two events occur at the *same position* in S′ and can be measured by *one* clock, the clock at the origin. Thus the time measured by the astronauts is the *proper time* $\Delta\tau$ between the two events. We can use Equation 36.9 with $\beta = 0.9$ to find

$$\Delta\tau = \sqrt{1 - \beta^2}\,\Delta t = \sqrt{1 - 0.9^2}\,(5300 \text{ s}) = 2310 \text{ s}$$

ASSESS The time interval measured between these two events by the astronauts is less than half the time interval measured by experimenters on earth. The difference has nothing to do with when earthbound astronomers *see* the rocket pass the sun and Saturn. Δt is the time interval from when the rocket actually passes the sun, as measured by a clock at the sun, until it actually passes Saturn, as measured by a synchronized clock at Saturn. The interval between *seeing* the events from earth, which would have to allow for light travel times, would be something other than 5300 s. Δt and $\Delta\tau$ are different because *time is different* in two reference frames moving relative to each other.

STOP TO THINK 36.6 Molly flies her rocket past Nick at constant velocity v. Molly and Nick both measure the time it takes the rocket, from nose to tail, to pass Nick. Which of the following is true?

a. Both Molly and Nick measure the same amount of time.
b. Molly measures a shorter time interval than Nick.
c. Nick measures a shorter time interval than Molly.

Experimental Evidence

Is there any evidence for the crazy idea that clocks moving relative to each other tell time differently? Indeed, there's plenty. An experiment in 1971 sent an atomic clock around the world on a jet plane while an identical clock remained in the laboratory. This was a difficult experiment because the traveling clock's speed was so small compared to c, but measuring the small differences between the time intervals was just barely within the capabilities of atomic clocks. It was also a more complex experiment than we've analyzed because the clock accelerated as it moved around a circle. The scientists found that, upon its return, the eastbound clock, traveling faster than the laboratory on a rotating earth, was 60 ns behind the stay-at-home clock, which was exactly as predicted by relativity.

Very detailed studies have been done on unstable particles called *muons* that are created at the top of the atmosphere, at a height of about 60 km, when high-energy cosmic rays collide with air molecules. It is well known, from laboratory studies, that stationary muons decay with a *half-life* of 1.5 μs. That is, half the muons decay within 1.5 μs, half of those remaining decay in the next 1.5 μs, and so on. The decays can be used as a clock.

The muons travel down through the atmosphere at very nearly the speed of light. The time needed to reach the ground, assuming $v \approx c$, is $\Delta t \approx (60,000 \text{ m})/(3 \times 10^8 \text{ m/s}) = 200 \ \mu$s. This is 133 half-lives, so the fraction of muons reaching the ground should be $\approx (\frac{1}{2})^{133} = 10^{-40}$. That is, only 1 out of every 10^{40} muons should reach the ground. In fact, experiments find that about 1 in 10 muons reach the ground, an experimental result that differs by a factor of 10^{39} from our prediction!

The discrepancy is due to time dilation. In **FIGURE 36.23**, the two events "muon is created" and "muon hits ground" take place at two different places in the earth's reference frame. However, these two events occur at the *same position* in the muon's reference frame. (The muon is like the rocket in Example 36.4.) Thus the muon's internal clock measures the proper time. The time-dilated interval $\Delta t = 200 \ \mu$s in the earth's reference frame corresponds to a proper time $\Delta \tau \approx 5 \ \mu$s in the muon's reference frame. That is, in the muon's reference frame it takes only 5 μs from creation at the top of the atmosphere until the ground runs into it. This is 3.3 half-lives, so the fraction of muons reaching the ground is $\left(\frac{1}{2}\right)^{3.3} = 0.1$, or 1 out of 10. We wouldn't detect muons at the ground at all if not for time dilation.

The details are beyond the scope of this textbook, but dozens of high-energy particle accelerators around the world that study quarks and other elementary particles have been designed and built on the basis of Einstein's theory of relativity. The fact that they work exactly as planned is strong testimony to the reality of time dilation.

FIGURE 36.23 We wouldn't detect muons at the ground if not for time dilation.

A muon travels ≈450 m in 1.5 μs. We would not detect muons at ground level if the half-life of a moving muon were 1.5 μs.

Muon is created.

Because of time dilation, the half-life of a muon is long enough in the earth's reference frame for 1 in 10 muons to reach the ground.

Muon hits ground.

The global positioning system (GPS), which allows you to pinpoint your location anywhere in the world to within a few meters, uses a set of orbiting satellites. Because of their motion, the atomic clocks on these satellites keep time differently from clocks on the ground. To determine an accurate position, the software in your GPS receiver must carefully correct for time-dilation effects.

The Twin Paradox

The most well-known relativity paradox is the twin paradox. George and Helen are twins. On their 25th birthday, Helen departs on a starship voyage to a distant star. Let's imagine, to be specific, that her starship accelerates almost instantly to a speed of 0.95c and that she travels to a star that is 9.5 light years (9.5 ly) from earth. Upon arriving, she discovers that the planets circling the star are inhabited by fierce aliens, so she immediately turns around and heads home at 0.95c.

A **light year,** abbreviated ly, is the distance that light travels in one year. A light year is vastly larger than the diameter of the solar system. The distance between two neighboring stars is typically a few light years. For our purpose, we can write the speed of light as $c = 1$ ly/year. That is, light travels 1 light year per year.

This value for c allows us to determine how long, according to George and his fellow earthlings, it takes Helen to travel out and back. Her total distance is 19 ly and, due to her rapid acceleration and rapid turnaround, she travels essentially the entire distance at speed $v = 0.95c = 0.95$ ly/year. Thus the time she's away, as measured by George, is

$$\Delta t_G = \frac{19 \text{ ly}}{0.95 \text{ ly/year}} = 20 \text{ years} \qquad (36.10)$$

George will be 45 years old when his sister Helen returns with tales of adventure.

While she's away, George takes a physics class and studies Einstein's theory of relativity. He realizes that time dilation will make Helen's clocks run more slowly than his clocks, which are at rest relative to him. Her heart—a clock—will beat fewer times and the minute hand on her watch will go around fewer times. In other words, she's aging more slowly than he is. Although she is his twin, she will be younger than he is when she returns.

Calculating Helen's age is not hard. We simply have to identify Helen's clock, because it's always with Helen as she travels, as the clock that measures proper time $\Delta \tau$. From Equation 36.9,

$$\Delta t_{\rm H} = \Delta \tau = \sqrt{1 - \beta^2}\, \Delta t_{\rm G} = \sqrt{1 - 0.95^2}\ (20 \text{ years}) = 6.25 \text{ years} \qquad (36.11)$$

George will have just celebrated his 45th birthday as he welcomes home his 31-year-and-3-month-old twin sister.

This may be unsettling because it violates our commonsense notion of time, but it's not a paradox. There's no logical inconsistency in this outcome. So why is it called "the twin paradox"?

Helen, knowing that she had quite of bit of time to kill on her journey, brought along several physics books to read. As she learns about relativity, she begins to think about George and her friends back on earth. Relative to her, they are all moving away at $0.95c$. Later they'll come rushing toward her at $0.95c$. Time dilation will cause their clocks to run more slowly than her clocks, which are at rest relative to her. In other words, as FIGURE 36.24 shows, Helen concludes that people on earth are aging more slowly than she is. Alas, she will be much older than they when she returns.

Finally, the big day arrives. Helen lands back on earth and steps out of the starship. George is expecting Helen to be younger than he is. Helen is expecting George to be younger than she is.

Here's the paradox! It's logically impossible for each to be younger than the other at the time they are reunited. Where, then, is the flaw in our reasoning? It seems to be a symmetrical situation—Helen moves relative to George and George moves relative to Helen—but symmetrical reasoning has led to a conundrum.

But are the situations really symmetrical? George goes about his business day after day without noticing anything unusual. Helen, on the other hand, experiences three distinct periods during which the starship engines fire, she's crushed into her seat, and free dust particles that had been floating inside the starship are no longer, in the starship's reference frame, at rest or traveling in a straight line at constant speed. In other words, George spends the entire time in an inertial reference frame, *but Helen does not*. The situation is *not* symmetrical.

The principle of relativity applies *only* to inertial reference frames. Our discussion of time dilation was for inertial reference frames. Thus George's analysis and calculations are correct. Helen's analysis and calculations are *not* correct because she was trying to apply an inertial reference frame result while traveling in a noninertial reference frame. (Or, alternatively, Helen was in two different inertial frames while George was only in one, and thus the situation is not symmetrical.)

Helen is younger than George when she returns. This is strange, but not a paradox. It is a consequence of the fact that time flows differently in two reference frames moving relative to each other.

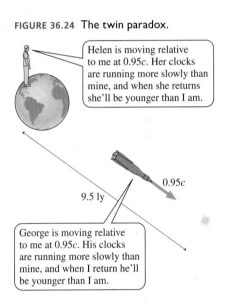

FIGURE 36.24 The twin paradox.

Helen is moving relative to me at $0.95c$. Her clocks are running more slowly than mine, and when she returns she'll be younger than I am.

$0.95c$

9.5 ly

George is moving relative to me at $0.95c$. His clocks are running more slowly than mine, and when I return he'll be younger than I am.

36.7 Length Contraction

We've seen that relativity requires us to rethink our idea of time. Now let's turn our attention to the concepts of space and distance. Consider the rocket that traveled from the sun to Saturn in Example 36.4. FIGURE 36.25a on the next page shows the rocket moving with velocity v through the solar system reference frame S. We define $L = \Delta x = x_{\rm Saturn} - x_{\rm sun}$ as the distance between the sun and Saturn in frame S or, more generally, the *length* of the spatial interval between two points. The rocket's speed is $v = L/\Delta t$, where Δt is the time measured in frame S for the journey from the sun to Saturn.

FIGURE 36.25 L and L' are the distances between the sun and Saturn in frames S and S'.

(a) Reference frame S: The solar system is stationary.   **(b)** Reference frame S': The rocket is stationary.

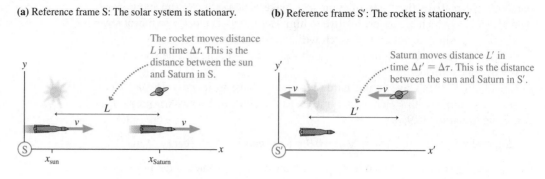

FIGURE 36.25b shows the situation in reference frame S', where the rocket is at rest. The sun and Saturn move to the left at speed $v = L'/\Delta t'$, where $\Delta t'$ is the time measured in frame S' for Saturn to travel distance L'.

Speed v is the relative speed between S and S' and is the same for experimenters in both reference frames. That is,

$$v = \frac{L}{\Delta t} = \frac{L'}{\Delta t'} \tag{36.12}$$

The time interval $\Delta t'$ measured in frame S' is the proper time $\Delta\tau$ because both events occur at the same position in frame S' and can be measured by one clock. We can use the time-dilation result, Equation 36.9, to relate $\Delta\tau$ measured by the astronauts to Δt measured by the earthbound scientists. Then Equation 36.12 becomes

$$\frac{L}{\Delta t} = \frac{L'}{\Delta\tau} = \frac{L'}{\sqrt{1 - \beta^2}\,\Delta t} \tag{36.13}$$

The Δt cancels, and the distance L' in frame S' is

$$L' = \sqrt{1 - \beta^2}\,L \tag{36.14}$$

Surprisingly, we find that **the distance between two objects in reference frame S' is *not the same* as the distance between the same two objects in reference frame S.**

Frame S, in which the distance is L, has one important distinction. It is the *one and only* inertial reference frame in which the objects are at rest. Experimenters in frame S can take all the time they need to measure L because the two objects aren't going anywhere. The distance L between two objects, or two points on one object, measured in the reference frame in which the objects are at rest is called the **proper length** ℓ. Only one inertial reference frame can measure the proper length.

We can use the proper length ℓ to write Equation 36.14 as

$$L' = \sqrt{1 - \beta^2}\,\ell \le \ell \tag{36.15}$$

The Stanford Linear Accelerator (SLAC) is a 2-mi-long electron accelerator. The accelerator's length is less than 1 m in the reference frame of the electrons.

This "shrinking" of the distance between two objects, as measured by an experiment moving with respect to the objects, is called **length contraction.** Although we derived length contraction for the distance between two distinct objects, it applies equally well to the length of any physical object that stretches between two points along the x- and x'-axes. **The length of an object is greatest in the reference frame in which the object is at rest.** The object's length is less (i.e., the length is contracted) when it is measured in any reference frame in which the object is moving.

EXAMPLE 36.5 | The distance from the sun to Saturn

In Example 36.4 a rocket traveled along a line from the sun to Saturn at a constant speed of $0.9c$ relative to the solar system. The Saturn-to-sun distance was given as 1.43×10^{12} m. What is the distance between the sun and Saturn in the rocket's reference frame?

MODEL Saturn and the sun are, at least approximately, at rest in the solar system reference frame S. Thus the given distance is the proper length ℓ.

SOLVE We can use Equation 36.15 to find the distance in the rocket's frame S':

$$L' = \sqrt{1 - \beta^2}\, \ell = \sqrt{1 - 0.9^2}\,(1.43 \times 10^{12}\ \text{m})$$
$$= 0.62 \times 10^{12}\ \text{m}$$

ASSESS The sun-to-Saturn distance measured by the astronauts is less than half the distance measured by experimenters on earth. L' and ℓ are different because *space is different* in two reference frames moving relative to each other.

The conclusion that space is different in reference frames moving relative to each other is a direct consequence of the fact that time is different. Experimenters in both reference frames agree on the relative velocity v, leading to Equation 36.12: $v = L/\Delta t = L'/\Delta t'$. We had already learned that $\Delta t' < \Delta t$ because of time dilation. Thus L' *has* to be less than L. That is the only way experimenters in the two reference frames can reconcile their measurements.

To be specific, the earthly experimenters in Examples 36.4 and 36.5 find that the rocket takes 5300 s to travel the 1.43×10^{12} m between the sun and Saturn. The rocket's speed is $v = L/\Delta t = 2.7 \times 10^8$ m/s $= 0.9c$. The astronauts in the rocket find that it takes only 2310 s for Saturn to reach them after the sun has passed by. But there's no conflict, because they also find that the distance is only 0.62×10^{12} m. Thus Saturn's speed toward them is $v = L'/\Delta t' = (0.62 \times 10^{12}\ \text{m})/(2310\ \text{s}) = 2.7 \times 10^8$ m/s $= 0.9c$.

Another Paradox?

Carmen and Dan are in their physics lab room. They each select a meter stick, lay the two side by side, and agree that the meter sticks are exactly the same length. Then, for an extra-credit project, they go outside and run past each other, in opposite directions, at a relative speed $v = 0.9c$. FIGURE 36.26 shows their experiment and a portion of their conversation.

Now, Dan's meter stick can't be both longer and shorter than Carmen's meter stick. Is this another paradox? No! Relativity allows us to compare the *same* events as they're measured in two different reference frames. This did lead to a real paradox when Peggy rolled past Ryan on the train. There the signal light on the box turns green (a single event) or it doesn't, and Peggy and Ryan have to agree about it. But the events by which Dan measures the length (in Dan's frame) of Carmen's meter stick are *not the same events* as those by which Carmen measures the length (in Carmen's frame) of Dan's meter stick.

There's no conflict between their measurements. In Dan's reference frame, Carmen's meter stick has been length contracted and is less than 1 m in length. In Carmen's reference frame, Dan's meter stick has been length contracted and is less than 1 m in length. If this weren't the case, if both agreed that one of the meter sticks was shorter than the other, then we could tell which reference frame was "really" moving and which was "really" at rest. But the principle of relativity doesn't allow us to make that distinction. Each is moving relative to the other, so each should make the same measurement for the length of the other's meter stick.

The Spacetime Interval

Forget relativity for a minute and think about ordinary geometry. FIGURE 36.27 shows two ordinary coordinate systems. They are identical except for the fact that one has been rotated relative to the other. A student using the xy-system would measure coordinates (x_1, y_1) for point 1 and (x_2, y_2) for point 2. A second student, using the $x'y'$-system, would measure (x'_1, y'_1) and (x'_2, y'_2).

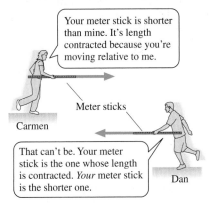

FIGURE 36.26 Carmen and Dan each measure the length of the other's meter stick as they move relative to each other.

Your meter stick is shorter than mine. It's length contracted because you're moving relative to me.

Meter sticks

Carmen

That can't be. Your meter stick is the one whose length is contracted. *Your* meter stick is the shorter one.

Dan

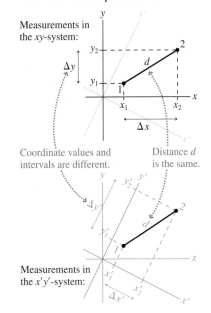

FIGURE 36.27 Distance d is the same in both coordinate systems.

Measurements in the xy-system:

Coordinate values and intervals are different.

Distance d is the same.

Measurements in the $x'y'$-system:

The students soon find that none of their measurements agree. That is, $x_1 \neq x'_1$ and so on. Even the intervals are different: $\Delta x \neq \Delta x'$ and $\Delta y \neq \Delta y'$. Each is a perfectly valid coordinate system, giving no reason to prefer one over the other, but each yields different measurements.

Is there *anything* on which the two students can agree? Yes, there is. The distance d between points 1 and 2 is independent of the coordinates. We can state this mathematically as

$$d^2 = (\Delta x)^2 + (\Delta y)^2 = (\Delta x')^2 + (\Delta y')^2 \tag{36.16}$$

The quantity $(\Delta x)^2 + (\Delta y)^2$ is called an **invariant** in geometry because it has the same value in any Cartesian coordinate system.

Returning to relativity, is there an invariant in the spacetime coordinates, some quantity that has the *same value* in all inertial reference frames? There is, and to find it let's return to the light clock of Figure 36.20. **FIGURE 36.28** shows the light clock as seen in reference frames S' and S". The speed of light is the same in both frames, even though both are moving with respect to each other and with respect to the clock.

Notice that the clock's height h is common to both reference frames. Thus

$$h^2 = \left(\tfrac{1}{2}c\,\Delta t'\right)^2 - \left(\tfrac{1}{2}\Delta x'\right)^2 = \left(\tfrac{1}{2}c\,\Delta t''\right)^2 - \left(\tfrac{1}{2}\Delta x''\right)^2 \tag{36.17}$$

The factor $\tfrac{1}{2}$ cancels, allowing us to write

$$c^2(\Delta t')^2 - (\Delta x')^2 = c^2(\Delta t'')^2 - (\Delta x'')^2 \tag{36.18}$$

Let us define the **spacetime interval** s between two events to be

$$s^2 = c^2(\Delta t)^2 - (\Delta x)^2 \tag{36.19}$$

What we've shown in Equation 36.18 is that **the spacetime interval s has the same value in all inertial reference frames**. That is, the spacetime interval between two events is an invariant. It is a value that all experimenters, in all reference frames, can agree upon.

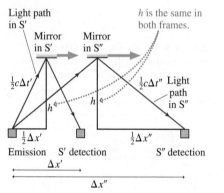

FIGURE 36.28 The light clock seen by experimenters in reference frames S' and S".

Light path in S'

Mirror in S' Mirror in S"

$\tfrac{1}{2}c\Delta t'$ $\tfrac{1}{2}c\Delta t''$ Light path in S"

h is the same in both frames.

h h

$\tfrac{1}{2}\Delta x'$ $\tfrac{1}{2}\Delta x''$

Emission S' detection S" detection

$\Delta x'$

$\Delta x''$

EXAMPLE 36.6 | Using the spacetime interval

A firecracker explodes at the origin of an inertial reference frame. Then, 2.0 μs later, a second firecracker explodes 300 m away. Astronauts in a passing rocket measure the distance between the explosions to be 200 m. According to the astronauts, how much time elapses between the two explosions?

MODEL The spacetime coordinates of two events are measured in two different inertial reference frames. Call the reference frame of the ground S and the reference frame of the rocket S'. The spacetime interval between these two events is the same in both reference frames.

SOLVE The spacetime interval (or, rather, its square) in frame S is

$$s^2 = c^2(\Delta t)^2 - (\Delta x)^2 = (600\text{ m})^2 - (300\text{ m})^2 = 270{,}000\text{ m}^2$$

where we used $c = 300$ m/μs to determine that $c\,\Delta t = 600$ m. The spacetime interval has the same value in frame S'. Thus

$$s^2 = 270{,}000\text{ m}^2 = c^2(\Delta t')^2 - (\Delta x')^2$$
$$= c^2(\Delta t')^2 - (200\text{ m})^2$$

This is easily solved to give $\Delta t' = 1.85$ μs.

ASSESS The two events are closer together in both space and time in the rocket's reference frame than in the reference frame of the ground.

Einstein's legacy, according to popular culture, was the discovery that "everything is relative." But it's not so. Time intervals and space intervals may be relative, as were the intervals Δx and Δy in the purely geometric analogy with which we opened this section, but some things are *not* relative. In particular, the spacetime interval s between two events is not relative. It is a well-defined number, agreed on by experimenters in each and every inertial reference frame.

STOP TO THINK 36.7 Beth and Charles are at rest relative to each other. Anjay runs past at velocity v while holding a long pole parallel to his motion. Anjay, Beth, and Charles each measure the length of the pole at the instant Anjay passes Beth. Rank in order, from largest to smallest, the three lengths L_A, L_B, and L_C.

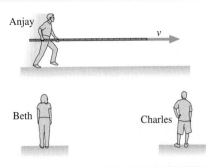

36.8 The Lorentz Transformations

The Galilean transformation $x' = x - vt$ of classical relativity lets us calculate the position x' of an event in frame S′ if we know its position x in frame S. Classical relativity, of course, assumes that $t' = t$. Is there a similar transformation in relativity that would allow us to calculate an event's spacetime coordinates (x', t') in frame S′ if we know their values (x, t) in frame S? Such a transformation would need to satisfy three conditions:

1. Agree with the Galilean transformations in the low-speed limit $v \ll c$.
2. Transform not only spatial coordinates but also time coordinates.
3. Ensure that the speed of light is the same in all reference frames.

We'll continue to use reference frames in the standard orientation of **FIGURE 36.29**. The motion is parallel to the x- and x'-axes, and we *define* $t = 0$ and $t' = 0$ as the instant when the origins of S and S′ coincide.

The requirement that a new transformation agree with the Galilean transformation when $v \ll c$ suggests that we look for a transformation of the form

$$x' = \gamma(x - vt) \quad \text{and} \quad x = \gamma(x' + vt') \tag{36.20}$$

where γ is a dimensionless function of velocity that satisfies $\gamma \rightarrow 1$ as $v \rightarrow 0$.

To determine γ, we consider the following two events:

Event 1: A flash of light is emitted from the origin of both reference frames $(x = x' = 0)$ at the instant they coincide $(t = t' = 0)$.

Event 2: The light strikes a light detector. The spacetime coordinates of this event are (x, t) in frame S and (x', t') in frame S′.

Light travels at speed c in both reference frames, so the positions of event 2 are $x = ct$ in S and $x' = ct'$ in S′. Substituting these expressions for x and x' into Equation 36.20 gives

$$ct' = \gamma(ct - vt) = \gamma(c - v)t$$
$$ct = \gamma(ct' + vt') = \gamma(c + v)t' \tag{36.21}$$

We solve the first equation for t', by dividing by c, then substitute this result for t' into the second:

$$ct = \gamma(c + v)\frac{\gamma(c - v)t}{c} = \gamma^2(c^2 - v^2)\frac{t}{c}$$

FIGURE 36.29 The spacetime coordinates of an event are measured in inertial reference frames S and S′.

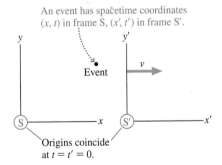

The t cancels, leading to

$$\gamma^2 = \frac{c^2}{c^2 - v^2} = \frac{1}{1 - v^2/c^2}$$

Thus the γ that "works" in the proposed transformation of Equation 36.20 is

$$\gamma = \frac{1}{\sqrt{1 - v^2/c^2}} = \frac{1}{\sqrt{1 - \beta^2}} \qquad (36.22)$$

You can see that $\gamma \rightarrow 1$ as $v \rightarrow 0$, as expected.

The transformation between t and t' is found by requiring that $x = x$ if you use Equation 36.20 to transform a position from S to S$'$ and then back to S. The details will be left for a homework problem. Another homework problem will let you demonstrate that the y and z measurements made perpendicular to the relative motion are not affected by the motion. We tacitly assumed this condition in our analysis of the light clock.

The full set of equations is called the **Lorentz transformations.** They are

$$
\begin{aligned}
x' &= \gamma(x - vt) & x &= \gamma(x' + vt') \\
y' &= y & y &= y' \\
z' &= z & z &= z' \\
t' &= \gamma(t - vx/c^2) & t &= \gamma(t' + vx'/c^2)
\end{aligned}
\qquad (36.23)
$$

The Lorentz transformations transform the spacetime coordinates of *one* event. Compare these to the Galilean transformation equations in Equations 36.1.

NOTE These transformations are named after the Dutch physicist H. A. Lorentz, who derived them prior to Einstein. Lorentz was close to discovering special relativity, but he didn't recognize that our concepts of space and time have to be changed before these equations can be properly interpreted.

Using Relativity

Relativity is phrased in terms of *events;* hence relativity problems are solved by interpreting the problem statement in terms of specific events.

PROBLEM-SOLVING STRATEGY 36.1 (MP)

Relativity

MODEL Frame the problem in terms of events, things that happen at a specific place and time.

VISUALIZE A pictorial representation defines the reference frames.

- Sketch the reference frames, showing their motion relative to each other.
- Show events. Identify objects that are moving with respect to the reference frames.
- Identify any proper time intervals and proper lengths. These are measured in an object's rest frame.

SOLVE The mathematical representation is based on the Lorentz transformations, but not every problem requires the full transformation equations.

- Problems about time intervals can often be solved using time dilation: $\Delta t = \gamma \Delta \tau$.
- Problems about distances can often be solved using length contraction: $L = \ell/\gamma$.

ASSESS Are the results consistent with Galilean relativity when $v \ll c$?

EXAMPLE 36.7 | Ryan and Peggy revisited

Peggy is standing in the center of a long, flat railroad car that has firecrackers tied to both ends. The car moves past Ryan, who is standing on the ground, with velocity $v = 0.8c$. Flashes from the exploding firecrackers reach him simultaneously 1.0 μs after the instant that Peggy passes him, and he later finds burn marks on the track 300 m to either side of where he had been standing.

a. According to Ryan, what is the distance between the two explosions, and at what times do the explosions occur relative to the time that Peggy passes him?

b. According to Peggy, what is the distance between the two explosions, and at what times do the explosions occur relative to the time that Ryan passes her?

MODEL Let the explosion on Ryan's right, the direction in which Peggy is moving, be event R. The explosion on his left is event L.

VISUALIZE Peggy and Ryan are in inertial reference frames. As FIGURE 36.30 shows, Peggy's frame S' is moving with $v = 0.8c$ relative to Ryan's frame S. We've defined the reference frames such that Peggy and Ryan are at the origins. The instant they pass, by definition, is $t = t' = 0$ s. The two events are shown in Ryan's reference frame.

FIGURE 36.30 A pictorial representation of the reference frames and events.

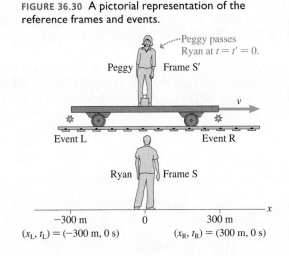

SOLVE a. The two burn marks tell Ryan that the distance between the explosions was $L = 600$ m. Light travels at $c = 300$ m/μs, and the burn marks are 300 m on either side of him, so Ryan can determine that each explosion took place 1.0 μs before he saw the flash. But this was the instant of time that Peggy passed him, so Ryan concludes that the explosions were simultaneous with each other and with Peggy's passing him. The spacetime coordinates of the two events in frame S are $(x_R, t_R) = (300 \text{ m}, 0 \text{ } \mu\text{s})$ and $(x_L, t_L) = (-300 \text{ m}, 0 \text{ } \mu\text{s})$.

b. We already know, from our qualitative analysis in Section 36.5, that the explosions are *not* simultaneous in Peggy's reference frame. Event R happens before event L in S', but we don't know how they compare to the time at which Ryan passes Peggy. We can now use the Lorentz transformations to relate the spacetime coordinates of these events as measured by Ryan to the spacetime coordinates as measured by Peggy. Using $v = 0.8c$, we find that γ is

$$\gamma = \frac{1}{\sqrt{1 - v^2/c^2}} = \frac{1}{\sqrt{1 - 0.8^2}} = 1.667$$

For event L, the Lorentz transformations are

$$x'_L = 1.667\big((-300 \text{ m}) - (0.8c)(0 \text{ } \mu\text{s})\big) = -500 \text{ m}$$

$$t'_L = 1.667\big((0 \text{ } \mu\text{s}) - (0.8c)(-300 \text{ m})/c^2\big) = 1.33 \text{ } \mu\text{s}$$

And for event R,

$$x'_R = 1.667\big((300 \text{ m}) - (0.8c)(0 \text{ } \mu\text{s})\big) = 500 \text{ m}$$

$$t'_R = 1.667\big((0 \text{ } \mu\text{s}) - (0.8c)(300 \text{ m})/c^2\big) = -1.33 \text{ } \mu\text{s}$$

According to Peggy, the two explosions occur 1000 m apart. Furthermore, the first explosion, on the right, occurs 1.33 μs before Ryan passes her at $t' = 0$ s. The second, on the left, occurs 1.33 μs after Ryan goes by.

ASSESS Events that are simultaneous in frame S are *not* simultaneous in frame S'. The results of the Lorentz transformations agree with our earlier qualitative analysis.

A follow-up discussion of Example 36.7 is worthwhile. Because Ryan moves at speed $v = 0.8c = 240$ m/μs relative to Peggy, he moves 320 m during the 1.33 μs between the first explosion and the instant he passes Peggy, then another 320 m before the second explosion. Gathering this information together, FIGURE 36.31 on the next page shows the sequence of events in Peggy's reference frame.

The firecrackers define the ends of the railroad car, so the 1000 m distance between the explosions in Peggy's frame is the car's length L' in frame S'. The car is at rest in frame S', hence length L' is the proper length: $\ell = 1000$ m. Ryan is measuring the length of a moving object, so he should see the car length contracted to

$$L = \sqrt{1 - \beta^2}\,\ell = \frac{\ell}{\gamma} = \frac{1000 \text{ m}}{1.667} = 600 \text{ m}$$

And, indeed, that is exactly the distance Ryan measured between the burn marks.

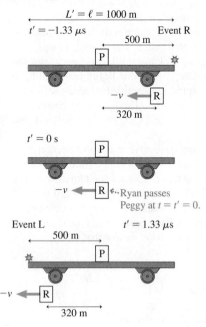

FIGURE 36.31 The sequence of events as seen in Peggy's reference frame.

Finally, we can calculate the spacetime interval s between the two events. According to Ryan,

$$s^2 = c^2(\Delta t^2) - (\Delta x)^2 = c^2(0 \ \mu\text{s})^2 - (600 \ \text{m})^2 = -(600 \ \text{m})^2$$

Peggy computes the spacetime interval to be

$$s^2 = c^2(\Delta t')^2 - (\Delta x')^2 = c^2(2.67 \ \mu\text{s})^2 - (1000 \ \text{m})^2 = -(600 \ \text{m})^2$$

Their calculations of the spacetime interval agree, showing that s really is an invariant, but notice that s itself is an imaginary number.

Length

We've already introduced the idea of length contraction, but we didn't precisely define just what we mean by the *length* of a moving object. The length of an object at rest is clear because we can take all the time we need to measure it with meter sticks, surveying tools, or whatever we need. But how can we give clear meaning to the length of a moving object?

A reasonable definition of an object's length is the distance $L = \Delta x = x_R - x_L$ between the right and left ends when the positions x_R and x_L are measured *at the same time t*. In other words, length is the distance spanned by the object at *one instant* of time. Measuring an object's length requires *simultaneous* measurements of two positions (i.e., two events are required); hence the result won't be known until the information from two spatially separated measurements can be brought together.

FIGURE 36.32 shows an object traveling through reference frame S with velocity v. The object is at rest in reference frame S′ that travels with the object at velocity v; hence the length in frame S′ is the proper length ℓ. That is, $\Delta x' = x_R' - x_L' = \ell$ in frame S′.

At time t, an experimenter (and his or her assistants) in frame S makes simultaneous measurements of the positions x_R and x_L of the ends of the object. The difference $\Delta x = x_R - x_L = L$ is the length in frame S. The Lorentz transformations of x_R and x_L are

$$
\begin{aligned}
x_R' &= \gamma(x_R - vt) \\
x_L' &= \gamma(x_L - vt)
\end{aligned}
\tag{36.24}
$$

where, it is important to note, t is the *same* for both because the measurements are simultaneous.

Subtracting the second equation from the first, we find

$$x_R' - x_L' = \ell = \gamma(x_R - x_L) = \gamma L = \frac{L}{\sqrt{1 - \beta^2}}$$

Solving for L, we find, in agreement with Equation 36.15, that

$$L = \sqrt{1 - \beta^2} \, \ell \tag{36.25}$$

This analysis has accomplished two things. First, by giving a precise definition of length, we've put our length-contraction result on a firmer footing. Second, we've had good practice at relativistic reasoning using the Lorentz transformation.

FIGURE 36.32 The length of an object is the distance between *simultaneous* measurements of the positions of the end points.

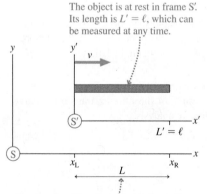

The object is at rest in frame S′. Its length is $L' = \ell$, which can be measured at any time.

$L' = \ell$

Because the object is moving in frame S, simultaneous measurements of its ends must be made to find its length L in frame S.

NOTE Length contraction does not tell us how an object would *look*. The visual appearance of an object is determined by light waves that arrive simultaneously at the eye. These waves left points on the object at different times (i.e., *not* simultaneously) because they had to travel different distances to the eye. The analysis needed to determine an object's visual appearance is considerably more complex. Length and length contraction are concerned only with the *actual* length of the object at one instant of time.

The Binomial Approximation

You've met the binomial approximation earlier in this text and in your calculus class. The binomial approximation is useful when we need to calculate a relativistic expression for a nonrelativistic velocity $v \ll c$. Because $v^2/c^2 \ll 1$ in these cases, we can write

The binomial approximation

If $x \ll 1$, then $(1 + x)^n \approx 1 + nx$.

$$\text{If } v \ll c: \begin{cases} \sqrt{1 - \beta^2} = \left(1 - \dfrac{v^2}{c^2}\right)^{1/2} \approx 1 - \dfrac{1}{2}\dfrac{v^2}{c^2} \\[3mm] \gamma = \dfrac{1}{\sqrt{1 - \beta^2}} = \left(1 - \dfrac{v^2}{c^2}\right)^{-1/2} \approx 1 + \dfrac{1}{2}\dfrac{v^2}{c^2} \end{cases} \quad (36.26)$$

The following example illustrates the use of the binomial approximation.

EXAMPLE 36.8 | **The shrinking school bus**

An 8.0-m-long school bus drives past at 30 m/s. By how much is its length contracted?

MODEL The school bus is at rest in an inertial reference frame S' moving at velocity $v = 30$ m/s relative to the ground frame S. The given length, 8.0 m, is the proper length ℓ in frame S'.

SOLVE In frame S, the school bus is length contracted to

$$L = \sqrt{1 - \beta^2}\,\ell$$

The bus's velocity v is much less than c, so we can use the binomial approximation to write

$$L \approx \left(1 - \frac{1}{2}\frac{v^2}{c^2}\right)\ell = \ell - \frac{1}{2}\frac{v^2}{c^2}\ell$$

The *amount* of the length contraction is

$$\ell - L = \frac{1}{2}\frac{v^2}{c^2}\ell = \frac{1}{2}\left(\frac{30 \text{ m/s}}{3.0 \times 10^8 \text{ m/s}}\right)^2 (8.0 \text{ m})$$

$$= 4.0 \times 10^{-14} \text{ m} = 40 \text{ fm}$$

where 1 fm = 1 femtometer = 10^{-15} m.

ASSESS The bus "shrinks" by only slightly more than the diameter of the nucleus of an atom. It's no wonder that we're not aware of length contraction in our everyday lives. If you had tried to calculate this number exactly, your calculator would have shown $\ell - L = 0$ because the difference between ℓ and L shows up only in the 14th decimal place. A scientific calculator determines numbers to 10 or 12 decimal places, but that isn't sufficient to show the difference. The binomial approximation provides an invaluable tool for finding the very tiny difference between two numbers that are nearly identical.

The Lorentz Velocity Transformations

FIGURE 36.33 shows an object that is moving in both reference frame S and reference frame S'. Experimenters in frame S determine that the object's velocity is u, while experimenters in frame S' find it to be u'. For simplicity, we'll assume that the object moves parallel to the x- and x'-axes.

The Galilean velocity transformation $u' = u - v$ was found by taking the time derivative of the position transformation. We can do the same with the Lorentz transformation if we take the derivative with respect to the time in each frame. Velocity u' in frame S' is

$$u' = \frac{dx'}{dt'} = \frac{d[\gamma(x - vt)]}{d[\gamma(t - vx/c^2)]} \quad (36.27)$$

where we've used the Lorentz transformations for position x' and time t'.

Carrying out the differentiation gives

$$u' = \frac{\gamma(dx - v\,dt)}{\gamma(dt - v\,dx/c^2)} = \frac{dx/dt - v}{1 - v(dx/dt)/c^2} \quad (36.28)$$

But dx/dt is u, the object's velocity in frame S, leading to

$$u' = \frac{u - v}{1 - uv/c^2} \quad (36.29)$$

You can see that Equation 36.29 reduces to the Galilean transformation $u' = u - v$ when $v \ll c$, as expected.

FIGURE 36.33 The velocity of a moving object is measured to be u in frame S and u' in frame S'.

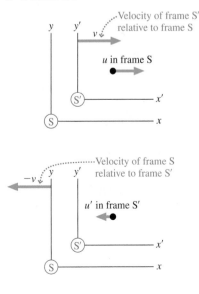

The transformation from S′ to S is found by reversing the sign of v. Altogether,

$$u' = \frac{u - v}{1 - uv/c^2} \quad \text{and} \quad u = \frac{u' + v}{1 + u'v/c^2} \tag{36.30}$$

Equations 36.30 are the Lorentz velocity transformation equations.

NOTE It is important to distinguish carefully between v, which is the relative velocity between two reference frames, and u and u', which are the velocities of an *object* as measured in the two different reference frames.

EXAMPLE 36.9 | **A really fast bullet**

A rocket flies past the earth at $0.90c$. As it goes by, the rocket fires a bullet in the forward direction at $0.95c$ with respect to the rocket. What is the bullet's speed with respect to the earth?

MODEL The rocket and the earth are inertial reference frames. Let the earth be frame S and the rocket be frame S′. The velocity of frame S′ relative to frame S is $v = 0.90c$. The bullet's velocity in frame S′ is $u' = 0.95c$.

SOLVE We can use the Lorentz velocity transformation to find

$$u = \frac{u' + v}{1 + u'v/c^2} = \frac{0.95c + 0.90c}{1 + (0.95c)(0.90c)/c^2} = 0.997c$$

The bullet's speed with respect to the earth is 99.7% of the speed of light.

NOTE Many relativistic calculations are much easier when velocities are specified as a fraction of c.

ASSESS In Newtonian mechanics, the Galilean transformation of velocity would give $u = 1.85c$. Now, despite the very high speed of the rocket and of the bullet with respect to the rocket, the bullet's speed with respect to the earth remains less than c. This is yet another indication that objects cannot travel faster than the speed of light.

Suppose the rocket in Example 36.9 fired a laser beam in the forward direction as it traveled past the earth at velocity v. The laser beam would travel away from the rocket at speed $u' = c$ in the rocket's reference frame S′. What is the laser beam's speed in the earth's frame S? According to the Lorentz velocity transformation, it must be

$$u = \frac{u' + v}{1 + u'v/c^2} = \frac{c + v}{1 + cv/c^2} = \frac{c + v}{1 + v/c} = \frac{c + v}{(c + v)/c} = c \tag{36.31}$$

Light travels at speed c in both frame S and frame S′. This important consequence of the principle of relativity is "built into" the Lorentz transformations.

36.9 Relativistic Momentum

In Newtonian mechanics, the total momentum of a system is a conserved quantity. Further, as we've seen, the law of conservation of momentum, $P_f = P_i$, is true in all inertial reference frames *if* the particle velocities in different reference frames are related by the Galilean velocity transformations.

The difficulty, of course, is that the Galilean transformations are not consistent with the principle of relativity. It is a reasonable approximation when all velocities are very much less than c, but the Galilean transformations fail dramatically as velocities approach c. We'll leave it as a homework problem to show that $P_f' \neq P_i'$ if the particle velocities in frame S′ are related to the particle velocities in frame S by the Lorentz transformations.

There are two possibilities:

1. The so-called law of conservation of momentum is not really a law of physics. It is approximately true at low velocities but fails as velocities approach the speed of light.
2. The law of conservation of momentum really is a law of physics, but the expression $p = mu$ is not the correct way to calculate momentum when the particle velocity u becomes a significant fraction of c.

Momentum conservation is such a central and important feature of mechanics that it seems unlikely to fail in relativity.

The classical momentum, for one-dimensional motion, is $p = mu = m(\Delta x/\Delta t)$. Δt is the time to move distance Δx. That seemed clear enough within a Newtonian framework, but now we've learned that experimenters in different reference frames disagree about the amount of time needed. So whose Δt should we use?

One possibility is to use the time measured *by the particle*. This is the proper time $\Delta \tau$ because the particle is at rest in its own reference frame and needs only one clock. With this in mind, let's redefine the momentum of a particle of mass m moving with velocity $u = \Delta x/\Delta t$ to be

$$p = m\frac{\Delta x}{\Delta \tau} \tag{36.32}$$

We can relate this new expression for p to the familiar Newtonian expression by using the time-dilation result $\Delta \tau = (1 - u^2/c^2)^{1/2}\Delta t$ to relate the proper time interval measured by the particle to the more practical time interval Δt measured by experimenters in frame S. With this substitution, Equation 36.32 becomes

$$p = m\frac{\Delta x}{\Delta \tau} = m\frac{\Delta x}{\sqrt{1 - u^2/c^2}\,\Delta t} = \frac{mu}{\sqrt{1 - u^2/c^2}} \tag{36.33}$$

You can see that Equation 36.33 reduces to the classical expression $p = mu$ when the particle's speed $u \ll c$. That is an important requirement, but whether this is the "correct" expression for p depends on whether the total momentum P is conserved when the velocities of a system of particles are transformed with the Lorentz velocity transformation equations. The proof is rather long and tedious, so we will assert, without actual proof, that the momentum defined in Equation 36.33 does, indeed, transform correctly. **The law of conservation of momentum is still valid in all inertial reference frames** *if* **the momentum of each particle is calculated with Equation 36.33.**

The factor that multiplies mu in Equation 36.33 looks much like the factor γ in the Lorentz transformation equations for x and t, but there's one very important difference. The v in the Lorentz transformation equations is the velocity of a *reference frame*. The u in Equation 36.33 is the velocity of a particle moving *in* a reference frame.

With this distinction in mind, let's define the quantity

$$\gamma_p = \frac{1}{\sqrt{1 - u^2/c^2}} \tag{36.34}$$

where the subscript p indicates that this is γ for a particle, not for a reference frame. In frame S', where the particle moves with velocity u', the corresponding expression would be called γ_p'. With this definition of γ_p, the momentum of a particle is

$$p = \gamma_p mu \tag{36.35}$$

EXAMPLE 36.10 | **Momentum of a subatomic particle**

Electrons in a particle accelerator reach a speed of $0.999c$ relative to the laboratory. One collision of an electron with a target produces a muon that moves forward with a speed of $0.95c$ relative to the laboratory. The muon mass is 1.90×10^{-28} kg. What is the muon's momentum in the laboratory frame and in the frame of the electron beam?

MODEL Let the laboratory be reference frame S. The reference frame S' of the electron beam (i.e., a reference frame in which the electrons are at rest) moves in the direction of the electrons at $v = 0.999c$. The muon velocity in frame S is $u = 0.95c$.

SOLVE γ_p for the muon in the laboratory reference frame is

$$\gamma_p = \frac{1}{\sqrt{1 - u^2/c^2}} = \frac{1}{\sqrt{1 - 0.95^2}} = 3.20$$

Thus the muon's momentum in the laboratory is

$$p = \gamma_p mu = (3.20)(1.90 \times 10^{-28}\text{ kg})(0.95 \times 3.00 \times 10^8\text{ m/s})$$
$$= 1.73 \times 10^{-19}\text{ kg m/s}$$

The momentum is a factor of 3.2 larger than the Newtonian momentum mu. To find the momentum in the electron-beam

Continued

reference frame, we must first use the velocity transformation equation to find the muon's velocity in frame S':

$$u' = \frac{u - v}{1 - uv/c^2} = \frac{0.95c - 0.999c}{1 - (0.95c)(0.999c)/c^2} = -0.962c$$

In the laboratory frame, the faster electrons are overtaking the slower muon. Hence the muon's velocity in the electron-beam frame is negative. γ'_p for the muon in frame S' is

$$\gamma'_p = \frac{1}{\sqrt{1 - u'^2/c^2}} = \frac{1}{\sqrt{1 - 0.962^2}} = 3.66$$

The muon's momentum in the electron-beam reference frame is

$$p' = \gamma'_p mu'$$

$$= (3.66)(1.90 \times 10^{-28}\,\text{kg})(-0.962 \times 3.00 \times 10^8\,\text{m/s})$$

$$= -2.01 \times 10^{-19}\,\text{kg m/s}$$

ASSESS From the laboratory perspective, the muon moves only slightly slower than the electron beam. But it turns out that the muon moves faster with respect to the electrons, although in the opposite direction, than it does with respect to the laboratory.

The Cosmic Speed Limit

FIGURE 36.34 The speed of a particle cannot reach the speed of light.

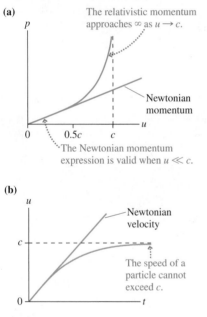

(a)

FIGURE 36.34a is a graph of momentum versus velocity. For a Newtonian particle, with $p = mu$, the momentum is directly proportional to the velocity. The relativistic expression for momentum agrees with the Newtonian value if $u \ll c$, but p approaches ∞ as $u \to c$.

The implications of this graph become clear when we relate momentum to force. Consider a particle subjected to a constant force, such as a rocket that never runs out of fuel. If F is constant, we can see from $F = dp/dt$ that the momentum is $p = Ft$. If Newtonian physics were correct, a particle would go faster and faster as its velocity $u = p/m = (F/m)t$ increased without limit. But the relativistic result, shown in **FIGURE 36.34b**, is that the particle's velocity asymptotically approaches the speed of light $(u \to c)$ as p approaches ∞. Relativity gives a very different outcome than Newtonian mechanics.

The speed c is a "cosmic speed limit" for material particles. A force cannot accelerate a particle to a speed higher than c because the particle's momentum becomes infinitely large as the speed approaches c. The amount of effort required for each additional increment of velocity becomes larger and larger until no amount of effort can raise the velocity any higher.

Actually, at a more fundamental level, c is a speed limit for *any* kind of **causal influence.** If I throw a rock and break a window, my throw is the *cause* of the breaking window and the rock is the *causal influence.* If I shoot a laser beam at a light detector that is wired to a firecracker, the light wave is the *causal influence* that leads to the explosion. A causal influence can be any kind of particle, wave, or information that travels from A to B and allows A to be the cause of B.

For two unrelated events—a firecracker explodes in Tokyo and a balloon bursts in Paris—the relativity of simultaneity tells us that they may be simultaneous in one reference frame but not in others. Or in one reference frame the firecracker may explode before the balloon bursts but in some other reference frame the balloon may burst first. These possibilities violate our commonsense view of time, but they're not in conflict with the principle of relativity.

For two causally related events—A *causes* B—it would be nonsense for an experimenter in any reference frame to find that B occurs before A. No experimenter in any reference frame, no matter how it is moving, will find that you are born before your mother is born. If A causes B, then it must be the case that $t_A < t_B$ in *all* reference frames.

Suppose there exists some kind of causal influence that *can* travel at speed $u > c$. **FIGURE 36.35** shows a reference frame S in which event A occurs at position $x_A = 0$. The faster-than-light causal influence—perhaps some yet-to-be-discovered "z ray"— leaves A at $t_A = 0$ and travels to the point at which it will cause event B. It arrives at x_B at time $t_B = x_B/u$.

How do events A and B appear in a reference frame S' that travels at an ordinary speed $v < c$ relative to frame S? We can use the Lorentz transformations to find out.

FIGURE 36.35 Assume that a causal influence can travel from A to B at a speed $u > c$.

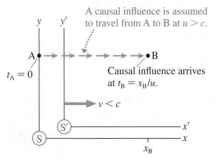

Because $x_A = 0$ and $t_A = 0$, it's easy to see that $x'_A = 0$ and $t'_A = 0$. That is, the origins of S and S′ overlap at the instant the causal influence leaves event A. More interesting is the time at which this influence reaches B in frame S′. The Lorentz time transformation for event B is

$$t'_B = \gamma\left(t_B - \frac{vx_B}{c^2}\right) = \gamma t_B\left(1 - \frac{v(x_B/t_B)}{c^2}\right) = \gamma t_B\left(1 - \frac{vu}{c^2}\right) \qquad (36.36)$$

where we first factored out t_B, then made use of the fact that $u = x_B/t_B$ in frame S.

We're assuming that $u > c$, so there exist ordinary reference frames, with $v < c$, for which $vu/c^2 > 1$. In that case, the term $(1 - vu/c^2)$ is negative and $t'_B < 0$. But if $t'_B < 0$, then event B happens *before* event A in reference frame S′. In other words, if a causal influence can travel faster than c, then there exist reference frames in which the effect happens before the cause. We know this can't happen, so our assumption $u > c$ must be wrong. **No causal influence of any kind—particle, wave, or yet-to-be-discovered z rays—can travel faster than** c.

The existence of a cosmic speed limit is one of the most interesting consequences of the theory of relativity. "Warp drive," in which a spaceship suddenly leaps to faster-than-light velocities, is simply incompatible with the theory of relativity. Rapid travel to the stars will remain in the realm of science fiction unless future scientific discoveries find flaws in Einstein's theory and open the doors to yet-undreamed-of theories. While we can't say with certainty that a scientific theory will never be overturned, there is currently not even a hint of evidence that disagrees with the special theory of relativity.

36.10 Relativistic Energy

Energy is our final topic in this chapter on relativity. Space, time, velocity, and momentum are changed by relativity, so it seems inevitable that we'll need a new view of energy.

In Newtonian mechanics, a particle's kinetic energy $K = \frac{1}{2}mu^2$ can be written in terms of its momentum $p = mu$ as $K = p^2/2m$. This suggests that a relativistic expression for energy will likely involve both the square of p and the particle's mass. We also hope that energy will be conserved in relativity, so a reasonable starting point is with the one quantity we've found that is the same in all inertial reference frames: the spacetime interval s.

Let a particle of mass m move through distance Δx during a time interval Δt, as measured in reference frame S. The spacetime interval is

$$s^2 = c^2(\Delta t)^2 - (\Delta x)^2 = \text{invariant}$$

We can turn this into an expression involving momentum if we multiply by $(m/\Delta\tau)^2$, where $\Delta\tau$ is the proper time (i.e., the time measured by the particle). Doing so gives

$$(mc)^2\left(\frac{\Delta t}{\Delta\tau}\right)^2 - \left(\frac{m\,\Delta x}{\Delta\tau}\right)^2 = (mc)^2\left(\frac{\Delta t}{\Delta\tau}\right)^2 - p^2 = \text{invariant} \qquad (36.37)$$

where we used $p = m(\Delta x/\Delta\tau)$ from Equation 36.32.

Now Δt, the time interval in frame S, is related to the proper time by the time-dilation result $\Delta t = \gamma_p \Delta\tau$. With this change, Equation 36.37 becomes

$$(\gamma_p mc)^2 - p^2 = \text{invariant}$$

Finally, for reasons that will be clear in a minute, we multiply by c^2, to get

$$(\gamma_p mc^2)^2 - (pc)^2 = \text{invariant} \qquad (36.38)$$

To say that the right side is an *invariant* means it has the same value in all inertial reference frames. We can easily determine the constant by evaluating it in the reference frame in which the particle is at rest. In that frame, where $p = 0$ and $\gamma_p = 1$, we find that

$$(\gamma_p mc^2)^2 - (pc)^2 = (mc^2)^2 \tag{36.39}$$

Let's reflect on what this means before taking the next step. The space-time interval s has the same value in all inertial reference frames. In other words, $c^2(\Delta t)^2 - (\Delta x)^2 = c^2(\Delta t')^2 - (\Delta x')^2$. Equation 36.39 was derived from the definition of the spacetime interval; hence the quantity mc^2 is also an invariant having the same value in all inertial reference frames. In other words, if experimenters in frames S and S' both make measurements on this particle of mass m, they will find that

$$(\gamma_p mc^2)^2 - (pc)^2 = (\gamma'_p mc^2)^2 - (p'c)^2 \tag{36.40}$$

Experimenters in different reference frames measure different values for the momentum, but experimenters in all reference frames agree that momentum is a conserved quantity. Equations 36.39 and 36.40 suggest that the quantity $\gamma_p mc^2$ is also an important property of the particle, a property that changes along with p in just the right way to satisfy Equation 36.39. But what is this property?

The first clue comes from checking the units. γ_p is dimensionless and c is a velocity, so $\gamma_p mc^2$ has the same units as the classical expression $\frac{1}{2}mv^2$—namely, units of energy. For a second clue, let's examine how $\gamma_p mc^2$ behaves in the low-velocity limit $u \ll c$. We can use the binomial approximation expression for γ_p to find

$$\gamma_p mc^2 = \frac{mc^2}{\sqrt{1 - u^2/c^2}} \approx \left(1 + \frac{1}{2}\frac{u^2}{c^2}\right)mc^2 = mc^2 + \frac{1}{2}mu^2 \tag{36.41}$$

The second term, $\frac{1}{2}mu^2$, is the low-velocity expression for the kinetic energy K. This is an energy associated with motion. But the first term suggests that the concept of energy is more complex than we originally thought. It appears that **there is an inherent energy associated with mass itself.**

Rest Energy and Total Energy

With that as a possibility, subject to experimental verification, let's define the **total energy** E of a particle to be

$$E = \gamma_p mc^2 = E_0 + K = \text{rest energy} + \text{kinetic energy} \tag{36.42}$$

This total energy consists of a **rest energy**

$$E_0 = mc^2 \tag{36.43}$$

and a relativistic expression for the *kinetic energy*

$$K = (\gamma_p - 1)mc^2 = (\gamma_p - 1)E_0 \tag{36.44}$$

This expression for the kinetic energy is very nearly $\frac{1}{2}mu^2$ when $u \ll c$ but, as **FIGURE 36.36** shows, differs significantly from the classical value for very high velocities.

Equation 36.43 is, of course, Einstein's famous $E = mc^2$, perhaps the most famous equation in all of physics. Before discussing its significance, we need to tie up some loose ends. First, we can use Equations 36.42 and 36.43 to rewrite Equation 36.39 as $E^2 - (pc)^2 = E_0^2$. This is easily rearranged to give the useful result

$$E = \sqrt{E_0^2 + (pc)^2} \tag{36.45}$$

The quantity E_0 is an *invariant* with the same value mc^2 in *all* inertial reference frames.

Second, notice that we can write

$$pc = (\gamma_p mu)c = \frac{u}{c}(\gamma_p mc^2)$$

FIGURE 36.36 The relativistic kinetic energy.

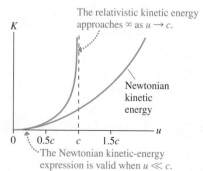

The relativistic kinetic energy approaches ∞ as $u \to c$.

Newtonian kinetic energy

The Newtonian kinetic-energy expression is valid when $u \ll c$.

But $\gamma_p mc^2$ is the total energy E and $u/c = \beta_p$, where the subscript p, as on γ_p, indicates that we're referring to the motion of a particle within a reference frame, not the motion of two reference frames relative to each other. Thus

$$pc = \beta_p E \qquad (36.46)$$

FIGURE 36.37 shows the "velocity-energy-momentum triangle," a convenient way to remember the relationships among the three quantities.

FIGURE 36.37 The velocity-energy-momentum triangle.

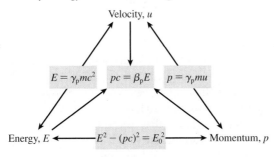

EXAMPLE 36.11 | Kinetic energy and total energy

Calculate the rest energy and the kinetic energy of (a) a 100 g ball moving with a speed of 100 m/s and (b) an electron with a speed of $0.999c$.

MODEL The ball, with $u \ll c$, is a classical particle. We don't need to use the relativistic expression for its kinetic energy. The electron is highly relativistic.

SOLVE a. For the ball, with $m = 0.10$ kg,

$$E_0 = mc^2 = 9.0 \times 10^{15} \text{ J}$$
$$K = \tfrac{1}{2}mu^2 = 500 \text{ J}$$

b. For the electron, we start by calculating

$$\gamma_p = \frac{1}{(1 - u^2/c^2)^{1/2}} = 22.4$$

Then, using $m_e = 9.11 \times 10^{-31}$ kg, we find

$$E_0 = mc^2 = 8.2 \times 10^{-14} \text{ J}$$
$$K = (\gamma_p - 1)E_0 = 170 \times 10^{-14} \text{ J}$$

ASSESS The ball's kinetic energy is a typical kinetic energy. Its rest energy, by contrast, is a staggeringly large number. For a relativistic electron, on the other hand, the kinetic energy is more important than the rest energy.

STOP TO THINK 36.8 An electron moves through the lab at 99% the speed of light. The lab reference frame is S and the electron's reference frame is S'. In which reference frame is the electron's rest mass larger?

a. In frame S, the lab frame
b. In frame S', the electron's frame
c. It is the same in both frames.

Mass-Energy Equivalence

Now we're ready to explore the significance of Einstein's famous equation $E = mc^2$. FIGURE 36.38 shows two balls of clay approaching each other. They have equal masses and equal kinetic energies, and they slam together in a perfectly inelastic collision to form one large ball of clay at rest. In Newtonian mechanics, we would say that the initial energy $2K$ is dissipated by being transformed into an equal amount of thermal energy, raising the temperature of the coalesced ball of clay. But Equation 36.42, $E = E_0 + K$, doesn't say anything about thermal energy. The total energy before the

FIGURE 36.38 An inelastic collision between two balls of clay does not seem to conserve the total energy E.

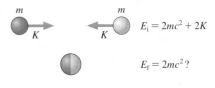

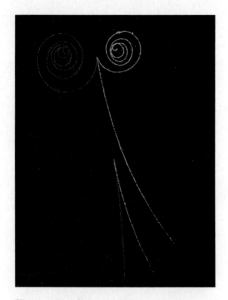

The tracks of elementary particles in a bubble chamber show the creation of an electron-positron pair. The negative electron and positive positron spiral in opposite directions in the magnetic field.

FIGURE 36.39 An inelastic collision between electrons can create an electron-positron pair.

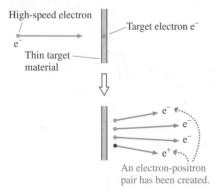

FIGURE 36.40 The annihilation of an electron-positron pair.

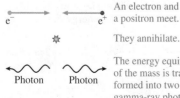

An electron and a positron meet.

They annihilate.

The energy equivalent of the mass is transformed into two gamma-ray photons.

collision is $E_i = 2mc^2 + 2K$, with the factor of 2 appearing because there are two masses. It seems like the total energy after the collision, when the clay is at rest, should be $2mc^2$, but this value doesn't conserve total energy.

There's ample experimental evidence that energy is conserved, so there must be a flaw in our reasoning. The statement of energy conservation is

$$E_f = Mc^2 = E_i = 2mc^2 + 2K \qquad (36.47)$$

where M is the mass of clay after the collision. But, remarkably, this requires

$$M = 2m + \frac{2K}{c^2} \qquad (36.48)$$

In other words, **mass is not conserved.** The mass of clay after the collision is larger than the mass of clay before the collision. Total energy can be conserved only if kinetic energy is transformed into an "equivalent" amount of mass.

The mass increase in a collision between two balls of clay is incredibly small, far beyond any scientist's ability to detect. So how do we know if such a crazy idea is true?

FIGURE 36.39 shows an experiment that has been done countless times in the last 50 years at particle accelerators around the world. An electron that has been accelerated to $u \approx c$ is aimed at a target material. When a high-energy electron collides with an atom in the target, it can easily knock one of the electrons out of the atom. Thus we would expect to see two electrons leaving the target: the incident electron and the ejected electron. Instead, *four* particles emerge from the target: three electrons and a positron. A *positron,* or positive electron, is the antimatter version of an electron, identical to an electron in all respects other than having charge $q = +e$.

In chemical-reaction notation, the collision is

$$e^-\,(\text{fast}) + e^-\,(\text{at rest}) \rightarrow e^- + e^- + e^- + e^+$$

An electron and a positron have been *created,* apparently out of nothing. Mass $2m_e$ before the collision has become mass $4m_e$ after the collision. (Notice that charge has been conserved in this collision.)

Although the mass has increased, it wasn't created "out of nothing." This is an inelastic collision, just like the collision of the balls of clay, because the kinetic energy after the collision is less than before. In fact, if you measured the energies before and after the collision, you would find that the decrease in kinetic energy is exactly equal to the energy equivalent of the two particles that have been created: $\Delta K = 2m_e c^2$. The new particles have been created *out of energy!*

Particles can be created from energy, and particles can return to energy. **FIGURE 36.40** shows an electron colliding with a positron, its antimatter partner. When a particle and its antiparticle meet, they *annihilate* each other. The mass disappears, and the energy equivalent of the mass is transformed into light. In Chapter 38, you'll learn that light is *quantized,* meaning that light is emitted and absorbed in discrete chunks of energy called *photons.* For light with wavelength λ, the energy of a photon is $E_{\text{photon}} = hc/\lambda$, where $h = 6.63 \times 10^{-34}$ J s is called *Planck's constant.* Photons carry momentum as well as energy. Conserving both energy and momentum in the annihilation of an electron and a positron requires the emission in opposite directions of two photons of equal energy.

If the electron and positron are fairly slow, so that $K \ll mc^2$, then $E_i \approx E_0 = mc^2$. In that case, energy conservation requires

$$E_f = 2E_{\text{photon}} = E_i \approx 2m_e c^2 \qquad (36.49)$$

Hence the wavelength of the emitted photons is

$$\lambda = \frac{hc}{m_e c^2} \approx 0.0024 \text{ nm} \qquad (36.50)$$

This is an extremely short wavelength, even shorter than the wavelengths of x rays. Photons in this wavelength range are called *gamma rays*. And, indeed, the emission of 0.0024 nm gamma rays is observed in many laboratory experiments in which positrons are able to collide with electrons and thus annihilate. In recent years, with the advent of gamma-ray telescopes on satellites, astronomers have found 0.0024 nm photons coming from many places in the universe, especially galactic centers—evidence that positrons are abundant throughout the universe.

Positron-electron annihilation is also the basis of the medical procedure known as positron-emission tomography, or a PET scan. A patient ingests a very small amount of a radioactive substance that decays by the emission of positrons. This substance is taken up by certain tissues in the body, especially those tissues with a high metabolic rate. As the substance decays, the positrons immediately collide with electrons, annihilate, and create two gamma-ray photons that are emitted back to back. The gamma rays, which easily leave the body, are detected, and their trajectories are traced backward into the body. The overlap of many such trajectories shows quite clearly the tissue in which the positron emission is occurring. The results are usually shown as false-color photographs, with redder areas indicating regions of higher positron emission.

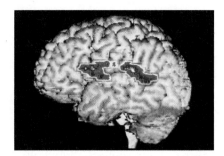

Positron-electron annihilation (a PET scan) provides a noninvasive look into the brain.

Conservation of Energy

The creation and annihilation of particles with mass, processes strictly forbidden in Newtonian mechanics, are vivid proof that neither mass nor the Newtonian definition of energy is conserved. Even so, the *total* energy—the kinetic energy *and* the energy equivalent of mass—remains a conserved quantity.

> **Law of conservation of total energy** The energy $E = \sum E_i$ of an isolated system is conserved, where $E_i = (\gamma_p)_i m_i c^2$ is the total energy of particle i.

Mass and energy are not the same thing, but, as the last few examples have shown, they are *equivalent* in the sense that mass can be transformed into energy and energy can be transformed into mass as long as the total energy is conserved.

Probably the most well-known application of the conservation of total energy is nuclear fission. The uranium isotope ^{236}U, containing 236 protons and neutrons, does not exist in nature. It can be created when a ^{235}U nucleus absorbs a neutron, increasing its atomic mass from 235 to 236. The ^{236}U nucleus quickly fragments into two smaller nuclei and several extra neutrons, a process known as **nuclear fission.** The nucleus can fragment in several ways, but one is

$$n + {}^{235}U \rightarrow {}^{236}U \rightarrow {}^{144}Ba + {}^{89}Kr + 3n$$

Ba and Kr are the atomic symbols for barium and krypton.

This reaction seems like an ordinary chemical reaction—until you check the masses. The masses of atomic isotopes are known with great precision from many decades of measurement in instruments called mass spectrometers. If you add up the masses on both sides, you find that the mass of the products is 0.185 u smaller than the mass of the initial neutron and ^{235}U, where, you will recall, $1\ u = 1.66 \times 10^{-27}$ kg is the atomic mass unit. In kilograms the mass loss is 3.07×10^{-28} kg.

Mass has been lost, but the energy equivalent of the mass has not. As **FIGURE 36.41** shows, the mass has been converted to kinetic energy, causing the two product nuclei and three neutrons to be ejected at very high speeds. The kinetic energy is easily calculated: $\Delta K = m_{lost}c^2 = 2.8 \times 10^{-11}$ J.

This is a very tiny amount of energy, but it is the energy released from *one* fission. The number of nuclei in a macroscopic sample of uranium is on the order of N_A, Avogadro's number. Hence the energy available if *all* the nuclei fission is enormous. This energy, of course, is the basis for both nuclear power reactors and nuclear weapons.

FIGURE 36.41 In nuclear fission, the energy equivalent of lost mass is converted into kinetic energy.

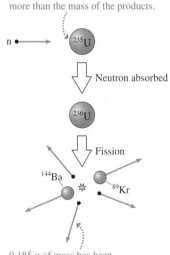

The mass of the reactants is 0.185 u more than the mass of the products.

n •⟶ ^{235}U

⬇ Neutron absorbed

^{236}U

⬇ Fission

^{144}Ba ^{89}Kr

0.185 u of mass has been converted into kinetic energy.

We started this chapter with an expectation that relativity would challenge our basic notions of space and time. We end by finding that relativity changes our understanding of mass and energy. Most remarkable of all is that each and every one of these new ideas flows from one simple statement: The laws of physics are the same in all inertial reference frames.

CHALLENGE EXAMPLE 36.12 | Goths and Huns

The rockets of the Goths and the Huns are each 1000 m long in their rest frame. The rockets pass each other, virtually touching, at a relative speed of 0.8c. The Huns have a laser cannon at the rear of their rocket that fires a deadly laser beam perpendicular to the rocket's motion. The captain of the Huns wants to send a threatening message to the Goths by "firing a shot across their bow." He tells his first mate, "The Goths' rocket is length contracted to 600 m. Fire the laser cannon at the instant the tail of their rocket passes the nose of ours. The laser beam will cross 400 m in front of them."

But things are different in the Goths' reference frame. The Goth captain muses, "The Huns' rocket is length contracted to 600 m, 400 m shorter than our rocket. If they fire as the nose of their ship passes the tail of ours, the lethal laser beam will pass right through our side."

The first mate on the Huns' rocket fires as ordered. Does the laser beam blast the Goths or not?

MODEL Both rockets are inertial reference frames. Let the Huns' rocket be frame S and the Goths' rocket be frame S′. S′ moves with velocity $v = 0.8c$ relative to S. We need to describe the situation in terms of events.

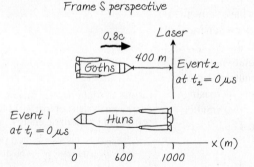

FIGURE 36.42 The situation seen by the Huns.

VISUALIZE Begin by considering the situation from the Huns' reference frame, as shown in FIGURE 36.42.

SOLVE The key to resolving the paradox is that two events simultaneous in one reference frame are not simultaneous in a different reference frame. The Huns do, indeed, see the Goths' rocket length contracted to $L_{Goths} = (1 - (0.8)^2)^{1/2}(1000 \text{ m}) = 600 \text{ m}$. Let event 1 be the tail of the Goths' rocket passing the nose of the Huns' rocket. Since we're free to define the origin of our coordinate system, we define this event to be at time $t_1 = 0$ μs and at position $x_1 = 0$ m. Then, in the Huns' reference frame, the spacetime coordinates of event 2, the firing of the laser cannon, are $(x_2, t_2) = (1000 \text{ m}, 0 \text{ μs})$. The nose of the Goths' rocket is at $x = 600$ m at $t = 0$ μs; thus the laser cannon misses the Goths by 400 m.

Now we can use the Lorentz transformations to find the spacetime coordinates of the events in the Goths' reference frame. The nose of the Huns' rocket passes the tail of the Goths' rocket at $(x_1', t_1') = (0 \text{ m}, 0 \text{ μs})$. The Huns fire their laser cannon at

$$x_2' = \gamma(x_2 - vt_2) = \tfrac{5}{3}(1000 \text{ m} - 0 \text{ m}) = 1667 \text{ m}$$

$$t_2' = \gamma\left(t_2 - \frac{vx_2}{c^2}\right) = \tfrac{5}{3}\left(0 \text{ μs} - (0.8)\frac{1000 \text{ m}}{300 \text{ m/μs}}\right) = -4.444 \text{ μs}$$

where we calculated $\gamma = 5/3$ for $v = 0.8c$. Events 1 and 2 are *not* simultaneous in S′. The Huns fire the laser cannon 4.444 μs *before* the nose of their rocket reaches the tail of the Goths' rocket. The laser is fired at $x_2' = 1667$ m, missing the nose of the Goths' rocket by 667 m. FIGURE 36.43 shows how the Goths see things.

FIGURE 36.43 The situation seen by the Goths.

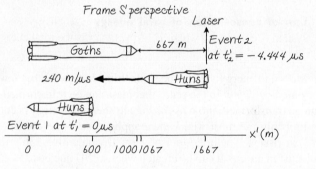

In fact, since the Huns' rocket is length contracted to 600 m, the nose of the Huns' rocket is at $x' = 1667 \text{ m} - 600 \text{ m} = 1067 \text{ m}$ at the instant they fire the laser cannon. At a speed of $v = 0.8c = 240 \text{ m/μs}$, in 4.444 μs the nose of the Huns' rocket travels $\Delta x' = (240 \text{ m/μs})(4.444 \text{ μs}) = 1067 \text{ m}$—exactly the right distance to be at the tail of the Goths' rocket at $t_1' = 0$ μs. We could also note that the 667 m "miss distance" in the Goths' frame is length contracted to $(1 - (0.8)^2)^{1/2}(667 \text{ m}) = 400 \text{ m}$ in the Huns' frame—exactly the amount by which the Huns think they miss the Goths' rocket.

ASSESS Thus we end up with a consistent explanation. The Huns miss the Goths' rocket because, to them, the Goths' rocket is length contracted. The Goths find that the Huns miss because event 2 (the firing of the laser cannon) occurs before event 1 (the nose of one rocket passing the tail of the other). The 400 m distance of the miss in the Huns' reference frame is the length-contracted miss distance of 667 m in the Goths' reference frame.

SUMMARY

The goal of Chapter 36 has been to learn how relativity changes our concepts of space and time.

GENERAL PRINCIPLES

Principle of Relativity

The laws of physics are the same in all inertial reference frames.

- The speed of light c is the same in all inertial reference frames.
- No particle or causal influence can travel at a speed greater than c.

Solving Relativity Problems

- Base the analysis on events.
- Time intervals can often be found using time dilation.
- Distances can often be found using length contraction.
- Use the **Lorentz transformations** for general problems.

IMPORTANT CONCEPTS

Space

Spatial measurements depend on the motion of the experimenter relative to the events. An object's length is the difference between *simultaneous* measurements of the positions of both ends.

Proper length ℓ is the length of an object measured in a reference frame in which the object is at rest. The object's length in a frame in which the object moves with velocity v is

$$L = \sqrt{1 - \beta^2}\, \ell \le \ell$$

This is called **length contraction.**

Time

Time measurements depend on the motion of the experimenter relative to the events. Events that are simultaneous in reference frame S are not simultaneous in frame S′ moving relative to S.

Proper time $\Delta\tau$ is the time interval between two events measured in a reference frame in which the events occur at the same position. The time interval between the events in a frame moving with relative velocity v is

$$\Delta t = \Delta\tau / \sqrt{1 - \beta^2} \ge \Delta\tau$$

This is called **time dilation.**

Momentum

The law of conservation of momentum is valid in all inertial reference frames if the momentum of a particle with velocity u is $p = \gamma_p m u$, where

$$\gamma_p = 1/\sqrt{1 - u^2/c^2}$$

The momentum approaches ∞ as $u \to c$.

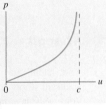

Energy

The law of conservation of energy is valid in all inertial reference frames if the energy of a particle with velocity u is $E = \gamma_p mc^2 = E_0 + K$.

Rest energy $E_0 = mc^2$

Kinetic energy $K = (\gamma_p - 1)mc^2$

Invariants

Invariants are quantities that have the same value in all inertial reference frames.

Spacetime interval: $s^2 = (c\,\Delta t)^2 - (\Delta x)^2$

Particle rest energy: $E_0^2 = (mc^2)^2 = E^2 - (pc)^2$

Mass-energy equivalence

Mass m can be transformed into energy $\Delta E = mc^2$.

Energy can be transformed into mass $m = \Delta E/c^2$.

APPLICATIONS

An **event** happens at a specific place in space and time. Spacetime coordinates are (x, t) in frame S and (x', t') in frame S′.

A **reference frame** is a coordinate system with meter sticks and clocks for measuring events.

The **Lorentz transformations** transform spacetime coordinates and velocities between reference frames S and S′.

$$x' = \gamma(x - vt) \qquad x = \gamma(x' + vt')$$
$$y' = y \qquad\qquad y = y'$$
$$z' = z \qquad\qquad z = z'$$
$$t' = \gamma(t - vx/c^2) \qquad t = \gamma(t' + vx'/c^2)$$
$$u' = \frac{u - v}{1 - uv/c^2} \qquad u = \frac{u' + v}{1 + u'v/c^2}$$

where u and u' are the x- and x'-components of an object's velocity.

$$\beta = v/c \quad \text{and} \quad \gamma = 1/\sqrt{1 - v^2/c^2} = 1/\sqrt{1 - \beta^2}$$

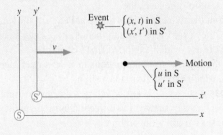

TERMS AND NOTATION

special relativity	spacetime coordinates,	time dilation	causal influence
reference frame	(x, y, z, t)	light year, ly	total energy, E
inertial reference frame	synchronized	proper length, ℓ	rest energy, E_0
Galilean principle of	simultaneous	length contraction	law of conservation of total
relativity	relativity of simultaneity	invariant	energy
principle of relativity	rest frame	spacetime interval, s	nuclear fission
event	proper time, $\Delta\tau$	Lorentz transformations	

CONCEPTUAL QUESTIONS

1. FIGURE Q36.1 shows two balls. What are the speed and direction of each (a) in a reference frame that moves with ball 1 and (b) in a reference frame that moves with ball 2?

FIGURE Q36.1

FIGURE Q36.2

2. Teenagers Sam and Tom are playing chicken in their rockets. As FIGURE Q36.2 shows, an experimenter on earth sees that each is traveling at $0.95c$ as he approaches the other. Sam fires a laser beam toward Tom.
   a. What is the speed of the laser beam relative to Sam?
   b. What is the speed of the laser beam relative to Tom?

3. Firecracker A is 300 m from you. Firecracker B is 600 m from you in the same direction. You see both explode at the same time. Define event 1 to be "firecracker A explodes" and event 2 to be "firecracker B explodes." Does event 1 occur before, after, or at the same time as event 2? Explain.

4. Firecrackers A and B are 600 m apart. You are standing exactly halfway between them. Your lab partner is 300 m on the other side of firecracker A. You see two flashes of light, from the two explosions, at exactly the same instant of time. Define event 1 to be "firecracker A explodes" and event 2 to be "firecracker B explodes." According to your lab partner, based on measurements he or she makes, does event 1 occur before, after, or at the same time as event 2? Explain.

5. FIGURE Q36.5 shows Peggy standing at the center of her railroad car as it passes Ryan on the ground. Firecrackers attached to the ends of the car explode. A short time later, the flashes from the two explosions arrive at Peggy at the same time.

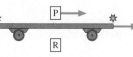

FIGURE Q36.5

   a. Were the explosions simultaneous in Peggy's reference frame? If not, which exploded first? Explain.
   b. Were the explosions simultaneous in Ryan's reference frame? If not, which exploded first? Explain.

6. FIGURE Q36.6 shows a rocket traveling from left to right. At the instant it is halfway between two trees, lightning simultaneously (in the rocket's frame) hits both trees.

   a. Do the light flashes reach the rocket pilot simultaneously? If not, which reaches her first? Explain.
   b. A student was sitting on the ground halfway between the trees as the rocket passed overhead. According to the student, were the lightning strikes simultaneous? If not, which tree was hit first? Explain.

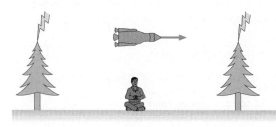

FIGURE Q36.6

7. Your friend flies from Los Angeles to New York. She carries an accurate stopwatch with her to measure the flight time. You and your assistants on the ground also measure the flight time.
   a. Identify the two events associated with this measurement.
   b. Who, if anyone, measures the proper time?
   c. Who, if anyone, measures the shorter flight time?

8. As the meter stick in FIGURE Q36.8 flies past you, you simultaneously measure the positions of both ends and determine that $L < 1$ m.

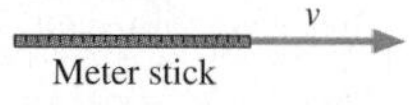

   Meter stick

   FIGURE Q36.8

   a. To an experimenter in frame S′, the meter stick's frame, did you make your two measurements simultaneously? If not, which end did you measure first? Explain.
   b. Can experimenters in frame S′ give an explanation for why your measurement is less than 1 m?

9. A 100-m-long train is heading for an 80-m-long tunnel. If the train moves sufficiently fast, is it possible, according to experimenters on the ground, for the entire train to be inside the tunnel at one instant of time? Explain.

10. Particle A has half the mass and twice the speed of particle B. Is the momentum p_A less than, greater than, or equal to p_B? Explain.

11. Event A occurs at spacetime coordinates (300 m, 2 μs).
    a. Event B occurs at spacetime coordinates (1200 m, 6 μs). Could A possibly be the cause of B? Explain.
    b. Event C occurs at spacetime coordinates (2400 m, 8 μs). Could A possibly be the cause of C? Explain.

EXERCISES AND PROBLEMS

Problems labeled ▨ integrate material from earlier chapters.

Exercises

Section 36.2 Galilean Relativity

1. ‖ A firecracker explodes in reference frame S at $t = 1.0$ s. A second firecracker explodes at the same position at $t = 3.0$ s. In reference frame S′, which moves in the x-direction at speed v, the first explosion is detected at $x' = 4.0$ m and the second at $x' = -4.0$ m.
   a. What is the speed of frame S′ relative to frame S?
   b. What is the position of the two explosions in frame S?

2. ‖ At $t = 1.0$ s, a firecracker explodes at $x = 10$ m in reference frame S. Four seconds later, a second firecracker explodes at $x = 20$ m. Reference frame S′ moves in the x-direction at a speed of 5.0 m/s. What are the positions and times of these two events in frame S′?

3. ‖ A newspaper delivery boy is riding his bicycle down the street at 5.0 m/s. He can throw a paper at a speed of 8.0 m/s. What is the paper's speed relative to the ground if he throws the paper (a) forward, (b) backward, and (c) to the side?

4. ‖ A baseball pitcher can throw a ball with a speed of 40 m/s. He is in the back of a pickup truck that is driving away from you. He throws the ball in your direction, and it floats toward you at a lazy 10 m/s. What is the speed of the truck?

Section 36.3 Einstein's Principle of Relativity

5. ‖ An out-of-control alien spacecraft is diving into a star at a speed of 1.0×10^8 m/s. At what speed, relative to the spacecraft, is the starlight approaching?

6. ‖ A starship blasts past the earth at 2.0×10^8 m/s. Just after passing the earth, it fires a laser beam out the back of the starship. With what speed does the laser beam approach the earth?

Section 36.4 Events and Measurements

Section 36.5 The Relativity of Simultaneity

7. ‖ Your job is to synchronize the clocks in a reference frame. You are going to do so by flashing a light at the origin at $t = 0$ s. To what time should the clock at $(x, y, z) = (30$ m, 40 m, 0 m$)$ be preset?

8. ‖ Bjorn is standing at $x = 600$ m. Firecracker 1 explodes at the origin and firecracker 2 explodes at $x = 900$ m. The flashes from both explosions reach Bjorn's eye at $t = 3.0$ μs. At what time did each firecracker explode?

9. ‖ Bianca is standing at $x = 600$ m. Firecracker 1, at the origin, and firecracker 2, at $x = 900$ m, explode simultaneously. The flash from firecracker 1 reaches Bianca's eye at $t = 3.0$ μs. At what time does she see the flash from firecracker 2?

10. ‖‖ You are standing at $x = 9.0$ km and your assistant is standing at $x = 3.0$ km. Lightning bolt 1 strikes at $x = 0$ km and lightning bolt 2 strikes at $x = 12.0$ km. You see the flash from bolt 2 at $t = 10$ μs and the flash from bolt 1 at $t = 50$ μs. According to your assistant, were the lightning strikes simultaneous? If not, which occurred first, and what was the time difference between the two?

11. ‖ You are standing at $x = 9.0$ km. Lightning bolt 1 strikes at $x = 0$ km and lightning bolt 2 strikes at $x = 12.0$ km. Both flashes reach your eye at the same time. Your assistant is standing at $x = 3.0$ km. Does your assistant see the flashes at the same time? If not, which does she see first, and what is the time difference between the two?

12. ‖ You are flying your personal rocketcraft at $0.90c$ from Star A toward Star B. The distance between the stars, in the stars' reference frame, is 1.0 ly. Both stars happen to explode simultaneously in your reference frame at the instant you are exactly halfway between them. Do you see the flashes simultaneously? If not, which do you see first, and what is the time difference between the two?

Section 36.6 Time Dilation

13. ‖ A cosmic ray travels 60 km through the earth's atmosphere in 400 μs, as measured by experimenters on the ground. How long does the journey take according to the cosmic ray?

14. ‖ At what speed, as a fraction of c, does a moving clock tick at half the rate of an identical clock at rest?

15. ‖ An astronaut travels to a star system 4.5 ly away at a speed of $0.90c$. Assume that the time needed to accelerate and decelerate is negligible.
   a. How long does the journey take according to Mission Control on earth?
   b. How long does the journey take according to the astronaut?
   c. How much time elapses between the launch and the arrival of the first radio message from the astronaut saying that she has arrived?

16. ‖ a. At what speed, as a fraction of c, must a rocket travel on a journey to and from a distant star so that the astronauts age 10 years while the Mission Control workers on earth age 120 years?
   b. As measured by Mission Control, how far away is the distant star?

17. ‖ At what speed, as a fraction of c, would a round-trip astronaut "lose" $\frac{1}{25}$ of the elapsed time shown on her watch?

18. ‖ At what speed, in m/s, would a moving clock lose 1.0 ns in 1.0 day according to experimenters on the ground?
   Hint: Use the binomial approximation.

19. ‖ You fly 5000 km across the United States on an airliner at 250 m/s. You return two days later at the same speed.
   a. Have you aged more or less than your friends at home?
   b. By how much?
   Hint: Use the binomial approximation.

Section 36.7 Length Contraction

20. ‖ Jill claims that her new rocket is 100 m long. As she flies past your house, you measure the rocket's length and find that it is only 80 m. What is Jill's speed, as a fraction of c?

21. ‖ At what speed, as a fraction of c, will a moving rod have a length 60% that of an identical rod at rest?

22. ‖ A cube has a density of 2000 kg/m³ while at rest in the laboratory. What is the cube's density as measured by an experimenter in the laboratory as the cube moves through the laboratory at 90% of the speed of light in a direction perpendicular to one of its faces?

23. ‖ A muon travels 60 km through the atmosphere at a speed of $0.9997c$. According to the muon, how thick is the atmosphere?

24. | Our Milky Way galaxy is 100,000 ly in diameter. A spaceship crossing the galaxy measures the galaxy's diameter to be a mere 1.0 ly.
    a. What is the spacecraft's speed, as a fraction of c, relative to the galaxy?
    b. How long is the crossing time as measured in the galaxy's reference frame?

25. ‖ A human hair is about 50 μm in diameter. At what speed, in m/s, would a meter stick "shrink by a hair"?
    Hint: Use the binomial approximation.

Section 36.8 The Lorentz Transformations

26. ‖ A rocket travels in the x-direction at speed $0.60c$ with respect to the earth. An experimenter on the rocket observes a collision between two comets and determines that the spacetime coordinates of the collision are $(x', t') = (3.0 \times 10^{10}$ m, 200 s$)$. What are the spacetime coordinates of the collision in earth's reference frame?

27. | An event has spacetime coordinates $(x, t) = (1200$ m, $2.0\ \mu$s$)$ in reference frame S. What are the event's spacetime coordinates (a) in reference frame S$'$ that moves in the positive x-direction at $0.80c$ and (b) in reference frame S$''$ that moves in the negative x-direction at $0.80c$?

28. ‖ In the earth's reference frame, a tree is at the origin and a pole is at $x = 30$ km. Lightning strikes both the tree and the pole at $t = 10\ \mu$s. The lightning strikes are observed by a rocket traveling in the x-direction at $0.50c$.
    a. What are the spacetime coordinates for these two events in the rocket's reference frame?
    b. Are the events simultaneous in the rocket's frame? If not, which occurs first?

29. | A rocket cruising past earth at $0.80c$ shoots a bullet out the back door, opposite the rocket's motion, at $0.90c$ relative to the rocket. What is the bullet's speed, as a fraction of c, relative to the earth?

30. ‖ A distant quasar is found to be moving away from the earth at $0.80c$. A galaxy closer to the earth and along the same line of sight is moving away from us at $0.20c$. What is the recessional speed of the quasar, as a fraction of c, as measured by astronomers in the other galaxy?

31. ‖ A laboratory experiment shoots an electron to the left at $0.90c$. What is the electron's speed, as a fraction of c, relative to a proton moving to the right at $0.90c$?

Section 36.9 Relativistic Momentum

32. | A proton is accelerated to $0.999c$.
    a. What is the proton's momentum?
    b. By what factor does the proton's momentum exceed its Newtonian momentum?

33. ‖ A 1.0 g particle has momentum 400,000 kg m/s. What is the particle's speed in m/s?

34. | At what speed, as a fraction of c, is a particle's momentum twice its Newtonian value?

35. | What is the speed, as a fraction of c, of a particle whose momentum is mc?

Section 36.10 Relativistic Energy

36. | A quarter-pound hamburger with all the fixings has a mass of 200 g. The food energy of the hamburger (480 food calories) is 2 MJ.
    a. What is the energy equivalent of the mass of the hamburger?
    b. By what factor does the energy equivalent exceed the food energy?

37. ‖ What are the rest energy, the kinetic energy, and the total energy of a 1.0 g particle with a speed of $0.80c$?

38. ‖ At what speed, as a fraction of c, must an electron move so that its total energy is 10% more than its rest mass energy?

39. ‖ At what speed, as a fraction of c, is a particle's kinetic energy twice its rest energy?

40. | At what speed, as a fraction of c, is a particle's total energy twice its rest energy?

41. | A modest *supernova* (the explosion of a massive star at the end of its life cycle) releases 1.5×10^{44} J of energy in a few seconds. This is enough to outshine the entire galaxy in which it occurs. Suppose a star with the mass of our sun collides with an antimatter star of equal mass, causing complete annihilation. What is the ratio of the energy released in this star-antistar collision to the energy released in the supernova?

42. ‖ One of the important ways in which the *Higgs boson* was detected at the Large Hadron Collider was by observing a type of decay in which the Higgs—which decays too quickly to be observed directly—is immediately transformed into two photons emitted back to back. Two photons, with momenta 3.31×10^{-17} kg m/s, were detected. What is the mass of the Higgs boson? Give your answer as a multiple of the proton mass.
    Hint: The relationship between energy and momentum applies to photons if you treat a photon as a massless particle.

Problems

43. ‖ The diameter of the solar system is 10 light hours. A spaceship crosses the solar system in 15 hours, as measured on earth. How long, in hours, does the passage take according to passengers on the spaceship?
    Hint: $c = 1$ light hour per hour.

44. | A 30-m-long rocket train car is traveling from Los Angeles to New York at $0.50c$ when a light at the center of the car flashes. When the light reaches the front of the car, it immediately rings a bell. Light reaching the back of the car immediately sounds a siren.
    a. Are the bell and siren simultaneous events for a passenger seated in the car? If not, which occurs first and by how much time?
    b. Are the bell and siren simultaneous events for a bicyclist waiting to cross the tracks? If not, which occurs first and by how much time?

45. ‖ The star Alpha goes supernova. Ten years later and 100 ly away, as measured by astronomers in the galaxy, star Beta explodes.
    a. Is it possible that the explosion of Alpha is in any way responsible for the explosion of Beta? Explain.
    b. An alien spacecraft passing through the galaxy finds that the distance between the two explosions is 120 ly. According to the aliens, what is the time between the explosions?

46. ‖ Two events in reference frame S occur 10 μs apart at the same point in space. The distance between the two events is 2400 m in reference frame S$'$.
    a. What is the time interval between the events in reference frame S$'$?
    b. What is the velocity of S$'$ relative to S?

47. ‖ A starship voyages to a distant planet 10 ly away. The explorers stay 1 year, return at the same speed, and arrive back on earth 26 years, as measured on earth, after they left. Assume that the time needed to accelerate and decelerate is negligible.
    a. What is the speed of the starship?
    b. How much time has elapsed on the astronauts' chronometers?

48. | The Stanford Linear Accelerator (SLAC) accelerates electrons to $v = 0.99999997c$ in a 3.2-km-long tube. If they travel the length of the tube at full speed (they don't, because they are accelerating), how long is the tube in the electrons' reference frame?

49. ‖ On a futuristic highway, a 15-m-long rocket travels so fast that a red stoplight, with a wavelength of 700 nm, appears to the pilot to be a green light with a wavelength of 520 nm. What is the length of the rocket to an observer standing at the intersection as the rocket speeds through?

    Hint: The Doppler effect for light was covered in Chapter 16.

50. ‖ In an attempt to reduce the extraordinarily long travel times for voyaging to distant stars, some people have suggested traveling at close to the speed of light. Suppose you wish to visit the red giant star Betelgeuse, which is 430 ly away, and that you want your 20,000 kg rocket to move so fast that you age only 20 years during the round trip.

    a. How fast, as a fraction of c, must the rocket travel relative to earth?

    b. How much energy is needed to accelerate the rocket to this speed?

    c. Compare this amount of energy to the total energy used by the United States in the year 2015, which was roughly 1.0×10^{20} J.

51. ‖ The quantity dE/dv, the rate of increase of energy with speed,
CALC is the amount of additional energy a moving object needs per 1 m/s increase in speed.

    a. A 25,000 kg truck is traveling at 30 m/s. How much additional energy is needed to increase its speed by 1 m/s?

    b. A 25,000 kg rocket is traveling at $0.90c$. How much additional energy is needed to increase its speed by 1 m/s?

52. ‖ A rocket traveling at $0.50c$ sets out for the nearest star, Alpha Centauri, which is 4.3 ly away from earth. It will return to earth immediately after reaching Alpha Centauri. What distance will the rocket travel and how long will the journey last according to (a) stay-at-home earthlings and (b) the rocket crew? (c) Which answers are the correct ones, those in part a or those in part b?

53. ‖ The star Delta goes supernova. One year later and 2.0 ly away, as measured by astronomers in the galaxy, star Epsilon explodes. Let the explosion of Delta be at $x_D = 0$ and $t_D = 0$. The explosions are observed by three spaceships cruising through the galaxy in the direction from Delta to Epsilon at velocities $v_1 = 0.30c$, $v_2 = 0.50c$, and $v_3 = 0.70c$. All three spaceships, each at the origin of its reference frame, happen to pass Delta as it explodes.

    a. What are the times of the two explosions as measured by scientists on each of the three spaceships?

    b. Does one spaceship find that the explosions are simultaneous? If so, which one?

    c. Does one spaceship find that Epsilon explodes before Delta? If so, which one?

    d. Do your answers to parts b and c violate the idea of causality? Explain.

54. ‖ Two rockets approach each other. Each is traveling at $0.75c$ in the earth's reference frame. What is the speed, as a fraction of c, of one rocket relative to the other?

55. ‖ Two rockets, A and B, approach the earth from opposite directions at speed $0.80c$. The length of each rocket measured in its rest frame is 100 m. What is the length of rocket A as measured by the crew of rocket B?

56. ‖ A rocket fires a projectile at a speed of $0.95c$ while traveling past the earth. An earthbound scientist measures the projectile's speed to be $0.90c$. What was the rocket's speed as a fraction of c?

57. ‖‖ Through what potential difference must an electron be accelerated, starting from rest, to acquire a speed of $0.99c$?

58. ‖ What is the speed, in m/s, of a proton after being accelerated from rest through a 50×10^6 V potential difference?

59. ‖ The half-life of a muon at rest is 1.5 μs. Muons that have been accelerated to a very high speed and are then held in a circular storage ring have a half-life of 7.5 μs.

    a. What is the speed, as a fraction of c, of the muons in the storage ring?

    b. What is the total energy of a muon in the storage ring? The mass of a muon is 207 times the mass of an electron.

60. ‖ This chapter has assumed that lengths perpendicular to the direction of motion are not affected by the motion. That is, motion in the x-direction does not cause length contraction along the y- or z-axes. To find out if this is really true, consider two spray-paint nozzles attached to rods perpendicular to the x-axis. It has been confirmed that, when both rods are at rest, both nozzles are exactly 1 m above the base of the rod. One rod is placed in the S reference frame with its base on the x-axis; the other is placed in the S′ reference frame with its base on the x'-axis. The rods then swoop past each other and, as **FIGURE P36.60** shows, each paints a stripe across the other rod.

    We will use proof by contradiction. Assume that objects perpendicular to the motion *are* contracted. An experimenter in frame S finds that the S′ nozzle, as it goes past, is less than 1 m above the x-axis. The principle of relativity says that an experiment carried out in two different inertial reference frames will have the same outcome in both.

    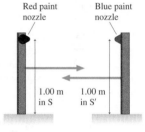

    Red paint nozzle — Blue paint nozzle

    1.00 m in S — 1.00 m in S′

    a. Pursue this line of reasoning and show that you end up with a logical contradiction, two mutually incompatible situations.

    b. What can you conclude from this contradiction?

    FIGURE P36.60

61. ‖ Derive the Lorentz transformations for t and t'.

    Hint: See the comment following Equation 36.22.

62. ‖ a. Derive a velocity transformation equation for u_y and u'_y. Assume that the reference frames are in the standard orientation with motion parallel to the x- and x'-axes.

    b. A rocket passes the earth at $0.80c$. As it goes by, it launches a projectile at $0.60c$ perpendicular to the direction of motion. What is the particle's speed, as a fraction of c, in the earth's reference frame?

63. ‖ A rocket is fired from the earth to the moon at a speed of $0.990c$. Let two events be "rocket leaves earth" and "rocket hits moon."

    a. In the earth's reference frame, calculate Δx, Δt, and the spacetime interval s for these events.

    b. In the rocket's reference frame, calculate $\Delta x'$, $\Delta t'$, and the spacetime interval s' for these events.

    c. Repeat your calculations of part a if the rocket is replaced with a laser beam.

64. ‖ Let's examine whether or not the law of conservation of momentum is true in all reference frames if we use the Newtonian definition of momentum: $p_x = mu_x$. Consider an object A of mass $3m$ at rest in reference frame S. Object A explodes into two pieces: object B, of mass m, that is shot to the left at a speed of $c/2$ and object C, of mass $2m$, that, to conserve momentum, is shot to the right at a speed of $c/4$. Suppose this explosion is observed in reference frame S′ that is moving to the right at half the speed of light.

    a. Use the Lorentz velocity transformation to find the velocity and the Newtonian momentum of A in S′.

    b. Use the Lorentz velocity transformation to find the velocities and the Newtonian momenta of B and C in S′.

    c. What is the total final momentum in S′?

    d. Newtonian momentum was conserved in frame S. Is it conserved in frame S′?

65. ‖ a. What are the momentum and total energy of a proton with speed 0.99c?

   b. What is the proton's momentum in a different reference frame in which $E' = 5.0 \times 10^{-10}$ J?

66. ‖‖ At what speed, as a fraction of c, is the kinetic energy of a particle twice its Newtonian value?

67. ‖ A typical nuclear power plant generates electricity at the rate of 1000 MW. The efficiency of transforming thermal energy into electrical energy is $\frac{1}{3}$ and the plant runs at full capacity for 80% of the year. (Nuclear power plants are down about 20% of the time for maintenance and refueling.)

   a. How much thermal energy does the plant generate in one year?

   b. What mass of uranium is transformed into energy in one year?

68. ‖ Many science fiction spaceships are powered by antimatter reactors. Suppose a 20-m-long spaceship, with a mass of 15,000 kg when empty, carries 2000 kg of fuel: 1000 kg each of matter and antimatter. The matter and antimatter are slowly combined, and the energy of their total annihilation is used to propel the ship. After consuming all the fuel and reaching top speed, the spaceship flies past a space station that is stationary with respect to the planet from which the ship was launched. What is the length of the spaceship as measured by astronauts on the space station?

69. ‖ The sun radiates energy at the rate 3.8×10^{26} W. The source of this energy is fusion, a nuclear reaction in which mass is transformed into energy. The mass of the sun is 2.0×10^{30} kg.

   a. How much mass does the sun lose each year?

   b. What percent is this of the sun's total mass?

   c. Fusion takes place in the core of a star, where the temperature and pressure are highest. A star like the sun can sustain fusion until it has transformed about 0.10% of its total mass into energy, then fusion ceases and the star slowly dies. Estimate the sun's lifetime, giving your answer in billions of years.

70. ‖ The radioactive element radium (Ra) decays by a process known as *alpha decay,* in which the nucleus emits a helium nucleus. (These high-speed helium nuclei were named alpha particles when radioactivity was first discovered, long before the identity of the particles was established.) The reaction is ^{226}Ra → ^{222}Rn + ^{4}He, where Rn is the element radon. The accurately measured atomic masses of the three atoms are 226.0254 u, 222.0176 u, and 4.0026 u. How much energy is released in each decay? (The energy released in radioactive decay is what makes nuclear waste "hot.")

71. ‖ The nuclear reaction that powers the sun is the fusion of four protons into a helium nucleus. The process involves several steps, but the net reaction is simply 4p → ^{4}He + energy. The mass of a proton, to four significant figures, is 1.673×10^{-27} kg, and the mass of a helium nucleus is known to be 6.644×10^{-27} kg.

   a. How much energy is released in each fusion?

   b. What fraction of the initial rest mass energy is this energy?

72. ‖ Consider the inelastic collision $e^- + e^- \rightarrow e^- + e^- + e^- + e^+$ in which an electron-positron pair is produced in a head-on collision between two electrons moving in opposite directions at the same speed. This is similar to Figure 36.39, but both of the initial electrons are moving.

   a. What is the threshold kinetic energy? That is, what minimum kinetic energy must each electron have to allow this process to occur?

   b. What is the speed of an electron with this kinetic energy?

Challenge Problems

73. ‖‖ An electron moving to the right at 0.90c collides with a positron moving to the left at 0.90c. The two particles annihilate and produce two gamma-ray photons. What is the wavelength of the photons?

74. ‖‖ Two rockets are each 1000 m long in their rest frame. Rocket Orion, traveling at 0.80c relative to the earth, is overtaking rocket Sirius, which is poking along at a mere 0.60c. According to the crew on Sirius, how long does Orion take to completely pass? That is, how long is it from the instant the nose of Orion is at the tail of Sirius until the tail of Orion is at the nose of Sirius?

75. ‖‖ Some particle accelerators allow protons (p^+) and antiprotons (p^-) to circulate at equal speeds in opposite directions in a device called a *storage ring.* The particle beams cross each other at various points to cause $p^+ + p^-$ collisions. In one collision, the outcome is $p^+ + p^- \rightarrow e^+ + e^- + \gamma + \gamma$, where γ represents a high-energy gamma-ray photon. The electron and positron are ejected from the collision at 0.9999995c and the gamma-ray photon wavelengths are found to be 1.0×10^{-6} nm. What were the proton and antiproton speeds, as a fraction of c, prior to the collision?

76. ‖‖ A ball of mass m traveling at a speed of 0.80c has a perfectly inelastic collision with an identical ball at rest. If Newtonian physics were correct for these speeds, momentum conservation would tell us that a ball of mass 2m departs the collision with a speed of 0.40c. Let's do a relativistic collision analysis to determine the mass and speed of the ball after the collision.

   a. What is γ_p, written as a fraction like a/b?

   b. What is the initial total momentum? Give your answer as a fraction times mc.

   c. What is the initial total energy? Give your answer as a fraction times mc^2. Don't forget that there are two balls.

   d. Because energy can be transformed into mass, and vice versa, you cannot assume that the final mass is 2m. Instead, let the final state of the system be an unknown mass M traveling at the unknown speed u_f. You have two conservation laws. Find M and u_f.

77. ‖‖ A very fast pole vaulter lives in the country. One day, while practicing, he notices a 10.0-m-long barn with the doors open at both ends. He decides to run through the barn at 0.866c while carrying his 16.0-m-long pole. The farmer, who sees him coming, says, "Aha! This guy's pole is length contracted to 8.0 m. There will be a short interval of time when the pole is entirely inside the barn. If I'm quick, I can simultaneously close both barn doors while the pole vaulter and his pole are inside." The pole vaulter, who sees the farmer beside the barn, thinks to himself, "That farmer is crazy. The barn is length contracted and is only 5.0 m long. My 16.0-m-long pole cannot fit into a 5.0-m-long barn. If the farmer closes the doors just as the tip of my pole reaches the back door, the front door will break off the last 11.0 m of my pole."

   Can the farmer close the doors without breaking the pole? Show that, when properly analyzed, the farmer and the pole vaulter agree on the outcome. Your analysis should contain both quantitative calculations and written explanation.

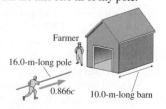

FIGURE CP36.77

Mathematics Review

Algebra

Using exponents:
$$a^{-x} = \frac{1}{a^x} \qquad a^x a^y = a^{(x+y)} \qquad \frac{a^x}{a^y} = a^{(x-y)} \qquad (a^x)^y = a^{xy}$$

$$a^0 = 1 \qquad a^1 = a \qquad a^{1/n} = \sqrt[n]{a}$$

Fractions:
$$\left(\frac{a}{b}\right)\left(\frac{c}{d}\right) = \frac{ac}{bd} \qquad \frac{a/b}{c/d} = \frac{ad}{bc} \qquad \frac{1}{1/a} = a$$

Logarithms:
If $a = e^x$, then $\ln(a) = x$ $\qquad \ln(e^x) = x$ $\qquad e^{\ln(x)} = x$

$$\ln(ab) = \ln(a) + \ln(b) \qquad \ln\left(\frac{a}{b}\right) = \ln(a) - \ln(b) \qquad \ln(a^n) = n\ln(a)$$

The expression $\ln(a + b)$ cannot be simplified.

Linear equations:
The graph of the equation $y = ax + b$ is a straight line. a is the slope of the graph. b is the y-intercept.

Proportionality:
To say that y is proportional to x, written $y \propto x$, means that $y = ax$, where a is a constant. Proportionality is a special case of linearity. A graph of a proportional relationship is a straight line that passes through the origin. If $y \propto x$, then

$$\frac{y_1}{y_2} = \frac{x_1}{x_2}$$

Quadratic equation:
The quadratic equation $ax^2 + bx + c = 0$ has the two solutions $x = \dfrac{-b \pm \sqrt{b^2 - 4ac}}{2a}$.

Geometry and Trigonometry

Area and volume:

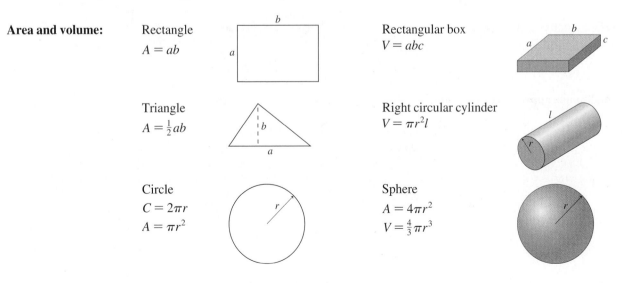

Rectangle
$A = ab$

Rectangular box
$V = abc$

Triangle
$A = \frac{1}{2}ab$

Right circular cylinder
$V = \pi r^2 l$

Circle
$C = 2\pi r$
$A = \pi r^2$

Sphere
$A = 4\pi r^2$
$V = \frac{4}{3}\pi r^3$

Arc length and angle:	The angle θ in radians is defined as $\theta = s/r$. The arc length that spans angle θ is $s = r\theta$. 2π rad $= 360°$	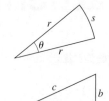

Right triangle: Pythagorean theorem $c = \sqrt{a^2 + b^2}$ or $a^2 + b^2 = c^2$

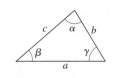

$$\sin\theta = \frac{b}{c} = \frac{\text{far side}}{\text{hypotenuse}} \qquad\qquad \theta = \sin^{-1}\!\left(\frac{b}{c}\right)$$

$$\cos\theta = \frac{a}{c} = \frac{\text{adjacent side}}{\text{hypotenuse}} \qquad\qquad \theta = \cos^{-1}\!\left(\frac{a}{c}\right)$$

$$\tan\theta = \frac{b}{a} = \frac{\text{far side}}{\text{adjacent side}} \qquad\qquad \theta = \tan^{-1}\!\left(\frac{b}{a}\right)$$

General triangle:

$$\alpha + \beta + \gamma = 180° = \pi \text{ rad}$$

Law of cosines $c^2 = a^2 + b^2 - 2ab\cos\gamma$

Identities:

$$\tan\alpha = \frac{\sin\alpha}{\cos\alpha} \qquad\qquad \sin^2\alpha + \cos^2\alpha = 1$$

$$\sin(-\alpha) = -\sin\alpha \qquad\qquad \cos(-\alpha) = \cos\alpha$$

$$\sin(\alpha \pm \beta) = \sin\alpha\cos\beta \pm \cos\alpha\sin\beta \qquad \cos(\alpha \pm \beta) = \cos\alpha\cos\beta \mp \sin\alpha\sin\beta$$

$$\sin(2\alpha) = 2\sin\alpha\cos\alpha \qquad\qquad \cos(2\alpha) = \cos^2\alpha - \sin^2\alpha$$

$$\sin(\alpha \pm \pi/2) = \pm\cos\alpha \qquad\qquad \cos(\alpha \pm \pi/2) = \mp\sin\alpha$$

$$\sin(\alpha \pm \pi) = -\sin\alpha \qquad\qquad \cos(\alpha \pm \pi) = -\cos\alpha$$

Expansions and Approximations

Binomial expansion: $(1 + x)^n = 1 + nx + \dfrac{n(n-1)}{2}x^2 + \cdots$

Binomial approximation: $(1 + x)^n \approx 1 + nx$ if $x \ll 1$

Trigonometric expansions: $\sin\alpha = \alpha - \dfrac{\alpha^3}{3!} + \dfrac{\alpha^5}{5!} - \dfrac{\alpha^7}{7!} + \cdots$ for α in rad

$$\cos\alpha = 1 - \frac{\alpha^2}{2!} + \frac{\alpha^4}{4!} - \frac{\alpha^6}{6!} + \cdots \text{ for } \alpha \text{ in rad}$$

Small-angle approximation: If $\alpha \ll 1$ rad, then $\sin\alpha \approx \tan\alpha \approx \alpha$ and $\cos\alpha \approx 1$.

The small-angle approximation is excellent for $\alpha < 5°$ (≈ 0.1 rad) and generally acceptable up to $\alpha \approx 10°$.

Calculus

The letters a and n represent constants in the following derivatives and integrals.

Derivatives

$$\frac{d}{dx}(a) = 0$$

$$\frac{d}{dx}(ax) = a$$

$$\frac{d}{dx}\left(\frac{a}{x}\right) = -\frac{a}{x^2}$$

$$\frac{d}{dx}(ax^n) = anx^{n-1}$$

$$\frac{d}{dx}\big(\ln(ax)\big) = \frac{1}{x}$$

$$\frac{d}{dx}(e^{ax}) = ae^{ax}$$

$$\frac{d}{dx}\big(\sin(ax)\big) = a\cos(ax)$$

$$\frac{d}{dx}\big(\cos(ax)\big) = -a\sin(ax)$$

Integrals

$$\int x\, dx = \frac{1}{2}x^2$$

$$\int x^2\, dx = \frac{1}{3}x^3$$

$$\int \frac{1}{x^2}\, dx = -\frac{1}{x}$$

$$\int x^n\, dx = \frac{x^{n+1}}{n+1} \qquad n \neq -1$$

$$\int \frac{dx}{x} = \ln x$$

$$\int \frac{dx}{a+x} = \ln(a+x)$$

$$\int \frac{x\, dx}{a+x} = x - a\ln(a+x)$$

$$\int \frac{dx}{\sqrt{x^2 \pm a^2}} = \ln\left(x + \sqrt{x^2 \pm a^2}\right)$$

$$\int \frac{x\, dx}{\sqrt{x^2 \pm a^2}} = \sqrt{x^2 \pm a^2}$$

$$\int \frac{dx}{x^2 + a^2} = \frac{1}{a}\tan^{-1}\left(\frac{x}{a}\right)$$

$$\int \frac{dx}{(x^2 + a^2)^2} = \frac{1}{2a^3}\tan^{-1}\left(\frac{x}{a}\right) + \frac{x}{2a^2(x^2 + a^2)}$$

$$\int \frac{dx}{(x^2 \pm a^2)^{3/2}} = \frac{\pm x}{a^2\sqrt{x^2 \pm a^2}}$$

$$\int \frac{x\, dx}{(x^2 \pm a^2)^{3/2}} = -\frac{1}{\sqrt{x^2 \pm a^2}}$$

$$\int e^{ax}\, dx = \frac{1}{a}e^{ax}$$

$$\int xe^{-x}\, dx = -(x+1)e^{-x}$$

$$\int x^2 e^{-x}\, dx = -(x^2 + 2x + 2)e^{-x}$$

$$\int \sin(ax)\, dx = -\frac{1}{a}\cos(ax)$$

$$\int \cos(ax)\, dx = \frac{1}{a}\sin(ax)$$

$$\int \sin^2(ax)\, dx = \frac{x}{2} - \frac{\sin(2ax)}{4a}$$

$$\int \cos^2(ax)\, dx = \frac{x}{2} + \frac{\sin(2ax)}{4a}$$

$$\int_0^\infty x^n e^{-ax}\, dx = \frac{n!}{a^{n+1}}$$

$$\int_0^\infty e^{-ax^2}\, dx = \frac{1}{2}\sqrt{\frac{\pi}{a}}$$

Periodic Table of Elements

Key:
27 — Atomic number
Co — Symbol
58.9 — Atomic mass

Period	1	2	3	4	5	6	7	8	9	10	11	12	13	14	15	16	17	18
1	H 1 1.0																	He 2 4.0
2	Li 3 6.9	Be 4 9.0											B 5 10.8	C 6 12.0	N 7 14.0	O 8 16.0	F 9 19.0	Ne 10 20.2
3	Na 11 23.0	Mg 12 24.3											Al 13 27.0	Si 14 28.1	P 15 31.0	S 16 32.1	Cl 17 35.5	Ar 18 39.9
4	K 19 39.1	Ca 20 40.1	Sc 21 45.0	Ti 22 47.9	V 23 50.9	Cr 24 52.0	Mn 25 54.9	Fe 26 55.8	Co 27 58.9	Ni 28 58.7	Cu 29 63.5	Zn 30 65.4	Ga 31 69.7	Ge 32 72.6	As 33 74.9	Se 34 79.0	Br 35 79.9	Kr 36 83.8
5	Rb 37 85.5	Sr 38 87.6	Y 39 88.9	Zr 40 91.2	Nb 41 92.9	Mo 42 95.9	Tc 43 [98]	Ru 44 101.1	Rh 45 102.9	Pd 46 106.4	Ag 47 107.9	Cd 48 112.4	In 49 114.8	Sn 50 118.7	Sb 51 121.8	Te 52 127.6	I 53 126.9	Xe 54 131.3
6	Cs 55 132.9	Ba 56 137.3	La 57 138.9	Hf 72 178.5	Ta 73 180.9	W 74 183.9	Re 75 186.2	Os 76 190.2	Ir 77 192.2	Pt 78 195.1	Au 79 197.0	Hg 80 200.6	Tl 81 204.4	Pb 82 207.2	Bi 83 209.0	Po 84 [209]	At 85 [210]	Rn 86 [222]
7	Fr 87 [223]	Ra 88 [226]	Ac 89 [227]	Rf 104 [265]	Db 105 [268]	Sg 106 [271]	Bh 107 [272]	Hs 108 [270]	Mt 109 [276]	Ds 110 [281]	Rg 111 [280]	Cn 112 [285]	113	Fl 114 [289]	115	Lv 116 [293]	117	118

Transition elements (groups 3–12). *Lu 71 175.0* and *Lr 103 [262]* head the inner-transition rows.

Lanthanides (Period 6):

La 57 138.9	Ce 58 140.1	Pr 59 140.9	Nd 60 144.2	Pm 61 144.9	Sm 62 150.4	Eu 63 152.0	Gd 64 157.3	Tb 65 158.9	Dy 66 162.5	Ho 67 164.9	Er 68 167.3	Tm 69 168.9	Yb 70 173.0

Actinides (Period 7):

Ac 89 [227]	Th 90 232.0	Pa 91 231.0	U 92 238.0	Np 93 [237]	Pu 94 [244]	Am 95 [243]	Cm 96 [247]	Bk 97 [247]	Cf 98 [251]	Es 99 [252]	Fm 100 [257]	Md 101 [258]	No 102 [259]

Inner transition elements

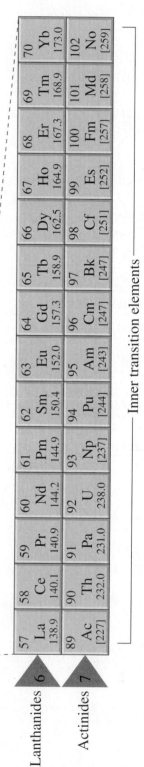

An atomic mass in brackets is that of the longest-lived isotope of an element with no stable isotopes.

Answers

Answers to Stop to Think Questions and Odd-Numbered Exercises and Problems

Chapter 22

Stop to Think Questions

1. **b.** Charged objects are attracted to neutral objects, so an attractive force is inconclusive. Repulsion is the only sure test.
2. $q_e(+3e) > q_a(+1e) > q_d(0) > q_b(-1e) > q_c(-2e)$.
3. **a.** The negative plastic rod will polarize the electroscope by pushing electrons down toward the leaves. This will partially neutralize the positive charge the leaves had acquired from the glass rod.
4. **b.** The two forces are an action/reaction pair, opposite in direction but *equal* in magnitude.
5. **c.** There's an electric field at *all* points, whether an $\vec{E}$ vector is shown or not. The electric field at the dot is to the right. But an electron is a negative charge, so the force of the electric field on the electron is to the left.
6. $E_b > E_a > E_d > E_c$.

Exercises and Problems

1. a Electrons removed   b. 5.0×10^{10}
3. a. Electrons transferred to the sphere   b. 3.1×10^{10}
5. -9.6×10^7 C
7. 2, negative
13. a. 0.90 N   b. 0.90 m/s^2
15. -10 nC
17. 0 N
19. 3.1×10^{-4} N, upward
21. $\vec{F}_{\text{B on A}} = +(7.2 \times 10^{-4}\,\text{N})\hat{\jmath}$, $\vec{F}_{\text{A on B}} = -(7.2 \times 10^{-4}\,\text{N})\hat{\jmath}$
23. a. 0.36 m/s^2 toward glass bead   b. 0.18 m/s^2 toward plastic bead
25. 8.9 kN/m
27. a. $(6.4\hat{\imath} + 1.6\hat{\jmath}) \times 10^{-17}$ N   b. $-(6.4\hat{\imath} + 1.6\hat{\jmath}) \times 10^{-17}$ N
   c. 4.0×10^{10} m/s^2 d. 7.3×10^{13} m/s^2
29. 0.11 nC
31. 12 nC
33. $-6.8 \times 10^4\hat{\imath}$ N/C, $3.0 \times 10^4\hat{\imath}$ N/C, $(8.1 \times 10^3\,\hat{\imath} - 4.1 \times 10^4\,\hat{\jmath})$ N/C
35. 1.4×10^5 C, -1.4×10^5 C
37. a. 5.0×10^2 N   b. 3.0×10^{29} m/s^2
39. 8.4×10^{21}
41. 4.3×10^{-3} N, 73° counterclockwise from the +x-axis
43. 2.0×10^{-4} N, 45° clockwise from the +x-axis
45. $1.1 \times 10^{-5}\hat{\jmath}$ N
47. a. -2.4 cm   b. Yes
49. -20 nC
51. $(2 - \sqrt{2})\dfrac{KQq}{L^2}$
53. a. 2.4×10^2 N   b. Yes
55. 8.1 nC
57. 7.3 m/s
59. 33 nC
61. a. 1.1×10^{18} m/s^2 b. 1.0×10^{-12} N   c. 6.3×10^6 N/C   d. 69 nC
63. $-1.0 \times 10^5\hat{\jmath}$ N/C, $(-2.9 \times 10^4\,\hat{\imath} - 2.2 \times 10^4\,\hat{\jmath})$ N/C, $-5.6 \times 10^4\,\hat{\imath}$ N/C
65. a. $(-1.0\,\text{cm}, 2.0\,\text{cm})$ b. $(3.0\,\text{cm}, 3.0\,\text{cm})$ c. $(4.0\,\text{cm}, -2.0\,\text{cm})$
67. 0.18 μC
69. b. 1.0×10^3
71. b. 5.0 cm
73. 0.75 μC
75. 41 g

77. a. $KQq\left(\dfrac{1}{(r - s/2)^2} - \dfrac{1}{(r + s/2)^2}\right)\hat{\imath}$ b. Toward Q

Chapter 23

Stop to Think Questions

1. **c.** From symmetry, the fields of the positive charges cancel. The net field is that of the negative charge, which is toward the charge.
2. $\eta_c = \eta_b = \eta_a$. All pieces of a uniformly charged surface have the same surface charge density.
3. **b, e, and h.** b and e both increase the linear charge density λ.
4. $E_a = E_b = E_c = E_d = E_e$. The field strength of a charged plane is the same at all distances from the plane. An electric field diagram shows the electric field vectors at only a few points; the field exists at all points.
5. $F_a = F_b = F_c = F_d = F_e$. The field strength inside a capacitor is the same at all points, hence the force on a charge is the same at all points. The electric field exists at all points whether or not a vector is shown at that point.
6. **c.** Parabolic trajectories require *constant* acceleration and thus a *uniform* electric field. The proton has an initial velocity component to the left, but it's being pushed back to the right.

Exercises and Problems

1. 7.6×10^3 N/C along the +x-axis
3. 7.6×10^3 N/C vertically downward
5. a. 2.0 nC   b. 180 N/C
7. 2.9×10^{-3} N
9. 2.3×10^5 N/C, 1.67×10^5 N/C, 2.3×10^5 N/C
11. 40 nC
13. 15 nC
15. a. 0 N/C   b. 4.1×10^3 N/C
17. 27 nC
19. 1.41×10^5 N/C
21. -0.35 pC
23. 25 nC, -25 nC
25. 0.9995 cm
27. a. 0.0023   b. 43 kN/C, up
29. 0.28 MN/C
31. 6.4 μC/m^2
33. a. $\dfrac{1}{4\pi\epsilon_0}\dfrac{qQs}{r^3}$ b. $\dfrac{1}{4\pi\epsilon_0}\dfrac{qQs}{r^2}$
35. a. $(1.0 \times 10^5\hat{\imath} - 3.6 \times 10^4\hat{\jmath})$ N/C
   b. 1.1×10^5 N/C, 19.4° ccw from the +x-axis
37. a. $(-4.7 \times 10^3\hat{\imath} + 8.6 \times 10^4\hat{\jmath})$ N/C
   b. 8.6×10^4 N/C, 93° cw from the +x-axis
39. $\dfrac{1}{4\pi\epsilon_0}\left(\dfrac{q}{5\sqrt{5}a^2}\right)(3\hat{\imath} + 2\hat{\jmath})$, $-\dfrac{1}{4\pi\epsilon_0}\left(\dfrac{17q}{9a^2}\right)\hat{\imath}$, $-\dfrac{1}{4\pi\epsilon_0}\left(\dfrac{7q}{9a^2}\right)\hat{\imath}$,
   and $\dfrac{1}{4\pi\epsilon_0}\left(\dfrac{q}{5\sqrt{5}a^2}\right)(3\hat{\imath} - 2\hat{\jmath})$
41. $\dfrac{1}{4\pi\epsilon_0}\dfrac{16\lambda y}{4y^2 + d^2}$
43. a. $\dfrac{1}{4\pi\epsilon_0}\dfrac{Q}{x^2 - L^2/4}$ b. $\dfrac{1}{4\pi\epsilon_0}\dfrac{Q}{x^2}$ c. 9.8×10^4 N/C

47. a. $\dfrac{1}{4\pi\epsilon_0}\left(\dfrac{2\pi Q}{L^2}\right)\hat{\imath}$ b. 1.70×10^5 N/C

49. 1.4×10^5 N/C

51. 2.2 mm

53. 1.19×10^7 m/s

55. 1.13×10^{14} Hz

57. a. $\dfrac{\frac{4}{3}\pi r^3\rho g + qE}{6\pi\eta r}$ b. 0.067 mm/s   c. 0.049 mm/s

59. 6.56×10^{15} Hz

63. b. 1.0 mm

65. b. $\dfrac{R}{\sqrt{3}}$

67. a.

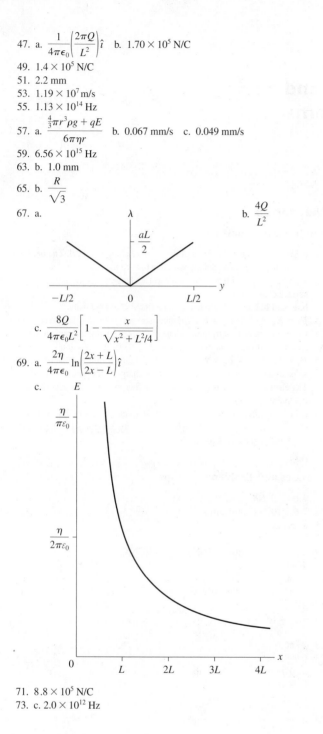

c. $\dfrac{8Q}{4\pi\epsilon_0 L^2}\left[1 - \dfrac{x}{\sqrt{x^2 + L^2/4}}\right]$

69. a. $\dfrac{2\eta}{4\pi\epsilon_0}\ln\left(\dfrac{2x+L}{2x-L}\right)\hat{\imath}$

c.

71. 8.8×10^5 N/C

73. c. 2.0×10^{12} Hz

Chapter 24

Stop to Think Questions

1. **a and d.** Symmetry requires the electric field to be unchanged if front and back are reversed, if left and right are reversed, or if the field is rotated about the wire's axis. Fields a and d both have the proper symmetry. Other factors would now need to be considered to determine the correct field.

2. **e.** The net flux is into the box.

3. **c.** There's no flux through the four sides. The flux is positive 1 N m²/C through both the top and bottom because $\vec{E}$ and $\vec{A}$ both point outward.

4. $\Phi_b = \Phi_e > \Phi_a = \Phi_c = \Phi_d$. The flux through a closed surface depends only on the amount of enclosed charge, not the size or shape of the surface.

5. **d.** A cube doesn't have enough symmetry to use Gauss's law. The electric field of a charged cube is *not* constant over the face of a cubic Gaussian surface, so we can't evaluate the surface integral for the flux.

Exercises and Problems

1.

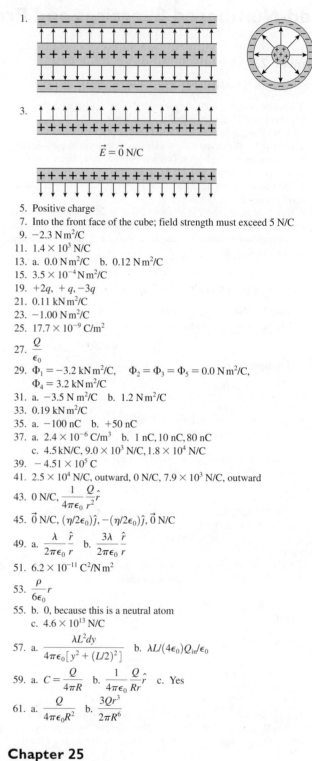

3.

$\vec{E} = \vec{0}$ N/C

5. Positive charge

7. Into the front face of the cube; field strength must exceed 5 N/C

9. -2.3 N m²/C

11. 1.4×10^3 N/C

13. a. 0.0 N m²/C   b. 0.12 N m²/C

15. 3.5×10^{-4} N m²/C

19. $+2q, +q, -3q$

21. 0.11 kN m²/C

23. -1.00 N m²/C

25. 17.7×10^{-9} C/m²

27. $\dfrac{Q}{\epsilon_0}$

29. $\Phi_1 = -3.2$ kN m²/C, $\Phi_2 = \Phi_3 = \Phi_5 = 0.0$ N m²/C, $\Phi_4 = 3.2$ kN m²/C

31. a. -3.5 N m²/C   b. 1.2 N m²/C

33. 0.19 kN m²/C

35. a. -100 nC   b. $+50$ nC

37. a. 2.4×10^{-6} C/m³   b. 1 nC, 10 nC, 80 nC
    c. 4.5 kN/C, 9.0×10^3 N/C, 1.8×10^4 N/C

39. -4.51×10^5 C

41. 2.5×10^4 N/C, outward, 0 N/C, 7.9×10^3 N/C, outward

43. 0 N/C, $\dfrac{1}{4\pi\epsilon_0}\dfrac{Q}{r^2}\hat{r}$

45. $\vec{0}$ N/C, $(\eta/2\epsilon_0)\hat{\jmath}, -(\eta/2\epsilon_0)\hat{\jmath}, \vec{0}$ N/C

49. a. $\dfrac{\lambda}{2\pi\epsilon_0}\dfrac{\hat{r}}{r}$ b. $\dfrac{3\lambda}{2\pi\epsilon_0}\dfrac{\hat{r}}{r}$

51. 6.2×10^{-11} C²/N m²

53. $\dfrac{\rho}{6\epsilon_0}r$

55. b. 0, because this is a neutral atom
    c. 4.6×10^{13} N/C

57. a. $\dfrac{\lambda L^2 dy}{4\pi\epsilon_0[y^2 + (L/2)^2]}$ b. $\lambda L/(4\epsilon_0)Q_{in}/\epsilon_0$

59. a. $C = \dfrac{Q}{4\pi R}$ b. $\dfrac{1}{4\pi\epsilon_0}\dfrac{Q}{Rr}\hat{r}$ c. Yes

61. a. $\dfrac{Q}{4\pi\epsilon_0 R^2}$ b. $\dfrac{3Qr^3}{2\pi R^6}$

Chapter 25

Stop to Think Questions

1. **Zero.** The motion is always perpendicular to the electric force.

2. $U_b = U_d > U_a = U_c$. The potential energy depends inversely on r. The effects of doubling the charge and doubling the distance cancel each other.

3. **c.** The proton gains speed by losing potential energy. It loses potential energy by moving in the direction of decreasing electric potential.
4. $V_a = V_b > V_c > V_d = V_e$. The potential decreases steadily from the positive to the negative plate. It depends only on the distance from the positive plate.
5. $\Delta V_{ac} = \Delta V_{bc} > \Delta V_{ab}$. The potential depends only on the *distance* from the charge, not the direction. $\Delta V_{ab} = 0$ because these points are at the same distance.

Exercises and Problems

1. 2.7×10^6 m/s
3. 2.1×10^6 m/s
5. -2.2×10^{-19} J
7. -4.7×10^{-6} J
9. 1.61×10^8 N/C
11. 4.38×10^5 m/s
13. 11.4 V
15. a. Higher   b. 3340 V
17. 1.0×10^{-5} V
19. a. 4.2×10^3 m/s   b. 1.8×10^6 m/s
23. a. 200 V   b. 6.3×10^{-10} C
25. a. 1000 V   b. 7.0×10^6 m/s
27. a. 1800 V, 1800 V, 900 V   b. 0 V, 900 V
29. a. 27 V   b. -4.3×10^{-18} J
31. -1600 V
33. a. No   b. Yes, at $x = 0$
   c.

35. 0 V
37. -10 nC, 40 nC
39. 3.0 cm, 6.0 cm
41. a. 0.72 J   b. 14 N   c. 2.0 g cube: 22 m/s, 4.0 g cube: 11 m/s
43. 1.0×10^5 m/s
45. 2.5 cm/s
47. a. 2.1×10^6 V/m   b. 9.4×10^7 m/s
49. 150 nC
51. 8.0×10^7 m/s
53. -5.1×10^{-19} J
55. 310 nC
57. 6.8 fm
59. a. Yes   c. 8.21×10^8 m/s
61. a. 15 V, 3.0 kV/m, 2.1×10^{-10} C   b. 15 V, 1/5 kV/m, 1.0×10^{-10} C
   c. 15 V, 3.0 kV/m, 8.3×10^{-10} C

63. a. $\dfrac{V_0}{R}$ b. 100 kV/m
65. a. 8.3 μC   b. 3.3×10^6 V/m
67. a. $\dfrac{2q}{4\pi\epsilon_0 x} \dfrac{1}{\sqrt{1 + s^2/4x^2}}$
   b.

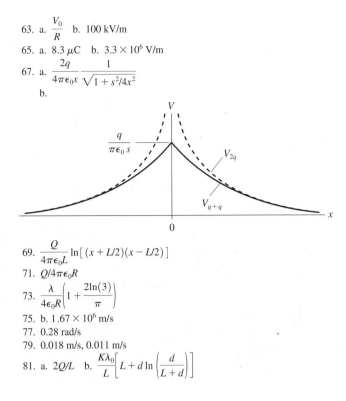

69. $\dfrac{Q}{4\pi\epsilon_0 L} \ln\left[(x + L/2)(x - L/2) \right]$
71. $Q/4\pi\epsilon_0 R$
73. $\dfrac{\lambda}{4\epsilon_0 R}\left(1 + \dfrac{2\ln(3)}{\pi}\right)$
75. b. 1.67×10^6 m/s
77. 0.28 rad/s
79. 0.018 m/s, 0.011 m/s
81. a. $2Q/L$ b. $\dfrac{K\lambda_0}{L}\left[L + d\ln\left(\dfrac{d}{L+d}\right)\right]$

Chapter 26

Stop to Think Questions

1. **c.** E_y is the negative of the slope of the V-versus-y graph. E_y is positive because $\vec{E}$ points up, so the graph has a negative slope. E_y has constant magnitude, so the slope has a constant value.
2. **c.** $\vec{E}$ points "downhill," so V must decrease from right to left. E is larger on the left than on the right, so the contour lines must be closer together on the left.
3. **b.** Because of the connecting wire, the three spheres form a single conductor in electrostatic equilibrium. Thus all points are at the same potential. The electric field of a sphere is related to the sphere's potential by $E = V/R$, so a smaller-radius sphere has a larger E.
4. **5.0 V.** The potentials add, but $\Delta V = -1.0$ V because the charge escalator goes *down* by 1.0 V.
5. $(C_{eq})_b > (C_{eq})_a = (C_{eq})_d > (C_{eq})_c$. $(C_{eq})_b = 3\,\mu\text{F} + 3\,\mu\text{F} = 6\,\mu\text{F}$. The equivalent capacitance of series capacitors is less than any capacitor in the group, so $(C_{eq})_c < 3\,\mu\text{F}$. Only d requires any real calculation. The two 4 μF capacitors are in series and are equivalent to a single 2 μF capacitor. The 2 μF equivalent capacitor is in parallel with 3 μF, so $(C_{eq})_d = 5\,\mu\text{F}$.

Exercises and Problems

1. -0.20 kV
3. -200 V
5. a. A   b. -70 V
7. $-(20\hat{j})$ kV/m
9. a. -5 V/m   b. 10 V/m   c. -5 V/m
11.

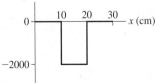

13. -1.0 kV/m
15. a. 27 V/m   b. 3.7 V/m

17. 1.5×10^{-6} J

19. 1.6×10^{-13} J

21. a. 13 pF   b. 1.3 nC

23. 4.8 cm

25. 3.0 μF

27. 7.5 μF

29. 20 μF, parallel

31. 1.4 kV

33. a. 1.1×10^{-7} J   b. 0.71 J/m^3

35. a. 0.15 nF   b. 12 kV

37. 89 pF

39. a.

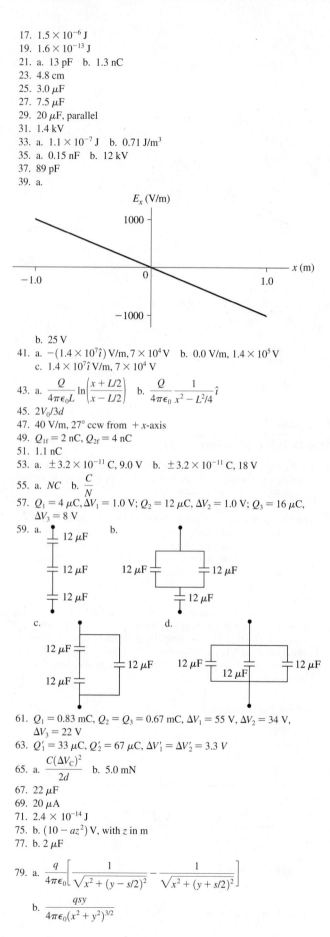

b. 25 V

41. a. $-(1.4 \times 10^7 \hat{\imath})$ V/m, 7×10^4 V   b. 0.0 V/m, 1.4×10^5 V

c. $1.4 \times 10^7 \hat{\imath}$ V/m, 7×10^4 V

43. a. $\dfrac{Q}{4\pi\epsilon_0 L} \ln\left(\dfrac{x + L/2}{x - L/2}\right)$ b. $\dfrac{Q}{4\pi\epsilon_0} \dfrac{1}{x^2 - L^2/4} \hat{\imath}$

45. $2V_0/3d$

47. 40 V/m, 27° ccw from $+x$-axis

49. $Q_{1f} = 2$ nC, $Q_{2f} = 4$ nC

51. 1.1 nC

53. a. $\pm 3.2 \times 10^{-11}$ C, 9.0 V   b. $\pm 3.2 \times 10^{-11}$ C, 18 V

55. a. NC b. $\dfrac{C}{N}$

57. $Q_1 = 4\ \mu$C, $\Delta V_1 = 1.0$ V; $Q_2 = 12\ \mu$C, $\Delta V_2 = 1.0$ V; $Q_3 = 16\ \mu$C,
$\Delta V_3 = 8$ V

59. a.   b.
c.   d.

```
a.  12 μF
    12 μF
    12 μF

b.  12 μF    12 μF
    12 μF

c.  12 μF
    12 μF    12 μF
    12 μF

d.  12 μF    12 μF
    12 μF    12 μF
```

61. $Q_1 = 0.83$ mC, $Q_2 = Q_3 = 0.67$ mC, $\Delta V_1 = 55$ V, $\Delta V_2 = 34$ V,
$\Delta V_3 = 22$ V

63. $Q'_1 = 33\ \mu$C, $Q'_2 = 67\ \mu$C, $\Delta V'_1 = \Delta V'_2 = 3.3$ V

65. a. $\dfrac{C(\Delta V_C)^2}{2d}$ b. 5.0 mN

67. 22 μF

69. 20 μA

71. 2.4×10^{-14} J

75. b. $(10 - az^2)$ V, with z in m

77. b. 2 μF

79. a. $\dfrac{q}{4\pi\epsilon_0}\left[\dfrac{1}{\sqrt{x^2 + (y - s/2)^2}} - \dfrac{1}{\sqrt{x^2 + (y + s/2)^2}}\right]$

b. $\dfrac{qsy}{4\pi\epsilon_0(x^2 + y^2)^{3/2}}$

c. $E_x = \dfrac{qs(3xy)}{4\pi\epsilon_0(x^2 + y^2)^{5/2}}$, $E_y = -\dfrac{qs(2y^2 - x^2)}{4\pi\epsilon_0(x^2 + y^2)^{5/2}}$

d. $\vec{E}_{\text{on-axis}} = \dfrac{2p}{4\pi\epsilon_0 r^3}\hat{\jmath}$, yes

e. $\vec{E}_{\text{bisecting axis}} = -\dfrac{p}{4\pi\epsilon_0 r^3}\hat{\jmath}$, yes

81. a. $\dfrac{1}{4\pi\epsilon_0}\dfrac{Q}{R}\left[\dfrac{3}{2} - \dfrac{r^2}{2R^2}\right]$ b. 3/2

c.

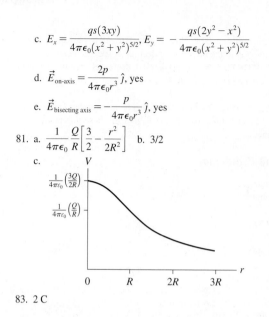

83. 2 C

Chapter 27

Stop to Think Questions

1. $i_c > i_b > i_a > i_d$. The electron current is proportional to $r^2 v_d$. Changing r by a factor of 2 has more influence than changing v_d by a factor of 2.

2. **The electrons don't have to move from the switch to the bulb, which could take hours.** Because the wire between the switch and the bulb is already full of electrons, a flow of electrons from the switch into the wire immediately causes electrons to flow from the other end of the wire into the lightbulb.

3. $E_d > E_b > E_e > E_a = E_c$. The electric field strength depends on the *difference* in the charge on the two wires. The electric fields of the rings in a and c are opposed to each other, so the net field is zero. The rings in d have the largest charge *difference*.

4. **1 A into the junction.** The total current entering the junction must equal the total current leaving the junction.

5. $J_b > J_a = J_d > J_c$. The current density $J = I/\pi r^2$ is independent of the conductivity σ, so a and d are the same. Changing r by a factor of 2 has more influence than changing I by a factor of 2.

6. $I_a = I_b = I_c = I_d$. Conservation of charge requires $I_a = I_b$. The current in each wire is $I = \Delta V_{\text{wire}}/R$. All the wires have the same resistance because they are identical, and they all have the same potential difference because each is connected directly to the battery, which is a *source of potential*.

Exercises and Problems

1. 75 μm/s

3. 0.93 mm

5. 0.023 V/m

7. a. 7.43×10^{-6} m/s   b. 2.1×10^{-14} s

9. a. 0.80 A   b. 7.0×10^7 A/m^2

11. 130 C

13. 1.8 μA

15. 4.2×10^6 A/m^2

17. a. 2.1×10^{-14} s   b. 4.3×10^{-15} s

19. a. 10 V/m   b. 6.7×10^6 A/m^2 c. 0.62 mm

21. Nichrome

23. a. 1.64×10^{-3} V/m   b. 1.10×10^{-5} m/s

25. a. 0.50 C/s   b. 1.5 J   c. 0.75 W

27. a. 1.5 Ω b. 3.5 Ω

29. 1.5 mV

31. 2.3 mA

33. 1.6 A

35. 50 Ω

37. 0.87 V

39. 0.10 V/m

41. 6.2×10^6
43. 23 mA
45. Yes, $2.2 \times 10^5 \ \Omega^{-1}\mathrm{m}^{-1}$
47. a. 120 C   b. 0.45 mm
49. 71°C
51. 100 V
53. a. $\dfrac{(\Delta V)A\left[1 - \alpha(T - T_0)\right]}{\rho_0 L}$ b. 4.4 A   c. −0.017 A/°C
55. 0.50 mm
57. a. $\dfrac{I}{4\pi\sigma r^2}$ b. $E_{\text{inner}} = 3.3 \times 10^{-4}$ V/m, $E_{\text{outer}} = 5.3 \times 10^{-5}$ V/m
59. a. $I(t) = (10 \text{ A})e^{-t/2.0 \text{ s}}$ b. 10 A
    c.

61. 1.01×10^{23}
63. 7.2 mm
65. $\dfrac{3}{2}\dfrac{I}{\pi R^3}$
67. 36 A
69. 1.80×10^3 C
71. $4R$
73. a. 9.4×10^{15} b. 115 A/m^2
75. 1.0 s

Chapter 28

Stop to Think Questions

1. **a, b, and d.** These three are the same circuit because the logic of the connections is the same. In c, the functioning of the circuit is changed by the extra wire connecting the two sides of the capacitor.
2. **ΔV increases by 2 V in the direction of I.** Kirchhoff's loop law, starting on the left side of the battery, is then $+12 \text{ V} + 2 \text{ V} - 8 \text{ V} - 6 \text{ V} = 0$.
3. **$P_b > P_d > P_a > P_c$.** The power dissipated by a resistor is $P_R = (\Delta V_R)^2/R$. Increasing R decreases P_R; increasing ΔV_R increases P_R. But the potential has a larger effect because P_R depends on the square of ΔV_R.
4. **$I = 2$ A for all. $V_a = 20$ V, $V_b = 16$ V, $V_c = 10$ V, $V_d = 8$ V, $V_e = 0$ V.** The potential is 0 V on the right and increases by IR for each resistor going to the left.
5. **A > B > C = D.** All the current from the battery goes through A, so it is brightest. The current divides at the junction, but not equally. Because B is in parallel with C + D but has half the resistance, twice as much current travels through B as through C + D. So B is dimmer than A but brighter than C and D. C and D are equal because current is the same through bulbs in series.
6. **b.** The two 2 Ω resistors are in series and equivalent to a 4 Ω resistor. Thus $\tau = RC = 4$ s.

Exercises and Problems

1.

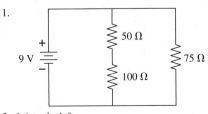

3. 1 A to the left
5. a. 0.5 A to the right
    b.

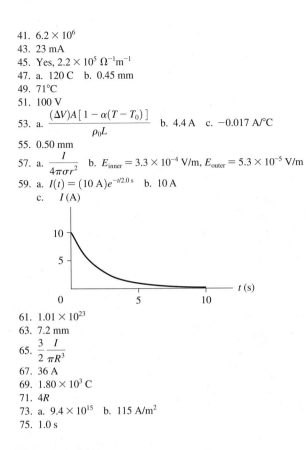

7. 9.60 Ω, 12.5 A
9. a. 60 W   b. 23 W, 14 W
11. D
13. 24 μm
15. $120
17. a. 0.65 Ω   b. 3.5 W
19. 1.2 Ω
21. 60 V, 10 Ω
23. 12 Ω
25. 14 Ω
27. 183 Ω
29. 20 W, 45 W
31. 5 V, −2 V
33. 8 ms
35. 69 ms
37. 0.87 kΩ
39. $65 for the incandescent bulb, $20 for the fluorescent tube
41. 19 W
43.

2.0 Ω

6.0 V, 3.0 Ω, 6.0 Ω

45. 7 Ω
47. 1.04
49. 9.5
51. a. $R = r$ b. 20 W
53. 3 A
55. a. 8 V   b. 0 V
57. a. 0.505 Ω   b. 0.500 Ω
59.

Resistor	Potential difference (V)	Current (A)
2 Ω	8	4
4 Ω	8	2
6 Ω	8	4/3
8 Ω	16	2
12 Ω	8	2/3

61. 2.0 A
63. 0.12 A, left to right
65. a. 200 A   b. 15 A   c. 200 A   d. 4 A
67. a. 6.9 ms   b. 3.5 ms
69. 89 Ω
71. 63 μJ
73. a. $\mathcal{E}$ b. $C\mathcal{E}$ c. $I = +dQ/dt$ d. $I = \dfrac{\mathcal{E}}{R}e^{-t/\tau}$

e. $I/(\mathcal{E}/R)$

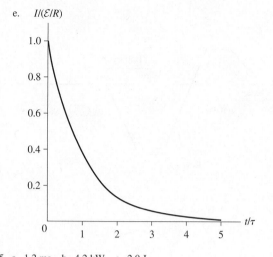

75. a. 1.2 ms b. 4.2 kW c. 2.9 J
77. 140 Ω, 72 Ω
79. 2.0 m, 0.49 mm
83. b. 5.1 kΩ

Chapter 29

Stop to Think Questions

1. **Not at all.** The charge exerts weak, attractive polarization forces on both ends of the compass needle, but in this configuration the forces will balance and have no net effect.
2. **d.** Point your right thumb in the direction of the current and curl your fingers around the wire.
3. **b.** Point your right thumb out of the page, in the direction of $\vec{v}$. Your fingers are pointing down as they curl around the left side.
4. **b.** The right-hand rule gives a downward $\vec{B}$ for a clockwise current. The north pole is on the side from which the field emerges.
5. **c.** For a field pointing into the page, $\vec{v} \times \vec{B}$ is to the right. But the electron is negative, so the force is in the direction of $-(\vec{v} \times \vec{B})$.
6. **b.** Repulsion indicates that the south pole of the loop is on the right, facing the bar magnet; the north pole is on the left. Then the right-hand rule gives the current direction.
7. **a or c.** Any magnetic field to the right, whether leaving a north pole or entering a south pole, will align the magnetic domains as shown.

Exercises and Problems

1. $B_2 = 40$ mT, $B_3 = 0$ T, $B_4 = 40$ mT
3. a. 0 T b. $1.60 \times 10^{-15}\hat{k}$ T c. $-4.0 \times 10^{-16}\hat{k}$ T
5. $2.83 \times 10^{-16}\hat{k}$ T
7. 2.5 A, 250 A, 5000 A to 50,000 A, 500,000 A
9. a. $2.0\ \mu$T b. 4.0%
11. a. 20 A b. 1.6×10^{-3} m
13. $\vec{0}$ T
15. 0.12 mT
17. a. 3.1×10^{-4} A m^2 b. 5.0×10^{-7} T
19. a. 6.2×10^{-5} T b. 6.3×10^{8} A
21. 0
23. 23.0 A, into the page
25. 2.4 kA
27. a. $8.0 \times 10^{-13}\hat{j}$ N b. $5.7 \times 10^{-13}(-\hat{j} - \hat{k})$ N
29. a. 1.4442 MHz b. 1.6450 MHz c. 1.6457 MHz
31. a. 86 mT b. 1.62×10^{-14} J
33. 0.131 T, out of page
35. 0.025 N, right
37. 240 A
39. 0.28 A m^2

41. a. $x = 0.50$ cm b. $x = 8.0$ cm
43. 4.1×10^{-4} T, into page
45. $\dfrac{\mu_0 I \theta}{4\pi R}$
47. 2.4×10^{-8} Ω m
49. #18, 4.1 A
51. a. 1.13×10^{10} A b. 0.014 A/m^2 c. 1.3×10^{6} A/m^2
53. a. 5.7×10^{-6} A b. 2.9×10^{-8} A m^2
55. a. Circular b. $\dfrac{\mu_0 NI}{2\pi r}$
57. 0; $\dfrac{\mu_0 I}{2\pi r}\left(\dfrac{r^2 - R_1^2}{R_2^2 - R_1^2}\right)$; $\dfrac{\mu_0 I}{2\pi r}$
59. 2.0 mT, into page
61. 2.4×10^{10} m/s^2, up
63. 16 cm
65. 0.82 mm, 3.0 mm
67. $\sqrt{v_0^2 + \left(\dfrac{qE_0}{m}t\right)^2}$
69. a. 0.45 s b. 2.8×10^{9} s
71. 0.12 T
73. 0.16 T
75. a. Into the page b. $\sqrt{\dfrac{2IlBd}{m}}$
77. a.

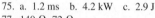

b. $\dfrac{1}{2}ILB_0\hat{j}$ c. $\dfrac{1}{3}IL^2 B_0$

79. 2.0 cm
81. $\dfrac{\mu_0 \omega Q}{2\pi R}$
83. a. Horizontal and to the left above the sheet; horizontal and to the right below the sheet b. $\frac{1}{2}\mu_0 J_s$

Chapter 30

Stop to Think Questions

1. **d.** According to the right-hand rule, the magnetic force on a positive charge carrier is to the right.
2. **No.** The charge carriers in the wire move parallel to $\vec{B}$. There's no magnetic force on a charge moving parallel to a magnetic field.
3. $F_b = F_d > F_a = F_c$. $\vec{F}_a$ is zero because there's no field. $\vec{F}_c$ is also zero because there's no current around the loop. The charge carriers in both the right and left edges are pushed to the bottom of the loop, creating a motional emf but no current. The currents at b and d are in opposite directions, but the forces on the segments in the field are both to the left and of equal magnitude.
4. **Clockwise.** The wire's magnetic field as it passes through the loop is into the page. The flux through the loop decreases into the page as the wire moves away. To oppose this decrease, the induced magnetic field needs to point into the page.
5. **d.** The flux is increasing into the loop. To oppose this increase, the induced magnetic field needs to point out of the page. This requires a ccw induced current. Using the right-hand rule, the magnetic force on the

current in the left edge of the loop is to the right, away from the field. The magnetic forces on the top and bottom segments of the loop are in opposite directions and cancel each other.

6. **b or f.** The potential decreases in the direction of increasing current and increases in the direction of decreasing current.

7. $\tau_c > \tau_a > \tau_b$, $\tau = L/R$, so smaller total resistance gives a larger time constant. The parallel resistors have total resistance $R/2$. The series resistors have total resistance $2R$.

Exercises and Problems

1. 2.0×10^4 m/s
3. a. 1.0 N   b. 2.2 T
5. 3.5×10^{-4} Wb
7. Bab
9. Decreasing
11. a. 8.7×10^{-4} Wb   b. Clockwise
13. 1.6 V
15. 4.7 T/s, increasing
17. a. $1.6 \times 10^{-3}(1 + t)$ A   b. 9.4×10^{-3} A, 1.7×10^{-2} A
19. a. 4.8×10^4 m/s², up   b. 0 m/s²   c. 4.8×10^4 m/s², down
    d. 9.6×10^4 m/s², down
21. 5.3 mT/s
23. 2400
25. 100 V, increases
27. 9.5×10^{-5} J
29. 0.25 μH
31. 3.8×10^{-18} F
33. a. 76 mA   b. 0.50 ms
35. 2.50×10^{-4} s
37. 1.6 A, 0.0 A, -1.6 A
39. 8.7 T/s
41. a. 5.0 mV   b. 10.0 mV
43. 44 mA
45. 44 μA
47. -15μA
49. 0.15 T
51. 4.0 nA
53. a. I (A)   b. 11 A

53. (graph) I (A), value 11 at t around 7 ms, triangular shape, t (ms) axis with marks at 10, 20

55. a. 0.20 A   b. 4.0 mN   c. 11 K
57. a. 6.3×10^{-4} N   b. 3.1×10^{-4} W   c. 1.3×10^{-2} A, ccw
    d. 3.1×10^{-4} W
59. a. $\dfrac{\mathcal{E}_{bat}}{Bl}$ b. 0.98 m/s
61. 3.9 V
63. $(R^2/2r)(dB/dt)$
65. 3.0 s
67. I (A)

67. (graph) I (A) axis with marks 0.20, 0.40; trapezoidal curve, t (ms) axis with marks 20, 40

69. a. $-L\omega I_0 \cos\omega t$ b. 1.3 mA
71. a. 6.3×10^{-7} s   b. 0.050 mA
73. 2.0 mH, 0.13 μF

75. a. 50 V   b. Close S_1 at $t = 0$ s, open S_1 and close S_2 at $t = 0.0625$ s, then open S_2 at $t = 0.1875$ s
77. a. $I_0 = \Delta V_{bat}/R$ b. $I = I_0\left(1 - e^{-t/(L/R)}\right)$
79. 0.72 mH
81. 0.50 m
83. a. $v_0 e^{-bt}$, where $b = l^2B^2/(mR)$
    b.

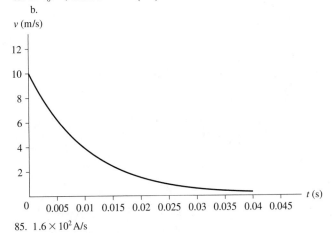

v (m/s)

85. 1.6×10^2 A/s

Chapter 31

Stop to Think Questions

1. **b.** $\vec{v}_{AB}$ is parallel to $\vec{B}_A$ hence $\vec{v}_{AB} \times \vec{B}_A$ is zero. Thus $\vec{E}_B = \vec{E}_A$ and points in the positive z-direction. $\vec{v}_{AB} \times \vec{E}_A$ points down, in the negative y-direction, so $-\vec{v}_{AB} \times \vec{E}_A/c^2$ points in the positive y-direction and causes $\vec{B}_B$ to be angled upward.

2. $B_c > B_a > B_d > B_b$. The induced magnetic field strength depends on the *rate dE/dt* at which the electric field is changing. Steeper slopes on the graph correspond to larger magnetic fields.

3. **e.** $\vec{E}$ is perpendicular to $\vec{B}$ and to $\vec{v}$, so it can only be along the z-axis. According to the Ampère-Maxwell law, $d\Phi_e/dt$ has the same sign as the line integral of $\vec{B} \cdot d\vec{s}$ around the closed curve. The integral is positive for a cw integration. Thus, from the right-hand rule, $\vec{E}$ is either into the page (negative z-direction) and increasing, or out of the page (positive z-direction) and decreasing. We can see from the figure that B is decreasing in strength as the wave moves from left to right, so E must also be decreasing. Thus $\vec{E}$ points along the positive z-axis.

4. **a.** The Poynting vector $\vec{S} = (\vec{E} \times \vec{B})/\mu_0$ points in the direction of travel, which is the positive y-direction. $\vec{B}$ must point in the positive x-direction in order for $\vec{E} \times \vec{B}$ to point upward.

5. **b.** The intensity along a line from the antenna decreases inversely with the square of the distance, so the intensity at 20 km is $\frac{1}{4}$ that at 10 km. But the intensity depends on the square of the electric field amplitude, or, conversely, E_0 is proportional to $I^{1/2}$. Thus E_0 at 20 km is $\frac{1}{2}$ that at 10 km.

6. $I_d > I_a > I_b = I_c$. The intensity depends on $\cos^2\theta$, where θ is the angle *between* the axes of the two filters. The filters in d have $\theta = 0°$. The two filters in both b and c are crossed ($\theta = 90°$) and transmit no light at all.

Exercises and Problems

1. a. Along the $-x$-axis   b. Along the y-axis ($+$ or $-$)
   c. Along the $+x$-axis
3. 16.3° above the x-axis
5. $-1.0 \times 10^6 \hat{k}$ V/m, $0.99998 \hat{j}$ T
9. 1.0 μF
11. a. 0 T   b. 1.67×10^{-13} T   c. 1.98×10^{-13} T
13. 6.0×10^5 V/m
15. a. 10.0 nm   b. 3.00×10^{16} Hz   c. 6.67×10^{-8} T
17. $-z$-direction

19. 980 V/M, 3.3 T
21. a. 5.1×10^{11} V/m   b. 0.43
23. 16 cm
25. a. 3.33×10^{-6} T   b. 1.67×10^{-6} T
27. 66 mW
29. 2.3×10^{-13} N, 45° ccw
31. a. (0.10 T, into page)
    b. 0 V/m, (0.10 T, into page)
33. a. $\vec{E} = \dfrac{\lambda}{2\pi\epsilon_0 r}$, away from wire; $\vec{B} = \vec{0}$ T

    b. $\vec{E} = \dfrac{\lambda}{2\pi\epsilon_0 r}$, away from wire; $\vec{B} = \dfrac{1}{c^2\epsilon_0}\dfrac{v\lambda}{2\pi r}$, into page at top

    d. $\vec{E} = \dfrac{\lambda}{2\pi\epsilon_0 r}$, away from the wire; $\vec{B} = \dfrac{1}{c^2\epsilon_0}\dfrac{v\lambda}{2\pi r}$, into page at top
35. a. 7.1 Vm, 0.17 A   b. 5.2 Vm and 0.044 A
37. a. 1.0 mT   b. 0.160 mT
39. a. $(2.8 \times 10^3 t^2)$ Vm   b.

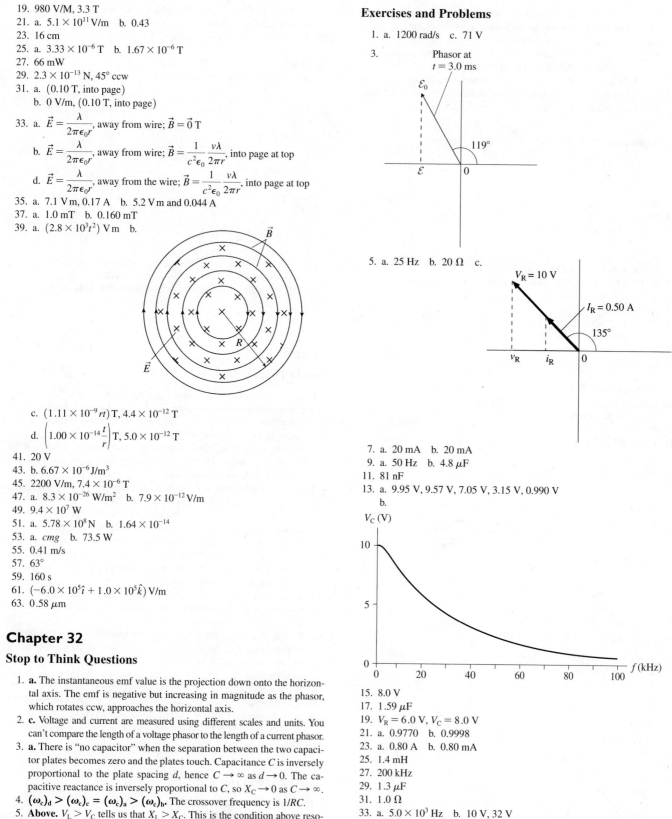

    c. $(1.11 \times 10^{-9} rt)$ T, 4.4×10^{-12} T
    d. $\left(1.00 \times 10^{-14}\dfrac{t}{r}\right)$ T, 5.0×10^{-12} T
41. 20 V
43. b. 6.67×10^{-6} J/m³
45. 2200 V/m, 7.4×10^{-6} T
47. a. 8.3×10^{-26} W/m²   b. 7.9×10^{-12} V/m
49. 9.4×10^7 W
51. a. 5.78×10^8 N   b. 1.64×10^{-14}
53. a. cmg b. 73.5 W
55. 0.41 m/s
57. 63°
59. 160 s
61. $(-6.0 \times 10^5\hat{i} + 1.0 \times 10^5\hat{k})$ V/m
63. 0.58 μm

Chapter 32

Stop to Think Questions

1. **a.** The instantaneous emf value is the projection down onto the horizontal axis. The emf is negative but increasing in magnitude as the phasor, which rotates ccw, approaches the horizontal axis.
2. **c.** Voltage and current are measured using different scales and units. You can't compare the length of a voltage phasor to the length of a current phasor.
3. **a.** There is "no capacitor" when the separation between the two capacitor plates becomes zero and the plates touch. Capacitance C is inversely proportional to the plate spacing d, hence $C \to \infty$ as $d \to 0$. The capacitive reactance is inversely proportional to C, so $X_C \to 0$ as $C \to \infty$.
4. $(\omega_c)_d > (\omega_c)_c = (\omega_c)_a > (\omega_c)_b.$ The crossover frequency is $1/RC$.
5. **Above.** $V_L > V_C$ tells us that $X_L > X_C$. This is the condition above resonance, where X_L is increasing with ω while X_C is decreasing.
6. **a, b, and f.** You can always increase power by turning up the voltage. The current leads the emf, telling us that the circuit is primarily capacitive. The current can be brought into phase with the emf, thus maximizing the power, by decreasing C or increasing L.

Exercises and Problems

1. a. 1200 rad/s   c. 71 V
3.
5. a. 25 Hz   b. 20 Ω   c.
7. a. 20 mA   b. 20 mA
9. a. 50 Hz   b. 4.8 μF
11. 81 nF
13. a. 9.95 V, 9.57 V, 7.05 V, 3.15 V, 0.990 V
    b.
15. 8.0 V
17. 1.59 μF
19. $V_R = 6.0$ V, $V_C = 8.0$ V
21. a. 0.9770   b. 0.9998
23. a. 0.80 A   b. 0.80 mA
25. 1.4 mH
27. 200 kHz
29. 1.3 μF
31. 1.0 Ω
33. a. 5.0×10^3 Hz   b. 10 V, 32 V
35. 22 V
37. 0.40 kW
39. 0.75
41. a. $\dfrac{\sqrt{3}}{RC}$ b. $\dfrac{\sqrt{3}}{2}\mathcal{E}_0$
43. 0.50 mm
45. a. 25 mA   b. 6.7 V

47. a. $I_R = \dfrac{\mathcal{E}_0}{R}, I_C = \dfrac{\mathcal{E}_0}{(\omega C)^{-1}}$ b. $\mathcal{E}_0\sqrt{(\omega C)^2 + \dfrac{1}{R^2}}$
49. a. $\mathcal{E}_0/\sqrt{R^2 + \omega^2 L^2}, \mathcal{E}_0 R/\sqrt{R^2 + \omega^2 L^2}, \mathcal{E}_0 \omega L/\sqrt{R^2 + \omega^2 L^2}$
     b. $V_R \rightarrow \mathcal{E}_0, V_R \rightarrow 0$ c. Low pass   d. R/L
51. a. 2.0 A   b. $-30°$ c. 150 W
53. 10 μT
55. 0.17 A
57. a. 3.6 V   b. 3.5 V   c. -3.6 V
61. a. 0.49 μH   b. 10.3 Ω
63. 24 W
65. a. 0.44 kA   b. 1.8×10^{-4} F   c. 7.4 MW
67. a. 0.83   b. 100 V   c. 13 Ω d. 3.2×10^{-4} F
71. b. $I = \mathcal{E}_0/R$ in both cases   c. 0

Chapter 33

Stop to Think Questions

1. The antinodal lines seen in Figure 33.4b are diverging.
2. **Smaller.** Shorter-wavelength light doesn't spread as rapidly as longer-wavelength light. The fringe spacing Δy is directly proportional to the wavelength λ.
3. **d.** Larger wavelengths have larger diffraction angles. Red light has a larger wavelength than violet light, so red light is diffracted farther from the center.
4. **b or c.** The width of the central maximum, which is proportional to λ/a, has increased. This could occur either because the wavelength has increased or because the slit width has decreased.
5. **d.** Moving M$_1$ in by λ decreases r_1 by 2λ. Moving M$_2$ out by λ increases r_2 by 2λ. These two actions together change the path length by $\Delta r = 4\lambda$.

Exercises and Problems

1. 470 nm
3. 1.2 mm
5. 1.3 m
7. 0.22 mm
9. 1.6°, 3.2°
11. 43.2°
13. 14.5 cm
15. 20 mm
17. 1.2 m
19. 2.9°
21. 633 nm
23. 9
25. 5.4 mm
27. 78 cm
29. 0.25 mm
31. 400 nm
33. a. Double slit   b. 0.16 mm
35. 0.40 mm
37. 500 nm
39. 500 nm
41. a. 9I_1 b. I_1
43. 1.3 m
45. 43 cm
47. a. $L\lambda/d$ b. $(L/d)\Delta\lambda$ c. 0.250 nm
49. 500 nm
51. 0.12 mm
53. 1.8 μm
55. $\dfrac{\sqrt{2}}{2}d$
57. 1.3 m
59. b. 50 μm
61. 0.88 mm
63. a. 550 nm   b. 0.40 mm
65. 50 cm

67. a. 22.3°   b. 16.6°
69. a. Dark   b. 1.597
71. 12.0 μm
73. a. $\Delta y = \dfrac{\Delta\lambda L}{d}$ c. $\Delta\lambda_{min} = \dfrac{\lambda}{N}$ d. 3646 lines
75. a. 0.52 mm   b. 0.074°   c. 1.3 m

Chapter 34

Stop to Think Questions

1. **c.** The light spreads vertically as it goes through the vertical aperture. The light spreads horizontally due to different points on the horizontal lightbulb.
2. **c.** There's one image behind the vertical mirror and a second behind the horizontal mirror. A third image in the corner arises from rays that reflect twice, once off each mirror.
3. **a.** The ray travels closer to the normal in both media 1 and 3 than in medium 2, so n_1 and n_3 are both larger than n_2. The angle is smaller in medium 3 than in medium 1, so $n_3 > n_1$.
4. **e.** The rays from the object are diverging. Without a lens, the rays cannot converge to form any kind of image on the screen.
5. **a, e, or f.** Any of these will increase the angle of refraction θ_2.
6. **Away from.** You need to decrease s' to bring the image plane onto the screen. s' is decreased by increasing s.
7. **c.** A concave mirror forms a real image in front of the mirror. Because the object distance is $s \approx \infty$, the image distance is $s' \approx f$.

Exercises and Problems

1. 8.0 cm
3. 5.4 m
5. 9.0 cm
9. 433 cm
11. 35°
13. 31°
15. 76.7°
17. 23 cm
19. 113 cm
21. 20 cm behind lens, inverted
23. 15 cm in front of lens, upright
25. 30 cm
27. -40 cm
29. 1.5 cm
31. 30 cm, 0.50 cm
33. b. 40 cm, 2.0 cm, agree
35. b. -60 cm, 8.0 cm, agree
37. b. -8.6 cm, 1.1 cm, agree
39. 30 cm, behind mirror, upright
41. 30 cm, 1.5 cm, behind, upright
43. 6.4 cm
45. 10 m
47. a. $2\cos^{-1}\left(\dfrac{n}{2n_{air}}\right)$ b. 82.8°
49. b. 60 cm
51. 4.0 m
53. 1.46
55. a. 5   b. 4   c. 3
57. 35°
59. 2.7 m/s
61. 15.1 cm
63. Concave, 3.6 cm
65. 2.8 cm
67. 93 cm
69. 0.67 m, 1.0 m
71. 0.014
73. 20 cm
75. 20 μm/s away from the lens

77. −100 cm

79. a. $\dfrac{(n_2 - n_1)}{n_1}\left(\dfrac{1}{R_1} - \dfrac{1}{R_2}\right)$ b. 40 cm, 1.6 m

81. a. 24 cm

b.

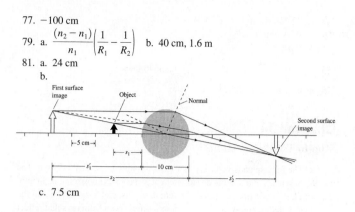

c. 7.5 cm

Chapter 35

Stop to Think Questions

1. **b.** A diverging lens refracts rays away from the optical axis, so the rays will travel farther down the axis before converging.
2. **a.** Because the shutter speed doesn't change, the f-number must remain unchanged. The f-number is f/D, so increasing f requires increasing D.
3. **a.** A magnifier is a converging lens. Converging lenses are used to correct hyperopia.
4. **b.** If the objective magnification is halved, the eyepiece magnification must be doubled. $M_{eye} = 25 \text{ cm}/f_{eye}$, so doubling M_{eye} requires halving f_{eye}.
5. $w_a > w_d > w_b = w_c$. The spot size is proportional to f/D.

Exercises and Problems

1. b. $s_2' = 49$ cm, $h_2' = 4.6$ cm
3. b. $s_2' = -30$ cm, $h_2' = 6.0$ cm
5. b. $s_2' = -3.33$ cm, $h_2' = 0.66$ cm
7. 5.0
9. 1/250 s
11. a. Hyperopia   b. 50 cm
13. 3.0
15. 2.0 mm
17. 9.2 mm
19. a. 40   b. 5.0
21. 1.48
23. 1600 nm
25. 55 km
27. Both images are 2.0 cm tall; one upright 10 cm left of lens, the other inverted 20 cm to right of lens
29. a. $f_2 + f_1$ b. $\dfrac{f_2}{|f_1|}w_1$
31. 16 cm placed 80 cm from screen
33. 23 cm
35. 5.0 cm
37. a. 2.5×   b. 2.0× lens   c. 17.5 cm
39. 4.6 mm
41. 1.0°
43. 2.6 cm
45. a. 1.3 cm   b. 1.2 cm   c. $f/1.7$
47. a. 3.8 cm   b. Sun is too bright
49. b. $\Delta n_2 = \frac{1}{2}\Delta n_1$ c. Crown converging, flint diverging   d. 4.18 cm

Chapter 36

Stop to Think Questions

1. **a, c, and f.** These move at constant velocity, or very nearly so. The others are accelerating.
2. **a.** $u' = u - v = -10$ m/s $-$ 6 m/s $= -16$ m/s. The *speed* is 16 m/s.
3. **c.** Even the light has a slight travel time. The event is the hammer hitting the nail, not your seeing the hammer hit the nail.

4. **At the same time.** Mark is halfway between the tree and the pole, so the fact that he *sees* the lightning bolts at the same time means they *happened* at the same time. It's true that Nancy *sees* event 1 before event 2, but the events actually occurred before she sees them. Mark and Nancy share a reference frame, because they are at rest relative to each other, and all experimenters in a reference frame, after correcting for any signal delays, *agree* on the spacetime coordinates of an event.

5. **After.** This is the same as the case of Peggy and Ryan. In Mark's reference frame, as in Ryan's, the events are simultaneous. Nancy *sees* event 1 first, but the time when an event is seen is not when the event actually happens. Because all experimenters in a reference frame agree on the spacetime coordinates of an event, Nancy's position in her reference frame cannot affect the order of the events. If Nancy had been passing Mark at the instant the lightning strikes occur in Mark's frame, then Nancy would be equivalent to Peggy. Event 2, like the firecracker at the front of Peggy's railroad car, occurs first in Nancy's reference frame.

6. **c.** Nick measures proper time because Nick's clock is present at both the "nose passes Nick" event and the "tail passes Nick" event. Proper time is the smallest measured time interval between two events.

7. $L_A > L_B = L_C$. Anjay measures the pole's proper length because it is at rest in his reference frame. Proper length is the longest measured length. Beth and Charles may *see* the pole differently, but they share the same reference frame and their *measurements* of the length agree.

8. **c.** The rest energy E_0 is an invariant, the same in all inertial reference frames. Thus $m = E_0/c^2$ is independent of speed.

Exercises and Problems

1. a. 4.0 m/s   b. $x_1 = 8.0$ m, $x_2 = 8.0$ m
3. a. 13 m/s   b. 3.0 m/s   c. 9.4 m/s
5. 3.0×10^8 m/s
7. 167 ns
9. 2.0 μs
11. Flash 2 is seen 40 μs after Flash 1.
15. a. 5.0 y   b. 2.2 y   c. 9.0 y
17. 0.28c
19. a. Aged less   b. 14 ns
21. 0.80c
23. 1.47 km
25. 3.0×10^6 m/s
27. a. 1200 m, -2.0 μs   b. 2800 m, 8.7 μs
29. 0.36c
31. 0.9944c
33. 240,000,000 m/s
35. 0.707c
37. 9.0×10^{13} J, 6.0×10^{13} J, 1.5×10^{14} J
39. 0.943c
41. 2400
43. 11.2 h
45. a. No   b. 67.1 y
47. a. 0.80c b. 16 y
49. 14 m
51. a. 750 kJ   b. 8.2×10^{13} J
53. a. $t_D' = t_D'' = t_D''' = 0$ y, $t_E' = 0.42$ y, $t_E'' = 0$ y, $t_E''' = -0.56$ y   b. Yes, spaceship 2   c. Yes, spaceship 3   d. No
55. 22 m
57. 3.1×10^6 V
59. a. 0.980c b. 8.5×10^{-11} J
63. a. 3.84×10^8 m, 1.29 s, 5.47×10^7 m   b. 0.00 m, 0.182 s, 5.47×10^7 m   c. 3.84×10^8 m, 1.28 s, 0.00 m
65. a. 3.5×10^{-18} kg m/s, 1.1×10^{-9} J   b. 1.6×10^{-18} kg m/s
67. a. 7.6×10^{16} J   b. 0.84 kg
69. a. 1.3×10^{17} kg   b. 6.7×10^{-12}%   c. 15 billion years
71. a. 4.3×10^{-12} J   b. 0.72%
73. 1.1 pm
75. 0.85c

Credits

Cover and title page: Thomas Vogel/Getty Images, Inc.
Page **vi:** Dominic Harrison/Alamy. Page **vii:** Mark Axcell/Alamy.

CHAPTER 22
Part VI Opener Page **601:** NASA Solar Dynamics Observatory. Page **602:** Lex Nichols/Alamy. Page **604:** George Resch/Fundamental Photographs. Page **617:** John Eisele/Colorado State University.

CHAPTER 23
Page **629:** Dominic Harrison/Alamy. Page **646:** Courtesy of Pacific Northwest National Laboratory.

CHAPTER 24
Page **658:** Alfred Benjamin/Science Source.

CHAPTER 25
Page **687:** NASA. Page **696:** Dave King/Dorling Kindersley Limited. Page **703:** Luc Diebold/Fotolia. Page **705:** Image provided by the Scientific Computing and Imaging Institute, University of Utah.

CHAPTER 26
Page **714:** Lightboxx/Alamy. Page **720:** Paul Silverman/Fundamental Photographs. Page **722:** AGE Fotostock America, Inc. Page **723:** Gusto/Science Source. Page **724:** Eric Schrader/Pearson Education, Inc. Page **726:** Pushish Images/Shutterstock. Page **729:** Adam Hart-Davis/Science Source.

CHAPTER 27
Page **742:** Tibip/Fotolia. Page **750:** Iphoto/Fotolia. Page **753:** May/Science Source. Page **754:** Ktsdesign/Fotolia. Page **758:** Tim Ridley/Dorling Kindersley, Ltd.

CHAPTER 28
Page **766:** Zelenenka/Getty Images. Page **773:** Chris Ehman/Getty Images. Page **782:** Francisco Cruz/Superstock. Page **784:** Randy Knight.

CHAPTER 29
Page **796:** Shin Okamoto/Getty Images. Page **798:** John Eisele/Colorado State University. Page **806:** John Eisele/Colorado State University. Page **813:** John Eisele/Colorado State University. Page **814:** Bradwieland/Getty Images. Page **817:** Richard Megna/Fundamental Photographs. Page **818:** Neutronman/Getty Images. Page **826:** Horizon International Images Limited/Alamy. Page **827:** Andrew Lambert/Science Source.

CHAPTER 30
Page **836:** Bai Heng-Yao/Thunderbolt _TW/Getty Images. Page **855:** Jeff T. Green/Reuters. Page **857:** Miss Kanithar Aiumla-or/Shutterstock. Page **862:** Alamy.

CHAPTER 31
Page **876:** Buccaneer/Alamy. Page **888:** USACE Engineer Research and Development Center. Page **895:** NASA. Page **897** All Photos: Richard Megna/Fundamental Photographs. Page **898** All Photos: Richard Megna/Fundamental Photographs. Page **900:** David Burton/Alamy.

CHAPTER 32
Page **905:** martinlisner/Fotolia. Page **919:** Bloomberg/Getty Images. Page **920:** Fotolia.

CHAPTER 33
Part VII Opener Page **929:** artpartner-images.com/Alamy, Page **930** Chapter Opener: Blickwinkel/Alamy; Top: Image reprinted courtesy of Dieter Zawischa; Center: Pearson Education, Inc.; Bottom: CENCO Physics/Fundamental Photographs. Page **931** Top: Richard Megna/Fundamental Photographs; Bottom: Todd Gipstein/Getty Images. Page **938:** Pran Mukherjee et al 2009 Nanotechnology 20 325301. doi:10.1088/0957-4484/20/32/325301. Page **940** Top: Devmarya/Fotolia; Top Insert: Jian Zi. Page **943:** Ken Kay/Fundamental Photographs. Page **947** Top Right: Randy Knight; Bottom: Pearson Education, Inc. Page **950:** CENCO Physics/Fundamental Photographs. Page **951:** Image reprinted courtesy of Rod Nave. Page **952:** Philippe Plailly/Science Source. Page **957:** Alekss/Fotolia.

CHAPTER 34
Page **960** Chapter Opener: Mark Axcell/Alamy; Bottom Left: sciencephotos/Alamy; Top Right: sciencephotos/Alamy. Page **966:** Pearson Education, Inc. Page **969:** Richard Megna/Fundamental Photographs. Page **973** All Photos: Richard Megna/Fundamental Photographs. Page **977:** Andrey Popov/Fotolia. Page **984:** Yaacov Dagan/Alamy. Page **987:** BSIP/Science Source.

CHAPTER 35
Page **995** Top: blickwinkel / Alamy Stock Photo; Bottom: Pearson Education, Inc. Page **999** All Photos: Richard Megna/Fundamental Photographs. Page **1000** Top: Underwater Photography/Canon U.S.A., Inc.; Center: NASA; Bottom: Randy Knight. Page **1002:** Tetra Images/Alamy. Page **1005:** Tetra Images/Alamy. Page **1006:** Biophoto Associates/Science Source. Page **1009:** Eric Schrader/Pearson Education, Inc. Page **1010:** Majeczka/Shutterstock. Page **1012:** Jeremy Burgess/Science Source. Page **1013:** Pearson Education, Inc.

CHAPTER 36
Part VIII Opener Page **1021:** Eric Toombs. Page **1022** Top: NG Images/Alamy; Bottom: 3d brained/Shutterstock. Page **1023:** TopFoto/The Image Works. Page **1038:** U.S. Department of Defense Visual Information Center. Page **1040:** Stanford Linear Accelerator Center/Science Source. Page **1054:** LBNL/Science Source. Page **1055:** Wellcome Dept. of Cognitive Neurology/ScienceSource.

Index

For users of the three-volume edition, pages 1–600 are in Volume 1, pages 601–1062 are in Volume 2, and pages 1021–1240 are in Volume 3.

Gaussian surfaces, 662
  calculating electric flux, 663–668
  electric field and, 662
  symmetry of, 662, 669
Gauss's law, 658–686, 809, 887, 889
  and conductors in electrostatic
    equilibrium, 676–679
  Coulomb's law versus, 669, 672
  for magnetic fields, 811, 882
Geiger counter, 734, 1221
Generator, 722, 841, 855–856, 906
Geomagnetism, 797–798
Geosynchronous orbit, 199, 348
Global warming, 540
Graphical addition of vectors, 67
Gravimeter, 405–406
Gravitational constant, 138, 339, 341–343
Gravitational field, 617, 699
Gravitational force, 113, 138–139, 339,
  341–343
  effective, 139, 192
  and weight, 339–340
Gravitational mass, 340
Gravitational potential energy, 233–239,
  343–346
Gravitational torque, 306–307, 407
Gravity, 616–617, 689
  little g (gravitational force) and big G
    (gravitational constant), 341–343
  motion with friction and, 238–239
  Newton's law of, 138–139
  on rotating earth, 192
  universal force, 338
Gray (Gy), 1230
Greenhouse effect, 540
Grounded circuits, 610, 782–784
Ground state, 1100, 1180, 1181, 1182,
  1193
Gyroscope, 324–326

H

Half-life, 410, 1038, 1222
Hall effect, 819–820
Hall voltage, 820
Harmonics, 461, 466–467
Hearing, threshold of, 444
Heat, 208, 222, 521–523
  heat-transfer mechanisms, 537–540
  in ideal-gas processes, 533–534
  specific heat and temperature change,
    526–527
  temperature and thermal energy
    versus, 523
  thermal interactions and, 517, 522,
    558–561
  units of, 523
  work and, 222, 517, 522, 536, 571–573
Heat engine, 570–600

Brayton cycle, 580–582
Carnot cycle, 587–591
ideal-gas, 578–582
perfect, 574, 577
perfectly reversible, 584–587
problem-solving strategy, 579
thermal efficiency of, 573–575, 581–582
Heat exchanger, 581
Heat of fusion, 528
Heat of transformation, 527–528
Heat of vaporization, 528
Heat pump, 596
Heat-transfer mechanisms, 537–540
Heisenberg uncertainty principle,
  1131–1134, 1152
Helium-neon laser, 1203–1204
Henry (H), 857
Hertz (Hz), 391, 394
Histogram, 549
History graph, 423–424, 426
Holography, 951–952
Hooke's law, 219–220, 241, 342–343,
  400–401
Horsepower, 224
Hot reservoir, 571
Huygens' principle, 940–941
Hydraulic lift, 367–369
Hydrogen atom
  angular momentum, 1180–1181
  Bohr's analysis of, 1103–1108, 1109,
    1184
  energy levels of, 1181–1182
  spectrum, 1108–1111
  stationary states of, 1179–1180
  wave functions and probabilities,
    1182–1185
Hydrogen-like ions, 1109–1110
Hydrostatic equilibrium, 364
Hydrostatic pressure, 363–365
Hyperopia, 1002, 1003

I

Ideal battery, 723, 775
Ideal-fluid model, 373
Ideal gas, 499–502
Ideal-gas heat engines, 578–582
Ideal-gas law, 500
Ideal-gas processes, 503–508, 578
  adiabatic process, 525, 534–535
  constant-pressure process, 504–506,
    520, 531
  constant-temperature process, 506–508,
    520–521, 525
  constant-volume process, 504, 520,
    525, 531
  first law of thermodynamics and, 524–525
  problem-solving strategy, 520
  pV diagram, 503

quasi-static processes, 503
  work in, 517–521
Ideal-gas refrigerators, 582–583
Ideal insulator, 758
Ideal spring, 219
Ideal wire, 758
Identical sources, 475–476
Image distance, 965, 971, 974, 978
Image formation
  by refraction, 971–972
  with spherical mirrors, 983–987
  with thin lenses, 982–983
Image plane, 974
Impedance, 915
Impulse, 262–266
Impulse approximation, 266
Impulsive force, 262
Inclined plane, motion on, 51–54
Independent particle approximation (IPA),
  1188
Index of refraction, 436–437, 473, 952,
  966–967, 969
Induced current, 837–851, 859
  applications of, 855–856
  in a circuit, 840–841
  eddy currents, 842
  Faraday's law, 848–851, 882
  Lenz's law, 845–848
  magnetic flux and, 842–845
  motional emf, 838–839
  non-Coulomb electric field and, 858
Induced electric dipole, 633
Induced electric field, 731, 852–855,
  882, 885
Induced emf, 848–851, 859
Induced magnetic dipole, 826
Induced magnetic field, 854–855, 858,
  885–886
Inductance, 857–858
Induction, charging by, 612
Inductive reactance, 913–914
Inductor circuits, 913–914
Inductor, 857–861, 919–920
Inelastic collisions, 272–274
Inertia, 119, 121, 527
Inertial mass, 119, 300, 340
Inertial reference frames, 122, 191, 1024
Instantaneous acceleration, 54–56, 84
Instantaneous velocity, 37–40
Insulators, 605, 758
  dielectrics, 731
  electric charges and forces, 608–612
Integrals, 42–43
Intensity, 443–445, 894, 897–898
  of double-slit interference pattern,
    936–937, 946–947
  of electromagnetic waves, 893–894
  of standing waves, 458